Physics Research and Technology

nova
science publishers

www.novapublishers.com

Physics Research and Technology

75 Years of the Pion
Igor Strakovsky, PhD, Alexey A. Petrov, PhD and Nicolai Popov, PhD (Editors)
2023. ISBN: 979-8-88697-917-6 (Hardcover)
2023. ISBN: 979-8-89113-164-4 (eBook)

Conjugate Construction of Quantum Optics: From Foundations to Applications
Jeffrey Zheng, PhD (Author)
2023. ISBN: 979-8-88697-651-9 (Hardcover)
2023. ISBN: 979-8-89113-150-7 (EPUB)

Four-Dimensional Atomic Structure and Law
Kunming Xu, PhD (Author)
2023. ISBN: 979-8-88697-989-3 (Hardcover)
2023. ISBN: 979-8-89113-062-3 (eBook)

Introduction to Multidisciplinary Science in an Artificial-Intelligence Age: Properties of Matter: Elasticity, Permeability, Porosity, Viscosity, and Wettability
Luc Thomas Ikelle, PhD (Editor)
2023. ISBN: 979-8-88697-522-2 (Hardcover)
2023. ISBN: (eBook)

An Introduction to Permittivity
Parveen Saini, PhD (Editor)
2022. ISBN: 979-8-88697-042-5 (Hardcover)
2022. ISBN: 979-8-88697-375-4 (eBook)

Electromagnetic Waves: Advances in Applications and Research
Manuel B. Hutchinson (Editor)
2022. ISBN: 979-8-88697-254-2 (Softcover)
2022. ISBN: 979-8-88697-265-8 (eBook)

More information about this series can be found at https://novapublishers.com/product-category/series/physics-research-and-technology/

Dr. Alina Ivashkevich
Dr. Nina Krylova
Dr. Elena Ovsiyuk
Dr. Vasiliy Kisel
Dr. Vladimir Balan
and
Dr. Viktor Red'kov

Fields of Particles with Spin, Theory and Applications

nova
science publishers
www.novapublishers.com

https://doi.org/10.52305/IBRV2431

NOTICE TO THE READER

Library of Congress Cataloging-in-Publication Data

ISBN: 979-8-89113-012-8

Published by Nova Science Publishers, Inc. † New York

Contents

Preface xi

Introduction xiii

1 Confluent Heun Functions and the Coulomb Problem for Spin 1/2 Particle 1
1.1 The Coulomb Problem: Solutions Constructed by Hypergeometricand Partially by Heun Functions 1
1.2 Standard Treatment of the Coulomb Problem 9
1.3 Solutions Constructed Completely in Terms of Heun Functions 12

2 Spin 1/2 Particle in 2D Spaces of Constant Curvature, in Presence of Magnetic Field 15
2.1 Cylindric and Conformal Coordinates in Lobachevsky Plane H_2 . . . 15
2.2 Landau Problem for a Scalar Particle in the Plane H_2 17
2.3 Dirac Particle in (x, y) Coordinates, Model H_2 20
2.4 Landau Problem in the Spherical Model S_2, Coordinates (r, ϕ) 22
2.5 Complex Poincaré Half-Plane for Spherical 2-Space 25

3 Hydrogen Atom in Static de Sitter Spaces 31
3.1 Separation of the Variables in *dS* Space 31
3.2 Qualitative Discussion 32
3.3 Reducing Radial Equation to the General Heun Equation 34
3.4 Semi-Classical Study 35
3.5 The Hydrogen Atom in *AdS* Space 37
3.6 Qualitative Study of the Problem in *AdS* Space 38
3.7 Semi-Classical Study for *AdS* Space 39
3.8 Spin 1/2 Particle in *dS* and *AdS* Spaces 40

4 Scalar Particle in Non-Static de Sitter Spaces 45
4.1 Nonrelativistic Approximation, and Riemann Geometry 45
4.2 Schrödinger Equation in de Sitter Non-Static Models 48
4.3 Solving Equations 51
4.4 Klein–Gordon–Fock Equation in Curved Space-Time 55
4.5 Solving Equation in Expanding de Sitter Metric 56

4.6 Solving Equation in Oscillating de Sitter Metric 62

5 Spin 1/2 Particle in Nonstatic de Sitter Spaces, Spherical Coordinates 65
5.1 Particle in Expanding de Sitter Model 65
5.2 Neutrino in Expanding de Sitter Space 77
5.3 Pauli Equation in Expanding de Sitter Space 78
5.4 Spin 1/2 Particle in Oscillating de Sitter Model 82
5.5 Pauli Equation in Oscillating de Sitter Space 90

6 Spin 1/2 Particle in Non-static de Sitter Models, Quasi-Cartesian Coordinates 95
6.1 Separation of the Variables in the Dirac Equation 95
6.2 Solving Equations in the Time Variable t 98
6.3 Behavior of Solutions in the Variable t Near the Points $\cos(t) = 0$ 101
6.4 Constructing Solutions in the Variable z 102
6.5 Majorana Spinor Field 108
6.6 Independent Majorana Components 109

7 The Fermion Doublet in Non-Abelian Monopole Field, Pauli Approximation, Geometry 111
7.1 Pauli Equation for Fermion Doublet, GeneralAnalysis 111
7.2 Non-Abelian Monopole in Schwinger Gauge 113
7.3 Separating the Variables 116
7.4 Nonrelativistic Approximation, the Case $j = 0$ 122
7.5 Nonrelativistic Approximation, the Case $j > 0$ 126
7.6 The Doublet in the Spaces of Constant Curvature 128
7.7 Geometrization of the Monopole Problem, KCC-Invariants 134
7.8 The Euclidean Space 138
7.9 Riemannian Space 139
7.10 Lobachevsky Space 140
7.11 Pure Monopole BPZ-Solution, Euclidean Space 142
7.12 Geometrizing the Doublet Problem, the Case $j = 0$ 146
7.13 Non-Relativistic Approximation, the Case $j > 0$ 149

8 To Analysis of the Dirac and Majorana Particle in Schwarzschild Field 151
8.1 Dirac and Weyl Equations in External Gravitational Field 151
8.2 Majorana Spinor Fields 152
8.3 Spin 1/2 Particle in Schwarzschild Field 153
8.4 Separation of the Variables 154
8.5 The Case of Majorana Particle 156
8.6 Qualitative Study 158
8.7 Analytical Treatment 161
8.8 Structure of the Power Series 165
8.9 General Study of the Tunneling Effect 167

8.10 Geometrization of the Maxwell and Dirac Theories in Schwarzschield Space-Time . . . 169

9 Dirac Particle in Cylindric Parabolic Coordinates and Spinor Space Structure 177
9.1 Spinor Structure and Solutions of the Klein–Gordon–Fock Equation 177
9.2 Solutions of the Klein–Gordon–Fock Equation and Spinors . . . 179
9.3 The Dirac Particle and the Space with Spinor Structure . . . 183

10 Maxwell Equations in Space with Spinor Structure 193
10.1 Spinor Form of Maxwell Equations . . . 193
10.2 Cylindrical Parabolic Coordinates . . . 195
10.3 Continuity and Spinor Space Structure . . . 201
10.4 Helicity Operator . . . 207

11 Geometrization of Maxwell Electrodynamics 217
11.1 Optics and Lagrange Formalism . . . 217
11.2 The Euler–Lagrange Equations . . . 220

12 Finslerian Geometrization for the Problem of a Vector Particle in External Coulomb Field 227
12.1 Setting the Problem . . . 227
12.2 KCC-Invariants . . . 231
12.3 Second KCC-Invariant . . . 232
12.4 Natural Splitting 4+4 . . . 238
12.5 Natural Splitting, Real-Valued Representation . . . 240
12.6 Projections of 8 Equations on Different Planes . . . 242

13 The Study of a Spin 1 Particle with Anomalous Magnetic Moment in the Coulomb Field 247
13.1 Separation of the Variables . . . 247
13.2 The Case of Minimal $j = 0$ . . . 250
13.3 The Non-Relativistic Approximation at $j = 0$ . . . 251
13.4 Non-Relativistic Equations, $j = 1, 2, 3, ...$ . . . 253
13.5 KCC-Geometrical Approach to the Problem . . . 259

14 Vector Particle with Electric Quadruple Moment in the Coulomb Field 265
14.1 Initial Equation . . . 265
14.2 Separating the Variables in the Relativistic Equation . . . 268
14.3 States with Parity $P = (-1)^{j+1}$ . . . 271
14.4 The Case of Minimal $j = 0$ . . . 274
14.5 Non-Relativistic Approximation, $P = (-1)^{j+1}, j = 1, 2, 3, ...$ . . . 276
14.6 Non-Relativistic Radial Equations, the Case of $j = 1, 2, 3, ...$ . . . 279
14.7 KCC-Geometrical Approach . . . 292

15 Massive and Massless Fields with Spin 3/2, Solutions and Helicity Operator 297
15.1 Massive and Massless Spin 3/2 Fields 297
15.2 Separating the Variables 300
15.3 Helicity Operator 303
15.4 Helicity Operator and Solutions of the Wave Equation 310
15.5 The Plane Wave Solutions in Massless Case 315
15.6 Relation to Initial Basis 320
15.7 Helicity Operator 322

16 Solutions with Spherical Symmetry for a Massive Spin 3/2 Particle 329
16.1 System of Equations and Spherical Symmetry 329
16.2 Separating the Variables 333
16.3 Separating the Variables and Additional Constraints 335
16.4 Solving Equations for Functions f_0, g_0 337
16.5 The Matrix Form of the Main System 338
16.6 The Case of Minimal Value $j = 1/2$ 339
16.7 Studying General Case $j = 3/2, 5/2, \ldots$ 343
16.8 Further Study of the Solutions 346
16.9 Accounting for Algebraic and Differential Constraints 349

17 Massless Spin 3/2 Field, Spherical Solutions, Exclusion of the Gauge Degrees of Freedom 355
17.1 Massless Spin 3/2 Particle, General Theory 355
17.2 Separation of the Variables 358
17.3 Gradient Type Solutions 362
17.4 Solving the System of Radial Equations 364
17.5 Solving the Homogeneous Equations 368

18 Spin 3/2 Massless Field, Cylindric Symmetry, Eliminating the Gauge Degrees of Freedom 371
18.1 Separating the Variables 371
18.2 Massless Field 375
18.3 Gauge Solutions 381
18.4 Solving the Second Order Equations, the First Order Constraints . 381

19 On the Matrix Equation for a Spin 2 Particle in Riemannian Space-Time, Tetrad Method 383
19.1 The Spin 2 Particle in Minkowski Space 383
19.2 Structure of the Matrices of the First Order System for a Spin 2 Field 384
19.3 Extension to Riemannian Space-Time Geometry 386
19.4 The Spin 2 Field in Cylindrical Coordinates 388
19.5 The Equation in Spherical Coordinates 389
19.6 The Structure of the Lorentzian Generators 391
19.7 Relativistic Invariance, Additional Checking 393
19.8 Matrix Blocks in the Theory of Spin 2 Particle 396

Conclusion 401

References 405

Index 433

About the Authors 435

Preface

The book is devoted to investigating the particles with spins in external fields and non-Euclidean space-time background. The key problems are: Coulomb task for spin 1/2 particle and Heun equation; the hydrogen atom in de Sitter space; fermion doublet in the non-Abelian monopole field and Pauli approximation; Pauli approximation for spin 1/2 and 1 particles in de Sitter space; the Dirac and Majorana particles in Schwarzschild space; the Dirac – Maxwell fields and spinor space structure; particles with spin 3/2, solutions with different symmetries and eliminating the gauge degrees of freedom in massless case; the matrix 30-component equation for a spin 2 field in Riemannian space-time; Finslerian geometzization of physical problems.

The book may be interesting for researchers, it may well serve as a pedagogical tool for either self study or in courses at both the undergraduate and graduate level. Bibliographies complete many chapters and an index covers the entire book.

Writing a handbook is a long process. It might not have been possible for us without the support and encouragement of the people whose names we want to mention here. We are greatly indebted to our teacher and colleague, Professor Andrey A. Bogush for his unwavering patience and support that have sustained us throughout many years. He helped us to a great extent, encouraged us to the very his end and made this book possible. We cannot thank him enough.

We are thankful to the staff of the Department "Fundamental Interactions and Astrophysics" of Institute of Physics of the National Academy of Sciences of Belarus, the staff of the Belarus State University, Mozyr State Pedagogical University, University Politehnica of Bucharest for many helpful discussions and useful comments. We would like to thank specially our colleagues, Prof. Ya.A. Kurochkin, Dr Yu.P. Vybly, V.V. Kudryashov, S.Yu. Sakovich, from Institute of Physics of the National Academy of Sciences of Belarus. We would like to our friends, Dr G.G. Krylov and Dr G.V. Grushevskaya from Belarus State University; Prof. V.A. Pletyukhov from Brest State University. Their help, stimulating suggestions, critics, interest and valuable hints helped us in writing this book.

Introduction

The present book is devoted to the study of certain problems for particles with different spins in presence of external electromagnetic fields and non-Euclidean space-time background.

In Chapter 1, we examine the possibility to apply the confluent Heun functions in studying the quantum mechanical problem for a spin 1/2 Dirac particle in external Coulomb field.

In Chapter 2, we study the scalar and spinor particles in 2D spaces of constant curvature (hyperbolic Lobachevsky and spherical Riemann models) in presence of the uniform magnetic field; in particular, the use of complex curvilinear coordinates in spherical plane is developed.

In Chapter 3, the quantum mechanical problem for the hydrogen atom is studied for de Sitter and anti de Sitter spaces in static coordinates. In both models, the problem is reduced to the general Heun equation. Qualitative examination shows that the energy spectrum for the hydrogen atom in the de Sitter space should be quasi-stationary, so the atom is not a stable system. A similar analysis shows that in the anti de Sitter model the hydrogen atom is a stable system. Also, we discuss the case of the Dirac spin 1/2 particle for both models.

In Chapter 4, for a scalar particle in spherical coordinates for expanding and oscillating de Sitter models, we develop the non-relativistic approach, and find the relevant exact solutions. In Chapter 5, we extend this analysis to the case of the Dirac particle.

In Chapter 6, we examine the property of the Lobachevsky geometry of acting on particles with different spins $S = 0, 1/2, 1$ as an ideal mirror distributed in the space. Since the Lobachevsky space is used in certain cosmological models, this property means that in such models it is necessary to take into account the effect of the presence of a "cosmological mirror"; it should effectively lead to a redistribution of the particle density in the whole space. We generalize this study for the oscillating de Sitter universe. The effect of the complete reflection of the particles from the effective potential barrier, is preserved.

In Chapter 7, we study the isotopic doublet of Dirac fermions in the presence of the non-Abelian monopole field. A special attention is given to the nonrelativistic approximation for the theory of isotopic doublet in the non-Abelian field. This analysis is detailed for the Bogomol'nyi – Prasad – Sommerfeld monopole field. Besides, we apply a special geometric method based on the geometrical Kosambi – Cartan – Chen invariants to study the behavior of solutions of the relevant differential equations from the point of view of the Jacobi stability.

In Chapter 8, we perform the mathematical study of the tunneling effect for the Dirac particle through the potential barrier generated by the Schwarzschild geometry. Our consideration is based on the use of Frobenius solutions of the relevant second order differential equations. We construct these solutions in explicit form and prove that the involved power series are convergent in all the physical region of the variable $r \in (1, +\infty)$. The results for the tunneling effect significantly differ for two situations: when the particle falls on the barrier from within, and when the particle falls from outside.

In Chapter 9, within the framework of applications of spinor theory to Quantum Mechanics, we study the role of spinor space structure in classifying solutions of scalar and spinor equations specified for the cylindric parabolic coordinates. The emphasis is put on doubling the set of space points, so that we get an extended space model. In such a space, instead of the 2π-rotation, the 4π-rotation is considered, which transfers the space into itself. We have constructed solutions of four types. The solutions of types $(--)$ and $(++)$ are single-valued in the spaces with vector structure, whereas the solutions of types $(-+)$ and $(+-)$ are not single-valued in spaces with vector structure. In Chapter 10, we extend the above analysis to the Maxwell theory. The final results turn out to be similar.

In Chapter 11, the equations of motion associated with a Lagrangian – inspired by relativistic optics in a nonuniform moving medium – are studied. The model describes optical effects in the nonuniform moving medium with special optical properties. We have established the Euler-Lagrange equations for the corresponding geodesics. We have specified the general model to the special case when the metric coefficient γ linearly increases along the direction Z. The exact analytical solutions of the Euler-Lagrange equations have been constructed. The study of the obtained solutions shows that the light ray bends to the axis along which the effective refractive index increases.

In Chapter 12, the quantum mechanical problem for the vector particle is studied. With the use of the space reflection operator, the derived radial system of 10 equations is split into independent subsystems, consisting of 4 and 6 equations. The last one reduces to a system of 4 linked first order differential equations for the complex radial functions $f^i(r)$, $i = 1, \ldots, 4$. We investigate this system by using the tools of the Jacobi stability theory, namely, the Kosambi – Cartan – Chen theory. In accordance with the general aproach, a pencil of geodesic curves from the point r_0 converges (or diverges) if the real parts of all eigenvalues of the 2-nd KCC-invariant $Pi^i{}_j$ are negative (or positive). Behavior of eigenvalues correlates with the existence of two solutions which may be associated with the bound states of a particle in the Coulomb field.

In Chapter 13, we study the generalized Duffin – Kemmer – Petiau equation for a spin 1 particle with anomalous magnetic moment in presence of the Coulomb field. In order to simplify the problem, the transition to the nonrelativistic approximation is performed. The states with minimal $j = 0$ are described by a 2-nd order equation of the double confluent Heun type. For states with $j = 1, 2, \ldots$, after performing the nonrelativistic approximation we have obtained a system of two 2-nd order linked differential equations for radial functions. This system gives a 4-th order equation for one function. Its Frobenius solutions are constructed, and the convergence of the involved power series with 8- and 9-terms recurrence relations, is studied. All these solutions are exact, but any quantization rules for energies of bound states are not known. In Chapter 14, the similar analysis will be performed for a spin 1 particle with electric quadruple moment.

In Chapter 15, the plane wave solutions for a massive spin 3/2 particle, are examined. The wave equation leads to 4 algebraic equations for 8 unknown variables, which assumes the existence of 4 independent solutions. In order to relate the choice of independent solutions to the quantum number of a certain physical operator, we study the problem of eigenvectors for the relevant helicity operator. Similar problems are studied for the massless case. The initial wave equation for vector bispinor $\Psi_a(x)$, which describes a massless spin 3/2 particle, is transformed to a new basis $\tilde{\Psi}_a(x)$, in which the presence of the gauge symmetry in the theory becomes evident: there exist solutions in the form of 4-gradient of an arbitrary bispinor $\tilde{\Psi}^0_a(x) = \partial_a \Psi(x)$. Finally, two independent solutions are explicitly constructed, which do not contain any gauge constituents.

In Chapter 16, the wave equation for a spin 3/2 particle, described by the 16-component vector-bispinor, is investigated in spherical coordinates. The complete equation is split into the main equation, and two additional (algebraic and differential) constraints. There are constructed the solutions, for which 4 operators are diagonalized: they correspond to the quantum numbers $\{\epsilon, j, m, P\}$. After separating the variables, we derive the main system of 8 radial first order equations and additional 2 algebraic and 2 differential constraints. The solutions of the radial equations are constructed as linear combinations of a number of Bessel functions. With the use of the known properties of the Bessel functions, the system of differential equations is transformed to an algebraic linear constraint for three numerical coefficients a_1, a_2, a_3. Its solutions may be chosen in various ways by resolving the simple linear condition $A_1 a_1 + A_2 a_2 + A_3 a_3 = 0$, where the coefficients A_i are expressed through the quantum numbers ϵ and j. Thus, at the fixed quantum numbers $\{\epsilon, j, m, P\}$ there exists a double-degeneration of the quantum states.

In Chapter 17, the system of equations for a massless spin 3/2 field has been studied in spherical coordinates. The general structure of the spherical gauge solutions is specified. It is proved that the general system reduces to two couples of independent 2-nd order nonhomogeneous differential equations; their particular solutions may be found with the use of gauge solutions of a special form. The corresponding homogeneous equations turn out to have the same form, and have three regular singularities and an irregular one, of the rank 2. The Frobenius type solutions for this equation are constructed, and the structure in power series expansion is studied. So we have obtained two types of solutions for the massless spin 3/2 field, which do not contain the gauge constituents.

In Chapter 18, we have extended the known first order matrix equation for a spin 2 particle in Minkowski space to pseudo-Riemannian space-time models, by applying the tetrad method. All the intrinsic constraints on the involved tensors $\Psi(x) = \{\Phi, \Phi_c, \Phi_{(ab)}, \Phi_{[ab]c}\}$ are contained in the structure of the basic matrices. This tetrad matrix equation is specified in cylindrical and spherical coordinates for the flat Minkowski space. The case of massless field is separately addressed, we focus in this theory on a detailed representation of the gauge symmetry.

Chapter 1

Confluent Heun Functions and the Coulomb Problem for Spin 1/2 Particle

1.1 The Coulomb Problem: Solutions Constructed by Hypergeometricand Partially by Heun Functions

The general Heun equation is a second-order linear differential equation which has four regular singularities and different confluent forms [2], [3], [4]. The general Heun equation and all its confluent forms turn out to be of primary significance in physical applications, for instance, in quantum mechanics and field theory on the background of curved space-time models, and in optics – see [5]–[65]. A more complete and comprehensive list of references can be found on the site The Heun Project [66].

In this book, the well-known quantum mechanical problem of a spin 1/2 particle in external Coulomb potential, reduced to a system of two first-order differential equations, is studied from the point of view of possible applications of the Heun function theory while addressing this system. It is shown that, in addition to the standard methods [70] used to solve the problem in terms of the confluent hypergeometric functions, there exist several other opportunities, which are based on applying confluent Heun functions. Namely, we elaborate two combined possibilities to construct solutions: one for the case when one equation of the pair of relevant functions is expressed trough hypergeometric functions, and the other relying on confluent Heun functions. We establish relations between these two classes of functions, and express both functions of the system in terms of confluent Heun functions. These approaches lead us to a unique energy spectrum, which confirms their correctness. In particular, there exist physical problems and, as well, possibilities to avoid (by using special tricks) Heun functions, and rely on hypergeometric functions only.

In spherical coordinates, a diagonal tetrad has the form [50, 64]

$$dS^2 = dt^2 - dr^2 - r^2(d\theta^2 + \sin^2 d\phi^2);, \quad e^{\alpha}_{(0)} = (1,0,0,0),$$
$$e^{\alpha}_{(3)} = (0,1,0,0), \quad e^{\alpha}_{(1)} = (0,0,r^{-1},0), \quad e^{\alpha}_{(2)} = (0,0,0,r^{-1}\sin^{-1}\theta). \tag{1.1}$$

We shall further use the notations from [50, 64]). The covariant Dirac equation

$$\Big[i\gamma^{c} \Big(e^{\alpha}_{(c)} \partial_{\alpha} + \frac{1}{2} j^{ab} \gamma_{abc} \Big) - m \Big] \Psi = 0$$

takes the form (let $\Psi = \frac{1}{r}\tilde{\Psi}$)

$$\Big(i\gamma^{0} \frac{\partial}{\partial t} + i\gamma^{3} \frac{\partial}{\partial r} + \frac{1}{r} \Sigma_{\theta\phi} - m \Big) \tilde{\Psi} = 0, \quad \Sigma_{\theta,\phi} = i\gamma^{1} \partial_{\theta} + \gamma^{2} \frac{i\partial_{\phi} + i\sigma^{12}}{\sin\theta}. \tag{1.2}$$

In order to diagonalize the operators $i\partial_t, \vec{J}^2, J_3$, one takes the wave function in the form [50]

$$\tilde{\Psi} = e^{-iEt} \left| \begin{array}{c} f_1(r)\, D_{-1/2} \\ f_2(r)\, D_{+1/2} \\ f_3(r)\, D_{-1/2} \\ f_4(r)\, D_{+1/2} \end{array} \right| , \tag{1.3}$$

where the Wigner functions [68] are denoted by $D_{\sigma} = D^{j}_{-m.\sigma}(\phi, \theta, 0)$. After separating the variables, we get four radial equations (let $\nu = j + 1/2$)

$$E f_3 - i\frac{d}{dr} f_3 - i\frac{\nu}{\sin r} f_4 - m f_1 = 0, \; E f_4 + i\frac{d}{dr} f_4 + i\frac{\nu}{\sin r} f_3 - m f_2 = 0,$$
$$E f_1 + i\frac{d}{dr} f_1 + i\frac{\nu}{\sin r} f_2 - m f_3 = 0, \; E f_2 - i\frac{d}{dr} f_2 - i\frac{\nu}{\sin r} f_1 - m f_4 = 0. \tag{1.4}$$

In the spherical tetrad, the space reflection operator is given by [50]

$$\hat{\Pi}_{sph} = \left| \begin{array}{cccc} 0 & 0 & 0 & -1 \\ 0 & 0 & -1 & 0 \\ 0 & -1 & 0 & 0 \\ -1 & 0 & 0 & 0 \end{array} \right| \otimes \hat{P} .$$

From the spectral equation $\hat{\Pi}_{sph}\, \Psi_{jm} = \Pi\, \Psi_{jm}$, we obtain

$$\Pi = \delta\, (-1)^{j+1}, \quad \delta = \pm 1, \quad f_4 = \delta\, f_1, \quad f_3 = \delta\, f_2, \tag{1.5}$$

which simplifies (1.4) to

$$\Big(\frac{d}{dr} + \frac{\nu}{r} \Big) f + (E + \delta m)\, g = 0, \quad \Big(\frac{d}{dr} - \frac{\nu}{r} \Big) g - (E - \delta m)\, f = 0, \tag{1.6}$$

where we use the new variables f and g given by

$$f = \frac{f_1 + f_2}{\sqrt{2}}, \quad g = \frac{f_1 - f_2}{i\sqrt{2}} .$$

For definiteness, let us consider the case when $\delta = 1$,

$$\Big(\frac{d}{dr} + \frac{\nu}{r} \Big) f + (E + m) g = 0, \quad \Big(\frac{d}{dr} - \frac{\nu}{r} \Big) g - (E - m) f = 0. \tag{1.7}$$

By performing the replacement $m \to -m$, we obtain the equations for the case $\delta = -1$.

The presence of the external Coulomb field is taken into account in (1.7) by the formal change $\epsilon \to \epsilon + e/r$. Thus, the quantum Coulomb problem for a Dirac particle is described by the following radial system

$$\Big(\frac{d}{dr}+\frac{\nu}{r}\Big)f+\Big(E+\frac{e}{r}+m\Big)g=0\,, \quad \Big(\frac{d}{dr}-\frac{\nu}{r}\Big)g-\Big(E+\frac{e}{r}-m\Big)f=0\,. \qquad (1.8)$$

Let us perform a linear transformation of the functions $f(r)$ and $g(r)$ (the coefficients of this transformation may depend on the radial variable; let its determinant obey the identity $a(r)b(r)-c(r)d(r)=1$)

$$\begin{aligned} f(r) &= aF(r)+cG(r)\,, \quad g(r)=dF(r)+bG(r)\,, \\ F(r) &= bf(r)-cg(r)\,, \quad G(r)=-df(r)+ag(r)\,. \end{aligned} \qquad (1.9)$$

Let us combine the equations (1.8) as follows: the first equation is multiplied by $+b$, the second by $-c$, and then sum the results; analogously, we add the first equation multiplied by $-d$ to the second multiplied by $+a$. Thus we arrive at

$$\Big[\frac{d}{dr}-b'a+c'd+\frac{\nu}{r}\,(ba+cd)+\Big(E+\frac{e}{r}+m\Big)\,bd+\Big(E+\frac{e}{r}-m\Big)\,ca\Big]F$$
$$=\Big[\,b'c-bc'-\frac{\nu}{r}2bc-\Big(E+\frac{e}{r}+m\Big)b^2-\Big(E+\frac{e}{r}-m\Big)c^2\Big]G,$$
$$\Big[\frac{d}{dr}+d'c-a'b-\frac{\nu}{r}\,(dc+ab)-\Big(E+\frac{e}{r}+m\Big)\,bd-\Big(E+\frac{e}{r}-m\Big)\,ca\Big]G$$
$$=\Big[-d'a+da'+\frac{\nu}{r}2ad+\Big(E+\frac{e}{r}+m\Big)d^2+\Big(E+\frac{e}{r}-m\Big)a^2\Big]F\,.$$

For simplicity, let us assume that the transformation (1.9) does not depend on r, and let it be orthogonal:

$$S=\begin{vmatrix} a & c \\ d & b \end{vmatrix}=\begin{vmatrix} \cos\gamma/2 & \sin\gamma/2 \\ -\sin\gamma/2 & \cos\gamma/2 \end{vmatrix}, \qquad (1.10)$$

which simplifies the equations (1.10) to

$$\begin{aligned} \Big(\frac{d}{dr}+\frac{\nu}{r}\cos\gamma-m\sin\gamma\Big)F &= \Big(-\frac{\nu}{r}\sin\gamma-\frac{e}{r}-E-m\cos\gamma\Big)G\,, \\ \Big(\frac{d}{dr}-\frac{\nu}{r}\cos\gamma+m\sin\gamma\Big)G &= \Big(-\frac{\nu}{r}\sin\gamma+\frac{e}{r}+E-m\cos\gamma\Big)F\,. \end{aligned} \qquad (1.11)$$

There exist four possibilities (only two of them are different in fact):

1)

$$-\frac{\nu}{r}\sin\gamma+\frac{e}{r}=0\,, \quad \sin\gamma=\frac{e}{\nu}\,, \quad \cos\gamma=\sqrt{1-e^2/\nu^2}\,,$$

$$\cos\frac{\gamma}{2}=\sqrt{\frac{\nu+\sqrt{\nu^2-e^2}}{2\nu}}\,, \quad \sin\frac{\gamma}{2}=\sqrt{\frac{\nu-\sqrt{\nu^2-e^2}}{2\nu}}\,;$$

1′)

$$
-\frac{\nu}{r}\sin\gamma-\frac{e}{r}=0,\quad \sin\gamma=-\frac{e}{\nu},\quad \cos\gamma=\sqrt{1-e^2/\nu^2},
$$
$$
\cos\frac{\gamma}{2}=\sqrt{\frac{\nu-\sqrt{\nu^2-e^2}}{2\nu}},\quad \sin\frac{\gamma}{2}=\sqrt{\frac{\nu+\sqrt{\nu^2-e^2}}{2\nu}}; \tag{1.12}
$$

2)

$$
E-m\cos\gamma=0,\quad \cos\gamma=+\frac{E}{m},\quad \sin\gamma=\sqrt{1-E^2/m^2},
$$
$$
\cos\frac{\gamma}{2}=\sqrt{\frac{m+E}{2m}},\qquad \sin\frac{\gamma}{2}=\sqrt{\frac{m-E}{2m}};
$$

2′)

$$
-E-m\cos\gamma=0,\quad \cos\gamma=-\frac{E}{m},\quad \sin\gamma=\sqrt{1-E^2/m^2},
$$
$$
\cos\frac{\gamma}{2}=\sqrt{\frac{m-E}{2m}},\quad \sin\frac{\gamma}{2}=\sqrt{\frac{m+E}{2m}}. \tag{1.13}
$$

First, consider the case 1). The equations (1.11) take the form

$$
\Big(\frac{d}{dr}+\frac{\nu}{r}\cos\gamma-m\sin\gamma\Big)F=\Big(-\frac{2e}{r}-E-m\cos\gamma\Big)G,
$$
$$
\Big(\frac{d}{dr}-\frac{\nu}{r}\cos\gamma+m\sin\gamma\Big)G=(E-m\cos\gamma)F. \tag{1.14}
$$

After eliminating the function F, we get a second order equation for G

$$
\Big(\frac{d^2}{dr^2}+E^2-m^2+\frac{\nu\cos\gamma-\nu^2\cos^2\gamma}{r^2}+\frac{2eE-2em\cos\gamma+2m\nu\sin\gamma\cos\gamma}{r}\Big)G=0. \tag{1.15}
$$

Having in mind the identity $\sin\gamma=e/\nu$, the last equation reduces to

$$
\Big(\frac{d^2}{dr^2}+E^2-m^2+\frac{\nu\cos\gamma-\nu^2\cos^2\gamma}{r^2}+\frac{2eE}{r}\Big)G=0. \tag{1.16}
$$

After changing the variable, $x=2\sqrt{m^2-E^2}\,r$, it reads

$$
\frac{d^2G}{dx^2}+\Big(-\frac{1}{4}-\frac{\nu\cos\gamma\,(\nu\cos\gamma-1)}{x^2}+\frac{eE}{\sqrt{m^2-E^2}\,x}\Big)G=0. \tag{1.17}
$$

With the use of the substitution $G(x)=x^A e^{Bx}\bar{G}(x)$, for $\bar{G}$ we get

$$
x\frac{d^2\bar{G}}{dx^2}+(2A+2Bx)\frac{d\bar{G}}{dx}
$$
$$
+\Big[(B^2-\frac{1}{4})x+\frac{A^2-A-\nu\cos\gamma\ (\nu\cos\gamma-1)}{x}+2AB+\frac{eE}{\sqrt{m^2-E^2}}\Big]\bar{G}=0. \tag{1.18}
$$

When

$$
A=+\nu\cos\gamma=\sqrt{\nu^2-e^2},\quad B=-\frac{1}{2}, \tag{1.19}
$$

this equation becomes simpler,

$$x\frac{d^2\bar{G}}{dx^2}+(2A-x)\frac{d\bar{G}}{dx}-\Big(A-\frac{eE}{\sqrt{m^2-E^2}}\Big)\bar{G}=0, \tag{1.20}$$

which is a confluent hypergeometric equation

$$x\,{}_1F_1''(x)+(c-x){}_1F_1'(x)-a\,{}_1F_1(x)=0\,, \quad a=A-\frac{eE}{\sqrt{m^2-E^2}}\,, \quad c=2a\,. \tag{1.21}$$

The known condition to have polynomial solutions, $a=-n$ $(n=0,1,2,...)$, gives the energy quantization rule

$$E=\frac{m}{\sqrt{1+e^2/(n+\sqrt{\nu^2-e^2})^2}}\,. \tag{1.22}$$

In turn, from (1.14), there follows a second-order equation for F:

$$\Big[\frac{d^2}{dr^2}+\Big(\frac{1}{r}-\frac{1}{r-R}\Big)\frac{d}{dr}+E^2-m^2+\frac{2eE}{r}+\frac{e^2-\nu^2}{r^2}+\frac{mR\,\sin\gamma-\nu\cos\gamma}{r\,(r-R)}\Big]F=0\,, \tag{1.23}$$

where

$$R=-\frac{2e}{E+m\cos\gamma}\,. \tag{1.24}$$

After changing the variable $z=r/R$, this reads

$$\frac{d^2F}{dz^2}+\Big(\frac{1}{z}-\frac{1}{z-1}\Big)\frac{dF}{dz}+\Big((E^2-m^2)R^2-\frac{\nu^2-e^2}{z^2}$$
$$+\frac{-\nu\cos\gamma+mR\,\sin\gamma}{z-1}+\frac{2eRE-mR\sin\gamma+\nu\cos\gamma}{z}\Big)F=0\,. \tag{1.25}$$

Let us search for solutions in the form $F=z^A\,e^{Bz}\,\bar{F}(z)$; the function $\bar{F}(z)$ satisfies the equation

$$\frac{d^2\bar{F}}{dz^2}+\Big(\frac{2A+1}{z}+2B-\frac{1}{z-1}\Big)\frac{d\bar{F}}{dz}$$
$$+\Big[B^2+(E^2-m^2)R^2+\frac{A^2-\nu^2+e^2}{z^2}-\frac{A+B+\nu\cos\gamma-mR\,\sin\gamma}{z-1}$$
$$+\frac{A+B+2AB+R(2eE-m\sin\gamma)+\nu\cos\gamma}{z}\Big]\bar{F}=0\,.$$

Using the underlined values for A,B:

$$A=\underline{+\sqrt{\nu^2-e^2}}\,,\ -\sqrt{\nu^2-e^2}\,,\quad B=+\sqrt{m^2-E^2}\,R\,,\underline{-\sqrt{m^2-E^2}\,R}\,, \tag{1.26}$$

the above equation becomes simpler:

$$\frac{d^2\bar{F}}{dz^2}+\Big(2B+\frac{2A+1}{z}-\frac{1}{z-1}\Big)\frac{d\bar{F}}{dz}$$
$$+\Big(\frac{A+B+2AB+2eRE-mR\sin\gamma+\nu\cos\gamma}{z}-\frac{A+B+\nu\cos\gamma-mR\,\sin\gamma}{z-1}\Big)\bar{F}=0\,. \tag{1.27}$$

This can be easily recognized as a confluent Heun equation (with two regular singularities $z=0,1$ and an irregular singularity of the rank 2 at the point $z=\infty$):

$$\frac{d^2H}{dz^2}+\Big(-t+\frac{c}{z}+\frac{d}{z-1}\Big)\frac{dH}{dz}-\frac{\lambda}{z}+\frac{\lambda-at}{z-1}H=0, \tag{1.28}$$

with the five parameters given by

$$\begin{gathered} t=-2B, \quad c=2A+1, \quad d=-1, \quad a=A+\frac{eER}{B}, \\ -\lambda=A+B+2AB+2eRE-mR\sin\gamma+\nu\cos\gamma, \end{gathered} \tag{1.29}$$

with

$$A=+\sqrt{\nu^2-e^2}, \; B=-\sqrt{m^2-E^2}\,R, \quad R=-\frac{2e}{E+m\cos\gamma}.$$

We consider the explicit form of a:

$$a=\sqrt{\nu^2-e^2}-\frac{eE}{\sqrt{m^2-E^2}}, \tag{1.30}$$

and use the known condition within the case of the so-called transcendental Heun functions, one of the two necessary conditions for getting polynomials. This leads to the energy quantization rule

$$a=-n, \quad n=0,1,2,\dots, \quad E=\frac{m}{\sqrt{1+e^2/(n+\sqrt{\nu^2-e^2})^2}}, \tag{1.31}$$

which coincides with the known formula for energy levels.

It should be emphasized that, as follows from (1.14), the function F (which is constructed in terms of confluent Heun functions) can be related to the function G (which is determined in terms of confluent hypergeometric functions) by means of the following differential operators:

$$G=\Big(-\frac{2e}{r}-E-m\cos\gamma\Big)^{-1}\Big(\frac{d}{dr}+\frac{\nu}{r}\cos\gamma-m\sin\gamma\Big)F, \tag{1.32}$$

$$F=\frac{1}{E-m\cos\gamma}\Big(\frac{d}{dr}-\frac{\nu}{r}\cos\gamma+m\sin\gamma\Big)G. \tag{1.33}$$

Let us consider the case 2) from (1.13). The equations (1.11) take the form

$$\begin{gathered} \Big(\frac{d}{dr}+\frac{\nu}{r}\cos\gamma-m\sin\gamma\Big)F=\Big(-\frac{\nu\sin\gamma+e}{r}-2m\cos\gamma\Big)G, \\ \Big(\frac{d}{dr}-\frac{\nu}{r}\cos\gamma+m\sin\gamma\Big)G=\frac{e-\nu\sin\gamma}{r}F. \end{gathered} \tag{1.34}$$

They lead to a second-order equation for $G(r)$ (using the identity $\cos\gamma=E/m$):

$$\Big(\frac{d^2}{dr^2}+\frac{1}{r}\frac{d}{dr}+E^2-m^2+\frac{e^2-\nu^2}{r^2}+\frac{2eE}{r}+\frac{\sqrt{m^2-E^2}}{r}\Big)G=0. \tag{1.35}$$

By making the change of the variable $x = 2\sqrt{m^2-E^2}\,r$, we get

$$\frac{d^2G}{dx^2} + \frac{1}{x}\frac{dG}{dx} + \Big(-\frac{1}{4} - \frac{\nu^2-e^2}{x^2} + \frac{1}{2}\frac{m^2-E^2+2Ee\sqrt{m^2-E^2}}{(m^2-E^2)x}\Big)G = 0 . \tag{1.36}$$

Let $G(x) = x^A e^{Bx}\bar{G}(x)$; the function $\bar{G}$ satisfies the equation

$$x\frac{d^2\bar{G}}{dx^2} + (2A+1+2Bx)\frac{d\bar{G}}{dx}$$
$$+\Big[\Big(B^2-\frac{1}{4}\Big)x + \frac{A^2-\nu^2+e^2}{x} + 2AB + B + \frac{1}{2}\frac{m^2-E^2+2Ee\sqrt{m^2-E^2}}{m^2-E^2}\Big]\bar{G} = 0 . \tag{1.37}$$

When $A = \sqrt{\nu^2-e^2}$, $B = -\frac{1}{2}$, we get

$$x\frac{d^2\varphi}{dx^2} + (2A+1-x)\frac{d\varphi}{dx} - \Big(A - \frac{Ee}{\sqrt{m^2-E^2}}\Big)\varphi = 0 , \tag{1.38}$$

which is a confluent hypergeometric equation,

$$x\,{}_1F_1'' + (c-x)\,{}_1F_1' - a\,{}_1F_1 = 0 , \quad a = A - \frac{Ee}{\sqrt{m^2-E^2}} , \quad x = 2A+1 . \tag{1.39}$$

One has polynomial solutions if $a = -n$, $n = 0, 1, 2, \ldots$; this provides the energy spectrum

$$E = \frac{m}{\sqrt{1+e^2/(n+\sqrt{\nu^2-e^2})^2}} . \tag{1.40}$$

In turn, from (1.34), the second-order equation for $F(r)$ follows:

$$\Big[\frac{d^2}{dr^2} + \Big(\frac{1}{r} - \frac{1}{r-D}\Big)\frac{d}{dr} + \frac{m\sin\gamma}{r-D} - \frac{\nu\cos\gamma}{D}\Big(\frac{1}{r-D} - \frac{1}{r}\Big)$$
$$+E^2 - m^2 + \frac{e^2-\nu^2}{r^2} + \frac{2eE - m\sin\gamma}{r}\Big]F = 0 , \tag{1.41}$$

where

$$D = -\frac{e+\nu\sin\gamma}{2m\cos\gamma} . \tag{1.42}$$

With the use of the variable $z = r/D$, this looks simpler,

$$\frac{d^2F}{dz^2} + \Big(\frac{1}{z} - \frac{1}{z-1}\Big)\frac{dF}{dz} + \Big[(E^2-m^2)D^2 - \frac{\nu^2-e^2}{z^2}$$
$$+\frac{-\nu\cos\gamma + mD\sin\gamma}{z-1} + \frac{D(2eE - m\sin\gamma) + \nu\cos\gamma}{z}\Big]F = 0 . \tag{1.43}$$

Let $F = y^A e^{By}\bar{F}(z)$; the function $\bar{F}(z)$ satisfies

$$\frac{d^2\bar{F}}{dz^2} + \Big(\frac{2A+1}{z} + 2B - \frac{1}{z-1}\Big)\frac{d\bar{F}}{dz}$$
$$+\Big[B^2 + (E^2-m^2)D^2 + \frac{A^2-\nu^2+e^2}{z^2} - \frac{A+B+\nu\cos\gamma - mD\sin\gamma}{z-1}$$
$$+\frac{A+B+2AB+D(2eE-m\sin\gamma)+\nu\cos\gamma}{z}\Big]\bar{F} = 0 . \tag{1.44}$$

Using underlined values, for A and B given by

$$A = \underline{+\sqrt{\nu^2 - e^2}}\,, \quad -\sqrt{\nu^2 - e^2}\,, \quad B = +\sqrt{m^2 - E^2}\,D\,, \quad \underline{-\sqrt{m^2 - E^2}\,D}\,, \tag{1.45}$$

we infer

$$\frac{d^2\bar{F}}{dz^2} + \Big(2B + \frac{2A+1}{y} - \frac{1}{y-1}\Big)\frac{d\bar{F}}{dz}$$
$$+\Big(\frac{A+B+2AB+D(2eE - m\sin\gamma) + \nu\cos\gamma}{z} - \frac{A+B+\nu\cos\gamma - mD\sin\gamma}{z-1}\Big)\bar{F} = 0, \tag{1.46}$$

which is a confluent Heun equation for H,

$$\frac{d^2H}{dz^2} + \Big(-t + \frac{c}{z} + \frac{d}{z-1}\Big)\frac{dH}{dz} - \frac{\lambda}{z} + \frac{\lambda - at}{z-1}\,H = 0,$$

with the five parameters given by

$$t = -2B\,, \quad c = 2A+1\,, \quad d = -1\,, \quad a = A + \frac{DeE}{B}\,,$$
$$-\lambda = A + B + 2AB + D(2eE - m\sin\gamma) + \nu\cos\gamma\,. \tag{1.47}$$

We use the following explicit form for a:

$$a = \sqrt{\nu^2 - e^2} - \frac{eE}{\sqrt{m^2 - E^2}}\,.$$

By using the above condition for transcendental Heun functions, we produce the energy quantization rule

$$a = -n\,, \quad n = 0,1,2,\ldots, \qquad E = \frac{m}{\sqrt{1 + e^2/(n + \sqrt{\nu^2 - e^2})^2}}\,,$$

which coincides with the known exact result.

It should be emphasized that two solutions in terms of Heun functions are similar, but still differ:

$$1) \qquad t = -2B\,, \quad c = 2A+1\,, \quad d = -1\,, \quad a = A + \frac{eER}{B}\,,$$
$$-\lambda = A + B + 2AB + 2eRE - mR\sin\gamma + \nu\cos\gamma\,;$$
$$2) \qquad t = -2B, \quad c = 2A+1\,, \quad d = -1\,, \quad a = A + \frac{eED}{B}\,,$$
$$-\lambda = A + B + 2AB + 2eDE - mD\sin\gamma + \nu\cos\gamma\,,$$

where

$$1) \quad R = -\frac{2e}{E + m\cos A}\,, \quad \sin\gamma = \frac{e}{\nu}\,, \quad \cos\gamma = \sqrt{1 - \frac{e^2}{\nu^2}}\,;$$

$$2) \quad D = -\frac{e + \nu\sin A}{2E}\,, \quad \cos\gamma = \frac{E}{m}\,, \quad \sin\gamma = \sqrt{1 - \frac{E^2}{m^2}}\,.$$

1.2 Standard Treatment of the Coulomb Problem

It should be noted that both proposed treatments of the Coulomb problem for the Dirac equation differ from the well known one – see [70]. Let us recall this standard approach. To this end, in the radial system (1.8)

$$\left(\frac{d}{dr}+\frac{\nu}{r}\right)f+\left(E+\frac{e}{r}+m\right)g=0, \quad \left(\frac{d}{dr}-\frac{\nu}{r}\right)g-\left(E+\frac{e}{r}-m\right)f=0, \tag{1.48}$$

one should introduce the new functions

$$f=\sqrt{m+E}\,(F_1+F_2), \quad g=\sqrt{m-E}\,(F_1-F_2); \tag{1.49}$$

this infers

$$r\left(\frac{d}{dr}+\frac{\nu}{r}\right)(F_1+F_2)+r\sqrt{m^2-E^2}(F_1-F_2)+e\frac{\sqrt{m-E}}{\sqrt{m+E}}\,(F_1-F_2)=0,$$
$$r\left(\frac{d}{dr}-\frac{\nu}{r}\right)(F_1-F_2)+r\sqrt{m^2-E^2}(F_1+F_2)-e\frac{\sqrt{m+E}}{\sqrt{m-E}}\,(F_1+F_2)=0.$$

By summing and subtracting the equations, we obtain

$$r\frac{d}{dr}F_1+\nu F_2+r\sqrt{m^2-E^2}F_1-\frac{eE}{\sqrt{m^2-E^2}}F_1-\frac{em}{\sqrt{m^2-E^2}}F_2=0,$$

$$r\frac{d}{dr}F_2+\nu F_1-r\sqrt{m^2-E^2}F_2+\frac{em}{\sqrt{m^2-E^2}}F_1+\frac{eE}{\sqrt{m^2-E^2}}F_2=0.$$

In the variables

$$\lambda=\sqrt{m^2-E^2}, \quad x=\lambda r, \quad \frac{em}{\lambda}=\mu, \quad \frac{eE}{\lambda}=\epsilon, \tag{1.50}$$

these equations read shorter

$$\left(x\frac{d}{dx}+x-\epsilon\right)F_1+\left(\nu-\mu\right)F_2=0, \quad \left(x\frac{d}{dx}-x+\epsilon\right)F_2+\left(\nu+\mu\right)F_1=0. \tag{1.51}$$

The system (1.51) can be solved in terms of hypergeometric functions [70]. For this, we translate (1.51) into the new variable $y=2x$:

$$\left(y\frac{d}{dy}+\frac{y}{2}-\epsilon\right)F_1+\left(\nu-\mu\right)F_2=0, \quad \left(y\frac{d}{dy}-\frac{y}{2}+\epsilon\right)F_2+\left(\nu+\mu\right)F_1=0, \tag{1.52}$$

whence there follow the following second-order differential equations for F_1 and F_2:

$$\left(y\frac{d^2}{dy^2}+\frac{d}{dy}+\epsilon+\frac{1}{2}-\frac{y}{4}+\frac{\mu^2-\nu^2-\epsilon^2}{y}\right)F_1=0, \tag{1.53}$$
$$\left(y\frac{d^2}{dy^2}+\frac{d}{dy}+\epsilon-\frac{1}{2}-\frac{y}{4}+\frac{\mu^2-\nu^2-\epsilon^2}{y}\right)F_2=0. \tag{1.54}$$

Let us study the equation (1.53). The substitution $F_1 = y^A e^{By} f_1$, leads to

$$\Big[y\frac{d^2}{dy^2} + \Big(2A+1+2By\Big)\frac{d}{dy}$$
$$+\epsilon + \frac{1}{2} + B\Big(1+2A\Big) - y\Big(\frac{1}{4} - B^2\Big) + \frac{A^2+\mu^2-\nu^2-\epsilon^2}{y}\Big] f_1 = 0 . \qquad (1.55)$$

The choice $A = +\sqrt{\epsilon^2 - \mu^2 + \nu^2}$, $B = -\frac{1}{2}$ turns (1.55) to the simpler confluent hypergeometric type

$$\Big[y\frac{d^2}{dy^2} + (2A+1-y)\frac{d}{dy} + \epsilon - A\Big] f_1 = 0 \qquad (1.56)$$

with the parameters $\alpha_1 = A - \epsilon$, $\gamma_1 = 2A+1$. By imposing the polynomial condition $\alpha_1 = -n_1$, we obtain the quantization rule for ϵ:

$$-n_1 = -\epsilon + \sqrt{\epsilon^2 - \mu^2 + \nu^2} . \qquad (1.57)$$

Taking into account the above definitions

$$\lambda = \sqrt{m^2 - E^2} , \qquad \frac{em}{\lambda} = \mu , \qquad \frac{eE}{\lambda} = \epsilon ,$$

we get

$$\sqrt{\epsilon^2 - \mu^2 + \nu^2} = \sqrt{\nu^2 - e^2} , \qquad A = +\sqrt{\nu^2 - e^2} , \qquad (1.58)$$

and further, from (1.58), we derive the formula for energy levels

$$\frac{eE}{\sqrt{m^2 - E^2}} = \sqrt{\nu^2 - e^2} + n_2 \equiv N \quad \Longrightarrow \quad E = \frac{m}{\sqrt{1 + e^2/N^2}} . \qquad (1.59)$$

Let us consider the second equation from (1.54). With the use of the substitution $F_2 = y^a e^{by} f_2$:

$$\Big[y\frac{d^2}{dy^2} + (2a+1+2by)\frac{d}{dy} + \epsilon - \frac{1}{2} + b(1+2a)$$
$$-y(\frac{1}{4} - b^2) + \frac{a^2+\mu^2-\nu^2-\epsilon^2}{y}\Big] f_2 = 0 ; \qquad (1.60)$$

at $a = \pm\sqrt{\epsilon^2 - \mu^2 + \nu^2}$, $b = -\frac{1}{2}$, we obtain an equation of confluent hypergeometric type

$$\Big[y\frac{d^2}{dy^2} + (2a+1-y)\frac{d}{dy} + \epsilon - a - 1\Big] f_2 = 0 \qquad (1.61)$$

with the parameters $\alpha_2 = a + 1 - \epsilon$, $\gamma_2 = 2a+1$. By imposing the polynomial restriction $\alpha_2 = -n_2$, we get the quantization rule for ϵ:

$$-n_2 = -\epsilon + 1 + \sqrt{\epsilon^2 - \mu^2 + \nu^2} ; \qquad (1.62)$$

whence we obtain the formula for energy levels

$$\frac{eE}{\sqrt{m^2-E^2}} = 1+\sqrt{\nu^2-e^2}+n_2 \equiv N \quad \Longrightarrow \quad E = \frac{m}{\sqrt{1+e^2/N^2}}. \tag{1.63}$$

Let us find the relative coefficient between the two functions:

$$\Big(y\frac{d}{dy}+\frac{y}{2}-\epsilon\Big)F_1+(\nu-\mu)F_2=0, \quad \Big(y\frac{d}{dy}-\frac{y}{2}+\epsilon\Big)F_2+(\nu+\mu)F_1=0, \tag{1.64}$$

where

$$\begin{gathered} F_1 = C_1 y^A e^{-y/2} F(-n_1,\gamma,y), \quad F_2 = C_2 y^A e^{-y/2} F(-n_2,\gamma,y), \\ A = +\sqrt{\epsilon^2-\mu^2+\nu^2}, \quad \gamma = 2A = 1, \quad -n_2 = -n_1+1. \end{gathered} \tag{1.65}$$

We plug in the expressions for F_1, F_2 into the first-order equation (1.64),

$$yC_1\frac{d}{dy}F(-n_1,\gamma,y)+C_1(A-\epsilon)F(-n_1,\gamma,y)+(\nu-\mu)C_2F(-n_1+1,\gamma,y)=0.$$

Then, the differentiation rule of the confluent hypergeometric function

$$\frac{d}{dy}F(-n_1,\gamma,y) = -\frac{n_1}{y}F(-n_1+1,\gamma,y)+\frac{n_1}{y}F(-n_1,\gamma,y),$$

leads to

$$\begin{gathered} -C_1 n_1 F(-n_1+1,\gamma,y)+C_1 n_1 F(-n_1,\gamma,y)+C_1(A-\epsilon)F(-n_1,\gamma,y) \\ +(\nu-\mu)C_2F(-n_1+1,\gamma,y)=0. \end{gathered}$$

Taking into account that $-n_1 = A-\epsilon$, we obtain

$$\frac{C_1}{C_2} = \frac{\nu-\mu}{n_1} = -\frac{\nu-\mu}{A-\epsilon}. \tag{1.66}$$

Now, we substitute the expressions for the functions F_1, F_2 into the first-order equation (1.65)

$$\begin{gathered} yC_2\frac{d}{dy}F(-n_1+1,\gamma,y)+C_2(A+\epsilon)F(-n_1+1,\gamma,y) \\ -yC_2F(-n_1+1,\gamma,y)+(\nu+\mu)C_1F(-n_1,\gamma,y)=0. \end{gathered}$$

We apply the rule of differentiation of the confluent hypergeometric functions,

$$\frac{d}{dy}F(-n_1+1,\gamma,y) = \Big(\frac{-n_1+1}{\gamma}-1\Big)F(-n_1+1,\gamma+1,y)+F(-n_1+1,\gamma,y),$$

and use the formula for contiguous confluent hypergeometric functions

$$yF(-n_1+1,\gamma+1,y) = \gamma F(-n_1+1,\gamma,y)-\gamma F(-n_1,\gamma,y),$$

and we derive

$$C_2\,(-n_1+1-\gamma)F(-n_1+1,\gamma,y)+C_2\,(A+\epsilon)\,F(-n_1+1,\gamma,y)$$
$$-C_2\,(-n_1+1-\gamma)F(-n_1,\gamma,y)+(\nu+\mu)C_1\,F(-n_1,\gamma,y)=0\,.$$

Allowing for the identity $-n_1=A-\epsilon$, we get

$$\frac{C_1}{C_2}=\frac{-n_1-2A}{\nu+\mu}=-\frac{A+\epsilon}{\nu+\mu}\,. \tag{1.67}$$

It can be easily checked that two expressions for relative coefficients, (1.66) and (1.67), coincide. Indeed, we have

$$\frac{C_1}{C_2}=-\frac{\nu-\mu}{A-\epsilon}=-\frac{A+\epsilon}{\nu+\mu}\quad\Longrightarrow\quad \nu^2-\mu^2=A^2-\epsilon^2\quad\Longleftrightarrow\quad e^2\equiv e^2\,. \tag{1.68}$$

1.3 Solutions Constructed Completely in Terms of Heun Functions

The idea to construct spectra in the framework of the Heun equation theory seems to be a very promising one. We shall consider, in this line of arguments, the known problem of a spin 1/2 particle in the presence of external Coulomb field. To this end, let us turn again to the equations in the presence of Coulomb potential (when $\delta=+1$)

$$\Big(\frac{d}{dr}+\frac{\nu}{r}\Big)f+\Big(E+\frac{e}{r}+m\Big)g=0,\quad \Big(\frac{d}{dr}-\frac{\nu}{r}\Big)g-\Big(E+\frac{e}{r}-m\Big)f=0\,. \tag{1.69}$$

After eliminating the function g, one gets

$$\frac{d^2f}{dr^2}+\frac{e}{r(Er+e+mr)}\frac{df}{dr}+\Big[\frac{e(e^2-\nu^2)}{r^2(Er+e+mr)}+\frac{E(3e^2-\nu^2)-\nu(m+E)+m(e^2-\nu^2)}{r(Er+e+mr)}$$
$$+\frac{e(E+m)\,(3E-m)}{Er+e+mr}+\frac{r(E-m)(E+m)^2}{Er+e+mr}\Big]f=0\,. \tag{1.70}$$

After changing the variable

$$z=-\frac{(E+m)\,r}{e}\,, \tag{1.71}$$

equation (1.70) takes the form

$$z\frac{d^2f}{dz^2}-\frac{1}{z-1}\frac{df}{dz}+\Big[\frac{e^2(Ez-mz-2E)}{E+m}+\frac{e^2-\nu^2}{z}-\frac{\nu}{z-1}\Big]f=0\,. \tag{1.72}$$

By separating the two factors $f(z)=x^Ae^{Cx}F(z)$, one derives for $F(z)$

$$\frac{d^2F}{dz^2}+\Big(2C+\frac{2A+1}{z}-\frac{1}{z-1}\Big)\frac{dF}{dz}+\Big[C^2+\frac{e^2(E-m)}{E+m}$$
$$+\frac{A^2+e^2-\nu^2}{z^2}+\frac{A+C+2AC-2Ee^2/(E+m)+\nu}{z}-\frac{A+C+\nu}{z-1}\Big]F=0\,. \tag{1.73}$$

When A and C are taken as (the bound states being of primary interest)

$$C^2+\frac{e^2(E-m)}{E+m}=0 \quad \Rightarrow \quad C=+e\sqrt{\frac{m-E}{m+E}}\,, \tag{1.74}$$

$$A^2+e^2-\nu^2=0 \quad \Rightarrow \quad A=+\sqrt{\nu^2-e^2}\,, \tag{1.75}$$

equation (1.73) becomes simpler, namely

$$\frac{d^2F}{dz^2}+\left(2C+\frac{2A+1}{z}-\frac{1}{z-1}\right)\frac{dF}{dz}$$
$$+\left[\frac{A+C+\nu+2AC-2Ee^2/(E+m)}{z}-\frac{A+C+\nu}{z-1}\right]F=0\,, \tag{1.76}$$

which is the confluent Heun equation

$$\frac{d^2H}{dz^2}+\left(-t+\frac{c}{z}+\frac{d}{z-1}\right)\frac{dH}{dz}-\frac{\lambda}{z}+\frac{\lambda-at}{z-1}H=0\,,$$

with the parameters determined by

$$t=-2C\,,\quad c=2A+1\,,\quad d=-1\,,\quad a=A-\frac{1}{C}\frac{e^2E}{E+m}\,,$$
$$-\lambda=A+C+\nu+2AC-2Ee^2/(E+m)\,. \tag{1.77}$$

We especially need an explicit form of a:

$$a=\sqrt{\nu^2-e^2}-\frac{eE}{\sqrt{m^2-E^2}}\,.$$

Having used the known transcendency condition, we produce the energy quantization rule

$$a=-n\,,\ n=0,1,2,\ldots\,,\quad E=\frac{m}{\sqrt{1+e^2/(n+\sqrt{\nu^2-e^2})^2}}\,,$$

which coincides with the known exact result.

Chapter 2

Spin 1/2 Particle in 2D Spaces of Constant Curvature, in Presence of Magnetic Field

2.1 Cylindric and Conformal Coordinates in Lobachevsky Plane H_2

In [74, 75], when describing magnetic field in hyperbolic plane, was used the coordinate system referring to the so called Poincaré half-plane. In [81], a similar analysis in cylindric coordinates for 3-dimensional Lobachevsky and Riemann models, was performed. Let us relate these two descriptions.

In the paper [94], it was defined the following coordinate system in the Lobachevsky plane

$$u_1 = \sinh\rho_1 + \frac{1}{2}\rho_2^2 e^{-\rho_1}, \qquad u_0 = \cosh\rho_1 + \frac{1}{2}\rho_2^2 e^{-\rho_1},$$
$$u_2 = \rho_2 e^{-\rho_1}, u_1^2 - u_2^2 - u_0^2 = -1, \; dl^2 = d\rho_1^2 + e^{-2\rho_1}\, d\rho_2^2. \tag{2.1}$$

By changing the variables, we arrive at the quasi-Cartesian coordinates $(x,\ y)$:

$$e^{+\rho_1} = y, \qquad \rho_2 = x, \qquad dl^2 = \frac{dx^2 + dy^2}{y^2}. \tag{2.2}$$

The coordinates (x, y) and u_i relate to each other by the formulas

$$u_1 = \frac{y^2 - 1 + x^2}{2y}, \; u_0 = \frac{y^2 + 1 + x^2}{2y}, \; u_2 = \frac{x}{y}. \tag{2.3}$$

Taking into account the formulas which define the cylindric coordinates [94]

$$u_1 = \sinh r \cos\phi, \quad u_2 = \sinh r \sin\phi, \quad u_0 = \cosh r,$$

we readily obtain the relations between (x,y) and (r,ϕ):

$$u_1 = \frac{y^2-1+x^2}{2y} = \sinh r \cos\phi\,, \quad u_2 = \frac{x}{y} = \sinh r \sin\phi\,,$$
$$u_0 = \frac{y^2+1+x^2}{2y} = \cosh r\,, \quad \tan\phi = \frac{2x}{x^2+y^2-1}\,. \tag{2.4}$$

Taking this in mind, let us express the potential of the magnetic field in the Lobachevsky space $A_\phi = -B(\cosh r - 1)$ in quasi-Cartesian coordinates (x,y):

$$A_x = \frac{\partial\phi}{\partial x}A_\phi\,, \quad A_y = \frac{\partial\phi}{\partial y}A_\phi\,;$$

this leads to

$$A_x = -B\frac{y^2-x^2-1}{x^2+(y+1)^2}\frac{1}{y}\,, \quad A_y = +B\frac{2x}{x^2+(y+1)^2}\,. \tag{2.5}$$

To this potential there corresponds the following electromagnetic tensor

$$F_{xy} = \frac{\partial}{\partial x}\frac{\partial\phi}{\partial y}A_\phi - \frac{\partial}{\partial y}\frac{\partial\phi}{\partial x}A_\phi = \frac{B}{y^2}\,. \tag{2.6}$$

Evidently, this tensor may be produced by a simpler potential,

$$\tilde{A}_x = \frac{B}{y} \quad \Longrightarrow \quad F_{xy} = -\partial_y A_x = \frac{B}{y^2}\,. \tag{2.7}$$

There must exist a corresponding gauge transformation

$$A_x + \frac{\partial}{\partial x}\Lambda = \tilde{A}_x = \frac{B}{y}\,, \qquad A_y + \frac{\partial}{\partial y}\Lambda = \tilde{A}_y = 0\,. \tag{2.8}$$

In order to find it, let us write down the equations which determine the gauge function,

$$\frac{\partial}{\partial y}\Lambda = -B\frac{2x}{x^2+(y+1)^2}\,, \qquad \frac{\partial}{\partial x}\Lambda = \frac{B}{y} + B\frac{y^2-x^2-1}{x^2+(y+1)^2}\frac{1}{y}\,.$$

From the first relation we deduce

$$\Lambda = f(x) - 2Bx\int\frac{dy}{x^2+(y+1)^2} = f(x) - 2B\arctan\frac{y+1}{x}\,. \tag{2.9}$$

Substituting this expression into the second equation, we get the needed result

$$\frac{df(x)}{dx} = 0\,, \qquad \Lambda = \text{const} - 2B\arctan\frac{y+1}{x}\,. \tag{2.10}$$

2.2 Landau Problem for a Scalar Particle in the Plane H_2

First we consider the problem in cylindric coordinates,

$$dS^2 = dt^2 - dr^2 - \sinh^2 r d\phi^2 . \tag{2.11}$$

The Schrödinger equation in presence of the magnetic field has the form

$$i\frac{\partial}{\partial t}\Psi = \frac{1}{2M}\Big[-\frac{1}{\sinh r}\frac{\partial}{\partial r}\sinh r\frac{\partial}{\partial r} + \frac{1}{\sinh^2 r}(i\frac{\partial}{\partial\phi} + eA_\phi)^2\Big]\Psi . \tag{2.12}$$

By applying the substitution $\Psi = e^{-i\epsilon t}e^{im\phi}R(r)$, we derive the equation in the variable r:

$$\frac{1}{\sinh r}\frac{d}{dr}\sinh r\frac{dR}{dr} - \frac{[-m - eB(\cosh r - 1)]^2}{\sinh^2 r}R + 2M\epsilon R = 0 . \tag{2.13}$$

This can be solved in terms of hypergeometric functions, and the corresponding energy spectrum is

$$-\Big(\frac{1}{2} + \sqrt{\frac{1}{4} + B^2 - 2M\epsilon} + n\Big) = \mid \frac{m}{2} \mid - \mid \frac{m}{2} - B \mid < 0 . \tag{2.14}$$

Now, let us consider the problem with the use of quasi-Cartesian coordinates (x,y). The Schrödinger equation in presence of the magnetic field

$$dS^2 = dt^2 - \frac{dx^2 + dy^2}{y^2}, \quad A_x = +\frac{B}{y}, \quad F_{xy} = \frac{B}{y^2},$$

takes on the form

$$i\partial_t \Psi = \frac{1}{2M}\Big[(i\frac{\partial}{\partial x} + eA_x)\,y^2\,(i\frac{\partial}{\partial x} + eA_x) - y^2\frac{\partial^2}{\partial y^2}\Big]\Psi$$

or,

$$i\partial_t \Psi = \frac{1}{2M}\Big[y^2\,(i\frac{\partial}{\partial x} + \frac{eB}{y})^2 - y^2\frac{\partial^2}{\partial y^2}\Big]\Psi . \tag{2.15}$$

We apply the substitution for the wave function $\Psi = e^{-i\epsilon t}e^{ikx}f(y)$, so producing the following equation in the y variable

$$y^2\frac{d^2}{dy^2}f + 2M\epsilon f - y^2\,(k - \frac{eB}{y})^2 f = 0 . \tag{2.16}$$

Let us introduce the new variable $z = 2ky$. It substantially depends on the sign of parameter k; first let us consider the case of positive k; for shortness let eB be noted as B. We search for solutions in the form $f(z) = z^A e^{Cz}F(z)$, and infer

$$z\frac{d^2F}{dz^2} + (2A + 2Cz)\frac{dF}{dz} + \Big[(C^2 - \frac{1}{4})z + 2AC + B + \frac{A(A-1) - B^2 + 2M\epsilon}{z}\Big]F = 0 .$$

We fix the values for A and C (to get solutions regular in $z=0$ we assume $A>0$)

$$A = \Big(\frac{1}{2} \pm \sqrt{\frac{1}{4} + B^2 - 2M\epsilon}\Big) > 0\,, \qquad C = -\frac{1}{2}\,, \tag{2.17}$$

and the above equation becomes simpler,

$$z\frac{d^2F}{dz^2} + (2A - z)\frac{dF}{dz} - (A-B)F = 0\,.$$

It is of hypergeometric type (for definiteness we assume $B>0$)

$$\alpha = A - B\,, \quad \gamma = 2A\,, \qquad F = \Phi(\alpha, \gamma, z)\,. \tag{2.18}$$

We need to clarify in detail the parameterizing of the hyperbolic plane by the coordinates (x,y). Taking in mind the relations

$$u_1 = \frac{x^2+y^2-1}{2y}\,, \quad u_0 = \frac{x^2+y^2+1}{2y}\,, \quad u_2 = \frac{x}{y}\,,$$

we conclude that $y \geq 0$. Besides, we notice that

$$x=0, y=+1 \qquad \Longrightarrow \qquad u_0 = +1\,, u_1 = 0, u_2 = 0\,, \tag{2.19}$$

and

$$q_1 = \frac{u_1}{u_0} = \frac{x^2+y^2-1}{x^2+y^2+1}\,, \qquad q_2 = \frac{u_2}{u_0} = \frac{2x}{x^2+y^2+1}\,, \tag{2.20}$$

which may be illustrated in Fig. 2.1 (the signs of q_1, q_2 are shown in the figure).

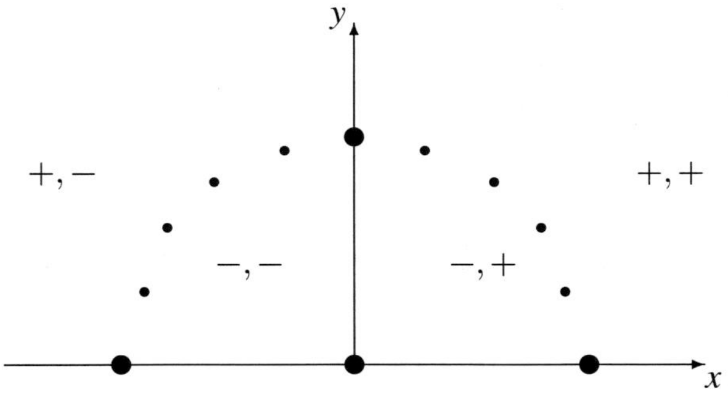

Fig. 2.1. Poincaré half-plane

In the half-plane $x \in (-\infty, +\infty), y \in [0, +\infty)$, the semi-circle $x^2+y^2=1, y>0$ is represented by the line $q_1 = 0, q_2 = x$. Similarly, the semi-axis $(x=0,y)$ is represented by the line $(q_1, q_2 = 0)$. The relations converse to (2.20) have the form

$$x = \frac{q_2}{1-q_1} \in (-\infty, +\infty)\,, \qquad y = +\frac{\sqrt{1-q_1^2-q_2^2}}{1-q_1} \in (0, +\infty)\,. \tag{2.21}$$

In particular, the boundary $x = 0, y$ determines the line

$$q_2 = 0,\ q_1 = \frac{y^2 - 1}{y^2 + 1} \Longrightarrow \begin{cases} (q_1 \to -1,\ q_2 = 0),y \to 0, \\ (q_1 \to +1, q_2 = 0),\ y \to +\infty. \end{cases} \tag{2.22}$$

Therefore, solutions which are finite at $q_1 \to \pm 1, q_2 = 0$ may have the structure (recall that $k > 0$):

$$f(z) = z^A e^{-z/2} \Phi(\alpha, \gamma, z),\ A > 0,\ z \in (0, +\infty),\ \Phi(\alpha, \gamma, z).$$

The quantization rule is

$$\begin{gathered} A = \frac{1}{2} + \sqrt{\frac{1}{4} + B^2 - 2M\epsilon} > 0, \quad \alpha = A - B, \\ \alpha = -n \Longrightarrow +\sqrt{\frac{1}{4} + B^2 - 2M\epsilon} = B - n - 1/2. \end{gathered} \tag{2.23}$$

The inequality $n < B - 1/2$ means that in the system there may exist only a finite number of bound states. If $0 < B < 1/2$, then no bound states exist; if $B = 1/2$, then there exists a single bound state. The formula for spectrum (2.23) can be written as

$$2M\epsilon = 2Bn - n(n+1) + B, \tag{2.24}$$

or,

$$2M\epsilon = \frac{1}{4} + B^2 - \left(n - (B - \frac{1}{2})\right)^2 > 0, \quad B - \frac{1}{2} \geq 0. \tag{2.25}$$

Let us consider the case of negative k. The equation

$$y^2 \frac{d^2 f}{dy^2} + \left(2M\epsilon - y^2 (k - \frac{B}{y})^2\right) f = 0 \tag{2.26}$$

is considered in the variable $z = -2ky$, $z > 0$. With the use of the substitution $f(z) = z^A e^{Cz} F(z)$, where

$$A = \frac{1}{2} \pm \sqrt{\frac{1}{4} + e^2 B^2 - 2M\epsilon}, \quad C = -\frac{1}{2}, \tag{2.27}$$

we derive the equation

$$z \frac{d^2 F}{dz^2} + (2A - z) \frac{dF}{dz} - (A + B) F = 0,$$

which is a hypergeometric equation with parameters

$$\alpha = A + B, \quad \gamma = 2A, \quad F = \Phi(\alpha, \gamma, z). \tag{2.28}$$

The admissible quantization rules take the form

$$A = \frac{1}{2} \pm \sqrt{\frac{1}{4} + e^2B^2 - 2M\epsilon} > 0\,,$$

$$\alpha = -n\,, \quad \pm\sqrt{\frac{1}{4} + B^2 - 2M\epsilon} = -n - B - \frac{1}{2}\,. \tag{2.29}$$

The choice of upper sign is forbidden (see (2.29)). The choice of the lower sign leads to

$$+\sqrt{\frac{1}{4} + B^2 - 2M\epsilon} = n + B + \frac{1}{2} > \frac{1}{2}\,,$$

$$A > 0 \quad \Longrightarrow \quad +\sqrt{\frac{1}{4} + B^2 - 2M\epsilon} < \frac{1}{2}\,,$$

and we readily see that these two inequalities are not consistent. Therefore, at $k < 0$ (and $B > 0$) any bound states do not exist.

2.3 Dirac Particle in (x, y) Coordinates, Model H_2

Let us study the Dirac equation in the 2-dimensional Lobachevsky space H_2 in presence of external magnetic field, using quasi-Cartesian coordinates. We start with the equation [88]

$$\Big\{\gamma^k [\, i\, (\, e^{\alpha}_{(k)} \partial_\alpha + B_k) + e e^{\alpha}_{(k)} A_\alpha\,] \, - M\Big\} \Psi = 0\,. \tag{2.30}$$

In the coordinates (x, y):

$$dS^2 = dt^2 - \frac{dx^2 + dy^2}{y^2} - dz^2\,, \quad dz = 0\,, \quad A_x = +\frac{B}{y}\,,$$

$$e^{\beta}_{(b)} = \left| \begin{array}{cccc} 1 & 0 & 0 & 0 \\ 0 & -y & 0 & 0 \\ 0 & 0 & -y & 0 \\ 0 & 0 & 0 & 0 \end{array} \right|, \qquad e_{(b)\beta} = \left| \begin{array}{cccc} 1 & 0 & 0 & 0 \\ 0 & 1/y & 0 & 0 \\ 0 & 0 & 1/y & 0 \\ 0 & 0 & 0 & 0 \end{array} \right|, \tag{2.31}$$

and in presence of magnetic field, it reads

$$\Big[\gamma^0 i\partial_t + \gamma^1 (ie^{x}_{(1)}\partial_x + e\, e^{x}_{(1)} A_x) + i\gamma^2 (e^{y}_{(2)}\partial_y + B_{(2)}) - M\Big] \Psi = 0\,,$$

$$B_{(2)} = \frac{1}{2} \frac{1}{\sqrt{-g}} \frac{\partial}{\partial y} \sqrt{-g} e^{y}_{(2)} = \frac{1}{2}\,.$$

We further obtain

$$\Big[\gamma^0 i\partial_t + \gamma^1 (-iy\partial_x - eB) + i\gamma^2 (-y\partial_y + 1/2) - M\Big] \Psi = 0\,. \tag{2.32}$$

The variables are separated by the following substitution

$$\Psi = e^{-i\epsilon t} e^{ipx} \left| \begin{array}{c} f_1(y) \\ f_2(y) \\ f_3(y) \\ f_4(y) \end{array} \right| ,$$

whence we derive the system for $f_a(y)$:

$$\begin{aligned} &\epsilon f_3 - (py - eB) f_4 - (-y\partial_y + 1/2) f_4 - M f_1 = 0, \\ &\epsilon f_4 - (py - eB) f_3 + (-y\partial_y + 1/2) f_3 - M f_2 = 0, \\ &\epsilon f_1 + (py - eB) f_2 + (-y\partial_y + 1/2) f_2 - M f_3 = 0, \\ &\epsilon f_2 + (py - eB) f_1 - (-y\partial_y + 1/2) f_1 - M f_4 = 0. \end{aligned} \tag{2.33}$$

These equations are consistent with the linear constraint $f_3 = K f_1$, $f_4 = K f_2$, when the parameter K takes two values

$$\epsilon - \frac{M}{K} = -\epsilon + MK \qquad \Longrightarrow \qquad K = K_{1,2} = \frac{\epsilon \pm \sqrt{\epsilon^2 - M^2}}{M} . \tag{2.34}$$

In this way, we derive two subsystems

$$\begin{aligned} &\Big(y \frac{\partial}{\partial y} - \frac{1}{2} - py + eB \Big) f_2 + (-\epsilon + MK) f_1 = 0, \\ &\Big(y \frac{\partial}{\partial y} - \frac{1}{2} + py - eB \Big) f_1 + (+\epsilon - MK) f_2 = 0. \end{aligned} \tag{2.35}$$

It is convenient to consider them separately (let it be $+\sqrt{\epsilon^2 - M^2} = \lambda$):

$$\begin{aligned} &MK = \epsilon + \sqrt{\epsilon^2 - M^2}, \\ &(y \frac{\partial}{\partial y} - \frac{1}{2} - py + eB) f_2 + \lambda f_1 = 0, \; (y \frac{\partial}{\partial y} - \frac{1}{2} + py - eB) f_1 - \lambda f_2 = 0; \end{aligned} \tag{2.36}$$

$$\begin{aligned} &MK = \epsilon - \sqrt{\epsilon^2 - M^2}, \\ &(y \frac{\partial}{\partial y} - \frac{1}{2} - py + eB) f_2 - \lambda f_1 = 0, \; (y \frac{\partial}{\partial y} - \frac{1}{2} + py - eB) f_1 + \lambda f_2 = 0. \end{aligned} \tag{2.37}$$

For definiteness, we will follow the case (2.36); so, for the function f_1 we have the equation

$$y \frac{d^2 f_1}{dy^2} + \Big(p - p^2 y + 2peB + \frac{1/4 - e^2 B^2 + \lambda^2}{y} \Big) f_1 = 0 . \tag{2.38}$$

Let us introduce the variable $z = 2py$, which substantially depends on the sign of p. First we consider the case of positive p (let eB be designated as B; also assume $B > 0$). Using the substitution $f_1(z) = z^A e^{Cz} F_1(z)$, we get

$$\begin{aligned} &z \frac{d^2 F_1}{dz^2} + (2A + 2Cz) \frac{dF_1}{dz} \\ &+ \Big[(C^2 - \frac{1}{4}) z + 2AC + B + \frac{1}{2} + \frac{A(A-1) - B^2 + 1/4 + \lambda^2}{z} \Big] F_1 = 0 . \end{aligned} \tag{2.39}$$

If

$$A = \frac{1}{2} + \sqrt{B^2 - \lambda^2} > 0\,, \quad C = -\frac{1}{2}\,,$$

Eq. (2.39) becomes of hypergeometric type with the parameters $\alpha = A - B - \frac{1}{2}$, $\gamma = 2A$. The possible quantization rule is

$$\alpha = -n \quad \Longrightarrow \quad \frac{1}{2} + \sqrt{B^2 - \lambda^2} - B - \frac{1}{2} = -n\,;$$

this provides us with the following spectrum

$$+\sqrt{B^2 - \lambda^2} = -n + B > 0\,, \tag{2.40}$$

and the number of bound states is finite and it is governed by the quantity B. For instance, if $B = 2/3$, then only one such a state when $n = 0$ exists.

Let us follow the case of negative p. In the variable $z = -2py$, $z > 0$, with the use of the substitution $f_1(z) = z^A e^{Cz} F_1(z)$, Eq. (2.38) reduces to

$$z\frac{d^2F_1}{dz^2} + (2A + 2Cz)\frac{dF_1}{dz} + \Big[(C^2 - \frac{1}{4})z$$
$$+2AC - B - \frac{1}{2} + \frac{A(A-1) - B^2 + 1/4 + \lambda^2}{z}\Big]F_1 = 0\,. \tag{2.41}$$

At the given A and C,

$$A = \frac{1}{2} \pm \sqrt{B^2 - \lambda^2} > 0\,, \quad C = -\frac{1}{2}\,,$$

Eq. (2.41) becomes of hypergeometric type with the parameters $\alpha = A + B + \frac{1}{2}$, $\gamma = 2A$. The quantization rule is

$$\alpha = -n\,, \quad \pm\sqrt{B^2 - \lambda^2} = -n - B - 1\,.$$

The choice of upper sign is forbidden. The choice of lower sign gives

$$0 < \sqrt{B^2 - \lambda^2} = B + n + 1 < \frac{1}{2}\,, \quad A > 0 \quad \Longrightarrow \quad \sqrt{B^2 - \lambda^2} > \frac{1}{2}\,.$$

We note that these inequalities are not consistent (at positive B).

2.4 Landau Problem in the Spherical Model S_2, Coordinates (r,ϕ)

In the 3-dimensional spherical Riemann space S_3, the following cylindric coordinate system is known

$$z \in [-\pi/2, +\pi/2]\,,\ r \in [0, +\pi],\ \phi \in [0, 2\pi]\,,$$
$$ds^2 = dt^2 - \cos^2 z(dr^2 + \sin^2 r\, d\phi^2) + dz^2\,, \tag{2.42}$$

they are linked to the coordinates (u_0, u_1, u_2, u_3) of the Euclidean 4-space by the formulas

$$u_1 = \cos z \sin r \cos\phi\,, \; u_2 = \cos z\; \sin r \sin\phi\,, \; u_3 = \sin z\,, \; u_0 = \cos z \cos r\,. \qquad (2.43)$$

For the plane S_2, the metric (2.42) simplifies

$$z = 0\,, \quad dz = 0\,, \qquad ds^2 = dt^2 - dr^2 - \sin^2 r\, d\phi^2\,. \qquad (2.44)$$

An analogue of the uniform magnetic field in the Riemannian space is determined by the potential

$$A_\phi = B\,(\cos r - 1)\,. \qquad (2.45)$$

The corresponding Schrödinger equation reads

$$i\frac{\partial}{\partial t}\Psi = \frac{1}{2M}\Big[-\frac{1}{\sin r}\frac{\partial}{\partial r}\sin r\frac{\partial}{\partial r} + \frac{1}{\sin^2 r}\Big(i\frac{\partial}{\partial\phi} + eA_\phi\Big)^2\Big]\Psi\,. \qquad (2.46)$$

With the use of the substitution $\Psi = e^{-i\epsilon t}e^{im\phi}R(r)$, we get

$$\frac{1}{\sin r}\frac{d}{dr}\sin r\frac{dR}{dr} - \frac{1}{\sin^2 r}\Big[-m + eB\,(\cos r - 1)\Big]^2 R + 2M\epsilon\, R = 0\,.$$

In the variable $z = (1 - \cos r)/2$, the last equation takes the form (for brevity let eB be designated as B; also assume $B > 0$)

$$z\,(1-z)\,\frac{d^2R}{dz^2} + (1-2z)\,\frac{dR}{dz} + \Big[2M\epsilon + B^2 - \frac{m^2}{4z} - \frac{(B+m/2)^2}{(1-z)}\Big]R = 0\,. \qquad (2.47)$$

With the substitution $R(z) = (1-z)^A z^C F(z)$, where

$$A = \pm(B + \frac{m}{2})\,, \quad C = \pm\frac{m}{2}\,,$$

for the function $F(z)$ we derive equation of hypergeometric type

$$z\,(1-z)\,\frac{d^2F}{dz^2} + (1 + 2C - 2(A+C+1)\,z)\,\frac{dF}{dz}$$
$$+ \big[2M\epsilon + B^2 - (A+C)\,(A+C+1)\big]\,F = 0\,,$$

with the following solutions

$$F = \,_2F_1\alpha, \beta, \gamma; z)\,, \quad \alpha = A + C + \frac{1}{2} - \sqrt{\frac{1}{4} + B^2 + 2M\epsilon}\,,$$
$$\beta = A + C + \frac{1}{2} + \sqrt{\frac{1}{4} + B^2 + 2M\epsilon}\,, \quad \gamma = 1 + 2C\,. \qquad (2.48)$$

In order to have solutions relevant to bound states, we require $A > 0$, $C > 0$. Besides, we shall use polynomials. There exist four possibilities to choose C and A:

$$1.\, A = -\Big(B + \frac{m}{2}\Big)\,, \quad C = -\frac{m}{2}\,; \quad 2.\, A = +\Big(B + \frac{m}{2}\Big)\,, \quad C = -\frac{m}{2}\,;$$

$$3.\, A=+\left(B+\frac{m}{2}\right), \quad C=+\frac{m}{2}; \quad 4.\, A=-\left(B+\frac{m}{2}\right), \quad C=+\frac{m}{2}.$$

Only three of them are appropriate to describe bound states (let $n \in \{0,1,2,...\}$):

$$\underline{1. \qquad m<-2B\,,} \qquad \alpha=-B-m+\frac{1}{2}-\sqrt{\frac{1}{4}+B^2+2M\epsilon}=-n\,,$$
$$\beta=-B-m+\frac{1}{2}+\sqrt{\frac{1}{4}+B^2+2M\epsilon}\,, \qquad \gamma=1-m\,, \tag{2.49}$$
$$\text{spectrum} \qquad \sqrt{\frac{1}{4}+B^2+2M\epsilon}=-B-m+\frac{1}{2}+n>0;$$

$$\underline{2. \qquad -2B<m<0\,,} \qquad \alpha=B+\frac{1}{2}-\sqrt{\frac{1}{4}+B^2+2M\epsilon}=-n\,,$$
$$\beta=B+\frac{1}{2}+\sqrt{\frac{1}{4}+B^2+2M\epsilon}\,, \qquad \gamma=1-m\,, \tag{2.50}$$
$$\text{spectrum} \qquad \sqrt{\frac{1}{4}+B^2+2M\epsilon}=B+\frac{1}{2}+n;$$

$$\underline{3. \qquad m>0\,,} \qquad \alpha=B+m+\frac{1}{2}-\sqrt{\frac{1}{4}+B^2+2M\epsilon}=-n\,,$$
$$\beta=B+m+\frac{1}{2}+\sqrt{\frac{1}{4}+B^2+2M\epsilon}\,, \qquad \gamma=1+m\,, \tag{2.51}$$
$$\text{spectrum} \qquad \sqrt{\frac{1}{4}+B^2+2M\epsilon}=B+m+\frac{1}{2}+n\,.$$

These three cases may by merged into one, by means of the notations

$$A=\mid B+\frac{m}{2}\mid, \quad C=\mid \frac{m}{2}\mid; \tag{2.52}$$

then the formula for the energy spectrum reads

$$2M\epsilon=-B^2-\frac{1}{4}+\left(\mid B+\frac{m}{2}\mid+\mid\frac{m}{2}\mid+\frac{1}{2}+n\right)^2. \tag{2.53}$$

This relation assumes restrictions on the possible values of quantum numbers m and n.

Let us detail the transition to the case of Euclidean plane. To this end, we shall express all the relations in terms of usual units:

$$B \;\Rightarrow\; \frac{eB\rho^2}{\hbar}, \quad 2M\epsilon \;\Rightarrow\; \frac{2ME\rho^2}{\hbar^2},$$

where ρ stands for the curvature radius of the spherical plane.

The case (2.49) at $m < -2eB\rho^2/\hbar$ is degenerated when $\rho \to \infty$, because it leads to the inequality $m < -\infty$.

The case (2.50)

$$\frac{2ME}{\hbar^2}\rho^2 = -\Big(\frac{eB\rho^2}{\hbar}\Big)^2 - \frac{1}{4} + \Big(\frac{eB\rho^2}{\hbar} + \frac{1}{2} + n\Big)^2$$

at the limit $\rho \to \infty$ gives

$$E = \frac{eB\hbar}{M}(n + \frac{1}{2}), \quad m < 0. \tag{2.54}$$

In the case (2.51), at the limit $\rho \to \infty$ we obtain

$$E = \frac{eB\hbar}{M}(n + m + \frac{1}{2}), \quad m > 0. \tag{2.55}$$

The last formulas may be merged into a single one,

$$E = \frac{eB\hbar}{M}(n + \frac{m + \mid m \mid}{2} + \frac{1}{2}), \tag{2.56}$$

which coincides with the known result.

2.5 Complex Poincaré Half-Plane for Spherical 2-Space

By analogy with the case of the 2-dimensional hyperbolic space, let us introduce quasi-Cartesian coordinates (x, y) by means of the relations

$$iu_1 = \frac{y^2 - 1 + x^2}{2y}, \; u_0 = \frac{y^2 + 1 + x^2}{2y}, \; iu_2 = \frac{x}{y}, \quad u_1^2 + u_2^2 + u_0^2 = +1. \tag{2.57}$$

The first three equations lead to

$$y = \frac{1}{u_0 - i\,u_1}, \quad x = \frac{i\,u_2}{u_0 - i\,u_1}. \tag{2.58}$$

Because the coordinates (x, y) should parameterize the real space, they must obey the additional constraints

$$xy^* + x^*y = 0, \quad yy^* - xx^* = 1. \tag{2.59}$$

Therefore we may introduce the following parametrization for the variables x, y:

$$\begin{aligned} x &= \sinh A i e^{i\alpha} = \sinh A(-\sin\alpha + i\cos\alpha), \\ y &= \cosh A e^{i\alpha} = \cosh A(\cos\alpha + i\sin\alpha); \end{aligned} \tag{2.60}$$

$$\begin{aligned} u_2 = \tanh A, \quad u_1 &= \frac{\sin\alpha}{\cosh A}, \quad u_0 = \frac{\cos\alpha}{\cosh A}; \\ A \in (-\infty, +\infty), &\qquad \alpha \in [0, 2\pi]. \end{aligned} \tag{2.61}$$

Taking in mind the formulas

$$u_1 = \sin r \cos\phi\,, \quad u_2 = \sin r \sin\phi\,, \quad u_0 = \cos r\,, \tag{2.62}$$

we derive the following two relations

$$\tan\phi = \frac{2x}{x^2+y^2-1}\,, \qquad \cos r = \frac{x^2+y^2+1}{2y}\,. \tag{2.63}$$

Let us find the explicit form of the metric in the spherical plane, while using the complex quasi-Cartesian coordinates (x,y). We start with the cylindric metric

$$dS^2 = dt^2 - (dr^2 + \sin^2 r\, d\phi^2)\,, \quad r \in [0,+\pi], \quad \phi \in [0,2\pi]\,. \tag{2.64}$$

From (2.64), and considering the identity

$$(1+\tan^2\phi)\, d\phi = \Big(\frac{2}{x^2+y^2-1} - \frac{4x^2}{(x^2+y^2-1)^2}\Big)\, dx - \frac{4xy}{(x^2+y^2-1)^2}\, dy\,,$$

$$-\sin r\, dr = \frac{x}{y}\, dx + \Big(1 - \frac{x^2+y^2+1}{2y^2}\Big)\, dy\,,$$

we obtain the metric for S_2 in the coordinates (x,y):

$$dl^2 = dr^2 + \sin^2 r\, d\phi^2 = -\frac{dx^2+dy^2}{y^2}\,. \tag{2.65}$$

Note that the metric (2.65) may be readily transformed to coordinates (A,α) from (2.61):

$$dl^2 = -\frac{dx^2+dy^2}{y^2} = \frac{dA^2+d\alpha^2}{\cosh^2 A}\,. \tag{2.66}$$

Now let us find expression for electromagnetic potential A_ϕ being transformed to coordinates $(x,\, y)$:

$$A_\phi = B(\cos r - 1), \quad A_x = \frac{\partial\phi}{\partial x} A_\phi, \quad A_y = \frac{\partial\phi}{\partial y} A_\phi,$$

whence it follows

$$A_x = -B\, \frac{x^2-y^2+1}{x^2+(y+1)^2}\, \frac{1}{y}\,, \quad A_y = -B\, \frac{2x}{x^2+(y+1)^2}\,. \tag{2.67}$$

The corresponding electromagnetic tensor has the form

$$F_{xy} = \frac{\partial}{\partial x}\frac{\partial\phi}{\partial y} A_\phi - \frac{\partial}{\partial y}\frac{\partial\phi}{\partial x} A_\phi = -\frac{B}{y^2}\,, \tag{2.68}$$

which, evidently, is a complex quantity. Besides, the same tensor follows from a simpler potential

$$\tilde{A}_x = -\frac{B}{y}\,,\ A_y = 0\,,\ A_z = 0 \implies F_{xy} = -\partial_y A_x = -\frac{B}{y^2}\,. \tag{2.69}$$

There must exist a gauge transformation

$$A_x + \frac{\partial}{\partial x}\Lambda = \tilde{A}_x = -\frac{B}{y}, \quad A_y + \frac{\partial}{\partial y}\Lambda = \tilde{A}_y = 0. \tag{2.70}$$

Such a gauge function Λ obeys two equations

$$\frac{\partial}{\partial x}\Lambda = -\frac{B}{y} + B\,\frac{x^2 - y^2 + 1}{x^2 + (y+1)^2}\,\frac{1}{y}, \quad \frac{\partial}{\partial y}\Lambda = B\,\frac{2x}{x^2 + (y+1)^2}. \tag{2.71}$$

From the second equation we infer

$$\Lambda = f(x) + 2Bx \int \frac{dy}{x^2 + (y+1)^2} = f(x) + 2B \arctan\frac{y+1}{x}. \tag{2.72}$$

Further, using the first equation we derive the needed explicit form for the gauge function

$$\frac{df(x)}{dx} = 0 \Longrightarrow \Lambda = \text{const} + 2B \arctan\frac{y+1}{x}. \tag{2.73}$$

Let us study the Schrödinger equation

$$i\partial_t \Psi = \frac{1}{2M}\Big[\Big(\frac{i}{\sqrt{-g}}\frac{\partial}{\partial x^k}\sqrt{-g} + eA_k\Big)(-g^{kl})\Big(i\frac{\partial}{\partial x^l} + eA_l\Big)\Big]\Psi;$$

in complex coordinates (x, y) and in presence of magnetic field. The explicit form of the equation is (note that $\sqrt{-g} = y^{-2}$):

$$i\partial_t \Psi = \frac{1}{2M}\Big[\Big(i\frac{\partial}{\partial x} + eA_x\Big)(-y^2)\Big(i\frac{\partial}{\partial x} + eA_x\Big) + y^2\frac{\partial^2}{\partial y^2}\Big]\Psi$$

or, differently,

$$i\partial_t \Psi = \frac{1}{2M}\Big[-y^2\Big(i\frac{\partial}{\partial x} - e\frac{B}{y}\Big)^2 + y^2\frac{\partial^2}{\partial y^2}\Big]\Psi. \tag{2.74}$$

By applying the substitution $\Psi = e^{-i\epsilon t}e^{ikx}f(y)$, after separating the variables, we get the following equation for $f(y)$

$$y^2\frac{d^2}{dy^2}f - 2M\epsilon\, f - y^2\Big(k + \frac{eB}{y}\Big)^2 f = 0. \tag{2.75}$$

In the new variable $z = 2ky$, with the substitution $f(z) = z^A e^{Cz} F(z)$, we get the equation (let eB be designated as B; we assume $B > 0$):

$$z\frac{d^2F}{dz^2} + (2A + 2Cz)\frac{dF}{dz}$$
$$+\Big[\Big(C^2 - \frac{1}{4}\Big)z + 2AC - B + \frac{A(A-1) - B^2 - 2M\epsilon}{z}\Big]F = 0. \tag{2.76}$$

Assume that

$$A = \frac{1}{2} \pm \sqrt{\frac{1}{4} + B^2 + 2M\epsilon}, \quad C = -\frac{1}{2};$$

then Eq. (2.76) takes the hypergeometric form

$$z\frac{d^2F}{dz^2} + (2A - z)\frac{dF}{dz} - (A+B)F = 0,$$

solutions which are finite at $y = 0$ are

$$F = {}_1F_1(\alpha, \gamma; z), \quad \alpha = A + B, \quad \gamma = 2A. \tag{2.77}$$

Let us examine the possibility to get polynomials. We consider the positive parameter A:

$$A = \frac{1}{2} + \sqrt{\frac{1}{4} + B^2 + 2M\epsilon},$$

and we arrive at the equation

$$\sqrt{\frac{1}{4} + B^2 + 2M\epsilon} = -n - B - \frac{1}{2},$$

which has no physically interpretable solutions. Consider the second possibility

$$A = \frac{1}{2} - \sqrt{\frac{1}{4} + e^2B^2 + 2M\epsilon}; \tag{2.78}$$

we can see that the functions ${}_1F_1(\alpha, \gamma; z)$ become polynomials if

$$\sqrt{\frac{1}{4} + B^2 + 2M\epsilon} = n + B + \frac{1}{2}. \tag{2.79}$$

The corresponding wave functions are determined by the relations

$$\Psi_{k,\epsilon} = e^{ikx} z^A e_1^{-z/2} F_1(\alpha, \gamma, z)$$

$$= e^{ikx}(2ky)^{-n-B} e^{-ky} {}_1F_1(-n, -2n - 2B, 2ky). \tag{2.80}$$

To the lowest energy level with $n = 0$ there corresponds the wave function

$$\Psi_{k,\epsilon^{min}} = e^{ikx}(2ky)^{-n-B} e^{-ky}. \tag{2.81}$$

The solutions (2.80) and (2.81) may be expressed in terms of the variables (A, α):

$$\Psi_{k,\epsilon} = \left(e^{-k}\right)^{\exp(A+i\alpha)} \left(2k \cosh A e^{i\alpha}\right)^{-n-B}$$

$$\times F(-n, -2n - 2B, 2k \cosh A\, e^{i\alpha}),$$

$$\Psi_{k,\epsilon^{min}} = \Psi_{k,\epsilon} = \left(e^{-k}\right)^{\exp(A+i\alpha)} \left(2k \cosh A e^{i\alpha}\right)^{-B};$$

the formulas include only independent real-valued variables; however, their structure is not of separable variables.

In the end, let us note the identities

$$x^* = -\frac{x}{x^2+y^2}\,, \quad y^* = \frac{y}{x^2+y^2}\,, \tag{2.82}$$

which allow us to derive the formulas

$$y^2 = -\frac{x}{x^*}(1+xx^*)\,, \quad x^2 = +\frac{y}{y^*}(1-yy^*)\,. \tag{2.83}$$

The last relations permit us to parameterize the spherical plane in two different ways: by coordinates (x,x^*) or by (y,y^*):

$$dl^2 = \frac{1}{(1+xx^*)^2}\Big[-\frac{1}{4}\Big(\frac{dx^2}{x^2}+\frac{dx^{*\,2}}{x^{*\,2}}\Big)+\Big(1+\frac{1}{2xx^*}\Big)dx\,dx^*\Big]\,, \tag{2.84}$$

$$dl^2 = -\frac{1}{yy^*(1-yy^*)}\Big[\frac{1}{4}\Big(\frac{dy^2}{y^2}+\frac{dy^{*\,2}}{y^{*\,2}}\Big)+\Big(1-\frac{1}{2\,yy^*}\Big)dy\,dy^*\Big]\,. \tag{2.85}$$

However, in these coordinates the variables in Schrödinger equations cannot be separated.

Chapter 3

Hydrogen Atom in Static de Sitter Spaces

3.1 Separation of the Variables in dS Space

Starting from the pioneering work by Dirac [97], a steady attention is devoted to the de Sitter geometrical models in the context of quantum theory for curved space-time (see, for instance, [98, 99]). In particular, the problem of description of the particles with different spins on such curved backgrounds has a long history (see [97]–[121]). The most of the work done, however, concerns the field theory on the background of such geometrical models. In the present chapter, we address the quantum mechanics and examine the influence of the de Sitter geometries on the behavior of the hydrogen atom.

If the origin of the coordinate system coincides with the location of the point electric charge, the Maxwell equations provide the 4-potential $A_\alpha = (e/r, 0, 0, 0)$ for the Coulomb field.

We consider the conventional generally covariant Klein–Gordon–Fock equation

$$\Big[\Big(i\hbar\nabla_\alpha + \frac{e}{c}A_\alpha\Big)\Big(i\hbar\nabla^\alpha + \frac{e}{c}A^\alpha\Big) - M^2c^2\Big]\Phi = 0$$

for the potential A_α in the de Sitter space model given by the static metrics [98]

$$dS^2 = \Big(1 - \frac{r^2}{\rho^2}\Big)dt^2 - \Big(1 - \frac{r^2}{\rho^2}\Big)^{-1}dr^2 - r^2\Big(d\theta^2 + \sin^2\theta d\phi^2\Big), \quad r \in [0, \rho). \tag{3.1}$$

Note that in these coordinates no nonrelativistic Schrödinger-type equation is known, so a relativistic description might be based on either the scalar Klein–Gordon–Fock equation or the spinor Dirac equation.

After the separation of the variables with the help of the substitution $\Phi = e^{-i\epsilon t/\hbar}Y_{lm}(\theta, \phi)f(r)$, we arrive at the radial equation

$$\frac{d^2}{dr^2}f + \frac{2(1 - 2r^2/\rho^2)}{r(1 - r^2/\rho^2)}\frac{d}{dr}f$$
$$+ \Big[\frac{(\epsilon + e^2/r)^2}{c^2\hbar^2}\frac{1}{(1 - r^2/\rho^2)^2} - \Big(\frac{M^2c^2}{\hbar^2} + \frac{l(l+1)}{r^2}\Big)\frac{1}{1 - r^2/\rho^2}\Big]f = 0. \tag{3.2}$$

Eq. (3.2) gives the following expression for the squared radial momentum:

$$p_r^2 = \frac{(\epsilon + e^2/r)^2}{c^2} \frac{1}{(1 - r^2/\rho^2)^2} - \left(M^2c^2 + \frac{L^2}{r^2}\right) \frac{1}{1 - r^2/\rho^2} . \tag{3.3}$$

3.2 Qualitative Discussion

To study the expression (3.3) for the squared radial momentum, it is convenient to compare with the flat Minkowski case, for which the squared momentum reads

$$p_r^2 = \frac{(\epsilon + e^2/r)^2}{c^2} - \left(M^2c^2 + \frac{L^2}{r^2}\right). \tag{3.4}$$

Consider first the free particle motion. In the Minkowski space p_r^2 vanishes at two points:

$$r_0 = \pm\sqrt{\frac{L^2c^2}{\epsilon^2 - M^2c^4}} . \tag{3.5}$$

The classical motion is possible in the domain where $p_r^2 > 0$, and at $\epsilon > Mc^2$ such a domain exists. For a free particle in dS-space, at the origin and at the horizon the momentum (3.3) behaves as

$$p_r^2(r \to 0) \sim -\frac{L^2}{r^2} \to -\infty, \quad p_r^2(r \to \rho) \sim \frac{\epsilon^2}{c^2(1 - \frac{r^2}{\rho^2})^2} \to +\infty. \tag{3.6}$$

We find the vanishing points of momentum from the fourth order polynomial equation $p_r^2 = 0$:

$$r_0 = \pm\rho\sqrt{-\frac{A}{2} \pm \sqrt{\frac{A^2}{4} + \frac{L^2}{M^2c^2\rho^2}}}, \quad A = \frac{\epsilon^2 - M^2c^4}{M^2c^4} + \frac{L^2}{M^2c^2\rho^2} . \tag{3.7}$$

Two of these roots are complex conjugate. The other two are real-valued, one of which is negative and another one is positive. We conclude that the character of the classical motion of a free particle in the de Sitter space is much the same as in the flat Minkowski space.

Now, let us consider the case of the Coulomb potential. For the particle on the background of the flat Minkowski space, the behavior of the momentum near the singular points is given by

$$p_r^2(r \to 0) \sim -\frac{L^2 - e^4/c^2}{r^2} \to -\infty , \quad p_r^2(r \to \infty) \sim \frac{\epsilon^2 - M^2c^4}{c^2} . \tag{3.8}$$

The turning points are given by

$$r_0 = \frac{c^2}{\epsilon^2 - M^2c^4}\left[-\frac{e^2\epsilon}{c^2} \pm \sqrt{\frac{e^4\epsilon^2}{c^4} + \left(L^2 - \frac{e^4}{c^2}\right)\frac{\epsilon^2 - M^2c^2}{c^2}}\right].$$

If $(L^2 - e^4/c^2) > 0$ and $\epsilon < Mc^2$, according to (3.7) we have two positive roots and two negative (nonphysical) ones. The plot of p_r^2 is schematically shown in Fig. 3.1.

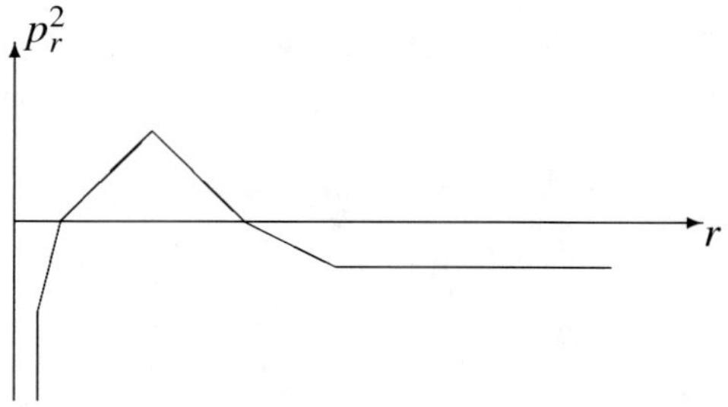

Fig. 3.1. Coulomb problem in the Minkowski space.

For the de Sitter space, at the origin and at the horizon, the momentum p_r^2 behaves as

$$p_r^2(r \to 0) \sim -\frac{L^2 - e^4/c^2}{r^2} \to -\infty, \quad p_r^2(r \to \rho) \sim \frac{(\epsilon + e^2/\rho)^2}{c^2\,(1 - r^2/\rho^2)^2} \to +\infty. \tag{3.9}$$

Furthermore, by Viete's theorem, for the product of the roots $r_1, ..., r_4$ of the equation, $p_r^2 = 0$ holds

$$r_1 r_2 r_3 r_4 = -\frac{(L^2 - e^4/c^2)\rho^2}{M^2 c^2} < 0. \tag{3.10}$$

From here, we conclude that the following three variants for the signs of the roots:

$$\text{sign}(r_1, r_2, r_3, r_4) = (-,-,+,+), \quad (-,-,-,-), \ (+,+,+,+), \tag{3.11}$$

are excluded. In turn, assuming two real-valued and two conjugate roots, we get the variants $\text{sign}(r_1, r_2, r_3, r_4) = (-,+,z,z^*)$, which has no physical sense. The relationship $r_1 r_2 r_3 r_4 < 0$ shows that, assuming all four roots are real, there are only two possibilities:

$$\text{sign}(r_1, r_2, r_3, r_4) = (-,-,-,+), \ (-,+,+,+). \tag{3.12}$$

Using physical arguments, the second case is more interesting. The schematic behavior of p_r^2 corresponding to this case is shown in Fig. 3.2. This behavior indicates that we should have a nonstable quantum-mechanical system.

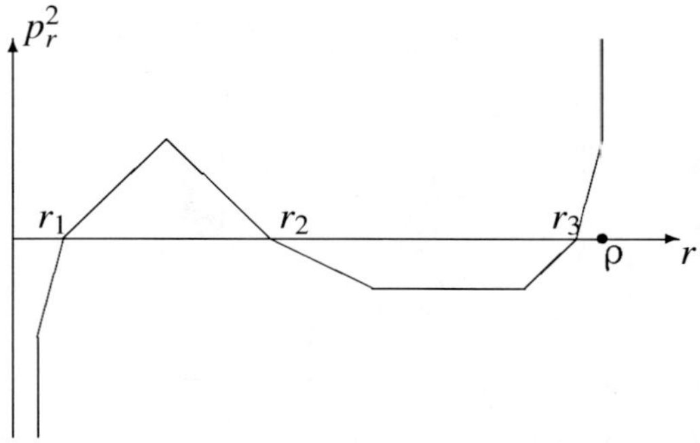

Fig. 3.2. Coulomb problem in dS space.

Let us discuss in more detail the behavior of p_r at the horizon $r \to \rho$. Because the coordinate r involved in the wave equation has no direct metrical sense, instead of it one can use any other coordinate. We consider the radial variable r^*, which is subject to:

$$\frac{d}{dr} = \frac{1}{1-r^2/\rho^2}\frac{d}{dr^*}, \quad r^* = \frac{\rho}{2}\ln\frac{1+r/\rho}{1-r/\rho}. \tag{3.13}$$

The result is

$$p_{r^*} \longrightarrow \frac{(\epsilon + e^2/\rho)^2}{c^2}. \tag{3.14}$$

The corresponding plot of $p^2_{r^*}$ is shown in Fig. 3.3. This picture is well understood from the physical point of view as representing a nonstable quantum-mechanical hydrogen atom model in the de Sitter space. We note that one may expect a similar behavior when considering the atom model on the basis of the Dirac equation.

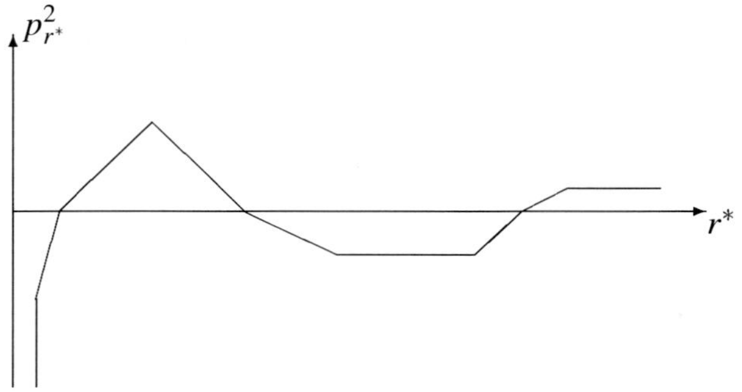

Fig. 3.3. Coulomb problem in dS space, the coordinate r^*.

3.3 Reducing Radial Equation to the General Heun Equation

The radial equation (3.2) is a Fuchsian ordinary linear differential equation having four regular singular points. Using the dimensionless quantities

$$x = \frac{r}{\rho}, \quad \frac{\epsilon\rho}{c\hbar} = E, \quad \frac{e^2}{c\hbar} = \alpha, \quad \frac{M^2c^2\rho^2}{\hbar^2} \Longrightarrow M^2, \tag{3.15}$$

and making the standard substitution $f = x^A(1-x)^B(1+x)^C H(x)$, it is reduced to the general Heun equation [3, 4]

$$\frac{d^2H}{dx^2} + \left(\frac{\gamma}{x} + \frac{\delta}{x-1} + \frac{\epsilon}{x+1}\right)\frac{dH}{dx} + \frac{\lambda\beta x - q}{x(x-1)(x+1)}H = 0, \tag{3.16}$$

where

$$\gamma = 2 + 2A\,, \quad \delta = 1 + 2B\,, \quad \epsilon = 1 + 2C,$$
$$\lambda, \beta = \frac{3}{2} + A + B + C \pm \sqrt{-M^2 + \frac{9}{4}}\,, \quad q = 2\,(E\alpha - (1+A)(B-C))$$

with the parameters A, B, C given as

$$A = -\frac{1}{2} \pm \sqrt{(l+1/2)^2 - \alpha^2}\,, \quad B = \pm\frac{i}{2}\,(E+\alpha)\,, \quad C = \pm\frac{i}{2}\,(E-\alpha)\,. \tag{3.17}$$

Note that the singular point $x = +1$ of the Heun equation (3.16) is physical, because it represents the de Sitter horizon, while the point $x = -1$ does not belong to the physical domain, the interval $x \in [0, +1)$.

To get solutions which are vanishing at the origin, we take

$$A = -\frac{1}{2} + \sqrt{(l+1/2)^2 - \alpha^2}\,. \tag{3.18}$$

Depending on the signs of B and C, we have four choices, $(+,+)$, $(-,-)$, $(+,-)$ and $(-,+)$, each leading to a different particular solution. The parameters λ, β and q for these solutions read:

$$\begin{array}{lll}
(+,+) & \lambda, \beta = \frac{3}{2} + A + iE \pm i\sqrt{M^2 - \frac{9}{4}}\,, & q = 2\alpha\,[\,E - i(A+1)\,]\,, \\
(-,-) & \lambda, \beta = \frac{3}{2} + A - iE \pm i\sqrt{M^2 - \frac{9}{4}}\,, & q = 2\alpha\,[\,E + i(A+1)\,]\,, \\
(+,-) & \lambda, \beta = \frac{3}{2} + A + i\alpha \pm i\sqrt{M^2 - \frac{9}{4}}\,, & q = 2E\,[\,\alpha - i(A+1)\,]\,, \\
(-,+) & \lambda, \beta = \frac{3}{2} + A - i\alpha \pm i\sqrt{M^2 - \frac{9}{4}}\,, & q = 2E\,[\,\alpha + i(A+1)\,]\,.
\end{array} \tag{3.19}$$

Note that by physical reason, we may assume the inequality $M^2 \to (M^2\rho^2c^2/\hbar^2) >> 1$.

3.4 Semi-Classical Study

To study the problem within the semi-classical approximation, we assume a large ρ, and consider the region $r \ll \rho$, where the potential well is located. Expanding the roots $r_1, \ldots, r_4$ of the equation $p_r^2 = 0$ (see 3.4) in terms of the small parameter ρ^{-1}, we get

$$r_i = r_{0i} + \frac{\Delta_i}{\rho^2}\,, \qquad i = 1, 2\,,$$

$$\begin{aligned}
r_{01,02} &= \frac{c^2}{M^2c^4 - \epsilon^2}\left[\,\frac{e^2\epsilon}{c^2} \pm M\sqrt{\frac{e^4\epsilon^2}{c^4} - \Big(L^2 - \frac{e^4}{c^2}\Big)\frac{M^2c^2 - \epsilon^2}{c^2}}\,\right] > 0\,, \\
\Delta_i &= -r_{0i}^2\,(L^2 + M^2c^2r_{0i}^2)\left[\,\frac{2e^2\epsilon}{c^2} + 2r_{0i}\Big(\frac{\epsilon^2}{c^2} - M^2c^2\Big)\right]^{-1}\,,
\end{aligned} \tag{3.20}$$

and

$$r_{3,4} = \pm\rho\sqrt{1-\frac{\epsilon^2}{M^2c^4}} - \frac{e^2\epsilon}{M^2c^4-\epsilon^2}, \qquad i=3,4. \tag{3.21}$$

It is immediately seen that for a large curvature radius ρ, the region between r_1 and r_2 lays far from $r_3 > 0$ and $r_4 < 0$. Then, rewriting

$$p_r^2 = \frac{1}{r^2\left(1-r^2/\rho^2\right)^2}\frac{M^2c^2}{\rho^2}(r-r_1)(r-r_2)(r-r_3)(r-r_4) \tag{3.22}$$

and expanding the product $(r-r_3)(r-r_4)$ for $r \ll r_{3,4}$, we obtain in virtue of Eq. (3.21), the approximate form for p_r in the region $r \ll \rho$, where the potential well is located:

$$p_r \sim Mc\frac{\sqrt{(r-r_1)(r-r_2)}}{r}\left(\frac{\epsilon^2}{M^2c^4}-1\right)^{1/2}\left(1+A\,\frac{r^2}{\rho^2}\right), \; A = 1-\frac{1}{2(1-\epsilon^2/M^2c^4)}. \tag{3.23}$$

The semi-classical quantization rule is

$$\int_L p_r dr = 2\pi\hbar(n+1/2). \tag{3.24}$$

From this, taking p_r according to (3.23) and choosing the path of integration surrounding the turning points r_1 and r_2, we reduce the problem to calculating the residues just at two points, $r=0$ and $r=\infty$. As a result, we arrive at the equation

$$\sum_{0,\infty} \text{res}\;\; p_r = i\hbar(n+1/2), \tag{3.25}$$

or

$$\sqrt{\frac{\epsilon^2}{c^2}-M^2c^2}\left[\left(\sqrt{r_1r_2}+\frac{r_1+r_2}{2}\right) - \frac{A}{\rho^2}r_1r_2\frac{r_1+r_2}{2}\right] = i\hbar(n+1/2). \tag{3.26}$$

Then, taking into account the relations (3.20), we get

$$\sqrt{\frac{\epsilon^2}{c^2}-m^2c^2}\left[\sqrt{r_{01}r_{02}}+\frac{r_{01}r_{02}}{2}+\frac{1}{2\rho^2}\left(\frac{\Delta_1+\Delta_2}{2}\right.\right.$$
$$\left.\left.+\frac{1}{2}\Delta_1\sqrt{\frac{r_{02}}{r_{01}}}+\frac{1}{2}\Delta_2\sqrt{\frac{r_{01}}{r_{02}}}-Ar_{01}r_{02}(r_{01}+r_{02})\right)\right] = i\hbar(n+1/2), \tag{3.27}$$

and we find the following approximate formula for energy levels in the de Sitter space:

$$\epsilon = \epsilon_0 + \frac{\Delta}{\rho^2}, \; \epsilon_0 = Mc^2\left[1+\frac{e^4/c^2}{[\hbar(n+1/2)+\sqrt{\hbar^2(l+1/2)^2-e^4/c^2}]^2}\right]^{-1/2} \tag{3.28}$$

with

$$\Delta = \frac{(M^2c^4-\epsilon_0^2)^2}{4e^2M^2c^4}\left[\Delta_1+\Delta_2+\Delta_1\sqrt{\frac{r_{02}}{r_{01}}}+\Delta_2\sqrt{\frac{r_{01}}{r_{02}}}-Ar_{01}r_{02}(r_{01}+r_{02})\right], \tag{3.29}$$

where $\Delta_{1,2}$, $r_{01,02}$ and A are given by the above formulas with ϵ replaced by ϵ_0.

In accordance with the behavior of the function p_r^2, the particle can move from the domain between r_1 and r_2 to the domain near the horizon $r \sim \rho$, due to the quantum-mechanical tunneling. The probability of this process is given as

$$W = \exp\Big(-\frac{2}{\hbar}\int_{r_2}^{r_3}\sqrt{-p_r^2}\,dr\Big), \tag{3.30}$$

from which we obtain a rough estimation $W \sim e^{-2\rho/\lambda}$, where λ is the Compton wavelength.

3.5 The Hydrogen Atom in *AdS* Space

We now consider the Coulomb problem in static coordinates of anti de Sitter space [98]

$$dS^2 = \Big(1+\frac{r^2}{\rho^2}\Big)dt^2 - \Big(1+\frac{r^2}{\rho^2}\Big)^{-1}dr^2 - r^2(d\theta^2+\sin^2\theta d\phi^2), \quad r\in[0,+\infty). \tag{3.31}$$

From the Klein–Gordon–Fock equation with the 4-potential $A_\alpha = (e/r,0,0,0)$, we get the radial equation

$$\frac{d^2}{dr^2}f + \frac{2(1+2r^2/\rho^2)}{r(1+r^2/\rho^2)}\frac{d}{dr}f$$
$$+\Big[\frac{(\epsilon+e^2/r)^2}{c^2\hbar^2}\frac{1}{(1+r^2/\rho^2)^2} - \Big(\frac{M^2c^2}{\hbar^2}+\frac{l(l+1)}{r^2}\Big)\frac{1}{1+r^2/\rho^2}\Big]f = 0. \tag{3.32}$$

In dimensionless parameters

$$x = \frac{ir}{\rho}, \quad \frac{\epsilon\rho}{c\hbar} = E, \quad \frac{e^2}{c\hbar} = \alpha, \quad \frac{M^2c^2\rho^2}{\hbar^2} \Longrightarrow M^2, \tag{3.33}$$

by applying the substitution $f = x^A\,(x-1)^B\,(x+1)^C\,H$ with

$$A = -\frac{1}{2} \pm \sqrt{(l+1/2)^2-\alpha^2}, \quad B = \pm\frac{1}{2}(E+i\alpha), \quad C = \pm\frac{1}{2}(E-i\alpha), \tag{3.34}$$

we reduce Eq. (3.32) to the general Heun equation

$$\frac{d^2H}{dx^2} + \Big(\frac{\gamma}{x}+\frac{\delta}{x-1}+\frac{\epsilon}{x+1}\Big)\frac{dH}{dx} + \frac{\lambda\beta x - q}{x(x-1)(x+1)}H = 0, \tag{3.35}$$

whose parameters are given as

$$\gamma = 2+2A, \quad \delta = 1+2B, \quad \epsilon = 1+2C,$$
$$\lambda,\beta = \frac{3}{2}+A+B+C \pm \sqrt{M^2+\frac{9}{4}}, \quad q = -2\,(iE\alpha + (1+A)(B-C)), \tag{3.36}$$

where the upper sign (+) conventionally stands for λ, and the lower sign (−) refers to β. Note that the variation region for the radial variable here is the interval

$x \in [0, +i\infty)$. To have finite solutions at the origin, we should use the positive value of A:

$$A = -\frac{1}{2} + \sqrt{(l+1/2)^2 - \alpha^2} > 0 . \tag{3.37}$$

Depending on the signs of B and C, we have four different possibilities: $(+,+)$, $(-,-)$, $(+,-)$ and $(-,+)$, each providing a different particular solution:

$$\begin{aligned}
(-,-): &\quad \lambda = 3/2 + A - E + \sqrt{M^2 + 9/4}, \quad q = -2i\alpha\,[\,E - (A+1)\,],\\
(+,+): &\quad \lambda = 3/2 + A + E + \sqrt{M^2 + 9/4}, \quad q = -2i\alpha\,[\,E + (A+1)\,],\\
(+,-): &\quad \lambda = 3/2 + A + i\alpha + \sqrt{M^2 + 9/4}, \quad q = -2E\,[\,i\alpha + (A+1)\,],\\
(-,+): &\quad \lambda = 3/2 + A - i\alpha + \sqrt{M^2 + 9/4}\,, \quad q = -2E\,[\,i\alpha - (A+1)\,].
\end{aligned} \tag{3.38}$$

To choose the appropriate branch, consider the situation without Coulomb field, $\alpha = 0$. The Heun equation (3.35), after transforming it to the variable $y = x^2$, becomes a hypergeometric equation

$$\begin{aligned}
&\frac{d^2 f}{dy^2} + \Big(\frac{3/2}{y} + \frac{1}{y-1}\Big)\frac{df}{dy} + \Big[-\frac{1}{4}\frac{l(l+1)}{y^2} - \frac{1}{4}\frac{E^2}{(y-1)^2}\\
&+\frac{1}{4}\frac{E^2 - M^2 + l(l+1)}{y-1} + \frac{1}{4}\frac{-E^2 + M^2 - l(l+1)}{y}\Big] f = 0 .
\end{aligned} \tag{3.39}$$

Using the substitution $f = y^A (y-1)^B \, {}_2F_1(a,b;c;y)$ with $A = +l/2$, $B = -E/2$, we construct the solution in terms of the hypergeometric function with the parameters

$$a = \frac{1}{2}\Big(\frac{3}{2} + l - E + \sqrt{9/4 + M^2}\,\Big), \ b = \frac{1}{2}\Big(\frac{3}{2} + l - E - \sqrt{9/4 + M^2}\ \Big), \ c = l + \frac{3}{2}. \tag{3.40}$$

3.6 Qualitative Study of the Problem in *AdS* Space

Consider the locations of the turning points of the classical momentum. By Viete's theorem, the roots of the equation $p_r^2 = 0$ obey the identities

$$r_1 + r_2 + r_3 + r_4 = 0 , \quad r_1 r_2 r_3 r_4 = + \frac{(L^2 - e^4/c^2)\rho^2}{M^2 c^2} > 0 . \tag{3.41}$$

The last relation shows that, assuming all four roots real-valued, the variants

$$\text{sign}(r_1, r_2, r_3, r_4) = (-,-,-,+)\,, \quad \text{sign}(r_1, r_2, r_3, r_4) = (+,+,+,-) \tag{3.42}$$

are forbidden. Besides, the variants

$$\text{sign}(r_1, r_2, r_3, r_4) = (-,-,-,-)\,, \quad \text{sign}(r_1, r_2, r_3, r_4) = (+,+,+,+) \tag{3.43}$$

are also forbidden because of the identity (3.41). So, we conclude that there exists only one possibility for the location of the real roots:

$$\mathrm{sign}(r_1, r_2, r_3, r_4) = (-,-,+,+)\,, \quad r_4 > r_3 > 0\,, \tag{3.44}$$

where r_4, r_3 stand for the physical turning points. Note also that the roots r_1, r_2 can be complex and conjugate:

$$\mathrm{sign}(r_1, r_2, r_3, r_4) \sim (z, z^*, +, +,)\,, \quad \mathrm{Re}\, z < 0\,. \tag{3.45}$$

Thus, the qualitative study shows that in the *AdS* space we can expect for the Coulomb problem the situation of a complicated potential well with normal (real) energy spectrum, so that the hydrogen atom is expected to be a stable system in the quantum-mechanical sense.

3.7 Semi-Classical Study for *AdS* Space

We use the formal method of analytical continuation, which gives the possibility to get results without tediously repeating the calculations:

$$dS \Longrightarrow AdS\,, \quad \rho \Longrightarrow i\rho\,. \tag{3.46}$$

The approximate roots r_1, r_2 of the equation $p^2(r) = 0$ in this case read

$$\begin{gathered} r_i = r_{0i} - \frac{\Delta_i}{\rho^2}\,, \qquad i = 1,2\,, \\ r_{01,02} = \frac{c^2}{M^2c^4 - \epsilon^2}\left[\frac{e^2\epsilon}{c^2} \pm M\sqrt{\frac{e^4\epsilon^2}{c^4} - \left(L^2 - \frac{e^4}{c^2}\right)\frac{M^2c^2 - \epsilon^2}{c^2}}\right] > 0\,, \\ \Delta_i = -r_{0i}^2\,(L^2 + M^2c^2r_{0i}^2)\left[\frac{2e^2\epsilon}{c^2} + 2r_{0i}\left(\frac{\epsilon^2}{c^2} - M^2c^2\right)\right]^{-1}, \end{gathered} \tag{3.47}$$

and for the roots r_3, r_4, we have

$$r_i = i\rho\left(b_{0i} + \frac{b_1}{i\rho}\right), \quad i = 3,4\,, \quad b_{03,04} = \pm\sqrt{1 - \frac{\epsilon^2}{M^2c^4}}\,, \quad b_1 = -\frac{e^2\epsilon}{M^2c^4 - \epsilon^2}\,. \tag{3.48}$$

Accordingly, the approximate form of p_r in the region $r \ll \rho$ (where the potential well is located) is written as

$$p_r \sim Mc\frac{\sqrt{(r-r_1)(r-r_2)}}{r}\left(\frac{\epsilon^2}{m^2c^4} - 1\right)^{1/2}\left(1 - A\frac{r^2}{\rho^2}\right), \; A = 1 - \frac{1}{2(1-\epsilon^2/M^2c^4)}\,. \tag{3.49}$$

The quantization condition is the same as Eq. (3.24). Choosing the contour of integration surrounding the points r_1 and r_2, we again reduce the problem to finding the residues at $r = 0$ and $r = \infty$ and arrive at the equation

$$\sqrt{\frac{\epsilon^2}{c^2} - M^2c^2}\left[\left(\sqrt{r_1r_2} + \frac{r_1+r_2}{2}\right) + \frac{A}{\rho^2}r_1r_2\frac{r_1+r_2}{2}\right] = i\hbar(n+1/2)\,. \tag{3.50}$$

From this, an approximation for the energy spectrum in AdS space follows:

$$\epsilon = \epsilon_0 - \frac{\Delta}{\rho^2}, \quad \epsilon_0 = Mc^2 \Big[1 + \frac{e^4/c^2}{[\hbar(n+1/2) + \sqrt{\hbar^2(l+1/2)^2 - e^4/c^2}]^2} \Big]^{-1/2},$$

$$\Delta = \frac{(M^2c^4 - \epsilon_0^2)^2}{4e^2M^2c^4} \Big[\Delta_1 + \Delta_2 + \Delta_1 \sqrt{\frac{r_{02}}{r_{01}}} + \Delta_2 \sqrt{\frac{r_{01}}{r_{02}}} + A r_{01} r_{02} (r_{01} + r_{02}) \Big], \quad (3.51)$$

where $\Delta_{1,2}$, $r_{01,02}$ and A are given by the above formulas with ϵ replaced by ϵ_0.

3.8 Spin 1/2 Particle in dS and AdS Spaces

To discuss the case of a spin 1/2 particle, we start with the Dirac free wave equation (the notations from [139] are used)

$$\Big[i\gamma^c \Big(e^{\alpha}_{(c)} \partial_\alpha + \frac{1}{2} \sigma^{ab} \gamma_{abc} \Big) - M \Big] \Psi = 0.$$

In terms of the static coordinates of the de Sitter space, using the spinor basis and the dimensionless coordinate $r \in [0,1)$, by applying the substitution $\Psi(x) = r^{-1} \Phi^{-1/4} F(x)$, $\Phi = 1 - r^2$, this equation reduces to the form

$$\Big[i \frac{\gamma^0}{\sqrt{\Phi}} \partial_t + i\sqrt{\Phi}\gamma^3 \partial_r + \frac{1}{r} \Sigma_{\theta,\phi} - M \Big] F(x) = 0, \quad (3.52)$$

$$\Sigma_{\theta,\phi} = i\gamma^1 \partial_\theta + \gamma^2 \frac{i\partial + i\sigma^{12}\cos\theta}{\sin\theta}.$$

The spherical waves are constructed as

$$\Psi_{\epsilon jm}(x) = \frac{e^{-i\epsilon t}}{r} \left| \begin{array}{c} f_1(r)\, D_{-1/2} \\ f_2(r)\, D_{+1/2} \\ f_3(r)\, D_{-1/2} \\ f_4(r)\, D_{+1/2} \end{array} \right|, \quad (3.53)$$

where the Wigner functions are denoted according to the rule $D^j_{-m,\sigma}(\phi,\theta,0) = D_\sigma$. With the use of the recurrence relations [68], we get

$$\Sigma_{\theta,\phi} \Psi_{\epsilon jm}(x) = i\nu \frac{e^{-i\epsilon t}}{r} \left| \begin{array}{c} - f_4(r)\, D_{-1/2} \\ + f_3(r)\, D_{+1/2} \\ + f_2(r)\, D_{-1/2} \\ - f_1(r)\, D_{+1/2} \end{array} \right|,$$

where $\nu = (j+1/2)$. Further we get the following radial equations

$$\begin{aligned} &\frac{\epsilon}{\sqrt{\Phi}} f_3 - i\sqrt{\Phi} \frac{d}{dr} f_3 - i\frac{\nu}{r} f_4 - M f_1 = 0, \\ &\frac{\epsilon}{\sqrt{\Phi}} f_4 + i\sqrt{\Phi} \frac{d}{dr} f_4 + i\frac{\nu}{r} f_3 - M f_2 = 0, \\ &\frac{\epsilon}{\sqrt{\Phi}} f_1 + i\sqrt{\Phi} \frac{d}{dr} f_1 + i\frac{\nu}{r} f_2 - M f_3 = 0, \\ &\frac{\epsilon}{\sqrt{\Phi}} f_2 - i\sqrt{\Phi} \frac{d}{dr} f_2 - i\frac{\nu}{r} f_1 - M f_4 = 0. \end{aligned} \quad (3.54)$$

To simplify the system, we diagonalize the P-operator in spherical tetrad

$$\hat{\Pi}_{sph} = \begin{vmatrix} 0 & 0 & 0 & -1 \\ 0 & 0 & -1 & 0 \\ 0 & -1 & 0 & 0 \\ -1 & 0 & 0 & 0 \end{vmatrix} \otimes \hat{P}.$$

From the equation $\hat{\Pi}_{sph} \Psi_{jm} = \Pi \Psi_{jm}$, it follows that $\Pi = \delta\,(-1)^{j+1}$, $\delta = \pm 1$ and

$$f_4 = \delta f_1, \; f_3 = \delta f_2, \; \Psi(x)_{\epsilon jm\delta} = \frac{e^{-i\epsilon t}}{r} \begin{vmatrix} f_1(r)\, D_{-1/2} \\ f_2(r)\, D_{+1/2} \\ \delta f_2(r)\, D_{-1/2} \\ \delta f_1(r)\, D_{+1/2} \end{vmatrix}.$$

The system (3.54) is then simplified as

$$\begin{aligned} &\Big(\sqrt{\Phi}\frac{d}{dr} + \frac{\nu}{r}\Big) f + \Big(\frac{\epsilon}{\sqrt{\Phi}} + \delta M\Big) g = 0, \\ &\Big(\sqrt{\Phi}\frac{d}{dr} - \frac{\nu}{r}\Big) g - \Big(\frac{\epsilon}{\sqrt{\Phi}} - \delta M\Big) f = 0, \end{aligned} \tag{3.55}$$

where the new functions

$$f = (f_1 + f_2)/\sqrt{2}, \quad g = (f_1 - f_2)/i\sqrt{2}$$

are used instead of f_1 and f_2. For definiteness, let us consider Eqs. (3.55) for $\delta = +1$

$$\Big(\sqrt{\Phi}\frac{d}{dr} + \frac{\nu}{r}\Big) f + \Big(\frac{\epsilon}{\sqrt{\Phi}} + M\Big) g = 0, \quad \Big(\sqrt{\Phi}\frac{d}{dr} - \frac{\nu}{r}\Big) g - \Big(\frac{\epsilon}{\sqrt{\Phi}} - M\Big) f = 0; \tag{3.56}$$

formally, the second case $\delta = -1$ corresponds to the change $M \to -M$.

To take into account the Coulomb field, it suffices to make the formal change $\epsilon \to \epsilon + e^2/r$. In this way we get the system

$$\begin{aligned} &\Big(\sqrt{1-r^2}\frac{d}{dr} + \frac{\nu}{r}\Big) f + \Big(\frac{1}{\sqrt{1-r^2}}\Big(\epsilon + \frac{e^2}{r}\Big) + M\Big) g = 0, \\ &\Big(\sqrt{1-r^2}\frac{d}{dr} - \frac{\nu}{r}\Big) g - \Big(\frac{1}{\sqrt{1-r^2}}\Big(\epsilon + \frac{e^2}{r}\Big) - M\Big) f = 0. \end{aligned} \tag{3.57}$$

Using the variable ρ ($\sin\rho = r$), these equations become shorter:

$$\begin{aligned} &\Big(\frac{d}{d\rho} + \frac{\nu}{\sin\rho}\Big) f + \Big(\frac{1}{\cos\rho}\Big(\epsilon + \frac{e^2}{\sin\rho}\Big) + M\Big) g = 0, \\ &\Big(\frac{d}{d\rho} - \frac{\nu}{\sin\rho}\Big) g - \Big(\frac{1}{\cos\rho}\Big(\epsilon + \frac{e^2}{\sin\rho}\Big) - M\Big) f = 0. \end{aligned} \tag{3.58}$$

Introducing two new functions:

$$f + g = e^{-i\rho/2}(F + G), \quad f - g = e^{+i\rho/2}(F - G),$$

and using a new variable $y = \tan(\rho/2)$ with the notations

$$y_{1,2} = -\frac{(\epsilon + M - i\nu - i/2) \pm \sqrt{(\epsilon + M - i\nu - i/2)^2 - e^4}}{e^2},$$
$$Y_{1,2} = -\frac{(\epsilon - M - i\nu + i/2) \pm \sqrt{(\epsilon - M - i\nu + i/2)^2 - e^4}}{e^2}, \tag{3.59}$$

we transform Eqs. (3.58) to the form

$$\Big[(1+y^2)\frac{d}{dy} - \nu y + \frac{\nu}{y} - \frac{a}{1-y} + \frac{b}{1+y} + \frac{a-b}{2}\Big] F + \frac{e^2}{y}(y-y_1)(y-y_2)\, G = 0,$$
$$\Big[(1+y^2)\frac{d}{dy} + \nu y - \frac{\nu}{y} + \frac{a}{1-y} - \frac{b}{1+y} - \frac{a-b}{2}\Big] G - \frac{e^2}{y}(y-Y_1)(y-Y_2)\, F = 0, \tag{3.60}$$

where $a = 2i(\epsilon + e^2)$, $b = 2i(\epsilon - e^2)$. Elimination of G (or F) from this system results in a second order differential equation for $F(y)$ (or $G(y)$), which has eight singular points: $0, \pm 1, \pm i, y_{1,2}, \infty$ (or $0, \pm 1, \pm i, Y_{1,2}, \infty$).

A similar situation is faced for the Dirac equation in the Coulomb potential on the background of the anti de Sitter space-time. The analysis here is much similar to that for the above de Sitter model. Using the variable $r = \sinh\rho$, instead of Eqs. (3.58) we have the equations

$$\Big(\frac{d}{d\rho} + \frac{\nu}{\sinh\rho}\Big) f + \Big(\frac{1}{\cosh\rho}(\epsilon + \frac{e^2}{\sinh\rho}) + M\Big) g = 0,$$
$$\Big(\frac{d}{d\rho} - \frac{\nu}{\sinh\rho}\Big) g - \Big(\frac{1}{\cosh\rho}(\epsilon + \frac{e^2}{\sinh\rho}) - M\Big) f = 0. \tag{3.61}$$

Introducing now the new functions F and G:

$$f + g = e^{-\rho/2}(F+G), \quad f - g = e^{+\rho/2}(F-G), \tag{3.62}$$

instead of Eq. (3.61) we have

$$\Big[\frac{d}{d\rho} + \nu\frac{\cosh\rho}{\sinh\rho} - \frac{\sinh\rho}{\cosh\rho}\Big(\epsilon + \frac{e^2}{\sinh\rho}\Big)\Big] F + \Big(\epsilon + \frac{e^2}{\sinh\rho} + M - \nu - \frac{1}{2}\Big) G = 0,$$
$$\Big[\frac{d}{d\rho} - \nu\frac{\cosh\rho}{\sinh\rho} + \frac{\sinh\rho}{\cosh\rho}\Big(\epsilon + \frac{e^2}{\sinh\rho}\Big)\Big] G + \Big(-\epsilon - \frac{e^2}{\sinh\rho} + M + \nu - \frac{1}{2}\Big) F = 0. \tag{3.63}$$

The coefficients of these equations become rational functions if one uses a new variable $y = \tanh(\rho/2)$; this system can be written as

$$\Big[(1-y^2)\frac{d}{dy} + \nu y + \frac{\nu}{y} - \frac{4(\epsilon y + e^2)}{1+y^2} + 2e^2\Big] F + \frac{e^2}{y}(y-\bar{y}_1)(y-\bar{y}_2) G = 0,$$
$$\Big[(1-y^2)\frac{d}{dy} - \nu y - \frac{\nu}{y} + \frac{4(\epsilon y + e^2)}{1+y^2} - 2e^2\Big] G + \frac{e^2}{y}(y-\bar{Y}_1)(y-\bar{Y}_2) F = 0, \tag{3.64}$$

where

$$\bar{y}_{1,2} = \frac{(\epsilon + M - \nu - 1/2) \pm \sqrt{(\epsilon + M - \nu - 1/2)^2 + e^4}}{e^2},$$
$$\bar{Y}_{1,2} = \frac{(\epsilon - M - \nu + 1/2) \pm \sqrt{(\epsilon - M - \nu + 1/2)^2 + e^4}}{e^2}. \tag{3.65}$$

Like in the case of the de Sitter space, the second order differential equations for $F(y)$ and $G(y)$ have eight singularities: $0, \pm 1, \pm i, \bar{y}_1, \bar{y}_2, \infty$ and $0, \pm 1, \pm i, \bar{Y}_1, \bar{Y}_2, \infty$, respectively.

However, the physical situation here is expected to be qualitatively similar to that arising in scalar models, due to the influence of any special space-time geometry being universal and, in principle, it should not depend on the spin of a quantum-mechanical particle. The dependence on the value of the spin should reveal only technical differences. So for the spin 1/2 particle, we can expect very complicated effective potential curves underlying the quasi-stationary quantum states in the dS space, and the discrete energy spectrum in the AdS space respectively.

Chapter 4

Scalar Particle in Non-Static de Sitter Spaces

4.1 Nonrelativistic Approximation, and Riemann Geometry

We start with the covariant first order tensor equations (for generality include additional interaction term through scalar curvature; see notations in [139]):

$$\Big(i\nabla_\alpha + \frac{e}{c\hbar}A_\alpha\Big)\Phi = \frac{Mc}{\hbar}\Phi_\alpha, \quad \Big(i\nabla_\alpha + \frac{e}{c\hbar}A_\alpha\Big)\Phi^\alpha = \frac{Mc}{\hbar}\Big(1+\frac{1}{6}\frac{R(x)}{m^2c^2/\hbar^2}\Big)\Phi. \tag{4.1}$$

We simplify the notations

$$1+\sigma\frac{R(x)}{M^2c^2/\hbar^2} = \Gamma(x), \quad \sigma = \frac{1}{6}, \tag{4.2}$$

then Eqs. (4.1) can be written as follows

$$\Big(i\partial_\alpha + \frac{e}{c\hbar}A_\alpha\Big)\Phi(x) = \frac{Mc}{\hbar}\,\Phi_\alpha, \quad \Big(\frac{i}{\sqrt{-g}}\frac{\partial}{\partial x^\alpha}\sqrt{-g} + \frac{e}{c\hbar}A_\alpha\Big)g^{\alpha\beta}\Phi_\beta = \frac{Mc}{\hbar}\Gamma\Phi. \tag{4.3}$$

Further, by considering space-times models with metrics of the form

$$dS^2 = c^2dt^2 + g_{kl}(x)dx^kdx^l\,,$$

we perform in (4.3) the (3+1)-spit:

$$\Big(i\,\partial_0 + \frac{e}{c\hbar}A_0\Big)\Phi = \frac{Mc}{\hbar}\,\Phi_0, \quad \Big(i\,\partial_l + \frac{e}{c\hbar}A_l\Big)\Phi(x) = \frac{Mc}{\hbar}\,\Phi_l\,,$$

$$\Big(\frac{i}{\sqrt{-g}}\frac{\partial}{\partial x^0}\sqrt{-g} + \frac{e}{c\hbar}A_0\Big)\Phi_0 + \Big(\frac{i}{\sqrt{-g}}\frac{\partial}{\partial x^k}\sqrt{-g} + \frac{e}{c\hbar}A_k\Big)g^{kl}\Phi_l = \frac{Mc}{\hbar}\Gamma\Phi,$$

or, differently,

$$\Big(i\partial_0 + \frac{e}{c\hbar}A_0\Big)\Phi = \frac{Mc}{\hbar}\,\Phi_0, \quad \Big(i\,\partial_l + \frac{e}{c\hbar}A_l\Big)\Phi = \frac{Mc}{\hbar}\Phi_l\,,$$

$$\Big(i\frac{\partial}{\partial x^0}+\frac{i}{\sqrt{-g}}\frac{\partial\sqrt{-g}}{\partial x^0}+\frac{e}{c\hbar}A_0\Big)\Phi_0+\Big(\frac{i}{\sqrt{-g}}\frac{\partial}{\partial x^k}\sqrt{-g}+\frac{e}{c\hbar}A_k\Big)g^{kl}\Phi_l=\frac{Mc}{\hbar}\Gamma\Phi\,.$$

We are to separate the rest energy by the special substitution

$$\Phi\Longrightarrow\exp(-i\frac{Mc^2t}{\hbar})\Phi,\quad \Phi_0\Longrightarrow\exp(-i\frac{Mc^2t}{\hbar})\Phi_0,\quad \Phi_l\Longrightarrow\exp(-i\frac{Mc^2t}{\hbar})\Phi_l\,. \tag{4.4}$$

This yields

$$\Big(\frac{i}{c}\,\partial_t+\frac{Mc}{\hbar}+\frac{e}{c\hbar}A_0\Big)\Phi(x)=\frac{Mc}{\hbar}\Phi_0(x)\,,$$

$$\Big(\frac{i}{c}\partial_t+\frac{Mc}{\hbar}+\frac{i}{\sqrt{-g}}\frac{\partial\sqrt{-g}}{\partial}+\frac{e}{c\hbar}A_0\Big)\Phi_0$$

$$+\Big(\frac{i}{\sqrt{-g}}\frac{\partial}{\partial x^k}\sqrt{-g}+\frac{e}{c\hbar}\,A_k\Big)g^{kl}\Phi_l=\frac{Mc}{\hbar}\Gamma\Phi(x)\,,$$

$$\Big(i\partial_l+\frac{e}{c\hbar}A_l\Big)\Phi(x)=\frac{Mc}{\hbar}\,\Phi_l(x)\,,$$

or

$$\Big(i\hbar\partial_t+Mc^2+eA_0\Big)\Phi(x)=Mc^2\Phi_0(x)\,, \tag{4.5}$$

$$\Big(i\hbar\partial_t+Mc^2+i\hbar\frac{1}{\sqrt{-g}}\frac{\partial\sqrt{-g}}{\partial t}+eA_0\Big)\Phi_0$$

$$+c\Big(\frac{i\hbar}{\sqrt{-g}}\frac{\partial}{\partial x^k}\sqrt{-g}+\frac{e}{c}A_k\Big)g^{kl}\Phi_l=Mc^2\Gamma\Phi(x)\,, \tag{4.6}$$

$$\Big(i\hbar\partial_l+\frac{e}{c}A_l\Big)\Phi(x)=Mc\Phi_l(x)\,. \tag{4.7}$$

After excluding non-dynamical variable Φ_l with the help of (4.7), we obtain

$$\Big(i\hbar\partial_t+Mc^2+eA_0\Big)\Phi(x)=Mc^2\Phi_0(x)\,, \tag{4.8}$$

$$\Big(i\hbar\partial_t+Mc^2+i\hbar\frac{1}{\sqrt{-g}}\frac{\partial\sqrt{-g}}{\partial t}+eA_0\Big)\Phi_0$$

$$+\frac{1}{M}\Big[(\frac{i\hbar}{\sqrt{-g}}\partial_k\sqrt{-g}+\frac{e}{c}A_k)g^{kl}(i\hbar\partial_l+\frac{e}{c}A_l)\Big]\Phi(x)=Mc^2\Gamma\Phi(x)\,. \tag{4.9}$$

Now we introduce the small φ and the big Ψ components [139],

$$\Phi-\Phi_0=\varphi\,,\quad \Phi+\Phi_0=\Psi,\quad \Phi=\frac{\Psi+\varphi}{2},\quad \Phi_0=\frac{\Psi-\varphi}{2}\,. \tag{4.10}$$

We infer

$$\Big(i\hbar\partial_t+Mc^2+eA_0\Big)\frac{\Psi+\varphi}{2}=Mc^2\frac{\Psi-\varphi}{2}\,,$$

$$\Big(i\hbar\partial_t+Mc^2+i\hbar\frac{1}{\sqrt{-g}}\frac{\partial\sqrt{-g}}{\partial t}+eA_0\Big)\frac{\Psi-\varphi}{2}$$

$$+\frac{1}{m}\Big[\Big(\frac{i\hbar}{\sqrt{-g}}\partial_k\sqrt{-g}+\frac{e}{c}A_k\Big)g^{kl}\Big(i\hbar\partial_l+\frac{e}{c}A_l\Big)\Big]\frac{\Psi+\varphi}{2}=Mc^2\Gamma\frac{\Psi+\varphi}{2},$$

whence it follows

$$(i\hbar\,\partial_t+eA_0)\,\frac{+\varphi+\Psi}{2}=-Mc^2\varphi\,, \tag{4.11}$$

$$\Big(i\hbar\partial_t+i\hbar\frac{1}{\sqrt{-g}}\frac{\partial\sqrt{-g}}{\partial t}+eA_0\Big)\frac{\Psi-\varphi}{2}$$
$$+\frac{1}{M}\Big[\Big(\frac{i\hbar}{\sqrt{-g}}\partial_k\sqrt{-g}+\frac{e}{c}A_k\Big)g^{kl}\Big(i\hbar\partial_l+\frac{e}{c}A_l\Big)\Big]\frac{\Psi-\varphi}{2}=Mc^2(\Gamma+1)\frac{\varphi}{2}+Mc^2(\Gamma-1)\frac{\Psi}{2}. \tag{4.12}$$

From this place, we distinguish two different cases. The first assumes $\Gamma=1$, and then the last equations lead to (we neglect the small component relative to the big one, Ψ)

$$(i\hbar\partial_t+eA_0)\frac{\Psi}{2}=-Mc^2\varphi\,, \tag{4.13}$$

$$\Big(i\hbar\partial_t+i\hbar\frac{1}{\sqrt{-g}}\frac{\partial\sqrt{-g}}{\partial t}+eA_0\Big)\frac{\Psi}{2}$$
$$+\frac{1}{M}\Big[\Big(\frac{i\hbar}{\sqrt{-g}}\partial_k\sqrt{-g}+\frac{e}{c}\,A_k\Big)g^{kl}\Big(i\hbar\partial_l+\frac{e}{c}A_l\Big)\Big]\frac{\Psi}{2}=Mc^2\varphi\,. \tag{4.14}$$

By excluding the small component, we derive the Schrödinger-like equation for the big component

$$\Big[i\hbar(\partial_t+\frac{1}{2\sqrt{-g}}\frac{\partial\sqrt{-g}}{\partial t})+eA_0\Big]\Psi$$
$$=\frac{1}{2M}\Big[\Big(\frac{i\hbar}{\sqrt{-g}}\partial_k\sqrt{-g}+\frac{e}{c}A_k\Big)(-g^{kl})\Big(i\hbar\partial_l+\frac{e}{c}A_l\Big)\Big]\Psi. \tag{4.15}$$

In static metrics, the determinant of the metrical tensor does not depend on time, so one term in the left hand side vanishes. Assuming the structure $\sqrt{-g}=\sqrt{g(t)}\sqrt{-g(x^1,x^2,x^3)}$, we reduce Eq. (4.15) to the form

$$\Big[i\hbar(\partial_t+\frac{1}{2\sqrt{g(t)}}\frac{\partial\sqrt{g(t)}}{\partial t})+eA_0\Big]\Psi$$
$$=\frac{1}{2M}\Big[\Big(\frac{i\hbar}{\sqrt{-g(x)}}\partial_k\sqrt{-g(x)}+\frac{e}{c}A_k\Big)(-g^{kl})\Big(i\hbar\,\partial_l+\frac{e}{c}A_l\Big)\Big]\Psi. \tag{4.16}$$

By the substitution $\Psi=g(t)^{-1/4}\Phi$, Eq. (4.16) is simplified,

$$(i\hbar\partial_t+eA_0)\Psi=\frac{1}{2M}\Big[\Big(\frac{i\hbar}{\sqrt{-g(x)}}\,\partial_k\sqrt{-g(x)}+\frac{e}{c}A_k\Big)(-g^{kl}(t,x))\Big(i\hbar\partial_l+\frac{e}{c}A_l\Big)\Big]\Psi. \tag{4.17}$$

The second possibility arises if $\Gamma\neq 1$. Correspondingly, from (4.11) and (4.12) it follows

$$(i\hbar\partial_t+eA_0)\frac{\Psi}{2}=-Mc^2\varphi\,, \tag{4.18}$$

$$\Big(i\hbar\partial_t + i\hbar\frac{1}{\sqrt{-g}}\frac{\partial\sqrt{-g}}{\partial t} + eA_0\Big)\frac{\Psi}{2} + \frac{1}{M}\Big[\Big(\frac{i\hbar}{\sqrt{-g}}\partial_k\sqrt{-g} + \frac{e}{c}A_k\Big)g^{kl}\Big(i\hbar\partial_l + \frac{e}{c}A_l\Big)\Big]\frac{\Psi}{2}$$
$$= Mc^2(\Gamma+1)\frac{\varphi}{2} + Mc^2(\Gamma-1)\frac{\Psi}{2}. \tag{4.19}$$

Therefore, the equation for the big component reads

$$\Big(i\hbar\,\partial_t + i\hbar\frac{1}{\sqrt{-g}}\frac{\partial\sqrt{-g}}{\partial t} + eA_0\Big)\frac{\Psi}{2} + \frac{(\Gamma+1)}{2}\,(i\hbar\,\partial_t + eA_0)\,\frac{\Psi}{2} - Mc^2(\Gamma-1)\frac{\Psi}{2}$$
$$= -\frac{1}{2M}\Big[\Big(\frac{i\hbar}{\sqrt{-g}}\partial_k\sqrt{-g} + \frac{e}{c}A_k\Big)\,g^{kl}\Big(i\hbar\partial_l + \frac{e}{c}A_l\Big)\Big]\Psi. \tag{4.20}$$

In may be written differently,

$$\Big[\Big(\frac{1}{2} + \frac{1}{2}\frac{\Gamma(x)+1}{2}\Big)(i\hbar\partial_t + eA_0) + \frac{i\hbar}{2\sqrt{-g}}\frac{\partial\sqrt{-g}}{\partial t}\Big]\Psi$$
$$= \frac{1}{2M}\Big[\Big(\frac{i\hbar}{\sqrt{-g}}\partial_k\sqrt{-g} + \frac{e}{c}A_k\Big)(-g^{kl})\Big(i\,\hbar\,\partial_l + \frac{e}{c}A_l\Big)\Big]\Psi + Mc^2\,\frac{(\Gamma(x)-1)}{2}\Psi.$$

Taking in mind the identity

$$\Gamma(x) = 1 + \frac{1}{6}\frac{\hbar^2 R(x)}{M^2c^2},$$

we reduce the last equation to the form

$$\Big[\Big(1 + \frac{1}{24}\frac{\hbar^2 R(x)}{M^2c^2}\Big)\Big(i\hbar\partial_t + eA_0\Big) + \frac{i\hbar}{2\sqrt{-g}}\frac{\partial\sqrt{-g}}{\partial t}\Big]\Psi$$
$$= \frac{1}{2M}\Big[\Big(\frac{i\hbar}{\sqrt{-g}}\partial_k\sqrt{-g} + \frac{e}{c}A_k\Big)(-g^{kl})\Big(i\hbar\,\partial_l + \frac{e}{c}A_l\Big)\Big] + \hbar^2\frac{R(x)}{6}\Big]\Psi. \tag{4.21}$$

It should be especially emphasized that in the nonrelativistic equation (4.16), the wave function is not a scalar function of the initial relativistic equation, but

$$\Psi(x) = \Phi(x) + \Phi_0(x), \qquad \Phi_0 \in \{\,\Phi_0, \Phi_1, \Phi_2, \Phi_3\,\}. \tag{4.22}$$

4.2 Schrödinger Equation in de Sitter Non-Static Models

In absence of external electromagnetic fields, Eq. (4.21) becomes simpler

$$i\hbar\Big[\Big(1 + \frac{1}{24}\lambda^2 R(x)\Big)\frac{\partial}{\partial t} + \frac{1}{2\sqrt{-g}}\frac{\partial\sqrt{-g}}{\partial t}\Big]\Psi$$
$$= -\frac{\hbar^2}{2M}\Big(-\frac{1}{\sqrt{-g}}\frac{\partial}{\partial x^k}\sqrt{-g}g^{kl}\frac{\partial}{\partial x^l} - \frac{1}{6}R(x)\Big)\Psi. \tag{4.23}$$

We shall consider in the following the de Sitter model (dS) and the anti de Sitter model (AdS), and employ non-static coordinates.

In the (dS)-case, the metric is given by the relations

$$dS, \qquad dS^2 = c^2 dt^2 - \rho^2 \cosh^2 \frac{ct}{\rho} [dr^2 + \sin^2 r (d\theta^2 + \sin^2\theta \, d\phi^2)],$$

$$\sqrt{-g} = \cosh^3 \frac{ct}{\rho} \sin^2 r \sin\theta, \quad R^2(x) = +\frac{1}{\rho^2}, \; (g_{ab}) = \mathrm{diag}(g_{00}, g_{11}, g_{22}, g_{33}),$$

$$g^{ab} = \mathrm{diag}\Big(c^{-1}, \frac{-1}{\rho^2 \cosh^2(ct/\rho)}, \frac{-1}{\rho^2 \cosh^2(ct/\rho)} \frac{1}{\sin^2 r}, \frac{-1}{\rho^2 \cosh^2(ct/\rho)} \frac{1}{\sin^2 r \sin^2\theta}\Big).$$

We need a 5D-description of de Sitter space. To this end, we consider 4-dimensional surface in 5-dimensional space with the symmetry group $SO(4,1)$; for simplicity we set $c = 1$ and $\rho = 1$):

$$(\xi^0)^2 - (\xi^1)^2 - (\xi^2)^2 - (\xi^3)^2 - (\xi^4)^2 = -\rho^2 = -1 . \tag{4.24}$$

It can be readily checked that the five coordinates ξ^A are linked to the four coordinates x^α by the formulas

$$\begin{aligned} &\xi^1 = \cosh t \sin r \sin\theta \cos\phi, \quad \xi^2 = \cosh t \sin r \sin\theta \sin\phi, \\ &\xi^3 = \cosh t \sin r \cos\theta, \quad \xi^0 = \sinh t, \quad \xi^4 = \cosh t \cos r, \end{aligned} \tag{4.25}$$

where

$$t \in (-\infty, +\infty), \quad r \in [0, \pi], \quad \theta \in [0, \pi], \quad \phi \in [0, 2\pi] .$$

Indeed, let us calculate the five differentials $d\xi^A$:

$$d\xi^0 = \cosh t dt, \quad d\xi^4 = \sinh t \cos r dt - \cosh t \sin r dr,$$

$$d\xi^1 = \sinh t \sin r \sin\theta \cos\phi dt + \cosh t \cos r \sin\theta \cos\phi dr$$

$$+ \cosh t \sin r \cos\theta \cos\phi d\theta - \cosh t \sin r \sin\theta \sin\phi d\phi,$$

$$d\xi^2 = \sinh t \sin r \sin\theta \sin\phi dt + \cosh t \cos r \sin\theta \sin\phi dr$$

$$+ \cosh t \sin r \cos\theta \sin\phi d\theta + \cosh t \sin r \sin\theta \cos\phi d\phi,$$

$$d\xi^3 = \sinh t \sin r \cos\theta dt + \cosh t \cos r \cos\theta dr - \cosh t \sin r \sin\theta d\theta;$$

then we derive the needed expression for the metric

$$dS^2 = (d\xi^0)^2 - (d\xi^1)^2 - (d\xi^2)^2 - (d\xi^3)^2 - (d\xi^4)^2 \tag{4.26}$$

in terms of the coordinates x^α:

$$dS^2 = dt^2 - \cosh^2 t \, [\, dr^2 + \sin^2 r (d\theta^2 + \sin^2\theta \, d\phi^2) \,] . \tag{4.27}$$

In the case of the (AdS)-space, we have the following non-static metric

$$AdS, \quad dS^2 = c^2 dt^2 - \rho^2 \cos^2 \frac{ct}{\rho} [dr^2 + \sinh^2 r (d\theta^2 + \sin^2\theta \, d\phi^2)],$$

$$\sqrt{-g} = \cos^3 \frac{ct}{\rho} \sinh^2 r \sin\theta, \quad R^2(x) = -\frac{1}{\rho^2},$$

$$g^{ab} = \text{diag}(c^{-1}, \frac{-1}{\rho^2 \cosh^2(ct/\rho)}, \frac{-1}{\rho^2 \cos^2(ct/\rho)} \frac{1}{\sinh^2 r}, \frac{-1}{\rho^2 \cos^2(ct/\rho)} \frac{1}{\sinh^2 r \sin^2\theta}).$$

Taking into account the formulas relating 5D-presentation of the model (with symmetry group $SO(3,2)$):

$$(\xi^0)^2 - (\xi^1)^2 - (\xi^2)^2 - (\xi^3)^2 + (\xi^4)^2 = +\rho^2 = +1,$$

$$\xi^1 = \cos t \sinh r \sin\theta \cos\phi, \quad \xi^2 = \cos t \sinh r \sin\theta \sin\phi,$$
$$\xi^3 = \cos t \sinh r \cos\theta, \quad \xi^0 = \sin t, \quad \xi^4 = \cos t \cosh r, \tag{4.28}$$

$$t \in (-\infty, +\infty), \qquad r \in [0, +\infty), \quad \theta \in [0, \pi], \quad \phi \in [0, 2\pi],$$

we find the five differentials $d\xi^A$:

$$d\xi^0 = \cos t dt, \quad d\xi^4 = -\sin t \coth r dt + \cos\, t \sinh r dr,$$
$$d\xi^1 = -\sin t \sinh r \sin\theta \cos\phi dt + \cos t \cosh r \sin\theta \cos\phi dr$$
$$+ \cos t \sinh r \cos\theta \cos\phi d\theta - \cos t \sinh r \sin\theta \sin\phi d\phi,$$
$$d\xi^2 = -\sin t \sinh r \sin\theta \sin\phi dt + \cos t \cosh r \sin\theta \sin\phi dr$$
$$+ \cos t \sinh r \cos\theta \sin\phi d\theta + \cos t \sinh r \sin\theta \cos\phi d\phi,$$
$$d\xi^3 = -\sin t \sinh r \cos\theta dt + \cos t \cosh r \cos\theta dr - \cos t \sinh r \sin\theta d\theta;$$

and the derive the needed identity

$$dS^2 = (d\xi^0)^2 - [(d\xi^1)^2 + (d\xi^2)^2 + (d\xi^3)^2] + (d\xi^4)^2$$
$$= dt^2 - \cos^2 t\, [\, dr^2 + \sinh^2 r (d\theta^2 + \sin^2\theta\, d\phi^2)\,]. \tag{4.29}$$

By using the notation for Compton wave length $\lambda = \hbar/Mc$, the Schrödinger-like equation in the de Sitter model is written as follows:

$$dS, \quad i\hbar\Big[\Big(1 + \frac{1}{24}\frac{\lambda^2}{\rho^2}\Big)\frac{\partial}{\partial t} + \frac{1}{4g}\frac{\partial g}{\partial t}\Big]\Psi = -\frac{\hbar^2}{2M}\Big(-\frac{1}{\sqrt{-g}}\partial_k \sqrt{-g} g^{kl} \partial_l - \frac{1}{6}\frac{1}{\rho^2}\Big)\Psi; \tag{4.30}$$

$$AdS, \quad i\hbar\Big[\Big(1 - \frac{1}{24}\frac{\lambda^2}{\rho^2}\Big)\frac{\partial}{\partial t} + \frac{1}{4g}\frac{\partial g}{\partial t}\Big]\Psi = -\frac{\hbar^2}{2M}\Big(-\frac{1}{\sqrt{-g}}\partial_k \sqrt{-g} g^{kl}\, \partial_l + \frac{1}{6}\frac{1}{\rho^2}\Big)\Psi. \tag{4.31}$$

We note the expressions of the metrical determinants:

$$dS, \quad g(t,x) = g(t)g(x),\ g(t) = \cosh^6 t,\ [-g(x)] = \sin^4 r \sin^2\theta,$$
$$AdS, \quad g(t,x) = g(t)g(x)\ g(t) = \cos^6 t,\ [-g(x)] = \sinh^4 r \sin^2\theta.$$

The structural equation

$$i\hbar\Big(\mu\frac{\partial}{\partial t} + \frac{1}{4g(t)}\frac{\partial g(t)}{\partial t}\Big)\Psi = H\Psi$$

may be simplified by the simple substitution

$$\Psi(x) = \varphi(t)\,\Phi(x) \equiv g(t)^{-\mu/4}\,\Phi\,, \qquad i\hbar\mu\frac{\partial}{\partial t}\Phi = H\Phi\,. \tag{4.32}$$

Thus, in the de Sitter model, we infer the equations:

$$dS, \qquad \mu = (1 + \frac{1}{24}\frac{\lambda^2}{\rho^2}), \quad \varphi(t) = \Big(\cosh^6\frac{ct}{\rho}\Big)^{-1/4\mu},$$

$$i\hbar\frac{\partial}{\partial t}\Phi = -\frac{\hbar^2}{2m\mu}\Big(-\frac{1}{\sqrt{-g(x)}}\,\partial_k\sqrt{-g(x)}g^{kl}\partial_l - \frac{1}{6}\frac{1}{\rho^2}\Big)\Phi\,; \tag{4.33}$$

$$AdS, \qquad \mu = (1 - \frac{1}{24}\frac{\lambda^2}{\rho^2}), \quad \varphi(t) = \Big(\cos^6\frac{ct}{\rho}\Big)^{-1/4\mu},$$

$$i\hbar\frac{\partial}{\partial t}\Phi = -\frac{\hbar^2}{2m\mu}\Big(-\frac{1}{\sqrt{-g(x)}}\partial_k\sqrt{-g(x)}g^{kl}\partial_l + \frac{1}{6}\frac{1}{\rho^2}\Big)\Phi\,. \tag{4.34}$$

4.3 Solving Equations

Let us derive the explicit form of the above equations in non-static coordinates
In the de Sitter space, we have

$$\Big(i\hbar\frac{\partial}{\partial t} - \frac{\hbar^2}{12M\mu\rho^2}\Big)\Phi = -\frac{\hbar^2}{2M\mu}\frac{1}{\rho^2\cosh^2(ct/\rho)}\Big[\frac{1}{\sin^2 r}\partial_r\sin^2 r\partial_r$$
$$+\frac{1}{\sin^2 r}\Big(\frac{1}{\sin\theta}\partial_\theta\sin\theta\partial_\theta + \frac{1}{\sin^2\theta}\partial_\phi\partial_\phi\Big)\Big]\Phi\,. \tag{4.35}$$

Further, taking in mind the form of squared orbital momentum

$$\mathbf{l}^2 = -\Big(\frac{1}{\sin\theta}\partial_\theta\sin\theta\partial_\theta + \frac{1}{\sin^2\theta}\partial_\phi\partial_\phi\Big),$$

we get

$$\Big(i\hbar\frac{\partial}{\partial t} - \frac{\hbar^2}{12M\mu\rho^2}\Big)\Phi = -\frac{\hbar^2}{2M\mu}\frac{1}{\rho^2\cosh^2(ct/\rho)}\Big(\frac{1}{\sin^2 r}\frac{\partial}{\partial r}\sin^2 r\frac{\partial}{\partial r} - \frac{\mathbf{l}^2}{\sin^2 r}\Big)\Phi\,. \tag{4.36}$$

In dimensionless coordinate $ct/\rho \Longrightarrow t$, this reads

$$\Big(i\frac{\hbar c}{\rho}\frac{\partial}{\partial t} - \frac{\hbar^2}{12M\mu\rho^2}\Big)\Phi = -\frac{\hbar^2}{2M\mu\rho^2}\frac{1}{\cosh^2\tau}\Big(\frac{1}{\sin^2 r}\frac{\partial}{partial r}\sin^2 r\frac{\partial}{\partial r} - \frac{\mathbf{l}^2}{\sin^2 r}\Big)\Phi\,.$$

In is convenient to employ a special unit for energy

$$\frac{\hbar^2}{2M\mu\rho^2}, \quad \frac{\hbar c}{\rho}\Big/\frac{\hbar^2}{2M\mu\rho^2} = 2\mu\frac{M\rho c}{\hbar} = \sigma,$$

and then the previous equation reads yet simpler

$$\cosh^2 t\Big(i\sigma\frac{\partial}{\partial t}-\frac{1}{6}\Big)\Phi = -\Big(\frac{1}{\sin^2 r}\frac{\partial}{\partial r}\sin^2 r\frac{\partial}{\partial r}-\frac{\mathbf{l}^2}{\sin^2 r}\Big)\Phi . \tag{4.37}$$

In this equation, the variables are separated by the substitution $\Phi = T(t)\,R(r)\,Y_{lm}(\theta,\phi)$, which yields

$$\frac{1}{T(t)}\cosh^2 t\Big(i\sigma\frac{\partial}{\partial t}-\frac{1}{6}\Big)T(t) = -\frac{1}{R(r)}\Big(\frac{1}{\sin^2 r}\frac{\partial}{\partial r}\sin^2 r\frac{\partial}{\partial r}-\frac{l(l+1)}{\sin^2 r}\Big)R(r)=\lambda ,$$

which results in two equations

$$\cosh^2 t\Big(i\sigma\frac{d}{dt}-\frac{1}{6}\Big)T(t)=\lambda T(t),$$
$$\Big(\frac{1}{\sin^2 r}\frac{d}{dr}\sin^2 r\frac{d}{dr}+\lambda-\frac{l(l+1)}{\sin^2 r}\Big)R(r)=0. \tag{4.38}$$

Similarly, we analyze the case of the AdS-space:

$$\Big(i\hbar\frac{\partial}{\partial t}+\frac{\hbar^2}{12M\mu\rho^2}\Big)\Phi = -\frac{\hbar^2}{2M\mu}\frac{1}{\rho^2\cos^2(ct/\rho)}\Big(\frac{1}{\sinh^2 r}\frac{\partial}{\partial r}\sinh^2 r\frac{\partial}{\partial r}-\frac{\mathbf{l}^2}{\sinh^2 r}\Big)\Phi ,$$

or shorter,

$$\cos^2 t\Big(i\sigma\frac{\partial}{\partial t}+\frac{1}{6}\Big)\Phi = -\Big(\frac{1}{\sinh^2 r}\frac{\partial}{\partial r}\sinh^2 r\frac{\partial}{\partial r}-\frac{\mathbf{l}^2}{\sinh^2 r}\Big)\Phi . \tag{4.39}$$

With the substitution $\Phi = T(t)\,R(r)\,Y_{lm}(\theta,\phi)$, we arrive at the equations

$$\cos^2 t\Big(i\sigma\frac{d}{dt}+\frac{1}{6}\Big)T(t)=\lambda T(t),$$
$$\Big(\frac{1}{\sinh^2 r}\frac{d}{dr}\sinh^2 r\frac{d}{dr}+\lambda-\frac{l(l+1)}{\sinh^2 r}\Big)R(r)=0. \tag{4.40}$$

Let us study Eqs. (4.38)

$$\cosh^2 t\Big(i\sigma\frac{d}{dt}+\frac{1}{6}\Big)T(t)=\lambda T(t), \tag{4.41}$$

$$\Big(\frac{1}{\sin^2 r}\frac{d}{dr}\sin^2 r\frac{d}{dr}+\lambda-\frac{l(l+1)}{\sin^2 r}\Big)R(r)=0. \tag{4.42}$$

Eq. (4.41) is solved by direct integrating:

$$i\sigma\frac{dT}{T}=\Big(\frac{\lambda}{\cosh^2 t}-\frac{1}{6}\Big)dt \qquad \Longrightarrow \qquad T(t)=e^{\frac{i}{\sigma}(t/6-\lambda\tanh t)} . \tag{4.43}$$

In absence of the non-minimal interaction term, we have a more simple result

$$T(t)=e^{-i\frac{\lambda}{\sigma}\tanh t}, \quad \text{or in ordinary units} \quad T=e^{-i\frac{\Lambda}{\hbar}\frac{\rho}{c}\tanh t}, \tag{4.44}$$

where we took into account the identities

$$\sigma = 2\frac{m\rho c}{\hbar}, \quad \lambda = \Lambda\frac{2m\rho^2}{\hbar^2}. \tag{4.45}$$

Under the restriction

$$\frac{ct}{\rho} << 1 \quad \text{or} \quad t << \frac{\rho}{c},$$

from $T(t)$ it follows the expected result $T = e^{-i\frac{\Lambda}{\hbar}t}$.

From the radial equation (4.42)

$$\Big(\frac{d^2}{dr^2} + 2\frac{\cos r}{\sin r}\frac{d}{dr} + \lambda - \frac{l(l+1)}{\sin^2 r}\Big)R(r) = 0, \tag{4.46}$$

by the substitution $R(r) = \sin^{-1} r\, F(r)$, we derive an equation without the first derivative

$$\frac{d^2}{dr^2}F + \Big(\lambda + 1 - \frac{l(l+1)}{\sin^2 r}\Big)F = 0. \tag{4.47}$$

In the variable

$$z = 1 - e^{-2ir} \quad (z = 0 \longleftrightarrow r = 0, \quad z = 1 \longleftrightarrow r = +\infty),$$

$$\frac{d}{dr} = 2ie^{-2ir}\frac{d}{dz} = 2i(1-z)\frac{d}{dz}, \quad \frac{d^2}{dr^2} = -4(1-z)^2\frac{d^2}{dz^2} + 4(1-z)\frac{d}{dz},$$

and the last equation reads

$$4(1-z)^2\frac{d^2F}{dz^2} - 4(1-z)\frac{dF}{dz} - \Big(\lambda + 1 + l(l+1)\frac{4(1-z)}{z^2}\Big)F = 0.$$

By using the substitution $F = z^a(1-z)^b f(z)$, we get

$$z(1-z)\frac{d^2f}{dz^2} + [2a - (2a+2b+1)z]\,\frac{df}{dz}$$
$$+\Big[\frac{\lambda+1}{4} - (a+b)^2 + \frac{a(a-1) - l(l+1)}{z} + (b^2 - \frac{\lambda+1}{4})\frac{1}{1-z}\Big]f = 0.$$

Consider now $a = l+1, -l$, $b = \pm\sqrt{\lambda+1}/2$; then we arrive at an equation of hypergeometric type

$$z(1-z)\frac{d^2f}{dz^2} + \Big[2a - (2a+2b+1)z\Big]\frac{df}{dz} - \Big[(a+b)^2 - \frac{\lambda+1}{4}\Big]f(z) = 0$$

with the parameters

$$\gamma = 2a, \quad \alpha = a + b - \frac{\sqrt{\lambda+1}}{2}, \quad \beta = a + b + \frac{\sqrt{\lambda+1}}{2}. \tag{4.48}$$

Let us fix the parameters:

$$a = l+1, \quad b = -\frac{\sqrt{\lambda+1}}{2}, \quad \gamma = 2l+2, \quad \alpha = l+1-\sqrt{\lambda+1}, \quad \beta = l+1. \quad (4.49)$$

We get polynomial solutions, by imposing the constraint $\alpha = -n$. This leads to the following spectrum for λ:

$$\lambda = (n+l+1)^2 - 1, \qquad n = 0, 1, 2, \dots; \qquad (4.50)$$

the corresponding solutions are given by the formula

$$\begin{aligned} F &= (1-e^{-2ir})^{l+1} e^{+i(n+l+1)r} F(-n, l+1, 2l+1, z) \\ &= (2i)^{l+1} (\sin r)^{l+1} e^{+inr} F(-n, l+1, 2l+1, 1-e^{-2ir}), \end{aligned} \qquad (4.51)$$

which vanish at the points $r = 0$ and $r = \pi$.

Now let us study the equations in the AdS-space (4.40):

$$\cos^2 t \left(i\sigma \frac{d}{dt} + \frac{1}{6} \right) T(t) = \lambda T(t), \qquad (4.52)$$

$$\left(\frac{1}{\sinh^2 r} \frac{d}{dr} \sinh^2 r \frac{d}{dr} + \lambda - \frac{l(l+1)}{\sinh^2 r} \right) R(r) = 0. \qquad (4.53)$$

Eq. (4.41) in terms of the time-variable is solved straightforwardly:

$$i\sigma \frac{dT}{T} = \left(\frac{\lambda}{\cos^2 t} - \frac{1}{6} \right) dt \quad \Longrightarrow \quad T(t) = e^{\frac{i}{\sigma}(t/6 - \lambda \tan t)}. \qquad (4.54)$$

In absence of the non-minimal interaction term, the result reads simpler,

$$T(t) = e^{-i\frac{\lambda}{\sigma}\tan t} \quad \text{or} \quad T = e^{-i\frac{\Lambda}{\hbar}\frac{\rho}{c}\tan t}. \qquad (4.55)$$

If $t << \frac{\rho}{c}$, we have the expected result $T(t) \approx e^{-i\frac{\lambda}{\sigma}\frac{ct}{\rho}}$.

The radial equation (4.53) leads to

$$\left(\frac{d^2}{dr^2} + 2\frac{\cosh r}{\sinh r}\frac{d}{dr} + \lambda - \frac{l(l+1)}{\sinh^2 r} \right) R(r) = 0. \qquad (4.56)$$

After the elementary transformation, $R(r) = \sinh^{-1} r F(r)$, we obtain

$$\frac{d^2}{dr^2} F + \left(\lambda - 1 - \frac{l(l+1)}{\sinh^2 r} \right) F = 0. \qquad (4.57)$$

In terms of the variable

$$z = 1 - e^{-2r}, \quad (z = 0 \longleftrightarrow r = 0, \quad z = 1 \longleftrightarrow r = +\infty),$$

$$\frac{d}{dr} = 2e^{-2r}\frac{d}{dz} = 2(1-z)\frac{d}{dz}, \quad \frac{d^2}{dr^2} = 4(1-z)^2\frac{d^2}{dz^2} - 4(1-z)\frac{d}{dz}$$

the above equation takes the form

$$4(1-z)^2\frac{d^2F}{dz^2}-4(1-z)\frac{dF}{dz}+\Big(\lambda-1-l(l+1)\frac{4(1-z)}{z^2}\Big)F=0 .$$

With the substitution $F=z^a(1-z)^bf(z)$, we get

$$z(1-z)\frac{d^2f}{dz^2}+[2a-(2a+2b+1)z]\,\frac{df}{dz}$$
$$+\Big[-\frac{\lambda-1}{4}-(a+b)^2+\frac{a(a-1)-l(l+1)}{z}+(b^2+\frac{\lambda-1}{4})\frac{1}{1-z}\Big]f=0 ,$$

whence, applying the restrictions

$$a=l+1,-l,\quad b=\pm\frac{i\sqrt{\lambda-1}}{2},$$

there follows hypergeometric equation

$$z(1-z)\frac{d^2f}{dz^2}+\Big[2a-(2a+2b+1)z\Big]\frac{df}{dz}-\Big[(a+b)^2+\frac{\lambda-1}{4}\Big]f=0 ,$$
$$\gamma=2a ,\quad \alpha=a+b-\frac{i\sqrt{\lambda-1}}{2},\quad \beta=a+b+\frac{i\sqrt{\lambda-1}}{2}. \tag{4.58}$$

We write down the two conjugate solutions (let $a=l+1$):

$$\text{I.}\quad b=+i\frac{\sqrt{\lambda-1}}{2},\quad F_I=z^a(1-z)^bf(z)$$
$$=(1-e^{-2r})^{l+1}(e^{-2r})^bF(l+1,l+1+i\sqrt{\lambda-1},2l+2,z), \tag{4.59}$$

$$\text{II.}\quad b'=-i\frac{\sqrt{\lambda-1}}{2},\quad F_{II}=z^a(1-z)^{b'}f(z)$$
$$=(1-e^{-2r})^{l+1}(e^{-2r})^{b'}F(l+1-i\sqrt{\lambda-1},l+1,2l+2,z). \tag{4.60}$$

4.4 Klein–Gordon–Fock Equation in Curved Space-Time

Let us consider the relativistic covariant scalar equation

$$\Big[\Big(i\hbar\nabla_\alpha+\frac{e}{c}A_\alpha\Big)\,g^{\alpha\beta}(x)\Big(i\hbar\nabla_\beta+\frac{e}{c}A_\beta\Big)\,-\frac{\hbar^2}{6}\,R\,-M^2c^2\Big]\Psi(x)=0. \tag{4.61}$$

Simplifying the notations, $\frac{e}{\hbar c}\Longrightarrow e$, $\frac{Mc}{h}\Longrightarrow M$, we have

$$\Big[(i\nabla_\alpha+eA_\alpha)g^{\alpha\beta}(x)(i\nabla_\beta+eA_\beta)-\frac{1}{6}R-M^2\Big]\Psi(x)=0. \tag{4.62}$$

This can be written differently

$$\Big[i^2\nabla_\alpha g^{\alpha\beta}(x)\nabla_\beta+ie(\nabla_\alpha g^{\alpha\beta}(x)A_\beta)$$

$$+2ieA_{\alpha}g^{\alpha\beta}(x)\nabla_{\beta}+e^{2}A_{\alpha}g^{\alpha\beta}(x)A_{\beta}-\frac{1}{6}R-M^{2}\Big]\Psi(x)=0\,,$$

and further with the use of the known relations [69]:

$$\nabla_{\alpha}g^{\alpha\beta}(x)\nabla_{\beta}\Phi=\frac{1}{\sqrt{-g}}\frac{\partial}{\partial x^{\alpha}}\sqrt{-g}g^{\alpha\beta}\frac{\partial}{\partial x^{\beta}}\Psi\,,$$
$$\nabla_{\alpha}g^{\alpha\beta}A_{\beta}=\frac{1}{\sqrt{-g}}\frac{\partial}{\partial x^{\alpha}}\sqrt{-g}g^{\alpha\beta}A_{\beta}\,, \tag{4.63}$$

we derive a more convenient form

$$\Big[i^{2}\frac{1}{\sqrt{-g}}\frac{\partial}{\partial x^{\alpha}}\sqrt{-g}g^{\alpha\beta}(x)\frac{\partial}{\partial x^{\beta}}+\frac{1}{\sqrt{-g}}\frac{\partial}{\partial x^{\alpha}}\sqrt{-g}g^{\alpha\beta}(x)A_{\beta}$$
$$+2ieA_{\alpha}g^{\alpha\beta}\frac{\partial}{\partial_{\beta}}+e^{2}A_{\alpha}g^{\alpha\beta}(x)A_{\beta}-\frac{1}{6}R-M^{2}\Big]\Psi(x)=0\,,$$

or, finally,

$$\Big[\frac{1}{\sqrt{-g}}\Big(i\frac{\partial}{\partial x^{\alpha}}+eA_{\alpha}\Big)\sqrt{-g}g^{\alpha\beta}(x)\Big(i\frac{\partial}{\partial x^{\beta}}+eA_{\beta}\Big)-\frac{1}{6}R-M^{2}\Big]\Psi(x)=0\,. \tag{4.64}$$

In ordinary units, this reads

$$\Big[\frac{1}{\sqrt{-g}}\Big(i\hbar\frac{\partial}{\partial x^{\alpha}}+\frac{e}{c}A_{\alpha}\Big)\sqrt{-g}g^{\alpha\beta}(x)\Big(i\hbar\frac{\partial}{\partial x^{\beta}}+\frac{e}{c}A_{\beta}\Big)-\frac{\hbar^{2}}{6}R-M^{2}c^{2}\Big]\Psi(x)=0\,. \tag{4.65}$$

4.5 Solving Equation in Expanding de Sitter Metric

Let us consider Eq. (4.64) in the non-static de Sitter metric

$$dS,\qquad dS^{2}=dt^{2}-\cosh^{2}t\,[\,dr^{2}+\sin^{2}r(d\theta^{2}+\sin^{2}\theta\,d\phi^{2})]\,,$$
$$\Big[\frac{1}{\sqrt{-g}}\frac{\partial}{\partial x^{\alpha}}\sqrt{-g}g^{\alpha\beta}(x)\frac{\partial}{\partial x^{\beta}}+\frac{1}{6}+M^{2}\Big]\Psi(x)=0\,, \tag{4.66}$$

or,

$$\cosh^{2}t\Big(\frac{1}{\cosh^{3}t}\partial_{t}\,\cosh^{3}t\partial_{t}+\frac{1}{6}\;+M^{2}\Big)\Psi(x)=\frac{1}{\sin^{2}r}\Big(\frac{\partial}{\partial r}\sin^{2}r\frac{\partial}{\partial r}-\mathbf{l}^{2}\Big)\Psi(x)\,. \tag{4.67}$$

The variables are separated by the evident substitution $\Phi(x)=T(t)R(r)Y_{lm}(\theta,\phi)$, so we derive ($K^2$ stands for the separating constant)

$$\frac{1}{T(t)}\cosh^{2}t\Big(\frac{1}{\cosh^{3}t}\frac{d}{dt}\cosh^{3}t\frac{d}{dt}+\frac{1}{6}+M^{2}\Big)T(t)$$
$$=\frac{1}{R(r)}\Big(\frac{1}{\sin^{2}r}\frac{d}{dr}\sin^{2}r\frac{d}{dr}-\frac{l(l+1)}{\sin^{2}r}\Big)R(r)=-K^{2}\,.$$

This results in two equations

$$\Big(\frac{d^2}{dr^2}+\frac{2}{\tan r}\frac{d}{dr}-\frac{l(l+1)}{\sin^2 r}+K^2\Big)R(r)=0, \tag{4.68}$$

$$\Big(\frac{d^2}{dt^2}+3\,\tanh t\frac{d}{dt}+\frac{1}{6}+M^2+\frac{K^2}{\cosh^2 t}\Big)T(t)=0. \tag{4.69}$$

In the limit of flat Minkowski space we get

$$\Big(\frac{d^2}{dt^2}+M^2+K^2\Big)T(t)=0, \quad T(t)=e^{-i\epsilon t}, \quad K^2=\epsilon^2-M^2>0. \tag{4.70}$$

Eq. (4.68) coincides (up to the replacement $\Lambda \Longleftrightarrow K^2$) with the yet studied equation. Hence we need to examine only Eq. (4.69). In Eq. (4.69), let us introduce the multiplier $T(t)=\varphi(t)f(t)$; then we get

$$f''+\Big(2\frac{\varphi'}{\varphi}+3\tanh t\Big)f'+\Big(\frac{\varphi''}{\varphi}+3\tanh t\frac{\varphi'}{\varphi}+M^2+\frac{1}{6}+\frac{K^2}{\coth^2 t}\Big)f=0.$$

Let the coefficient at the first derivative vanish,

$$\varphi=\cosh^{-\frac{3}{2}}t, \quad f''+(\frac{\varphi''}{\varphi}+3\tanh t\frac{\varphi'}{\varphi}+M^2+\frac{1}{6}+\frac{K^2}{\cosh^2 t})f=0,$$

$$\varphi'(t)=-\frac{3}{2}\frac{\sinh t}{\cosh^{\frac{5}{2}}t}, \quad \frac{\varphi'}{\varphi}=-\frac{3}{2}\tanh t, \quad \frac{\varphi''}{\varphi}=\frac{3}{4}\frac{3\sinh^2 t-2}{\cosh^2 t}.$$

Explicitly, the equation for $f(t)$ reads

$$\frac{d^2f}{dt^2}+\Big(\frac{9}{4}\tanh^2 t-\frac{3}{2\cosh^2 t}-\frac{9}{2}\tanh^2 t+M^2+\frac{1}{6}+\frac{K^2}{\cosh^2 t}\Big)f=0,$$

or,

$$\frac{d^2f}{dt^2}+\Big((M^2-\frac{25}{12})+\frac{(K^2+3/4)}{\cosh^2 t}\Big)f(t)=0. \tag{4.71}$$

In the variable $y=\tanh t$, this equation takes the form

$$(1-y^2)\frac{d^2f}{dy^2}-2y\frac{df}{dy}+\Big(K^2+\frac{3}{4}-\frac{25/12-M^2}{1-x^2}\Big)f=0, \tag{4.72}$$

which may be identified with the Legendre equation [1]

$$(1-y^2)\frac{d^2W}{dy^2}-2y\frac{dW}{dy}+\Big(\nu(\nu+1)-\frac{\mu^2}{1-y^2}\Big)W=0 \tag{4.73}$$

with the parameters

$$\nu=-\frac{1}{2}\pm\sqrt{1+K^2}, \quad \mu=\pm i\sqrt{M^2-25/12}. \tag{4.74}$$

For a more detailed study of Eq. (4.71), let us use the alternative variable

$$x = 1 + e^{+2t}, \quad e^{2t} = x - 1,$$

$$\frac{1}{\cosh^2 t} = 1 - \tanh^2 t = 1 - \frac{(x-2)^2}{x^2} = \frac{4x-4}{x^2},$$

$$\frac{d}{dt} = 2e^{2t}\frac{d}{dx} = 2(x-1)\frac{d}{dx}, \quad \frac{d^2}{dt^2} = 4(x-1)^2\frac{d^2}{dx^2} + 4(x-1)\frac{d}{dx}.$$

Then Eq. (4.71) can be written as

$$x(1-x)\frac{d^2 f}{dx^2} - x\frac{df}{dx} - \Big(\frac{1}{4}M^2 - \frac{25}{48} + \frac{K^2 + 3/4}{x} - \frac{1}{48}\frac{12M^2 - 25}{1-x}\Big) f = 0.$$

With the substitution $f = x^a (1-x)^b \bar{f}(x)$, we get

$$x(1-x)\frac{d^2 \bar{f}}{dx^2} + \Big[2a - (2a+2b+1)x\Big]\frac{d\bar{f}}{dx}$$

$$-\Big[(a+b)^2 + \frac{1}{4}M^2 - \frac{25}{48} + \frac{a(1-a) + K^2 + 3/4}{x} - \frac{1}{48}\frac{48b^2 + 12M^2 - 25}{1-x}\Big]\bar{f} = 0.$$

We further consider

$$a(1-a) + K^2 + 3/4 = 0 \quad \Longrightarrow \quad a = \frac{1}{2} \pm \sqrt{1+K^2},$$

$$48b^2 + 12M^2 - 25 = 0 \quad \Longrightarrow \quad b = \pm\frac{i}{2}\sqrt{M^2 - 25/12} \equiv \pm\frac{i}{2}\Gamma,$$

and then the previous equation becomes of hypergeometric type

$$x(1-x)\frac{d^2 \bar{f}}{dx^2} + [2a - (2a+2b+1)x]\frac{d\bar{f}}{dx} - \Big[(a+b)^2 + \frac{1}{4}(M^2 - \frac{25}{12)}\Big]\bar{f} = 0$$

with the parameters

$$\alpha = a + b + \frac{i}{2}\Gamma, \quad \beta = a + b - \frac{i}{2}\Gamma, \quad \gamma = 2a. \tag{4.75}$$

Note that the two different values $b = +i\Gamma/2, -i\Gamma/2$ lead to the same solution $f(x)$; this follows from the known identity [1]

$$F(\alpha, \beta, \gamma; x) = (1-x)^{\gamma-\alpha-\beta} F(\gamma - a, \gamma - \beta, \gamma; x). \tag{4.76}$$

Indeed, with the identities in mind

$$\gamma - \alpha - \beta = -2b, \quad \gamma - \alpha = a - b - i\Gamma/2, \quad \gamma - \beta = a - b + i\Gamma/2,$$

we obtain the needed result

$$x^a (1-x)^b F(\alpha, \beta, \gamma; x) = x^a (1-x)^{-b} F(a - b + i\Gamma/2, a - b - i\Gamma/2, \gamma; x).$$

Let us study the behavior of the constructed solutions in the regions $t \to \pm\infty$:

$$t \to +\infty \, (x \to +\infty), \quad t \to -\infty \; (x \to +1).$$

To describe the asymptotic behavior at $r \to \infty$, we make use of the Kummer relation [1]

$$U_1 = \frac{\Gamma(\gamma)\Gamma(\beta-\alpha)}{\Gamma(\gamma-\alpha)\Gamma(\beta)} U_3 + \frac{\Gamma(\gamma)\Gamma(\alpha-\beta)}{\Gamma(\gamma-\beta)\Gamma(\alpha)} U_4, \tag{4.77}$$

where

$$U_1 = F(\alpha, \beta, \gamma; x), \quad U_3 = (-x)^{-\alpha} F(\alpha, \alpha+1-\gamma, \alpha+1-\beta; \frac{1}{x}),$$

$$U_4 = (-x)^{-\beta} F(\beta, \beta+1-\gamma, \beta+1-\alpha; \frac{1}{x}).$$

At $x \to +\infty$, the formula (4.77) gives

$$F(a,b,c; x \to \infty) = \frac{\Gamma(\gamma)\Gamma(\beta-\alpha)}{\Gamma(\gamma-\alpha)\Gamma(\beta)} (-x)^{-\alpha} + \frac{\Gamma(\gamma)\Gamma(\alpha-\beta)}{\Gamma(\gamma-\beta)\Gamma(\alpha)} (-x)^{-\beta}.$$

For the complete function

$$f(x) = x^a (1-x)^b F(\alpha, \beta, \gamma; x)$$

at this limit we get expression

$$f(x \to +\infty) \sim (-1)^a (-x)^{a+b} \Big[\frac{\Gamma(\gamma)\Gamma(\beta-\alpha)}{\Gamma(\gamma-\alpha)\Gamma(\beta)} (-x)^{-\alpha} + \frac{\Gamma(\gamma)\Gamma(\alpha-\beta)}{\Gamma(\gamma-\beta)\Gamma(\alpha)} (-x)^{-\beta} \Big],$$

whence it follows

$$f(t \to +\infty) \left[\frac{\Gamma(\gamma)\Gamma(\beta-\alpha)}{\Gamma(\gamma-\alpha)\Gamma(\beta)} (-1)^{-i\Gamma/2} e^{-i\Gamma t} + \frac{\Gamma(\gamma)\Gamma(\alpha-\beta)}{\Gamma(\gamma-\beta)\Gamma(\alpha)} (-1)^{+i\Gamma/2} e^{+i\Gamma t} \right]; \tag{4.78}$$

we recall that

$$\alpha = a+b+\frac{i}{2}\Gamma, \quad \beta = a+b-\frac{i}{2}\Gamma, \quad \gamma = 2a.$$

The first linearly independent solution is

$$\begin{gathered} a = \frac{1}{2} + \sqrt{K^2+1}, \quad b = +\frac{i}{2}\Gamma, \\ \alpha = \frac{1}{2} + \sqrt{K^2+1} + i\Gamma, \quad \beta = \frac{1}{2} + \sqrt{K^2+1}, \quad \gamma = 1 + 2\sqrt{K^2+1}, \\ \beta - \alpha = -i\Gamma, \quad \alpha - \beta = +i\Gamma, \\ \gamma - \alpha = \frac{1}{2} + \sqrt{K^2+1} - i\Gamma = \alpha^*, \quad \gamma - \beta = \frac{1}{2} + \sqrt{K^2+1} = \beta; \end{gathered} \tag{4.79}$$

the correspondent asymptotic (4.78) takes the form a of standing wave

$$f(t \to +\infty) \sim \Big[\text{const}e^{-i\Gamma t} + \text{const}^* e^{+i\Gamma t}\Big]_{t\to+\infty}$$
$$= 2\text{Re}\left[\frac{\Gamma(1+2\sqrt{K^2+1})\Gamma(-i\Gamma)}{\Gamma(\frac{1}{2}+\sqrt{K^2+1}-i\Gamma)\Gamma(\frac{1}{2}+\sqrt{K^2+1})}(-1)^{-i\Gamma/2}e^{-i\Gamma t}\right]. \tag{4.80}$$

The second independent solution is

$$a = \frac{1}{2} - \sqrt{K^2+1}\,, \quad b = +\frac{i}{2}\Gamma,$$
$$\alpha = \frac{1}{2} - \sqrt{K^2+1} + i\Gamma, \quad \beta = \frac{1}{2} - \sqrt{K^2+1}\,, \quad \gamma = 1 - 2\sqrt{K^2+1}\,,$$
$$\beta - \alpha = -i\Gamma, \quad \alpha - \beta = +i\Gamma,$$
$$\gamma - \alpha = \frac{1}{2} - \sqrt{K^2+1} - i\Gamma = \alpha^*, \quad \gamma - \beta = \frac{1}{2} - \sqrt{K^2+1} = \beta, \tag{4.81}$$

and its asymptotic also represent a standing wave (with a different phase shift)

$$f(t \to +\infty) \sim \Big[\text{const}\, e^{-i\Gamma t} + \text{const}^* e^{+i\Gamma t}\Big]_{t\to+\infty}$$
$$= 2\text{Re}\left[\frac{\Gamma(1-2\sqrt{K^2+1})\,\Gamma(-i\Gamma)}{\Gamma(\frac{1}{2}-\sqrt{K^2+1}-i\Gamma)\Gamma(\frac{1}{2}-\sqrt{K^2+1})}(-1)^{-i\Gamma/2}e^{-i\Gamma t}\right]. \tag{4.82}$$

Taking as solutions the Kummer functions U_3 and U_4, we provide solutions $f(t)$ describing the running waves in the region $t \to +\infty$. These solutions are

$$f_3(x) = x^a(1-x)^b U_3 = x^a(1-x)^b(-x)^{-\alpha}F(\alpha, \alpha+1-\gamma, \alpha+1-\beta; \frac{1}{x}),$$
$$f_4(x) = x^a(1-x)^b U_4 = x^a(1-x)^b(-x)^{-\beta}F(\beta, \beta+1-\gamma, \beta+1-\alpha; \frac{1}{x});$$

at $t \to +\infty$, they behave as follows

$$f_3(t \to +\infty) = x^a(1-x)^b(-x)^{-(a+b+i\Gamma/2)} \sim e^{-i\Gamma t},$$
$$f_4(t \to +\infty) = x^a(1-x)^b(-x)^{-(a+b-i\Gamma/2)} \sim e^{+i\Gamma t}. \tag{4.83}$$

Now, we are to study behavior of solutions in the region $r \to -\infty$. Here we need the following Kummer relations

$$U_1 = \frac{\Gamma(\gamma)\Gamma(\gamma-\beta-\alpha)}{\Gamma(\gamma-\alpha)\Gamma(\gamma-\beta)}U_2 + \frac{\Gamma(\gamma)\Gamma(\alpha+\beta-\gamma)}{\Gamma(\beta)\Gamma(\alpha)}U_6, \tag{4.84}$$

where

$$U_1 = F(\alpha, \beta, \gamma; x), \quad U_2 = F(\alpha, \beta, \alpha+\beta+1-\gamma; 1-x),$$
$$U_6 = (1-x)^{\gamma-\alpha-\beta}F(\gamma-\alpha, \gamma-\beta, \gamma+1-\alpha-\beta; 1-x).$$

In the region $t \to -\infty$, it gives

$$U_1 = \frac{\Gamma(\gamma)\Gamma(\gamma-\beta-\alpha)}{\Gamma(\gamma-\alpha)\Gamma(\gamma-\beta)} + \frac{\Gamma(\gamma)\Gamma(\alpha+\beta-\gamma)}{\Gamma(\beta)\Gamma(\alpha)}(1-x)^{\gamma-\alpha-\beta}, \tag{4.85}$$

and for the complete function $f(t)$ we derive the asymptotic

$$f(t \to -\infty) = \frac{\Gamma(\gamma)\Gamma(\gamma-\beta-\alpha)}{\Gamma(\gamma-\alpha)\Gamma(\gamma-\beta)}(1-x)^b + \frac{\Gamma(\gamma)\Gamma(\alpha+\beta-\gamma)}{\Gamma(\beta)\Gamma(\alpha)}(1-x)^b(1-x)^{\gamma-\alpha-\beta}. \quad (4.86)$$

Taking in mind the identities

$$\alpha = a+b+\frac{i}{2}\Gamma, \quad \beta = a+b-\frac{i}{2}\Gamma, \quad \gamma = 2a, \quad \gamma-\alpha-\beta = -2b,$$

we write the last relation in the form

$$f(t \to -\infty) = \frac{\Gamma(\gamma)\Gamma(\gamma-\beta-\alpha)}{\Gamma(\gamma-\alpha)\Gamma(\gamma-\beta)}(-e^{2t})^b + \frac{\Gamma(\gamma)\Gamma(\alpha+\beta-\gamma)}{\Gamma(\beta)\Gamma(\alpha)}(-e^{2t})^{-b}.$$

Finally, by using the relations

$$a = \frac{1}{2} \pm \sqrt{K^2+1}, \qquad b = +\frac{i}{2}\Gamma, \quad \gamma = 2a,$$

and their consequences

$$\alpha = \frac{1}{2} \pm \sqrt{K^2+1} + i\Gamma, \quad \beta = \frac{1}{2} \pm \sqrt{K^2+1},$$

$$\gamma = 1 \pm 2\sqrt{K^2+1}, \quad \gamma-\alpha-\beta = -i\Gamma,$$

$$\gamma-\alpha = \frac{1}{2} \pm \sqrt{K^2+1} - i\Gamma = \alpha^*, \quad \gamma-\beta = \frac{1}{2} \pm \sqrt{K^2+1} = \beta = \beta^*,$$

we conclude that in this region the solutions behave as standing waves:

$$a = \tfrac{1}{2} + \sqrt{K^2+1}, \qquad f(t \to -\infty)$$

$$= 2\mathrm{Re}\Big[\frac{\Gamma(1+2\sqrt{K^2+1})\Gamma(+i\Gamma)}{\Gamma(\frac{1}{2}+\sqrt{K^2+1}+i\Gamma)\Gamma(\frac{1}{2}+\sqrt{K^2+1})}(-1)^{-i\Gamma/2}e^{-i\Gamma t}\Big]; \quad (4.87)$$

$$a = \tfrac{1}{2} - \sqrt{K^2+1}, \qquad f(t \to -\infty)$$

$$= 2\mathrm{Re}\Big[\frac{\Gamma(1-2\sqrt{K^2+1})\Gamma(+i\Gamma)}{\Gamma(\frac{1}{2}-\sqrt{K^2+1}+i\Gamma)\Gamma(\frac{1}{2}-\sqrt{K^2+1})}(-1)^{-i\Gamma/2}e^{-i\Gamma t}\Big]. \quad (4.88)$$

The complete functions $f(x)$, which refer to the Kummer solutions U_2 and U_6:

$$f_2 = x^a(1-x)^b F(\alpha,\beta,\alpha+\beta+1-\gamma;1-x),$$

$$f_6 = x^a(1-x)^b(1-x)^{\gamma-\alpha-\beta} F(\gamma-\alpha,\gamma-\beta,\gamma+1-\alpha-\beta;1-x),$$

represent the running waves

$$f_2 \sim e^{+i\Gamma t}, \quad f_4 \sim e^{-i\Gamma t}. \quad (4.89)$$

The two pairs of independent solutions

$$U_3(x),\ U_4(x) \quad \text{and} \quad U_2(x),\ U_6(x)$$

are linked to the Kummer formulas [1], as well.

4.6 Solving Equation in Oscillating de Sitter Metric

Let us discuss the solutions of Eq. (4.65) in the anti de Sitter metric

$$AdS, \quad dS^2 = dt^2 - \cos^2 t\,[dr^2 + \sinh^2 r(d\theta^2 + \sin^2\theta d\phi^2)]\,, \tag{4.90}$$

$$\Big[\frac{1}{\sqrt{-g}}\frac{\partial}{\partial x^\alpha}\sqrt{-g}g^{\alpha\beta}(x)\frac{\partial}{\partial x^\beta} - \frac{1}{6} + M^2\Big]\Psi(x) = 0. \tag{4.91}$$

We further get

$$\cos^2 t\Big(\frac{1}{\cos^3 t}\partial_t \cos^3 t\partial_t - \frac{1}{6} + M^2\Big)\Psi(x) = \frac{1}{\sinh^2 r}\Big(\frac{\partial}{\partial r}\sinh^2 r\frac{\partial}{\partial r} - \mathbf{l}^2\Big)\Psi(x)\,. \tag{4.92}$$

After the separation of variables, we find the two equations

$$\Big(\frac{d^2}{dr^2} + \frac{2}{\tanh r}\frac{d}{dr} - \frac{l(l+1)}{\sinh^2 r} + K^2\Big)R(r) = 0\,, \tag{4.93}$$

$$\Big(\frac{d^2}{dt^2} - 3\tanh t\frac{d}{dt} - \frac{1}{6} + M^2 + \frac{K^2}{\cos^2 t}\Big)T(t) = 0. \tag{4.94}$$

We need to study only Eq. (4.94). After separating the simple multiplier $T(t) = (\cos^{-\frac{3}{2}} t)f(t)$, we get

$$\frac{d^2 f}{dt^2} + \Big[(M^2 + \frac{25}{12}) + \frac{(K^2 - 3/4)}{\cos^2 t}\Big]f(t) = 0. \tag{4.95}$$

In the variable $ix = \tan t$, this reads

$$(1-x^2)\frac{d^2 f}{dx^2} - 2x\frac{df}{dx} + \Big(-K^2 + \frac{3}{4} - \frac{25/12 + M^2}{1-x^2}\Big)f = 0\,,$$

which can be identified with the Legendre equation

$$(1-z^2)\frac{d^2\omega}{dz^2} - 2z\frac{d\omega}{dz} + \Big[\nu(\nu+1) - \frac{\mu^2}{1-z^2}\Big]\omega = 0\,,$$

$$\nu = -\frac{1}{2} \pm \sqrt{1-K^2}\,, \quad \mu = \pm\frac{1}{6}\sqrt{75 + 36M^2}\,.$$

We can use another variable,

$$x = 1 + e^{+2it}\,, \quad e^{2it} = x - 1\,, \quad \tan t = -i\frac{e^{2it} - 1}{e^{2it} + 1} = -i\frac{x-2}{x}\,,$$

$$\frac{1}{\cos^2 t} = 1 + \tan^2 t = 1 - \frac{(x-2)^2}{x^2} = \frac{4x-4}{x^2}\,,$$

$$\frac{d}{dt} = 2ie^{2it}\frac{d}{dx} = 2i(x-1)\frac{d}{dx}, \quad \frac{d^2}{dt^2} = -4(x-1)^2\frac{d^2}{dx^2} - 4(x-1)\frac{d}{dx}\,,$$

and then we obtain

$$x(1-x)\frac{d^2f}{dx^2} - x\frac{df}{dx} + \Big(\frac{1}{4}M^2 + \frac{25}{48} + \frac{K^2 - 3/4}{x} - \frac{1}{48}\frac{12M^2+25}{1-x}\Big)f = 0.$$

By using the substitution $f = x^a(1-x)^b \bar{f}(x)$, we infer

$$x(1-x)\frac{d^2\bar{f}}{dx^2} + \Big[2a - (2a+2b+1)x\Big]\frac{d\bar{f}}{dx}$$

$$+\Big[-(a+b)^2 + \frac{1}{4}M^2 + \frac{25}{48} + \frac{a(a-1)+K^2-3/4}{x} - \frac{1}{48}\frac{-48b^2+12M^2+25}{1-x}\Big]\bar{f} = 0.$$

Further, by considering

$$a(a-1) + K^2 - 3/4 = 0 \quad \Longrightarrow \quad a = \frac{1}{2} \pm \sqrt{1-K^2},$$

$$-48b^2 + 12M^2 + 25 = 0 \quad \Longrightarrow \quad b = \pm\frac{1}{12}\sqrt{75+36M^2},$$

we obtain an equation of hypergeometric type

$$x(1-x)\frac{d^2\bar{f}}{dx^2} + \Big[2a - (2a+2b+1)x\Big]\frac{d\bar{f}}{dx} - \Big[(a+b)^2 - \frac{1}{4}M^2 - \frac{25}{48}\Big]\bar{f} = 0,$$

$$\alpha = a + b - \frac{1}{12}\sqrt{75+36M^2}, \quad \beta = a + b + \frac{1}{12}\sqrt{75+36M^2}, \quad \gamma = 2a.$$

Obviously, in this space model, neither bound states, nor quantization for the parameter K, exist.

Chapter 5

Spin 1/2 Particle in Nonstatic de Sitter Spaces, Spherical Coordinates

5.1 Particle in Expanding de Sitter Model

We start with the Dirac equation

$$\Big[i\gamma^c(e^{\alpha}_{(c)}\partial_\alpha + \frac{1}{2}\sigma^{ab}\gamma_{abc}) - M\Big]\Psi = 0\,, \tag{5.1}$$

where γ_{abc} stand for the Ricci rotation coefficients, $\gamma_{bac} = -\gamma_{abc} = -e_{(b)\beta;\alpha}\, e^{\beta}_{(a)} e^{\alpha}_{(c)}$. In curvilinear orthogonal coordinates Eq. (5.1) reads

$$\Big\{i\gamma^k(\, e^{\alpha}_{(k)}\partial_\alpha + B_k\,) - M\Big\}\,\Psi = 0\,, \tag{5.2}$$

where

$$B_k(x) = \frac{1}{2}\, e^{\alpha}_{(k)\,;\alpha}(x) = \frac{1}{2}\frac{1}{\sqrt{-g}}\frac{\partial}{\partial x^\alpha}\sqrt{-g}e^{\alpha}_{(k)}\,.$$

In the de Sitter space, we employ the nonstatic coordinates $x^\alpha = (t, r, \theta, \phi)$

$$dS^2 = dt^2 - \cosh^2 t[\, dr^2 + \sin^2 r(d\theta^2 + \sin^2\theta d\phi^2)\,] \tag{5.3}$$

and the following tetrad

$$e^{\alpha}_{(0)} = (1,0,0,0)\,, \quad e^{\alpha}_{(3)} = (0, \frac{1}{\cosh t}, 0, 0)\,,$$
$$e^{\alpha}_{(1)} = (0,0,\frac{1}{\cosh t \sin r},0)\,, \; e^{\alpha}_{(2)} = (0,0,0,\frac{1}{\cosh t \sin r\, \sin\theta})\,. \tag{5.4}$$

Taking into account the expressions for B_k

$$B_0 = \frac{3}{2}\tanh t\,, \quad B_1 = +\frac{1}{2}\frac{1}{\cosh t}\frac{1}{\sin r}\frac{\cos\theta}{\sin\theta}\,, \quad B_2 = 0\,, \quad B_3 = \frac{1}{\cosh t}\frac{1}{\tan r}\,,$$

Eq. (5.2)

$$\Big[i\gamma^0(e^t_{(0)}\partial_t + B_0) + i\gamma^3(e^r_{(3)}\partial_r + B_3)$$
$$+ i\gamma^1(e^\theta_{(1)}\partial_\theta + B_1) + i\gamma^2(e^\phi_{(2)}\partial_\phi + B_2) - M\Big]\Psi = 0$$

takes the form

$$\Big[i\gamma^0\Big(\partial_t + \frac{3}{2}\tanh t\Big) + i\gamma^3\Big(\frac{1}{\cosh t}\partial_r + \frac{1}{\cosh t}\frac{1}{\tan r}\Big)$$
$$+ i\gamma^1\Big(\frac{1}{\cosh t \sin r}\partial_\theta + \frac{1}{2}\frac{1}{\cosh t}\frac{1}{\sin r}\frac{\cos\theta}{\sin\theta}\Big) + i\gamma^2\frac{1}{\cosh t \sin r \sin\theta}\partial_\phi - M\Big]\Psi = 0.$$

It may be re-written as follows

$$\Big\{i\gamma^0\cosh t\Big(\partial_t + \frac{3}{2}\tanh t\Big) + i\gamma^3\Big(\partial_r + \frac{1}{\tan r}\Big)$$
$$+ \frac{1}{\sin r}\Big[i\gamma^1\Big(\partial_\theta + \frac{1}{2}\frac{\cos\theta}{\sin\theta}\Big) + i\gamma^2\frac{1}{\sin\theta}\partial_\phi\Big] - M\cosh t\Big\}\Psi = 0. \tag{5.5}$$

Taking in mind the identity $\gamma^2\sigma^{12} = +\gamma^1/2$, we arrive at

$$\Big[i\gamma^0\cosh t\Big(\partial_t + \frac{3}{2}\tanh t) + i\gamma^3(\partial_r + \frac{1}{\tan r}\Big) + \frac{1}{\sin r}\Sigma_{\theta\phi} - M\cosh t\Big]\Psi = 0, \tag{5.6}$$

where

$$\Sigma_{\theta,\phi} = i\gamma^1\partial_\theta + \gamma^2\frac{i\partial_\phi + i\sigma^{12}\cos\theta}{\sin\theta}.$$

Having separated in the wave function two multipliers

$$\Psi(x) = \frac{1}{\sin r}\frac{1}{\cosh^{3/2} t}\,\varphi(x), \tag{5.7}$$

we get the simpler form

$$\Big(i\gamma^0\cosh t\partial_t + i\gamma^3\partial_r + \frac{1}{\sin r}\Sigma_{\theta\phi} - M\cosh t\Big)\varphi = 0. \tag{5.8}$$

Let us use the following substitution (where $D^j_{-m,\sigma}(\phi,\theta,0) \equiv D_\sigma$):

$$\varphi_{jm}(x) = \begin{vmatrix} f_1(t,r)\,D_{-1/2} \\ f_2(t,r)\,D_{+1/2} \\ f_3(t,r)\,D_{-1/2} \\ f_4(t,r)\,D_{+1/2} \end{vmatrix}. \tag{5.9}$$

By applying the needed recurrent formulas [68]

$$\begin{gathered} \partial_\theta D_{+1/2} = a\,D_{-1/2} - b\,D_{+3/2}, \\ \frac{-m - 1/2\,\cos\theta}{\sin\theta}D_{+1/2} = -a\,D_{-1/2} - b\,D_{+3/2}, \\ \partial_\theta\,D_{-1/2} = b\,D_{-3/2} - a\,D_{+1/2}, \\ \frac{-m + 1/2\,\cos\theta}{\sin\theta}D_{-1/2} = -b\,D_{-3/2} - a\,D_{+1/2}, \\ a = \frac{j+1/2}{2}, \quad b = \frac{1}{2}\sqrt{(j-1/2)(j+3/2)}, \end{gathered} \tag{5.10}$$

we infer (let $\nu = j + 1/2$)

$$\Sigma_{\theta,\phi}\, \varphi_{jm}(x) = i\,\nu \left| \begin{array}{c} -\, f_4(t,r)\, D_{-1/2} \\ +\, f_3(t,r)\, D_{+1/2} \\ +\, f_2(t,r)\, D_{-1/2} \\ -\, f_1(t,r)\, D_{+1/2} \end{array} \right| , \qquad (5.11)$$

and further we obtain the radial equations

$$\begin{aligned} & i\cosh t \frac{\partial}{\partial t} f_3 - i \frac{\partial}{\partial r} f_3 - i \frac{\nu}{\sin r} f_4 - M\, \cosh t\, f_1 = 0\,, \\ & i\cosh t \frac{\partial}{\partial t} f_4 + i \frac{\partial}{\partial r} f_4 + i \frac{\nu}{\sin r} f_3 - M\, \cosh t\, f_2 = 0\,, \\ & i\cosh t \frac{\partial}{\partial t} f_1 + i \frac{\partial}{\partial r} f_1 + i \frac{\nu}{\sin r} f_2 - M\, \cosh t\, f_3 = 0\,, \\ & i\cosh t \frac{\partial}{\partial t} f_2 - i \frac{\partial}{\partial r} f_2 - i \frac{\nu}{\sin r} f_1 - M\, \cosh t\, f_4 = 0\,. \end{aligned} \qquad (5.12)$$

In spherical tetrad basis, the spatial reflection operator is given by the formula

$$\hat{\Pi}_{sph} = \left| \begin{array}{cccc} 0 & 0 & 0 & -1 \\ 0 & 0 & -1 & 0 \\ 0 & -1 & 0 & 0 \\ -1 & 0 & 0 & 0 \end{array} \right| \hat{P}\,. \qquad (5.13)$$

From the eigenvalue equation

$$\hat{\Pi}_{sph} \Psi_{jm} = \Pi\, \Psi_{jm}\,, \quad \text{note the rule} \quad \hat{P}\, D^{j}_{-m,\sigma}(\phi,\theta,0) = (-1)^{j}\, D^{j}_{-m,-\sigma}(\phi,\theta,0)\,,$$

we obtain two eigenvalues and corresponding restrictions on the radial functions $f_i(r)$:

$$\Pi = \delta\, (-1)^{j+1}, \quad \delta = \pm 1\,, \quad f_4 = \delta\, f_1\,, \quad f_3 = \delta\, f_2\,. \qquad (5.14)$$

Thus, in the following, we employ the more simple form for the wave functions

$$\varphi(x)_{jm\delta} = \left| \begin{array}{c} f_1(t,r)\, D_{-1/2} \\ f_2(t,r)\, D_{+1/2} \\ \delta\, f_2(t,r)\, D_{-1/2} \\ \delta\, f_1(t,r)\, D_{+1/2} \end{array} \right| . \qquad (5.15)$$

Correspondingly, the system (5.12) gives two equations (we have two variants depending on δ)

$$\begin{aligned} & \left(\frac{\partial}{\partial r} + \frac{\nu}{\sin r} \right) f + \left(i\cosh t\, \frac{\partial}{\partial t} + \delta\, M\, \cosh t \right) g = 0\,, \\ & \left(\frac{\partial}{\partial r} - \frac{\nu}{\sin r} \right) g - \left(i\cosh t\, \frac{\partial}{\partial t} - \delta\, M\, \cosh t \right) f = 0\,, \end{aligned} \qquad (5.16)$$

where instead of $f_1(t,r)$ and $f_2(t,r)$, we use their combinations

$$f(t,r)=\frac{f_1+f_2}{\sqrt{2}}, \quad g(t,r)=\frac{f_1-f_2}{i\sqrt{2}}.$$

The variables in the system (5.16) can be separated, if one searches for solutions in the form

$$f(t,r)=f(t)f(r), \quad g(t,r)=g(t)g(r),$$

this results in

$$\frac{1}{g(r)}\Big(\frac{d}{dr}+\frac{\nu}{\sin r}\Big)f(r)=-\frac{1}{f(t)}\Big(i\cosh t\frac{d}{dt}+\delta\, M\,\cosh t\Big)\,g(t)=\lambda,$$

$$\frac{1}{f(r)}\Big(\frac{d}{dr}-\frac{\nu}{\sin r}\Big)g(r)=\frac{1}{g(t)}\Big(i\cosh t\frac{d}{dt}-\delta\, M\,\cosh t\Big)f(t)=\mu.$$

Therefore, we arrive at at two subsystems in variables t and r:

$$\Big(\frac{d}{dr}+\frac{\nu}{\sin r}\Big)f(r)=\lambda\, g(r), \quad \Big(\frac{d}{dr}-\frac{\nu}{\sin r}\Big)g(r)=\mu\, f(r); \tag{5.17}$$

$$\Big(i\cosh t\frac{d}{dt}+\delta M\cosh t\Big)g(t)=-\lambda f(t), \quad \Big(i\cosh t\frac{d}{dt}-\delta M\cosh t\Big)f(t)=\mu g(t). \tag{5.18}$$

For definiteness, in (5.18) we shall consider the case of $\delta=+1$:

$$\Big(i\cosh t\frac{d}{dt}+M\cosh t\Big)g(t)=-\lambda f(t), \quad \Big(i\cosh t\frac{d}{dt}-M\cosh t\Big)f(t)=\mu g(t); \tag{5.19}$$

the variant with $\delta=-1$ follows from the previous one (5.19) by means of the formal change $M\Longrightarrow -M$.

From (5.17), we exclude the function $g(r)$:

$$g(r)=\frac{1}{\lambda}\Big(\frac{d}{dr}+\frac{\nu}{\sin r}\Big)f(r), \quad \Big(\frac{d^2}{dr^2}-\nu\frac{\cos r}{\sin^2 r}-\frac{\nu^2}{\sin^2 r}-\lambda\mu\Big)f(r)=0. \tag{5.20}$$

Similarly, we derive

$$f(r)=\frac{1}{\mu}\Big(\frac{d}{dr}-\frac{\nu}{\sin r}\Big)g(r), \qquad \Big(\frac{d^2}{dr^2}+\nu\frac{\cos r}{\sin^2 r}-\frac{\nu^2}{\sin^2 r}-\lambda\mu\Big)g(r)=0. \tag{5.21}$$

Note the symmetry between the equations (5.20) and (5.21): they follow from each other by the formal replacement $\nu\Longrightarrow-\nu$; therefore, we do not need follow in detail both variants. Let us focus on Eqs. (5.19). By excluding the function $f(t)$, we get

$$\begin{gathered} f(t)=-\frac{1}{\lambda}\Big(i\cosh t\frac{d}{dt}+M\,\cosh t\Big)g(t), \\ \Big(\frac{d^2}{dt^2}+\frac{\sinh t}{\cosh t}\frac{d}{dt}+M^2-iM\frac{\sinh t}{\cosh t}-\frac{\lambda\mu}{\cosh^2 t}\Big)g(t)=0. \end{gathered} \tag{5.22}$$

By excluding the function $g(t)$, we get

$$g(t) = +\frac{1}{\lambda}\Big(i\cosh t \frac{d}{dt} - M \cosh t\Big) f(t),$$
$$\Big(\frac{d^2}{dt^2} + \frac{\sinh t}{\cosh t}\frac{d}{dt} + M^2 + iM\frac{\sinh t}{\cosh t} - \frac{\lambda\mu}{\cosh^2 t}\Big) f(t) = 0. \tag{5.23}$$

Let us study the equation

$$\Big(\frac{d^2}{dr^2} - \nu\frac{\cos r}{\sin^2 r} - \frac{\nu^2}{\sin^2 r} - \lambda\mu\Big) f(r) = 0. \tag{5.24}$$

Because the 3-space is compact, the motion in the radial variable must be quantized; besides, the inequality $(-\nu\mu) > 0$ should hold true. Let us make change of the variable, $z = \cos r$, $z \in (-1, +1)$:

$$z = \cos r, \quad \frac{d}{dr} = -\sqrt{1-z^2}\frac{d}{dz}, \quad \frac{d^2}{dr^2} = (1-z^2)\frac{d^2}{dz^2} - z\frac{d}{dz};$$

then Eq. (5.24) reads

$$(1-z^2)\frac{d^2 f}{dz^2} - z\frac{df}{dz} - \Big(\nu\frac{z}{1-z^2} + \frac{\nu^2}{1-z^2} + \lambda\mu\Big) f = 0. \tag{5.25}$$

In the variable

$$y = \frac{1-z}{2} = \frac{1-\cos r}{2}, \quad (1-y) = \frac{1+\cos r}{2},$$

the above equation becomes

$$y(1-y)\frac{d^2}{dy^2} + \Big(\frac{1}{2} - y\Big)\frac{d}{dy} - \Big[\frac{\nu}{4y} - \frac{\nu}{4(1-y)} + \frac{\nu^2}{4}\frac{1}{y} + \frac{\nu^2}{4}\frac{1}{1-y} + \lambda\mu\Big] f = 0. \tag{5.26}$$

By using the substitution $f = y^A(1-y)^B F$, we arrive at

$$y(1-y)F'' + [\, 2A(1-y) - 2By + \frac{1}{2} - y\,]\, F'$$
$$+A(A-1)\frac{1}{y}F - A(A-1)F - 2ABF - B(B-1)F + B(B-1)\frac{1}{1-y}F$$
$$+\frac{A}{2}\frac{1}{y}F - \frac{B}{2}\frac{1}{1-y}F - AF + B\frac{1}{1-y}F - BF$$
$$-\frac{\nu}{4y}F + \frac{\nu}{4(1-y)}F - \frac{\nu^2}{4}\frac{1}{y}F - \frac{\nu^2}{4}\frac{1}{1-y}F - \lambda\mu F = 0.$$

Let the coefficients at y^{-1} and $(1-y)^{-1}$ vanish; then

$$\frac{1}{y}, \quad 2A = \nu + 1, -\nu; \quad \frac{1}{1-y}, \quad 2B = -\nu + 1, +\nu, \tag{5.27}$$

and then we get a simpler equation

$$y(1-y)F'' + [\,(2A + 1/2) - (2A + 2B + 1)y\,]F' - [(A+B)^2 + \lambda\mu] = 0.$$

This is a hypergeometric equation,

$$y(1-y)F'' + [c-(a+b+1)y]F' - abF = 0$$

with the parameters

$$c = 2A + 1/2\,, \quad a = A + B - \sqrt{-\lambda\mu}\,, \quad b = A + B + \sqrt{-\lambda\mu}\,. \tag{5.28}$$

By physical considerations, we should expect discrete values for Λ^2:

$$\Lambda^2 = -\lambda\mu > 0\,. \tag{5.29}$$

The needed solutions are then constructed as follows

$$\begin{gathered} 2A = +\nu + 1 = j + 3/2\,, \quad c = j + 2\,, \\ 2B = +\nu = j + 1/2\,, \quad A + B = j + 1\,, \\ a = j + 1 - \Lambda = -n\,, \quad n = 0, 1, 2, ..., \quad \Lambda = j + 1 + n\,, \\ b = 2(j+1) + n\,, \quad f(r) = C y^{(\nu+1)/2}(1-y)^{\nu/2} F(a,b,c,y)\,. \end{gathered} \tag{5.30}$$

In accordance with the mentioned symmetry, we have similar solutions for the function g:

$$\begin{gathered} g = C'\, y^{A'}(1-y)^{B'} G\,, \\ 2A' = \nu = j + 1/2\,, \quad c' = j + 1 = c - 1\,, \\ 2B' = +\nu + 1 = j + 3/2\,, \quad A' + B' = j + 1\,, \\ a' = j + 1 - \Lambda = -n = a\,, \quad n = 0, 1, 2, ..., \quad \Lambda = j + 1 + n\,, \\ b' = j + 1 + \Lambda = 2(j+1) + n = b\,, \\ g(r) = C' y^{\nu/2}(1-y)^{(\nu+1)/2} F(a,b,c-1,y)\,. \end{gathered} \tag{5.31}$$

To find a relative coefficient for C and C', we make use of the first order equation from (5.20):

$$g(r) = \frac{1}{\lambda}\Big(\frac{d}{dr} + \frac{\nu}{\sin r}\Big) f(r)\,, \tag{5.32}$$

whence, taking into account (5.30) and (5.31), we derive

$$\begin{gathered} \lambda\, C'\, (\sin\frac{r}{2})^{\nu}(\cos\frac{r}{2})^{\nu+1} F(a,b,c-1,y) \\ = \Big(\frac{d}{dr} + \frac{\nu}{\sin r}\Big) C\, (\sin\frac{r}{2})^{\nu+1}(\cos\frac{r}{2})^{\nu} F(a,b,c,y)\,, \end{gathered} \tag{5.33}$$

or,

$$\lambda\,\frac{C'}{C}(\sin\frac{r}{2})^{\nu}(\cos\frac{r}{2})^{\nu+1} F(a,b,c-1,y)$$

$$= (\sin\frac{r}{2})^{\nu+1}(\cos\frac{r}{2})^{\nu}\frac{d}{dr}F(a,b,c,y) + \frac{\nu+1}{2}(\sin\frac{r}{2})^{\nu}(\cos\frac{r}{2})^{\nu+1} F(a,b,c,y)$$

$$-\frac{\nu}{2}(\sin\frac{r}{2})^{\nu+2}(\cos\frac{r}{2})^{\nu-1}F(a,b,c,y)+\frac{\nu}{2}(\sin\frac{r}{2})^{\nu}(\cos\frac{r}{2})^{\nu-1}F(a,b,c,y).$$

Further, with the use of the identity

$$\frac{d}{dr}=\frac{1}{2}\sin r\frac{d}{dy}=\sin\frac{r}{2}\cos\frac{r}{2}\frac{d}{dy}=\frac{\cos(r/2)}{\sin(r/2)}y\frac{d}{dy},$$

we produce

$$\lambda\frac{C'}{C}(\sin\frac{r}{2})^{\nu}(\cos\frac{r}{2})^{\nu+1}F(a,b,c-1,y)$$
$$=(\sin\frac{r}{2})^{\nu}(\cos\frac{r}{2})^{\nu+1}\Big[y\frac{d}{dy}F(a,b,c,y)+\frac{\nu+1}{2}F(a,b,c,y)$$
$$-\frac{\nu}{2}\frac{1-\cos^2(r/2)}{\cos^2(r/2)}F(a,b,c,y)+\frac{\nu}{2}\frac{1}{\cos^2(r/2)}F(a,b,c,y)\Big],$$

which after an elementary manipulation (recall $\nu+\frac{1}{2}=j+1=c-1$), yields

$$\lambda\frac{C'}{C}F(a,b,c-1,y)=y\frac{d}{dy}F(a,b,c,y)+(c-1)F(a,b,c,y). \tag{5.34}$$

Moreover, making use of the known relation for hypergeometric functions [1]

$$y\frac{d}{dy}F(a,b,c,y)+(c-1)F(a,b,c,y)=(c-1)F(a,b,c-1,y),$$

we arrive at the needed linear relation

$$\lambda C'=(j+1)C. \tag{5.35}$$

Now, let us study the relevant equation in the variable t:

$$\Big(\frac{d^2}{dt^2}+\frac{\sinh t}{\cosh t}\frac{d}{dt}+M^2-iM\frac{\sinh t}{\cosh t}-\frac{\lambda\mu}{\cosh^2}\Big)g(t)=0. \tag{5.36}$$

In the variable $x=\tanh t$, we get

$$\frac{d}{dt}=(1-x^2)\frac{d}{dx},\quad \frac{d^2}{dt^2}=(1-x^2)^2\frac{d^2}{dx^2}-2x(1-x^2)\frac{d}{dx}.$$

Eq. (5.36) takes the form

$$\Big[(1-x^2)\frac{d^2}{dx^2}-x\frac{d}{dx}+\frac{M^2}{1-x^2}-\frac{iMx}{1-x^2}+\Lambda^2\Big]g=0. \tag{5.37}$$

Let us change the variable once again,

$$y=\frac{1-x}{2}=\frac{1-\tanh t}{2},\quad (1-y)=\frac{1+\tanh t}{2},\quad y\in(0,+1);$$

$$x=1-2y,\quad 1-x^2=4y(1-y),\qquad \frac{d}{dx}=-\frac{1}{2}\frac{d}{dy},\quad \frac{d^2}{dx^2}=\frac{1}{4}\frac{d^2}{dy^2};$$

then we arrive at

$$\Big[y(1-y)\frac{d^2}{dy^2}+(\frac{1}{2}-y)\frac{d}{dy}-\frac{im(1-2y)}{4y(1-y)}+\frac{m^2}{4y(1-y)}+\Lambda^2\Big]g=0\,.$$

With the substitution $g=y^A(1-y)^B G$, we obtain

$$y(1-y)G''+[\,2A(1-y)-2By+\frac{1}{2}-y\,]G'$$

$$+A(A-1)\frac{1}{y}G-A(A-1)G-2ABG-B(B-1)G+B(B-1)\frac{1}{1-y}G$$

$$+\frac{A}{2}\frac{1}{y}G-\frac{B}{2}\frac{1}{1-y}G-AG+B\frac{1}{1-y}G-BG$$

$$-\frac{iM}{4y}G+\frac{iM}{4(1-y)}G+\frac{M^2}{4}\frac{1}{y}G+\frac{M^2}{4}\frac{1}{1-y}G+\Lambda^2G=0\,.$$

Let the coefficients at y^{-1} and at $(1-y)^{-1}$ both vanish; this yields:

$$\frac{1}{y},\qquad A(A-1)+\frac{A}{2}-\frac{iM}{4}+\frac{M^2}{4}=0\implies 2A=1+iM,-iM\,;$$

$$\frac{1}{1-y},\qquad B(B-1)-\frac{B}{2}+B+\frac{iM}{4}+\frac{M^2}{4}=0\implies 2B=1-iM,iM\,.$$

Thus, we obtain the simpler equation

$$y(1-y)G''+[\,(2A+\frac{1}{2})-(2A+2B+1)y\,]G'-[(A+B)^2-\Lambda^2]=0;$$

which can be identified with hypergeometric one

$$y(1-y)F''+[c-(a+b+1)y]F'-abF=0\,.$$

Thus we get

$$\begin{gathered}2A=1+iM,\ -iM,\quad 2B=1-iM,\ iM,\\ c=2A+\frac{1}{2},\quad a=A+B-\Lambda,\quad b=A+B+\Lambda,\\ g(y)=Ly^A(1-y)^BG=Ly^A(1-y)^B\,F(a,b,c,y)\,,\end{gathered}\tag{5.38}$$

where $2A=1+iM,-iM$, $2B=1-iM,iM$.

Let us consider the behavior of these solutions at large values of time, $t\to+\infty$:

$$\begin{gathered}t\to+\infty,\quad y=\frac{1-\tanh t}{2}\to 0,\quad F(a,b,c,0)=1,\quad g\sim y^A,\\ 2A=1+iM\implies g(t)\sim(\frac{1-\tanh t}{2})^A=e^{-2At}=e^{-t}e^{-iMt},\\ 2A=-iM\implies g(t)\sim(\frac{1-\tanh t}{2})^A=e^{-2At}=e^{+iMt}\,.\end{gathered}\tag{5.39}$$

In order to find the description of solutions in the region $t \to -\infty$ $(y \to 1)$, we shall use the Kummer relation [1]

$$U_1 = \frac{\Gamma(c)\Gamma(c-a-b)}{\Gamma(c-a)\Gamma(c-b)} U_2 + \frac{\Gamma(c)\Gamma(-c+a+b)}{\Gamma(a)\Gamma(b)} U_6 ,$$
$$U_1 = F(a,b,c,y), \qquad U_2 = F(a,b,a+b+1-c,1-y),$$
$$U_6 = (1-y)^{c-a-b} F(c-a,c-b,c+1-a-b,1-y). \tag{5.40}$$

When $y \to 1 (t \to -\infty)$, this Kummer formula yields (note that $c-a-b = 1/2 - 2B$):

$$F(a,b,c,y) = \frac{\Gamma(c)\Gamma(c-a-b)}{\Gamma(c-a)\Gamma(c-b)} + \frac{\Gamma(c)\Gamma(-c+a+b)}{\Gamma(a)\Gamma(b)} (1-y)^{1/2-2B};$$

therefore for the function $g(y)$, we get the following expression

$$g = (1-y)^B \frac{\Gamma(c)\Gamma(c-a-b)}{\Gamma(c-a)\Gamma(c-b)} + \frac{\Gamma(c)\Gamma(-c+a+b)}{\Gamma(a)\Gamma(b)} (1-y)^{1/2-B}. \tag{5.41}$$

Correspondingly, we have two possibilities:

$$t \to -\infty, \qquad 1-y \to 0, \qquad B = \frac{1}{2} - \frac{iM}{2},$$
$$g = (1-y)^{1/2-iM/2} \frac{\Gamma(c)\Gamma(c-a-b)}{\Gamma(c-a)\Gamma(c-b)} + \frac{\Gamma(c)\Gamma(-c+a+b)}{\Gamma(a)\Gamma(b)} (1-y)^{+iM/2}; \tag{5.42}$$
$$t \to -\infty, \quad 1-y \to 0, \qquad B = +\frac{iM}{2},$$
$$g = (1-y)^{iM/2} \frac{\Gamma(c)\Gamma(c-a-b)}{\Gamma(c-a)\Gamma(c-b)} + \frac{\Gamma(c)\Gamma(-c+a+b)}{\Gamma(a)\Gamma(b)} (1-y)^{1/2-iM/2}. \tag{5.43}$$

For estimating the contribution of the terms in (5.42) and (5.43) into the total asymptotic, we need to take into account the following relations

$$(1-y)^{1/2-iM/2} = e^{(1-iM)t} = [e^{\ln(1-y)}]^{1/2-iM/2} = e^{(1/2)\ln(1-y)} e^{-i(M/2)\ln(1-y)}$$

$$= e^{(1/2)\ln(1-y)} \left[\cos\frac{M}{2} \ln(1-y) - i \sin\frac{M}{2} \ln(1-y) \right] \to 0;$$

$$(1-y)^{+iM/2} = e^{iMt} = \left(e^{\ln(1-y)} \right)^{+iM/2} = \cos\frac{M}{2} \ln(1-y) + i \sin\frac{M}{2} \ln(1-y).$$

By remembering the above mentioned symmetry, through the formal change $M \Longrightarrow = -M$, we find the behavior of the function $f(t)$:

$$c' = 2A' + \frac{1}{2}, \quad a' = A' + B' - \Lambda, \quad b' = A' + B' + \Lambda,$$
$$f(y) = L' y^{A'} (1-y)^{B'} G' = L' y^{A'} (1-y)^{B'} F(a',b',c',y), \tag{5.44}$$

where $2A' = 1 - iM, +iM$, $2B' = 1 + iM, -iM$.

Let $t \to +\infty$:

$$
\begin{aligned}
t \to +\infty, \quad y = \frac{1-\tanh t}{2} \to 0, \quad F(a',b',c',0) = 1, \quad f \approx L' y^{A'}, \\
2A' = 1 - iM, \quad f(t) \approx L' (\frac{1-\tanh t}{2})^{A'} = L' e^{-2A't} = L' e^{-t} e^{+iMt}, \\
2A' = +iM, \quad f(t) \approx L' (\frac{1-\tanh t}{2})^{A'} = L' e^{-2A't} = L' e^{-iMt}.
\end{aligned} \tag{5.45}
$$

Let $t \to -\infty$; then we may use the Kummer formula

$$
f = (1-y)^{B'} \frac{\Gamma(c')\Gamma(c'-a'-b')}{\Gamma(c'-a')\Gamma(c'-b')} + \frac{\Gamma(c')\Gamma(-c'+a'+b')}{\Gamma(a')\Gamma(b')} (1-y)^{1/2-B'}. \tag{5.46}
$$

Here there arise two possibilities

$$
t \to -\infty, \quad 1-y \to 0, \quad B' = \frac{1}{2} + \frac{iM}{2},
$$

$$
f = (1-y)^{1/2+iM/2} \frac{\Gamma(c')\Gamma(c'-a'-b')}{\Gamma(c'-a')\Gamma(c'-b')} + \frac{\Gamma(c')\Gamma(-c'+a'+b')}{\Gamma(a')\Gamma(b')} (1-y)^{-iM/2}; \tag{5.47}
$$

$$
t \to -\infty, \quad 1-y \to 0, \quad B' = -\frac{iM}{2},
$$

$$
f = (1-y)^{-iM/2} \frac{\Gamma(c')\Gamma(c'-a'-b')}{\Gamma(c'-a')\Gamma(c'-b')} + \frac{\Gamma(c')\Gamma(-c'+a'+b')}{\Gamma(a')\Gamma(b')} (1-y)^{1/2+iM/2}. \tag{5.48}
$$

In estimating the contribution of the terms from (5.47) and (5.48) into the total asymptotic, we are to take into account the following relations

$$
(1-y)^{1/2+iM/2} = e^{(1+iM)t} = e^{(1/2)\ln(1-y)} \Big[\cos(-M/2)\ln(1-y) + i \sin\frac{M}{2}\ln(1-y) \Big] \to 0,
$$

$$
(1-y)^{-iM/2} = e^{iMt} = \left(e^{\ln(1-y)}\right)^{-iM/2} = \cos(-M/2)\ln(1-y) - i\sin(M/2)\ln(1-y).
$$

Let us find the relative coefficients for the functions $f(t)$ and $g(t)$. We start with the first equation of the system (5.18) and we find function $f(t)$ by a fixed function $g(t)$:

$$
f(t) = -\frac{\cosh t}{\lambda} \Big(i \frac{d}{dt} + M \Big) g(t). \tag{5.49}
$$

With the use of the identities

$$
\frac{d}{dt} = -\frac{1}{2} \frac{1}{\cosh^2 t} \frac{d}{dy} = -2y(1-y)\frac{d}{dy}, \quad \cosh t = \frac{1}{\sqrt{1-\tanh^2 t}} = \frac{1}{2} \frac{1}{\sqrt{y(1-y)}},
$$

the equation (5.49) takes the form

$$
f(t) = -\frac{1}{2\lambda} \frac{L}{\sqrt{y(1-y)}} \Big(-2iy(1-y)\frac{d}{dy} + M \Big) y^{A} (1-y)^{B} F(a,b,c,y),
$$

and we further get

$$f(t) = -\frac{(-i)}{2\lambda}\frac{L}{\sqrt{y(1-y)}}y^A(1-y)^B\Big[2A(1-y)F - 2ByF + 2y(1-y)\frac{d}{dy}F + iMF\Big]. \qquad (5.50)$$

We shall prove that there exist pairs of solutions $f(y), g(y)$, which obey the equation (5.50).

The first pair is

$$g(y) = Ly^A(1-y)^B F(a,b,c,y),$$
$$A = \frac{1+iM}{2}, \quad B = \frac{iM}{2}, \quad c = 2A + \frac{1}{2} = iM + \frac{3}{2}, \qquad (5.51)$$
$$a = A + B - \Lambda = iM + \frac{1}{2} - \Lambda, \quad b = A + B + \Lambda = iM + \frac{1}{2} + \Lambda;$$

$$f(y) = L'y^{A'}(1-y)^{B'}F(a',b',c',y), \quad A' = \frac{iM}{2} = A - \frac{1}{2},$$
$$B' = \frac{1+iM}{2} = B + \frac{1}{2}, \; c' = 2A' + \frac{1}{2} = iM + \frac{1}{2} = c - 1, \qquad (5.52)$$
$$a' = A' + B' - \Lambda = iM + \frac{1}{2} - \Lambda = a, \; b' = A' + B' + \Lambda = iM + \frac{1}{2} + \Lambda = b.$$

The equation (5.50) for these functions yields

$$f(t) = -\frac{(-i)}{2\lambda}\frac{L}{\sqrt{y(1-y)}}y^A(1-y)^B\Big[2A(1-y)F + 2y(1-y)\frac{d}{dy}F + iM(1-y)F\Big], \qquad (5.53)$$

or,

$$f(t) = -\frac{(-i)}{2\lambda}Ly^{A-1/2}(1-y)^{B+1/2}\Big[2AF + 2y\frac{d}{dy}F + iMF\Big],$$

or, differently,

$$f(t) = \frac{i}{\lambda}Ly^{A'}(1-y)^{B'}\Big[(c-1)F + y\frac{d}{dy}F\Big]. \qquad (5.54)$$

The left side of the relation can be taken in the form

$$f(y) = L'y^{A'}(1-y)^{B'}F(a',b',c',y) = L'y^{A'}(1-y)^{B'}F(a,b,c-1,y);$$

so we get

$$L'F(a,b,c-1,y) = \frac{i}{\lambda}L\Big[(c-1)F(a,b,c,y) + y\frac{d}{dy}F(a,b,c,y)\Big],$$

whence, by applying a known property of hypergeometric functions[1]

$$y\frac{d}{dy}F(a,b,c,y) + (c-1)F(a,b,c,y) = (c-1)F(a,b,c-1,y),$$

we derive the needed linear identity which involves the coefficients L and L':

$$L' = \frac{i}{\lambda} L(c-1)\,. \tag{5.55}$$

The second pair of related solutions is

$$\begin{gathered} g(y) = L\, y^{A}(1-y)^{B} F(a,b,c,y)\,, \\ A = -\frac{iM}{2} = A' - \frac{1}{2}, \quad B = \frac{1-iM}{2} = B' + \frac{1}{2}, \quad c = 2A + \frac{1}{2} = -iM + \frac{1}{2} = c' - 1\,, \\ a = A + B - \Lambda = -iM + \frac{1}{2} - \Lambda = a', \;\; b = A + B + \Lambda = -iM + \frac{1}{2} + \Lambda = b'; \end{gathered} \tag{5.56}$$

$$\begin{gathered} f(y) = L' y^{A'}(1-y)^{B'} F(a',b',c',y)\,, \\ A' = \frac{1-iM}{2}, \quad B' = -\frac{iM}{2}, c' = 2A' + \frac{1}{2} = -iM + \frac{3}{2}, \\ a' = A' + B' - \Lambda = -iM + \frac{1}{2} - \Lambda = a, \;\; b' = A' + B' + \Lambda = -iM + \frac{1}{2} + \Lambda = b\,. \end{gathered} \tag{5.57}$$

Here it is more convenient to employ the second equation in the system (5.18):

$$g(t) = \frac{1}{\mu} \cosh t \left(i\frac{d}{dt} - M \right) f(t)\,; \tag{5.58}$$

it is equivalent to

$$g(y) = \frac{1}{2\mu} \frac{L'}{\sqrt{y(1-y)}} \Big(-2iy(1-y)\frac{d}{dy} - M \Big) y^{A'}(1-y)^{B'} F(a',b',c',y)\,,$$

or,

$$g(t) = -\frac{i}{2\mu} \frac{L'}{\sqrt{y(1-y)}} y^{A'}(1-y)^{B'} \Big[2A'(1-y)F - 2B' yF + 2y(1-y)\frac{d}{dy}F - iMF \Big], \tag{5.59}$$

or, differently,

$$g(t) = -\frac{i}{2\mu} L'\, y^{A'-1/2}(1-y)^{B'+1/2} \Big[2A'F + 2y\frac{d}{dy}F - iMF \Big],$$

whence it follows

$$g(t) = -\frac{i}{2\mu} L'\, y^{A}(1-y)^{B} \Big[(c'-1)F + y\frac{d}{dy}F \Big]. \tag{5.60}$$

The left hand side of this relation can be taken in the form

$$g(y) = Ly^{A}(1-y)^{B} F(a,b,c,y) = Ly^{A}(1-y)^{B} F(a',b',c'-1,y)\,,$$

which leads to

$$LF(a',b',c'-1,y) = -\frac{i}{2\mu} L' \left[(c'-1)F(a',b',c',y) + y\frac{d}{dy}F(a',b',c',y) \right].$$

Finally, using the known property of hypergeometric functions [1]

$$y\frac{d}{dy}F(a',b',c',y) + (c'-1)F(a',b',c',y) = (c'-1)F(a',b',c'-1,y)\,,$$

we arrive at the needed relationship involving the coefficients L and L':

$$L = -\frac{i}{2\mu} L'(c'-1)\,. \tag{5.61}$$

5.2 Neutrino in Expanding de Sitter Space

Let us start with the massless Dirac equation written in the form (see (5.8))

$$\Big(i\gamma^0 \cosh t\, \partial_t + i\gamma^3 \partial_r + \frac{1}{\sin r} \Sigma_{\theta\phi}\Big)\, \varphi = 0\,. \tag{5.62}$$

We take in mind the structure of the 4-spinor wave function

$$\varphi(x) = \left| \begin{array}{c} \xi(x) \\ \eta(x) \end{array} \right| = \varphi_{jm}(x) = \left| \begin{array}{c} f_1(t,r)\, D_{-1/2} \\ f_2(t,r)\, D_{+1/2} \\ f_3(t,r)\, D_{-1/2} \\ f_4(4,r)\, D_{+1/2} \end{array} \right| , \tag{5.63}$$

where the component $\eta(x)$ relates to the neutrino and the component $\xi(x)$ refers to the antineutrino. Instead of four linked radial equations, we have now two unlinked subsystems. The first one is for the 2-spinor $\eta(x)$:

$$\cosh t \frac{\partial}{\partial t} f_3 - \frac{\partial}{\partial r} f_3 - \frac{\nu}{\sin r} f_4 = 0\,, \quad \cosh t \frac{\partial}{\partial t} f_4 + \frac{\partial}{\partial r} f_4 + \frac{\nu}{\sin r} f_3 = 0\,; \tag{5.64}$$

by summing and subtracting, we derive (let $f_3 + f_4 = f,\ f_3 - f_4 = g$)

$$\cosh t \frac{\partial}{\partial t} f - (\frac{\partial}{\partial r} - \frac{\nu}{\sin r}) g = 0\,, \quad \cosh t \frac{\partial}{\partial t} g - (\frac{\partial}{\partial r} + \frac{\nu}{\sin r}) f = 0\,. \tag{5.65}$$

The second pair of equations for the spinor $\xi(x)$ is

$$\cosh t \frac{\partial}{\partial t} f_1 + \frac{\partial}{\partial r} f_1 + \frac{\nu}{\sin r} f_2 = 0\,, \quad \cosh t \frac{\partial}{\partial t} f_2 - \frac{\partial}{\partial r} f_2 - \frac{\nu}{\sin r} f_1 = 0\,. \tag{5.66}$$

Let $f_1 + f_2 = F,\ f_1 - f_2 = G$; then the last subsystem yields

$$\cosh t \frac{\partial}{\partial t} F + (\frac{\partial}{\partial r} - \frac{\nu}{\sin r}) G = 0\,, \quad \cosh t \frac{\partial}{\partial t} G + (\frac{\partial}{\partial r} F + \frac{\nu}{\sin r}) F = 0\,. \tag{5.67}$$

It suffices to solve the system (5.65), whence we produce solutions for the subsystem (5.67):

$$F(t,r) = +g(t,r)\,, \quad G(t,r) = -g(t,r)\,. \tag{5.68}$$

Let us study the system(5.65):

$$\cosh t \frac{\partial}{\partial t} f - (\frac{\partial}{\partial r} - \frac{\nu}{\sin r}) g = 0\,, \quad \cosh t \frac{\partial}{\partial t} g - (\frac{\partial}{\partial r} + \frac{\nu}{\sin r}) f = 0\,. \tag{5.69}$$

Using the substitutions

$$f(t,r) = f(t) f(r)\,, \quad g(t,r) = g(t) g(r)\,,$$

we derive

$$f(r) \cosh t \frac{d}{dt} f(t) = g(t) (\frac{d}{dr} - \frac{\nu}{\sin r}) g(r)\,,$$
$$g(r) \cosh t \frac{d}{dt} g(t) = f(t) (\frac{d}{dr} + \frac{\nu}{\sin r}) f(r)\,,$$

or,

$$\frac{1}{g(t)}\cosh t\frac{d}{dt}f(t) = \frac{1}{f(r}(\frac{d}{dr} - \frac{\nu}{\sin r})g(r) = \lambda ,$$
$$\frac{1}{f(t)}\cosh t\frac{d}{dt}g(t) = \frac{1}{g(r)}(\frac{d}{dr} + \frac{\nu}{\sin r})f(r) = \mu .$$

Thus, we have two subsystems:

$$(\frac{d}{dr} + \frac{\nu}{\sin r})f(r) = \lambda g(r) , \quad (\frac{d}{dr} - \frac{\nu}{\sin r})g(r) = \mu f(r) , \tag{5.70}$$

and

$$\cosh t\frac{d}{dt}f(t) = \mu g(t) , \quad \cosh t\frac{d}{dt}g(t) = \lambda f(t) . \tag{5.71}$$

By comparing these equation with the above established ones for the massive Dirac field,

$$\Big(\frac{d}{dr} + \frac{\nu}{\sin r}\Big)f(r) = \lambda g(r) , \quad \Big(\frac{d}{dr} - \frac{\nu}{\sin r}\Big)g(r) = \mu f(r) ,$$

and

$$\Big(\cosh t\frac{d}{dt} - iM\cosh t\Big)g(t) = i\lambda f(t) , \quad \Big(\cosh t\frac{d}{dt} + iM\cosh t\Big)f(t) = -i\mu g(t) ,$$

we conclude that the solution of the system (5.70) coincides with the solution for (5.17). Also, we see that the system (5.71), with the formal change in notations

$$i\lambda \Longrightarrow \lambda, \quad -i\mu \Longrightarrow \mu, \quad M = 0, \quad \Lambda^2 = -\lambda\mu > 0$$

coincides with the one previously solved.

5.3 Pauli Equation in Expanding de Sitter Space

It is convenient to perform a restriction to nonrelativistic limit just in the radial system:

$$\begin{aligned} &\frac{1}{\cosh t}\Big(\frac{\partial}{\partial r} + \frac{\nu}{\sin r}\Big)f + \Big(i\frac{\partial}{\partial t} + \delta\, M\Big)g = 0 , \\ &\frac{1}{\cosh t}\Big(\frac{\partial}{\partial r} - \frac{\nu}{\sin r}\Big)g - \Big(i\frac{\partial}{\partial t} - \delta\, M\Big)f = 0 . \end{aligned} \tag{5.72}$$

First, we separate the rest energy:

$$i\frac{\partial}{\partial t} \quad \Longrightarrow \quad M + i\frac{\partial}{\partial t} .$$

Further, the two cases $\delta=+1$ and $\delta=-1$ will be separately considered:

$$\delta=+1,\quad \begin{array}{l} \dfrac{1}{\cosh t}\Big(\dfrac{\partial}{\partial r}+\dfrac{\nu}{\sin r}\Big)f+\Big(M+i\dfrac{\partial}{\partial t}+M\Big)g=0, \\ \dfrac{1}{\cosh t}\Big(\dfrac{\partial}{\partial r}-\dfrac{\nu}{\sin r}\Big)g-\Big(M+i\dfrac{\partial}{\partial t}-M\Big)f=0; \end{array} \tag{5.73}$$

$$\delta=-1,\quad \begin{array}{l} \dfrac{1}{\cosh t}\Big(\dfrac{\partial}{\partial r}+\dfrac{\nu}{\sin r}\Big)f+\Big(M+i\dfrac{\partial}{\partial t}-M\Big)g=0, \\ \dfrac{1}{\cosh t}\Big(\dfrac{\partial}{\partial r}-\dfrac{\nu}{\sin r}\Big)g-\Big(M+i\dfrac{\partial}{\partial t}-M\Big)f=0, \end{array} \tag{5.74}$$

whence, by neglecting the term $i\partial_t$ (which is small in comparison with the term $2M$), we obtain

$$\delta=+1,\quad \frac{1}{\cosh t}\Big(\frac{\partial}{\partial r}+\frac{\nu}{\sin r}\Big)f+2Mg=0,\quad \frac{1}{\cosh t}\Big(\frac{\partial}{\partial r}-\frac{\nu}{\sin r}\Big)g-i\frac{\partial}{\partial t}f=0; \tag{5.75}$$

$$\delta=-1,\quad \frac{1}{\cosh t}\Big(\frac{\partial}{\partial r}+\frac{\nu}{\sin r}\Big)f+i\frac{\partial}{\partial t}g=0,\quad \frac{1}{\cosh t}\Big(\frac{\partial}{\partial r}-\frac{\nu}{\sin r}\Big)g-2Mf=0. \tag{5.76}$$

In each variant, we can derive the nonrelativistic equation for the big component:

$$\delta=+1,\quad i\frac{\partial}{\partial t}f=-\frac{1}{2M}\frac{1}{\cosh^2 t}\Big(\frac{\partial}{\partial r}-\frac{\nu}{\sin r}\Big)\Big(\frac{\partial}{\partial r}+\frac{\nu}{\sin r}\Big)f, \tag{5.77}$$

$$\delta=-1,\quad i\frac{\partial}{\partial t}g=-\frac{1}{2M}\frac{1}{\cosh t}\Big(\frac{\partial}{\partial r}+\frac{\nu}{\sin r}\Big)\Big(\frac{\partial}{\partial r}-\frac{\nu}{\sin r}\Big)g. \tag{5.78}$$

Let us recall the constructing of nonrelativistic wave function from the initial relativistic one:

$$\varphi(x)_{jm\delta}=\left|\begin{array}{c} f_1D_{-1/2} \\ f_2D_{+1/2} \\ \delta f_2D_{-1/2} \\ \delta f_1D_{+1/2} \end{array}\right| \quad\Longrightarrow\quad \Psi^{Pauli}=\left|\begin{array}{c} f_1D_{-1/2}+\delta f_2D_{-1/2} \\ f_2D_{+1/2}+\delta f_1\,D_{+1/2} \end{array}\right|. \tag{5.79}$$

For a state with different parities, we have

$$\delta=+1,\quad \Psi^{Pauli}=\left|\begin{array}{c} f_1D_{-1/2}+f_2D_{-1/2} \\ f_2D_{+1/2}+f_1D_{+1/2} \end{array}\right|=\left|\begin{array}{c} fD_{-1/2} \\ fD_{+1/2} \end{array}\right|, \tag{5.80}$$

$$\delta=-1,\quad \Psi^{Pauli}=\left|\begin{array}{c} f_1D_{-1/2}-f_2D_{-1/2} \\ f_2D_{+1/2}-f_1D_{+1/2} \end{array}\right|=\left|\begin{array}{c} -igD_{-1/2} \\ -igD_{+1/2} \end{array}\right|. \tag{5.81}$$

The respective functions $f(t,r)$ and $g(t,r)$ obey slightly different equations:

$$\delta=+1,\quad i\frac{\partial}{\partial t}f=-\frac{1}{2M}\frac{1}{\cosh^2 t}\Big(\frac{\partial^2}{\partial r^2}-\frac{\nu^2+\nu\cos r}{\sin^2 r}\Big)f, \tag{5.82}$$

$$\delta = -1, \quad i\frac{\partial}{\partial t}g = -\frac{1}{2M}\frac{1}{\cosh^2 t}\Big(\frac{\partial^2}{\partial r^2} - \frac{\nu^2 - \nu\cos r}{\sin^2 r}\Big)g\,. \tag{5.83}$$

This means that the degeneration in parity is preserved at the nonrelativistic limit.

With the help of the substitutions

$$f(t,r) = f(t)f(r)\,, \quad g(t,r) = g(t)g(r)\,,$$

the variables are separated, so we obtain

$\delta = +1,$

$$i\cosh^2 t\frac{1}{f(t)}\frac{d}{dt}f(t) = -\frac{1}{2M}\frac{1}{f(r)}\Big(\frac{d^2}{dr^2} - \frac{\nu^2 - \nu\cos r}{\sin^2 r}\Big)f(r) = E\,; \tag{5.84}$$

$\delta = -1,$

$$i\cosh^2 t\frac{1}{g(t)}\frac{d}{dt}g(t) = -\frac{1}{2M}\frac{1}{g(r)}\Big(\frac{d^2}{dr^2} - \frac{\nu^2 - \nu\cos r}{\sin^2 r}\Big)g(r) = E\,. \tag{5.85}$$

Thus we arrive at

$$\delta = +1, \quad \begin{array}{l} i\cosh^2 t\dfrac{1}{f(t)}\dfrac{d}{dt}f(t) = E \quad \Longrightarrow \quad f(t) = e^{-iE\tanh t}\,, \\ -\dfrac{1}{2M}\dfrac{1}{f(r)}\Big(\dfrac{d^2}{dr^2} - \dfrac{\nu^2 + \nu\cos r}{\sin^2 r}\Big)f(r) = E\,; \end{array}$$

$$\delta = -1, \quad \begin{array}{l} i\cosh^2 t\dfrac{1}{g(t)}\dfrac{d}{dt}g(t) = E \quad \Longrightarrow \quad g(t) = e^{-iE\tanh t}\,, \\ -\dfrac{1}{2M}\dfrac{1}{g(r)}\Big(\dfrac{d^2}{dr^2} - \dfrac{\nu^2 - \nu\cos r}{\sin^2 r}\Big)g(r) = E\,. \end{array}$$

Due to the symmetry $\nu \Longrightarrow -\nu$, it suffices to study in detail only one case; for definiteness, let it be $\delta = +1$:

$$\Big(\frac{d^2}{dr^2} - \frac{\nu^2 + \nu\cos r}{\sin^2 r} + 2ME\Big)f(r) = 0\,. \tag{5.86}$$

Because the 3-space is compact, the motion must be quantized. Besides, the inequality $2ME > 0$ should hold. In Eq. (5.86), we change the variable $z = \cos r$, $z \in (-1,+1)$:

$$(1-z^2)\frac{d^2 f}{dz^2} - z\frac{df}{dz} - \Big(\nu\frac{z}{1-z^2} + \frac{\nu^2}{1-z^2} - 2ME\Big)f = 0\,, \tag{5.87}$$

and change the variable

$$y = \frac{1-z}{2} = \frac{1-\cos r}{2}\,,$$

$$y(1-y)\,\frac{d^2f}{dy^2}+\Big(\frac{1}{2}-y\Big)\,\frac{df}{dy}-\Big[\frac{\nu(\nu+1)}{4y}+\frac{\nu(\nu-1)}{4(1-y)}-2ME\Big]f=0\,.$$

Now, using the substitution $f=y^A(1-y)^BF$, we obtain

$$y(1-y)\frac{d^2F}{dy^2}+\Big[2A+\frac{1}{2}-(2A+2B+1)\,y\Big]\frac{dF}{dy}$$

$$-\Big[(A+B)^2-\frac{2A\,(2A-1)-\nu(\nu+1)}{4y}-\frac{2B\,(2B-1)-\nu(\nu-1)}{4(1-y)}-2ME\Big]F=0\,.$$

Let the coefficients at y^{-1} and $(1-y)^{-1}$ vanish; this yields

$$2A=\nu+1,\ -\nu,\qquad 2B=-\nu+1,\ +\nu\,. \tag{5.88}$$

The above equation becomes simpler

$$y(1-y)\frac{d^2F}{dy^2}+\Big[2A+\frac{1}{2}-(2A+2B+1)\,y\Big]\frac{dF}{dy}-\Big[(A+B)^2-2ME\Big]F=0\,,$$

and it can be identified as being of hypergeometric type

$$y(1-y)F''+[c-(a+b+1)y]F'-abF=0\,,$$
$$c=2A+\frac{1}{2}\,,\quad a=A+B-\sqrt{2ME}\,,\quad b=A+B+\sqrt{2ME}\,,$$

where we recall that $2ME>0$. The needed solutions are constructed as follows

$$2A=+\nu+1=j+3/2\,,\quad 2B=+\nu=j+1/2\,,\quad A+B=j+1\,,$$

from condition $a=-n$ we derive

$$2ME=(j+1+n)^2\,;$$

corresponding wave functions are

$$f(y)=C\,y^{(\nu+1)/2}(1-y)^{\nu/2}F(a,b,c,y)\,,$$
$$b=2(j+1)+n\,,\quad c=j+2\quad(\nu=j+1/2)\,. \tag{5.89}$$

Taking in mind the mentioned symmetry between equations for $f(y)$ and $g(y)$, we obtain the needed solutions for the function $g(y)$:

$$g(y)=C'\,y^{A'}(1-y)^{B'}G,\quad 2A'=-\nu+1,\nu,\quad 2B'=\nu+1,-\nu;$$

at positive A' and B':

$$2A'=\nu=j+1/2\,,\quad c'=j+1=c-1\,,$$
$$2B'=+\nu+1=j+3/2\,,\quad A'+B'=j+1\,,$$
$$a'=-n=a\,,\quad n=0,1,2,\ldots,\quad b'=2(j+1)+n=b\,,$$
$$g(r)=C'\,y^{\nu/2}(1-y)^{(\nu+1)/2}F(a,b,c-1,y)\,. \tag{5.90}$$

Thus, the solutions of the Pauli equations in an expanding de Sitter space have the form

$\delta = +1$

$$\Psi^{Pauli}_{Ejm,+1}(t,r,\theta,\phi) = e^{-iE\tanh t} f(r) \left| \begin{array}{c} D^{j}_{-m,-1/2}(\phi,\theta,0) \\ D^{j}_{-m,+1/2}(\phi,\theta,0) \end{array} \right| ,$$

$$f(r) = C \left(\sin\frac{r}{2}\right)^{j+3/2} \left(\cos\frac{r}{2}\right)^{j+1/2} F(-n, n+2j+2, j+2, \sin^2\frac{r}{2}); \tag{5.91}$$

$\delta = -1$

$$\Psi^{Pauli}_{Ejm,-1}(t,r,\theta,\phi) = e^{-iE\tanh t} g(r) \left| \begin{array}{c} D^{j}_{-m,-1/2}(\phi,\theta,0) \\ D^{j}_{-m,+1/2}(\phi,\theta,0) \end{array} \right| ,$$

$$g(r) = C' \left(\sin\frac{r}{2}\right)^{j+1/2} \left(\cos\frac{r}{2}\right)^{j+3/2} F(-n, n+2j+2, j+1, \sin^2\frac{r}{2}). \tag{5.92}$$

The parameter E, analogue to the energy, is quantized, and for both classes of states with opposite parities, it is given by the same formula

$$2ME = (j+1+n)^2. \tag{5.93}$$

5.4 Spin 1/2 Particle in Oscillating de Sitter Model

We start with the Dirac equation

$$\left\{ i\gamma^k (e^{\alpha}_{(k)}\partial_\alpha + B_k) - M \right\}\Psi = 0, \quad B_k(x) = \frac{1}{2}\frac{1}{\sqrt{-g}}\frac{\partial}{\partial x^\alpha}\sqrt{-g}\, e^{\alpha}_{(k)}. \tag{5.94}$$

In nonstatic coordinates $x^\alpha = (t, r, \theta, \phi)$, we have

$$dS^2 = dt^2 - \cos^2 t\,[dr^2 + \sinh^2 r(d\theta^2 + \sin^2\theta\, d\phi^2)]$$

and the corresponding tetrad

$$e^{\alpha}_{(0)} = (1,0,0,0), \quad e^{\alpha}_{(3)} = (0, \frac{1}{\cos t}, 0, 0),$$

$$e^{\alpha}_{(2)} = (0,0,0, \frac{1}{\cos t\, \sinh r \sin\theta}), \quad e^{\alpha}_{(1)} = (0,0, \frac{1}{\cos t\, \sinh r}, 0).$$

For B_k, we get expressions

$$B_0 = -\frac{3}{2}\tan t, \quad B_1 = \frac{1}{2}\frac{\cos\theta}{\cos t\, \sinh r \sin\theta}, \quad B_2 = 0, \quad B_3 = \frac{1}{\cos t\, \tanh r}.$$

Therefore, Eq. (5.94) takes the form

$$\left[i\gamma^0 \cos t \left(\partial_t - \frac{3}{2}\tan t\right) + i\gamma^3 \left(\partial_r + \frac{1}{\tanh r}\right) + \frac{1}{\sinh r}\Sigma_{\theta\phi} - M\cos t \right]\Psi = 0, \tag{5.95}$$

where

$$\Sigma_{\theta,\phi} = i\gamma^1\partial_\theta + \gamma^2 \frac{i\partial_\phi + i\,\sigma^{12}\cos\theta}{\sin\theta}.$$

By separating the simple factors:

$$\Psi(x) = \frac{1}{\sinh r}\frac{1}{\cos^{3/2}t}\,\varphi(x), \tag{5.96}$$

we reduce the last equation to a simpler form

$$\left(i\gamma^0 \cos t\partial_t + i\gamma^3\partial_r + \frac{1}{\sinh r}\Sigma_{\theta\phi} - M\cos t\right)\varphi = 0. \tag{5.97}$$

We consider the following substitution for the wave function

$$\varphi_{jm}(x) = \left|\begin{array}{c} f_1(t,r)\,D_{-1/2} \\ f_2(t,r)\,D_{+1/2} \\ f_3(t,r)\,D_{-1/2} \\ f_4(t,r)\,D_{+1/2} \end{array}\right|. \tag{5.98}$$

Allowing for the relation (let $\nu = j+1/2$),

$$\Sigma_{\theta,\phi}\,\varphi_{jm}(x) = i\nu\left|\begin{array}{c} -\,f_4(t,r)\,D_{-1/2} \\ +\,f_3(t,r)\,D_{+1/2} \\ +\,f_2(t,r)\,D_{-1/2} \\ -\,f_1(t,r)\,D_{+1/2} \end{array}\right|, \tag{5.99}$$

we derive 4 equations

$$\begin{aligned} &i\cos t\frac{\partial}{\partial t}f_3 - i\frac{\partial}{\partial r}f_3 - i\frac{\nu}{\sinh r}f_4 - M\,\cos t f_1 = 0, \\ &i\cos t\frac{\partial}{\partial t}f_4 + i\frac{\partial}{\partial r}f_4 + i\frac{\nu}{\sinh r}f_3 - M\,\cos t f_2 = 0, \\ &i\cos t\frac{\partial}{\partial t}f_1 + i\frac{\partial}{\partial r}f_1 + i\frac{\nu}{\sinh r}f_2 - M\,\cos t f_3 = 0, \\ &i\cos t\frac{\partial}{\partial t}f_2 - i\frac{\partial}{\partial r}f_2 - i\frac{\nu}{\sinh r}f_1 - M\cos t f_4 = 0. \end{aligned} \tag{5.100}$$

Diagonalizing the spatial reflection operator, we find the additional constraints

$$\Pi = \delta\,(-1)^{j+1},\ \delta = \pm 1, \quad f_4 = \delta\,f_1,\ f_3 = \delta\,f_2, \tag{5.101}$$

which simplify the structure of wave functions

$$\varphi(x)_{jm\delta} = \left|\begin{array}{c} f_1(t,r)\,D_{-1/2} \\ f_2(t,r)\,D_{+1/2} \\ \delta\,f_2(t,r)\,D_{-1/2} \\ \delta\,f_1(t,r)\,D_{+1/2} \end{array}\right|. \tag{5.102}$$

Taking in mind (5.101), from Eqs. (5.100) we derive two more simple subsystems:

$$\begin{gathered}\Big(\frac{\partial}{\partial r}+\frac{\nu}{\sinh r}\Big)f+\Big(i\cos t\,\frac{\partial}{\partial t}+\delta M\,\cos t\Big)g=0\,,\\ \Big(\frac{\partial}{\partial r}-\frac{\nu}{\sinh r}\Big)g-\Big(i\cos t\frac{\partial}{\partial t}-\delta M\,\cos t\Big)f=0\,,\end{gathered} \tag{5.103}$$

where

$$f(t,r)=\frac{f_1+f_2}{\sqrt{2}},\quad g(t,r)=\frac{f_1-f_2}{i\sqrt{2}}\,.$$

From the system (5.103), by employing the substitutions

$$f(t,r)=f(t)f(r)\,,\quad g(t,r)=g(t)g(r)\,,$$

we produce

$$\begin{gathered}\frac{1}{g(r)}\Big(\frac{d}{dr}+\frac{\nu}{\sinh r}\Big)f(r)=-\frac{1}{f(t)}\Big(i\cos t\frac{d}{dt}+\delta M\cos t\Big)g(t)=\lambda\,,\\ \frac{1}{f(r)}\Big(\frac{d}{dr}-\frac{\nu}{\sinh r}\Big)g(r)=\frac{1}{g(t)}\Big(i\cos t\frac{d}{dt}-\delta M\cos t\Big)f(t)=\mu\,.\end{gathered} \tag{5.104}$$

Thus we have two separated subsystems:

$$\Big(\frac{d}{dr}+\frac{\nu}{\sinh r}\Big)f(r)=\lambda\,g(r)\,,\quad \Big(\frac{d}{dr}-\frac{\nu}{\sinh r}\Big)g(r)=\mu\,f(r)\,; \tag{5.105}$$

$$\Big(i\cos t\frac{d}{dt}+M\cos t\Big)g(t)=-\lambda f(t)\,,\ \Big(i\cos t\frac{d}{dt}-M\cos t\Big)f(t)=\mu g(t)\,. \tag{5.106}$$

For definiteness we shall follow the case of $\delta=+1$. Excluding from (5.105) the function $g(r)$:

$$\begin{gathered}g(r)=\frac{1}{\lambda}\Big(\frac{d}{dr}+\frac{\nu}{\sinh r}\Big)f(r)\,,\\ \Big(\frac{d^2}{dr^2}-\nu\frac{\cosh r}{\sinh^2 r}-\frac{\nu^2}{\sinh^2 r}-\lambda\mu\Big)f(r)=0\,.\end{gathered} \tag{5.107}$$

Similarly, we may exclude the function $f(r)$:

$$\begin{gathered}f(r)=\frac{1}{\mu}\Big(\frac{d}{dr}-\frac{\nu}{\sinh r}\Big)g(r)\,,\\ \Big(\frac{d^2}{dr^2}+\nu\frac{\cosh r}{\sinh^2 r}-\frac{\nu^2}{\sinh^2 r}-\lambda\mu\Big)g(r)=0\,.\end{gathered} \tag{5.108}$$

In a similar way, the system in the variable t may reduce to

$$\begin{gathered}f(t)=-\frac{1}{\lambda}\Big(i\cos t\frac{d}{dt}+M\,\cos t\Big)g(t)\,,\\ \Big(\frac{d^2}{dt^2}-\frac{\sin t}{\cos t}\frac{d}{dt}+M^2+iM\frac{\sin t}{\cos t}-\frac{\lambda\mu}{\cos^2 t}\Big)g(t)=0\,,\end{gathered} \tag{5.109}$$

or,

$$g(t) = \frac{1}{\mu}\Big(i\cos t\,\frac{d}{dt} - M\cos t\Big)f(t),$$
$$\Big(\frac{d^2}{dt^2} - \frac{\sin t}{\cos t}\frac{d}{dt} + M^2 - iM\frac{\sin t}{\cos t} - \frac{\lambda\mu}{\cos^2 t}\Big)f(t) = 0. \tag{5.110}$$

The equations (5.109) and (5.110) are symmetric under the change $M \Longrightarrow -M$.

Let us study the equation

$$\Big(\frac{d^2}{dr^2} - \nu\frac{\cosh r}{\sinh^2 r} - \frac{\nu^2}{\sinh^2 r} - \lambda\mu\Big)f(r) = 0. \tag{5.111}$$

By physical reason, we assume $\nu\mu < 0$. In terms of the variable $z = \cosh r$, $z \in (+1, +\infty)$, the equation reads

$$(1-z^2)\frac{d^2 f}{dz^2} - z\frac{df}{dz} - \Big(\nu\frac{z}{1-z^2} + \frac{\nu^2}{1-z^2} - \lambda\mu\Big)f = 0. \tag{5.112}$$

However, more convenient is the variable

$$y = \frac{1-z}{2} = \frac{1-\cosh r}{2},$$

which leads to

$$y(1-y)\frac{d^2 f}{dy^2} + (\frac{1}{2} - y)\frac{df}{dy} - \Big[\frac{\nu(\nu+1)}{4y} + \frac{\nu(\nu-1)}{4(1-y)} - \lambda\mu\Big]f = 0.$$

With the substitution $f = y^A(1-y)^B F$, we get

$$y(1-y)\frac{d^2F}{dy^2} + \Big[\frac{1}{2} + 2A - (1+2A+2B)y\Big]\frac{dF}{dy}$$
$$-\Big[(A+B)^2 - \lambda\mu - \frac{2A(2A-1) - \nu(\nu+1)}{4y} - \frac{2B(2B-1) - \nu(\nu-1)}{4(1-y)}\Big]F = 0.$$

Let the coefficients at y^{-1} and at $(1-y)^{-1}$ both vanish:

$$A = -\frac{1}{2}\nu,\ \frac{1}{2} + \frac{1}{2}\nu; \quad B = \frac{1}{2} - \frac{1}{2}\nu,\ \frac{1}{2}\nu; \tag{5.113}$$

then the above equation becomes simpler

$$y(1-y)\frac{d^2F}{dy^2} + \Big[\frac{1}{2} + 2A - (1+2A+2B)y\Big]\frac{dF}{dy} - \Big[(A+B)^2 - \lambda\mu\Big]F = 0,$$

and it is of hypergeometric type

$$y(1-y)F'' + [c - (a+b+1)y]F' - abF = 0,$$
$$c = 2A + \frac{1}{2}, \quad a+b = 2A+2B, \quad ab = (A+B)^2 - \lambda\mu, \tag{5.114}$$
$$a = A+B - i\sqrt{-\lambda\mu}, \quad b = A+B + i\sqrt{-\lambda\mu}.$$

Thus, the complete solutions are defined by the formulas

$$
\begin{gathered}
a=\frac{1}{2}+\nu-i\sqrt{-\lambda\mu}, \; b=\frac{1}{2}+\nu+i\sqrt{-\lambda\mu}, \quad c=\frac{3}{2}+\nu, \; \nu=j+\frac{1}{2}, \\
a=1+j-i\sqrt{-\lambda\mu}, \quad b=1+j+i\sqrt{-\lambda\mu}\,, \quad c=j+2, \\
f(r)=C\,y^{(\nu+1)/2}(1-y)^{\nu/2}F(a,b,c,y)\,.
\end{gathered} \tag{5.115}
$$

Taking in mind the mentioned symmetry, we can get the similar functions $g(r)$:

$$
\begin{gathered}
g=C'\,y^{A'}(1-y)^{B'}G, \quad A'=\frac{1}{2}\nu, \; \frac{1}{2}-\frac{1}{2}\nu, \quad B'=\frac{1}{2}+\frac{1}{2}\nu, \; -\frac{1}{2}\nu, \\
a'=\frac{1}{2}+\nu-i\sqrt{-\lambda\mu}, \quad b'=\frac{1}{2}+\nu+i\sqrt{-\lambda\mu}\,, \qquad c'=\frac{1}{2}+\nu, \\
a'=1+j-i\sqrt{-\lambda\mu}, \quad b'=1+j+i\sqrt{-\lambda\mu}\,, \quad c'=j+1=c-1, \\
g(r)=C'\,y^{\nu/2}(1-y)^{(\nu+1)/2}F(a,b,c-1,y)\,.
\end{gathered} \tag{5.116}
$$

In order to find a relative coefficient between C and C', we turn to the differential equation from (5.107), written in the variable $y=(1-\cosh r)/2$:

$$
g(y)=\frac{1}{\lambda}\Big(-\sqrt{y(y-1)}\frac{d}{dy}+\frac{1}{2}\frac{\nu}{\sqrt{y(y-1)}}\Big)f(y)\,. \tag{5.117}
$$

Taking into account (5.115) and (5.116), we reduce it to the form

$$
i\lambda\frac{C'}{C}\,y^{\nu/2}(1-y)^{(\nu+1)/2}F(a,b,c-1,y)
$$

$$
=\left(y^{1/2}(1-y)^{1/2}\frac{d}{dy}+\frac{\nu}{2}y^{-1/2}(1-y)^{-1/2}\right)y^{(\nu+1)/2}(1-y)^{\nu/2}F(a,b,c,y)\,,
$$

or,

$$
i\lambda\frac{C'}{C}\,F(a,b,c-1,y)=\Big(y\frac{d}{dy}+\nu+\frac{1}{2}\Big)F(a,b,c,y)\,.
$$

Taking in mind the identity $\nu+1/2=j+1=c-1$; we obtain

$$
\lambda\,\frac{C'}{C}\,F(a,b,c-1,y)=y\frac{d}{dy}F(a,b,c,y)+(c-1)F(a,b,c,y),
$$

whence, by applying the known relation for hypergeometric functions [1]

$$
y\frac{d}{dy}F(a,b,c,y)+(c-1)F(a,b,c,y)=(c-1)F(a,b,c-1,y)\,,
$$

we arrive at the needed relation

$$
i\lambda\,\frac{C'}{C}\,F(a,b,c-1,y)=(c-1)F(a,b,c-1,y) \quad \Longrightarrow \quad i\lambda\,C'=(j+1)\,C\,. \tag{5.118}
$$

Now, let us consider the equation for $g(t)$:

$$\left(\frac{d^2}{dt^2} - \frac{\sin t}{\cos t}\frac{d}{dt} + M^2 + iM\frac{\sin t}{\cos t} - \frac{\lambda\mu}{\cos^2 t}\right)g(t) = 0. \tag{5.119}$$

In the variable $x = i\tan t$, it reads

$$\left[(1-x^2)\frac{d^2}{dx^2} - x\frac{d}{dx} - \frac{M^2}{1-x^2} - \frac{Mx}{1-x^2} + \lambda\mu\right]g = 0.$$

Let us change the variable yet another time

$$y = \frac{1-x}{2} = \frac{1-i\tan t}{2} = \frac{e^{-it}}{2\cos t}, \quad 1-y = \frac{1+i\tan t}{2} = \frac{e^{+it}}{2\cos t};$$

then we get

$$\left[y(1-y)\frac{d^2}{dy^2} + (\frac{1}{2}-y)\frac{d}{dy} - \frac{1}{4}\frac{M(M+1)}{y} - \frac{1}{4}\frac{M(M-1)}{1-y} + \lambda\mu\right]g = 0. \tag{5.120}$$

With the use of the substitution $g(y) = y^A(1-y)^B G(y)$, we derive

$$y(1-y)\frac{d^2G}{dy^2} + \left[\frac{1}{2} + 2A - (2A+2B+1)y\right]\frac{dG}{dy}$$

$$+\left[\lambda\mu - (A+B)^2 - \frac{1}{4}\frac{2A(1-2A)+M(M+1)}{y} - \frac{1}{4}\frac{2B(1-2B)+M(M-1)}{1-y}\right]G = 0.$$

Let the coefficients at y^{-1} and at $(1-y)^{-1}$ both vanish:

$$A = -\frac{1}{2}M, \ \frac{1}{2} + \frac{1}{2}M, \quad B = -\frac{1}{2}M + \frac{1}{2}, \ \frac{1}{2}M, \tag{5.121}$$

and then the above equation simplifies

$$y(1-y)\frac{d^2G}{dy^2} + \left[\frac{1}{2} + 2A - (2A+2B+1)y\right]\frac{dG}{dy} - \left[(A+B)^2 - \lambda\mu\right]G = 0;$$

and it may be identified with one of hypergeometric type

$$y(1-y)F'' + [c - (a+b+1)y]F' - abF = 0, \quad c = \frac{1}{2} + 2A,$$
$$a = A + B - i\sqrt{-\lambda\mu}, \quad b = A + B + i\sqrt{-\lambda\mu}.$$

Thus, we have the solutions

$$A = \frac{1}{2} + \frac{1}{2}M, \quad B = \frac{1}{2}M,$$
$$a = \frac{1}{2} + M - i\sqrt{-\lambda\mu}, \quad b = \frac{1}{2} + M + i\sqrt{-\lambda\mu}, \quad c = \frac{3}{2} + M, \tag{5.122}$$
$$g(y) = Ly^A(1-y)^B G = Ly^A(1-y)^B F(a,b,c,y).$$

Taking in mind the mentioned symmetry between $f(t)$ and $g(t)$ $(M \Longrightarrow -M)$, we write down similar expressions for f:

$$A' = \frac{1}{2}M, \quad B' = \frac{1}{2}M + \frac{1}{2},$$
$$a' = \frac{1}{2} + M - i\sqrt{-\lambda\mu} = a\,,\; b' = \frac{1}{2} + M + i\sqrt{-\lambda\mu} = b\,,\; c' = \frac{1}{2} + M = c - 1\,, \tag{5.123}$$
$$f(y) = L'\, y^{A'}(1-y)^{B'} G' = L' y^{A'}(1-y)^{B'}\, F(ab, c-1, y)\,.$$

To find the relative coefficient between L and L' in the expression for f and g we use the first order relationship (expressed in terms of the variable $y = (1 - i\tan t)/2$)

$$f(y) = -\frac{1}{\lambda}\Big(\sqrt{y(1-y)}\frac{d}{dy} + \frac{1}{2}\frac{M}{\sqrt{y(1-y)}}\Big)g(y)\,. \tag{5.124}$$

By allowing for (5.122) and (5.123), we transform this to the form

$$-\lambda\frac{L'}{L}\, y^{M/2}(1-y)^{(M+1)/2}F(a,b,c-1,y)$$
$$= \Big(y^{1/2}(1-y)^{1/2}\frac{d}{dy} + \frac{M}{2}y^{-1/2}(1-y)^{-1/2}\Big)y^{(M+1)/2}(1-y)^{M/2}F(a,b,c,y)\,,$$

or,

$$-\lambda\frac{L'}{L}\, F(a,b,c-1,y) = \Big(y\frac{d}{dy} + M + \frac{1}{2}\Big)F(a,b,c,y)\,.$$

Taking in mind the identity $M + 1/2 = c - 1$, we get

$$-\lambda\,\frac{L'}{L}\, F(a,b,c-1,y) = y\frac{d}{dy}F(a,b,c,y) + (c-1)F(a,b,c,y)\,,$$

whence by applying the known relation for hypergeometric functions

$$y\frac{d}{dy}F(a,b,c,y) + (c-1)F(a,b,c,y) = (c-1)F(a,b,c-1,y)\,,$$

we arrive at the needed result

$$-\lambda\,\frac{L'}{L}\, F(a,b,c-1,y) = (c-1)F(a,b,c-1,y) \quad\Longrightarrow\quad -\lambda\, L' = \Big(m + \frac{1}{2}\Big)L\,. \tag{5.125}$$

Choosing in different ways the two parameters A and B, we may construct four solutions

$$\text{I.}\quad A = \frac{M}{2} + \frac{1}{2}, \quad B = \frac{M}{2}, \quad A + B = M + \frac{1}{2},$$
$$f = y^{\frac{M}{2}+\frac{1}{2}}(1-y)^{\frac{M}{2}}, \quad a(b) = M + \frac{1}{2} \mp i\sqrt{-\lambda\mu}, \quad c = M + \frac{1}{2}; \tag{5.126}$$

$$\text{II.}\quad A = \frac{M}{2}, \quad B = -\frac{M}{2} + \frac{1}{2}, \quad A + B = \frac{1}{2},$$
$$f = y^{\frac{M}{2}}(1-y)^{-\frac{M}{2}+\frac{1}{2}}, \quad a(b) = 1 \mp i\sqrt{-\lambda\mu}\,, \quad c = M + \frac{1}{2}; \tag{5.127}$$

$$\begin{gathered}\text{III.} \quad A = -\frac{M}{2}, \quad B = -\frac{M}{2} + \frac{1}{2}, \quad A + B = -M + \frac{1}{2}, \\ f = y^{-\frac{M}{2}} (1-y)^{-\frac{M}{2}+\frac{1}{2}}, \ a(b) = -M + \frac{1}{2} \mp i\sqrt{-\lambda\mu}, \ c = -M + \frac{1}{2};\end{gathered} \tag{5.128}$$

$$\begin{gathered}\text{IV.} \quad A = -\frac{M}{2}, \quad B = \frac{M}{2}, \quad A + B = 0, \\ f = y^{-\frac{m}{2}} (1-y)^{\frac{M}{2}}, \quad a(b) = \mp i\sqrt{-\lambda\mu}, \quad c = -M + \frac{1}{2}.\end{gathered} \tag{5.129}$$

Let us recall the formulas (5.120)

$$y = \frac{1-x}{2} = \frac{1 - i\tan t}{2} = \frac{e^{-it}}{2\cos t}, \quad 1 - y = \frac{1 + i\tan t}{2} = \frac{e^{+it}}{2\cos t}. \tag{5.130}$$

We see that at $\cos t = 0$, the variable y tends to infinity. In order to describe behavior of solutions near this point

$$\cos t = 0, \quad t = \frac{\pi}{2} + \pi N, \quad y = \infty, \tag{5.131}$$

we make use of the Kummer relation [1]

$$U_1 = \frac{\Gamma(c)\Gamma(b-a)}{\Gamma(c-a)\Gamma(b)} U_3 + \frac{\Gamma(c)\Gamma(a-b)}{\Gamma(c-b)\Gamma(a)} U_4, \tag{5.132}$$

where

$$U_1 = F(a,b,c;y), \quad U_3 = (-y)^{-a} F(a, a+1-c, a+1-b; \frac{1}{y}),$$

$$U_4 = (-y)^{-b} F(b, b+1-c, b+1-a; \frac{1}{y}).$$

When $y \to \infty$, the formula (5.130) reads

$$F(a,b,c; y \to \infty) = \frac{\Gamma(c)\Gamma(b-a)}{\Gamma(c-a)\Gamma(b)} (-y)^{-a} + \frac{\Gamma(c)\Gamma(a-b)}{\Gamma(c-b)\Gamma(a)} (-y)^{-b}. \tag{5.133}$$

Therefore, the complete solution $f(y)$ in the vicinity of the infinite point, by taking in mind the formulas

$$c = 2A + \frac{1}{2}, \quad a = A + B - i\sqrt{-\lambda\mu}, \quad b = A + B + i\sqrt{-\lambda\mu},$$

is given by the relation (the total multiplier $\Gamma(c)$ is omitted)

$$f \sim y^A (-y)^B \Big[\frac{\Gamma(b-a)}{\Gamma(c-a)\Gamma(b)} (-y)^{-A-B+i\sqrt{-\lambda\mu}} + \frac{\Gamma(a-b)}{\Gamma(c-b)\Gamma(a)} (-y)^{-A-B-i\sqrt{-\lambda\mu}} \Big],$$

or,

$$f(y \to \infty) \sim \Big[\frac{\Gamma(b-a)}{\Gamma(c-a)\Gamma(b)} (-y)^{+i\sqrt{-\lambda\mu}} + \frac{\Gamma(a-b)}{\Gamma(c-b)\Gamma(a)} (-y)^{-i\sqrt{-\lambda\mu}} \Big]. \tag{5.134}$$

All the four solutions behave themselves similarly as linear combinations of two terms:

$$f(y \to \infty) \sim \text{const}(-y)^{-A-B+i\sqrt{-\lambda\mu}} + \text{const}'(-y)^{-A-B-i\sqrt{-\lambda\mu}}. \tag{5.135}$$

Let us take into account the following two identities (when $e^{2it} \to -1$)

$$(-y)^{+i\sqrt{-\lambda\mu}} = \Big(-\frac{1}{e^{2it}+1}\Big)^{+i\sqrt{-\lambda\mu}}, \quad (-y)^{-i\sqrt{-\lambda\mu}} = \Big(-\frac{1}{e^{2it}+1}\Big)^{-i\sqrt{-\lambda\mu}}.$$

We consider the approaching to the singular point from different sides:
let $\delta \to +\infty$,

$$\begin{gathered}(a) \qquad e^{2it} \to -1, \quad e^{2it} = -1 - \frac{1}{\delta}, \quad \delta \to +\infty, \\ (-y)^{+i\sqrt{-\lambda\mu}} = \Big(+\delta\Big)^{+i\sqrt{-\lambda\mu}} = e^{+i\sqrt{-\lambda\mu}\ln\delta}, \\ (-y)^{-i\sqrt{-\lambda\mu}} = \Big(-\delta\Big)^{-i\sqrt{-\lambda\mu}} = e^{-i\sqrt{-\lambda\mu}\ln(-\delta)};\end{gathered} \tag{5.136}$$

let $\delta \to +\infty$,

$$\begin{gathered}(b) \qquad e^{2it} \to -1, \quad e^{2it} = -1 + \frac{1}{\delta}, \quad \delta \to +\infty, \\ (-y)^{+i\sqrt{-\lambda\mu}} = \Big(-\delta\Big)^{+i\sqrt{-\lambda\mu}} = e^{+i\sqrt{-\lambda\mu}\ln(-\delta)}, \\ (-y)^{-i\sqrt{-\lambda\mu}} = \Big(+\delta\Big)^{-i\sqrt{-\lambda\mu}} = e^{-i\sqrt{-\lambda\mu}\ln\delta}.\end{gathered} \tag{5.137}$$

5.5 Pauli Equation in Oscillating de Sitter Space

Let us perform the nonrelativistic approximation in the system (5.103):

$$\begin{gathered}\frac{1}{\cos t}\Big(\frac{\partial}{\partial r} + \frac{\nu}{\sinh r}\Big) f + \Big(i\frac{\partial}{\partial t} + \delta\, M\Big) g = 0, \\ \frac{1}{\cos t}\Big(\frac{\partial}{\partial r} - \frac{\nu}{\sinh r}\Big) g - \Big(i\frac{\partial}{\partial t} - \delta\, M\Big) f = 0.\end{gathered} \tag{5.138}$$

We separate the rest energy by the formal change

$$i\frac{\partial}{\partial t} \quad \Longrightarrow \quad \Big(M + i\frac{\partial}{\partial t}\Big).$$

Further, the two cases $\delta = \pm 1$ are considered separately:

$$\delta = +1, \quad \begin{gathered}\frac{1}{\cos t}\Big(\frac{\partial}{\partial r} + \frac{\nu}{\sinh r}\Big) f + \Big(M + i\frac{\partial}{\partial t} + M\Big) g = 0, \\ \frac{1}{\cos t}\Big(\frac{\partial}{\partial r} - \frac{\nu}{\sinh r}\Big) g - \Big(M + i\frac{\partial}{\partial t} - M\Big) f = 0;\end{gathered} \tag{5.139}$$

$$\delta=-1,\quad \begin{aligned}&\frac{1}{\cos t}\Big(\frac{\partial}{\partial r}+\frac{\nu}{\sinh r}\Big)f+\Big(M+i\frac{\partial}{\partial t}-M\Big)g=0,\\ &\frac{1}{\cos t}\Big(\frac{\partial}{\partial r}-\frac{\nu}{\sinh r}\Big)g-\Big(M+i\frac{\partial}{\partial t}+M\Big)f=0.\end{aligned} \tag{5.140}$$

By neglecting the small terms $i\partial_t$ (in comparison with $2M$), we get

$$\delta=+1,\quad \frac{1}{\cos t}\Big(\frac{\partial}{\partial r}+\frac{\nu}{\sinh r}\Big)f+2Mg=0,\quad \frac{1}{\cos t}\Big(\frac{\partial}{\partial r}-\frac{\nu}{\sinh r}\Big)g-i\frac{\partial}{\partial t}f=0;$$

$$\delta=-1,\quad \frac{1}{\cos t}\Big(\frac{\partial}{\partial r}+\frac{\nu}{\sinh r}\Big)f+i\frac{\partial}{\partial t}g=0,\quad \frac{1}{\cos t}\Big(\frac{\partial}{\partial r}-\frac{\nu}{\sinh r}\Big)g-2Mf=0.$$

In both cases, we derive the nonrelativistic equations for the big components (the second equation permits us to find the concomitant small component):

$$\delta=+1,\quad \begin{aligned}&i\frac{\partial}{\partial t}f=-\frac{1}{2M}\frac{1}{\cos^2 t}\Big(\frac{\partial}{\partial r}-\frac{\nu}{\sinh r}\Big)\Big(\frac{\partial}{\partial r}+\frac{\nu}{\sinh r}\Big)f,\\ &g=-\frac{1}{2M}\frac{1}{\cos t}\Big(\frac{\partial}{\partial r}+\frac{\nu}{\sinh r}\Big)f;\end{aligned} \tag{5.141}$$

$$\delta=-1,\quad \begin{aligned}&i\frac{\partial}{\partial t}g=-\frac{1}{2M}\frac{1}{\cos^2 t}\Big(\frac{\partial}{\partial r}+\frac{\nu}{\sinh r}\Big)\Big(\frac{\partial}{\partial r}-\frac{\nu}{\sinh r}\Big)g,\\ &f=\frac{1}{2M}\frac{1}{\cos t}\Big(\frac{\partial}{\partial r}-\frac{\nu}{\sinh r}\Big)g.\end{aligned} \tag{5.142}$$

Let us recall the structure of the nonrelativistic wave functions for a spinor particle:

$$\varphi(x)_{jm\delta}=\left|\begin{array}{c} f_1D_{-1/2}\\ f_2D_{+1/2}\\ \delta f_2D_{-1/2}\\ \delta f_1D_{+1/2}\end{array}\right| \Longrightarrow \Psi^{Pauli}=\left|\begin{array}{c} f_1D_{-1/2}+\delta f_2D_{-1/2}\\ f_2D_{+1/2}+\delta f_1\,D_{+1/2}\end{array}\right|. \tag{5.143}$$

At different parities, we respectively obtain

$$\delta=+1,\quad \Psi^{Pauli}=\left|\begin{array}{c} f_1D_{-1/2}+f_2D_{-1/2}\\ f_2D_{+1/2}+f_1D_{+1/2}\end{array}\right|=\left|\begin{array}{c} fD_{-1/2}\\ fD_{+1/2}\end{array}\right|, \tag{5.144}$$

$$\delta=-1,\quad \Psi^{Pauli}=\left|\begin{array}{c} f_1D_{-1/2}-f_2D_{-1/2}\\ f_2D_{+1/2}-f_1D_{+1/2}\end{array}\right|=\left|\begin{array}{c} -igD_{-1/2}\\ -igD_{+1/2}\end{array}\right|. \tag{5.145}$$

The functions $f(t,r)$ and $g(t,r)$ satisfy slightly different equations:

$$\delta=+1,\quad i\frac{\partial}{\partial t}f=-\frac{1}{2M}\frac{1}{\cos^2 t}\Big(\frac{\partial^2}{\partial r^2}-\frac{\nu^2+\nu\cosh r}{\sinh^2 r}\Big)f, \tag{5.146}$$

$$\delta=-1,\quad i\frac{\partial}{\partial t}g=-\frac{1}{2M}\frac{1}{\cos^2 t}\Big(\frac{\partial^2}{\partial r^2}-\frac{\nu^2-\nu\cosh r}{\sinh^2 r}\Big)g. \tag{5.147}$$

The variables are separated by the substitutions

$$f(t,r) = f(t)f(r), \quad g(t,r) = g(t)g(r).$$

Thus, we obtain

$$\delta = +1, \quad i\cos^2 t \frac{1}{f(t)}\frac{d}{dt}f(t) = -\frac{1}{2M}\frac{1}{f(r)}\Big(\frac{d^2}{dr^2} - \frac{\nu^2+\nu\cosh r}{\sinh^2 r}\Big)f(r) = E; \quad (5.148)$$

$$\delta = -1, \quad i\cos^2 t \frac{1}{g(t)}\frac{d}{dt}g(t) = -\frac{1}{2M}\frac{1}{g(r)}\Big(\frac{d^2}{dr^2} - \frac{\nu^2-\nu\cosh r}{\sinh^2 r}\Big)g(r) = E. \quad (5.149)$$

The separated equations are

$$\delta = +1, \quad \begin{array}{l} i\cos^2 t \dfrac{1}{f(t)}\dfrac{d}{dt}f(t) = E \quad \Longrightarrow \quad f(t) = e^{-iE\tan t}, \\ -\dfrac{1}{2M}\dfrac{1}{f(r)}\Big(\dfrac{d^2}{dr^2} - \dfrac{\nu^2+\nu\cosh r}{\sinh^2 r}\Big)f(r) = E; \end{array} \quad (5.150)$$

$$\delta = -1, \quad \begin{array}{l} i\cos^2 t \dfrac{1}{g(t)}\dfrac{d}{dt}g(t) = E \quad \Longrightarrow \quad g(t) = e^{-iE\tan t}, \\ -\dfrac{1}{2M}\dfrac{1}{g(r)}\Big(\dfrac{d^2}{dr^2} - \dfrac{\nu^2-\nu\cosh r}{\sinh^2 r}\Big)g(r) = E. \end{array} \quad (5.151)$$

The equations for $f(r)$ and $g(r)$ are symmetrical under the replacement $\nu \Longrightarrow -\nu$. For definiteness, let us study the equation for $\delta = +1$:

$$\Big(\frac{d^2}{dr^2} - \frac{\nu^2+\nu\cosh r}{\sinh^2 r} + 2ME\Big)f(r) = 0. \quad (5.152)$$

In terms of the variable $y = \frac{1-\cosh r}{2}$, we get

$$y(1-y)\frac{d^2 f}{dy^2} + \Big(\frac{1}{2} - y\Big)\frac{df}{dy} - \Big[\frac{\nu(\nu+1)}{4y} + \frac{\nu(\nu-1)}{4(1-y)} + 2ME\Big]f = 0.$$

With the substitution $f(y) = y^A(1-y)^B F(y)$, we get

$$y(1-y)\frac{d^2 F}{dy^2} + \Big[\frac{1}{2} + 2A - (1+2A+2B)y\Big]\frac{dF}{dy}$$

$$-\Big[(A+B)^2 + 2ME - \frac{2A(2A-1)-\nu(\nu+1)}{4y} - \frac{2B(2B-1)-\nu(\nu-1)}{4(1-y)}\Big]F = 0.$$

Let it be

$$A = -\frac{1}{2}\nu, \ \frac{1}{2}+\frac{1}{2}\nu; \quad B = \frac{1}{2}-\frac{1}{2}\nu, \ \frac{1}{2}\nu,$$

and then the above equation simplifies

$$y(1-y)\frac{d^2F}{dy^2}+\Big[\frac{1}{2}+2A-(1+2A+2B)y\Big]\frac{dF}{dy}-\Big[(A+B)^2+2ME\Big]F=0;$$

it is identified as a hypergeometric equation

$$y(1-y)F''+[c-(a+b+1)y]F'-abF=0,$$
$$c=2A+\frac{1}{2},\quad a=A+B-i\sqrt{2ME},\quad b=A+B+i\sqrt{2ME}. \tag{5.153}$$

Its solutions are

$$a=\frac{1}{2}+\nu-i\sqrt{2ME},\quad b=\frac{1}{2}+\nu+i\sqrt{2ME},$$
$$c=\frac{3}{2}+\nu,\quad \nu=j+\frac{1}{2},\quad c=j+2, \tag{5.154}$$
$$a=1+j-i\sqrt{2ME},\quad b=1+j+i\sqrt{2ME},$$
$$f(r)=C\,y^{(\nu+1)/2}(1-y)^{\nu/2}F(a,b,c,y).$$

Similar solutions for $g(y)$ are

$$g(y)=C'y^{A'}(1-y)^{B'}G(y),\quad A'=\frac{1}{2}\nu,\ \frac{1}{2}-\frac{1}{2}\nu,\quad B'=\frac{1}{2}+\frac{1}{2}\nu,\ -\frac{1}{2}\nu,$$
$$a'=\frac{1}{2}+\nu-i\sqrt{2ME},\quad b'=\frac{1}{2}+\nu+i\sqrt{2ME},\quad c'=\frac{1}{2}+\nu, \tag{5.155}$$
$$a'=1+j-i\sqrt{2ME},\quad b'=1+j+i\sqrt{2ME},\quad c'=j+1=c-1,$$
$$g(r)=C'y^{\nu/2}(1-y)^{(\nu+1)/2}F(a,b,c-1,y).$$

Thus, the nonrelativistic solutions in the nonstatic anti de Sitter model are

$$\delta=+1,\quad \Psi^{Pauli}_{Ejm,+1}(t,r,\theta,\phi)=e^{-iE\tan t}f(r)\begin{vmatrix} D^j_{-m,-1/2}(\phi,\theta,0)\\ D^j_{-m,+1/2}(\phi,\theta,0)\end{vmatrix},$$
$$f(r)=C\Big(\frac{1-\cosh r}{2}\Big)^{(\nu+1)/2}\Big(\frac{1+\cosh r}{2}\Big)^{\nu/2}F\Big(-n,n+2\nu+1,\nu+\frac{3}{2},\frac{1-\cosh r}{2}\Big); \tag{5.156}$$

$$\delta=-1,\quad \Psi^{Pauli}_{Ejm,-1}(t,r,\theta,\phi)=e^{-iE\tan t}g(r)\begin{vmatrix} D^j_{-m,-1/2}(\phi,\theta,0)\\ D^j_{-m,+1/2}(\phi,\theta,0)\end{vmatrix},$$
$$g(r)=C'\Big(\frac{1-\cosh r}{2}\Big)^{\nu/2}\Big(\frac{1+\cosh r}{2}\Big)^{(\nu+1)/2}F\Big(-n,n+2\nu+1,\nu+\frac{1}{2},\frac{1-\cosh r}{2}\Big). \tag{5.157}$$

Chapter 6

Spin 1/2 Particle in Non-static de Sitter Models, Quasi-Cartesian Coordinates

6.1 Separation of the Variables in the Dirac Equation

It is known that the geometry of the Lobachevsky space acts on particles with spins $0, 1/2, 1$ as an ideal mirror distributed in the space. The depth of penetration of the field in such an effective medium increases with increasing field energy; also this penetration depends on the radius of curvature of the Lobachevsky space. Since the Lobachevsky model is a constituent element in some cosmological models, this property means that in such models it is necessary to take into account the effect of the presence of a "cosmological mirror"; it must effectively lead to a redistribution of the particle density in the space. The earlier analysis assumed the static nature of the space-time geometry. In this chapter we generalize the investigation for the spin 1/2 field in the case of the oscillating de Sitter universe. The Dirac equation is solved in non-static quasi-Cartesian coordinates. At this we substantially use a generalized helicity operator. The wave functions of the particle depend nontrivially on the time, however the effect of the complete reflection of the particles from the effective potential barrier is preserved. For the real Majorana 4-spinor field, similar results are valid as well.

We will use the following system of coordinates in the anti de Sitter space-time

$$dS^2 = dt^2 - \cos^2 t \left[e^{-2z}(dx^2 + dy^2) + dz^2 \right]. \tag{6.1}$$

Starting from the Dirac equation in the form

$$\left[i\gamma^a \left(e^{\alpha}_{(a)} \partial_\alpha + \frac{1}{2} \left(\frac{1}{\sqrt{-g}} \partial_\alpha \sqrt{-g}\, e^{\alpha}_{(a)} \right) \right) - M \right] \Psi = 0,$$

we use a diagonal tetrad; then equation takes the form

$$\left[i\gamma^0 \cos t \left(\frac{\partial}{\partial t} - \frac{3}{2} \tan t \right) + i\gamma^1 e^z \frac{\partial}{\partial x} + i\gamma^2 e^z \frac{\partial}{\partial y} + i\gamma^3 \left(\frac{\partial}{\partial z} - 1 \right) - M \cos t \right] \Psi = 0. \tag{6.2}$$

By separating the simple multiplier in the wave function, $\Psi = (e^z/\cos^{3/2} t)\,\psi$, we reduce the above equation to a simpler form

$$\left[i\gamma^0 \cos t \frac{\partial}{\partial t} + i\gamma^1 e^z \frac{\partial}{\partial x} + i\gamma^2 e^z \frac{\partial}{\partial y} + i\gamma^3 \frac{\partial}{\partial z} - M \cos t\right]\psi = 0. \tag{6.3}$$

The solutions are searched as

$$\psi = e^{ik_1 x}\, e^{ik_2 y} \left| \begin{array}{c} f_1(t,z) \\ f_2(t,z) \\ f_3(t,z) \\ f_4(t,z) \end{array} \right|. \tag{6.4}$$

Using the Dirac matrices in spinor basis, we derive the equations for $f_i(t,z)$ (let $k_1 = a, k_2 = b$):

$$\begin{aligned}
& i\cos t \frac{\partial}{\partial t} f_3 - M \cos t\, f_1 + e^z(a-ib) f_4 - i\frac{\partial}{\partial z} f_3 = 0, \\
& i\cos t \frac{\partial}{\partial t} f_4 - M \cos t\, f_2 + e^z(a+ib) f_3 + i\frac{\partial}{\partial z} f_4 = 0, \\
& i\cos t \frac{\partial}{\partial t} f_1 - M \cos t\, f_3 - e^z(a-ib) f_2 + i\frac{\partial}{\partial z} f_1 = 0, \\
& i\cos t \frac{\partial}{\partial t} f_2 - M \cos t\, f_4 - e^z(a+ib) f_1 - i\frac{\partial}{\partial z} f_2 = 0.
\end{aligned} \tag{6.5}$$

There exists a generalized helicity operator, commuting with the operator in (6.3):

$$\Sigma = \frac{1}{2}\left(e^z\gamma^2\gamma^3 \frac{\partial}{\partial x} + e^z\gamma^3\gamma^1 \frac{\partial}{\partial y} + \gamma^1\gamma^2 \frac{\partial}{\partial z}\right). \tag{6.6}$$

From the eigenvalue equation $\Sigma\Psi = p\,\Psi$, we derive

$$\begin{aligned}
e^z(a-ib) f_2(t,z) &= \left(+i\frac{\partial}{\partial z} + p\right) f_1(t,z), \\
e^z(a+ib) f_1(t,z) &= \left(-i\frac{\partial}{\partial z} + p\right) f_2(t,z), \\
e^z(a-ib) f_4(t,z) &= \left(+i\frac{\partial}{\partial z} + p\right) f_3(t,z), \\
e^z(a+ib) f_3(t,z) &= \left(-i\frac{\partial}{\partial z} + p\right) f_4(t,z).
\end{aligned} \tag{6.7}$$

In the case of a static metric, we know the eigenvalues of the helicity (the index 0 serves to distinguish this case from the one studied below)

$$p_0 = \pm\sqrt{\epsilon^2 - M^2}, \quad f_3(z) = \lambda_0 f_1(z), \quad f_4(z) = \lambda_0 f_2(z), \quad \lambda_0 = \frac{\epsilon - p}{M}. \tag{6.8}$$

From Eq. (6.7) we may derive the second order equations for the separate functions:

$$\begin{aligned}
&\Big(\frac{d^2}{dz^2}-\frac{d}{dz}-e^{2z}(a^2+b^2)+p(i+p)\Big)f_1=0\,,\\
&\Big(\frac{d^2}{dz^2}-\frac{d}{dz}-e^{2z}(a^2+b^2)-p(i-p)\Big)f_2=0\,,\\
&\Big(\frac{d^2}{dz^2}-\frac{d}{dz}-e^{2z}(a^2+b^2)+p(i+p)\Big)f_3=0\,,\\
&\Big(\frac{d^2}{dz^2}-\frac{d}{dz}-e^{2z}(a^2+b^2)-p(i-p)\Big)f_4=0\,,
\end{aligned} \tag{6.9}$$

whence one may conclude that, together with the solution for the eigenvalue $(+p)$, there exists a solution for the eigenvalue $(-p)$; by physical reason we assume the real-valuedness of these $\pm p$.

The form of Eqs. (6.7) assumes that the pairs of functions

$$f_1(t,z)-f_2(t,z)\,,\quad f_3(t,z)-f_4(t,z)$$

should have a quite definite structure

$$\begin{aligned}
&f_1(t,z)=F(t)f_1(z)\,,\ f_2(t,z)=F(t)f_2(z)\,,\\
&f_3(t,z)=G(t)f_3(z)\,,\ f_4(t,z)=G(t)f_4(z)\,;
\end{aligned} \tag{6.10}$$

which reduces the Eqs. (6.7) to two sub-systems in the variable z:

$$\begin{aligned}
e^z(a-ib)f_2(z)&=\Big(+i\frac{d}{dz}+p\Big)f_1(z)\,,\\
e^z(a+ib)f_1(z)&=\Big(-i\frac{d}{dz}+p\Big)f_2(z)\,;\\
e^z(a-ib)f_4(z)&=\Big(+i\frac{d}{dz}+p\Big)f_3(z)\,,\\
e^z(a+ib)f_3(z)&=\Big(-i\frac{d}{dz}+p\Big)f_4(z)\,.
\end{aligned} \tag{6.11}$$

Taking into account (6.11) in Eqs. (6.5), we get two sub-systems in the variable t:

$$\begin{aligned}
&f_3(z)\Big(i\cos t\frac{d}{dt}+p\Big)G(t)-M\cos t F(t)f_1(z)=0\,,\\
&f_1(z)\Big(i\cos t\frac{d}{dt}-p\Big)F(t)-M\cos t G(t)f_3(z)=0;\\
&f_4(z)\Big(i\cos t\frac{d}{dt}+p\Big)G(t)-M\cos t G(t)f_2(z)=0\,,\\
&f_2(z)\Big(i\cos t\frac{d}{dt}-p\Big)F(t)-M\cos t G(t)f_4(z)=0\,.
\end{aligned} \tag{6.12}$$

The form of these equations assumes the relationships $f_3(z) = \lambda f_1(z)$, $f_4(z) = \lambda f_2(z)$; then we arrive at one the same equations in the variable t:

$$\Big(i\cos t\frac{d}{dt}+p\Big)G(t)-\frac{M}{\lambda}\cos t F(t)=0\,,\qquad \Big(i\cos t\frac{d}{dt}-p\Big)F(t)-\lambda M\cos t G(t)=0\,. \tag{6.13}$$

Evidently, the parameter λ may be inserted into the notation $F(t)/\lambda \longrightarrow F(t)$, and therefore the parameter λ may be excluded from Eqs.(6.13). In fact, without loss of generality, we may set $\lambda = 1$ in Eqs. (6.13). Also, it is understandable that the functions $F(t)$ and $G(t)$ depend on the sign of $\pm p$.

Thus we have two systems, in variables t and z, respectively:

$$\Big(i\cos t\frac{d}{dt}+p\Big)G(t)-M\cos t F(t)=0\,,\qquad \Big(i\cos t\frac{d}{dt}-p\Big)F(t)-M\cos t\, G(t)=0; \tag{6.14}$$

$$e^{z}(a-ib)f_2(z)=\Big(+i\frac{d}{dz}+p\Big)f_1(z)\,,\qquad e^{z}(a+ib)f_1(z)=\Big(-i\frac{d}{dz}+p\Big)f_2(z)\,. \tag{6.15}$$

Their solutions determine the wave functions

$$\psi = e^{ik_1x}e^{ik_2y}\left|\begin{array}{c} F(t)\,f_1(z)\\ F(t)\,f_2(z)\\ \lambda G(t)\,f_1(z)\\ \lambda G(t)\,f_2(z)\end{array}\right|. \tag{6.16}$$

6.2 Solving Equations in the Time Variable t

From (6.15) we obtain the second order equations for f and g:

$$\Big(\frac{d^2}{dt^2}+M^2+\frac{ip\,(\sin t-ip)}{\cos^2 t}\Big)F=0,\quad \Big(\frac{d^2}{dt^2}+M^2-\frac{ip\,(\sin t+ip)}{\cos^2 t}\Big)G=0. \tag{6.17}$$

In the variable $\tau = (1+\sin t)/2$, they read

$$\tau(1-\tau)\frac{d^2F}{d\tau^2}+\Big(\frac{1}{2}-\tau\Big)\frac{dF}{d\tau}+\Big(M^2+\frac{1}{4}\frac{p\,(p+i)}{1-\tau}+\frac{1}{4}\frac{p\,(p-i)}{\tau}\Big)F=0,$$
$$\tau(1-\tau)\frac{d^2G}{d\tau^2}+\Big(\frac{1}{2}-\tau\Big)\frac{dG}{d\tau}+\Big(M^2+\frac{1}{4}\frac{p\,(p-i)}{1-\tau}+\frac{1}{4}\frac{p\,(p+i)}{\tau}\Big)G=0. \tag{6.18}$$

Due to the symmetry $p \rightarrow -p$, it suffices to study one equation. In fact, two last equations relate to each other by the complex conjugation: $G = \text{const}\ F^*$; let us follow the function F. With the substitution $F = \tau^A (1-\tau)^B f(\tau)$, for $f(\tau)$ we get the following equation

$$\tau(1-\tau)\frac{d^2 f}{d\tau^2} + \Big[2A + \frac{1}{2} - (2A+2B+1)\tau\Big]\frac{df}{d\tau}$$

$$+\Big[M^2 - (A+B)^2 + \frac{1}{4}\frac{2B(2B-1)+p(p+i)}{1-\tau} + \frac{1}{4}\frac{2A(2A-1)+p(p-i)}{\tau}\Big] f = 0.$$

At A and B chosen as

$$A = \frac{1}{2}(ip+1),\ -\frac{1}{2}ip, \quad B = \frac{1}{2}(-ip+1),\ \frac{1}{2}ip,$$

the above equation becomes simpler

$$\tau(1-\tau)\frac{d^2 f}{d\tau^2} + \Big[2A + \frac{1}{2} - (2A+2B+1)\tau\Big]\frac{df}{d\tau} + \Big[M^2 - (A+B)^2\Big] f = 0,$$

and it is recognized as a hypergeometric equation with the parameters α, β, γ:

$$\alpha = A+B-M,\ \beta = A+B+M,\ \gamma = 2A + \frac{1}{2}.$$

Below, we shall use two independent solutions of this equation (see [1])

$$A = \frac{-ip}{2}, \quad B = \frac{ip}{2}, \quad \alpha = -M, \quad \beta = +M, \quad \gamma = -ip + \frac{1}{2},$$
$$F^{(1)}(\tau) = \tau^A (1-\tau)^B F(\alpha,\beta,\gamma;\tau) = \tau^A (1-\tau)^B U_1(\tau); \tag{6.19}$$

$$A' = \frac{ip+1}{2}, \quad B' = \frac{-ip+1}{2}, \quad \alpha' = 1-M,\ \beta' = 1+M,\ \gamma' = ip + \frac{3}{2},$$
$$F^{(2)}(\tau) = \tau^{A'} (1-\tau)^{B'} F(\alpha',\beta',\gamma';\tau) = \tau^A (1-\tau)^B U_5(\tau); \tag{6.20}$$

here we have used two Kummer's solutions [1]

$$U_1(\tau) = F(a,b,c;\tau) \quad \Longrightarrow \quad F^{(1)}(\tau),$$

$$U_5(\tau) = y^{1-c}(1-y)^{c-a-b} F(1-a, 1-b, 2-c;\tau) \quad \Longrightarrow \quad F^{(2)}(\tau).$$

By the formal change $p \to -p$, we readily obtain similar results for the functions $G(\tau)$:

$$A_* = \frac{ip}{2}, \quad B_* = \frac{-ip}{2}, \quad \alpha = -M, \quad \beta = +M, \quad \gamma_* = ip + \frac{1}{2},$$
$$G^{(1)} = \tau^{A_*} (1-\tau)^{B_*} F(\alpha,\beta,\gamma_*;\tau) = \tau^{A_*} (1-\tau)^{B_*} U_1(\tau); \tag{6.21}$$

$$A'_* = \frac{-ip+1}{2}, \quad B'_* = \frac{ip+1}{2}, \quad \alpha' = 1-M,\ \beta' = 1+M,\ \gamma'_* = -ip + \frac{3}{2},$$
$$G^{(2)} = \tau^{A'_*} (1-\tau)^{B'_*} F(\alpha',\beta',\gamma'_*;\tau) = \tau^{A_*} (1-\tau)^{B_*} U_5(\tau). \tag{6.22}$$

Now, let us relate these 4 functions pairwise. First, we consider (6.19) and (6.22):

$$F^{(1)}(\tau) = \nu \tau^{A}(1-\tau)^{B} F(\alpha,\beta,\gamma;\tau),$$
$$A = \frac{-ip}{2},\ B = \frac{ip}{2}, \quad \alpha = -M,\ \beta = +M,\ \gamma = -ip + \frac{1}{2}, \tag{6.23}$$

and

$$G^{(2)}(\tau) = \nu' \tau^{A'_*}(1-\tau)^{B'_*} F(\alpha',\beta',\gamma'_*;\tau) = \nu'(1-\tau)^{B'_*} F(1+\alpha,1+\beta,1+\gamma;\tau),$$
$$A'_* = \frac{-ip+1}{2},\ B'_* = \frac{ip+1}{2}, \quad \alpha' = 1-M,\ \beta' = 1+M,\ \gamma'_* = -ip + \frac{3}{2}. \tag{6.24}$$

To find a relative coefficient for ν and ν', we turn to the 1-st order equation in (6.14), written in the variable $\tau = (1+\sin t)/2$:

$$G^{(2)}(\tau) = \frac{1}{M}\Big(i\sqrt{\tau(1-\tau)}\frac{d}{d\tau} - \frac{1}{2}\frac{p}{\sqrt{\tau(1-\tau)}}\Big) F^{(1)}(\tau); \tag{6.25}$$

we get

$$M\frac{\nu'}{\nu}\tau^{\frac{-ip+1}{2}}(1-\tau)^{\frac{ip+1}{2}} F(1+\alpha,1+\beta,1+\gamma;\tau)$$
$$= \Big(i\tau^{1/2}(1-\tau)^{1/2}\frac{d}{d\tau} - \frac{p}{2}\tau^{-1/2}(1-\tau)^{-1/2}\Big)\tau^{\frac{-ip}{2}}(1-\tau)^{\frac{ip}{2}} F(\alpha,\beta,\gamma;\tau),$$

or,

$$M\frac{\nu'}{\nu} F(1+\alpha,1+\beta,1+\gamma;\tau) = i\frac{d}{d\tau} F(\alpha,\beta,\gamma;\tau),$$

whence we find the needed relative coefficient

$$M\frac{\nu'}{\nu} = i\frac{\alpha\beta}{\gamma} \quad \Longrightarrow \quad \nu' = \frac{-iM}{-ip+1/2}\nu. \tag{6.26}$$

Now, consider the solutions (6.20) and (6.21):

$$F^{(2)} = \mu' \tau^{A'}(1-\tau)^{B'} F(\alpha',\beta',\gamma';\tau) = \mu' \tau^{A'}(1-\tau)^{B'} F(1+\alpha,1+\beta,1+\gamma_*;\tau),$$

$$A' = \frac{ip+1}{2},\ B' = \frac{-ip+1}{2},\ \alpha' = 1-M,\ \beta' = 1+M,\ \gamma' = ip + \frac{3}{2}; \tag{6.27}$$

$$G^{(1)} = \mu L_* \tau^{A_*}(1-\tau)^{B_*} F(\alpha,\beta,\gamma_*;\tau),$$
$$A_* = \frac{ip}{2},\ B_* = \frac{-ip}{2},\ \alpha = -M, \quad \beta = +M,\ \gamma_* = ip + \frac{1}{2}. \tag{6.28}$$

To find the relative coefficient for μ' and μ, we turn to the 1-st order differential equation in (6.14)), written as

$$F^{(2)}(\tau) = \frac{1}{M}\Big(i\sqrt{\tau(1-\tau)}\frac{d}{d\tau} + \frac{1}{2}\frac{p}{\sqrt{\tau(1-\tau)}}\Big) G^{(1)}(\tau);$$

we yield

$$M\frac{\mu'}{\mu}\tau^{\frac{ip+1}{2}}(1-\tau)^{\frac{-ip+1}{2}}F(1+\alpha,1+\beta,1+\gamma_*;\tau)$$
$$=\left(i\tau^{1/2}(1-\tau)^{1/2}\frac{d}{d\tau}+\frac{p}{2}\tau^{-1/2}(1-\tau)^{-1/2}\right)\tau^{\frac{ip}{2}}(1-\tau)^{\frac{-ip}{2}}F(\alpha,\beta,\gamma_*;\tau),$$

or,

$$M\frac{\mu'}{\mu}F(1+\alpha,1+\beta,1+\gamma_*;\tau)=i\frac{d}{d\tau}F(\alpha,\beta,\gamma_*;\tau),$$

whence, we get the needed coefficient (we may set $\nu=\mu=1$)

$$M\frac{\mu'}{\mu}=i\frac{\alpha\beta}{\gamma_*}\quad\Longrightarrow\quad\mu'=\frac{iM}{ip+1/2}\mu. \tag{6.29}$$

6.3 Behavior of Solutions in the Variable t Near the Points $\cos(t)=0$

The metric of the space-time model is singular when $\cos t=0$. Let us consider the singular point $t=-\pi/2$, to which there corresponds $\tau=0$; the solutions behave as follows:

the pair

$$F^{(1)}\sim\tau^{A}=\left(\frac{1+\sin t}{2}\right)^{-ip/2},G^{(2)}\sim\frac{+iM}{1/2-ip}\tau^{A'_*}=\frac{-iM}{1/2-ip}\left(\frac{1+\sin t}{2}\right)^{1/2-ip/2}; \tag{6.30}$$

the pair

$$F^{(2)}\sim\frac{+iM}{1/2+ip}\tau^{A'}=\frac{+iM}{1/2+ip}\left(\frac{1+\sin t}{2}\right)^{1/2+ip/2},G^{(1)}\sim\tau^{A'_*}=\left(\frac{1+\sin t}{2}\right)^{+ip/2}. \tag{6.31}$$

Taking in mind the formula $t\approx-\pi/2+2\Delta,\Delta\to 0$, we get

$$F^{(1)}\sim e^{-ip/2\ln\Delta},\quad G^{(2)}\sim\frac{-iM}{1/2-ip}\sqrt{\Delta}e^{-ip/2\ln\Delta}\Longrightarrow 0; \tag{6.32}$$

$$F^{(2)}\sim\frac{+iM}{1/2+ip}\sqrt{\Delta}e^{+ip/2\ln\Delta}\quad\Longrightarrow 0,\quad G^{(1)}\sim e^{+ip/2\ln\Delta}. \tag{6.33}$$

At the singular point $t=+\pi/2$, the asymptotic behavior is similar. To prove this, let us write U_1 and U_5 as combinations of Kummer solutions [1] in the argument $(1-\tau)$:

$$U_2=F(\alpha,\beta,\alpha+\beta+1-c;1-\tau),$$
$$U_6=(1-\tau)^{\gamma-\alpha-\beta}F(\gamma-\alpha,\gamma-\beta,\gamma+1-\alpha-\beta;1-\tau).$$

Hence, near the point $\tau \to 1$, we obtain

$$U_1 = \frac{\Gamma(\gamma)\Gamma(\gamma-\alpha-\beta)}{\Gamma(\gamma-\alpha)\Gamma(\gamma-\beta)} \cdot 1 + \frac{\Gamma(\gamma)\Gamma(\alpha+\beta-\gamma)}{\Gamma(\alpha)\Gamma(\beta)} (1-\tau)^{\gamma-\alpha-\beta},$$

$$U_5 = \frac{\Gamma(2-\gamma)\Gamma(\gamma-\alpha-\beta)}{\Gamma(1-\alpha)\Gamma(1-\beta)} \cdot 1 + \frac{\Gamma(2-\gamma)\Gamma(\alpha+\beta-\gamma)}{\Gamma(\alpha+1-\gamma)\Gamma(\beta+1-\gamma)} (1-\tau)^{\gamma-\alpha-\beta}.$$

Whence, taking in mind identity $\gamma-\alpha-\beta = 1/2 - ip$, we derive asymptotic for $F^{(1)}, F^{(2)}$:

$$F^{(1)} \sim \frac{\Gamma(\gamma)\Gamma(\gamma-\alpha-\beta)}{\Gamma(\gamma-\alpha)\Gamma(\gamma-\beta)} (1-\tau)^{+ip/2}, \quad F^{(2)} \sim \frac{\Gamma(2-\gamma)\Gamma(\gamma-\alpha-\beta)}{\Gamma(1-\alpha')\Gamma(1-\beta)} (1-\tau)^{+ip/2}.$$

In the vicinity of the point $t = +\pi/2$:

$$t = +\frac{\pi}{2} + 2\Delta, \quad (1-\tau) = \frac{1-\sin t}{2} = \Delta, \quad (1-\tau)^{+ip/2} = e^{+ip/2 \ln \Delta}. \tag{6.34}$$

6.4 Constructing Solutions in the Variable z

Let us turn to Eqs. (6.15):

$$e^{z}(a-ib) f_2(z) = \Big(+i\frac{d}{dz} + p\Big) f_1(z), \quad e^{z}(a+ib) f_1(z) = \Big(-i\frac{d}{dz} + p\Big) f_2(z); \tag{6.35}$$

whence we derive 2-nd order equations

$$\begin{aligned} &\Big[\frac{d^2}{dz^2} - \frac{d}{dz} - e^{2z}(a^2+b^2) + p(i+p)\Big] f_1 = 0, \\ &\Big[\frac{d^2}{dz^2} - \frac{d}{dz} - e^{2z}(a^2+b^2) - p(i-p)\Big] f_2 = 0. \end{aligned} \tag{6.36}$$

The interpretation of this situation in terms of barrier and reflection is troublesome, because Eqs. (6.36) involve complex terms. To overcome this, we first write the equations (6.35) and (6.36) in the variable $Z = e^{z}$, $Z \in (0, +\infty)$:

$$\Big(\frac{d}{dZ} - \frac{ip}{Z}\Big) f_1 + i(a-ib) f_2 = 0, \quad \Big(\frac{d}{dZ} + \frac{ip}{Z}\Big) f_2 - i(a+ib) F_1 = 0; \tag{6.37}$$

$$\Big(\frac{d^2}{dZ^2} + \frac{p^2+ip}{Z^2} - a^2 - b^2\Big) f_1 = 0, \quad \Big(\frac{d^2}{dZ^2} + \frac{p^2-ip}{Z^2} - a^2 - b^2\Big) f_2 = 0. \tag{6.38}$$

The form of these equations assumes the relationship $f_2 \sim f_1^*$; taking this in mind, we combine the two functions as follows

$$H = c f_1 + c^* f_2, \quad G = c f_1 - c^* f_2; \tag{6.39}$$

the coefficient c may be arbitrary; in particular, we may choose $c = 1$. Further, we readily obtain the following two second order equations

$$\Big[\frac{d^2}{dZ^2} + \frac{p^2}{Z^2} - (a^2+b^2)\Big] H + \frac{ip}{Z^2} G = 0, \quad \Big[\frac{d^2}{dZ^2} + \frac{p^2}{Z^2} - (a^2+b^2)\Big] G + \frac{ip}{Z^2} H = 0. \tag{6.40}$$

In the variable $z = \ln Z$, the equations (6.40) read (it is convenient to make the substitutions $H \Longrightarrow e^{z/2}H, G \Longrightarrow e^{z/2}G$):

$$\Big[\frac{d^2}{dz^2} + p^2 - \frac{1}{4} - (a^2+b^2)e^{2z}\Big]H + ipG = 0, \quad \Big[\frac{d^2}{dz^2} + p^2 - \frac{1}{4} - (a^2+b^2)e^{2z}\Big]G + ipH = 0.$$

The point at which function decreases is determined by the condition

$$p^2 - 1/4 = (k_1^2 + k_2^2)\, e^{2z_0} \quad \Longrightarrow \quad z_0 = \ln\sqrt{\frac{p^2-1/4}{a^2+b^2}}. \tag{6.41}$$

In the vicinity of the point z_0, the equations become simpler

$$\frac{d^2}{dz^2}H + ipG = 0, \quad \frac{d^2}{dz^2}G + ipH = 0. \tag{6.42}$$

These solutions should be of exponential form:

$$H = H_0 e^{\nu(z-z_0)}, \quad G = G_0 e^{\nu(z-z_0)};$$

we easily derive the algebraic system

$$\begin{vmatrix} \nu^2 & ip \\ ip & \nu^2 \end{vmatrix} \begin{vmatrix} H_0 \\ G_0 \end{vmatrix} = 0 \quad \Longrightarrow \quad \nu = -\frac{1 \pm i}{\sqrt{2}}\sqrt{p}, +\frac{1 \pm i}{\sqrt{2}}\sqrt{p}. \tag{6.43}$$

We see that the two possibilities correspond to the needed solutions. Note that in usual units, the point z_0 is defined by the formula

$$z_0 = \rho \ln\sqrt{\frac{(E^2 - M^2c^4)/c^2\hbar^2 - 1/4\rho^2}{(K_1^2 + K_2^2)}}, \tag{6.44}$$

where ρ stands for the curvature radius of the Lobachebsky space. We see that if K_1, K_2 tend to zero, then the depth of penetration of the wave into the space increases to infinity. This agrees with the fact that if $a^2 + b^2 = 0$, then the effective potential vanishes.

The above qualitative analysis shows that the solutions which correspond to the reflections of the particles from the potential barrier are constructed by combining solutions with opposite values of the helicity. We further prove this statement, more accurately.

Let us turn to constructing exact solutions of the equations (6.38). It suffices to study the function f_1. Using the substitution $f_1(Z) = Z^A e^{BZ} \bar{f}_1(Z)$, we get

$$\bar{f}_1'' + \Big(\frac{2A}{Z} + 2B\Big)\bar{f}_1' + \frac{2AB}{Z}\bar{f}_1 + \Big(\frac{A(A-1)}{Z^2} + \frac{p^2+ip}{Z^2}\Big)\bar{f}_1 + [B^2 - (a^2+b^2)]\bar{f}_1 = 0.$$

If $A = +ip,\ 1-ip$, $B = \pm\sqrt{a^2+b^2}$, then the above equation, for $\bar{F}_1$, becomes simpler

$$Z\bar{f}_1'' + \Big(2A + 2BZ\Big)\bar{f}_1' + 2AB\bar{f}_1 = 0.$$

Without loss of generality, we fix the values

$$A = +ip, \quad B = -\sqrt{a^2 + b^2}\,; \tag{6.45}$$

the transition to the case of opposite helicity is reached by the change $A \Longrightarrow -A$.

In the variable $y = -2BZ = 2\sqrt{a^2+b^2}e^z$, we arrive at the equation

$$y\frac{d^2}{dy^2}\bar{f}_1 + (2A - y)\frac{d}{dy}\bar{f}_1 - A\bar{f}_1 = 0\,,$$

which is of hypergeometric type

$$\Phi'' + (\gamma - y)\Phi' - \alpha\Phi = 0\,, \quad \alpha = A = ip\,, \quad \gamma = 2A = 2ip\,. \tag{6.46}$$

As linearly independent solutions we may take the following two ones:

$$\begin{aligned} \bar{f}_1^{(1)}(y) &= \Phi(\alpha, \gamma; y) = \Phi(ip, 2ip; y)\,, \\ \bar{f}_1^{(2)}(y) &= y^{1-\gamma}\Phi(\alpha - \gamma + 1, 2 - \gamma; y) = y^{1-2ip}\Phi(1 - ip, 2 - 2ip; y)\,, \end{aligned} \tag{6.47}$$

this leads to expressions for complete functions $f_1(Z) = Z^A e^{BZ}\bar{f}_1$:

$$f_1^{(1)} = y^{ip}e^{-y/2}\,\Phi(ip, 2ip; y)\,, \quad f_1^{(2)} = y^{1-ip}e^{-y/2}\,\Phi(1 - ip, 2 - 2ip; y)\,. \tag{6.48}$$

With the use of the above noted symmetry, we may get the following two solutions for F_2:

$$f_2^{(1)} = y^{1+ip}e^{-y/2}\,\Phi(1 + ip, 2 + 2ip; y)\,, \quad f_2^{(2)} = y^{-ip}e^{-y/2}\,\Phi(-ip, -2ip; y)\,. \tag{6.49}$$

Now let us pairwise relate the functions from the set $\{f_1^{(1)}, f_1^{(2)};\ f_2^{(1)}, f_2^{(2)}\}$ (the question of calculating the relative coefficients in each pair can be solved with the use of the 1-st order equations (6.35)). We state result, and then prove it.

Let us write down the needed solutions

$$\begin{aligned} f_1^{+(1)} &= e^{-y/2}y^A\,\Phi(A, 2A, y) = f\,, \\ f_2^{+(1)} &= Le^{-y/2}y^{1+A}\,\Phi(1 + A, 2 + 2A, y) = g\,, \\ f_1^{+(2)} &= L^*e^{-y/2}y^{1-A}\Phi(1 - A, 2 - 2A, y) = g^*\,, \\ f_2^{+(2)} &= e^{-y/2}y^{-A}\Phi(-A, -2A, y) = f^*\,; \end{aligned}$$

$A \Longrightarrow -A$

$$\begin{aligned} f_1^{-(1)} &= e^{-y/2}y^{-A}\,\Phi(-A, -2A, y) = f^*\,, \\ f_2^{-(1)} &= L^*e^{-y/2}y^{1-A}\,\Phi(1 - A, 2 - 2A, y) = g^*\,, \\ f_1^{-(2)} &= Le^{-y/2}y^{1+A}\Phi(1 + A, 2 + 2A, y) = g\,, \\ f_2^{-(2)} &= e^{-y/2}y^{A}\Phi(A, 2A, y) = f\,. \end{aligned}$$

To get the coefficient L, we relate the functions f_1, f_2 by the first order equations

$$\Big(\frac{d}{dy}-\frac{A}{y}\Big)f_1+\frac{e^{i\alpha}}{2}f_2=0, \quad \Big(\frac{d}{dy}+\frac{A}{y}\Big)f_2+\frac{e^{-i\alpha}}{2}f_1=0;$$

for brevity we use notations

$$e^{i\alpha}=i\frac{a-ib}{\sqrt{a^2+b^2}}, \qquad e^{-i\alpha}=-i\frac{a+ib}{\sqrt{a^2+b^2}}.$$

We will prove the existence of the two pairs:

$$f_1^{+(1)}=e^{-y/2}y^A\,\Phi(A,2A,y)=f, \quad f_2^{+(1)}=Le^{-y/2}y^{1+A}\,\Phi(1+A,2+2A,y)=g,$$

and

$$f_2^{+(2)}=e^{-y/2}y^{-A}\Phi(-A,-2A,y)=f^*, \quad f_1^{+(2)}=L^*e^{-y/2}y^{1-A}\Phi(1-A,2-2A,y)=g^*.$$

In fact, it suffices to prove only one relation, from the first line. The substitution of the two functions in the 1-st equation

$$\Big(\frac{d}{dy}-\frac{A}{y}\Big)e^{-y/2}y^A\,\Phi(A,2A,y)+\frac{e^{i\alpha}}{2}Le^{-y/2}y^{1+A}\,\Phi(1+A,2+2A,y)=0,$$

leads to

$$-\Phi(A,2A)+\Phi(A+1,2A+1)+e^{i\alpha}Ly\Phi(1+A,2+2A)=0,$$

or,

$$-\Phi(A,2A)+\Phi(A+1,2A+1)=x\frac{1}{2(2A+1)}\,\Phi(A+1,2A+2).$$

Therefore we get the needed coefficient

$$\frac{1}{2(2A+1)}+e^{i\alpha}L=0 \quad\Longrightarrow\quad L=-\frac{e^{-i\alpha}}{2(2A+1)}=\frac{i/2}{2A+1}\frac{a+ib}{\sqrt{a^2+b^2}}. \qquad (6.50)$$

Now let us find the asymptotic behavior of the functions $f(z),g(z)$ in the region $z\to-\infty$ (at $y\to 0$),

$$f\sim y^A=(2\sqrt{a^2+b^2})^{ip}e^{ipz}, \quad g\sim L(2\sqrt{a^2+b^2})^{1+ip}e^{(1+ip)z}\to 0. \qquad (6.51)$$

With the use of the known formula [1]:

$$\Phi(\alpha,\gamma,x)=\frac{\Gamma(\gamma)}{\Gamma(\alpha)}\,e^x(x)^{\alpha-\gamma},\ \mathrm{Re}\,x\to+\infty,$$

we derive $z\to+\infty$ $(y=\to+\infty)$,

$$f\sim e^{-y/2}y^{ip}\,\frac{\Gamma(2ip)}{\Gamma(ip)}\,e^y y^{-ip}\to\infty, \quad g\sim Le^{-y/2}y^{1-ip}\,\frac{\Gamma(2ip+2)}{\Gamma(ip+1)}\,e^y y^{1-ip}\to\infty. \qquad (6.52)$$

The relations (6.52) mean that the solutions based on f, g do not have the needed behavior at $z \to +\infty$. Therefore, we should search other solutions; at this we will follow only the function f_1, considering it as being the main one.

First, let us recall the procedure for finding the needed solutions in the case of static Lobachevsky space, and then extend this procedure to non-static space.

We have used above only two solutions of the hypergeometric equation (with two possibilities for A: $+A$ and $-A$)

$$Y^{+(1)} = \Phi(A, 2A, y), \quad Y^{+(2)} = y^{1-2A}\Phi(1-A, 2-2A, y);$$
$$Y^{-(1)} = \Phi(-A, -2A, y), \quad Y^{-(2)} = y^{1+2A}\Phi(1+A, 2+2A, y). \tag{6.53}$$

Now we need yet another pair of solutions [1] (also with two possibilities for A: $+A$ and $-A$):

$$Y^{+(5)} = \Psi(A, 2A, y), \quad Y^{+(7)} = e^{y}\Psi(A, 2A, -y);$$
$$Y^{-(5)} = \Psi(-A, -2A, y), \quad Y^{-(7)} = e^{y}\Psi(-A, -2A, -y). \tag{6.54}$$

The solutions (6.53) and (6.54) are related by the linear Kummer's formulas [1]

$$Y^{+(5)} = \frac{\Gamma(1-2A)}{\Gamma(1-A)} Y^{+(1)} + \frac{\Gamma(2A-1)}{\Gamma(A)} Y^{+(2)},$$
$$Y^{+(7)} = \frac{\Gamma(1-2A)}{\Gamma(1-A)} Y^{+(1)} - \frac{\Gamma(2A-1)}{\Gamma(A)} Y^{+(2)};$$

$$Y^{-(5)} = \frac{\Gamma(1+2A)}{\Gamma(1+A)} Y^{-(1)} + \frac{\Gamma(-2A-1)}{\Gamma(-A)} Y^{-(2)},$$
$$Y^{-(7)} = \frac{\Gamma(1+2A)}{\Gamma(1+A)} Y^{-(1)} - \frac{\Gamma(-2A-1)}{\Gamma(-A)} Y^{-(2)};$$

after multiplying them by $y^{A}e^{-y/2}$ and with $y^{-A}e^{-y/2}$ respectively, they take the form

$$f_1^{+(5)} = \frac{\Gamma(1-2A)}{\Gamma(1-A)} f + \frac{\Gamma(2A-1)}{\Gamma(A)} \frac{1}{L^*} g^*,$$
$$f_1^{+(7)} = \frac{\Gamma(1-2A)}{\Gamma(1-A)} f - \frac{\Gamma(2A-1)}{\Gamma(A)} \frac{1}{L^*} g^*; \tag{6.55}$$

$$f_1^{-(5)} = \frac{\Gamma(1+2A)}{\Gamma(1+A)} f^* + \frac{\Gamma(-2A-1)}{\Gamma(-A)} \frac{1}{L} g,$$
$$f_1^{-(7)} = \frac{\Gamma(1+2A)}{\Gamma(1+A)} f^* - \frac{\Gamma(-2A-1)}{\Gamma(-A)} \frac{1}{L} g. \tag{6.56}$$

The functions $f_1^{\pm(5)}$ and $f_1^{\pm(7)}$, at $z \longrightarrow -\infty$ behave as a plane wave

$$f_1^{+(5)} = \frac{\Gamma(1-2A)}{\Gamma(1-A)} \left(2\sqrt{a^2+b^2}\right)^{+ip} e^{+ipz},$$

$$f_1^{+(7)} = \frac{\Gamma(1-2A)}{\Gamma(1-A)} \left(2\sqrt{a^2+b^2}\right)^{+ip} e^{+ipz}. \tag{6.57}$$

Due to the above stated symmetry, the solutions with opposite helicity behave as follows ($p \Longrightarrow -p$, $a = ip \Longrightarrow -a$) $z \to -\infty$ ($y \to +\infty$),

$$f_1^{-(5)} = \frac{\Gamma(1+2A)}{\Gamma(1+A)} \left(2\sqrt{a^2+b^2}\right)^{-ip} e^{-ipz},$$

$$f_1^{-(7)} = \frac{\Gamma(1+2A)}{\Gamma(1+A)} \left(2\sqrt{a^2+b^2}\right)^{-ip} e^{-ipz}. \qquad (6.58)$$

Now we are to study behavior of solutions $f_1^{\pm(5)}z$ at $z \to +\infty$. Making use of the known formula [1]

$$Y_5 = \Psi(A, 2A, y) \sim y^{-A},$$

we get
$y \to +\infty$ ($z \to +\infty$),

$$f_1^{+(5)} = y^A e^{-y/2} y^{-A} \sim e^{-y/2} \sim \exp\left(-\sqrt{a^2+b^2 e^z}\right) \longrightarrow 0,$$
$$f_1^{-(5)} = y^{-A} e^{-y/2} y^{+A} \sim e^{-y/2} \sim \exp\left(-\sqrt{a^2+b^2 e^z}\right) \longrightarrow 0. \qquad (6.59)$$

Similarly, by applying the formulas

$$Y_7 = e^y \Psi(A, 2A, -y) \sim e^y y^{-A},$$

we get
$y \to +\infty$ ($z \to +\infty$),

$$f_1^{+(7)} \sim y^A e^{-y/2} e^y y^{-A} \sim e^{+y/2} \sim \exp\left(+\sqrt{a^2+b^2 e^z}\right) \longrightarrow \infty,$$
$$f_1^{-(7)} \sim y^{-A} e^{-y/2} e^y y^{+A} \sim e^{+y/2} \sim \exp\left(+\sqrt{a^2+b^2 e^z}\right) \longrightarrow \infty. \qquad (6.60)$$

The most interesting are solutions of the type $\pm(5)$ with behavior (6.59), because they vanish at $z \to$ and look as plane waves at $z \to -\infty$. Let us introduce the following special linear combinations of $F_1^{+(5)}$ and $F_1^{-(5)}$:

$$H_1 = f_1^{+(5)} + f_1^{-(5)}, \; H^* = H;$$
$$G_1 = f_1^{+(5)} - f_1^{-(5)}, \; G^* = -G. \qquad (6.61)$$

These new functions behave at $z \to -\infty$ as follows

$$H_1(z \to -\infty) \sim M_+ \, e^{+ipz} + M_- \, e^{-ipz},$$
$$G_1(z \to -\infty) \sim M_+ \, e^{+ipz} - M_- \, e^{-ipz}, \qquad (6.62)$$

where $M_- = (M_+)^*$,

$$M_+ = \frac{\Gamma(1-2A)}{\Gamma(1-A)} (2\sqrt{a^2+b^2})^{+ip}, \quad M_- = \frac{\Gamma(1+2A)}{\Gamma(1+A)} (2\sqrt{a^2+b^2})^{-ip}. \qquad (6.63)$$

By determining the reflection coefficient in the common manner, we readily obtain the needed result,

$$H_{\pm}(z) \sim M_{-}e^{-ipz} \pm M_{+}e^{+ipz}, \quad R = \left| \frac{M_{-}}{M_{+}} \right|^2 = 1. \tag{6.64}$$

Now, having constructed solutions referred to the reflection effect, we can easily obtain the explicit form for the concomitant function (we omit the details, see [309]).

Now we are ready to extend the above procedure to the non-static case. It suffices to follow only the first component of the bispinor wave function. Taking in mind the explicit form of the 2-nd order equation for $F(t)$

$$\left(\frac{d^2}{dt^2} + M^2 + \frac{ip(\sin t - ip)}{\cos^2 t} \right) F = 0, \quad \left(\frac{d^2}{dt^2} + M^2 - \frac{ip(\sin t + ip)}{\cos^2 t} \right) G = 0, \tag{6.65}$$

we immediately conclude that to opposite values of helicity $+p$ and $(-p)$ correspond to the conjugate functions $F_{-p}(t) = [F_{+p}(t)]^*$. Therefore, the needed solutions in non static case are constructed according to the rule

$$H_{\pm}(t,z) \sim \left\{ F_{+p}(t) f_1^{+(5)}(z) \pm F_{-p}(t) f_1^{-(5)}(z) \right\}. \tag{6.66}$$

Though the wave functions depend on the time variable in a nontrivial manner, the effect of reflection is still valid, like in the static space model.

6.5 Majorana Spinor Field

Let us fix a Majorana basis by the following transformation from the spinor one [139]:

$$\begin{gathered} \Psi_M = A\,\Psi,\, \Gamma^a_M = A\gamma^a A^{-1}, \quad A = \frac{1-\gamma^2}{\sqrt{2}},\, A^{-1} = \frac{1+\gamma^2}{\sqrt{2}}, \\ \gamma^0_M = \gamma^0\gamma^2, \quad \gamma^1_M = \gamma^1\gamma^2, \quad \gamma^2_M = \gamma^2, \quad \gamma^3_M = \gamma^3\gamma^2. \end{gathered} \tag{6.67}$$

The explicit Dirac matrices in this new basis are:

$$\gamma^0_M = \begin{vmatrix} 0 & -i & 0 & 0 \\ i & 0 & 0 & 0 \\ 0 & 0 & 0 & i \\ 0 & 0 & -i & 0 \end{vmatrix}, \gamma^1_M = \begin{vmatrix} -i & 0 & 0 & 0 \\ 0 & i & 0 & 0 \\ 0 & 0 & -i & 0 \\ 0 & 0 & 0 & i \end{vmatrix},$$

$$\gamma^2_M = \begin{vmatrix} 0 & 0 & 0 & i \\ 0 & 0 & -i & 0 \\ 0 & -i & 0 & 0 \\ i & 0 & 0 & 0 \end{vmatrix}, \gamma^3_M = \begin{vmatrix} 0 & i & 0 & 0 \\ i & 0 & 0 & 0 \\ 0 & 0 & 0 & i \\ 0 & 0 & i & 0 \end{vmatrix}.$$

We note the property $(\gamma^a_M)^* = -\gamma^a_M$. This means that in the Majorana basis, the Dirac operator from the equation $(i\gamma^a\partial_a - m)\Psi_M = 0$ is real; therefore the real and the imaginary parts of the complex bispinor function obey the independent wave equations

$$\Psi_M = \text{Re}\,\Psi + i\,\text{Im}\Psi = \Psi_+ + \Psi_- , \quad (i\gamma^a \partial_a - m)\Psi_+ = 0, \quad (i\gamma^a \partial_a - m)\Psi_- = 0. \quad (6.68)$$

The extension to General relativity case is straightforward. By turning to tetrad-based Dirac equation in Riemannian space-time [139]:

$$\left\{\gamma^\alpha(x)[i(\partial_\alpha + \Gamma_\alpha(x)) - eA_\alpha] - m\right\}\Psi(x) = 0,$$

and by specifying this equation in the Majorana basis with the properties

$$(i\gamma^a_M)^* = +\gamma^a_M, \quad (\sigma^{ab}_M)^* = +\sigma^{ab}_M, \quad (i\gamma^5_M)^* = +\gamma^5_M,$$

we conclude that in absence of external electromagnetic fields, the generally covariant Dirac operator is real. Therefore, in any Riemannian space-time there exist wave equations for Majorana particles with the wave functions Ψ_+ and Ψ_-: $\Psi_M = \Psi_+ + i\Psi_-$.

6.6 Independent Majorana Components

The decomposition of quasi-plane waves into sum of Majorana components $\Psi = \Psi_+ + \Psi_-$ in spinor basis is done according to the general formulas

$$\Psi^c = \left| \begin{array}{c} -\sigma^2\eta^* \\ \sigma^2\xi^* \end{array} \right|, \quad \Psi = \frac{\Psi + \Psi^c}{2} + \frac{\Psi - \Psi^c}{2},$$

and reads as follows

$$\Psi_+ = \left| \begin{array}{c} (\xi - \sigma_2\eta^*)/2 \\ (\eta + \sigma_2\xi^*)/2 \end{array} \right|, \quad \Psi_- = \left| \begin{array}{c} (\xi + \sigma_2\eta^*)/2 \\ (\eta - \sigma_2\xi^*)/2 \end{array} \right|.$$

We use the following shortening notations in spinor basis

$$\varphi = e^{iax}\, e^{iby}, \quad \varphi^* = e^{-iax}\, e^{-iby},$$

$$\xi = \varphi F(t) \left| \begin{array}{c} f_1(z) \\ f_2(z) \end{array} \right|, \quad \xi^* = \varphi^* F^*(t) \left| \begin{array}{c} f_1^*(z) \\ f_2^*(z) \end{array} \right|,$$

$$\eta = \varphi G(t) \left| \begin{array}{c} Kf_1(z) \\ Kf_2(z) \end{array} \right|, \quad \eta^* = \varphi^* G^*(t) \left| \begin{array}{c} Kf_1^*(z) \\ Kf_2^*(z) \end{array} \right|.$$

We further we the expressions for the two Majorana components (still in the spinor basis)

$$\Psi_+ = \left| \begin{array}{c} \varphi F(t) f_1(z) + i\varphi^* G^*(t) K f_2^*(z) \\ \varphi F(t) f_2(z) - i\varphi^* G^*(t) K f_1^*(z) \\ \varphi G(t) K f_1(z) - i\varphi^* F^*(t) f_2^*(z) \\ \varphi G(t) K f_2(z) + i\varphi F^*(t) f_1^*(z) \end{array} \right|, \Psi_- = \left| \begin{array}{c} \varphi F(t) f_1(z) - i\varphi^* G^*(t) K f_2^*(z) \\ \varphi F(t) f_2(z) + i\varphi^* G^*(t) K f_1^*(z) \\ \varphi G(t) K f_1(z) + i\varphi^* F^*(t) f_2^*(z) \\ \varphi G(t) K f_2(z) - i\varphi^* F^*(t) f_1^*(z) \end{array} \right|.$$

The transition to Majorana basis in these formulas is done according to the rule

$$\Psi^M_\pm = \frac{1-\gamma^2}{\sqrt{2}}\Psi_\pm = \frac{1}{\sqrt{2}}\begin{vmatrix} 1 & 0 & 0 & -i \\ 0 & 1 & i & 0 \\ 0 & i & 1 & 0 \\ -i & 0 & 0 & 1 \end{vmatrix}\Psi_\pm\,. \tag{6.69}$$

This leads to the wave functions having the needed properties $\Psi_+ = +(\Psi_+)^*$, $\Psi_- = -(\Psi_-)^*$,

$$\Psi_+ = \begin{vmatrix} [\varphi F(t)f_1(z) + i\varphi^* G^*(t)K f_2^*(z)] - i[\varphi G(t)K f_2(z) + i\varphi F^*(t)f_1^*(z)] \\ [\varphi F(t)f_2(z) - i\varphi^* G^*(t)K f_1^*(z)] + i[\varphi G(t)K f_1(z) - i\varphi^* F^*(t)f_2^*(z)] \\ i[\varphi F(t)f_1(z) + i\varphi^* G^*(t)K f_2^*(z)] + [\varphi G(t)K f_1(z) - i\varphi^* F^*(t)f_2^*(z)] \\ -i[\varphi F(t)f_1(z) + i\varphi^* G^*(t)K f_2^*(z)] + [\varphi G(t)K f_2(z) + i\varphi F^*(t)f_1^*(z)] \end{vmatrix},$$

$$\Psi_- = \begin{vmatrix} [\varphi F(t)f_1(z) - i\varphi^* G^*(t)K f_2^*(z)] - i[\varphi G(t)K f_2(z) - i\varphi^* F^*(t)f_1^*(z)] \\ [\varphi F(t)f_2(z) + i\varphi^* G^*(t)K f_1^*(z)] + i[\varphi G(t)K f_1(z) + i\varphi^* F^*(t)f_2^*(z)] \\ i[\varphi F(t)f_2(z) + i\varphi^* G^*(t)K f_1^*(z)] + [\varphi G(t)K f_1(z) + i\varphi^* F^*(t)f_2^*(z)] \\ -i[\varphi G(t)K f_2(z) - i\varphi^* F^*(t)f_1^*(z)] + [\varphi G(t)K f_2(z) - i\varphi^* F^*(t)f_1^*(z)] \end{vmatrix}.$$

Evidently, for both Majorana cases, the conclusion on the complete reflection of the particles from the effective geometrical barrier is valid, as well.

Chapter 7

The Fermion Doublet in Non-Abelian Monopole Field, Pauli Approximation, Geometry

7.1 Pauli Equation for Fermion Doublet, GeneralAnalysis

Since the non-Abelian monopole was introduced (by t'Hooft [146], Polyakov [147], Julia, Zee [148], and Bais-Russel [149]), in scientific practice its basic properties have been studied in detail. There exist several ways to solve the monopole problems. In particular, the method which is based on the investigation of the monopole-caused physical phenomena, when the monopoles are considered as external potentials, can be applied to monopole problem analysis along with geometrical and topological methods (see [150] – [154]). It this chapter we study the behavior of an isotopic doublet of Dirac fermions on the background of non-Abelian monopole potential. A special attention is considered for the unsolved problem of the nonrelativistic approximation in the theory of isotopic multiplets in non-Abelian fields. In this approximation, the analysis of the problem is significantly simplified due to the twice reduction of the variables. The general Pauli-like equation which accounts for the presence of the external non-Abelian monopole field, has been derived. It is detailed for the case of non-Abelian monopole potentials, particularly, for the Bogomol'nyi – Prasad – Sommerfeld exact solution [155, 156]. We apply a special geometric method based on the KCC-invariants [163] – [166] to study the behavior of solutions of differential equations from the point of view of Jacobi stability.

Let us consider an isotopic doublet of Dirac particles on the background of an external Yang-Mills field [159] – [162]:

$$\Big[\, i\gamma^{\alpha}(x)\Big(\frac{\partial}{\partial x^{\alpha}} + \Gamma_{\alpha}(x) - iet^{a}W^{a}_{\alpha}(x)\Big) - M\Big]\, \Psi(x) = 0 . \tag{7.1}$$

In the spaces admitting nonrelativistic approximation [139]

$$dS^2 = (dx^0)^2 + g_{ij}(x^0, x^1, x^2, x^3)\, dx^i\, dx^j ,$$

the equation (7.1) takes the form

$$\Big\{\gamma^0\Big[i\Big(\frac{\partial}{\partial t}+\Gamma_t\Big)-et^aW_0^a\Big]+\gamma^j(x)\Big[i\Big(\frac{\partial}{\partial x_j}+\Gamma_j\Big)-et^aW_j^a\Big]-M\Big\}\Psi(x)=0\,, \tag{7.2}$$

where the generalized Dirac matrices and connection are defined by the equalities

$$\begin{gathered}\gamma^0(x)=\gamma^0, \quad \gamma^j(x)=\gamma^k e^j_{(k)}(x)\,,\\ \Gamma_t(x)=\Gamma_0(x)=\frac{1}{4}\gamma^k(x)\gamma_{k;0}(x)\,, \quad \Gamma_j(x)=\frac{1}{4}\gamma^0\gamma_{0;j}(x)+\frac{1}{4}\gamma^k(x)\gamma_{k;j}(x)\,.\end{gathered} \tag{7.3}$$

The nonrelativistic approximation for the generally covariant Dirac equation can be performed in any basis of Dirac matrices. The large and the small components in an arbitrary basis of Dirac matrices are given by two projection operators, according to the rule

$$\Psi_+=\frac{1+\gamma^0}{2}\Psi, \quad \Psi_-=\frac{1+\gamma^0}{2}\Psi.$$

Acting by these operators on the left hand side of the equation (7.2), one gets

$$\begin{gathered}\Big[i\Big(\frac{\partial}{\partial t}+\Gamma_t\Big)-e\,t^a\,W_t^a\Big]\Psi_+ +\gamma^j(x)\Big[i\Big(\frac{\partial}{\partial x^j}+\Gamma_j\Big)-et^aW_j^a\Big]\Psi_- -M\Psi_+(x)=0\,,\\ -\Big[i\Big(\frac{\partial}{\partial t}+\Gamma_t\Big)-et^aW_t^a\Big]\Psi_- +\gamma^j(x)\Big[i\Big(\frac{\partial}{\partial x_j}+\Gamma_j\Big)-e\,t^aW_j^a\Big]\Psi_+ -M\Psi_-(x)=0\,.\end{gathered} \tag{7.4}$$

We separates the rest energy by the formal substitution $\Psi_\pm(x)=\exp(-iMt)\Phi_\pm(x)$. Then, from the equations (7.4), one finds

$$\Big[i\Big(\frac{\partial}{\partial t}+\Gamma_t\Big)-et^aW_t^a\Big]\Phi_+ +\gamma^j(x)\Big[i\Big(\frac{\partial}{\partial x_j}+\Gamma_j\Big)-et^aW_j^a\Big]\Phi_- =0\,, \tag{7.5}$$

$$-\Big[i(\frac{\partial}{\partial t}+\Gamma_t)-et^aW_t^a\Big]\Phi_- +\gamma^j(x)\Big[i(\frac{\partial}{\partial x^j}+\Gamma_j)-et^aW_j^a\Big]\Phi_+ -2M\Phi_-(x)=0\,. \tag{7.6}$$

By replacing the last equation by its approximation (it should be noted that by this operation we assume the neglecting of the time components of the Yang-Mills field, in comparison with $2M\Phi_-(x)$)

$$\Phi_-(x)=\frac{1}{2M}\,\gamma^j(x)\left[i\hbar(\frac{\partial}{\partial x^j}+\Gamma_j)-e\,t^a\,W_j^a\right]\Phi_+\,,$$

and by eliminating the small component from equation (7.5), we get

$$\begin{gathered}\Big[i(\frac{\partial}{\partial t}+\Gamma_t)-et^aW_0^a\Big]\Phi_+\\ =-\frac{1}{2M}\gamma^j(x)\Big[i(\frac{\partial}{\partial x^j}+\Gamma_j)-t^aW_j^a\Big]\gamma^l(x)\Big[i(\frac{\partial}{\partial x^l}+\Gamma_l)-et^aW_l^a\Big]\Phi_+\,.\end{gathered} \tag{7.7}$$

The obtained equation is the generally covariant Pauli equation for the Dirac fermion doublet on the background of external Yang – Mills field. This is valid for an arbitrary basis of Dirac matrices and for the class of Riemannian spaces (7.2) admitting nonrelativistic approximation. The wave function of the doublet obeys the identity $\gamma^0 \Phi_+ = \Phi_+$. The existence of this additional condition is provided by the fact that the Pauli wave function for the Dirac particle contains only two independent components (they can be established explicitly at the choice of Dirac matrices in the conventional basis). Respectively, the doublet wave-function in nonrelativistic approximation has to contain only 4 components. Let us recall that the Weyl (spinor) basis is determined by the relations

$$
(i\gamma^a \partial_a - m)\,\Psi = 0, \quad \Psi = \begin{vmatrix} \xi \\ \eta \end{vmatrix}, \quad \gamma^0 = \begin{vmatrix} 0 & I \\ I & 0 \end{vmatrix},
$$
$$
\gamma^k = \begin{vmatrix} 0 & -\sigma_k \\ \sigma_k & 0 \end{vmatrix}, \quad \gamma^5 = -i\gamma^0\gamma^1\gamma^2\gamma^3 = \begin{vmatrix} -I & 0 \\ 0 & I \end{vmatrix}. \tag{7.8}
$$

The Pauli basis is defined by the formulas

$$
(i\gamma^a_{st}\partial_a - m)\Psi_{st} = 0, \quad \Psi_{st} = \begin{vmatrix} \varphi \\ \chi \end{vmatrix}, \quad \Psi_{st} = \frac{1}{\sqrt{2}} \begin{vmatrix} I & I \\ I & -I \end{vmatrix} \begin{vmatrix} \xi \\ \eta \end{vmatrix},
$$
$$
\gamma^0_{st} = \begin{vmatrix} I & 0 \\ 0 & -I \end{vmatrix}, \quad \gamma^k_{st} = \begin{vmatrix} 0 & \sigma_k \\ -\sigma_k & 0 \end{vmatrix}, \quad \gamma^5_{st} = -\gamma^0 = \begin{vmatrix} 0 & -I \\ -I & 0 \end{vmatrix}. \tag{7.9}
$$

The transformation which describes the transition from spinor (7.8) to basic (7.9) is determined by the following relations (here we use the Dirac matrices in the representation (7.8)):

$$
\Psi_{st} = S\,\Psi, \quad S = S^{-1} = \frac{\gamma^0 - \gamma^5}{\sqrt{2}}. \tag{7.10}
$$

7.2 Non-Abelian Monopole in Schwinger Gauge

It is known that the ordinary Abelian potential generates some non-Abelian potential, which is a solution of the Yang-Mills equations. Such a special non-Abelian solution has been found out firstly in [149] during the study of the Yang-Mills equation on the background of curved space-time with spherical symmetry.

The procedure of embedding the 4-vector $A_\mu(x)$ in the non-Abelian scheme

$$
A_\mu(x) \Longrightarrow A^{(a)}_\mu(x) \equiv (0,\, 0,\, A^{(3)}_\mu = A_\mu(x)) \tag{7.11}
$$

leads to the vector $A^{(a)}_\mu(x)$, which satisfies the free Yang-Mills equations. Indeed, it is easy to prove that the vector $A_\mu(x) = (0,0,0,A_\phi = g\cos\theta)$ obeys the generally covariant Maxwell equations in a space-time with spherical symmetry

$$
dS^2 = e^{2\nu}dt^2 - e^{2\mu}dr^2 - r^2(d\theta^2 + \sin^2\theta d\phi^2), \quad A_\phi = g\cos\theta, \quad F_{\theta\phi} = -g\sin\theta,
$$

while the Maxwell equations

$$
\frac{1}{\sqrt{-g}} \frac{\partial}{\partial x^\alpha} \sqrt{-g} F^{\alpha\beta} = 0
$$

can be reduced to one equation

$$\frac{1}{e^{\nu+\mu}r^2\sin\theta}\frac{\partial}{\partial\theta}e^{\nu+\mu}r^2\sin\theta\frac{1}{r^4\sin^2\theta}(-g\sin\theta)\equiv 0\,,$$

that is identically satisfied. In turn, the non-Abelian tensor

$$F^{(a)}_{\mu\nu}(x)=\nabla_\mu A^{(a)}_\nu-\nabla_\nu A^{(a)}_\mu+e\epsilon_{abc}A^{(b)}_\mu A^{(c)}_\nu\,,$$

corresponding to the introduced isotopic vector $A^{(a)}_\mu$ has a simple isotopic structure: $F^{(3)}_{\theta\phi}=-g\,\sin\theta$, and all the other components $F^{(a)}_{\nu\mu}$ are equal to zero. This substitution

$$F^{(a)}_{\nu\mu}=(0,0,F^{(3)}_{\theta\phi}=-g\,\sin\theta)$$

leads to the fact that the Yang-Mills equations are reduced to only one equation, which coincides with the Abelian one. Consequently, the potential (7.11) can be considered as (a trivially non-Abelian) solution of the Yang-Mills equations.

One can assume that such type of trivial solution should be presented as a part in the known monopole solution obtained by t'Hooft and Polyakov [146, 147], Julia and Zee [148]:

$$\Phi^{(a)}(x)=x^a\Phi(r)\,,\quad W^{(a)}_0(x)=x^aF(r)\,,\quad W^{(a)}_i(x)=\epsilon_{iab}x^bK(r)\,. \tag{7.12}$$

We shall further explicitly detail this embedding. To this aim, we shall introduce three isotopic gauges: Cartesian, Dirac and Schwinger ones. To avoid any confusion, we use the corresponding marks $S., D., C.$ for Schwinger, Dirac and Cartesian gauges. The abbreviations sph and Cart correspond to spherical and Cartesian tetrads, respectively.

Let us consider a special representation of the non-Abelian monopole potential that is the most convenient in analyzing the problem of isotopic particle multiplet in this field. The known monopole solution found by t'Hooft and Polyakov is assumed to be known. The field $W^{(a)}_\alpha$ is a covariant vector with the ordinary transformational properties: $W^{(a)}_\beta=(\partial x^i/\partial x^\beta)W^{(a)}_i$.

We start with the structure of this solution in Cartesian coordinates. Then the potential ($\Phi^{(a)}(x)$, $W^{(a)}_\alpha$) is transformed to the spherical coordinate system:

$$\Phi^{(a)}(x),\quad W^{(a)}_t,\,W^{(a)}_r,\,W^{(a)}_\theta,\,W^{(a)}_\phi\,.$$

Now we can perform the special local transformation in the isotopic space. To do this, we determine the gauge matrix from the condition

$$(O_{ab}\Phi^b(x))=(0,0,r\Phi(r))\,.$$

This equation has a set of solutions, because any rotation around the third isotopic axis $(0,0,1)$ does not change the finite vector $(0,0,r\Phi(r))$. While working with $SO(3.R)$, it is convenient to apply the parametrization of rotations by the three-dimensional Gibbs vector [157]

$$O=O(\vec{c})=I+2\,\frac{\vec{c}^{\,\times}+(\vec{c}^{\,\times})^2}{1+\vec{c}^{\,2}}\,,\quad (\vec{c}^{\,\times})_{ac}=-\epsilon_{acb}\,c_b\,. \tag{7.13}$$

The simplest rotation transferring one vector to the another is given by the formulas [157]

$$\vec{B} = O(\vec{c})\,\vec{A}\,, \quad \vec{c} = \frac{\vec{B}\times\vec{A}}{(\vec{A}+\vec{B})\,\vec{A}}\,;$$

whence one finds

$$\vec{A} = r\,\Phi(r)\,\vec{n}_{\theta,\phi}\,, \quad \vec{B} = r\,\Phi(r)(0,0,1) \Longrightarrow \vec{c} = \frac{\sin\theta}{1+\cos\theta}\,(+\sin\phi, -\cos\phi, 0)\,.$$

The gauge transformation of a scalar field $\Phi^a(x)$ is accompanied by the transformation of the vector triplet $W^{(a)}_{\beta}(x)$ [158]:

$$W'^{(a)}_{\alpha}(x) = O_{ab}(\vec{c})\,W^{(b)}_{\alpha} + \frac{1}{e}f_{ab}(\vec{c})\,\frac{\partial c_b}{\partial x^{\alpha}}\,, \quad f(\vec{c}) = -2\,\frac{1+\vec{c}^{\,\times}}{1+\vec{c}^{\,2}}\,. \qquad (7.14)$$

Using (7.14), we get new representation for the non-Abelian potentials

$$\Phi^{D.(a)} = r\Phi(r)\begin{vmatrix} 0 \\ 0 \\ 1 \end{vmatrix}, \quad W^{D.(a)}_{\theta} = (r^2K+1/e)\begin{vmatrix} -\sin\phi \\ +\cos\phi \\ 0 \end{vmatrix}, \quad W^{D.(a)}_{r} = \begin{vmatrix} 0 \\ 0 \\ 0 \end{vmatrix},$$

$$W^{D.(a)}_{t} = \begin{vmatrix} 0 \\ 0 \\ rF(r) \end{vmatrix}, \quad W^{D.(a)}_{\phi} = \begin{vmatrix} -(r^2K+1/e)\sin\theta\cos\phi \\ -(r^2K+1/e)\sin\theta\sin\phi \\ \frac{1}{e}(\cos\theta-1) \end{vmatrix}. \qquad (7.15)$$

It should be noted that the factor $r^2K(r)+1/e$ vanishes at $K=-1/er^2$. Thus, the special choice of the function $K(r)$ leads to a significant simplification of the structure of the non-Abelian monopole potentials to the form:

$$\Phi^{D.(a)} = r\Phi(r)\begin{vmatrix} 0 \\ 0 \\ 1 \end{vmatrix}, \quad W^{D.(a)}_{\theta} = \begin{vmatrix} 0 \\ 0 \\ 0 \end{vmatrix}, \quad W^{D.(a)}_{r} = \begin{vmatrix} 0 \\ 0 \\ 0 \end{vmatrix},$$

$$W^{D.(a)}_{t} = \begin{vmatrix} 0 \\ 0 \\ rF(r) \end{vmatrix}, \quad W^{D.(a)}_{\phi} = \begin{vmatrix} 0 \\ 0 \\ \frac{1}{e}(\cos\theta-1) \end{vmatrix}. \qquad (7.16)$$

There exists a close relation between $W^{D.(a)}_{\phi}$ in (7.16) and the Dirac expression for the Abelian monopole potential (let $\vec{n}=(0,0,+1)$):

$$A^{\beta}_{D.} = g\Big(0, \frac{[\,\vec{n}\,\vec{r}\,]}{(r+\vec{r}\,\vec{n})\,r}\Big), \quad A^{D.}_{\phi} = g\,(\cos\theta-1)\,. \qquad (7.17)$$

In other words, $W^{triv}_{(a)\alpha}(x)$ in (7.16) can be considered as a result of enclosure of the Abelian potential (7.17) in the non-Abelian scheme: $W^{(a)D.}_{\alpha}(x) \equiv (0,0,A^{D.}_{\alpha}(x))$. The quantity $W^{(a)D.}_{\alpha}(x)$ designed by the symbol $D.$ will be further interpreted as related to the Dirac non-Abelian gauge in isotopic space. In Abelian case, the Dirac potential $A^{D.}_{\alpha}(x)$ can be easy transformed into a Schwinger form

$$A^{S.}_{\alpha} = \Big(0,\, g\,\frac{[\,\vec{r}\,\vec{n}\,]\,(\vec{r}\,\vec{n})}{(r^2-(\vec{r}\,\vec{n})^2)r}\Big), \quad A^{S.}_{\phi} = g\,\cos\theta \qquad (7.18)$$

by the transformation

$$A^{S.}_{\alpha} = A^{D.}_{\alpha} + \frac{\hbar c}{ie} S \frac{\partial}{\partial x^{\alpha}} S^{-1}, \quad S(x) = \exp(-i\frac{eg}{\hbar c}\phi).$$

By analogy, we can introduce a Schwinger basis in the isotopic space:

$$(\Phi^{D.(a)}, W^{D.(a)}_{\alpha}) \quad \stackrel{\vec{c}\,'}{\longrightarrow} \quad (\Phi^{S.(a)}, W^{S.(a)}_{\alpha}), \quad \vec{c}\,' = (0, 0, -\text{tg}\,\phi/2),$$

where

$$O(\vec{c}\,') = \left| \begin{array}{ccc} \cos\phi & \sin\phi & 0 \\ -\sin\phi & \cos\phi & 0 \\ 0 & 0 & 1 \end{array} \right| .$$

In this gauge, the explicit form of the non-Abelian potentials is simplified (comparing with (7.16)):

$$W^{S.(a)}_{\theta} = \left| \begin{array}{c} 0 \\ (r^2 K + 1/e) \\ 0 \end{array} \right|, \quad W^{S.(a)}_{\phi} = \left| \begin{array}{c} -(r^2 K + 1/e) \\ 0 \\ \frac{1}{e}\cos\theta \end{array} \right|,$$
$$W^{S.(a)}_{r} = \left| \begin{array}{c} 0 \\ 0 \\ 0 \end{array} \right|, \quad W^{S.(a)}_{t} = \left| \begin{array}{c} 0 \\ 0 \\ rF(r) \end{array} \right|, \quad \Phi^{S.(a)} = \left| \begin{array}{c} 0 \\ 0 \\ r\Phi(r) \end{array} \right|, \qquad (7.19)$$

where the symbol *S.* designates the Schwinger gauge. Both gauges (7.15)–(7.16) and (7.19) are unitary, because the corresponding scalar field $\Phi^{D.}_{(a)}(x)$ has only the third isotopic component nonvanishing; however, the Schwinger gauge seems to be more simple.

7.3 Separating the Variables

We shall use in the tetrad formalism. In the basis of spherical tetrad and in the Schwinger unitary gauge of monopole potential, the basic equation takes the form

$$\Big[\gamma^0(i\partial_t + erF(r)t^3) + i\gamma^3(\partial_r + \frac{1}{r}) + \frac{1}{r}\Sigma^{S.}_{\theta,\phi}$$
$$+ \frac{er^2K(r)+1}{r}(\gamma^1 \otimes t^2 - \gamma^2 \otimes t^1) - (M + \kappa r\Phi(r)t^3)\Big]\Psi^{S.} = 0, \qquad (7.20)$$

$$\Sigma^{S.}_{\theta,\phi} = i\gamma^1\partial_\theta + \gamma^2 \frac{i\partial_\phi + (i\sigma^{12} + t^3)\cos\theta}{\sin\theta}, \quad t^j = \frac{1}{2}\sigma^j.$$

The special choice of the basis automatically leads to the necessary rearrangement of summands in the wave equation (7.20). In particular, only the term which is proportional to $(er^2K(r)+1)$ mixes up the doublet components, and it disappears at applying the simplest monopole potential of special type.

In the representation (7.20), the components of the total conserved angular momentum are determined according to

$$J^{S.}_1 = l_1 + \frac{(i\sigma^{12} + t^3)\cos\phi}{\sin\theta}, \quad J^{S.}_2 = l_2 + \frac{(i\sigma^{12} + t^3)\sin\phi}{\sin\theta}, \quad J^{S.}_3 = l_3, \qquad (7.21)$$

and then the substitution for doublet wave function $\Psi_{\epsilon jm}(x)$ should be taken in the form

$$\Psi_{\epsilon jm}(x) = \frac{e^{-i\epsilon t}}{r}\Big[T_{+1/2}\otimes F(r,\theta,\phi) + T_{-1/2}\otimes G(r,\theta,\phi)\Big],$$

$$T_{+1/2} = \left|\begin{array}{c} 1 \\ 0 \end{array}\right|, \; F = \left|\begin{array}{c} f_1(r)D_{-1} \\ f_2(r)D_0 \\ f_3(r)D_{-1} \\ f_4(r)D_0 \end{array}\right|, \; T_{-1/2} = \left|\begin{array}{c} 0 \\ 1 \end{array}\right|, \; G = \left|\begin{array}{c} g_1(r)D_0 \\ g_2(r)D_{+1} \\ g_3(r)D_0 \\ g_4(r)D_{+1} \end{array}\right|, \tag{7.22}$$

$D_\sigma \equiv D^j_{-m,\sigma}(\phi,\theta,0)$. The quantum number j takes the values $j = 0,1,2,3,\ldots$

The important case in the electron-monopole quantum problem is for the minimal value of quantum number j. If $j = 0$, then the used symbols $D^0_{0,\pm 1}$ in (7.22) have no sense, and the wave function $\Psi_{\epsilon 0}(x)$ has to be constructed as follows

$$\Psi_{\epsilon 0} = T_{+1/2}\otimes \left|\begin{array}{c} 0 \\ f_2(r) \\ 0 \\ f_4(r) \end{array}\right| + T_{-1/2}\otimes \left|\begin{array}{c} g_1(r) \\ 0 \\ g_3(r) \\ 0 \end{array}\right|. \tag{7.23}$$

Hereafter, the factor $e^{-i\epsilon t}/r$ is omitted.

Using the recurrent relations for Wigner functions [68]

$$\nu = \sqrt{j(j+1)}, \quad \omega = \sqrt{(j-1)(j+2)}, \quad j \neq 0,$$

$$\partial_\theta D_{-1} = \frac{1}{2}(\omega D_{-2} - \nu D_0), \quad \frac{m-\cos\theta}{\sin\theta}D_{-1} = \frac{1}{2}(\omega D_{-2} + \nu D_0),$$

$$\partial_\theta D_0 = \frac{1}{2}(\nu D_{-1} - \nu D_{+1}), \quad \frac{m}{\sin\theta}D_0 = \frac{1}{2}(\nu D_{-1} + \nu D_0),$$

$$\partial_\theta D_{+1} = \frac{1}{2}(\nu D_0 - \omega D_{+2}), \quad \frac{m+\cos\theta}{\sin\theta}D_{+1} = \frac{1}{2}(\nu D_0 + \omega D_{+2}),$$

we find

$$\Sigma^{S.}_{\theta,\phi}\,\Psi^{S.}_{jm} = i\nu\left[T_{+1/2}\otimes \left|\begin{array}{c} -f_4\,D_{-1} \\ +f_3\,D_0 \\ +f_2\,D_{-1} \\ -f_1\,D_0 \end{array}\right| + T_{-1/2}\otimes \left|\begin{array}{c} -g_4\,D_0 \\ +g_3\,D_{+1} \\ +g_2\,D_0 \\ -g_1\,D_{+1} \end{array}\right|\right]. \tag{7.24}$$

The term with mixed isotopic components has the form

$$\frac{er^2K(r)+1}{r}\,(\gamma^1\otimes t^2 - \gamma^2\otimes t^1)\,\Psi_{jm}$$

$$= i\,\frac{er^2K(r)+1}{r}\Big[\,T_{+1/2}\otimes \left|\begin{array}{c} 0 \\ +g_3D_0 \\ 0 \\ -g_1D_0 \end{array}\right| + T_{-1/2}\otimes \left|\begin{array}{c} -f_4D_0 \\ 0 \\ +f_2D_0 \\ 0 \end{array}\right|\Big]. \tag{7.25}$$

For brevity, we introduce the following notations

$$W \equiv e\,r^2\,K(r) + 1, \quad \tilde{F} \equiv \frac{e\,r\,F(r)}{2}, \quad \tilde{\Phi} \equiv \frac{\kappa\,r\,\Phi(r)}{2}. \tag{7.26}$$

After simple calculations we find the radial system

$$
\begin{aligned}
(-i\frac{d}{dr}+\epsilon+\tilde{F})f_3-i\frac{\nu}{r}f_4-(M+\tilde{\Phi})f_1=0,\\
(+i\frac{d}{dr}+\epsilon+\tilde{F})f_4+i\frac{\nu}{r}f_3+i\frac{W}{r}g_3-(M+\tilde{\Phi})f_2=0,\\
(+i\frac{d}{dr}+\epsilon+\tilde{F})f_1+i\frac{\nu}{r}f_2-(M+\tilde{\Phi})f_3=0,\\
(-i\frac{d}{dr}+\epsilon+\tilde{F})f_2-i\frac{\nu}{r}f_1-i\frac{W}{r}g_1-(M+\tilde{\Phi})f_4=0,\\
(-i\frac{d}{dr}+\epsilon-\tilde{F})g_3-i\frac{\nu}{r}g_4-i\frac{W}{r}f_4-(M-\tilde{\Phi})g_1=0,\\
(+i\frac{d}{dr}+\epsilon-\tilde{F})g_4+i\frac{\nu}{r}g_3-(M-\tilde{\Phi})g_2=0,\\
(+i\frac{d}{dr}+\epsilon-\tilde{F})g_1+i\frac{\nu}{r}g_2+i\frac{W}{r}f_2-(M-\tilde{\Phi})g_3=0,\\
(-i\frac{d}{dr}+\epsilon-\tilde{F})g_2-i\frac{\nu}{r}g_1-(M-\tilde{\Phi})g_4=0.
\end{aligned}
\tag{7.27}
$$

In the case when $j=0$ (at that $\Sigma_{\theta,\phi}\Psi_{\epsilon 0}\equiv 0$), the radial equations are simplified

$$
\begin{aligned}
(+i\frac{d}{dr}+\epsilon+\tilde{F})f_4+i\frac{W}{r}g_3-(M+\tilde{\Phi})f_2=0,\\
(-i\frac{d}{dr}+\epsilon+\tilde{F})f_2-i\frac{W}{r}g_1-(M+\tilde{\Phi})f_4=0,\\
(-i\frac{d}{dr}+\epsilon-\tilde{F})g_3-i\frac{W}{r}f_4-(M-\tilde{\Phi})g_1=0,\\
(+i\frac{d}{dr}+\epsilon-\tilde{F})g_1+i\frac{W}{r}f_2-(M-\tilde{\Phi})g_3=0.
\end{aligned}
\tag{7.28}
$$

Let us introduce an additional diagonalizing operator. The ordinary operator of P-inversion for the bispinor field is not suitable for this purpose; instead, the required operator can be constructed as a combination of a bispinor P-reflection and some discrete transformation of the isotopic space. Indeed, we take into account the bispinor P-reflection in Cartesian tetrad basis

$$
\hat{P}^{Cart}_{bisp}\otimes\hat{P}=i\gamma^0\otimes\hat{P},
$$

where the P-reflection of space coordinates is determined in spherical basis as

$$
\hat{P}^{sph}_{bisp}\otimes\hat{P}=\left|\begin{array}{cccc}0&0&0&-1\\0&0&-1&0\\0&-1&0&0\\-1&0&0&0\end{array}\right|\otimes\hat{P}=-(\gamma^5\gamma^1)\otimes\hat{P}.
\tag{7.29}
$$

This operator acts on the wave function $\Psi_{jm}(x)$ in the following way

$$
(\hat{P}^{sph}_{bisp}\otimes\hat{P})\Psi_{\epsilon jm}(x)=(-1)^{j+1}\Big[T_{+1/2}\otimes\left|\begin{array}{c}f_4D_0\\f_3D_{+1}\\f_2D_0\\f_1D_{+1}\end{array}\right|+T_{-1/2}\otimes\left|\begin{array}{c}g_4D_{-1}\\g_3D_0\\g_2D_{-1}\\g_1D_0\end{array}\right|\Big].
$$

The given relation indicates the way to construct the quantity with the required properties. Its structure can be of the form

$$\hat{N}^{S.}_{sph} \equiv \hat{\pi}^{S.} \otimes \hat{P}^{sph}_{bisp} \otimes \hat{P}, \quad \hat{\pi}^{S.} = a\sigma^1 + b\sigma^2,$$
$$\hat{\pi}^{S.} T_{+1/2} = (a+ib) T_{-1/2}, \quad \hat{\pi}^{S.} T_{-1/2} = (a-ib) T_{+1/2}. \tag{7.30}$$

The general factor at the quantity $\hat{\pi}^{S.}$ does not matter in separating the variables; hence further we shall suppose $(\hat{\pi}^{S.})^2 = a^2 + b^2 = +1$. From the equation $\hat{N}^{S.}_{sph} \Psi_{jm} = N \Psi_{jm}$, one finds the two eigenvalues $N = \delta(-1)^{j+1}$, $\delta = \pm 1$ and the corresponding constraints on the functions,

$$g_1 = \delta(a+ib) f_4, \quad g_2 = \delta(a+ib) f_3,$$
$$g_3 = \delta(a+ib) f_2, \quad g_4 = \delta(a+ib) f_1. \tag{7.31}$$

Taking into consideration (7.31), the following equations are obtained (we introduce the notation $\Delta = e^{iA} = a + ib$)

$$\begin{aligned}
&(-i\frac{d}{dr} + \epsilon + \tilde{F}) f_3 - \frac{i\nu}{r} f_4 - (M + \tilde{\Phi}) f_1 = 0, \\
&(+i\frac{d}{dr} + \epsilon + \tilde{F}) f_4 + \frac{i\nu}{r} f_3 + i\frac{W}{r} \delta \Delta f_2 - (M + \tilde{\Phi}) f_2 = 0, \\
&(+i\frac{d}{dr} + \epsilon + \tilde{F}) f_1 + \frac{i\nu}{r} f_2 - (M + \tilde{\Phi}) f_3 = 0, \\
&(-i\frac{d}{dr} + \epsilon + \tilde{F}) f_2 - \frac{i\nu}{r} f_1 - i\frac{W}{r} \delta \Delta f_4 - (M + \tilde{\Phi}) f_4 = 0, \\
&(-i\frac{d}{dr} + \epsilon - \tilde{F}) f_2 - \frac{i\nu}{r} f_1 - i\frac{W}{r} \Delta^{-1} \delta f_4 - (M - \tilde{\Phi}) f_4 = 0, \\
&(+i\frac{d}{dr} + \epsilon - \tilde{F}) f_1 + \frac{i\nu}{r} f_2 - (M - \tilde{\Phi}) f_3 = 0, \\
&(+i\frac{d}{dr} + \epsilon - \tilde{F}) f_4 + \frac{i\nu}{r} f_3 + i\frac{W}{r} \Delta^{-1} \delta f_2 - (M - \tilde{\Phi}) f_2 = 0, \\
&(-i\frac{d}{dr} + \epsilon - \tilde{F}) f_3 - \frac{i\nu}{r} f_4 - (M - \tilde{\Phi}) f_1 = 0.
\end{aligned} \tag{7.32}$$

It is obvious that this system can be self-consistent only if $\tilde{F}(r) = 0$ and $\tilde{\Phi}(r) = 0$. In other words, the operator $\hat{N}^{S.}$ can be diagonalized on the functions $\Psi_{\epsilon jm}(x)$ only when $W_t^{(a)} = 0$ and $\kappa = 0$. We further assume that these conditions are fulfilled. It should be noted that the condition $\kappa = 0$ is necessary to produce the nonrelativistic approximation, while the condition $W_t^{(a)} = 0$ just simplifies the Pauli equation due to the operator Δ.

We will restrict our considerations to the pure monopole potential and exclude the additional interaction of the doublet with Higgs scalar fields. At that, the system (7.32) takes the simpler form

$$
\begin{gathered}
(-i\frac{d}{dr}+\epsilon)f_3-\frac{i\nu}{r}f_4-Mf_1=0\,,\\
(+i\frac{d}{dr}+\epsilon)f_4+\frac{i\nu}{r}f_3+i\frac{W}{r}\delta\Delta f_2-Mf_2=0\,,\\
(+i\frac{d}{dr}+\epsilon)f_1+\frac{i\nu}{r}f_2-Mf_3=0\,,\\
(-i\frac{d}{dr}+\epsilon)f_2-\frac{i\nu}{r}f_1-i\frac{W}{r}\delta\Delta f_4-Mf_4=0\,,\\
(-i\frac{d}{dr}+\epsilon)f_2-\frac{i\nu}{r}f_1-i\frac{W}{r}\Delta^{-1}\delta f_4-M-\tilde{\Phi})f_4=0\,,\\
(+i\frac{d}{dr}+\epsilon)f_1+\frac{i\nu}{r}f_2-Mf_3=0\,,\\
(+i\frac{d}{dr}+\epsilon)f_4+\frac{i\nu}{r}f_3+i\frac{W}{r}\Delta^{-1}\delta f_2-Mf_2=0\,,\\
(-i\frac{d}{dr}+\epsilon)f_3-\frac{i\nu}{r}f_4-Mf_1=0\,.
\end{gathered}
\tag{7.33}
$$

In the system (7.33), two cases have to be distinguished in dependence on the explicit expression for $W(r)$.

If $W(r)=0$, then the difference between Δ and Δ^{-1} in the equations (7.32) is not significant, because the corresponding terms disappear from the equation. For this case, the system (7.32) transforms to the following one (a symbol A at $\hat{N}$ designates the dependence on the parameters a and b):

$W(r)=0,\quad \hat{N}^{S.}_{A}=(a\sigma^1+b\sigma^2)\otimes\hat{P}_{bisp}\otimes\hat{P}\,,$

$$
\begin{gathered}
(-i\frac{d}{dr}+\epsilon)f_3-\frac{i\nu}{r}f_4-Mf_1=0,\quad (+i\frac{d}{dr}+\epsilon)f_4+\frac{i\nu}{r}f_3-Mf_2=0\,,\\
(+i\frac{d}{dr}+\epsilon)f_1+\frac{i\nu}{r}f_2-Mf_3=0,\quad (-i\frac{d}{dr}+\epsilon)f_2-\frac{i\nu}{r}f_1-Mf_4=0\,.
\end{gathered}
\tag{7.34}
$$

The analysis of these radial equations can be reduced to find explicit solutions. Indeed, the equations (7.34) allow the following simplification due to diagonalization of the operator

$$\hat{K}_{\theta,\phi}=-i\gamma^0\gamma^5\Sigma_{\theta,\phi}\,.$$

From the equation $\hat{K}_{\theta,\phi}\Psi_{jm}=\lambda\Psi_{jm}$, one obtains $\lambda=-\mu\sqrt{j(j+1)}$ and

$$f_4=\mu f_1\,,\quad f_3=\mu f_2\,,\quad g_4=\mu g_1\,,\quad g_3=\mu g_2\,,\quad \mu=\pm1\,. \tag{7.35}$$

Correspondingly, the system (7.34) leads to

$$(+i\frac{d}{dr}+\epsilon)f_1+i\frac{\nu}{r}f_2-\mu Mf_2=0,\quad (-i\frac{d}{dr}+\epsilon)f_2-i\frac{\nu}{r}f_1-\mu Mf_1=0\,. \tag{7.36}$$

These equations are solved in terms of Bessel functions. The wave functions with the quantum numbers $(\epsilon,j,m,\delta,\mu)$ have the structure

$$\Psi^{\Delta}_{\epsilon jm\delta\mu}(x)=T_{+1/2}\otimes\left|\begin{array}{c} f_1\,D_{-1}\\ f_2\,D_0\\ \mu f_3\,D_{-1}\\ \mu f_4\,D_0\end{array}\right|+\mu\delta e^{iA}T_{-1/2}\otimes\left|\begin{array}{c} f_4\,D_0\\ f_3\,D_{+1}\\ \mu f_2\,D_0\\ \mu f_1\,D_{+1}\end{array}\right|. \tag{7.37}$$

We can associate these non-Abelian functions with the wave functions of the corresponding Abelian problem, so we get the following expansion

$$\text{at } j>0, \quad \Psi^{\Delta\delta\mu}_{\epsilon jm} = T_{+1/2}\otimes\Phi^{eg=-1/2}_{\epsilon jm\mu} + \mu\,\delta\,\Delta\,T_{-1/2}\otimes\Phi^{eg=+1/2}_{\epsilon jm\mu}\,, \tag{7.38}$$

$$\text{at } j=0, \quad \Psi^{\Delta}_{\epsilon 0\delta} = T_{+1/2}\otimes\Phi^{eg=-1/2}_{\epsilon 0} + \delta\,\Delta\,T_{-1/2}\otimes\Phi^{eg=+1/2}_{\epsilon 0}\,. \tag{7.39}$$

A completely different situation arises at $W \neq 0$. In this case, the equations (7.33) are consistent only if $\Delta = \Delta^{-1}$; consequently, $\Delta = (a+ib) = \pm 1$. Combining this relation with the normalizing condition $(a+ib)(a-ib) = 1$, one can find that $a = \pm 1$ and $b = 0$ (for definiteness we choose the parameter a as being equal to $+1$). The corresponding radial system takes the form

$$\begin{aligned}
(-i\frac{d}{dr}+\epsilon)f_3 - \frac{i\nu}{r}f_4 - Mf_1 = 0\,,\\
(+i\frac{d}{dr}+\epsilon)f_1 + \frac{i\nu}{r}f_2 - Mf_3 = 0\,,\\
(+i\frac{d}{dr}+\epsilon)f_4 + \frac{i\nu}{r}f_3 + i\frac{W}{r}\delta f_2 - Mf_2 = 0\,,\\
(-i\frac{d}{dr}+\epsilon)f_2 - \frac{i\nu}{r}f_1 - i\frac{W}{r}\delta f_4 - Mf_4 = 0\,.
\end{aligned} \tag{7.40}$$

By analogy, one can consider the case $j = 0$. Here the eigenvalues and the restrictions on the wave function are

$$N = -\delta\,,\ \delta = \pm 1\,, \qquad g_1(r) = \delta\,\Delta\,f_4(r)\,,\ g_3(r) = \delta\,\Delta\,f_2(r)\,. \tag{7.41}$$

Further, the quantities $\tilde{F}$ and $\tilde{\Phi}$ have to be equal to zero. As previously, there exist two cases depending on $W(r)$:

$W(r) = 0$,

$$(i\frac{d}{dr}+\epsilon)f_4 - Mf_2 = 0\,, \quad (-i\frac{d}{dr}+\epsilon)f_2 - Mf_4 = 0\,; \tag{7.42}$$

$W(r) \neq 0$,

$$(i\frac{d}{dr}+\epsilon)f_4 - (M - i\frac{\delta}{r}W)f_2 = 0\,, \quad (-i\frac{d}{dr}+\epsilon)f_2 - (M + i\frac{\delta}{r}W)f_4 = 0\,. \tag{7.43}$$

The explicit expressions for the wave functions $\Psi_{\epsilon jm\delta}(x)$ and $\Psi_{\epsilon 0\delta}(x)$ are
$W(r) \neq 0$, $j > 0$,

$$\Psi_{\epsilon jm\delta} = T_{+1/2}\otimes\left|\begin{array}{c} f_1\,D_{-1}\\ f_2\,D_0\\ f_3\,D_{-1}\\ f_4\,D_0 \end{array}\right| + \delta T_{-1/2}\otimes\left|\begin{array}{c} f_4\,D_0\\ f_3\,D_{+1}\\ f_2\,D_0\\ f_1\,D_{+1} \end{array}\right|; \tag{7.44}$$

$W(r) \neq 0, \ j = 0,$

$$\Psi_{\epsilon 0 \delta} = T_{+1/2} \otimes \begin{vmatrix} 0 \\ f_2 \\ 0 \\ f_4 \end{vmatrix} + \delta \, T_{-1/2} \otimes \begin{vmatrix} f_4 \\ 0 \\ f_2 \\ 0 \end{vmatrix} . \qquad (7.45)$$

When $W = 0$, in the last two relations the quantity $\delta T_{-1/2}$ is changed to $\delta e^{iA} T_{-1/2}$. These formulas mean a special way of combining one isotopic component with the other, in the complete doublet function.

7.4 Nonrelativistic Approximation, the Case $j = 0$

In the system (7.43), at $\delta = +1$:

$$\Big(i\frac{d}{dr} + \epsilon\Big) f_4 - \Big(M - i\frac{W}{r}\Big) f_2 = 0, \quad \Big(-i\frac{d}{dr} + \epsilon\Big) f_2 - \Big(M + i\frac{W}{r}\Big) f_4 = 0; \qquad (7.46)$$

let us sum and subtract the equations. This results in

$$\begin{aligned} &\Big(\frac{d}{dr} + \frac{W}{r}\Big) f + (\epsilon + M) g = 0, \\ &\Big(\frac{d}{dr} - \frac{W}{r}\Big) g - (\epsilon - M) f = 0, \end{aligned} \qquad (7.47)$$

where

$$f_2(r) + f_4(r) = f(r), \quad i[f_2(r) - f_4(r)] = g(r) .$$

Now, separating the rest energy by the formal substitution $\epsilon = M + E$, one gets (in the first equation we neglect the nonrelativistic energy compared with the rest energy)

$$\Big(\frac{d}{dr} + \frac{W}{r}\Big) f + 2M \, g = 0, \quad \Big(\frac{d}{dr} - \frac{W}{r}\Big) g - E \, f = 0 .$$

Using the first equation, we exclude the small component $g(r)$ and obtain the nonrelativistic equation for the big component $f(r)$:

$$\Big[\Big(\frac{d}{dr} - \frac{W(r)}{r}\Big)\Big(\frac{d}{dr} + \frac{W(r)}{r}\Big) + 2ME\Big] f(r) = 0 .$$

This equation can be written as one-dimensional Schrödinger equation with an effective potential

$$\Big[\frac{d^2}{dr^2} + 2ME + \Big(\frac{W}{r}\Big)' - \frac{W^2}{r^2}\Big] f(r) = 0 . \qquad (7.48)$$

We shall use the exact solution known as Bogomol'nyi – Prasad – Sommerfeld monopole solution [155, 156]; in Cartesian isotopic gauge it has the structure

$$W_i^a(x) = \epsilon_{iab} x^b K(r), \quad \text{let } 1 + er^2 K(r) = W(r) .$$

The explicit form of the function $W(r)$ is given by six variants:

$$W = \pm 1\,, \quad W = \pm\,\frac{Ar}{\sinh Ar}\,, \quad W = \pm\,\frac{Ar}{\sin Ar}\,, \tag{7.49}$$

where A is an arbitrary constant.

We start by considering the first two cases from (7.49). The first one,

$$W(r) = +1\,, \quad \Big(\frac{d^2}{dr^2} + 2ME - \frac{2}{r^2}\Big) f(r) = 0\,; \tag{7.50}$$

leads to one-dimensional Schrödinger problem with effective centrifugal repulsion potential. In the variable $x = 2\sqrt{-2ME}\,r$, the solutions of the equation (7.50) are constructed in terms of Bessel functions. For the second case we have

$$W(r) = -1\,, \quad \Big(\frac{d^2}{dr^2} + 2ME\Big) f(r) = 0\,, \quad f = e^{\pm i\sqrt{2ME}\,r}\,. \tag{7.51}$$

Now let us consider two solutions known as referring to a non-Abelian monopole. The first of them is

$$W = +\frac{Ar}{\sinh Ar}\,, \quad \Big(\frac{d^2}{dr^2} + 2ME - \frac{A^2}{\cosh Ar - 1}\Big) f = 0\,. \tag{7.52}$$

This is the Schrödinger equation in the effective repulsion field. Such a system has no bound states. For the second solution, we have

$$W = -\frac{Ar}{\sinh Ar}\,, \quad \Big(\frac{d^2}{dr^2} + 2ME + \frac{A^2}{\cosh Ar + 1}\Big) f = 0\,; \tag{7.53}$$

This is the Schrödinger equation in the effective attracting field

$$U = -\frac{A^2}{\cosh Ar + 1}\,, \quad F_r = -\frac{dU}{dr} = -A^3 \frac{\sinh Ar}{(\cosh Ar + 1)^2}\,,$$

and for this problem the bounded states can exist.

Let us consider the equation (7.52) in detail. By changing the variable

$$x = \frac{\cosh Ar + 1}{2} \qquad r \to 0 (x \to 1)\,, \quad r \to +\infty (x \to +\infty)\,,$$

we get the equation

$$\Big[x(1-x)\frac{d^2}{dx^2} + \Big(\frac{1}{2} - x\Big)\frac{d}{dx} - \frac{2ME}{A^2} - \frac{1}{2(1-x)}\Big] f(x) = 0\,,$$

and applying the substitution $f(x) = x^a (1-x)^b F(x)$:

$$\Big[x(1-x)\frac{d^2}{dx^2} + \big(2a + \frac{1}{2} - (2a + 2b + 1)x\big)\frac{d}{dx}$$
$$-(a+b)^2 - \frac{2ME}{A^2} + \frac{a(2a-1)}{2x} - \frac{1 + b - 2b^2}{2(1-x)}\Big] F(x) = 0$$

at $a=0,\ 1/2,\ b=-1/2,\ 1$, the equation for $F(x)$ reduces to the hypergeometric type

$$\Big[x(1-x)\frac{d^2}{dx^2}+(2a+\frac{1}{2}-(2a+2b+1)x)\frac{d}{dx}-(a+b)^2-\frac{2ME}{A^2}\Big]F(x)=0$$

with the parameters

$$\alpha=a+b+\frac{\sqrt{-2ME}}{A}, \quad \beta=a+b-\frac{\sqrt{-2mE}}{A}, \quad \gamma=2a+\frac{1}{2}. \tag{7.54}$$

To have the required bounded state behavior, we have to choose $a=0, b=+1/2$; at that $a+b$ cannot be negative, and consequently $\alpha>0$. This means that bound states do not exist in the system.

Now we consider the equation (7.53) in the variable

$$x=\frac{\cosh Ar+1}{2}, \quad r\to 0(x\to 1)\,, \quad r\to+\infty(x\to+\infty)\,;$$

we get

$$\Big[x(1-x)\frac{d^2}{dx^2}+(\frac{1}{2}-x)\frac{d}{dx}-\frac{2ME}{A^2}-\frac{1}{2x}\Big]f(x)=0;$$

making the substitution $f(x)=x^a(1-x)^bF(x)$, we infer

$$\Big[x(1-x)\frac{d^2}{dx^2}+(2a+\frac{1}{2}-(2a+2b+1)x)\frac{d}{dx}$$
$$-(a+b)^2-\frac{2ME}{A^2}-\frac{1+a-2a^2}{2x}+\frac{b(2b-1)}{2(1-x)}\Big]F(x)=0.$$

Let $a=-1/2, 1\,,\ b=0,\ 1/2$; then the equation reduces to the hypergeometric type

$$\Big[x(1-x)\frac{d^2}{dx^2}+(2a+\frac{1}{2}-(2a+2b+1)x)\frac{d}{dx}-(a+b)^2-\frac{2ME}{A^2}\Big]F(x)=0$$

with the parameters

$$\alpha=a+b+\frac{\sqrt{-2ME}}{A}, \quad \beta=a+b-\frac{\sqrt{-2ME}}{A}, \quad \gamma=2a+\frac{1}{2}. \tag{7.55}$$

We shall construct the solutions which turn to zero at infinity. Such a behavior is supported by two solutions of this hypergeometric equation

$$\begin{aligned} U_3(x)&=(-x)^{-\alpha}F(\alpha,\alpha+1-\gamma,\alpha+1-\beta;\frac{1}{x}), \quad U_3(\infty)\sim(-x)^{-\alpha}, \quad \alpha>0;\\ U_4(x)&=(-x)^{-\beta}F(\beta+1-\gamma,\beta,\beta+1-\alpha;\frac{1}{x}), \quad U_4(\infty)\sim(-x)^{-\beta} \quad \beta>0. \end{aligned} \tag{7.56}$$

Accounting for the identities

$$-\alpha=-a-b-\frac{\sqrt{-2ME}}{A}, \quad -\beta=-a-b+\frac{\sqrt{-2ME}}{A},$$

we conclude that only the solution U_3 is suitable, and the corresponding full function $f(x)$ vanishes at infinity

$$f(x\to+\infty)\sim x^{a}(-x)^{b}U_3(x\to+\infty)\sim x^{-\frac{\sqrt{-2ME}}{A}}\,. \tag{7.57}$$

One can find the behavior of this solution near the point $x=1$ $(r=0)$. To do this, we use Kummer's relation [1] (the explicit form of the expansion coefficients K,L is not needed)

$$U_3(x)=KU_2(x)+LU_6(x)\,,$$
$$U_2(x)=F(\alpha,\beta,\alpha+\beta+1-\gamma;1-x)\,,$$
$$U_6(x)=(1-x)^{\gamma-\alpha-\beta}F(\gamma-\alpha,\gamma-\beta,\gamma+1-\alpha-\beta;1-x)\,,$$
$$U_3(x\to1)=K+L(1-x)^{\gamma-\alpha-\beta},\quad \gamma-\alpha-\beta=\frac{1}{2}-2b\,.$$

Thus, we have

$$f(x\to1)\sim(1-x)^{b}[K+L(1-x)^{\gamma-\alpha-\beta}]\sim(1-x)^{b}K+(1-x)^{1/2-b}L.$$

Two possibilities arise (both of them lead to finite values of the function $f(r\to0)$):

$$b=0\,,\qquad f(x\to1)\sim K;\quad b=\frac{1}{2}\,,\qquad f(x\to1)\sim L\,. \tag{7.58}$$

Let us find the quantization rule. Applying the ordinary polynomial condition

$$\alpha=a+b+\frac{\sqrt{-2ME}}{A}=-n\,.$$

we obtain the only nontrivial possibility

$$a=-\frac{1}{2}\,,\quad b=0\,,\quad \frac{\sqrt{-2ME}}{|A|}=-n+\frac{1}{2}\,,\quad n=0\,,$$

so there exists a single bound state

$$\frac{\sqrt{-2ME}}{|A|}=\frac{1}{2}\,,\quad E=-\frac{A^2}{8M}\,. \tag{7.59}$$

It is easy to verify that the relevant function

$$f(x)=x^{-1/2}=\sqrt{\frac{2}{\cosh Ar+1}}$$

is a solution of the equation (7.54) at the energy value (7.59). Also, it should be noted that this function is a square-integrable function. Indeed, taking into account that the factor r^{-1} has been already separated in the wave function (see (7.22)), the normalization integral has the form

$$I=\int_0^{\infty}f^2(r)dr\,,$$

and by changing to the variable $x=(\cosh Ar+1)/2$, we get

$$I=\frac{1}{A}\int_1^{\infty}\frac{dx}{x\sqrt{x^2-x}}=\frac{1}{A}\frac{2}{x}\sqrt{x^2-x}\Big|_1^{+\infty}=\frac{2}{A}\,.$$

7.5 Nonrelativistic Approximation, the Case $j>0$

We shall now consider the system at $j>0$ (see (7.40)) for $\delta=+1$:

$$\begin{aligned}
(-i\frac{d}{dr}+\epsilon)f_3-\frac{i\nu}{r}f_4-Mf_1=0\,,\\
(+i\frac{d}{dr}+\epsilon)f_1+\frac{i\nu}{r}f_2-Mf_3=0\,,\\
(-i\frac{d}{dr}+\epsilon)f_2-\frac{i\nu}{r}f_1-i\frac{W}{r}f_4-Mf_4=0\,,\\
(+i\frac{d}{dr}+\epsilon)f_4+\frac{i\nu}{r}f_3+i\frac{W}{r}f_2-Mf_2=0\,.
\end{aligned} \tag{7.60}$$

First of all, we find the explicit expressions for big and small components of the doublet wave function. The action of the projection operator on the functions (7.22) gives

$$\Psi_{\pm}^{\epsilon,j,m}=T_{+1/2}\otimes\left|\begin{array}{c}(f_1\pm f_3)\,D_{-1}\\(f_2\pm f_4)\,D_0\\(f_3\pm f_1)\,D_{-1}\\(f_4\pm f_2)\,D_0\end{array}\right|+T_{-1/2}\otimes\left|\begin{array}{c}(g_1\pm g_3)\,D_0\\(g_2\pm g_4)\,D_{+1}\\(g_3\pm g_1)\,D_0\\(g_4\pm g_2)\,D_{+1}\end{array}\right|.$$

We take into account the restrictions arisen from the diagonalization of the parity operator

$$\Psi_{\pm}^{\epsilon,j,m}=T_{+1/2}\otimes\left|\begin{array}{c}(f_1\pm f_3)\,D_{-1}\\(f_2\pm f_4)\,D_0\\(f_3\pm f_1)\,D_{-1}\\(f_4\pm f_2)\,D_0\end{array}\right|+\delta\,T_{-1/2}\otimes\left|\begin{array}{c}(f_4\pm f_2)\,D_0\\(f_3\pm f_1)\,D_{+1}\\(f_2\pm f_4)\,D_0\\(f_1\pm f_3)\,D_{+1}\end{array}\right|. \tag{7.61}$$

For each pair in (7.60), we sum and subtract the equations: with the use of notations

$$f_1+f_3=F\,,\quad i(f_1-f_3)=f\,,\quad (f_2+f_4)=G\,,\quad i(f_2-f_4)=g\,, \tag{7.62}$$

we can write the obtained equations in the form:

$$\frac{d}{dr}f+\epsilon F+\frac{\nu}{r}g-MF=0\,,\quad -\frac{d}{dr}F+\epsilon f-\frac{\nu}{r}G+Mf=0\,,$$

$$-\frac{d}{dr}g+\epsilon G-\frac{\nu}{r}f+\frac{W}{r}g-MG=0\,,\quad -\frac{d}{dr}G-\epsilon g-\frac{\nu}{r}F-\frac{W}{r}G-Mg=0\,.$$

Separating the rest energy by the substitution $\epsilon=E+M$ and neglecting the nonrelativistic energy E compared to the double rest energy $2M$, one gets

$$\frac{d}{dr}f+EF+\frac{\nu}{r}g=0\,,\quad \frac{d}{dr}F+\frac{\nu}{r}G-2Mf=0\,,$$
$$-\frac{d}{dr}g+EG-\frac{\nu}{r}f+\frac{W}{r}g=0\,,\quad \frac{d}{dr}G+\frac{\nu}{r}F+\frac{W}{r}G+2Mg=0\,.$$

Finally, by excluding the small components

$$2Mf = \frac{d}{dr}F + \frac{\nu}{r}G, \quad 2Mg = -\frac{d}{dr}G - \frac{\nu}{r}F - \frac{W}{r}G,$$

we find two equations for the big components:

$$\Delta F = \left(\frac{\nu}{r^2} + \frac{\nu W}{r^2}\right)G, \quad \Delta G = -\left(\frac{d}{dr}\frac{W}{r} - \frac{W^2}{r^2}\right)G + \left(\frac{\nu}{r^2} + \frac{\nu W}{r^2}\right)F, \tag{7.63}$$

where Δ denotes the operator

$$\Delta = \frac{d^2}{dr^2} + 2ME - \frac{\nu^2}{r^2}, \quad \nu = \sqrt{j(j+1)}. \tag{7.64}$$

Let us consider the case $W(r) = +1$:

$$\Delta F = \frac{2\nu}{r^2}G, \quad \Delta G = +\frac{2}{r^2}G + \frac{2\nu}{r^2}F, \quad \frac{1}{2}r^2\Delta \begin{vmatrix} F \\ G \end{vmatrix} = \begin{vmatrix} 0 & \nu \\ \nu & 1 \end{vmatrix} \begin{vmatrix} F \\ G \end{vmatrix}.$$

We need to find such a transformation which transforms the right matrix to its diagonal form

$$\frac{1}{2}r^2\Delta\Psi = A\psi, \quad [S\,\frac{1}{2}r^2\Delta\, S^{-1}](S\Psi) = (SAS^{-1})\,(S\psi),$$

$$S\psi = \psi' = \begin{vmatrix} F' \\ G' \end{vmatrix}, \quad \frac{1}{2}r^2\Delta\Psi' = A'\psi', \quad A' = SAS^{-1} = \begin{vmatrix} \lambda_1 & 0 \\ 0 & \lambda_2 \end{vmatrix}.$$

The solution is

$$A = \begin{vmatrix} 0 & \nu \\ \nu & 1 \end{vmatrix}, \quad S^{-1}AS = \begin{vmatrix} \lambda_1 & 0 \\ 0 & \lambda_2 \end{vmatrix},$$

$$S^{-1} = \begin{vmatrix} 1 & \frac{1}{2\nu} + \frac{\sqrt{1+4\nu^2}}{2\nu} \\ 1 & \frac{1}{2\nu} - \frac{\sqrt{1+4\nu^2}}{2\nu} \end{vmatrix}, \quad S = \begin{vmatrix} \frac{1}{2} - \frac{1}{2\sqrt{1+4\nu^2}} & \frac{1}{2} + \frac{1}{2\sqrt{1+4\nu^2}} \\ \frac{\nu}{\sqrt{1+4\nu^2}} & -\frac{\nu}{\sqrt{1+4\nu^2}} \end{vmatrix},$$

$$\lambda_1 = \frac{1}{2} - \frac{1}{2}\sqrt{1+4\nu^2}, \quad \lambda_2 = \frac{1}{2} + \frac{1}{2}\sqrt{1+4\nu^2},$$

Taking into account that $\sqrt{1+4\nu^2} = 2j+1$, one finds the expressions for λ_1, λ_2:

$$\lambda_1 = -j, \quad \lambda_2 = j+1.$$

We further obtain the differential equations

$$\left(\frac{d^2}{dr^2} + 2ME + \frac{j(j+3)}{r^2}\right)F' = 0, \quad \left(\frac{d^2}{dr^2} + 2ME + \frac{(j-2)(j+1)}{r^2}\right)G' = 0. \tag{7.65}$$

In the second case $W(r) = -1$ from (7.63), we get two separate equations

$$\left(\frac{d^2}{dr^2} + 2ME + \frac{j(j+1)}{r^2}\right)F = 0, \quad \left(\frac{d^2}{dr^2} + 2ME + \frac{j(j+1)-1}{r^2}\right)G = 0. \tag{7.66}$$

All the four equations are solved in terms of Bessel functions and do not have any solutions corresponding to the bound states.

In the cases

$$W = \pm \frac{Ar}{\sinh Ar}$$

it is obvious that the corresponding systems of equations for $F(r), G(r)$:

$$\Delta F = \Big(\frac{\nu}{r^2} + \frac{\nu W}{r^2}\Big) G, \qquad \Delta G = -\Big(\frac{d}{dr}(\frac{W}{r}) - \frac{W^2}{r^2}\Big) G + \Big(\frac{\nu}{r^2} + \frac{\nu W}{r^2}\Big) F, \tag{7.67}$$

where

$$\Delta = \Big(\frac{d^2}{dr^2} + 2ME + \frac{\nu^2}{r^2}\Big), \quad \nu = \sqrt{j(j+1)},$$

cannot be studied analytically, because of the simultaneous presence of the variable r and of the transcendental function $\sinh Ar$ in the equations.

7.6 The Doublet in the Spaces of Constant Curvature

Let us generalize the above analysis to consider the doublet of Dirac particles in three spaces of constant curvature: Euclidean space E_3, Lobachevsky space H_3 and Riemann space S_3. This gives the opportunity to formulate additional arguments concerning to problem of deciding which Yang – Mills equation solutions are interesting from the physical and theoretical points of view.

In a spherical coordinate system of the space S_3, the metric and tetrad are determined by the formulas

$$dS^2 = dt^2 - dr^2 - \sin^2 r(\, d\theta^2 + \sin^2\theta d\phi^2\,),$$
$$e^{\alpha}_{(0)} = (1,\, 0,\, 0,\, 0),\ e^{\alpha}_{(1)} = (0,\, 0,\, \sin^{-1} r,\, 0),$$
$$e^{\alpha}_{(2)} = (0,\, 0,\, 0,\, \sin^{-1} r \sin^{-1}\theta),\ e^{\alpha}_{(3)} = (0,1,0,0).$$

In Schwinger unitary gauge, the generalized Dirac equation for the particles doublet takes the following form

$$\Big[\gamma^0(i\partial_t + erF(r)t^3) + i\gamma^3(\partial_r + \frac{1}{\tan r}) + \frac{1}{\sin r}\Sigma^{S.}_{\theta,\phi}$$
$$+ \frac{er^2K+1}{\sin r}(\gamma^1 \otimes t^2 \; - \; \gamma^2 \otimes t^1) - (M + \kappa r \Phi(r) t^3)\Big]\Psi^{S.} = 0. \tag{7.68}$$

At this special choice of tetrad and isotopic bases for all three geometrical models, the explicit form of the symmetry operators for total orbital, spin and isotopic momenta, is

$$J^{S.}_1 = l_1 + \frac{(i\sigma^{12} + t^3)\cos\phi}{\sin\theta}, \quad J^{S.}_2 = l_2 + \frac{(i\sigma^{12} + t^3)\sin\phi}{\sin\theta}, \quad J^{S.}_3 = l_3.$$

The dependence of the wave functions Ψ_{jm} is described by Wigner's functions, $D^j_{-m,\sigma}(\phi,\theta,0)$ (in the same way for all three models)

$$\Psi_{\epsilon jm}(x) = \frac{e^{-i\epsilon t}}{\sin r}\,\big[T_{+1/2} \otimes F(r,\theta,\phi) + T_{-1/2} \otimes G(r,\theta,\phi)\big],$$

where

$$F = \begin{vmatrix} f_1(r)D_{-1} \\ f_2(r)D_0 \\ f_3(r)D_{-1} \\ f_4(r)D_0 \end{vmatrix}, \quad G = \begin{vmatrix} g_1(r)D_0 \\ g_2(r)D_{+1} \\ g_3(r)D_0 \\ g_4(r)D_{+1} \end{vmatrix}, \quad T_{+1/2} = \begin{vmatrix} 1 \\ 0 \end{vmatrix}, T_{-1/2} = \begin{vmatrix} 0 \\ 1 \end{vmatrix};$$

the factor $e^{-i\epsilon t}/\sin r$ is omitted for shortness. It should noted that the term mixing isotopic components of the doublet is proportional to (er^2K+1) and for all three model spaces this term vanishes identically for the trivial monopole solution of Yang – Mills equation.

For the case $j=0$, the wave function is constructed as follows

$$\Psi_{\epsilon 0} = T_{+1/2} \otimes \begin{vmatrix} 0 \\ f_2(r) \\ 0 \\ f_4(r) \end{vmatrix} + T_{-1/2} \otimes \begin{vmatrix} g_1(r) \\ 0 \\ g_3(r) \\ 0 \end{vmatrix}.$$

After calculations (which mainly coincide with the ones performed for flat space) we get the radial system

$$\begin{aligned}
(-i\frac{d}{dr} + \epsilon + \tilde{F})f_3 - i\frac{\nu}{\sin r} f_4 - (M + \tilde{\Phi})f_1 = 0, \\
(+i\frac{d}{dr} + \epsilon + \tilde{F})f_4 + i\frac{\nu}{\sin r} f_3 + i\frac{W}{\sin r} g_3 - (M + \tilde{\Phi})f_2 = 0, \\
(+i\frac{d}{dr} + \epsilon + \tilde{F})f_1 + i\frac{\nu}{\sin r} f_2 - (M + \tilde{\Phi})f_3 = 0, \\
(-i\frac{d}{dr} + \epsilon + \tilde{F})f_2 - i\frac{\nu}{\sin r} f_1 - i\frac{W}{\sin r} g_1 - (M + \tilde{\Phi})f_4 = 0, \\
(-i\frac{d}{dr} + \epsilon - \tilde{F})g_3 - i\frac{\nu}{\sin r} g_4 - i\frac{W}{\sin r} f_4 - (M - \tilde{\Phi})g_1 = 0, \\
(+i\frac{d}{dr} + \epsilon - \tilde{F})g_4 + i\frac{\nu}{\sin r} g_3 - (M - \tilde{\Phi})g_2 = 0, \\
(+i\frac{d}{dr} + \epsilon - \tilde{F})g_1 + i\frac{\nu}{\sin r} g_2 + i\frac{W}{\sin r} f_2 - (M - \tilde{\Phi})g_3 = 0, \\
(-i\frac{d}{dr} + \epsilon - \tilde{F})g_2 - i\frac{\nu}{\sin r} g_1 - (M - \tilde{\Phi})g_4 = 0,
\end{aligned} \tag{7.69}$$

where

$$W \equiv (er^2K+1)/2, \quad \tilde{F} \equiv erF/2, \quad \tilde{\Phi} \equiv \kappa r\Phi/2.$$

At $j=0$, the angular operator acts as the null one $\Sigma_{\theta,\phi}\Psi_{\epsilon 0} \equiv 0$ and the radial system simplifies:

$$\begin{aligned}
(+i\frac{d}{dr} + \epsilon + \tilde{F})f_4 + i\frac{W}{\sin r} g_3 - (M + \tilde{\Phi})f_2 = 0, \\
(-i\frac{d}{dr} + \epsilon + \tilde{F})f_2 - i\frac{W}{\sin r} g_1 - (M + \tilde{\Phi})f_4 = 0, \\
(-i\frac{d}{dr} + \epsilon - \tilde{F})g_3 - i\frac{W}{\sin r} f_4 - (M - \tilde{\Phi})g_1 = 0, \\
(+i\frac{d}{dr} + \epsilon - \tilde{F})g_1 + i\frac{W}{\sin r} f_2 - (M - \tilde{\Phi})g_3 = 0.
\end{aligned} \tag{7.70}$$

The obtained radial equations can be significantly simplified by diagonalizing the discrete operator which acts in bispinor and isotopic spaces. The final results are

at $W(r)=0,\ j=0$

$$(i\frac{d}{dr}+\epsilon)f_4-Mf_2=0, \quad (-i\frac{d}{dr}+\epsilon)f_2-Mf_4=0; \tag{7.71}$$

at $W(r)=0,\ j\neq 0$

$$\begin{gathered}(-i\frac{d}{dr}+\epsilon)f_3-\frac{i\nu}{\sin r}f_4-Mf_1=0, \quad +i\frac{d}{dr}+\epsilon)f_4+\frac{i\nu}{\sin r}f_3-Mf_2=0,\\ (+i\frac{d}{dr}+\epsilon)f_1+\frac{i\nu}{\sin r}f_2-Mf_3=0, \quad (-i\frac{d}{dr}+\epsilon)f_2-\frac{i\nu}{\sin r}f_1-Mf_4=0.\end{gathered} \tag{7.72}$$

The equations (7.71) coincide with the ones which arise in flat space, and are solved in the same manner. Equations (7.72) can be solved in the same way as for the flat space model. Indeed, these equations are simplified by the diagonalization of the operator $\hat{K}_{\theta,\phi}=-i\gamma^0\gamma^5\Sigma_{\theta,\phi}$. From the equation $\hat{K}_{\theta,\phi}\Psi_{jm}=\lambda\Psi_{jm}$, it follows

$$\lambda=-\mu\sqrt{j(j+1)}(\mu=\pm 1), \quad f_4=\mu f_1, \quad f_3=\mu f_2, g_4=\mu g_1, \quad g_3=\mu g_2. \tag{7.73}$$

Correspondingly, from the system (7.72), we derive

$$\begin{gathered}(+i\frac{d}{dr}+\epsilon)f_1+i\frac{\nu}{\sin r}f_2-\mu M f_2=0,\\ (-i\frac{d}{dr}+\epsilon)f_2-i\frac{\nu}{\sin r}f_1-\mu M f_1=0.\end{gathered} \tag{7.74}$$

These equations have solutions in terms of hypergeometric functions (see [50]). The wave function with quantum numbers $(\epsilon, j, m, \delta, \mu)$ has the structure in the form (7.37); at that, the construction of non-Abelian solutions is performed in terms of Abelian ones, according to (7.38), (7.39). The solutions are suitable for spherical Riemannian space (but in contrast to flat space, for Riemann space the energy spectra are discrete because of the compactness of this geometrical model).

The situation is quite different for the nontrivial non-Abelian potential:

$W(r)\neq 0,\ j=0,$

$$(i\frac{d}{dr}+\epsilon)f_4-(M-i\delta\frac{W}{\sin r})f_2=0, \quad (-i\frac{d}{dr}+\epsilon)f_2-(M+i\delta\frac{W}{\sin r})f_4=0; \tag{7.75}$$

$W(r)\neq 0,\ j\neq 0,$

$$\begin{gathered}(-i\frac{d}{dr}+\epsilon)f_3-\frac{i\nu}{\sin r}f_4-Mf_1=0,\\ (+i\frac{d}{dr}+\epsilon)f_4+\frac{i\nu}{\sin r}f_3+i\frac{W}{\sin r}\delta f_2-Mf_2=0,\\ (+i\frac{d}{dr}+\epsilon)f_1+\frac{i\nu}{\sin r}f_2-Mf_3=0,\\ (-i\frac{d}{dr}+\epsilon)f_2-\frac{i\nu}{\sin r}f_1-i\frac{W}{\sin r}\delta f_4-Mf_4=0.\end{gathered} \tag{7.76}$$

The transition from the radial system in Riemann space to the analogues for spaces E_3 and H_3 is performed by means of the formal changes $\sin r \Longrightarrow r$, $\sin r \Longrightarrow \sinh r$ respectively.

Let us recall the explicit form of the radial function $W(r)$, found for all three spaces:

$$S_3\,, \quad \frac{W}{\sin r} = \varphi(r)\,, \quad r \in [0,\pi]\,,$$
$$H_3\,, \quad \frac{W}{\sinh r} = \varphi(r)\,, \quad r \in [0,+\infty)\,,$$
$$E_3\,, \quad \frac{W}{r} = \varphi(r)\,, \quad r \in [0,+\infty)\,. \tag{7.77}$$

There exist six solutions for the function $\varphi(r)$:

$$\varphi(r) = \pm\frac{a}{ar}\,, \quad \pm\frac{a}{\sinh ar}\,, \quad \pm\frac{a}{\sin ar}\,. \tag{7.78}$$

It is evident that among the three pairs of monopole solutions there is only one pair which is distinguished by its obvious linking with a corresponding geometry E_3, H_3, or S_3. This situation can be illustrated by the scheme

$$\begin{array}{cccc} & E_3 & H_3 & S_3 \\ ar & * & - & - \\ \sinh ar & - & * & - \\ \sin ar & - & - & * \end{array} \tag{7.79}$$

It should be emphasized that the known non-singular monopole solution by Bogomol'ny – Prasad – Sommerfeld in Minkowski space is, in some sense, an artificial combination of flat space geometry with the possibility of being associated with the geometry of Lobachevsky space according to (7.79).

In the three geometries, at the zero quantum number j, the radial systems are

$$E_3 \qquad \begin{array}{l} (i\dfrac{d}{dr} + \epsilon) f_4 - (M - \delta\mu\dfrac{i}{r}) f_2 = 0\,, \\ (-i\dfrac{d}{dr} + \epsilon) f_2 - (M + \delta\mu\dfrac{i}{r}) f_4 = 0\,; \end{array} \tag{7.80}$$

$$H_3 \qquad \begin{array}{l} (i\dfrac{d}{dr} + \epsilon) f_4 - (M - \delta\mu\dfrac{i}{\sinh r}) f_2 = 0\,, \\ (-i\dfrac{d}{dr} + \epsilon) f_2 - (M + \delta\mu\dfrac{i}{\sinh r}) f_4 = 0\,; \end{array} \tag{7.81}$$

$$S_3 \qquad \begin{array}{l} (i\dfrac{d}{dr} + \epsilon) f_4 - (M - \delta\mu\dfrac{i}{\sin r}) f_2 = 0\,, \\ (-i\dfrac{d}{dr} + \epsilon) f_2 - (M + \delta\mu\dfrac{i}{\sin r}) f_4 = 0\,. \end{array} \tag{7.82}$$

For the three geometry models, at the non-zero quantum number j, the radial systems are

$$E_3 \qquad \begin{aligned} &(-i\frac{d}{dr}+\epsilon)f_3-\frac{i\nu}{r}f_4-Mf_1=0,\\ &(i\frac{d}{dr}+\epsilon)f_4+\frac{i\nu}{r}f_3+\delta\mu\frac{i}{r}f_2-Mf_2=0,\\ &(i\frac{d}{dr}+\epsilon)f_1+\frac{i\nu}{\sin r}f_2-Mf_3=0,\\ &(-i\frac{d}{dr}+\epsilon)f_2-\frac{i\nu}{r}f_1-\delta\mu\frac{i}{r}f_4-Mf_4=0; \end{aligned} \qquad (7.83)$$

$$H_3 \qquad \begin{aligned} &(-i\frac{d}{dr}+\epsilon)f_3-\frac{i\nu}{\sinh r}f_4-Mf_1=0,\\ &(i\frac{d}{dr}+\epsilon)f_4+\frac{i\nu}{\sinh r}f_3+\delta\mu\frac{i}{\sinh r}f_2-mf_2=0,\\ &(+i\frac{d}{dr}+\epsilon)f_1+\frac{i\nu}{\sinh r}f_2-Mf_3=0,\\ &(-i\frac{d}{dr}+\epsilon)f_2-\frac{i\nu}{\sinh r}f_1-\delta\mu\frac{i}{\sinh r}f_4-mf_4=0; \end{aligned} \qquad (7.84)$$

$$S_3 \qquad \begin{aligned} &(-i\frac{d}{dr}+\epsilon)f_3-\frac{i\nu}{\sin r}f_4-Mf_1=0,\\ &(i\frac{d}{dr}+\epsilon)f_4+\frac{i\nu}{\sin r}f_3+\delta\mu\frac{i}{\sin r}f_2-Mf_2=0,\\ &(i\frac{d}{dr}+\epsilon)f_1+\frac{i\nu}{\sin r}f_2-Mf_3=0,\\ &(-i\frac{d}{dr}+\epsilon)f_2-\frac{i\nu}{\sin r}f_1-\delta\mu\frac{i}{\sin r}f_4-Mf_4=0. \end{aligned} \qquad (7.85)$$

We sum and subtract the equations from each other in the systems (7.80), (7.81), (7.82). As a result, we have:

$$\begin{aligned} &E_3, \quad (\frac{d}{dr}+\frac{\delta\mu}{r})f+(\epsilon+M)g=0, \quad (\frac{d}{dr}-\frac{\delta\mu}{r})g-(\epsilon-M)f=0,\\ &H_3, \quad (\frac{d}{dr}+\frac{\delta\mu}{\sinh r})f+(\epsilon+M)g=0, \quad (\frac{d}{dr}-\frac{\delta\mu}{\sinh r})g-(\epsilon-M)f=0,\\ &S_3, \quad (\frac{d}{dr}+\frac{\delta\mu}{\sin r})f+(\epsilon+M)g=0, \quad (\frac{d}{dr}-\frac{\delta\mu}{\sin r})g-(\epsilon-M)f=0, \end{aligned} \qquad (7.86)$$

where the following notations are used: $f_2(r)+f_4(r)=f(r)$, $i\,[f_2(r)-f_4(r)]=g(r)$. Let us separate the rest energy by the formal substitution $\epsilon=M+E$. After that we neglect the nonrelativistic energy, comparing with the rest energy. So we obtain

$$\begin{aligned} &E_3, \quad \left(\frac{d}{dr}+\frac{\delta\mu}{r}\right)f+2Mg=0, \quad \left(\frac{d}{dr}-\frac{\delta\mu}{r}\right)g-Ef=0,\\ &H_3, \quad \left(\frac{d}{dr}+\frac{\delta\mu}{\sinh r}\right)f+2Mg=0, \quad \left(\frac{d}{dr}-\frac{\delta\mu}{\sinh r}\right)g-Ef=0,\\ &S_3, \quad \left(\frac{d}{dr}+\frac{\delta\mu}{\sin r}\right)f+2Mg=0, \quad \left(\frac{d}{dr}-\frac{\delta\mu}{\sin r}\right)g-Ef=0. \end{aligned} \qquad (7.87)$$

Using the fist equations, we exclude the small components $g(r)$ and obtain the following nonrelativistic equations for the big components $f(r)$:

$$
\begin{aligned}
E_3, &\quad \Big(\frac{d^2}{dr^2}+2ME-\delta\mu\frac{1}{r^2}-\frac{1}{r^2}\Big)f=0,\\
H_3, &\quad \Big(\frac{d^2}{dr^2}+2ME-\delta\mu\frac{\cosh r}{\sinh^2 r}-\frac{1}{\sinh^2 r}\Big)f=0,\\
S_3, &\quad \Big(\frac{d^2}{dr^2}+2ME-\delta\mu\frac{\cos r}{\sin^2 r}-\frac{1}{\sin^2 r}\Big)f=0.
\end{aligned}
\tag{7.88}
$$

All three equations are solved in hypergeometric functions; at that, there are no solutions corresponding to the bound states.

Now we find the nonrelativistic approximation at $j>0$. We perform the calculations for the space S_3 (see (7.85)). We sum and subtract the equations in each pair

$$
\begin{aligned}
(-i\frac{d}{dr}+\epsilon)f_3-\frac{i\nu}{\sin r}f_4-Mf_1=0,\\
(i\frac{d}{dr}+\epsilon)f_1+\frac{i\nu}{\sin r}f_2-Mf_3=0,\\
(i\frac{d}{dr}+\epsilon)f_4+\frac{i\nu}{\sin r}f_3+\delta\mu\frac{i}{\sin r}f_2-Mf_2=0,\\
(-i\frac{d}{dr}+\epsilon)f_2-\frac{i\nu}{\sin r}f_1-\delta\mu\frac{i}{\sin r}f_4-Mf_4=0,
\end{aligned}
$$

whence using the notations

$$
f_1+f_3=F, \quad i(f_1-f_3)=f, \quad (f_2+f_4)=G, \quad i(f_2-f_4)=g, \tag{7.89}
$$

we derive

$$
\frac{d}{dr}f+\epsilon F+\frac{\nu}{\sin r}g-MF=0, \quad \frac{d}{dr}F-\epsilon f+\frac{\nu}{r}G-Mf=0,
$$

$$
\frac{d}{dr}g-\epsilon G+\frac{\nu}{\sin r}f-\frac{\delta\mu}{\sin r}g+MG=0,
$$

$$
\frac{d}{dr}G+\epsilon g+\frac{\nu}{\sin r}F+\frac{\delta\mu}{\sin r}G+Mg=0.
$$

Then we separate the rest energy by the substitution $\epsilon=E+M$ and neglect nonrelativistic energy E comparing with the double rest energy $2M$:

$$
\frac{d}{dr}f+\frac{\nu}{\sin r}g+EF=0, \quad \frac{d}{dr}F+\frac{\nu}{r}G-2Mf=0,
$$

$$
\frac{d}{dr}g+\frac{\nu}{\sin r}f-\frac{\delta\mu}{\sin r}g-EG=0, \quad \frac{d}{dr}G+\frac{\nu}{\sin r}F+\frac{\delta\mu}{\sin r}G+2Mg=0.
$$

Excluding the small components

$$
2Mf=\Big(\frac{d}{dr}F+\frac{\nu}{\sin r}G\Big), \quad 2Mg=-\Big(\frac{d}{dr}G+\frac{\nu}{\sin r}F+\frac{\delta\mu}{\sin r}G\Big),
$$

we find two equations for the big components:

$$\frac{d}{dr}\Big(\frac{d}{dr}F+\frac{\nu}{\sin r}G\Big)-\frac{\nu}{\sin r}\Big(\frac{d}{dr}G+\frac{\nu}{\sin r}F+\frac{\delta\mu}{\sin r}G\Big)+2MEF=0,$$

$$-\frac{d}{dr}\Big(\frac{d}{dr}G+\frac{\nu}{\sin r}F+\frac{\delta\mu}{\sin r}G\Big)+\frac{\delta\mu}{\sin r}\Big(\frac{d}{dr}G+\frac{\nu}{\sin r}F+\frac{\delta\mu}{\sin r}G\Big)$$

$$+\frac{\nu}{\sin r}\Big(\frac{d}{dr}F+\frac{\nu}{\sin r}G\Big)-2MEG=0,$$

which after some transformation give

$$\Big(\frac{d^2}{dr^2}+2ME-\frac{\nu^2}{\sin^2 r}\Big)F+\Big(-\frac{\nu\cos r}{\sin^2 r}-\delta\mu\frac{\nu}{\sin^2 r}\Big)G=0,$$

$$\Big(-\frac{d^2}{dr^2}+\delta\mu\frac{\cos r}{\sin^2 r}+\frac{1+\nu^2}{\sin^2 r}-2ME\Big)G+\Big(\frac{\nu\cos r}{\sin^2 r}+\delta\mu\frac{\nu}{\sin^2 r}\Big)F=0.$$

This system can be written as follows

$$\Big(\frac{d^2}{dr^2}+2ME-\frac{\nu^2}{\sin^2 r}\Big)F=\frac{\cos r+\delta\mu}{\sin^2 r}\Big(0\cdot G+\nu G\Big),$$

$$\Big(\frac{d^2}{dr^2}+2ME-\frac{\nu^2}{\sin^2 r}\Big)G=\frac{\cos r+\delta\mu}{\sin^2 r}\Big(\nu F+\delta\mu\, G\Big),$$

or in matrix form,

$$\Delta\begin{vmatrix} F \\ G \end{vmatrix}=\begin{vmatrix} 0 & \nu \\ \nu & \delta\mu \end{vmatrix}\begin{vmatrix} F \\ G \end{vmatrix},\quad \Delta=\Big(\frac{\cos r+\delta\mu}{\sin^2 r}\Big)^{-1}\Big(\frac{d^2}{dr^2}+2ME-\frac{\nu^2}{\sin^2 r}\Big). \tag{7.90}$$

An analog of this system in the Lobachevsky space has the form

$$\Delta\begin{vmatrix} F \\ G \end{vmatrix}=\begin{vmatrix} 0 & \nu \\ \nu & \delta\mu \end{vmatrix}\begin{vmatrix} F \\ G \end{vmatrix},\quad \Delta=\Big(\frac{\cosh r+\delta\mu}{\sinh^2 r}\Big)^{-1}\Big(\frac{d^2}{dr^2}+2ME-\frac{\nu^2}{\sinh^2 r}\Big). \tag{7.91}$$

In both systems, the right matrices are diagonalized by linear transformations. As a result, two uncoupled equations arise. In the case of spherical Riemannian space, these equations lead to discrete energy values because of the compactness of the space. In Lobachevsky space, the differential equations do not have any solutions corresponding to the bound states.

7.7 Geometrization of the Monopole Problem, KCC-Invariants

We apply a special geometric method based on the KCC-invariants [163] – [166] to study the behavior of solutions of differential equations from the point of view of Jacobi stability. In this approach, one considers a system of second order differential equations

$$\dot{y}^i(r)+2Q^i(r,x,y)=0, \tag{7.92}$$

which corresponds to the the Euler – Lagrange equations for some dynamical system with a corresponding Lagrangian L. In (7.92), the symbol x^i designates the so called coordinates; their derivatives in argument r are $y^i = dx^i/dr = \dot{x}^i$, and the quantities Q_i are determined by some Lagrangian L in accordance with the formulas

$$Q^i = \frac{1}{4} g^{il} \Big(\frac{\partial^2 L}{\partial x^k \partial y^l} y^k - \frac{\partial L}{\partial x^l} + \frac{\partial^2 L}{\partial y^l \partial r} \Big), \qquad g_{ij} = \frac{1}{2} \frac{\partial^2 L}{\partial y^i \partial y^j} . \tag{7.93}$$

The first and second invariants, $\varepsilon^i(r,x,y)$ and $P^i_{\ j}$, are introduced by the following definitions

$$\varepsilon^i = \frac{\partial Q^i}{\partial y^j} y^j - 2Q^i, \qquad P^i_{\ j} = 2\frac{\partial Q^i}{\partial x^j} + 2Q^s \frac{\partial^2 Q^i}{\partial y^j \partial y^s} - \frac{\partial^2 Q^i}{\partial y^j \partial x^s} y^s - \frac{\partial Q^i}{\partial y^s} \frac{\partial Q^s}{\partial y^j} - \frac{\partial^2 Q^i}{\partial y^j \partial r} . \tag{7.94}$$

The second invariant $P^i_{\ j}$ relates to the Jacobi stability of the dynamical system. There is an analogy between the equations of geodesic deviation expressed in terms of the Riemannian curvature, and the equations of geodesic deviation expressed in terms of the second KCC-invariant:

$$\frac{D^2 \xi^i}{Ds^2} = R^i_{kjl} \frac{dx^k}{ds} \frac{dx^l}{ds} \xi^j = -K^i_j \xi^j , \qquad \frac{D^2 \xi^i}{Dr^2} = P^i_j \xi^j . \tag{7.95}$$

It is known that a pencil of geodesic curves from the some point r_0 converges (or diverges) if the real parts of all eigenvalues of the invariant $P^i_{\ j}$ are negative (or positive) ones.

We start with the system of radial Yang – Mills equations in the Prasad – Sommerfeld – Bogomol'nyi [155], [156] limit, that can be presented as follows

$$\begin{gathered}
\frac{d^2\Phi}{dr^2} + \frac{4}{r}\frac{d\Phi}{dr} - 2e\Phi(2+er^2K)K - \frac{\dot{\Sigma}}{\Sigma}\Big(\frac{d\Phi}{dr} + \frac{\Phi}{r}\Big) = 0, \\
\frac{d^2 f}{dr^2} + \frac{4}{r}\frac{df}{dr} - 2e\Phi(2+er^2K)K - \frac{\dot{\Sigma}}{\Sigma}\Big(\frac{df}{dr} + \frac{f}{r}\Big) = 0, \\
\frac{d^2K}{dr^2} + \frac{4}{r}\frac{dK}{dr} - e(3+er^2K)K^2 + \frac{\dot{\Sigma}}{\Sigma}\Big(\frac{dK}{dr} + \frac{2K}{r}\Big) + e\frac{(f^2-\Phi^2)(1+er^2K)}{\Sigma^2} = 0,
\end{gathered} \tag{7.96}$$

where the quantity Σ determines the metric of the geometrical model:

$$dS^2 = dt^2 + \frac{1}{\Sigma^2(r)}[dx_1^2 + dx_2^2 + dx_3^2], \quad r = \sqrt{x_1^2 + x_2^2 + x_3^2} .$$

The quantity $\Sigma(r) = 1$ corresponds to the Euclidean space, $\Sigma(r) = 1 + r^2/4$ corresponds to Riemannian space, and $\Sigma(r) = 1 - r^2/4$ is associated with the Lobachevsky space; the point over Σ designates the derivative d/dr. Applying the following notations

$$x^i = \{\Phi(r), \quad f(r), \; K(r)\}, \quad y^i = \frac{d}{dr} x^i(r) = \{\dot{\Phi}, \; \dot{f}, \; \dot{K}\}$$

and comparing equations (7.96) with (7.92), one finds the relevant expressions for quantities Q^i:

$$Q^1 = -eK\Phi\left(eKr^2+2\right) - \frac{\dot{\Sigma}}{2\Sigma}\left(\frac{\Phi}{r}+\dot{\Phi}\right) + \frac{2\dot{\Phi}}{r},$$
$$Q^2 = -efK\left(eKr^2+2\right) - \frac{\dot{\Sigma}}{2\Sigma}\left(\dot{f}+\frac{f}{r}\right) + \frac{2\dot{f}}{r}, \tag{7.97}$$
$$Q^3 = \frac{e\left(f^2-\Phi^2\right)\left(eKr^2+1\right)}{2\Sigma^2} - \frac{1}{2}eK^2\left(eKr^2+3\right) + \frac{\dot{\Sigma}}{2\Sigma}\left(\dot{K}+\frac{2K}{r}\right) + \frac{2\dot{K}}{r}.$$

By the direct calculation, we find the components of the first KCC-invariant:

$$\varepsilon^1 = 2eK\Phi\left(eKr^2+2\right) + \frac{1}{2}\dot{\Phi}\left(\frac{\dot{\Sigma}}{\Sigma}-\frac{4}{r}\right) + \frac{\Phi}{r}\frac{\dot{\Sigma}}{\Sigma},$$
$$\varepsilon^2 = 2eKf\left(eKr^2+2\right) + \frac{1}{2}\dot{f}\left(\frac{\dot{\Sigma}}{\Sigma}-\frac{4}{r}\right) + \frac{f}{r}\frac{\dot{\Sigma}}{\Sigma}, \tag{7.98}$$
$$\varepsilon^3 = \frac{e\left(\Phi^2-f^2\right)\left(eKr^2+1\right)}{\Sigma^2} + eK^2\left(eKr^2+3\right) - \frac{\dot{\Sigma}}{2r\Sigma}\left(4K+r\dot{K}\right) - \frac{2}{r}\dot{K},$$

and the components of the second invariant (elements of the matrix are presented by columns)

$$P^i_{\ j} = \left| \begin{array}{c} \frac{\ddot{\Sigma}}{2\Sigma} - \frac{3\left(\dot{\Sigma}\right)^2}{4\Sigma^2} + \frac{\dot{\Sigma}}{r\Sigma} - \frac{2\left(eKr^2+1\right)^2}{r^2} \\ 0 \\ -\frac{2e\Phi\left(eKr^2+1\right)}{\Sigma^2} \end{array} \quad \begin{array}{c} 0 \\ \frac{\ddot{\Sigma}}{2\Sigma} - \frac{3\left(\dot{\Sigma}\right)^2}{4\Sigma^2} + \frac{\dot{\Sigma}}{r\Sigma} - \frac{2\left(eKr^2+1\right)^2}{r^2} \\ \frac{2ef\left(eKr^2+1\right)}{\Sigma^2} \end{array} \right.$$
$$\left. \begin{array}{c} -4e\Phi\left(eKr^2+1\right) \\ -4ef\left(eKr^2+1\right) \\ -\frac{2e\Phi\left(eKr^2+1\right)}{\Sigma^2}\frac{e^2r^2\left(f^2-\Phi^2\right)}{\Sigma^2} + \frac{\dot{\Sigma}^2}{4\Sigma^2} - 3eK\left(eKr^2+2\right) - \frac{2}{r^2} - \frac{\ddot{\Sigma}}{2\Sigma} \end{array} \right| . \tag{7.99}$$

The eigenvalues Λ_i $(i=1,2,3)$ of the second invariant are given by the formulas

$$\Lambda_1 = -\frac{2\left(eKr^2+1\right)^2}{r^2} + \frac{\dot{\Sigma}}{r\Sigma} + \frac{\ddot{\Sigma}}{2\Sigma} - \frac{3\dot{\Sigma}^2}{4\Sigma^2}, \tag{7.100}$$

$$\Lambda_{2,3} = -\frac{1}{4\Sigma^2}\left(2e^2r^2\left(\Phi^2-f^2\right)+\dot{\Sigma}^2\right) - \frac{1}{2r^2}\left(5eKr^2\left(eKr^2+2\right)+4\right) + \frac{\dot{\Sigma}}{2r\Sigma}$$
$$\pm\frac{1}{2r\Sigma^2}\Big[\left(e^2r^3\left(f^2-\Phi^2\right)+r\dot{\Sigma}^2\right)^2 + r^2\Sigma^2\left(2e^2\left(\Phi^2-f^2\right)\left(17eKr^2\left(eKr^2+2\right)+16\right)+\ddot{\Sigma}^2\right)$$
$$+\Sigma^2\left(\dot{\Sigma}^2\left(1-2eKr^2\left(eKr^2+2\right)\right)+2r\dot{\Sigma}\ddot{\Sigma}\right) + 2r\Sigma\left(r\ddot{\Sigma}+\dot{\Sigma}\right)\left(e^2r^2\left(\Phi^2-f^2\right)-\left(\dot{\Sigma}\right)^2\right)$$
$$+e^2K^2r^2\Sigma^4\left(eKr^2+2\right)^2 + 2eKr\Sigma^3\left(eKr^2+2\right)\left(r\ddot{\Sigma}+\dot{\Sigma}\right)\Big]^{1/2}. \tag{7.101}$$

The next step is to construct a Lagrangian function L for the phase space $\{y_i,\, x_i\}$, such that the formulas for coefficients Q^i (7.93) hold true, and the dynamics of the system is defined by the equations (7.96). We will search for the function L in the form

$$L = g_{ij}(r)y^i y^j + b_j(r,x)y^j; \tag{7.102}$$

assuming the diagonal structure of the metrical tensor:

$$g_{ij}(r) = \begin{vmatrix} g_{11}(r) & 0 & 0 \\ 0 & g_{22}(r) & 0 \\ 0 & 0 & g_{33}(r) \end{vmatrix}. \tag{7.103}$$

In this case, by substituting (7.102)–(7.103) into (7.93), we derive

$$\begin{aligned} Q^1 &= \frac{1}{4g_{11}}\Big(2\dot{g}_{11}y^1 + \frac{\partial b_1}{\partial r} + \Big(\frac{\partial b_1}{\partial x^2} - \frac{\partial b_2}{\partial x^1}\Big)y^2 + \Big(\frac{\partial b_1}{\partial x^3} - \frac{\partial b_3}{\partial x^1}\Big)y^3\Big), \\ Q^2 &= \frac{1}{4g_{22}}\Big(2\dot{g}_{22}y^2 + \frac{\partial b_2}{\partial r} + \Big(\frac{\partial b_2}{\partial x^1} - \frac{\partial b_1}{\partial x^2}\Big)y^1 + \Big(\frac{\partial b_2}{\partial x^3} - \frac{\partial b_3}{\partial x^2}\Big)y^3\Big), \\ Q^3 &= \frac{1}{4g_{33}}\Big(2\dot{g}_{33}y^3 + \frac{\partial b_3}{\partial r} + \Big(\frac{\partial b_3}{\partial x^1} - \frac{\partial b_1}{\partial x^3}\Big)y^1 + \Big(\frac{\partial b_3}{\partial x^2} - \frac{\partial b_2}{\partial x^3}\Big)y^2\Big). \end{aligned} \tag{7.104}$$

Equating the terms from (7.97) to the corresponding terms from (7.104) and taking in mind that $x^1 = \Phi$, $x^2 = f$, $x^3 = K$, we obtain the system of equations with respect to the variables $g_{ij}(r)$ and $b_j(r,x)$:

$$\begin{aligned} &\frac{\dot{g}_{11}}{g_{11}} = \frac{4}{r} - \frac{\dot{\Sigma}}{\Sigma}, \quad \frac{\dot{g}_{22}}{g_{22}} = \frac{4}{r} - \frac{\dot{\Sigma}}{\Sigma}, \quad \frac{\dot{g}_{33}}{g_{33}} = \frac{4}{r} - \frac{\dot{\Sigma}}{\Sigma}, \\ &\frac{\partial b_1}{\partial f} - \frac{\partial b_2}{\partial \Phi} = 0, \quad \frac{\partial b_1}{\partial K} - \frac{\partial b_3}{\partial \Phi} = 0, \quad \frac{\partial b_2}{\partial K} - \frac{\partial b_3}{\partial f} = 0, \\ &\frac{1}{4g_1}\frac{\partial b_1}{\partial r} = -e\Phi K\left(2 + er^2K\right) - \frac{1}{2r}\frac{\dot{\Sigma}}{\Sigma}\Phi, \\ &\frac{1}{4g_2}\frac{\partial b_2}{\partial r} = -efK\left(2 + er^2K\right) - \frac{1}{2r}\frac{\dot{\Sigma}}{\Sigma}f, \\ &\frac{1}{2g_3}\frac{\partial b_3}{\partial r} = -eK^2\left(3 + er^2K\right) + \frac{2}{r}\frac{\dot{\Sigma}}{\Sigma}K - \frac{e(\Phi^2 - f^2)(1 + er^2K)}{\Sigma^2}. \end{aligned} \tag{7.105}$$

Its solution is given by the formulas

$$\begin{aligned} &g_{11} = \frac{C_1 r^4}{\Sigma}, \quad g_{22} = -\frac{C_1 r^4}{\Sigma}, \quad g_{33} = 2C_1 r^4 \Sigma, \\ &b_1 = B_1(\Phi, f, K) - 2C_1\Phi \int \frac{4eKr^4\Sigma(r) + 2e^2K^2r^6\Sigma(r) + r^3\dot{\Sigma}(r)}{\Sigma(r)^2} dr, \\ &b_2 = B_2(\Phi, f, K) + 2C_1 f \int \frac{4eKr^4\Sigma(r) + 2e^2K^2r^6\Sigma(r) + r^3\dot{\Sigma}(r)}{\Sigma(r)^2} dr, \\ &b_3 = B_3(\Phi, f, K) - 4C_1 \\ &\times \int \frac{1}{\Sigma(r)^2}\left(e(\Phi^2 - f^2)r^4(1 + eKr^2) + eK^2\Sigma(r)^2 r^4(3 + eKr^2) - 2Kr^3\Sigma(r)\dot{\Sigma}(r)\right) dr, \end{aligned} \tag{7.106}$$

where C_1 is an arbitrary constant, and the functions $B_i(\Phi, f, K)$ obey the following constraints

$$\frac{\partial B_1}{\partial f} - \frac{\partial B_2}{\partial \Phi} = 0, \qquad \frac{\partial B_2}{\partial K} - \frac{\partial B_3}{\partial f} = 0, \qquad \frac{\partial B_1}{\partial K} - \frac{\partial B_3}{\partial \Phi} = 0. \tag{7.107}$$

In accordance with the known theorem from (7.107), we conclude that this 3-dimensional vector field B_j can be written as a gradient of some scalar function $\varphi(x^1, x^2, x^3)$:

$$B_1 = \frac{\partial \varphi}{\partial x^1}, \quad B_2 = \frac{\partial \varphi}{\partial x^2}, \quad B_3 = \frac{\partial \varphi}{\partial x^3}, \quad B_i = \operatorname{grad} \varphi . \tag{7.108}$$

Therefore, there exists some gauge freedom in choosing the Lagrange function L referring to the dynamical system under consideration.

7.8 The Euclidean Space

Now, we consider the case of the Euclidean space, $\Sigma(r) = 1$, $\dot{\Sigma}(r) = 0$. In this case, the expressions for Q^i become simpler

$$Q^1 = -eK\Phi\left(eKr^2 + 2\right) + \frac{2\dot{\Phi}}{r}, \quad Q^2 = -efK\left(eKr^2 + 2\right) + \frac{2\dot{f}}{r},$$
$$Q^3 = \frac{1}{2}e\left(f^2 - \Phi^2\right)\left(eKr^2 + 1\right) - \frac{1}{2}eK^2\left(eKr^2 + 3\right) + \frac{2\dot{K}}{r}. \tag{7.109}$$

The first and the second invariants are simplified as well:

$$\varepsilon^1 = 2eK\Phi\left(eKr^2 + 2\right) - \frac{2}{r}\dot{\Phi}, \quad \varepsilon^2 = 2eKf\left(eKr^2 + 2\right) - \frac{2}{r}\dot{f},$$
$$\varepsilon^3 = e\left(\Phi^2 - f^2\right)\left(eKr^2 + 1\right) + eK^2\left(eKr^2 + 3\right) - \frac{2}{r}\dot{K}; \tag{7.110}$$

$$P^i{}_j = \left| \begin{array}{ccc} -\frac{2\left(eKr^2+1\right)^2}{r^2} & 0 & -4e\Phi\left(eKr^2 + 1\right) \\ 0 & -\frac{2\left(eKr^2+1\right)^2}{r^2} & -4ef\left(eKr^2 + 1\right) \\ -2e\Phi\left(eKr^2 + 1\right) & 2ef\left(eKr^2 + 1\right) & e^2r^2(f^2 - \Phi^2 - 3K^2) - 6eK - \frac{2}{r^2} \end{array} \right| . \tag{7.111}$$

The eigenvalues Λ_i of the second invariant are given by the formulas

$$\Lambda_1 = -\frac{2\left(eKr^2 + 1\right)^2}{r^2},$$
$$\Lambda_{2,3} = -\frac{2}{r^2} - 5eK + \frac{1}{2}e^2r^2(-f^2 + 5K^2 + \Phi^2) \pm \left[e^2f^4r^3 + K^2(2 + er^2K)^2\right.$$
$$\left. - 2f^2(16 + 17er^2K(2 + er^2K)) + \Phi^2(32 + 68er^2K - 2e^2r^4(f^2 - 17K^2)) + e^2r^4\Phi^4\right]^{1/2}. \tag{7.112}$$

The Lagrangian function L has the following explicit form

$$L = g_{ij}(r)y^iy^j + b_j(r,x)y^j, \tag{7.113}$$

where the metric tensor is

$$g_{ij} = \begin{vmatrix} C_1 r^4 & 0 & 0 \\ 0 & -C_1 r^4 & 0 \\ 0 & 0 & 2C_1 r^4 \end{vmatrix},$$

and the functions $b_j(r,x)$ are determined by the expressions

$$\begin{aligned} b_1 &= B_1(\Phi, f, K) - 4eC_1\Phi K r^5\Big(\frac{2}{5} + \frac{1}{7}er^2K\Big), \\ b_2 &= B_2(\Phi, f, K) + 4eC_1 f K r^5\Big(\frac{2}{5} + \frac{1}{7}er^2K\Big), \\ b_3 &= B_3(\Phi, f, K) - 4eC_1 r^5\Big(\Big(\frac{1}{5} + \frac{1}{7}er^2K\Big)\Big(\Phi^2 - f^2\Big) + K^2\Big(\frac{1}{7}er^2K + \frac{3}{5}\Big)\Big). \end{aligned} \tag{7.114}$$

7.9 Riemannian Space

Let us consider the spherical Riemannian space, where

$$\Sigma(r) = 1 + r^2/4, \quad \dot{\Sigma}(r) = r/2. \tag{7.115}$$

In this case, the quantities Q^i are given by the formulas

$$\begin{aligned} Q^1 &= -eK\Phi\left(eKr^2 + 2\right) + \frac{2\dot{\Phi}}{r} - \frac{\Phi + r\dot{\Phi}}{4 + r^2}, \\ Q^2 &= -efK\left(eKr^2 + 2\right) + \frac{2\dot{f}}{r} - \frac{f + r\dot{f}}{4 + r^2}, \\ Q^3 &= \frac{8e\left(f^2 - \Phi^2\right)\left(eKr^2 + 1\right)}{(4 + r^2)^2} - \frac{1}{2}eK^2\left(eKr^2 + 3\right) + \frac{2\dot{K}}{r} + \frac{2K + r\dot{K}}{4 + r^2}. \end{aligned} \tag{7.116}$$

The first and the second invariants are

$$\begin{aligned} \varepsilon^1 &= 2eK\Phi\left(eKr^2 + 2\right) + \frac{2\Phi}{4 + r^2} - \frac{8 + r^2}{r(4 + r^2)}\dot{\Phi}, \\ \varepsilon^2 &= 2eKf\left(eKr^2 + 2\right) + \frac{2f}{4 + r^2} - \frac{8 + r^2}{r(4 + r^2)}\dot{f}, \\ \varepsilon^3 &= \frac{16e\left(\Phi^2 - f^2\right)\left(eKr^2 + 1\right)}{(4 + r^2)^2} + eK^2\left(eKr^2 + 3\right) - \frac{4K}{4 + r^2} - \frac{8 + 3r^2}{r(4 + r^2)}\dot{K}; \end{aligned} \tag{7.117}$$

$$P^i{}_j =$$

$$\begin{vmatrix} -\frac{2\left(eKr^2+1\right)^2}{r^2} + \frac{12}{(4+r^2)^2} & 0 & -4e\Phi\left(eKr^2+1\right) \\ 0 & -\frac{2\left(eKr^2+1\right)^2}{r^2} + \frac{12}{(4+r^2)^2} & -4ef\left(eKr^2+1\right) \\ -32e\Phi\frac{eKr^2+1}{(4+r^2)^2} & 32ef\frac{eKr^2+1}{(4+r^2)^2} & -\frac{2}{r^2} - \frac{4}{(4+r^2)^2} - 6eK - 3e^2r^2K^2 + \frac{16e^2r^2(f^2-\Phi^2)}{(4+r^2)^2} \end{vmatrix}.$$

The eigenvalues Λ_i of the second invariant are given by the formulas

$$\begin{aligned} \Lambda_1 &= -\frac{2\left(eKr^2+1\right)^2}{r^2} + \frac{12}{(4+r^2)^2}, \\ \Lambda_{2,3} &= \frac{1}{2r^2(r^2+4)^2}\{r^2e^2\left(16r^2(f^2-\Phi^2)-5K^2(r^3+4r)^2\right) \\ &\quad -10eKr^2\left(r^2+4\right)^2 - 4r^2\left(r^2+6\right)-64 \\ &\quad \pm r^2\left[e^4K^4(r^3+4r)^4+64Ke(r^2+4)^2(17e^2r^2(\Phi^2-f^2)+1)\right. \\ &\quad +4e^2K^2(r^2+4)^2(r^4(136e^2(\Phi^2-f^2)+1) \\ &\quad +16(r^2+1))+256e^2(\Phi^2-f^2)(r^4(e^2(\Phi^2-f^2)+2)+18r^2+32) \\ &\quad \left.+4e^3K^3r^2(r^2+4)^4+256\right]^{1/2}\}. \end{aligned} \tag{7.118}$$

The Lagrangian function L can be found in the form

$$L = g_{ij}(r)y^iy^j + b_j(r,x)y^j, \tag{7.119}$$

where the metric tensor is

$$g_{ij} = C_1\begin{vmatrix} \frac{r^4}{4+r^2} & 0 & 0 \\ 0 & -\frac{r^4}{4+r^2} & 0 \\ 0 & 0 & \frac{1}{8}r^4(4+r^2) \end{vmatrix},$$

and the functions $b_j(r,x)$ are determined by the following expressions

$$\begin{aligned} b_1 &= B_1(\Phi,f,K) - 4C_1\Phi\Big\{\frac{2r}{4+r^2}+\frac{1}{5}e^2r^5K^2-\frac{2}{3}er^3K(2eK-1) \\ &\quad +r(4eK-1)^2+(3-16eK+32e^2K^2)\arctan\frac{2}{r}\Big\}, \\ b_2 &= B_2(\Phi,f,K) + 4C_1f\Big\{\frac{2r}{4+r^2}+\frac{1}{5}e^2r^5K^2-\frac{2}{3}er^3K(2eK-1) \\ &\quad +r(4eK-1)^2+(3-16eK+32e^2K^2)\arctan\frac{2}{r}\Big\}, \\ b_3 &= B_3(\Phi,f,K) - \frac{1}{4}C_1\Big\{\frac{1}{9}e^2r^9K^3+\frac{4}{5}r^5K(3eK-1)+\frac{1}{7}er^7K^2(3+4eK) \\ &\quad +\frac{16}{15}e(\Phi^2-f^2)\Big(3er^5K-5\Big(-12r+r^3-24\arctan\frac{2}{r}\Big)\Big(4eK-1\Big)\Big)\Big\}. \end{aligned} \tag{7.120}$$

7.10 Lobachevsky Space

Let us consider the space of negative constant curvature (the Lobachevsky model):

$$\Sigma(r) = 1 - r^2/4, \quad \dot{\Sigma}(r) = -r/2. \tag{7.121}$$

In this case, the quantities Q^i equal

$$
\begin{aligned}
Q^1 &= -eK\Phi\left(eKr^2+2\right)+\frac{2\dot{\Phi}}{r}-\frac{\Phi+r\dot{\Phi}}{r^2-4}\,,\\
Q^2 &= -efK\left(eKr^2+2\right)+\frac{2\dot{f}}{r}-\frac{f+r\dot{f}}{r^2-4}\,,\\
Q^3 &= \frac{8e\left(f^2-\Phi^2\right)\left(eKr^2+1\right)}{(r^2-4)^2}-\frac{1}{2}eK^2\left(eKr^2+3\right)+\frac{2\dot{K}}{r}+\frac{2K+r\dot{K}}{r^2-4}\,.
\end{aligned}
\tag{7.122}
$$

The first and the second invariants are

$$
\begin{aligned}
\varepsilon^1 &= 2eK\Phi\left(eKr^2+2\right)+\frac{2\Phi}{r^2-4}-\frac{r^2-8}{r(r^2-4)}\dot{\Phi}\,,\\
\varepsilon^2 &= 2eKf\left(eKr^2+2\right)+\frac{2f}{r^2-4}-\frac{r^2-8}{r(r^2-4)}\dot{f}\,,\\
\varepsilon^3 &= \frac{16e\left(\Phi^2-f^2\right)\left(eKr^2+1\right)}{(r^2-4)^2}+eK^2\left(eKr^2+3\right)-\frac{4K}{r^2-4}-\frac{3r^2-8}{r(r^2-4)}\dot{K}\,;
\end{aligned}
\tag{7.123}
$$

$$
P^i{}_j =
$$

$$
\begin{vmatrix}
-\frac{2\left(eKr^2+1\right)^2}{r^2}-\frac{12}{(r^2-4)^2} & 0 & -4e\Phi\left(eKr^2+1\right)\\
0 & -\frac{2\left(eKr^2+1\right)^2}{r^2}-\frac{12}{(r^2-4)^2} & -4ef\left(eKr^2+1\right)\\
-32e\Phi\frac{eKr^2+1}{(r^2-4)^2} & 32ef\frac{eKr^2+1}{(r^2-4)^2} & -\frac{2}{r^2}+\frac{4}{(r^2-4)^2}-6eK-3e^2r^2K^2+\frac{16e^2r^2(f^2-\Phi^2)}{(r^2-4)^2}
\end{vmatrix}.
$$

The eigenvalues Λ_i of the second invariant are given by the formulas

$$
\Lambda_1 = -\frac{2(eKr^2+1)^2}{r^2}-\frac{12}{(r^2-4)^2}\,,
$$

$$
\begin{aligned}
\Lambda_{2,3} = \frac{1}{2r^2(r^2-4)^2}\Big\{&r^2e^2(16r^2(f^2-\Phi^2)-5K^2(r^3-4r)^2)-10eKr^2(r^2-4)^2-4r^2(r^2-6)-64\\
&\pm r^2\Big[e^4K^4(r^3-4r)^4+64Ke(r^2-4)^2(17e^2r^2(\Phi^2-f^2)-1)+4e^2K^2(r^2-4)^2(r^4(136e^2(\Phi^2-f^2)+1)\\
&-16(r^2-1))+256e^2(\Phi^2-f^2)(r^4(e^2(\Phi^2-f^2)+2)-18r^2+32)+4e^3K^3r^2(r^2-4)^4+256\Big]^{1/2}\Big\}.
\end{aligned}
$$

The Lagrangian function L is

$$
L = g_{ij}(r)y^iy^j+b_j(r,x)y^j\,,
\tag{7.124}
$$

where the metric tensor is

$$
g_{ij} = C_1\begin{vmatrix}
-\frac{r^4}{r^2-4} & 0 & 0\\
0 & \frac{r^4}{r^2-4} & 0\\
0 & 0 & -\frac{1}{8}r^4(r^2-4)
\end{vmatrix},
$$

and the functions $b_j(r,x)$ are determined as

$$
\begin{aligned}
b_1 &= B_1(\Phi,f,K)+4C_1\Phi\Big\{-\frac{2r}{r^2-4}+\frac{1}{5}e^2r^5K^2+\frac{2}{3}er^3K(2eK+1) \\
&\qquad +r(4eK+1)^2-(3+16eK+32e^2K^2)\operatorname{arctanh}\frac{r}{2}\Big\}, \\
b_2 &= B_2(\Phi,f,K)-4C_1f\Big\{-\frac{2r}{r^2-4}+\frac{1}{5}e^2r^5K^2+\frac{2}{3}er^3K(2eK+1) \\
&\qquad +r(4eK+1)^2-(3+16eK+32e^2K^2)\operatorname{arctanh}\frac{r}{2}\Big\}, \\
b_3 &= B_3(\Phi,f,K)+\frac{1}{4}C_1\Big\{\frac{1}{9}e^2r^9K^3-\frac{4}{5}r^5K(3eK+1)+\frac{1}{7}er^7K^2(3-4eK) \\
&+\frac{16}{15}e(\Phi^2-f^2)\Big(3er^5K+5\Big(12r+r^3-24\operatorname{arctanh}\frac{r}{2}\Big)\Big(4eK+1\Big)\Big)\Big\}.
\end{aligned} \tag{7.125}
$$

7.11 Pure Monopole BPZ-Solution, Euclidean Space

The exact solutions of the radial Yang – Mills equations in the BPZ-limit are known (for simplicity we restrict ourselves to the pure monopole case, setting $f(r)=0$):

$$
\begin{aligned}
\ddot{\Phi}+\frac{4}{r}\dot{\Phi}-2e\Phi\,(2+er^2K)\,K=0, \\
\ddot{K}+\frac{4\dot{K}}{r}-e\,\Phi^2\,(1+er^2K)-eK^2\,(3+er^2K)=0.
\end{aligned} \tag{7.126}
$$

Instead of $\Phi(r)$ and $K(r)$, it is more convenient to use the variables f_1 and f_2:

$$
1+er^2K(r)=rf_1(r), \qquad 1+er^2\Phi(r)=rf_2(r); \tag{7.127}
$$

in this way we transform equations (7.126) to a simpler form

$$
\begin{aligned}
2(\dot{f}_2+f_1^2)+r(\ddot{f}_2-2f_1^2f_2)=0, \\
2(\dot{f}_1+f_1f_2)+r(\ddot{f}_1-f_1f_2^2-f_1^3)=0.
\end{aligned} \tag{7.128}
$$

In these variables, the quantities Q^i (7.93) and the first KCC-invariants ε (7.94) read

$$
Q_f^1=-\frac{1}{2}f_1(f_1^2+f_2^2)+\frac{1}{r}(\dot{f}_1+f_1f_2), \quad Q_f^2=\frac{1}{r}(f_1^2+\dot{f}_2)-f_1^2f_2; \tag{7.129}
$$

$$
\varepsilon_f^1=-\frac{2f_2f_1}{r}+f_1^3+f_2^2f_1-\frac{\dot{f}_1}{r}, \quad \varepsilon_f^2=-\frac{2f_1^2}{r}+2f_2f_1^2-\frac{\dot{f}_2}{r}. \tag{7.130}
$$

The second invariant P_j^i (7.94) is given by the formula

$$
P_j{}^i=\begin{vmatrix} (\frac{2}{r}f_2-f_2^2-3f_1^2) & (\frac{2}{r}f_1-2f_1f_2) \\ (\frac{4}{r}f_1-4f_1f_2) & -2f_1^2 \end{vmatrix}. \tag{7.131}
$$

The eigenvalues of the second invariant are

$$\Lambda_{f\,(1,2)} = \frac{1}{2r}\Big(\pm\sqrt{f_1^4 r^2 + f_1^2\,(34 f_2 r\,(f_2 r - 2) + 32) + f_2^2\,(f_2 r - 2)^2}$$
$$- (5 f_1^2 + f_2^2)\, r + 2 f_2\Big)\,. \qquad (7.132)$$

To analyze the behavior of these eigenvalues, we will apply the known solutions of equations (7.128) in BPZ-limit. To this end, we search for the functions f_1 and f_2, which obey four equations:

$$\frac{df_2}{dr} + f_1^2 = 0\,, \qquad \frac{df_2}{dr} - 2\, f_1^2\, f_2 = 0\,,$$
$$\frac{df_1}{dr} + f_1\, f_2 = 0\,, \qquad \frac{df_1}{dr} - f_1\, f_2^2 - f_1^3 = 0\,. \qquad (7.133)$$

It is readily seen that in (7.133), the second and the fourth equations are consequences of the first and of the third ones, so we have only two independent equations

$$\frac{df_1}{dr} = -f_1\, f_2\,, \quad \frac{df_2}{dr} = -f_1^2 \;\Longrightarrow\; f_2 = -\frac{1}{f_1}\frac{df_1}{dr}\,, \quad \frac{d}{dr}\Big(\frac{1}{f_1}\frac{df_1}{dr}\Big) = f_1^2\,. \qquad (7.134)$$

Therefore, the problem reduces to a single equation

$$\Big(\frac{f_1'}{f_1}\Big)' = f_1^2 \qquad \Longrightarrow \qquad (\ln f_1)'' = f_1^2\,,$$

the last being equivalent to

$$\frac{d}{dr}\,\big[\,(\ln f_1)'\,\big]^2 = \frac{d}{dr}\, f_1^2\,,$$

whence we obtain

$$(\ln f_1)' = \pm\sqrt{f_1^2 + c} \;\Longrightarrow\; \int \frac{d\,f_1}{f_1\sqrt{c + f_1^2}} = \pm\,(r + \text{const})\,.$$

Depending on the sign of the constant c, we have three different types of solutions:

$$I \quad c = 0 \;\Longrightarrow\; f_1 = \pm\frac{A}{Ar + B}\,, \quad f_2 = \frac{A}{Ar + B}\,;$$
$$II \quad c < 0 \;\Longrightarrow\; f_1 = \pm\frac{A}{\text{sh}\,(Ar + B)}\,, \quad f_2 = \frac{A}{\text{th}\,(Ar + B)}\,; \qquad (7.135)$$
$$III \quad c > 0 \;\Longrightarrow\; f_1 = \pm\frac{A}{\sin\,(Ar + B)}\,, \quad f_2 = \frac{A}{\text{tg}\,(Ar + B)}\,,$$

where A and B are arbitrary constants. By physical reasons, we should set $B = 0$, and then the solution will contain only one singular point, $r = 0$. For generality, we shall follow below all the three possibilities: $B = 0, B < 0, B > 0$.

It should be noted only the relative sign for the parameters A and B is significant; for this reason, we will consider only two cases:

$$A > 0, B > 0, \quad \text{and} \quad A > 0, B < 0.$$

In the case I, the eigenvalues (7.132) take the form

$$\Lambda_{f\,1} = -\frac{2A}{Ar^2 + Br}, \quad \Lambda_{f\,2} = -\frac{2A(Ar - 2B)}{r(Ar+B)^2}; \tag{7.136}$$

$$B \neq 0, \qquad \begin{array}{c} r \to 0, \quad \Lambda_{f\,1} \to -2A/Br, \; \Lambda_{f\,2} \to +4A/Br; \\ r \to \infty, \quad \Lambda_{f\,1} \to -2/r^2, \; \Lambda_{f\,2} \to -2/r^2. \end{array} \tag{7.137}$$

When $B \neq 0$, the behavior of $\Lambda_{f\,(1,2)}(r)$ (7.138) is illustrated in figure 7.1a,c.

When $B = 0$, the solutions and the eigenvalues simplify,

$$f_1 = \pm\frac{1}{r}, \quad f_2 = \frac{1}{r}, \quad \Lambda_{f\,1} = \Lambda_{f\,2} = -\frac{2}{r^2} < 0.$$

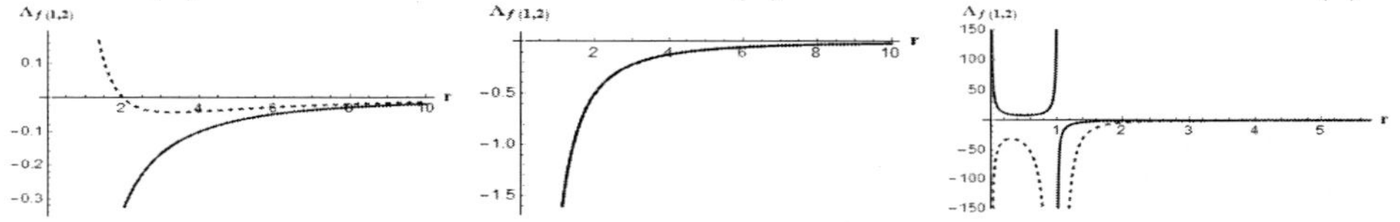

Figure 7.1: The dependencies of eigenvalues $\Lambda_{f\,1}(r)$ (solid line) and $\Lambda_{f\,2}$ (dashed line) in Euclidean space for the solution of the type I at $B = 1$ (a), $B = 0$ (b) and $B = -1$ (c). $A = 1$.

Now, let us consider solutions of the type II. In this case, the eigenvalues (7.132) take the form (recall the notation csch $= 1/\sinh$)

$$\Lambda_{f\,(1,2)} = \pm\frac{A}{2r}\Big[A^2r^2\text{csch}^4(Ar+B) + \coth^2(Ar+B)\,[Ar\coth(Ar+B) - 2]^2$$
$$+ [34Ar\coth(Ar+B)\,[Ar\coth(Ar+B) - 2] + 32]\,\text{csch}^2(Ar+B)\Big]^{1/2} \tag{7.138}$$
$$-\frac{1}{2}A^2\left[6\text{csch}^2(Ar+B) + 1\right] + \frac{A\coth(Ar+B)}{r}.$$

They become much simpler when $B = 0$:

$$\Lambda_{f\,(1,2)} = \pm\frac{A}{2r}\Big[A^2r^2 + \text{csch}^2(Ar)[2Ar\coth(Ar)[-\cosh(2Ar)$$

$$+18Ar\coth(Ar) - 35] + 36] + 4\Big]^{1/2} - 3A^2\text{csch}^2(Ar) - \frac{A^2}{2} + \frac{A\coth(Ar)}{r}.$$

Let us specify their behavior near the points $r=0,\infty$. When $B=0$, we have

$$r\to 0,\quad \Lambda_{f\,1}=\Lambda_{f\,2}=-\frac{2}{r^2},\qquad r\to\infty,\quad \Lambda_{f\,1}=-A^2,\quad \Lambda_{f\,2}=\frac{A}{r}. \tag{7.139}$$

If $B\neq 0$, we get

$$\begin{gathered} r\to 0,\quad \Lambda_{f\,(1,2)}=\frac{A}{r}\left(\coth B\pm\sqrt{\coth^2 B+8\operatorname{csch}^2 B}\right),\\ r\to\infty,\quad \Lambda_{f\,1}=-A^2,\quad \Lambda_{f\,2}=\frac{A}{r}. \end{gathered} \tag{7.140}$$

In the case III, the eigenvalues (7.132) are

$$\begin{aligned} \Lambda_{f\,(1,2)}=\pm\frac{A}{2r}\Big[&A^2r^2\csc^4(Ar+B)+\cot^2(Ar+B)\,[Ar\cot(Ar+B)-2]^2\\ &+[34Ar\cot(Ar+B)\,[Ar\cot(Ar+B)-2]+32]\csc^2(Ar+B)\Big]^{1/2}\\ &-3A^2\csc^2(Ar+B)+\frac{A^2}{2}+\frac{A\cot(Ar+B)}{r}; \end{aligned} \tag{7.141}$$

when $B=0$, they become simpler

$$\Lambda_{f\,(1,2)}=\pm\frac{A}{2r}\Big[A^2r^2\csc^4 Ar+\cot^2 Ar\,[Ar\cot Ar-2]^2$$

$$+[34Ar\cot Ar\,[Ar\cot Ar-2]+32]\csc^2 Ar\Big]^{1/2}-3A^2\csc^2 Ar+\frac{A^2}{2}+\frac{A\cot Ar}{r}.$$

The expression of the Lagrangian is as follows (the arbitrary constant has been taken as 1):

$$\begin{aligned} L=\left(2\dot f_1^2+\dot f_2^2\right)r^2+\frac{2}{3}f_1r^2\left(-2\left(f_1^2+f_2^2\right)\dot f_1 r+f_1 f_2\left(3-2f_2r\right)+6f_2\dot f_1\right)\\ +\dot f_1 B_1\left(f_1,f_2\right)+\dot f_2 B_2\left(f_1,f_2\right); \end{aligned} \tag{7.142}$$

the functions B_1 and B_2 obey the following restriction

$$\frac{\partial B_1}{\partial f_2}-\frac{\partial B_2}{\partial f_1}=0.$$

So, the vector field B_1,B_2 can be presented as a gradient of some scalar function $B_i=\operatorname{grad}\varphi(f_1,f_2)$.

The substitution of the explicit solutions (7.135) leads to the following results:

$$I\quad L=\frac{A^4r^2(Ar-3B)}{(Ar+B)^5},$$

$$II\quad L=\frac{1}{6}A^4r^2\operatorname{csch}^5(Ar+B)(2Ar(11\cosh(Ar+B)$$

$$+\cosh(3(Ar+B)))-3(3\sinh(Ar+B)+\sinh(3(Ar+B)))),$$

$$III \;\; L = \frac{1}{6} A^4 r^2 \csc^5(Ar+B)(2Ar(11\cos(Ar+B) + \cos(3(Ar+B))) - 3(3\sin(Ar+B) + \sin(3(Ar+B)))) .$$

When $B = 0$, the formulas become simpler

$$I \;\; L = \frac{1}{r^2},$$

$$II \;\; L = \frac{1}{6} A^4 r^2 \mathrm{csch}^5(Ar)(2Ar(11\cosh(Ar) + \cosh(3Ar)) - 3(3\sinh(Ar) + \sinh(3Ar))),$$

$$III \;\; L = \frac{1}{6} A^4 r^2 \csc^5(Ar)(2Ar(11\cos(Ar) + \cos(3Ar)) - 3(3\sin(Ar) + \sin(3Ar))) .$$

Their behavior in the physically meaningful case I is illustrated by figure 7.2.

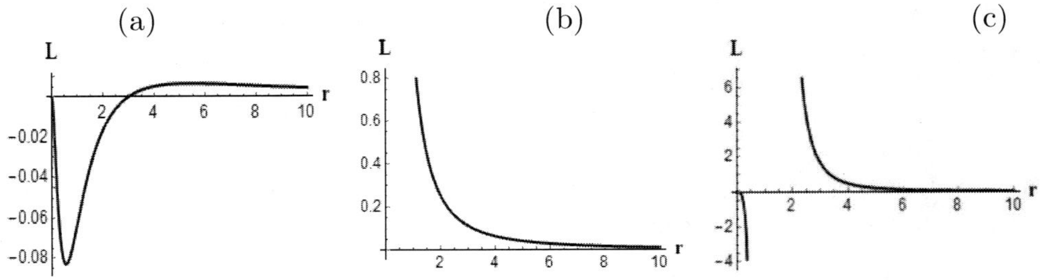

Figure 7.2: The dependencies of Lagrangians on radial coordinate r, in Euclidean space for the solution of the type I; $A = 1$, $B = +1$ (a), $B = 0$ (b) and $B = -1$ (c).

7.12 Geometrizing the Doublet Problem, the Case $j = 0$

We start with the system of equations (7.48):

$$\left(\frac{d}{dr} + \frac{W}{r}\right) f + 2M\, g = 0, \quad \left(\frac{d}{dr} - \frac{W}{r}\right) g - E\, f = 0 .$$

We differentiate them, and substitute the expressions for the derivatives f' and g' from the initial equations. In this way, we get

$$\left[\frac{d^2}{dr^2} + 2ME + \left(\frac{W}{r}\right)' - \frac{W^2}{r^2}\right] f(r) = 0,$$
$$\left[\frac{d^2}{dr^2} + 2ME - \left(\frac{W}{r}\right)' - \frac{W^2}{r^2}\right] g(r) = 0. \quad (7.143)$$

It is easy to find the explicit form of the coefficients Q^i (we apply the notations $x^1 \equiv f$, $x^2 \equiv g$):

$$Q^1 = \left[ME + \frac{1}{2}\left(\frac{W}{r}\right)' - \frac{1}{2}\frac{W^2}{r^2}\right] x^1 , \quad (7.144)$$

$$Q^2 = \left[ME - \frac{1}{2}\left(\frac{W}{r}\right)' - \frac{1}{2}\frac{W^2}{r^2}\right] x^2 . \quad (7.145)$$

The direct calculation of the second invariant gives

$$P^i_j = \left| \begin{array}{cc} 2ME + (\frac{W}{r})' - \frac{W^2}{r^2} & 0 \\ 0 & 2ME - (\frac{W}{r})' - \frac{W^2}{r^2} \end{array} \right| . \tag{7.146}$$

The diagonal elements of the matrix (7.146) are the eigenvalues of the second invariant. Let us study the behavior of the eigenvalue in the singular points $r = 0$, $r = \infty$ for the six different variants of the function $W(r)$:

$$W = \pm 1 , \quad W = \pm \frac{Ar}{\operatorname{sh} Ar} , \quad W = \pm \frac{Ar}{\sin Ar} . \tag{7.147}$$

Each pair of the solution for $W(r)$ has the same two eigenvalues, so the corresponding eigenvalues are

$$\begin{array}{ll} W = \pm 1 , & \Lambda_1 = 2EM , \quad \Lambda_2 = 2EM - \dfrac{2}{r^2} ; \\ W = \pm \dfrac{Ar}{\operatorname{sh} Ar} , & \Lambda_{1,2} = 2EM + \dfrac{A^2}{1 \pm \cosh(Ar)} ; \\ W = \pm \dfrac{Ar}{\sin Ar} , & \Lambda_{1,2} = 2EM - \dfrac{A^2}{1 \pm \cos(Ar)} . \end{array} \tag{7.148}$$

Near the singular points, in the case $W = \pm 1$, the eigenvalues behave as follows:

$$r \to 0, \ \Lambda_1 = 2EM , \ \Lambda_2 = -\frac{2}{r^2} ; \qquad r \to \infty, \ \Lambda_{1,2} = 2EM . \tag{7.149}$$

In the case $W = \pm \frac{Ar}{\operatorname{sh} Ar}$, the behavior of eigenvalues is given by:

$$r \to 0 , \ \Lambda_1 = 2EM + \frac{1}{2} A^2 , \ \Lambda_2 = -\frac{2}{r^2} ; \qquad r \to \infty , \ \Lambda_{1,2} = 2EM . \tag{7.150}$$

Now we will construct a Lagrangian function for the dynamical system (7.143), in the following form:

$$L = g_{ij}(r) y^i y^j + b_j(r,x) y^j , \quad g_{ij}(r) = \left| \begin{array}{cc} g_{11}(r) & 0 \\ 0 & g_{22}(r) \end{array} \right| . \tag{7.151}$$

The coefficients Q^i take the form

$$\begin{aligned} Q^1 &= \frac{1}{2g_{11}} \frac{dg_{11}}{dr} y^1 + \frac{1}{4g_{11}} \Big(\frac{\partial b_1}{\partial x^2} - \frac{\partial b_2}{\partial x^1} \Big) y^2 + \frac{1}{4g_{11}} \frac{\partial b_1}{\partial r} , \\ Q^2 &= \frac{1}{2g_{22}} \frac{dg_{22}}{dr} y^2 - \frac{1}{4g_{22}} \Big(\frac{\partial b_1}{\partial x^2} - \frac{\partial b_2}{\partial x^1} \Big) y^1 + \frac{1}{4g_{22}} \frac{\partial b_2}{\partial r} . \end{aligned} \tag{7.152}$$

Following the same considerations as previously performed for the monopole problem, we equate the coefficients (7.152) and (7.145), producing the equations:

$$\begin{aligned} & \frac{dg_{11}}{dr} = 0, \quad \frac{dg_{22}}{dr} = 0 , \quad \frac{\partial b_1}{\partial x^2} - \frac{\partial b_2}{\partial x^1} = 0 , \\ & \frac{1}{4g_{11}} \frac{\partial b_1}{\partial r} = x^1 \Big[ME + \frac{1}{2} \Big(\frac{W}{r} \Big)' - \frac{1}{2} \frac{W^2}{r^2} \Big] , \\ & \frac{1}{4g_{22}} \frac{\partial b_2}{\partial r} = x^2 \Big[ME - \frac{1}{2} \Big(\frac{W}{r} \Big)' - \frac{1}{2} \frac{W^2}{r^2} \Big] . \end{aligned} \tag{7.153}$$

The solution of this system is

$$
\begin{aligned}
g_{11} &= C_1, \qquad b_1 = B_1(x^1,x^2) + 2C_1 x^1 \Big(2MEr + \frac{W}{r} - \int \frac{W^2}{r^2} dr\Big), \\
g_{22} &= C_2, \qquad b_2 = B_2(x^1,x^2) + 2C_2 x^2 \Big(2MEr - \frac{W}{r} - \int \frac{W^2}{r^2} dr\Big)
\end{aligned}
\tag{7.154}
$$

with the restriction

$$
\frac{\partial B_1}{\partial x^2} - \frac{\partial B_2}{\partial x^1} = 0.
$$

For the particular case of the pure monopole solutions, we get the Lagrangians (we choose the constants C_1, C_2 equal to 1):

$W = \pm 1$,

$$
L = (y^1)^2 + (y^2)^2 + 2\Big(2EMr + \frac{1\pm 1}{r}\Big)x^1 y^1
$$
$$
+2\Big(2EMr + \frac{1\mp 1}{r}\Big)x^2 y^2 + B_1 y^1 + B_2 y^2;
$$

$W = \frac{Ar}{\sinh Ar}$,

$$
L = (y^1)^2 + (y^2)^2 + 2\Big(2EMr + 2A \coth \frac{AM}{2}\Big)x^1 y^1
$$
$$
+2\Big(2EMr + 2A \tanh \frac{AM}{2}\Big)x^2 y^2 + B_1 y^1 + B_2 y^2;
$$

$W = -\frac{Ar}{\sinh Ar}$,

$$
L = (y^1)^2 + (y^2)^2 + 2\Big(2EMr + 2A \tanh \frac{AM}{2}\Big)x^1 y^1
$$
$$
+2\Big(2EMr + 2A \coth \frac{AM}{2}\Big)x^2 y^2 + B_1 y^1 + B_2 y^2;
$$

$W = \frac{Ar}{\sin Ar}$,

$$
L = (y^1)^2 + (y^2)^2 + 2\Big(2EMr + 2A \cot \frac{AM}{2}\Big)x^1 y^1
$$
$$
+2\Big(2EMr + 2A \tan \frac{AM}{2}\Big)x^2 y^2 + B_1 y^1 + B_2 y^2;
$$

$W = -\frac{Ar}{\sin Ar}$,

$$
L = (y^1)^2 + (y^2)^2 + 2\Big(2EMr + 2A \tan \frac{AM}{2}\Big)x^1 y^1
$$
$$
+2\Big(2EMr + 2A \cot \frac{AM}{2}\Big)x^2 y^2 + B_1 y^1 + B_2 y^2.
$$

The Hamiltonian can be found from the Lagrangian according the formula

$$
H = y^i \frac{\partial L}{\partial y^i} - L. \tag{7.155}
$$

Using the explicit form of Lagrangian, we obtain:

$$H = C_1(y^1)^2 + C_2(y^2)^2 = C_1\left(\frac{df}{dr}\right)^2 + C_2\left(\frac{dg}{dr}\right)^2 \tag{7.156}$$

for all the monopole solutions. This structure (7.156) is formally similar to the expression for kinetic energy of a free particle in classical mechanics.

7.13 Non-Relativistic Approximation, the Case $j > 0$

Now we consider the system of equation (7.63)

$$\frac{d}{dr}f + EF + \frac{\nu}{r}g = 0\,, \quad \frac{d}{dr}F + \frac{\nu}{r}G - 2Mf = 0\,,$$
$$-\frac{d}{dr}g + EG - \frac{\nu}{r}f + \frac{W}{r}g = 0, \quad \frac{d}{dr}G + \frac{\nu}{r}F + \frac{W}{r}G + 2Mg = 0\,.$$

Using the new notations $x^1 \equiv f$, $x^2 \equiv g$, $x^3 \equiv F$, $x^4 \equiv G$, and performing the same procedure as in the previous case, we find the coefficients Q^i:

$$\begin{aligned} Q^1 &= -\left(\frac{\nu^2}{2r^2} - EM\right)x^1 + \frac{\nu(W-1)}{2r^2}x^2\,,\\ Q^2 &= -\left(\frac{\nu^2}{2r^2} - EM + \frac{W^2}{2r^2} + \frac{1}{2}\left(\frac{W}{r}\right)'\right)x^2 + \frac{\nu(W-1)}{2r^2}x^1\,,\\ Q^3 &= -\left(\frac{\nu^2}{2r^2} - EM\right)x^3 - \frac{\nu(W+1)}{2r^2}x^4\,,\\ Q^4 &= -\left(\frac{\nu^2}{2r^2} - EM + \frac{W^2}{2r^2} - \frac{1}{2}\left(\frac{W}{r}\right)'\right)x^4 - \frac{\nu(W+1)}{2r^2}x^3\,. \end{aligned} \tag{7.157}$$

The second invariant reads

$$P^i_j = \begin{vmatrix} 2EM - \frac{\nu^2}{r^2} & \frac{\nu(W-1)}{r^2} & 0 & 0 \\ \frac{\nu(W-1)}{r^2} & 2EM - \frac{\nu^2}{r^2} - \frac{W^2}{r^2} - \left(\frac{W}{r}\right)' & 0 & 0 \\ 0 & 0 & 2EM - \frac{\nu^2}{r^2} & -\frac{\nu(W+1)}{r^2} \\ 0 & 0 & -\frac{\nu(W+1)}{r^2} & 2EM - \frac{\nu^2}{r^2} - \frac{W^2}{r^2} + \left(\frac{W}{r}\right)' \end{vmatrix}.$$

The four eigenvalues of the invariant are

$$\Lambda_{1,2} = 2EM + \frac{1}{2}\left(\frac{W}{r}\right)' - \frac{\nu^2}{r^2} - \frac{W^2}{2r^2} \pm \sqrt{\frac{\nu^2(W+1)^2}{r^4} + \frac{1}{4}\left(\frac{W^2}{r^2} - \left(\frac{W}{r}\right)'\right)^2}\,,$$

$$\Lambda_{3,4} = 2EM - \frac{1}{2}\left(\frac{W}{r}\right)' - \frac{\nu^2}{r^2} - \frac{W^2}{2r^2} \pm \sqrt{\frac{\nu^2(W-1)^2}{r^4} + \frac{1}{4}\left(\frac{W^2}{r^2} + \left(\frac{W}{r}\right)'\right)^2}\,.$$

For the solution $W = \pm 1$, the eigenvalues take the form:

$$\Lambda_{1,2} = 2EM - \frac{\nu^2 + 1 \pm \sqrt{4\nu^2+1}}{r^2}\,, \qquad \Lambda_{3,4} = 2EM - \frac{\nu^2}{r^2}\,. \tag{7.158}$$

Near the singular points $r=0$ and $r=\infty$, they behave as follows:

$$r\to 0,\ \Lambda_{1,2}=-\frac{\nu^2+1\pm\sqrt{4\nu^2+1}}{r^2}\,,\ \Lambda_{3,4}=-\frac{\nu^2}{r^2}\,;\quad r\to\infty,\ \Lambda_{1,2,3,4}=2EM\,. \tag{7.159}$$

The other solutions for W lead to rather complicated expressions for eigenvalues, that cannot be analytically analyzed.

The Lagrangian function is searched in the form

$$L=g_{ij}(r)y^iy^j+b_j(r,x)y^j\,,\quad g_{ij}(r)=\left|\begin{array}{cccc} g_{11}(r) & 0 & 0 & 0\\ 0 & g_{22}(r) & 0 & 0\\ 0 & 0 & g_{33}(r) & 0\\ 0 & 0 & 0 & g_{44}(r)\end{array}\right|\,. \tag{7.160}$$

Equating the coefficients Q^i obtained from the formulas (7.93) and (7.157), we get the coefficients of the Lagrangian:

$$g_{11}=C_1\,,\quad g_{22}=C_2\,,\quad g_{33}=C_3\,,\quad g_{44}=C_4\,,$$

$$b_1=B_1(x^i)+2C_1\Big(2EMrx^1+\frac{\nu^2}{r}x^1+\frac{\nu}{r}x^2+\nu x^2\int\frac{W}{r^2}dr\Big)\,,$$

$$b_2=B_2(x^i)+2C_2\Big(2EMrx^2+\frac{\nu^2}{r}x^2+\frac{\nu}{r}x^1-\frac{W}{r}x^2+\nu x^1\int\frac{W}{r^2}dr-x^2\int\frac{W^2}{r^2}dr\Big)\,,$$

$$b_3=B_3(x^i)+2C_3\Big(2EMrx^3+\frac{\nu^2}{r}x^3+\frac{\nu}{r}x^4-\nu x^4\int\frac{W}{r^2}dr\Big)\,,$$

$$b_4=B_4(x^i)+2C_4\Big(2EMrx^4+\frac{\nu^2}{r}x^4+\frac{\nu}{r}x^3+\frac{W}{r}x^4+\nu x^3\int\frac{W}{r^2}dr-x^4\int\frac{W^2}{r^2}dr\Big)\,.$$

The gauge condition reads

$$\frac{\partial B_i}{\partial x^j}-\frac{\partial B_j}{\partial x^i}=0\,.$$

For the monopole solution $W=\pm1$ (16.52), we obtain the following Lagrangians (by setting $C_1=C_2=C_3=C_4=1$):

$W=1$,

$$L=(y^1)^2+(y^2)^2+(y^3)^2+(y^4)^2+\frac{2}{r}\Big(2EMr^2+\nu^2\Big)\Big(x^1y^1+x^2y^2+x^3y^3+x^4y^4\Big)$$

$$+\frac{4\nu}{r}\Big(x^4y^3+x^3y^4\Big)+\frac{4x^4y^4}{r}+B_1y^1+B_2y^2+B_3y^3+B_4y^4\,;$$

$W=-1$,

$$L=(y^1)^2+(y^2)^2+(y^3)^2+(y^4)^2+\frac{2}{r}\Big(2EMr^2+\nu^2\Big)\Big(x^1y^1+x^2y^2+x^3y^3+x^4y^4\Big)$$

$$+\frac{4\nu}{r}\Big(x^2y^1+x^1y^2\Big)+\frac{4x^2y^2}{r}+B_1y^1+B_2y^2+B_3y^3+B_4y^4\,.$$

The corresponding Hamiltonian is the same for all the monopole solutions:

$$H=C_1\Big(\frac{df}{dr}\Big)^2+C_2\Big(\frac{dg}{dr}\Big)^2+C_3\Big(\frac{dF}{dr}\Big)^2+C_4\Big(\frac{dG}{dr}\Big)^2\,. \tag{7.161}$$

Chapter 8

To Analysis of the Dirac and Majorana Particle in Schwarzschild Field

8.1 Dirac and Weyl Equations in External Gravitational Field

The initial idea in the subject under consideration appeared many years ago in the famous paper by T. Regge and John A. Wheeler [167].

The abstract of the paper is: It is shown that a Schwarzschild singularity, spherically symmetrical and endowed with mass, will undergo small vibrations about the spherical form and will therefore remain stable if subjected to a small nonspherical perturbation. In fact, this paper contains an (accessory) by-product: namely, the wave equation describing the linearized spin 2 field on the background of undisturbed Schwarzschild background [170] (see also the classical papers [168] and [169]). Linear differential equations were derived for the small first-order departures from the Schwarzschild metric; these equations ensure that the perturbed field will satisfy the Einstein equations and represent a mass-free space of nearly spherical symmetry. It turns out that the equations for disturbances can be presented in the form of Schrödinger-lire equation with effective potential of a barrier type. A much more detailed development of the subject was done by Chandrasekhar[171]. Renewed interest and attention were given to this problem after establishing the black hole radiation by [172–175] and many other authors. The Bibliography is enormous, see more references in [176, 177]. At present time, the subject has become much more clear. In particular, for scalar, spin 1/2 and electromagnetic fields, the relevant mathematical tasks are reduced to an ordinary 2-nd order differential equation of the Heun class, or more complicated ones [3, 4].

In the present chapter we turn to the case of the spinor field (mainly the massless one), and perform the general mathematical study of the tunneling process through effective potential barrier generated by Schwarzschild black hole background. The study will be based on the use of 8 Frobenius solutions of the related 2-nd order differential equations with nonregular singularities of the rank 2. The obtained results are valid for massless Dirac and Majorana particles.

The Dirac equation in presence of external gravitational fields takes the form (we use the notations from [139])

$$[i\gamma^{\alpha}(x)\,(\partial_{\alpha}+\Gamma_{\alpha}(x)\,)-m]\,\Psi(x)=0\,, \tag{8.1}$$

where

$$\gamma^{\alpha}(x)=\gamma^{a}\,e^{\alpha}_{(a)}(x)\,,\quad \Gamma_{\alpha}(x)=\frac{1}{2}\,\sigma^{ab}\,e^{\beta}_{(a)}\,\nabla_{\alpha}\,(\,e_{(b)\beta}\,)\,; \tag{8.2}$$

where $e^{\alpha}_{(a)}(x)$ is the tetrad, $\Gamma_{\alpha}(x)$ is the bispinor connection, and ∇_{α} stands for the covariant derivative. In the spinor basis of Dirac matrices

$$\psi(x)=\left|\begin{array}{c}\xi(x)\\ \eta(x)\end{array}\right|,\quad \gamma^{a}=\left|\begin{array}{cc}0 & \bar{\sigma}^{a}\\ \sigma^{a} & 0\end{array}\right|,\quad \xi(x)=\left|\begin{array}{c}\xi^{1}(x)\\ \xi^{2}(x)\end{array}\right|,\quad \eta(x)=\left|\begin{array}{c}\eta_{\dot{1}}(x)\\ \eta_{\dot{2}}(x)\end{array}\right|, \tag{8.3}$$

where the Pauli matrices are $\sigma^{a}=(I,+\sigma^{k})$, $\bar{\sigma}^{a}=(I,-\sigma^{k})$, equation (8.1) is written in 2-component form:

$$i\,\sigma^{\alpha}(x)\,[\,\partial_{\alpha}+\Sigma_{\alpha}(x)\,]\,\xi(x)=m\,\eta(x)\,,\quad i\,\bar{\sigma}^{\alpha}(x)\,[\,\partial_{\alpha}+\bar{\Sigma}_{\alpha}(x)\,]\,\eta(x)=m\,\xi(x)\,; \tag{8.4}$$

where the following notations are used:

$$\begin{gathered}\sigma^{\alpha}(x)=\sigma^{a}e^{\alpha}_{(a)}(x),\quad \bar{\sigma}^{\alpha}(x)=\bar{\sigma}^{a}\,e^{\alpha}_{(a)}(x)\,,\\ \Sigma_{\alpha}=\frac{1}{2}\Sigma^{ab}e^{\beta}_{(a)}\nabla_{\alpha}(e_{(b)\beta}),\quad \bar{\Sigma}_{\alpha}=\frac{1}{2}\bar{\Sigma}^{ab}e^{\beta}_{(x)}\nabla_{\alpha}(e_{(b)\beta})\,,\\ \Sigma^{ab}=\frac{1}{4}\,(\,\bar{\sigma}^{a}\,\sigma^{b}-\bar{\sigma}^{b}\,\sigma^{a}\,),\quad \bar{\Sigma}^{ab}=\frac{1}{4}\,(\,\sigma^{a}\,\bar{\sigma}^{b}-\sigma^{b}\,\bar{\sigma}^{a}\,)\,.\end{gathered} \tag{8.5}$$

From (8.4), by setting $m=0$, we readily derive the 2-component Weyl equations for the neutrino $\eta(x)$ and the anti-neutrino $\xi(x)$

$$\begin{gathered}i\,\sigma^{\alpha}(x)\,[\,\partial_{\alpha}+\Sigma_{\alpha}(x)\,]\,\xi(x)=m\,\eta(x)\,,\\ i\,\bar{\sigma}^{\alpha}(x)\,[\,\partial_{\alpha}+\bar{\Sigma}_{\alpha}(x)\,]\,\eta(x)=m\,\xi(x)\,.\end{gathered} \tag{8.6}$$

8.2 Majorana Spinor Fields

Let us specify the Majorana basis, having applied the following transformation to the spinor basis

$$\begin{gathered}\Psi_{M}=A\,\Psi,\quad A=\frac{1-\gamma^{2}}{\sqrt{2}}\,,\ \gamma^{a}_{M}=A\gamma^{a}A^{-1}\,,\\ \gamma^{0}_{M}=\gamma^{0}\gamma^{2}\,,\ \gamma^{1}_{M}=\gamma^{1}\gamma^{2}\,,\ \gamma^{2}_{M}=\gamma^{2}\,,\ \gamma^{3}_{M}=\gamma^{3}\gamma^{2}\,.\end{gathered}$$

In the new basis, the explicit form of Dirac matrices is

$$\gamma^{2}_{M}=\left|\begin{array}{cccc}0&0&0&i\\ 0&0&-i&0\\ 0&-i&0&0\\ i&0&0&0\end{array}\right|,\quad \gamma^{0}_{M}=\left|\begin{array}{cccc}0&-i&0&0\\ i&0&0&0\\ 0&0&0&i\\ 0&0&-i&0\end{array}\right|,$$

$$\gamma^{1}_{M}=\left|\begin{array}{cccc}-i&0&0&0\\ 0&i&0&0\\ 0&0&-i&0\\ 0&0&0&i\end{array}\right|,\quad \gamma^{3}_{M}=\left|\begin{array}{cccc}0&i&0&0\\ i&0&0&0\\ 0&0&0&i\\ 0&0&i&0\end{array}\right|;$$

we note that the Dirac matrices are now purely imaginary. Therefore, in this representation the Dirac operator is real-valued, so that real and imaginary parts of the complex wave function obey the independent Dirac equations

$$\Psi_M = \text{Re}\,\Psi + i\,\text{Im}\,\Psi = \Psi_- + \Psi_- \,,$$

$$\Big(i\gamma^a \frac{\partial}{\partial x^a} - m\Big)\Psi_+ = 0\,, \quad \Big(i\gamma^a \frac{\partial}{\partial x^a} - m\Big)\Psi_- = 0\,. \tag{8.7}$$

Both equations are invariant under the local Lorentz gauge transformations related to the freedom in choosing tetrads in the covariant Dirac equation. These equations describe the Majorana fermions Ψ_+ and Ψ_- with the charge parity $+1$ and -1, respectively.

To describe the interaction of the Majorana particles with the external gravitational field, we should specify the covariant Dirac equation in the Majorana basis from above

$$\Big[i\gamma^\alpha(x)(\partial_\alpha + \Gamma_\alpha(x)) - m\Big]\Psi(x) = 0\,.$$

Due to the properties $(i\gamma^a_M)^* = +\gamma^a_M$, $(\sigma^{ab}_M)^* = +\sigma^{ab}_M$, the generators $\sigma^{ab}_M = \frac{1}{4}(\gamma^a_M\gamma^b_M - \gamma^b_M\gamma^a_M)$ are real in any Majorana basis, and the covariant Dirac wave operator is real-valued. This means that the real and the imaginary parts of the wave function interact with non-Euclidean structure of space-time independently; in other words, there exist separate wave equations for the different Majorana fermions ψ_+, Ψ_-:

$$\Big[i\gamma^\alpha(x)\,(\partial_\alpha + \Gamma_\alpha(x)\,) - m\Big]\Psi_+ = 0\,, \quad \Big[i\gamma^\alpha(x)\,(\partial_\alpha + \Gamma_\alpha(x)\,) - m\Big]\Psi_- = 0\,. \tag{8.8}$$

8.3 Spin 1/2 Particle in Schwarzschild Field

In static Schwarzschild coordinates $x^\alpha = (t, \theta, \phi, r)$ (note $r \in (1, +\infty)$)

$$dS^2 = \Phi dt^2 - r^2 d\theta^2 - r^2 \sin^2\theta d\phi^2 - \frac{1}{\Phi}dr^2, \quad \Phi = 1 - \frac{1}{r}\,,$$

we use the following tetrad

$$e^\beta_{(a)} = g^{\beta\alpha} e_{(a)\alpha} = \left|\begin{array}{cccc} \Phi^{-1/2} & 0 & 0 & 0 \\ 0 & 1/r & 0 & 0 \\ 0 & 0 & 1/r\sin\theta & 0 \\ 0 & 0 & 0 & \Phi^{1/2} \end{array}\right|.$$

The Dirac equation takes the form

$$\Big[i\frac{\gamma^0}{\sqrt{\Phi}}\partial_t + i\sqrt{\Phi}\gamma^3\Big(\partial_r + \frac{1}{r} + \frac{\Phi'}{4\Phi}\Big) + \frac{1}{r}\Sigma_{\theta,\phi} - M\Big]\Psi(x) = 0\,, \tag{8.9}$$

where the angular operator $\Sigma_{\theta,\phi}$ is given by

$$\Sigma_{\theta,\phi} = i\gamma^1\partial_\theta + \gamma^2\frac{i\partial_\phi + i\sigma^{12}\cos\theta}{\sin\theta}\,. \tag{8.10}$$

8.4 Separation of the Variables

Equation (8.9), with the help of the substitution $\Psi(x) = r^{-1}\Phi^{-1/4}(r)\,\psi(x)$, can be simplified:

$$\Big(i\frac{\gamma^0}{\sqrt{\Phi}}\partial_t + i\sqrt{\Phi}\gamma^3\partial_r + \frac{1}{r}\Sigma_{\theta,\phi} - M\Big)\psi(x) = 0\,.$$

We shall further use the spinor representation for Dirac matrices [70]:

$$\gamma^0 = \left|\begin{array}{cc} 0 & I \\ I & 0 \end{array}\right|, \; \gamma^j = \left|\begin{array}{cc} 0 & -\sigma_j \\ \sigma_j & 0 \end{array}\right|, \; i\sigma^{12} = \left|\begin{array}{cc} \sigma_3 & 0 \\ 0 & \sigma_3 \end{array}\right|.$$

The solutions with spherical symmetry are constructed by using the following substitution

$$\psi_{\epsilon jm}(x) \;=\; e^{-i\epsilon t}\left|\begin{array}{c} f_1(r)\,D_{-1/2} \\ f_2(r)\,D_{+1/2} \\ f_3(r)\,D_{-1/2} \\ f_4(r)\,D_{+1/2} \end{array}\right|; \tag{8.11}$$

here we use Wigner D-functions [68] $D_\sigma = D^j_{-m,\sigma}(\phi,\theta,0)$; the fixed parameters $j = 1/2, 3/2, ..$ and $m \in \{-j,...,+j\}$ will be omitted for brevity. In accordance with the structure of the wave functions in spinor representation

$$\psi(x) = \left|\begin{array}{c} \xi(x) \\ \eta(x) \end{array}\right|, \tag{8.12}$$

one immediately obtains the general substitution for the wave function of the two Weyl particles:

$$\xi_{\epsilon jm}(x) = e^{-i\epsilon t}\left|\begin{array}{c} f_1(r)\,D_{-1/2} \\ f_2(r)\,D_{+1/2} \end{array}\right|, \quad \eta_{\epsilon jm}(x) = e^{-i\epsilon t}\left|\begin{array}{c} f_3(r)\,D_{-1/2} \\ f_4(r)\,D_{+1/2} \end{array}\right|. \tag{8.13}$$

Using the recurrent formulas for Wigner functions, $D^j_{m,\sigma} = e^{im\phi}d^j_{-m,\sigma}(\theta)$, $\sigma = \pm 1/2$:

$$\begin{aligned} \partial_\theta\,D_{+1/2} &= a\,D_{-1/2} - b\,D_{+3/2}\,, \; \frac{-m - 1/2\cos\theta}{\sin\theta}D_{+1/2} = -aD_{-1/2} - bD_{+3/2}\,, \\ \partial_\theta\,D_{-1/2} &= b\,D_{-3/2} - a\,D_{+1/2}\,, \; \frac{-m + 1/2\cos\theta}{\sin\theta}D_{-1/2} = -bD_{-3/2} - aD_{+1/2}\,, \end{aligned}$$

where $a = (j+1)/2$, $b = (1/2)\sqrt{(j-1/2)(j+3/2)}$, we find the action of the angular operator on the wave function (note that $\nu = j + 1/2$)

$$\Sigma_{\theta,\phi}\,\psi_{\epsilon jm}(x) = i\nu\frac{e^{-i\epsilon t}}{r}\left|\begin{array}{c} -f_4(r)D_{-1/2} \\ +f_3(r)D_{+1/2} \\ +f_2(r)D_{-1/2} \\ -f_1(r)D_{+1/2} \end{array}\right|. \tag{8.14}$$

Then we obtain the radial equations

$$\begin{gathered}
\frac{\epsilon}{\sqrt{\Phi}} f_3 - i\sqrt{\Phi}\frac{d}{dr} f_3 - i\frac{\nu}{r} f_4 - M f_1 = 0, \; \frac{\epsilon}{\sqrt{\Phi}} f_4 + i\sqrt{\Phi}\frac{d}{dr} f_4 + i\frac{\nu}{r} f_3 - M f_2 = 0, \\
\frac{\epsilon}{\sqrt{\Phi}} f_1 + i\sqrt{\Phi}\frac{d}{dr} f_1 + i\frac{\nu}{r} f_2 - M f_3 = 0, \; \frac{\epsilon}{\sqrt{\Phi}} f_2 - i\sqrt{\Phi}\frac{d}{dr} f_2 - i\frac{\nu}{r} f_1 - M f_4 = 0.
\end{gathered} \tag{8.15}$$

In massless case, the system of radial equations consists of two subsystems which can be considered as related to neutrino and anti-neutrino, respectively:

$$\left(\sqrt{\Phi}\frac{d}{dr} + i\frac{\epsilon}{\sqrt{\Phi}}\right) f_3 + \frac{\nu}{r} f_4 = 0, \quad \left(\sqrt{\Phi}\frac{d}{dr} - i\frac{\epsilon}{\sqrt{\Phi}}\right) f_4 + \frac{\nu}{r} f_3 = 0; \tag{8.16}$$

and

$$\left(\sqrt{\Phi}\frac{d}{dr} - i\frac{\epsilon}{\sqrt{\Phi}}\right) f_1 + \frac{\nu}{r} f_2 = 0, \quad \left(\sqrt{\Phi}\frac{d}{dr} + i\frac{\epsilon}{\sqrt{\Phi}}\right) f_2 + \frac{\nu}{r} f_1 = 0. \tag{8.17}$$

Also, we can consider these four equations as being concerned with the massless Dirac field.

Turning again to the system (8.15), we may simplify it by using the discrete P-reflection operator (see in [140])

$$\hat{\Pi}_{sph.} = \begin{vmatrix} 0 & 0 & 0 & -1 \\ 0 & 0 & -1 & 0 \\ 0 & -1 & 0 & 0 \\ -1 & 0 & 0 & 0 \end{vmatrix} \otimes \hat{P}.$$

From the eigenvalue equation $\hat{\Pi}_{sph.}\,\Psi_{jm} = \Pi\,\Psi_{jm}$, it follows $\Pi = \delta\,(-1)^{j+1}, \delta = \pm 1$, and

$$f_4 = \delta\, f_1, \quad f_3 = \delta\, f_2, \quad \psi(x)_{\epsilon jm\delta} = e^{-i\epsilon t} \begin{vmatrix} f_1(r)\, D_{-1/2} \\ f_2(r)\, D_{+1/2} \\ \delta\, f_2(r)\, D_{-1/2} \\ \delta\, f_1(r)\, D_{+1/2} \end{vmatrix}. \tag{8.18}$$

Taking into account (8.18) in (8.15), we get only two equations

$$\begin{gathered}
\frac{\epsilon}{\sqrt{\Phi}} f_2 - i\sqrt{\Phi}\frac{d}{dr} f_2 - i\frac{\nu}{r} f_1 - M\delta f_1 = 0, \\
\frac{\epsilon}{\sqrt{\Phi}} f_1 + i\sqrt{\Phi}\frac{d}{dr} f_1 + i\frac{\nu}{r} f_2 - M\delta f_2 = 0.
\end{gathered} \tag{8.19}$$

They are transformed to equations without complex unit i:

$$\left(\sqrt{\Phi}\frac{d}{dr} + \frac{\nu}{r}\right) f + \left(\frac{\epsilon}{\sqrt{\Phi}} + \delta M\right) g = 0, \; \left(\sqrt{\Phi}\frac{d}{dr} - \frac{\nu}{r}\right) g - \left(\frac{\epsilon}{\sqrt{\Phi}} - \delta M\right) f = 0, \tag{8.20}$$

where $f = (f_1 + f_2)$, $g = -i(f_1 - f_2)$.

For definiteness we consider the case $\delta = +1$ (the transition to $\delta = -1$ is reached by the formal change $M \Longrightarrow -M$):

$$\Big(\sqrt{\Phi}\frac{d}{dr} + \frac{\nu}{r}\Big) f + \Big(\frac{\epsilon}{\sqrt{\Phi}} + M\Big) g = 0, \; \Big(\sqrt{\Phi}\frac{d}{dr} - \frac{\nu}{r}\Big) g - \Big(\frac{\epsilon}{\sqrt{\Phi}} - M\Big) f = 0. \qquad (8.21)$$

These are symmetrical under the change: $\nu \Longrightarrow -\nu$, $\epsilon \Longrightarrow -\epsilon$.

For Weyl particles we have the following equations:

$$\Big(\sqrt{\Phi}\frac{d}{dr} - i\frac{\epsilon}{\sqrt{\Phi}}\Big) f_1 + \frac{\nu}{r} f_2 = 0, \quad \Big(\sqrt{\Phi}\frac{d}{dr} + i\frac{\epsilon}{\sqrt{\Phi}}\Big) f_2 + \frac{\nu}{r} f_1 = 0; \qquad (8.22)$$

and

$$\Big(\sqrt{\Phi}\frac{d}{dr} + i\frac{\epsilon}{\sqrt{\Phi}}\Big) f_3 + \frac{\nu}{r} f_4 = 0, \quad \Big(\sqrt{\Phi}\frac{d}{dr} - i\frac{\epsilon}{\sqrt{\Phi}}\Big) f_4 + \frac{\nu}{r} f_3 = 0. \qquad (8.23)$$

They are related to each other by the formal transformation $\{f_3, f_4, \epsilon\} \iff \{f_1, f_2, -\epsilon\}$. Eqs. (8.22) and (8.23) may be transformed into (let $f = f_1 + f_2$, $g = -i(f_1 - f_2)$)

$$\Big(\sqrt{\Phi}\frac{d}{dr} + \frac{\nu}{r}\Big) f + \frac{\epsilon}{\sqrt{\Phi}} g = 0, \quad \Big(\sqrt{\Phi}\frac{d}{dr} - \frac{\nu}{r}\Big) g - \frac{\epsilon}{\sqrt{\Phi}} f = 0; \qquad (8.24)$$

and (let $F = f_3 + f_4$, $G = -i(f_3 - f_4)$)

$$\Big(\sqrt{\Phi}\frac{d}{dr} + \frac{\nu}{r}\Big) F - \frac{\epsilon}{\sqrt{\Phi}} G = 0, \quad \Big(\sqrt{\Phi}\frac{d}{dr} - \frac{\nu}{r}\Big) G + \frac{\epsilon}{\sqrt{\Phi}} F = 0. \qquad (8.25)$$

8.5 The Case of Majorana Particle

Now we turn to resolving Dirac wave functions into the Majorana components $\Psi = \Psi_+ + \Psi_-$ in spinor basis; this means

$$\Psi^c = \gamma^2 \Psi^* = \left| \begin{array}{c} -\sigma^2 \eta^* \\ \sigma^2 \xi^* \end{array} \right|, \quad \Psi = \Psi_+ + \Psi_- = \frac{\Psi + \Psi^c}{2} + \frac{\Psi - \Psi^c}{2}, \qquad (8.26)$$

or,

$$\Psi_+ = \left| \begin{array}{c} \xi_+ = (\xi - \sigma_2 \eta^*)/2 \\ \eta_+ = (\eta + \sigma_2 \xi^*)/2 \end{array} \right|, \quad \Psi_- = \left| \begin{array}{c} \xi_- = (\xi + \sigma_2 \eta^*)/2 \\ \eta_- = (\eta - \sigma_2 \xi^*)/2 \end{array} \right|. \qquad (8.27)$$

Applying these formulas to spherical waves, we get

$$\Psi_+ = \frac{1}{2r} \left| \begin{array}{c} e^{-i\epsilon t} f_1 D_{-1/2} + i e^{i\epsilon t} f_4^* D^*_{+1/2} \\ e^{-i\epsilon t} f_2 D_{+1/2} - i e^{i\epsilon t} f_3^* D^*_{-1/2} \\ e^{-i\epsilon t} f_3 D_{-1/2} - i e^{i\epsilon t} f_2^* D^*_{+1/2} \\ e^{-i\epsilon t} f_4 D_{+1/2} + i e^{i\epsilon t} f_1^* D^*_{-1/2} \end{array} \right|, \Psi_- = \frac{1}{2r} \left| \begin{array}{c} e^{-i\epsilon t} f_1 D_{-1/2} - i e^{i\epsilon t} f_4^* D^*_{+1/2} \\ e^{-i\epsilon t} f_2 D_{+1/2} + i e^{i\epsilon t} f_3^* D^*_{-1/2} \\ e^{-i\epsilon t} f_3 D_{-1/2} + i e^{i\epsilon t} f_2^* D^*_{+1/2} \\ e^{-i\epsilon t} f_4 D_{+1/2} - i e^{i\epsilon t} f_1^* D^*_{-1/2} \end{array} \right|. \qquad (8.28)$$

In order to have the spherical Majorana waves explicitly real (or imaginary), we translate these formulas to the Majorana basis in accordance with the rule

$$\Psi^M_{\pm} = \frac{1-\gamma^2}{\sqrt{2}}\Psi_{\pm} = \frac{1}{\sqrt{2}}\left|\begin{array}{cccc} 1 & 0 & 0 & -i \\ 0 & 1 & i & 0 \\ 0 & i & 1 & 0 \\ -i & 0 & 0 & 1 \end{array}\right|\Psi_{\pm}\,.$$

In this way we obtain

$$\Psi^M_{+} = \frac{1}{\sqrt{2}}\left|\begin{array}{c} (e^{-i\epsilon t} f_1 D_{-1/2} + i e^{i\epsilon t} f_4^* D^*_{+1/2}) - i(e^{-i\epsilon t} f_4 D_{+1/2} + i e^{i\epsilon t} f_1^* D^*_{-1/2}) \\ (e^{-i\epsilon t} f_2 D_{+1/2} - i e^{i\epsilon t} f_3^* D^*_{-1/2}) + i(e^{-i\epsilon t} f_3 D_{-1/2} - i e^{i\epsilon t} f_2^* D^*_{+1/2}) \\ i(e^{-i\epsilon t} f_2 D_{+1/2} - i e^{i\epsilon t} f_3^* D^*_{-1/2}) + (e^{-i\epsilon t} f_3 D_{-1/2} - i e^{i\epsilon t} f_2^* D^*_{+1/2}) \\ -i(e^{-i\epsilon t} f_1 D_{-1/2} + i e^{i\epsilon t} f_4^* D^*_{+1/2}) + (e^{-i\epsilon t} f_4 D_{+1/2} + i e^{i\epsilon t} f_1^* D^*_{-1/2}) \end{array}\right|,$$

$$\Psi^M_{-} = \frac{1}{\sqrt{2}}\left|\begin{array}{c} (e^{-i\epsilon t} f_1 D_{-1/2} - i e^{i\epsilon t} f_4^* D^*_{+1/2}) - i(e^{-i\epsilon t} f_4 D_{+1/2} - i e^{i\epsilon t} f_1^* D^*_{-1/2}) \\ (e^{-i\epsilon t} f_2 D_{+1/2} + i e^{i\epsilon t} f_3^* D^*_{-1/2}) + i(e^{-i\epsilon t} f_3 D_{-1/2} + i e^{i\epsilon t} f_2^* D^*_{+1/2}) \\ i(e^{-i\epsilon t} f_2 D_{+1/2} + i e^{i\epsilon t} f_3^* D^*_{-1/2}) + (e^{-i\epsilon t} f_3 D_{-1/2} + i e^{i\epsilon t} f_2^* D^*_{+1/2}) \\ -i(e^{-i\epsilon t} f_1 D_{-1/2} - i e^{i\epsilon t} f_4^* D^*_{+1/2}) + (e^{-i\epsilon t} f_4 D_{+1/2} - i e^{i\epsilon t} f_1^* D^*_{-1/2}) \end{array}\right|,$$

note identities

$$\Psi^M_{+} = (\Psi^M_{+})^*, \quad \Psi^M_{+} = -(\Psi^M_{+})^*. \tag{8.29}$$

Finally, let us write down the Majorana wave functions for fixed values of parity:
in spinor basis

$$\Psi^{\delta}_{+} = \frac{1}{2}\left|\begin{array}{c} e^{-i\epsilon t} f_1 D_{-1/2} + i e^{i\epsilon t} f_4^* D^*_{+1/2} \\ e^{-i\epsilon t} f_2 D_{+1/2} - i e^{i\epsilon t} f_3^* D^*_{-1/2} \\ e^{-i\epsilon t} f_3 D_{-1/2} - i e^{i\epsilon t} f_2^* D^*_{+1/2} \\ e^{-i\epsilon t} f_4 D_{+1/2} + i e^{i\epsilon t} f_1^* D^*_{-1/2} \end{array}\right|, \Psi^{\delta}_{-} = \frac{1}{2}\left|\begin{array}{c} e^{-i\epsilon t} f_1 D_{-1/2} - i e^{i\epsilon t} f_4^* D^*_{+1/2} \\ e^{-i\epsilon t} f_2 D_{+1/2} + i e^{i\epsilon t} f_3^* D^*_{-1/2} \\ e^{-i\epsilon t} f_3 D_{-1/2} + i e^{i\epsilon t} f_2^* D^*_{+1/2} \\ e^{-i\epsilon t} f_4 D_{+1/2} - i e^{i\epsilon t} f_1^* D^*_{-1/2} \end{array}\right|; \tag{8.30}$$

in Majorana basis

$$\Psi^{M,\delta}_{+} = \sqrt{2}\left|\begin{array}{c} \text{Re}\, e^{-i\epsilon t} f_1 D_{-1/2} + \delta\, \text{Im}\, e^{-i\epsilon t} f_1 D_{+1/2} \\ \text{Re}\, e^{-i\epsilon t} f_2 D_{+1/2} - \delta\, \text{Im}\, e^{-i\epsilon t} f_2 D_{-1/2} \\ -\text{Im}\, e^{-i\epsilon t} f_2 D_{+1/2} + \delta\, \text{Re}\, e^{-i\epsilon t} f_2 D_{-1/2} \\ \delta\, \text{Re}\, e^{-i\epsilon t} f_1 D_{+1/2} + \text{Im}\, e^{-i\epsilon t} f_1 D_{-1/2} \end{array}\right|, \tag{8.31}$$

$$\Psi^{M,\delta}_{-} = i\sqrt{2}\left|\begin{array}{c} \text{Im}\, e^{-i\epsilon t} f_1 D_{-1/2} - \delta\, \text{Re}\, e^{-i\epsilon t} f_1 D_{+1/2} \\ \text{Im}\, e^{-i\epsilon t} f_2 D_{+1/2} + \delta\, \text{Re}\, e^{-i\epsilon t} f_2 D_{-1/2} \\ \text{Re}\, e^{-i\epsilon t} f_2 D_{+1/2} + \delta\, \text{Im}\, e^{-i\epsilon t} f_2 D_{-1/2} \\ -\ \text{Re}\, e^{-i\epsilon t} f_1 D_{-1/2} + \delta\, \text{Im}\, e^{-i\epsilon t} f_1 D_{+1/2} \end{array}\right|. \tag{8.32}$$

8.6 Qualitative Study

Let us start with the equations (recall that $\Phi = \sqrt{1-1/r}$)

$$\Big(\Phi\frac{d}{dr} + \frac{\nu\sqrt{\Phi}}{r}\Big)\, f = -(\epsilon + M\sqrt{\Phi})\, g\,, \quad \Big(\Phi\frac{d}{dr} - \frac{\nu\sqrt{\Phi}}{r}\Big)\, g = +(\epsilon - M\sqrt{\Phi}) f\,. \tag{8.33}$$

We transform equations to the alternative variable r_* defined by relations

$$\Phi\frac{d}{dr} = \frac{d}{dr_*}\,, \quad dr_* = \frac{dr}{1-1/r}\,,$$

$$r_* \in (-\infty, +\infty)\,, \quad r \to 1,\; r_* \to -\infty \quad r \to \infty,\; r_* \to +\infty\,.$$

In this way we obtain

$$\Big(\frac{d}{dr_*} + \nu\varphi(r_*)\Big) f = -(\epsilon + M\sqrt{\Phi})\, g\,, \quad \Big(\frac{d}{dr_*} - \nu\varphi(r_*)\Big) g = +(\epsilon - M\sqrt{\Phi})\, f\,, \tag{8.34}$$

where

$$\varphi(r_*) = \frac{\sqrt{\Phi}}{r} = \frac{\sqrt{1-1/r}}{r}\,, \quad r_* \to -\infty,\; \varphi(r_*) \to 0\,, \quad r_* \to +\infty,\; \varphi(r_*) \to 0\,. \tag{8.35}$$

We can derive the 2-nd order equations for the separate functions

$$\Big\{(\epsilon - M\sqrt{\Phi})\Big(\frac{d}{dr_*} + \nu\varphi(r_*)\Big)\frac{1}{(\epsilon - M\sqrt{\Phi})}\Big\}\Big(\frac{d}{dr_*} - \nu\varphi(r_*)\Big) g + (\epsilon^2 - M^2\Phi)\, g = 0\,,$$

$$\Big\{(\epsilon + M\sqrt{\Phi})\Big(\frac{d}{dr_*} - \nu\varphi(r_*)\Big)\frac{1}{(\epsilon + M\sqrt{\Phi})}\Big\}\Big(\frac{d}{dr_*} + \nu\varphi(r_*)\Big)\, f + (\epsilon^2 - M^2\Phi)\, f = 0\,.$$

They are symmetric, so it is enough to perform the calculations only for one equation. Consider the terms (let the prime designates derivative d/dr^*)

$$(\epsilon - M\sqrt{\Phi})\Big(\frac{d}{dr_*} + \nu\varphi(r_*)\Big)\frac{1}{(\epsilon - M\sqrt{\Phi})} = \frac{d}{dr_*} + \nu\varphi(r_*) - \frac{(\epsilon - M\sqrt{\Phi})'}{(\epsilon - M\sqrt{\Phi})}\,,$$

$$(\epsilon + M\sqrt{\Phi})\Big(\frac{d}{dr_*} - \nu\varphi(r_*)\Big)\frac{1}{(\epsilon + M\sqrt{\Phi})} = \frac{d}{dr_*} - \nu\varphi(r_*) - \frac{(\epsilon + M\sqrt{\Phi})'}{(\epsilon + M\sqrt{\Phi})}\,.$$

With the notations

$$A = \frac{1}{(\epsilon - M\sqrt{\Phi})}\frac{d}{dr_*}(\epsilon - M\sqrt{\Phi})\,, \qquad B = \frac{1}{(\epsilon + M\sqrt{\Phi})}\frac{d}{dr_*}(\epsilon + M\sqrt{\Phi})\,,$$

the second order equations can be written as follows

$$\Big(\frac{d}{dr_*} + \nu\varphi(r_*) - A\Big)\Big(\frac{d}{dr_*} - \nu\varphi(r_*)\Big) g + (\epsilon^2 - M^2\Phi)\, g = 0\,,$$

$$\Big(\frac{d}{dr_*} - \nu\varphi(r_*) - B\Big)\Big(\frac{d}{dr_*} + \nu\varphi(r_*)\Big) f + (\epsilon^2 - M^2\Phi)\, f = 0\,,$$

or, differently,

$$\begin{aligned}
&\Big[\frac{d^2}{dr_*^2} - A\frac{d}{dr_*} + \epsilon^2 - \nu^2\varphi^2 - \nu\frac{d\varphi}{dr_*} + \nu A\varphi - M^2\Phi\Big] g = 0,\\
&\Big[\frac{d^2}{dr_*^2} - B\frac{d}{dr_*} + \epsilon^2 - \nu^2\varphi^2 + \nu\frac{d\varphi}{dr_*} - \nu B\varphi - M^2\Phi\Big] f = 0.
\end{aligned} \tag{8.36}$$

Let us eliminate the first derivative terms. For definiteness, let us follow the equation for f. With the substitution

$$f(r_*) = \beta(r_*)F(r_*); \tag{8.37}$$

we derive (the prime designates derivative on the variable r_*)

$$\Big[\frac{d^2}{dr_*^2} + \Big(2\frac{\beta'}{\beta} - B\Big)\frac{d}{dr_*} + \Big(\frac{\beta''}{\beta} - A\frac{\alpha'}{\beta} + \epsilon^2 - \nu^2\varphi^2 + \nu\frac{d\varphi}{dr_*} - \nu B\varphi - M^2\Phi\Big)\Big] F = 0.$$

Let us fix the function $\beta(x)$:

$$2\frac{\beta'}{\beta} = B = \frac{(\epsilon + M\sqrt{\Phi})'}{\epsilon + M\sqrt{\Phi}} \quad \Longrightarrow \quad \beta(r_*) = \sqrt{\epsilon + M\sqrt{\Phi}}; \tag{8.38}$$

then we find

$$\beta' = \frac{B}{2}\beta, \quad \beta'' = \frac{B'}{2}\beta + \frac{B^2}{4}\beta \quad \Longrightarrow \quad \frac{\beta''}{\beta} = \frac{B'}{2} + \frac{B^2}{4}.$$

Therefore, the above equation reads

$$\Big[\frac{d^2}{dr_*^2} + \Big(\frac{1}{2}\frac{dB}{dr_*} + \frac{B^2}{4} - \frac{AB}{2} + \epsilon^2 - \nu^2\varphi^2 + \nu\frac{d\varphi}{dr_*} - \nu B\varphi - M^2\Phi\Big)\Big] F = 0. \tag{8.39}$$

Similarly, we consider the function G:

$$\begin{aligned}
&g(r_*) = \alpha(r_*)G(r_*), \qquad \alpha(r_*) = \sqrt{\epsilon - M\sqrt{\Phi}},\\
&\Big[\frac{d^2}{dr_*^2} + \Big(\frac{1}{2}\frac{dA}{dr_*} + \frac{A^2}{4} - \frac{AB}{2} + \epsilon^2 - \nu^2\varphi^2 - \nu\frac{d\varphi}{dr_*} + \nu A\varphi - M^2\Phi\Big)\Big] G = 0.
\end{aligned} \tag{8.40}$$

Considering the definitions

$$\varphi = \frac{1}{r}\sqrt{1 - 1/r}, \quad \frac{d}{dr_*} = (1 - 1/r)\frac{d}{dr}, \quad \frac{d}{dr_*}\varphi = \frac{1/2r - (1 - 1/r)}{r^2}\sqrt{1 - 1/r},$$

we obtain

$$(\epsilon^2 - \nu^2\varphi^2 - \nu\frac{d\varphi}{dr_*}) = \epsilon^2 - \nu^2\frac{1 - 1/r}{r^2} - \nu\,\frac{1/2r - (1 - 1/r)}{r^2}\sqrt{1 - 1/r},$$

$$(\epsilon^2 - \nu^2\varphi^2 + \nu\frac{d\varphi}{dr_*}) = \epsilon^2 - \nu^2\frac{1 - 1/r}{r^2} + \nu\,\frac{1/2r - (1 - 1/r)}{r^2}\sqrt{1 - 1/r},$$

$$A = \frac{1}{(\epsilon - M\sqrt{\Phi})} \frac{d}{dr_*}(\epsilon - M\sqrt{\Phi}) = -\frac{M\sqrt{1-1/r}}{2r^2\,(\epsilon - M\sqrt{1-1/r})},$$

$$B = \frac{1}{(\epsilon + M\sqrt{\Phi})} \frac{d}{dr_*}(\epsilon + M\sqrt{\Phi}) = +\frac{M\sqrt{1-1/r}}{2r^2\,(\epsilon + M\sqrt{1-1/r})},$$

$$A|_{r\to 1} = 0, \quad B|_{r\to 1} = 0, \quad A|_{r\to\infty} = 0, \quad B|_{r\to\infty} = 0,$$

$$A \cdot \varphi = -\frac{M(1-1/r)}{2r^3(\epsilon - M\sqrt{1-1/r})}, \qquad B \cdot \varphi = +\frac{M(1-1/r)}{2r^3(\epsilon + M\sqrt{1-1/r})}.$$

Note that

$$A\varphi|_{r\to 1} = 0, \quad B\varphi|_{r\to 1} = 0, \quad A\varphi|_{r\to\infty} = 0, \quad B\varphi|_{r\to\infty} = 0,$$

$$\frac{1}{2}\frac{dA}{dr_*} = -\frac{M}{4}(1-1/r)\frac{d}{dr}\Big\{\sqrt{1-1/r}\cdot\frac{1}{r^2}\cdot\frac{1}{\epsilon - M\sqrt{1-1/r}}\Big\}$$

$$= -\frac{M}{4}(1-1/r)\Big\{\frac{1}{2r^2\sqrt{1-1/r}}\cdot\frac{1}{r^2}\cdot\frac{1}{\epsilon - M\sqrt{1-1/r}}$$

$$-\sqrt{1-1/r}\cdot\frac{2}{r^3}\cdot\frac{1}{\epsilon - M\sqrt{1-1/r}} + \sqrt{1-1/r}\cdot\frac{1}{r^2}\cdot\frac{1}{2(\epsilon - M\sqrt{1-1/r})^2}\frac{M}{2\sqrt{1-1/r}}\frac{1}{r^2}\Big\}.$$

Also note that

$$\frac{1}{2}\frac{dA}{dr_*}\Big|_{r\to 1} = 0, \qquad \frac{1}{2}\frac{dA}{dr_*}\Big|_{r\to\infty} = 0, \qquad \frac{1}{2}\frac{dB}{dr_*}\Big|_{r\to 1} = 0, \qquad \frac{1}{2}\frac{dB}{dr_*}\Big|_{r\to\infty} = 0,$$

$$\frac{1}{2}\frac{dB}{dr_*} = +\frac{1}{2}(1-1/r)\frac{d}{dr}\frac{M\sqrt{1-1/r}}{2r^2\,(\epsilon + M\sqrt{1-1/r})},$$

$$\frac{1}{4}B^2 = \frac{1}{4}\,\frac{M^2(1-1/r)}{4r^4(\epsilon + M\sqrt{1-1/r})^2}, \qquad \frac{1}{4}A^2 = \frac{1}{4}\,\frac{M^2(1-1/r)}{4r^4(\epsilon - M\sqrt{1-1/r})^2}$$

$$-\frac{AB}{2} = \frac{1}{2}\,\frac{M^2(1-1/r)}{4r^4(\epsilon + M\sqrt{1-1/r})(\epsilon - M\sqrt{1-1/r})},$$

$$M^2\Phi = M^2(1-1/r), \quad M^2\Phi|_{r\to 1} = 0, \quad M^2\Phi|_{r\to\infty} = M^2.$$

Thus, the asymptotic form of the 2-nd order equations in the regions $r \to 1$ and $r \to +\infty$ is as follows

$$r \to +1, \qquad \Big(\frac{d^2}{dr_*^2} + \epsilon^2\Big)f = 0, \quad \Big(\frac{d^2}{dr_*^2} + \epsilon^2\Big)g = 0; \tag{8.41}$$

$$r \to \infty, \qquad \Big(\frac{d^2}{dr_*^2} + \epsilon^2 - M^2\Big)f = 0, \quad \Big(\frac{d^2}{dr_*^2} + \epsilon^2 - M^2\Big)g = 0. \tag{8.42}$$

8.7 Analytical Treatment

Let us introduce the variable in which we eliminate the square root from equation (8.24):

$$\sqrt{\Phi} = \sqrt{\frac{r-1}{r}} = x, \quad \frac{1}{r} = 1-x^2, \quad 2x\frac{dx}{dr} = \frac{1}{r^2} = (1-x^2)^2,$$
$$\Phi\frac{d}{dr} = x^2\frac{dx}{dr}\frac{d}{dx} = \frac{1}{2}x(1-x^2)^2\frac{d}{dx}, \quad \frac{\sqrt{\Phi}}{r} = x(1-x^2), \quad \frac{1}{r} = 1-x^2, \tag{8.43}$$
$$r \to 1, x \to 0; \quad r \to +\infty, x \to \pm 1; \quad r \to 0, \quad x \to \pm i\infty .$$

The physical region for x is the interval $x \in (0,1)$. The equations (8.24) take the form

$$\Big(\frac{1}{2}x(1-x^2)^2\frac{d}{dx} + \nu x(1-x^2)\Big)f + \epsilon g = 0,$$
$$\Big(\frac{1}{2}x(1-x^2)^2\frac{d}{dx} - \nu x(1-x^2)\Big)g - \epsilon f = 0. \tag{8.44}$$

From (8.44) we derive the 2-nd order equations for g and f:

$$\Big[\frac{d^2}{dx^2} + \frac{(1-5x^2)}{x(1-x^2}\frac{d}{dx} - 2\nu\frac{(1-3x^2)}{x(1-x^2)^2} - \frac{4\nu^2}{(1-x^2)^2} + \frac{4\epsilon^2}{x^2(1-x^2)^4}\Big]g = 0,$$
$$\Big[\frac{d^2}{dx^2} + \frac{(1-5x^2)}{x(1-x^2}\frac{d}{dx} + 2\nu\frac{(1-3x^2)}{x(1-x^2)^2} - \frac{4\nu^2}{(1-x^2)^2} + \frac{4\epsilon^2}{x^2(1-x^2)^4}\Big]f = 0. \tag{8.45}$$

These equations may be presented in a more symmetrical form:

$$\frac{d^2g}{dx^2} + \Big[\frac{2}{x+1} + \frac{2}{x-1} + \frac{1}{x}\Big]\frac{dg}{dx} + \Big[\frac{-8\nu^2+35\epsilon^2+8\nu}{8(x+1)}$$
$$+\frac{-8\nu^2+19\epsilon^2-8\nu}{8(x+1)^2} + \frac{\epsilon^2}{(x+1)^3} + \frac{\epsilon^2}{4(x+1)^4} + \frac{8\nu^2-35\epsilon^2+8\nu}{8(x-1)}$$
$$+\frac{-8\nu^2+19\,\epsilon^2+8\nu}{8\,(x-1)^2} - \frac{\epsilon^2}{(x-1)^3} + \frac{\epsilon^2}{4\,(x-1)^4} - \frac{2\nu}{x} + \frac{4\,\epsilon^2}{x^2}\Big]g = 0, \tag{8.46}$$

$$\frac{d^2f}{dx^2} + \Big[\frac{2}{x+1} + \frac{2}{x-1} + \frac{1}{x}\Big]\frac{df}{dx} + \Big[\frac{-8\nu^2+35\,\epsilon^2-8\nu}{8\,(x+1)}$$
$$+\frac{-8\nu^2+19\,\epsilon^2+8\nu}{8\,(x+1)^2} + \frac{\epsilon^2}{(x+1)^3} + \frac{\epsilon^2}{4\,(x+1)^4} + \frac{8\nu^2-35\,\epsilon^2-8\nu}{8\,(x-1)}$$
$$+\frac{-8\nu^2+19\,\epsilon^2-8\nu}{8\,(x-1)^2} - \frac{\epsilon^2}{(x-1)^3} + \frac{\epsilon^2}{4\,(x-1)^4} + \frac{2\nu}{x} + \frac{4\,\epsilon^2}{x^2}\Big]f = 0. \tag{8.47}$$

They relate to each other by the formal change $\nu \Longrightarrow -\nu$; hence we may detail only one case, let it be the function $g(x)$. The points $x = 0, \infty$ are regular singularities; the points $x = \pm 1$ are irregular of the rank 2 – see [3, 4]. In order to consider the point $x=\infty$, we should translate equation to the variable $y=x^{-1}$ and look at the resulting equation in vicinity of the point $y=0$:

$$\Big(\frac{d^2}{dy^2}+\frac{2}{y}\frac{d}{dy}-(\frac{2}{y}+\frac{2}{y}+\frac{1}{y})\frac{d}{dy}$$
$$+\frac{-8\nu^2+35\epsilon^2+8\nu}{8y^3}+\frac{8\nu^2-35\epsilon^2+8\nu}{8y^3}-\frac{2\nu}{y^3}+\frac{\dots}{y^2}+\dots\Big)g=0.$$

Because the total coefficient at y^3 equals to zero, we conclude that the singular point $y=0$ ($x=\infty$) is regular. The equation (8.47) has the same singular points.

It is convenient to present Eq. (8.46) in a shorter form,

$$\frac{d^2g}{dx^2}+\Big(\frac{2}{x+1}+\frac{2}{x-1}+\frac{1}{x}\Big)\frac{dg}{dx}+\Big[-\frac{2\nu}{x}+\frac{4\epsilon^2}{x^2}+\frac{A}{(x+1)}+\frac{B}{(x+1)^2}$$
$$+\frac{\epsilon^2}{(x+1)^3}+\frac{\epsilon^2/4}{(x+1)^4}+\frac{A'}{(x-1)}+\frac{B'}{(x-1)^2}-\frac{\epsilon^2}{(x-1)^3}+\frac{\epsilon^2/4}{(x-1)^4}\Big]g=0, \qquad (8.48)$$

where

$$A=\frac{-8\nu^2+35\epsilon^2+8\nu}{8}, B=\frac{-8\nu^2+19\epsilon^2-8\nu}{8},$$
$$A'=\frac{8\nu^2-35\epsilon^2+8\nu}{8}, \quad B'=\frac{-8\nu^2+19\epsilon^2+8\nu}{8}.$$

It should be stressed that the last equation does not contain complex quantities, therefore with any complex-valued solution, its conjugate function will be solution of the equation as well, $g(x)$, $g^*(x)$.

Near the point $x=0$, the solutions behave in accordance with

$$\Big(\frac{d^2}{dx^2}+\frac{1}{x}\frac{d}{dx}+\frac{4\epsilon^2}{x^2}\Big)g=0, \quad g(x)\approx x^\gamma\sim(1-r)^{\gamma/2}, \quad \gamma=\pm 2i\epsilon.$$

This asymptotic formula can be readily translated to the variable r_*:

$$r_*-1=r-1+\ln(r-1), \quad r\to 1, \quad g(r_*)\sim e^{\pm i\epsilon r_*}, \quad r^*\to-\infty.$$

Now consider the solutions in the vicinity of the point $x=+1$; Eq. (8.46) simplifies

$$\frac{d^2g}{dx^2}+\frac{2}{x-1}\frac{dg}{dx}+\Big[\frac{A'}{(x-1)}+\frac{B'}{(x-1)^2}-\frac{\epsilon^2}{(x-1)^3}+\frac{\epsilon^2/4}{(x-1)^4}\Big]g=0,$$

and its solution can be searched in the form $g\approx(x-1)^{\alpha'}\exp\frac{\beta'}{(x-1)}\,G(x)$. This results in

$$\frac{d^2G}{dx^2}+\Big[\frac{2+2\alpha'}{x-1}-\frac{2\beta'}{(x-1)^2}\Big]\frac{dG}{dx}$$
$$+\Big[\frac{A'}{x-1}+\frac{\alpha'(\alpha'+1)+B'}{(x-1)^2}-\frac{2\alpha'\beta'+\epsilon^2}{(x-1)^3}+\frac{\beta'^2+\epsilon^2/4}{(x-1)^4}\Big]G=0.$$

Let the coefficient at the main terms $(x-1)^{-3}, (x-1)^{-4}$ vanish – this gives two possibilities: $2\beta' = +i\epsilon$, $\alpha' = +i\epsilon$ and $2\beta' = -i\epsilon$, $\alpha' = -i\epsilon$. Thus, near the point $x = +1$, the main asymptotic is given by $x \to +1$, $\quad g(x) \approx (x-1)^{\alpha'} \exp \frac{\beta'}{(x-1)}$. After translating it to the variable r_*, we get

$$\begin{aligned}
2\beta' &= \pm i\epsilon, \quad r_* = r + \ln(r-1) \approx r, \\
x - 1 &= \sqrt{1 - \frac{1}{r}} - 1 \approx -\frac{1}{2r} \approx -\frac{1}{2r_*}, \\
g(x) &\approx (x-1)^{\alpha'} \exp \frac{\beta'}{(x-1)} \approx (-2r_*)^{-\alpha'} e^{-2\beta' r_*} \\
&\sim e^{-\alpha' \ln r_* - 2\beta' r_*} \approx e^{\mp i\epsilon \ln r_* \mp i\epsilon r_*} \approx e^{\mp i\epsilon r_*}.
\end{aligned} \tag{8.49}$$

Similarly, near the point $x = -1$, we have

$$\frac{d^2 g}{dx^2} + \frac{2}{x+1}\frac{dg}{dx} + \Big[\frac{A}{(x+1)} + \frac{B}{(x+1)^2} + \frac{\epsilon^2}{(x+1)^3} + \frac{\epsilon^2/4}{(x+1)^4}\Big] g = 0.$$

The substitution $g \approx (x+1)^{\alpha} \exp \frac{\beta}{(x+1)} G(x)$ leads to

$$\frac{d^2 G}{dx^2} + \Big[\frac{2(\alpha+1)}{x+1} - \frac{2\beta}{(x+1)^2}\Big]\frac{dG}{dx}$$
$$+ \Big[\frac{A}{x+1} + \frac{\alpha(\alpha+1) + B}{(x+1)^2} - \frac{2\alpha\beta - \epsilon^2}{(x+1)^3} + \frac{\beta^2 + \epsilon^2/4}{(x+1)^4}\Big] G = 0.$$

Hence we obtain two sets of parameters: $2\beta = +i\epsilon$, $\alpha = -i\epsilon$, and $2\beta = -i\epsilon$, $\alpha = +i\epsilon$. The main asymptotic term is $x \to -1$, $\quad g(x) \approx (x+1)^{\alpha} \exp \frac{\beta}{(x+1)}$; in the variable r_* it looks as

$$\begin{aligned}
\beta &= \pm i\epsilon/2, \quad r_* = r + \ln(r-1) \approx r, \\
x + 1 &= -\sqrt{1 - \frac{1}{r}} + 1 \approx +\frac{1}{2r} \approx +\frac{1}{2r_*}, \\
g(x) &\approx (x+1)^{\alpha} \exp \frac{\beta}{(x+1)} \approx (-2r_*)^{-\alpha} e^{-2\beta r_*} \\
&\sim e^{-\alpha \ln r_* - 2\beta r_*} \approx e^{\pm i\epsilon \ln r_* \mp i\epsilon r_*} \approx e^{\mp i\epsilon r_*};
\end{aligned} \tag{8.50}$$

compare (8.50) and (8.49).

Now we turn to the main equation again

$$\begin{aligned}
&\frac{d^2 g}{dx^2} + \Big(\frac{2}{x+1} + \frac{2}{x-1} + \frac{1}{x}\Big)\frac{dg}{dx} + \Big[-\frac{2\nu}{x} + \frac{4\epsilon^2}{x^2} + \frac{A}{(x+1)} + \frac{B}{(x+1)^2} \\
&+ \frac{\epsilon^2}{(x+1)^3} + \frac{\epsilon^2/4}{(x+1)^4} + \frac{A'}{(x-1)} + \frac{B'}{(x-1)^2} - \frac{\epsilon^2}{(x-1)^3} + \frac{\epsilon^2/4}{(x-1)^4}\Big] g = 0,
\end{aligned} \tag{8.51}$$

and search for its exact solution in the form

$$g = x^{\gamma}(x+1)^{\alpha}(x-1)^{\alpha'} e^{\frac{\beta}{x+1}} e^{\frac{\beta'}{x-1}} G(x). \tag{8.52}$$

We derive

$$G'' + \Big(\frac{2+2\alpha}{x+1} + \frac{2+2\alpha'}{x-1} + \frac{1+2\gamma}{x} - \frac{2\beta}{(x+1)^2} - \frac{2\beta'}{(x-1)^2} \Big) G'$$
$$+ \Big\{ \frac{\gamma(\gamma-1)+\gamma+4\epsilon^2}{x^2} + \frac{\beta^2+\epsilon^2/4}{(x+1)^4} + \frac{\beta'^2+\epsilon^2/4}{(x-1)^4}$$
$$+ \frac{-\alpha\beta-\beta(\alpha-2)-2\beta+\epsilon^2}{(x+1)^3} + \frac{-\alpha'\beta'-\beta'(\alpha'-2)-2\beta'-\epsilon^2}{(x-1)^3} \Big\} G$$
$$+ \Big[\frac{2\gamma\alpha}{x(x+1)} + \frac{2\gamma\alpha'}{x(x-1)} - \frac{2\gamma\beta}{x(x+1)^2} - \frac{2\gamma\beta'}{x(x-1)^2} + \frac{\alpha(\alpha-1)}{(x+1)^2} + \frac{2\alpha\alpha'}{(x+1)(x-1)}$$
$$- \frac{2\alpha\beta'}{(x+1)(x-1)^2} + \frac{\alpha'(\alpha'-1)}{(x-1)^2} - \frac{2\alpha'\beta}{(x+1)^2(x-1)} + \frac{2\beta\beta'}{(x+1)^2(x-1)^2} + \frac{2\gamma}{x(x+1)}$$
$$+ \frac{2\alpha}{(x+1)^2} + \frac{2\alpha'}{(x+1)(x-1)} - \frac{2\beta'}{(x+1)(x-1)^2} + \frac{2\gamma}{x(x-1)} + \frac{2\alpha}{(x-1)(x+1)}$$
$$+ \frac{2\alpha'}{(x-1)^2} - \frac{2\beta}{(x-1)(x+1)^2} + \frac{\alpha}{x(x+1)} + \frac{\alpha'}{x(x-1)} - \frac{\beta}{x(x+1)^2} - \frac{\beta'}{x(x-1)^2}$$

$$- \frac{2\nu}{x} + \frac{A}{(x+1)} + \frac{B}{(x+1)^2} + \frac{A'}{x-1} + \frac{B'}{(x-1)^2} \Big] G = 0. \qquad (8.53)$$

Now we impose the restrictions

$$\gamma^2 + 4\epsilon^2 = 0, \; -2\alpha\beta + \epsilon^2 = 0, \; 4\beta^2 + \epsilon^2 = 0, \; -\epsilon^2 - 2\alpha'\beta' = 0, \; \epsilon^2 + 4\beta'^2 = 0;$$

whence it follows

$$\gamma = \pm 2i\epsilon; \quad (\beta = \pm i\epsilon/2, \quad \alpha = \mp i\epsilon); \quad (\beta' = \pm i\epsilon/2, \quad \alpha' = \pm i\epsilon). \qquad (8.54)$$

So we get

$$G'' + \Big(\frac{1+2\gamma}{x} + \frac{2+2\alpha}{x+1} + \frac{2+2\alpha'}{x-1} - \frac{2\beta}{(x+1)^2} - \frac{2\beta'}{(x-1)^2} \Big) G'$$
$$+ \Big\{ -\frac{2\nu}{x} + \frac{A}{(x+1)} + \frac{A'}{x-1} + \frac{\alpha(\alpha-1)+2\alpha+B}{(x+1)^2} + \frac{\alpha'(\alpha'-1)+2\alpha'+B'}{(x-1)^2}$$
$$+ \frac{2\gamma\alpha+2\gamma+\alpha}{x(x+1)} + \frac{2\gamma\alpha'+2\gamma+\alpha'}{x(x-1)} + \frac{-2\gamma\beta-\beta}{x(x+1)^2} + \frac{-2\gamma\beta'-\beta'}{x(x-1)^2}$$
$$+ \frac{-2\alpha\beta'-2\beta'}{(x+1)(x-1)^2} + \frac{-2\alpha'\beta-2\beta}{(x-1)(x+1)^2} + \frac{2\alpha\alpha'+2\alpha+2\alpha'}{(x+1)(x-1)} + \frac{2\beta\beta'}{(x+1)^2(x-1)^2} \Big\} G = 0.$$

The last equation can be presented in the more symmetrical form

$$\frac{d^2G}{dx^2}+\Big[\frac{2\gamma+1}{x}+\frac{2(\alpha+1)}{x+1}-\frac{2\beta}{(x+1)^2}+\frac{2(\alpha'+1)}{x-1}-\frac{2\beta'}{(x-1)^2}\Big]\frac{dG}{dx}$$
$$+\Big[\frac{2\gamma\alpha-2\gamma\beta-2\gamma\alpha'-2\gamma\beta'-2\nu+\alpha-\beta-\alpha'-\beta'}{x}$$
$$+\frac{-4\gamma\alpha-2\alpha\alpha'-\alpha\beta'+4\gamma\beta+\alpha'\beta+\beta\beta'+2A-4\alpha+3\beta-4\gamma-2\alpha'-\beta'}{2(x+1)}$$
$$+\frac{2\alpha^2+4\gamma\beta+2\alpha'\beta+\beta\beta'+2B+2\alpha+4\beta}{2(x+1)^2}$$
$$+\frac{2\alpha\alpha'+\alpha\beta'-\alpha'\beta-\beta\beta'+4\gamma\alpha'+4\gamma\beta'+2\alpha-\beta+4\gamma+2A'+4\alpha'+3\beta'}{2(x-1)}$$
$$+\frac{-2\alpha\beta'+\beta\beta'-4\gamma\beta'+2\alpha'^2+2B'+2\alpha'-4\beta'}{2(x-1)^2}\Big]G=0. \tag{8.55}$$

The general structure of the equation may be presented in a shorter form

$$G''+\Big(\frac{n}{x}+\frac{n_1}{x-1}+\frac{n_2}{(x-1)^2}+\frac{n_3}{x+1}+\frac{n_4}{(x+1)^2}\Big)G'$$
$$+\Big(\frac{m}{x}+\frac{m_1}{x-1}+\frac{m_2}{(x-1)^2}+\frac{m_3}{x+1}+\frac{m_4}{(x+1)^2}\Big)G=0. \tag{8.56}$$

8.8 Structure of the Power Series

Let us start with

$$G''+\Big(\frac{n}{x}+\frac{n_1}{x-1}+\frac{n_2}{(x-1)^2}+\frac{n_3}{x+1}+\frac{n_4}{(x+1)^2}\Big)G'$$
$$+\Big(\frac{m}{x}+\frac{m_1}{x-1}+\frac{m_2}{(x-1)^2}+\frac{m_3}{x+1}+\frac{m_4}{(x+1)^2}\Big)G=0, \tag{8.57}$$

where the following designations are used:

$$n=2\gamma+1,\quad n_1=2\alpha'+2,\quad n_2=-2\beta',\quad n_3=2\alpha+2,\quad n_4=-2\beta,$$

$$m=2\gamma\alpha-2\gamma\beta-2\gamma\alpha'-2\gamma\beta'-2\nu+\alpha-\beta-\alpha'-\beta',$$

$$m_1=\frac{2\alpha\alpha'+\alpha\beta'-\alpha'\beta-\beta\beta'+4\gamma\alpha'+4\gamma\beta'+2\alpha-\beta+4\gamma+2A'+4\alpha'+3\beta'}{2},$$

$$m_2=\frac{-2\alpha\beta'+\beta\beta'-4\gamma\beta'+2\alpha'^2+2B'+2\alpha'-4\beta'}{2},$$

$$m_3=\frac{-4\gamma\alpha-2\alpha\alpha'-\alpha\beta'+4\gamma\beta+\alpha'\beta+\beta\beta'+2A-4\alpha+3\beta-4\gamma-2\alpha'-\beta'}{2},$$

$$m_4=\frac{2\alpha^2+4\gamma\beta+2\alpha'\beta+\beta\beta'+2B+2\alpha+4\beta}{2}.$$

Equation (8.57) can be written in the more tractable form

$$x^5G'' - 2x^3G'' + xG'' + (n+n_1+n_3)x^4G' + (n_1+n_2-n_3+n_4)x^3G'$$
$$+(-2n-n_1+2n_2-n_3-2n_4)x^2G' + (-n_1+n_2+n_3+n_4)xG' + nG'$$
$$+(m+m_1+m_3)x^4G + (m_1+m_2-m_3+m_4)x^3G$$
$$+(-2m-m_1+2m_2-m_3-2m_4)x^2G + (-m_1+m_2+m_3+m_4)xG + mG = 0.$$

Let us introduce the shortening notations:

$$n = e\,, \quad (n+n_1+n_3) = a, \quad (n_1+n_2-n_3+n_4) = b,$$
$$(-2n-n_1+2n_2-n_3-2n_4) = c, \quad (-n_1+n_2+n_3+n_4) = d,$$
$$m = E\,, \quad (m+m_1+m_3) = A\,, \quad (m_1+m_2-m_3+m_4) = B,$$
$$(-2m-m_1+2m_2-m_3-2m_4) = C, \quad (-m_1+m_2+m_3+m_4) = D, \tag{8.58}$$

then we have equation in the form

$$x^5G'' - 2x^3G'' + xG'' + ax^4G' + bx^3G'$$
$$+cx^2G' + dxG' + eG' + Ax^4G + Bx^3G + Cx^2G + DxG + EG = 0.$$

Searching for the solutions $G(x)$ as power series

$$G = \sum_{k=0}^{\infty} c_k x^k, \quad G' = \sum_{k=1}^{\infty} k c_k x^{k-1}, \quad G'' = \sum_{k=2}^{\infty} k(k-1) c_k x^{k-2},$$

leads to

$$\sum_{k=5}^{\infty}(k-3)(k-4)c_{k-3}x^k - 2\sum_{k=3}^{\infty}(k-1)(k-2)c_{k-1}x^k + \sum_{k=1}^{\infty}(k+1)kc_{k+1}x^k$$
$$+a\sum_{k=4}^{\infty}(k-3)c_{k-3}x^k + b\sum_{k=3}^{\infty}(k-2)c_{k-2}x^k + c\sum_{k=2}^{\infty}(k-1)c_{k-1}x^k$$
$$+d\sum_{k=1}^{\infty}kc_kx^k + e\sum_{k=0}^{\infty}(k+1)c_{k+1}x^k + A\sum_{k=4}^{\infty}c_{k-4}x^k + B\sum_{k=3}^{\infty}c_{k-3}x^k$$
$$+C\sum_{k=2}^{\infty}c_{k-2}x^k + D\sum_{k=1}^{\infty}c_{k-1}x^k + E\sum_{k=0}^{\infty}c_kx^k = 0. \tag{8.59}$$

By collecting the coefficients at all the powers of x, we arrive at the recurrent formulas.
The main 6-term recurrent relationship is $\underline{k = 4,5,6,7,...}$

$$Ac_{k-4} + [B + (k-3)(k-4) + (k-3)a]c_{k-3} + [\,C + (k-2)b\,]\,c_{k-2}$$
$$+[\,D + (k-1)c - 2(k-1)(k-2)\,]\,c_{k-1} + (E+kd)\,c_k + (k+1)(k+e)\,c_{k+1} = 0. \tag{8.60}$$

To solve the question on convergence of the series, we use the known method by Poincaré–Perrone [3, 4]. To this end, we should divide the relation (8.60) by c_{k-4}:

$$A+[B+(k-3)(k-4)+(k-3)a]\frac{c_{k-3}}{c_{k-4}}+[C+(k-2)b]\frac{c_{k-2}}{c_{k-3}}\frac{c_{k-3}}{c_{k-4}}$$

$$+[D+(k-1)c-2(k-1)(k-2)]\frac{c_{k-1}}{c_{k-2}}\frac{c_{k-2}}{c_{k-3}}\frac{c_{k-3}}{c_{k-4}}+(E+kd)\frac{c_k}{c_{k-1}}\frac{c_{k-1}}{c_{k-2}}\frac{c_{k-2}}{c_{k-3}}\frac{c_{k-3}}{c_{k-4}}$$

$$+(k+1)(k+e)\frac{c_{k+1}}{c_k}\frac{c_k}{c_{k-1}}\frac{c_{k-1}}{c_{k-2}}\frac{c_{k-2}}{c_{k-3}}\frac{c_{k-3}}{c_{k-4}}=0. \qquad (8.61)$$

The convergence radius is determined as follows

$$r=\lim_{k\to\infty}\frac{c_{k+1}}{c_n}, \qquad R_{conv}=\frac{1}{|r|}. \qquad (8.62)$$

To get the algebraic equation for r, let us multiply relation (8.60) by k^{-2}, and let k tend to infinity. This results in the simple equation

$$r-2r^3+r^5=0 \quad \Longrightarrow \quad r(r^2-1)^2=0, \quad r=0,\ \pm 1.$$

Thus, the minimal convergence radius is $R_{conv}=1$, and it covers all the physical region for the variable $x\in(0,+1)$.

8.9 General Study of the Tunneling Effect

The above specified possibilities permit to define 8 different solutions (we take into account the convergence of all power series):

$$\begin{aligned}
g_1(x)&=e^{+2i\epsilon\ln x}\,e^{+i\epsilon\ln(x-1)}e^{\frac{+i\epsilon/2}{x-1}}\,e^{-i\epsilon\ln(x+1)}e^{\frac{+i\epsilon/2}{x+1}}\,(R_1(x)+iI_1(x)),\\
g_2(x)&=e^{-2i\epsilon\ln x}\,e^{-i\epsilon\ln(x-1)}e^{\frac{-i\epsilon/2}{x-1}}\,e^{+i\epsilon\ln(x+1)}e^{\frac{-i\epsilon/2}{x+1}}\,(R_1(x)-iI_1(x));\\
g_3(x)&=e^{+2i\epsilon\ln x}\,e^{-i\epsilon\ln(x-1)}e^{\frac{-i\epsilon/2}{x-1}}\,e^{+i\epsilon\ln(x+1)}e^{\frac{-i\epsilon/2}{x+1}}\,(R_3(x)+iI_3(x)),\\
g_4(x)&=e^{-2i\epsilon\ln x}\,e^{+i\epsilon\ln(x-1)}e^{\frac{+i\epsilon/2}{x-1}}\,e^{-i\epsilon\ln(x+1)}e^{\frac{+i\epsilon/2}{x+1}}\,(R_3(x)-iI_3(x));\\
g_5(x)&=e^{+2i\epsilon\ln x}\,e^{+i\epsilon\ln(x-1)}e^{\frac{+i\epsilon/2}{x-1}}\,e^{+i\epsilon\ln(x+1)}e^{\frac{-i\epsilon/2}{x+1}}\,(R_5(x)+iI_5(x));\\
g_6(x)&=e^{-2i\epsilon\ln x}\,e^{-i\epsilon\ln(x-1)}e^{\frac{-i\epsilon/2}{x-1}}\,e^{-i\epsilon\ln(x+1)}e^{\frac{+i\epsilon/2}{x+1}}\,(R_5(x)-iI_5(x));\\
g_7(x)&=e^{+2i\epsilon\ln x}\,e^{-i\epsilon\ln(x-1)}e^{\frac{-i\epsilon/2}{x-1}}\,e^{-i\epsilon\ln(x+1)}e^{\frac{+i\epsilon/2}{x+1}}\,(R_7(x)+iI_7(x));\\
g_8(x)&=e^{-2i\epsilon\ln x}\,e^{+i\epsilon\ln(x-1)}e^{\frac{+i\epsilon/2}{x-1}}\,e^{+i\epsilon\ln(x+1)}e^{\frac{-i\epsilon/2}{x+1}}\,(R_7(x)-iI_7(x)).
\end{aligned} \qquad (8.63)$$

Note that the factor $e^{+i\epsilon\ln(x-1)}$, $x\in(0,1)$ leads to a multi-valued function,

$$e^{\pm i\epsilon\ln(x-1)}=e^{\pm i\epsilon[\ln|x-1|+i(\pi+2\pi n)]}=e^{\mp\epsilon(\pi+2\pi n)}\,e^{\pm i\epsilon\ln|x-1|}. \qquad (8.64)$$

The different branches of the function differ only in multipliers, so in (8.64) for definiteness we take $n=0$.

Relations (8.63) determine the exact solutions in the whole physical region $x \in (0,+1)$, because the involved series converges for $x \in (0,+1)$. Note that the solutions are divided in conjugate pairs. Applying the numerical method, it is easily to show that all series have monotonic behavior with a well-marked asymptotic depending on the values ϵ and j.

Let us write down the asymptotic behavior of g_1, g_3 and g_2, g_4:

$x \to 0 (r_* \to -\infty)$,

$$g_1 = [e^{-\epsilon\pi}]\, e^{+i\epsilon r_*}, g_2 = [e^{+\epsilon\pi}]\, e^{-i\epsilon r_*}, \qquad g_3 = [e^{+\epsilon\pi}]\, e^{+i\epsilon r_*}, g_4 = [e^{-\epsilon\pi}]\, e^{-i\epsilon r_*}; \tag{8.65}$$

$x \to +1 (r_* \to +\infty)$,

$$g_1 = [e^{-i\epsilon \ln 2} e^{+i\epsilon/4}\, (R_1 + iI_1)]\, e^{-i\epsilon r_*}, \quad g_2 = [e^{+i\epsilon \ln 2} e^{-i\epsilon/4}\, (R_1 - iI_1)]\, e^{+i\epsilon r_*},$$

$$g_3 = [e^{+i\epsilon \ln 2} e^{-i\epsilon/4}\, (R_3 + iI_3)]\, e^{+i\epsilon r_*}, \quad g_4 = [e^{-i\epsilon \ln 2} e^{+i\epsilon/4}\, (R_3 - iI_3)]\, e^{-i\epsilon r_*}. \tag{8.66}$$

The linear combination of the above solutions

$$H_+ = \frac{g_3 + g_1}{2}, \; H_- = \frac{g_3 - g_1}{2}, \quad F_+ = \frac{g_2 + g_4}{2}, \; F_- = \frac{g_2 - g_4}{2} \tag{8.67}$$

behaves as

$$r_* \to -\infty, \quad H_+ = \cosh \epsilon\pi\, e^{+i\epsilon r_*}, \qquad H_- = \sinh \epsilon\pi\, e^{+i\epsilon r_*};$$

$$r_* \to -\infty, \quad F_+ = \cosh \epsilon\pi\, e^{-i\epsilon r_*}, \quad F_- = \sinh \epsilon\pi\, e^{-i\epsilon r_*}; \tag{8.68}$$

$r_* \to +\infty$,

$$H_\pm = \frac{1}{2}[e^{+i\epsilon \ln 2} e^{-i\epsilon/4} (R_3 + iI_3))] e^{+i\epsilon r_*} \pm \frac{1}{2}[e^{-i\epsilon \ln 2} e^{+i\epsilon/4} (R_1 + iI_1)] e^{-i\epsilon r_*},$$
$$F_\pm = \frac{1}{2}[e^{+i\epsilon \ln 2} e^{-i\epsilon/4} (R_1 - iI_1))]\, e^{+i\epsilon r_*} \pm \frac{1}{2}[e^{-i\epsilon \ln 2} e^{+i\epsilon/4} (R_3 - iI_3)] e^{-i\epsilon r_*}. \tag{8.69}$$

Thus, we have two pairs of solutions with the following asymptotic (on the left $\leftrightarrow$ on the right):

$$H_+, \; \cosh \epsilon\pi e^{+i\epsilon r_*} \;\leftrightarrow\; A e^{+i\epsilon r_*} + B e^{-i\epsilon r_*},$$

$$H_-, \; \sinh \epsilon\pi e^{+i\epsilon r_*} \;\leftrightarrow\; A e^{+i\epsilon r_*} - B e^{-i\epsilon r_*}; \tag{8.70}$$

$$F_+, \; \cosh \epsilon\pi e^{-i\epsilon r_*} \;\leftrightarrow\; A^* e^{-i\epsilon r_*} + B^* e^{+i\epsilon r_*},$$

$$F_-, \; \sinh \epsilon\pi e^{-i\epsilon r_*} \;\leftrightarrow\; A^* e^{-i\epsilon r_*} - B^* e^{+i\epsilon r_*}. \tag{8.71}$$

The function (8.71) seems to be more simple for interpretation: it corresponds to the case when a particle falls from the right on the gravitational barrier, is partially reflected and partially passes through this barrier.

The physical information on tunneling process (8.71)

$$\frac{C_\pm}{A^*} e^{-i\epsilon r_*} \; < - - - \; e^{-i\epsilon r_*} \pm \frac{B^*}{A^*} e^{+-i\epsilon r_*} \tag{8.72}$$

is determined by the reflection R and the transmission D coefficients:

$$D = \mid \frac{C_\pm}{A^*} \mid^2, \qquad R = 1 - D = \mid \frac{B^*}{A^*} \mid^2 . \tag{8.73}$$

If we combine the functions within the pairs $H_+ - -H_-$ or $F_+ - -F_-$, then we obtain the solutions describing the tunneling effect for Majorana particles.

Similarly, we can consider the other physical case when the particle falls from the left on the gravitational barrier, is partially reflected and partially passes through this barrier. At this, we employ the solutions $g_5, ..., g_8$ with the asymptotic behavior

$$x \to +1, \; g_5 = e^{+i\epsilon/2} e^{-i\epsilon/4} (R_5 + iI_5) e^{-i\epsilon r_*}, \quad g_8 = e^{+i\epsilon/2} e^{-i\epsilon/4} (R_7 - iI_7) e^{-i\epsilon r_*},$$

$$x \to 0, \; g_5 = e^{-\epsilon\pi} e^{i\epsilon r_*}, \; g_8 = e^{+\epsilon\pi} e^{-i\epsilon r_*}; \tag{8.74}$$

$$x \to +1, \; g_6 = e^{-i\epsilon/2} e^{+i\epsilon/4} (R_5 - iI_5) e^{+i\epsilon r_*}, \quad g_7 = e^{-i\epsilon/2} e^{+i\epsilon/4} (R_7 + iI_7) e^{+i\epsilon r_*},$$

$$x \to 0, \; g_6 = e^{+\epsilon\pi} e^{-i\epsilon r_*}, \; g_7 = e^{-\epsilon\pi} e^{+i\epsilon r_*}. \tag{8.75}$$

Let us introduce the new combinations $\bar{H}_\pm = g_5 \pm g_8,$; $\bar{F}_\pm = g_6 \pm g_7$; with the following asymptotic

$$\bar{H}_\pm, \; e^{-\epsilon\pi} e^{+i\epsilon r_*} \pm e^{+\epsilon\pi} e^{-i\epsilon r_*} <--> e^{+i\epsilon/2} e^{-i\epsilon/4} [(R_5 + iI_5) \pm (R_7 - iI_7)] \, e^{-i\epsilon r_*}; \tag{8.76}$$

$$\bar{F}_\pm, \; e^{-\epsilon\pi} e^{+i\epsilon r_*} \pm e^{+\epsilon\pi} e^{-i\epsilon r_*} <--> e^{+i\epsilon/2} e^{-i\epsilon/4} [(R_5 - iI_5) \pm (R_7 + iI_7)] \, e^{+i\epsilon r_*}. \tag{8.77}$$

The functions (8.77) may be associated with the physical situation when the particle falls from the left on the gravitational barrier, is partially reflected and partially passes through this barrier:

$$\bar{A} e^{+i\epsilon r_*} \pm \bar{B} e^{-i\epsilon r_*} ---> \quad \bar{C}_\pm \, e^{+i\epsilon r_*} . \tag{8.78}$$

The physics of tunneling process (8.78) is determined by the reflection $\bar{R}$ and the transmission $\bar{D}$ coefficients:

$$\bar{D} = \mid \frac{\bar{C}_\pm}{\bar{A}} \mid^2, \qquad \bar{R} = \mid \frac{\bar{B}}{\bar{A}} \mid^2 . \tag{8.79}$$

8.10 Geometrization of the Maxwell and Dirac Theories in Schwarzschield Space-Time

The behavior of material fields in the vicinity of cosmological objects such as black holes or neutron stars is of great interest. Relevant space-time models describe the gravitational potentials of these objects. However, searching the analytical solutions under the background of curved space-time remains a complicated problem that stipulates the development of other methods to analyze the behavior of the corresponding dynamical systems. For instance, the method of KCC-invariants is known [163, 179].

We consider the Maxwell equations on the background of Schwarzschild space-time

$$dS^2 = \Phi dt^2 - \frac{dr^2}{\Phi} - r^2\left(d\theta^2 + \sin^2\theta d\varphi^2\right), \tag{8.80}$$

where the function $\Phi = 1 - \frac{M}{r}$, and the differential equation system for the radial components were found after separating the variables in initial Maxwell equations both in 3-dimensional Majorana–Oppenheimer and in 10-demensional Duffin–Kemmer–Petiau approaches. We start with the second order differential equation for the primary radial function G that was obtained in the Majorana–Oppenheimer formalism:

$$\frac{d^2F}{dx^2} + \left(\frac{1}{x-1} - \frac{1}{x}\right)\frac{dF}{dx}$$
$$+\left(M^2\omega^2 + \frac{j(j+1)}{x} + \frac{2M^2\omega^2 - j(j+1)}{x-1} + \frac{M^2\omega^2}{(x-1)^2}\right)F = 0, \tag{8.81}$$

where $x = r/M$ is a dimensionless coordinate. The spray coefficient G equals

$$G = \frac{1}{2}\left(\frac{1}{x-1} - \frac{1}{x}\right)\frac{dF}{dx}$$
$$+\frac{1}{2}\left(M^2\omega^2 + \frac{j(j+1)}{x} + \frac{2M^2\omega^2 - j(j+1)}{x-1} + \frac{M^2\omega^2}{(x-1)^2}\right)F. \tag{8.82}$$

The first KCC invariant is found in the form

$$\varepsilon = -\frac{1}{2}\left(\frac{1}{x-1} - \frac{1}{x}\right)\frac{dF}{dx} - \left(M^2\omega^2 + \frac{j\big(j+1\big)}{x} + \frac{2M^2\omega^2 - j\big(j+1\big)}{x-1} + \frac{M^2\omega^2}{(x-1)^2}\right)F,$$

the second KCC invariant is

$$P = \frac{-3 + 4x\Big(1 - j(j+1)(x-1) + M^2\omega^2x^3\Big)}{4(x-1)^2x^2}. \tag{8.83}$$

In the Duffin–Kemmer formalism, the equation for the other primary radial function f is

$$\frac{d^2f}{dr^2} + \left(\frac{\Phi'}{\Phi} + \frac{2}{r}\right)\frac{df}{dr} + \left(\frac{\omega^2}{\Phi^2} - \frac{2\nu^2}{r^2\Phi} + \frac{\Phi'}{\Phi}\frac{1}{r}\right)f = 0, \tag{8.84}$$

where the derivative over r is denoted by a prime. The coefficient G equals

$$G = \frac{(M-2r)}{2(M-r)r}\frac{df}{dr} + \frac{\left(-M^2 - 2\nu^2r^2 + M(r + 2\nu^2r) + r^4\omega^2\right)}{2(M-r)^2r^2}f. \tag{8.85}$$

The first and the second KCC-invariants take the form

$$\varepsilon = -\frac{(M-2r)}{2(M-r)r}\frac{df}{dr} + \frac{\left(M^2 + 2\nu^2r^2 - M(r + 2\nu^2r) - r^4\omega^2\right)}{(M-r)^2r^2}f,$$
$$P = \frac{-3M^2 - 8\nu^2r^2 + 4M(r + 2\nu^2r) + 4\omega^2r^4}{4(1 - r/M)^2r^2}. \tag{8.86}$$

Utilizing the notations $x = r/M$, one finds the second KCC-invariant in the form

$$P = \frac{-3+4x\left(1-2\nu^2(x-1)+M^2\omega^2x^3\right)}{4(x-1)^2x^2}, \quad 2\nu^2 = j(j+1),$$

which coincides with (8.83). Near the singular points, the eigenvalue of the second invariant $(\Lambda \equiv P)$ behaves as follows:

$$x = \frac{r}{M} \to 0, \quad \Lambda \to -\frac{3}{4x^2} < 0; \quad x \to \infty, \quad \Lambda \to \omega^2 > 0; \tag{8.87}$$

$$x \to 1 \quad (r \to M), \quad \Lambda \to \frac{1+4M^2\omega^2}{4(x-1)^2} > 0. \tag{8.88}$$

It indicates that in the vicinity of $x = 0$ (this point is nonphysical) the geodesics converge. Vice versa, near the physical points, at the Schwarzschild horizon $x = 1$ and at $x \to \infty$, the Jacobi instability exists and the geodesics diverge. The typical behavior of the eigenvalue as function of the radial coordinate is shown in fig. 8.1.

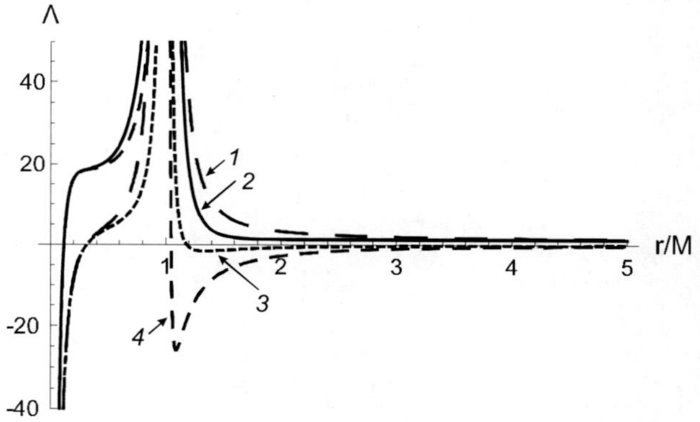

Figure 8.1: Behavior of the real parts of the second invariant eigenvalues for the geometrized problem of the electromagnetic field on the background of the Schwarzschild spacetime. The values of parameters used are: $M = 1$, $\omega = 0.0001$ (curves 3 and 4) and 1.0001 (curves 1 and 2), $j = 1$ (curves 1 and 3) and 2 (curves 2 and 4).

Now let us consider the spin 1/2 particle on the background of Schwarzschild space-time. Starting from a generally covariant form of the Dirac equation

$$\left[i\gamma^a\left(e^{\alpha}_{(a)}\frac{\partial}{\partial x^\alpha}+\frac{1}{2}\left(\frac{1}{\sqrt{-g}}\frac{\partial}{\partial x^\alpha}\sqrt{-g}e^{\alpha}_{(a)}\right)\right)-m\right]\Psi(x) = 0 \tag{8.89}$$

in orthogonal static coordinates $x^\alpha = (t, \theta, \varphi, r)$ and tetrad in Schwarzschild space-time

$$e^{\beta}_{(a)} = g^{\beta\alpha}e_{(a)\alpha} = \begin{vmatrix} \Phi^{-1/2} & 0 & 0 & 0 \\ 0 & 1/r & 0 & 0 \\ 0 & 0 & 1/r\sin\theta & 0 \\ 0 & 0 & 0 & \Phi^{1/2} \end{vmatrix}, \tag{8.90}$$

after separating the variables with the diagonalization of the total angular momentum we derive the system of equations

$$
\begin{aligned}
&\frac{\varepsilon}{\sqrt{1-\frac{M}{r}}} f_3 - i\sqrt{1-\frac{M}{r}}\frac{df_3}{dt} - i\frac{\nu}{r} f_4 - M f_1 = 0 ,\\
&\frac{\varepsilon}{\sqrt{1-\frac{M}{r}}} f_4 + i\sqrt{1-\frac{M}{r}}\frac{df_4}{dt} + i\frac{\nu}{r} f_3 - M f_2 = 0 ,\\
&\frac{\varepsilon}{\sqrt{1-\frac{M}{r}}} f_1 + i\sqrt{1-\frac{M}{r}}\frac{df_1}{dt} - i\frac{\nu}{r} f_2 - M f_3 = 0 ,\\
&\frac{\varepsilon}{\sqrt{1-\frac{M}{r}}} f_2 - i\sqrt{1-\frac{M}{r}}\frac{df_2}{dt} - i\frac{\nu}{r} f_1 - M f_4 = 0 .
\end{aligned}
\qquad (8.91)
$$

Each complex function $f_i(r)$ we will resolve into a sum of real and imaginary parts:

$$
f_1 = f_{11} + i f_{12} , \quad f_2 = f_{21} + i f_{22} , \quad f_3 = f_{31} + i f_{32} \quad f_4 = f_{41} + i f_{42} . \qquad (8.92)
$$

Substituting the expressions (8.92) into the system (8.91), one gets the system of 8 connected differential equations of the first order:

$$
\begin{aligned}
&\frac{df_{11}}{dr}\sqrt{1-\frac{M}{r}} + \frac{f_{12}\varepsilon}{\sqrt{1-\frac{M}{r}}} + \frac{f_{21}\nu}{r} - f_{32} m = 0 ,\\
&-\frac{df_{12}}{dr}\sqrt{1-\frac{M}{r}} + \frac{f_{11}\varepsilon}{\sqrt{1-\frac{M}{r}}} - \frac{f_{22}\nu}{r} - f_{31} m = 0 ,\\
&-\frac{df_{21}}{dr}\sqrt{1-\frac{M}{r}} - \frac{f_{11}\nu}{r} + \frac{f_{22}\varepsilon}{\sqrt{1-\frac{M}{r}}} - f_{42} m = 0 ,\\
&\frac{df_{22}}{dr}\sqrt{1-\frac{M}{r}} + \frac{f_{12}\nu}{r} + \frac{f_{21}\varepsilon}{\sqrt{1-\frac{M}{r}}} - f_{41} m = 0 ,\\
&-\frac{df_{31}}{dr}\sqrt{1-\frac{M}{r}} - f_{12} m + \frac{f_{32}\varepsilon}{\sqrt{1-\frac{M}{r}}} - \frac{f_{41}\nu}{r} = 0 ,\\
&\frac{df_{32}}{dr}\sqrt{1-\frac{M}{r}} - f_{11} m + \frac{\varepsilon f_{31}}{\sqrt{1-\frac{M}{r}}} + \frac{f_{42}\nu}{r} = 0 ,\\
&\frac{df_{41}}{dr}\sqrt{1-\frac{M}{r}} - f_{22} m + \frac{f_{31}\nu}{r} + \frac{f_{42}\varepsilon}{\sqrt{1-\frac{M}{r}}} = 0 ,\\
&-\frac{df_{42}}{dr}\sqrt{1-\frac{M}{r}} - f_{21} m - \frac{f_{32}\nu}{r} + \frac{f_{41}\varepsilon}{\sqrt{1-\frac{M}{r}}} = 0 .
\end{aligned}
\qquad (8.93)
$$

We differentiate each equation over the radial variable and introduce the notations:

$$f_{11}=x^1,\ f_{12}=x^2,\ f_{21}=x^3,\ f_{22}=x^4,\ f_{31}=x^5,\ f_{32}=x^6,\ f_{41}=x^7,\ f_{42}=x^8,$$

$$\frac{df_{11}}{dr}=y^1,\ \frac{df_{12}}{dr}=y^2,\ \frac{df_{21}}{dr}=y^3,\ \frac{df_{22}}{dr}=y^4,$$

$$\frac{df_{31}}{dr}=y^5,\ \frac{df_{32}}{dr}=y^6,\ \frac{df_{41}}{dr}=y^7,\ \frac{df_{42}}{dr}=y^8,$$

this results in

$$\frac{dy^1}{dr}-\frac{my^6}{\sqrt{1-\frac{M}{r}}}-\frac{\nu x^3}{r^2\sqrt{1-\frac{M}{r}}}-\frac{Mx^2\varepsilon}{2r^2(1-\frac{M}{r})^2}+\frac{My^1}{2r^2(1-\frac{M}{r})}+\frac{\nu y^3}{r\sqrt{1-\frac{M}{r}}}+\frac{y^2\varepsilon}{(1-\frac{M}{r})}=0,$$

$$\frac{dy^2}{dr}+\frac{my^5}{\sqrt{1-\frac{M}{r}}}-\frac{\nu x^4}{r^2\sqrt{1-\frac{M}{r}}}+\frac{Mx^1\varepsilon}{2r^2(1-\frac{M}{r})^2}+\frac{My^2}{2r^2(1-\frac{M}{r})}+\frac{\nu y^4}{r\sqrt{1-\frac{M}{r}}}-\frac{y^1\varepsilon}{(1-\frac{M}{r})}=0,$$

$$\frac{dy^3}{dr}+\frac{my^8}{\sqrt{1-\frac{M}{r}}}+\frac{Mx^4\varepsilon}{2r^2(1-\frac{M}{r})^2}-\frac{\nu x^1}{r^2\sqrt{1-\frac{M}{r}}}+\frac{My^3}{2r^2(1-\frac{M}{r})}-\frac{y^4\varepsilon}{(1-\frac{M}{r})}+\frac{\nu y^1}{r\sqrt{1-\frac{M}{r}}}=0,$$

$$\frac{dy^4}{dr}-\frac{my^7}{\sqrt{1-\frac{M}{r}}}-\frac{Mx^3\varepsilon}{2r^2(1-\frac{M}{r})^2}-\frac{\nu x^2}{r^2\sqrt{1-\frac{M}{r}}}+\frac{My^4}{2r^2(1-\frac{M}{r})}+\frac{y^3\varepsilon}{(1-\frac{M}{r})}+\frac{\nu y^2}{r\sqrt{1-\frac{M}{r}}}=0,$$

$$\frac{dy^5}{dr}+\frac{my^2}{\sqrt{1-\frac{M}{r}}}-\frac{\nu x^7}{r^2\sqrt{1-\frac{M}{r}}}+\frac{Mx^6\varepsilon}{2r^2(1-\frac{M}{r})^2}+\frac{My^5}{2r^2(1-\frac{M}{r})}+\frac{\nu y^7}{r\sqrt{1-\frac{M}{r}}}-\frac{y^6\varepsilon}{(1-\frac{M}{r})}=0,$$

$$\frac{dy^6}{dr}-\frac{my^1}{\sqrt{1-\frac{M}{r}}}-\frac{\nu x^8}{r^2\sqrt{1-\frac{M}{r}}}-\frac{Mx^5\varepsilon}{2r^2(1-\frac{M}{r})^2}+\frac{My^6}{2r^2(1-\frac{M}{r})}+\frac{\nu y^8}{r\sqrt{1-\frac{M}{r}}}+\frac{y^5\varepsilon}{(1-\frac{M}{r})}=0,$$

$$\frac{dy^7}{dr}-\frac{my^4}{\sqrt{1-\frac{M}{r}}}-\frac{Mx^8\varepsilon}{2r^2(1-\frac{M}{r})^2}-\frac{\nu x^5}{r^2\sqrt{1-\frac{M}{r}}}+\frac{My^7}{2r^2(1-\frac{M}{r})}+\frac{y^8\varepsilon}{(1-\frac{M}{r})}+\frac{\nu y^5}{r\sqrt{1-\frac{M}{r}}}=0,$$

$$\frac{dy^8}{dr}+\frac{my^3}{\sqrt{1-\frac{M}{r}}}+\frac{Mx^7\varepsilon}{2r^2(1-\frac{M}{r})^2}-\frac{\nu x^6}{r^2\sqrt{1-\frac{M}{r}}}+\frac{My^8}{2r^2(1-\frac{M}{r})}-\frac{y^7\varepsilon}{(1-\frac{M}{r})}+\frac{\nu y^6}{r\sqrt{1-\frac{M}{r}}}=0.$$

One can find the coefficients Q^i in explicit form. Then the first invariant is determined as (here $\zeta^i\equiv\varepsilon^i$):

$$\zeta^1=\frac{My^1-2\sqrt{1-\frac{M}{r}}\,(mr^2y^6-\nu ry^3+2\nu x^3)}{4r(M-r)}+\frac{\varepsilon\left(ry^2(M-r)+Mx^2\right)}{2(M-r)^2},$$

$$\zeta^2=\frac{2\sqrt{1-\frac{M}{r}}\,(mr^2y^5+\nu ry^4-2\nu x^4)+My^2}{4r(M-r)}+\frac{\varepsilon(ry^1(r-M)-Mx^1)}{2(M-r)^2},$$

$$\zeta^3=\frac{1}{4r(M-r)}\left(2mr^2y^8\sqrt{1-\frac{M}{r}}+My^3-2\nu\sqrt{1-\frac{M}{r}}(2x^1-ry^1)\right)$$

$$+\frac{M\varepsilon(x^4-ry^4)+r^2y^4\varepsilon}{2(M-r)^2},$$

$$\zeta^4=\frac{1}{4r(M-r)}\Big(-2mr^2y^7\sqrt{1-\frac{M}{r}}+My^4+2\nu\sqrt{1-\frac{M}{r}}(ry^2-2x^2)\Big)$$

$$+\frac{M\varepsilon(ry^3+x^3)-r^2y^3\varepsilon}{2r(M-r)^2},$$

$$\zeta^5=\frac{2\sqrt{1-\frac{M}{r}}\left(mr^2y^2+\nu ry^7-2\nu x^7\right)+My^5}{4r(M-r)}+\frac{\varepsilon\left(ry^6(r-M)-Mx^6\right)}{2(M-r)^2},$$

$$\zeta^6=\frac{My^6-2\sqrt{1-\frac{M}{r}}\left(mr^2y^1-\nu ry^8+2\nu x^8\right)}{4r(M-r)}+\frac{\varepsilon\left(ry^5(M-r)+Mx^5\right)}{2(M-r)^2},$$

$$\zeta^7=\frac{M^2(2mr^2y^4+ry^7\sqrt{1-\frac{M}{r}}-2\nu(ry^5-2x^5))}{4r^2\sqrt{1-\frac{M}{r}}(M-r)^2}$$

$$-\frac{r^2(y^8\varepsilon\sqrt{1-\frac{M}{r}}-my^4)+\nu\left(ry^5-2x^5\right)}{2\sqrt{1-\frac{M}{r}}(M-r)^2}$$

$$+\frac{Mr(\sqrt{1-\frac{M}{r}}\left(2\varepsilon\left(ry^8+x^8\right)-y^7\right)-4mry^4)+4M\nu(ry^5-2x^5)}{4r\sqrt{1-\frac{M}{r}}(M-r)^2},$$

$$\zeta^8=\frac{M^2(-2mr^2y^3+ry^8\sqrt{1-\frac{M}{r}}-2\nu ry^6+4\nu x^6)}{4r^2\sqrt{1-\frac{M}{r}}(M-r)^2}$$

$$-\frac{r^2(my^3-y^7\varepsilon\sqrt{1-\frac{M}{r}})+\nu(ry^6-2x^6)}{2\sqrt{1-\frac{M}{r}}(M-r)^2}$$

$$-\frac{Mr(\sqrt{1-\frac{M}{r}}(2\varepsilon(ry^7+x^7)+y^8)-4mry^3)+4M\nu(2x^6-ry^6)}{4r\sqrt{1-\frac{M}{r}}(M-r)^2},$$

or, in complex form:

$$\Omega^1=\zeta^1+i\zeta^2$$

$$=\frac{1}{4r(M-r)}\Big(\frac{df_1}{dr}(M+2i\varepsilon r^2)+2\sqrt{1-\frac{M}{r}}\Big(\frac{df_2}{dr}\nu r+i\frac{df_3}{dr}mr^2-2f_2\nu\Big)\Big)-\frac{2i\varepsilon M}{4(M-r)^2}f_1,$$

$$\Omega^2=\zeta^3+i\zeta^4$$

$$=\frac{1}{4r(M-r)}\Big(\frac{df_2}{dr}(M+2i\varepsilon r^2)+2\sqrt{1-\frac{M}{r}}\Big(-\frac{df_1}{dr}\nu+i\frac{df_4}{dr}mr+2f_2\frac{\nu}{r}\Big)\Big)-\frac{i\varepsilon M}{2(M-r)^2}f_2,$$

$$\Omega^3 = \zeta^5 + i\zeta^6$$

$$= \frac{1}{4r(M-r)}\Big(\frac{df_3}{dr}(M+2i\varepsilon r^2)+2\sqrt{1-\frac{M}{r}}\Big(\frac{df_4}{dr}\nu r - i\frac{df_1}{dr}mr^2 - 2f_4\nu\Big)\Big) + \frac{i\varepsilon M}{2(M-r)^2}f_3,$$

$$\Omega^4 = \zeta^7 + i\zeta^8$$

$$= \frac{1}{4r(M-r)}\Big(\frac{df_4}{dr}(M-2i\varepsilon r^2)+2\sqrt{1-\frac{M}{r}}\Big(-\frac{df_3}{dr}\nu - i\frac{df_2}{dr}mr + 2f_3\frac{\nu}{r}\Big)\Big) - \frac{i\varepsilon M}{2(M-r)^2}f_4.$$

The second invariant P^i_j can be directly calculated according to the the known formula, which gives

$$P^i_j = \begin{vmatrix} \alpha & -\beta & \gamma & 0 & 0 & 0 & 0 & 0 \\ \beta & \alpha & 0 & \gamma & 0 & 0 & 0 & 0 \\ \gamma & 0 & \alpha & \beta & 0 & 0 & 0 & 0 \\ 0 & \gamma & -\beta & \alpha & 0 & 0 & 0 & 0 \\ 0 & 0 & 0 & 0 & \alpha & \beta & \gamma & 0 \\ 0 & 0 & 0 & 0 & -\beta & \alpha & 0 & \gamma \\ 0 & 0 & 0 & 0 & \gamma & 0 & \alpha & -\beta \\ 0 & 0 & 0 & 0 & 0 & \gamma & \beta & \alpha \end{vmatrix}, \qquad (8.94)$$

where

$$\alpha = \frac{4r^4(\varepsilon^2-m^2)+4Mr\left(m^2r^2+\nu^2+2\right)-5M^2-4\nu^2r^2}{16r^2(M-r)^2},$$

$$\beta = \frac{\varepsilon M}{4(M-r)^2}, \qquad \gamma = -\frac{\nu}{2r^2\sqrt{1-\frac{M}{r}}}.$$

The eigenvalues Λ_i of the second invariant P^i_j are four-fold degenerate, and the different two are determined as

$$\Lambda_{1,2} = \alpha \pm \sqrt{\gamma^2-\beta^2}$$

$$= \frac{1}{16}\Big(\frac{4m^2r^2+4\nu^2-2}{r(M-r)} \pm 4\sqrt{-\frac{M^2\varepsilon^2}{(M-r)^4} - \frac{4\nu^2}{r^3(M-r)}} + \frac{4r^2\varepsilon^2+3}{(M-r)^2} - \frac{5}{r^2}\Big).$$

The dependencies of the two different eigenvalues $\Lambda_{1,2}$ of the second invariant on the radial variable have been analyzed for different values of energy ϵ (see fig. 8.2) Near the singular points, the eigenvalues behave as follows:

$$r \to 0, \qquad \Lambda_{1,2} \to -\frac{5}{16r^2} < 0,$$

$$r \to M, \qquad \Lambda_{1,2} \to \frac{r^2\varepsilon^2+3}{4(M-r)^2} \pm \frac{iM\varepsilon}{4(M-r)^2}, \quad \mathrm{Re}(\Lambda_{1,2}) \to \frac{r^2\varepsilon^2+3}{4(M-r)^2} > 0,$$

$$r \to \infty, \qquad \Lambda_{1,2} \to \frac{1}{4}\left(\varepsilon^2-m^2\right).$$

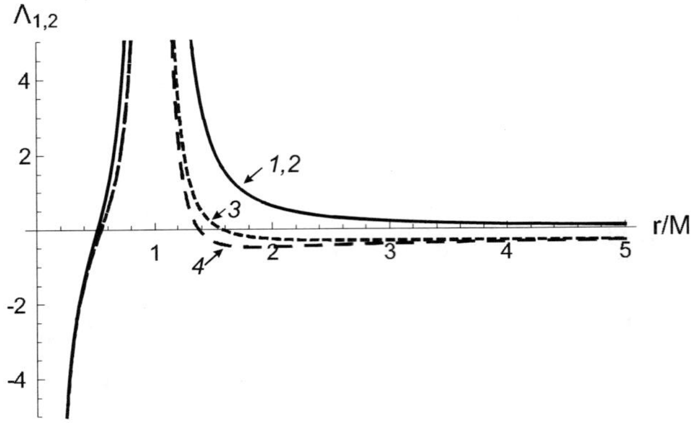

Figure 8.2: Behavior of the real parts of the second invariant eigenvalues for the geometrized problem of the spinor field on the background of the Schwarzschild spacetime. The used values of parameters are: $m = 1$, $M = 1$, $\epsilon = 0.0001$ (curves 3 and 4) and 1.0001 (curves 1 and 2), $j = 1/2$; curves 1 and 3 and curves 2 and 4 correspond to two different eigenvalues Λ_1 and Λ_2, correspondingly.

So, in the vicinity of the Schwarzschild horizon, the geodesics diverge at any energy ϵ, while at $r \rightarrow \infty$ the geodesics diverge at $\epsilon > m$ and converge at $\epsilon < m$; however, the second possibility may be ignored as being nonphysical. Contradictory results have been found out for the problem of the existence of bound states for the Dirac equation in the presence of a black hole [180].Our results on the KCC-analysis support the conclusion that the fermion bound state on the background of the Schwarzschild black hole are absent as the bound state solution has to be characterized by the convergence of the geodesics flow near the extreme points $r \rightarrow M$ and $r \rightarrow \infty$.

Chapter 9

Dirac Particle in Cylindric Parabolic Coordinates and Spinor Space Structure

9.1 Spinor Structure and Solutions of the Klein–Gordon–Fock Equation

In the literature [182]–[253], there exist three different approaches, whose intrinsic essence is similar: a space-time spinor structure (Penrose, Rindler et al.); the Hopf bundle and the Kustaanheimo – Stiefel bundles. In Hopf's technique, one uses only complex 2-spinors (ξ) and their conjugates (ξ^*), instead of real-valued 4-vector (tensor) objects. In the Kustaanheimo–Stiefel approach, there are used four real-valued coordinates, the real and imaginary parts of 2-spinor components. The formalism developed in the present work exploits as well the possibilities given by spinors to construct 3-vectors; however, the emphasis is put on doubling the set of spacial points, so that we get an extended space model. In such a space, instead of the 2π-rotation, the 4π-rotation is considered, which transfers the space into itself. The procedure of extending the set of manifold points is achieved much easier by using curvilinear coordinates.

Within the framework of applications of spinor theory to Quantum Mechanics, we discuss examine the role of spinor space structure in classifying solutions of scalar and spinor wave equations specified for cylindric parabolic coordinates.

Let us start with the parabolic cylindrical coordinates

$$x = (u^2 - v^2)/2\,, \qquad y = u\,v\,, \qquad z = z\,.$$

In order to cover the vector space (x, y, z), it suffices to make a choice out of the following four possibilities:

$$
\begin{array}{ll}
v = +\sqrt{-x+\sqrt{x^2+y^2}}\,, & u = \pm\sqrt{+x+\sqrt{x^2+y^2}}\,, \\
v = -\sqrt{-x+\sqrt{x^2+y^2}}\,, & u = \pm\sqrt{+x+\sqrt{x^2+y^2}}\,, \\
v = \pm\sqrt{-x+\sqrt{x^2+y^2}}\,, & u = +\sqrt{+x+\sqrt{x^2+y^2}}\,, \\
v = \pm\sqrt{-x+\sqrt{x^2+y^2}}\,, & u = -\sqrt{+x+\sqrt{x^2+y^2}}\,.
\end{array}
$$

For definiteness, let us use the first variant from the above:

$$
v = +\sqrt{-x+\sqrt{x^2+y^2}}\,, \qquad u = \pm\sqrt{+x+\sqrt{x^2+y^2}}\,.
$$

which is illustrated in Fig. 9.1.

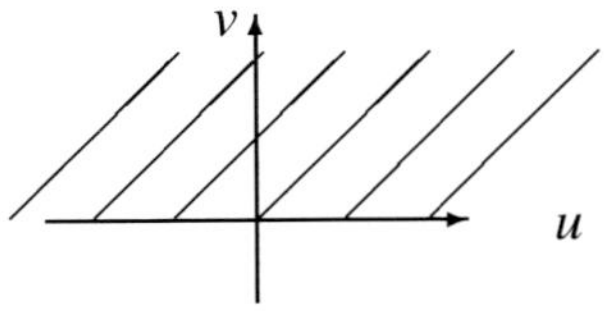

Fig. 9.1. The region $G(u,v)$, used to parameterize the vector model.

The correspondence between the points (x,y) and (u,v) can be illustrated by the following formulas and Fig. 9.2:

$$
u = k\,\cos\phi\,, \quad v = k\,\sin\phi\,, \quad \phi \in [\,0,\,\pi\,]\,;
$$
$$
x = (k^2/2)\,\cos 2\phi\,, \quad y = (k^2/2)\,\sin 2\phi\,, \quad 2\phi \in [0,\,2\pi]\,.
$$

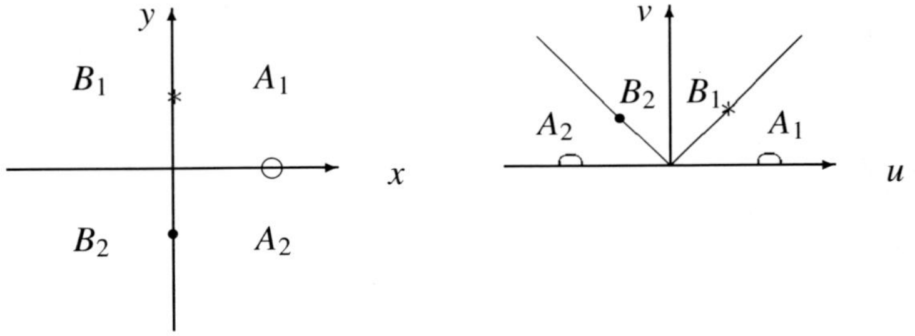

Fig. 9.2. The mapping $G(x,y) \Longrightarrow G(u,v)$; identification rules

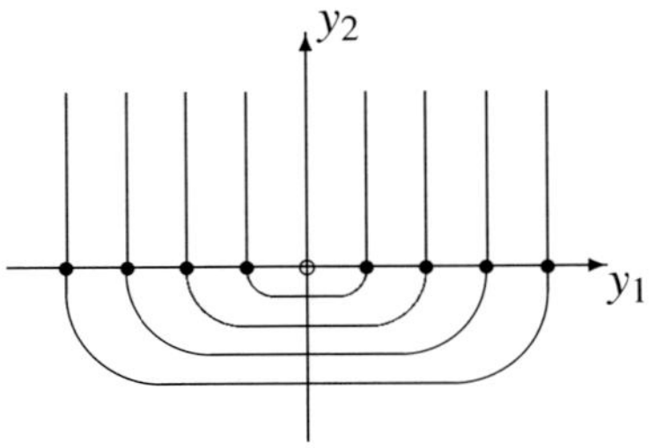

Fig. 9.3. Parabolic cylindrical coordinates

In Fig. 9.3, the identified points on the boundary are connected by lines, and the domain $G(y_1,y_2)^{y_3}$ (at arbitrary y_3) ranging in the half-plane (y_1,y_2) covers the whole vector plane $(x_1,x_2)^{x_3}$. When turning to the case of spinor space, we shall note the complete symmetry between the coordinates u and v; they relate to the Cartesian coordinates of the extended model $(x,y,z)\oplus(x',y',z')$ through the formulas (see Fig. 9.4.):

$$v=\pm\sqrt{-x+\sqrt{x^2+y^2}}, \qquad u=\pm\sqrt{+x+\sqrt{x^2+y^2}}.$$

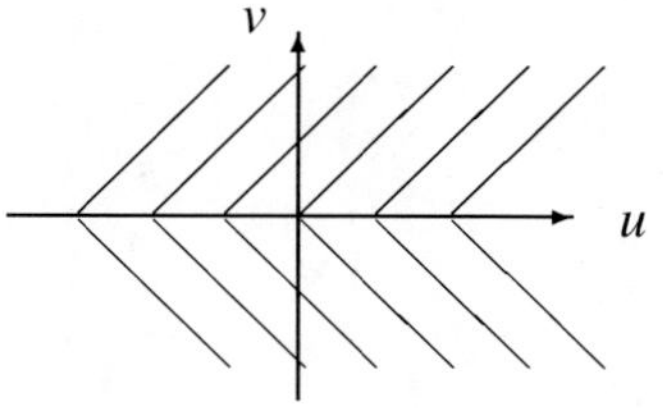

Fig. 9.4. $\tilde{G}(u,v)$ covering the spinor space

The metric of space-time in parabolic cylindrical coordinates has the form

$$dS^2=c^2dt^2-(u^2+v^2)(du^2+dv^2)-dz^2.$$

9.2 Solutions of the Klein–Gordon–Fock Equation and Spinors

Let us consider the equation

$$\Big[-\frac{1}{c^2}\frac{\partial^2}{\partial t^2}+\frac{\partial^2}{\partial z^2}+\frac{1}{u^2+v^2}\Big(\frac{\partial^2}{\partial u^2}+\frac{\partial^2}{\partial v^2}\Big)-\frac{m^2c^2}{\hbar^2}\Big]\Psi=0. \tag{9.1}$$

After separating the variables by the substitution

$$\Psi(t,u,v,\phi)=e^{-i\epsilon t/\hbar}\,e^{ipz/\hbar}\,U(u)\,V(v),$$

one gets

$$\Big[\frac{1}{U}\frac{d^2U}{du^2}+\Big(\frac{\epsilon^2}{\hbar^2c^2}-\frac{m^2c^2}{\hbar^2}-\frac{p^2}{\hbar^2}\Big)u^2\Big]+\Big[\frac{1}{V}\frac{d^2V}{dv^2}+\Big(\frac{\epsilon^2}{\hbar^2c^2}-\frac{m^2c^2}{\hbar^2}-\frac{p^2}{\hbar^2}\Big)v^2\Big]=0. \tag{9.2}$$

In the following, we shall use the notation

$$\lambda^2=\Big(\frac{\epsilon^2}{\hbar^2c^2}-\frac{m^2c^2}{\hbar^2}-\frac{p^2}{\hbar^2}\Big), \qquad [\lambda]=\frac{1}{\text{meter}}. \tag{9.3}$$

By introducing two separation constants a and b (satisfying $a+b=0$), we can derive from (9.2) two distinct equations

$$\frac{d^2U}{du^2}+(\lambda^2u^2-a)\,U=0, \qquad \frac{d^2V}{dv^2}+(\lambda^2v^2-b)\,V=0. \tag{9.4}$$

Transition in eqs. (9.4) to the canonical form is obtained by using dimensionless variables

$$\sqrt{2\lambda}\,u \;\to\; u\,, \quad \frac{a}{2\lambda} \;\to\; a\,, \quad \sqrt{2\lambda}\,v \;\to\; v\,, \quad \frac{b}{2\lambda} \;\to\; b\,. \tag{9.5}$$

The equations (9.4) will take the form

$$\frac{d^2U}{du^2} + \Big(\frac{u^2}{4} - a\Big)\,U = 0\,, \qquad \frac{d^2V}{dv^2} + \Big(\frac{v^2}{4} + a\Big)\,V = 0\,. \tag{9.6}$$

The solutions of these similar equations can be found in series form:

$$F(\xi) = c_0 + c_1\xi + c_2\xi^2 + \sum_{k=1,2,...} c_{2k+1}\xi^{2k+1} + \sum_{k=1,2,...} c_{2k+2}\xi^{2k+2}\,; \tag{9.7}$$

we note that in (9.7) the terms of even and odd powers of ξ are separated.

After tedious calculations, one derives two independent groups of recurrent relations:

for even powers

$$\begin{aligned} \xi^0 :& \quad 2\,c_2 - \alpha\,c_0 = 0\,, \\ \xi^2 :& \quad c_4\,4\times 3 + \tfrac{c_0}{4} - \alpha\,c_2 = 0\,, \\ \xi^4 :& \quad c_6\,6\times 5 + \tfrac{c_2}{4} - \alpha\,c_4 = 0\,, \\ n = 3,4,...,\; \xi^{2n} :& \quad c_{2n+2}(2n+2)(2n+1) + \tfrac{1}{4}\,c_{2n-2} - \alpha\,c_{2n} = 0\,; \end{aligned} \tag{9.8}$$

for odd powers

$$\begin{aligned} \xi^1 :& \quad c_3\,3\times 2 - \alpha\,c_1 = 0\,, \\ \xi^3 :& \quad c_5\,5\times 4 + \tfrac{c_1}{4} - \alpha\,c_3 = 0\,, \\ n = 3,4,...,\; \xi^{2n-1} :& \quad c_{2n+1}(2n+1)(2n) + \tfrac{1}{4}\,c_{2n-3} - \alpha\,c_{2n-1} = 0\,. \end{aligned} \tag{9.9}$$

So one can construct the following two linearly independent solutions

even

$$F_1(\xi^2) = 1 + a_2\frac{\xi^2}{2!} + a_4\frac{\xi^4}{4!} + \ldots\,, \quad a_2 = \alpha\,, \qquad a_4 = \alpha^2 - \frac{1}{2}\,, \qquad c_6 = \alpha^3 - \frac{7}{2}\alpha\,,$$

$$n = 3,4,\ldots : \qquad a_{2n+2} = \alpha\,a_{2n} - \frac{(2n)(2n-1)}{4}\,a_{2n-2}\,; \tag{9.10}$$

odd

$$F_2(\xi) = \xi + a_3\,\frac{\xi^3}{3!} + a_5\,\frac{\xi^5}{5!} + \ldots\,, \quad a_3 = \alpha\,, \qquad a_5 = \alpha^2 - \frac{3}{2}\,, \tag{9.11}$$

$$n = 3,4,\ldots : \qquad a_{2n+1} = \alpha\,a_{2n-1} - \frac{(2n-1)(2n-2)}{4}\,a_{2n-3}\,. \tag{9.12}$$

Having combined the two previous solutions F_1 and F_2, we can obtain four types of wave functions[1]

$$
\begin{array}{ll}
(\text{even} \otimes \text{even}), & \Phi_{++} = E(a,u^2)\, E(-a,v^2)\,; \\
(\text{odd} \otimes \text{odd}), & \Phi_{--} = O(a,u)\, O(-a,v\,)\,; \\
(\text{even} \otimes \text{odd}), & \Phi_{+-} = E(a,u^2)\, O(-a,v\,)\,; \\
(\text{odd} \otimes \text{even}), & \Phi_{-+} = O(a,u)\, E(-a,v^2)\,.
\end{array}
\tag{9.13}
$$

We note the behavior of the constructed wave functions:

$$
\begin{array}{ll}
\Phi_{++}(x=0,y=0) \neq 0\,, & \Phi_{--}(x=0,y=0) = 0\,, \\
\Phi_{+-}(x>0,y=0) = 0\,, & \Phi_{-+}(x<0,y=0) = 0\,.
\end{array}
\tag{9.14}
$$

Now let us find which restrictions for the wave functions Ψ follow from the requirement of single-valuedness. To this aim, two properties of the parametrization are essential (see Fig. 9.5):

$$
\underline{v=0}\ :\ x = +\frac{u^2}{2} \geq 0\,,\ y=0; \quad \underline{u=0}\ :\ x = -\frac{v^2}{2} \leq 0\,,\ y=0\,.
$$

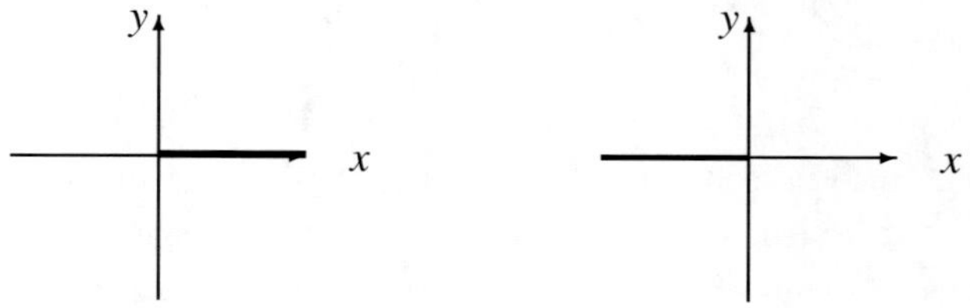

Fig. 9.5. The peculiarities of the parametrization

The four solutions behave in special regions, as follows:

$$
\begin{array}{l}
\Phi_{++}(a;u=0,v) = +\,\Phi_{++}(a;u=0,-v)\,, \\
\Phi_{++}(a;+u,v=0) = +\,\Phi_{++}a;-u,v=0)\,, \\
\Phi_{--}(a;u=0,+v) = +\,\Phi_{--}(a;u=0,-v) = 0\,, \\
\Phi_{--}(a;u,v=0) = +\,\Phi_{--}(a;-u,v=0) = 0\,,
\end{array}
\tag{9.15}
$$

$$
\begin{array}{l}
\Phi_{+-}(a;u=0,+v) = -\,\Phi_{+-}(a;u=0,-v)\,, \\
\Phi_{+-}(a;u,v=0) = \Phi_{+-}(a;-u,v=0) = 0\,, \\
\Phi_{-+}(a;u=0,+v) = \Phi_{-+}(a;u=0,-v) = 0\,, \\
\Phi_{-+}(a;+u,v=0) = -\,\Phi_{-+}(a;-u,v=0)\,.
\end{array}
\tag{9.16}
$$

The boundary properties of the constructed wave functions can be illustrated by the schemes described in Fig. 9.6. So we conclude that the solutions Ψ of the types $(++)$ and $(--)$ are single-valued in the space with vector structure, whereas

[1] We will change the notation: $F_1 \Longrightarrow E$; $F_2 \Longrightarrow O$.

the solutions of the types (+−) and (−+) are not single-valued in such a space, so these latter types (+−) and (−+) must be discarded. However, these solutions ((+−) and (−+)) must be retained in the space with spinor structure.

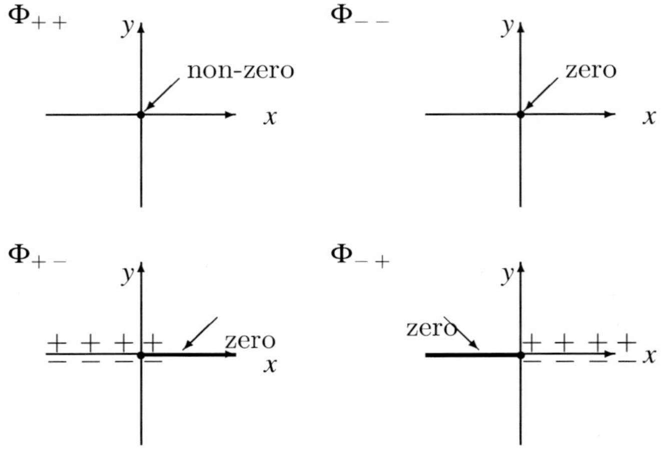

Fig 9.6. Boundary behavior of the wave functions in the (x,y)-plane

When using the spinor space model, the two sets (u,v) and $(-u,-v)$ represent different geometrical points in the spinor space, so the requirement of single valuedness, like applied in the case of the spinor space does not assume that the values of the wave functions must be equal at the points (u,v) and $(-u,-v)$:

$$\Phi(u,v) = \Phi(x,y) \neq \Phi(-u,-v) = \Phi(x',y').$$

Now let us consider the scalar multiplication

$$\int \Psi^*_{\mu'}\, \Psi_{\mu}\, \sqrt{-g}\; dtdzdudv.$$

of the basic constructed wave functions:

$$\begin{aligned} &\Psi_{++}(\epsilon,p,a) = e^{i\epsilon t}\, e^{ipz}\, \Phi_{++}(a;u,v), \quad \Psi_{--}(\epsilon,p,a) = e^{i\epsilon t}\, e^{ipz}\, \Phi_{--}(a;u,v), \\ &\Psi_{+-}(\epsilon,p,a) = e^{i\epsilon t}\, e^{ipz}\, \Phi_{+-}(a;u,v), \quad \Psi_{-+}(\epsilon,p,a) = e^{i\epsilon t}\, e^{ipz}\, \Phi_{-+}(a;u,v), \end{aligned} \tag{9.17}$$

where μ and μ' stand for generalized quantum numbers. We should study the integrals [2]:

in the vector space

$$I_0 = \int_0^{+\infty} dv \int_{-\infty}^{+\infty} du\, \Phi^*_{++}\, \Phi_{--}\, (u^2+v^2),$$

[2]The arguments $(a;u,v)$ are omitted here.

in the spinor space

$$I_1 = \int_{-\infty}^{+\infty} dv \int_{-\infty}^{+\infty} du\, \Phi^*_{++}\ \Phi_{--}\ (u^2+v^2)\,,$$
$$I_2 = \int_{-\infty}^{+\infty} dv \int_{-\infty}^{+\infty} du\, \Phi^*_{+-}\ \Phi_{-+}\ (u^2+v^2)\,,$$
$$I_3 = \int_{-\infty}^{+\infty} dv \int_{-\infty}^{+\infty} du\, \Phi^*_{++}\ \Phi_{+-}\ (u^2+v^2)\,,$$
$$I_4 = \int_{-\infty}^{+\infty} dv \int_{-\infty}^{+\infty} du\, \Phi^*_{++}\ \Phi_{-+}\ (u^2+v^2)\,,$$
$$I_5 = \int_{-\infty}^{+\infty} dv \int_{-\infty}^{+\infty} du\, \Phi^*_{--}\ \Phi_{+-}\ (u^2+v^2)\,,$$
$$I_6 = \int_{-\infty}^{+\infty} dv \int_{-\infty}^{+\infty} du\, \Phi^*_{--}\ \Phi_{-+}\ (u^2+v^2)\,.$$

All these seven integrals $I_0, I_1 ... I_6$ are equal to zero, which means that the constructed functions provide us with an orthogonal basis for the Hilbert space $\Psi(t,z,u,v)$, where (u,v,z) belong to the spinor space model.

9.3 The Dirac Particle and the Space with Spinor Structure

We shall apply a tetrad based form of the Dirac equation to construct solutions of the Dirac equation in parabolic cylindric coordinates. The last ones are defined by the formulas

$$x_1 = \frac{u^2-v^2}{2}\,, \qquad x_2 = uv\,, \qquad x_3 = z\,.$$

To parameterize the space with vector structure (x,y,z), it is enough to make use of the following formulas

$$v = +\sqrt{-x_1 + \sqrt{x_1^2+x_1^2}}\,, \qquad u = \pm\sqrt{+x_1 + \sqrt{x_1^2+x_2^2}}\,,$$

However, to parameterize the space $(x,y,z) \oplus (x,y,z)$ with spinor structure, one must use for u and v more symmetrical, substantially different formulas:

$$v = \pm\sqrt{-x_1 + \sqrt{x_1^2+x_2^2}}\,, \qquad u = \pm\sqrt{+x_1 + \sqrt{x_1^2+x_2^2}}\,.$$

We shall further denote the parabolic cylindric coordinates by $(u,v,z) = (y^1,y^2,y^3)$, and use the explicit formulas

$$dx^i = \frac{\partial x^i}{\partial y^j} dy^j\,, \quad (S^i{}_j) = \frac{\partial x^i}{\partial y^j} = \begin{vmatrix} u & -v & 0 \\ v & u & 0 \\ 0 & 0 & 1 \end{vmatrix}\,, \quad (S^{-1})^j_{\ k} = \frac{\partial y^j}{\partial x^k} = \frac{1}{u^2+v^2} \begin{vmatrix} u & v & 0 \\ -v & u & 0 \\ 0 & 0 & 1 \end{vmatrix}\,.$$

The metric of the Minkowski 4-space in cylindric parabolic coordinates is given by

$$g_{\alpha\beta} = \begin{vmatrix} 1 & 0 & 0 & 0 \\ 0 & -(u^2+v^2) & 0 & 0 \\ 0 & 0 & -(u^2+v^2) & 0 \\ 0 & 0 & 0 & -1 \end{vmatrix}\,,$$

and the most simple tetrad to use is

$$e_{(k)}^{\alpha} = \left| \begin{array}{cccc} 1 & 0 & 0 & 0 \\ 0 & (u^2+v^2)^{-1/2} & 0 & 0 \\ 0 & 0 & (u^2+v^2)^{-1/2} & 0 \\ 0 & 0 & 0 & 1 \end{array} \right| .$$

Let us specify the Dirac equation in this tetrad. It is convenient to start with the general form of the equation in any orthogonal coordinate system

$$\left[i\gamma^k \left(e_{(k)}^{\alpha}(y) \frac{\partial}{\partial y^{\alpha}} + \frac{1}{2} e_{(k);\alpha}^{\alpha}(y) \right) - M \right] \Psi(y) = 0. \tag{9.18}$$

By using the known formula

$$A_{;\alpha}^{\alpha} = \frac{1}{\sqrt{-g}} \frac{\partial \sqrt{-g} A^{\alpha}}{\partial y^{\alpha}}, \qquad g = \det(g_{\alpha\beta}),$$

we get

$$e_{(0);\alpha}^{\alpha} = 0, \quad e_{(3);\alpha}^{\alpha} = 0, \quad e_{(1);\alpha}^{\alpha}(y) = \frac{u}{(u^2+v^2)^{3/2}}, \quad e_{(2);\alpha}^{\alpha}(y) = \frac{v}{(u^2+v^2)^{3/2}}.$$

Therefore, (9.18) reads

$$\Big\{ i\gamma^0 \frac{\partial}{\partial t} + i\gamma^3 \frac{\partial}{\partial z} - M + \frac{i}{\sqrt{u^2+v^2}}$$
$$\times \Big[\gamma^1 \Big(\frac{\partial}{\partial u} + \frac{u}{2(u^2+v^2)} \Big) + \gamma^2 \Big(\frac{\partial}{\partial v} + \frac{v}{2(u^2+v^2)} \Big) \Big] \Big\} \Psi(y) = 0.$$

In spinor basis, this is written in 2-block form as follows:

$$\Big\{ i\frac{\partial}{\partial t} + i\sigma^3 \frac{\partial}{\partial z} + \frac{i}{\sqrt{u^2+v^2}} \Big[\sigma^1 \Big(\frac{\partial}{\partial u} + \frac{u}{2(u^2+v^2)} \Big) + \sigma^2 \Big(\frac{\partial}{\partial v} + \frac{v}{2(u^2+v^2)} \Big) \Big] \Big\} E = MH,$$
$$\Big\{ i\frac{\partial}{\partial t} - i\sigma^3 \frac{\partial}{\partial z} - \frac{i}{\sqrt{u^2+v^2}} \Big[\sigma^1 \Big(\frac{\partial}{\partial u} + \frac{u}{2(u^2+v^2)} \Big) + \sigma^2 \Big(\frac{\partial}{\partial v} + \frac{v}{2(u^2+v^2)} \Big) \Big] \Big\} H = ME.$$

By using the substitutions

$$E(t,z,u,v) = e^{-i\epsilon t} e^{ikz} E(u,v), \quad H(t,z,u,v) = e^{-i\epsilon t} e^{ikz} H(u,v),$$

we further obtain

$$\Big\{ \epsilon - k\sigma^3 + \frac{i}{\sqrt{u^2+v^2}} \Big[\sigma^1 \Big(\frac{\partial}{\partial u} + \frac{u}{2(u^2+v^2)} \Big) + \sigma^2 \Big(\frac{\partial}{\partial v} + \frac{v}{2(u^2+v^2)} \Big) \Big] \Big\} E = MH,$$
$$\Big\{ \epsilon + k\sigma^3 - \frac{i}{\sqrt{u^2+v^2}} \Big[\sigma^1 \Big(\frac{\partial}{\partial u} + \frac{u}{2(u^2+v^2)} \Big) + \sigma^2 \Big(\frac{\partial}{\partial v} + \frac{v}{2(u^2+v^2)} \Big) \Big] \Big\} H = ME.$$

It is convenient to make the following substitutions for the following 2-component entities

$$E(u,v) = (u^2+v^2)^{-1/4} e(u,v), \quad H(u,v) = (u^2+v^2)^{-1/4} h(u,v).$$

This yields

$$\Big[(\epsilon - k\sigma^3) + \frac{i}{\sqrt{u^2+v^2}}\Big(\sigma^1\frac{\partial}{\partial u} + \sigma^2\frac{\partial}{\partial v}\Big)\Big]e(y) = Mh(y),$$

$$\Big[(\epsilon + k\sigma^3) - \frac{i}{\sqrt{u^2+v^2}}\Big(\sigma^1\frac{\partial}{\partial u} + \sigma^2\frac{\partial}{\partial v}\Big)\Big]h(y) = Me(y). \quad (9.19)$$

The equations (9.19) are rather complicated, and to proceed with them we shall diagonalize the helicity operator $(\mathbf{sp})_0$. We shall translate it to cylindric coordinates and then translate it to the cylindric parabolic tetrad

$$(\mathbf{sp}) = S(\mathbf{sp})_0 S^{-1}, \; S = \left| \begin{array}{cc} B(y) & 0 \\ 0 & B(y) \end{array} \right|,$$

$$B(y) = \frac{1}{(u^2+v^2)^{1/4}} \left| \begin{array}{cc} \sqrt{u+iv} & 0 \\ 0 & \sqrt{u-iv} \end{array} \right|.$$

In this way, we get

$$(\,\vec{\sigma}\vec{p}\,) = B(\,\vec{\sigma}\vec{p}\,)_0 B^{-1}$$

$$= -i\Big\{\sigma^3\frac{\partial}{x^3} + \frac{1}{\sqrt{u^2+v^2}}\Big[\sigma^1\Big(\frac{\partial}{\partial u} + \frac{u}{2(u^2+v^2)}\Big) + \sigma^2\Big(\frac{\partial}{\partial v} + \frac{v}{2(u^2+v^2)}\Big)\Big]\Big\}.$$

The explicit form of the eigenvalue equation for the 4-spinor wave function can be simplified by the following substitution

$$(\,\vec{\sigma}\vec{p}\,)\,E(y) = \lambda E(y), \quad E(y) = (u^2+v^2)^{-1/4}e(y),$$

$$(\,\vec{\sigma}\vec{p}\,)\,H(y) = \lambda H(y), \quad H(y) = (u^2+v^2)^{-1/4}h(y); \quad (9.20)$$

so we obtain

$$\Big[\sigma^3 k - \frac{i}{\sqrt{u^2+v^2}}\Big(\sigma^1\frac{\partial}{\partial u} + \sigma^2\frac{\partial}{\partial v}\Big)\Big]e(u,v) = \lambda\, e(u,v),$$

$$\Big[\sigma^3 k - \frac{i}{\sqrt{u^2+v^2}}\Big(\sigma^1\frac{\partial}{\partial u} + \sigma^2\frac{\partial}{\partial v}\Big)\Big]h(u,v) = \lambda\, h(u,v). \quad (9.21)$$

Let us turn back to Eqs. (9.19). With the help of (9.21), from (9.19) we infer the algebraic system

$$(\epsilon - \lambda)e(u,v) = Mh(u,v), \quad (\epsilon + \lambda)h(u,v) = Me(u,v),$$

which leads to the following solutions

$$\lambda = \pm\sqrt{\epsilon^2 - M^2}, \quad \mu = \frac{(\epsilon-\lambda)}{M} = \frac{M}{(\epsilon+\lambda)},$$

$$h_j = \mu\, e_j, \quad \Psi = \frac{e^{-i\epsilon t}e^{ikz}}{\sqrt{u^2+v^2}} \left| \begin{array}{c} e_1(u,v) \\ e_2(u,v) \\ \mu e_1(u,v) \\ \mu e_2(u,v) \end{array} \right|. \quad (9.22)$$

Thus, we have two independent variables, $e_1(u,v)$ and $e_2(u,v)$, which obey the following equations

$$(\lambda - k)\, e_1 + \frac{i}{\sqrt{u^2+v^2}}(\partial_u - i\partial_v)\, e_2 = 0\,,$$
$$(\lambda + k)\, e_2 + \frac{i}{\sqrt{u^2+v^2}}(\partial_u + i\partial_v)\, e_1 = 0\,. \qquad (9.23)$$

Using the substitutions $e_1 = \sqrt{u+iv}\, F$, $e_2 = \sqrt{u-iv}\, G$, and taking into account the identities

$$(\partial_u + i\partial_v)\, \sqrt{u+iv}\, F = \sqrt{u+iv}\, (\partial_u + i\partial_v)\, F\,,$$
$$(\partial_u - i\partial_v)\, \sqrt{u-iv}\, G = \sqrt{u-iv}\, (\partial_u - i\partial_v)\, G\,,$$

we reduce (9.23) to the simpler form

$$(\lambda - k)\; (u+iv)\, F + i\, (\partial_u - i\partial_v)\, G = 0\,,$$
$$(\lambda + k)\; (u-iv)\, G + i\, (\partial_u + i\partial_v)\, F = 0\,. \qquad (9.24)$$

These equations are associated with the wave functions of the form

$$\Psi = \frac{e^{-i\epsilon t} e^{ikz}}{\sqrt{u^2+v^2}} \left| \begin{array}{c} \sqrt{u+iv}\, F(u,v) \\ \sqrt{u-iv}\, G(u,v) \\ \mu\, \sqrt{u+iv}\, F(u,v) \\ \mu\, \sqrt{u-iv}\, G(u,v) \end{array} \right| ; \qquad (9.25)$$

the transition of (9.25) to Cartesian tetrad is made according to the rule

$$\Psi_0(x) = \frac{1}{(u^2+v^2)^{1/4}} \left| \begin{array}{cccc} \sqrt{u-iv} & 0 & 0 & 0 \\ 0 & \sqrt{u+iv} & 0 & 0 \\ 0 & 0 & \sqrt{u-iv} & 0 \\ 0 & 0 & 0 & \sqrt{u+iv} \end{array} \right| \Psi(x)\,,$$

whence it follows

$$\Psi_0(x) = \frac{e^{-i\epsilon t} e^{ikz}}{(u^2+v^2)^{1/4}} \left| \begin{array}{c} F(u,v) \\ G(u,v) \\ \mu\, F(u,v) \\ \mu\, G(u,v) \end{array} \right| .$$

Note that after introducing the two complex variables

$$u + iv = x\,,\; u - iv = y\,, \quad (\partial_u + i\partial_v) = 2\partial_y\,,\; (\partial_u - i\partial_v) = 2\partial_x\,,$$

the system (9.24) takes the form

$$(\lambda - k)\, x\, F + 2i\partial_x\, G = 0\,, \quad (\lambda + k)\, y\, G + 2i\partial_y\, F = 0\,;$$

by eliminating the function G, we get

$$G = -\frac{2i}{(\lambda+k)}\, \frac{1}{y} \partial_y\, F\,, \quad (\lambda - k)\, x\, F - 2i\partial_x \frac{2i}{(\lambda+k)}\, \frac{1}{y} \partial_y\, F = 0\,.$$

Searching for solutions $F(x,y)$ of the form $F = X(x)Y(y)$, we infer:

$$-\frac{\lambda^2-k^2}{4} = \frac{X'}{X}\frac{1}{x}\frac{Y'}{Y}\frac{1}{y} \quad \Longrightarrow \quad \frac{X'}{X}\frac{1}{x} = \alpha, \quad \frac{Y'}{Y}\frac{1}{y} = \beta, \quad \alpha\beta = -\frac{\lambda^2-k^2}{4}.$$

The separated equations can be readily integrated,

$$X = C_1 e^{\alpha x^2/2}, \quad Y = C_2 e^{\beta y^2/2};$$

to which there corresponds a quite definite function $G(x,y)$:

$$G = -\frac{2i}{\lambda+k}\frac{1}{y}\frac{\partial}{\partial y}XY = -\frac{2i}{\lambda+k}C_1 e^{\alpha x^2/2}\beta C_2 e^{\beta y^2/2}.$$

In fact, the constructed solutions represent the well known plane waves. Indeed, assuming that $x_1 = (u^2-v^2)/2$, $x_2 = uv$, we get:

$$X(x)Y(y) \sim e^{\alpha(u^2-v^2+2iub)/2} e^{\beta(u^2-v^2-2iub)/2} = e^{(\alpha+\beta)x_1} e^{(\alpha-\beta)x_2} = e^{ik_1x_1} e^{ik_2x_2},$$

whence it follows

$$\alpha+\beta = ik_1, \quad \alpha-\beta = k_2 \quad \Longrightarrow$$

$$\alpha = \frac{ik_1+k_2}{2}, \quad \beta = \frac{ik_1-k_2}{2}, \quad \alpha\beta = -\frac{k_1^2+k_2^2}{4} = -\frac{\lambda^2-k_3^2}{4};$$

so we get the claimed identity $\epsilon^2 = M^2+k_1^2+k_2^2+k_3^2$.

Let us turn again to the system (9.24). First, by eliminating the function G from (9.24), we derive

$$G = -\frac{i}{\lambda+k}\frac{1}{u-iv}\Big(\frac{\partial}{\partial u}+i\frac{\partial}{\partial v}\Big)F,$$

$$\Big[\frac{\partial^2}{\partial u^2}+\frac{\partial^2}{\partial v^2}+(\lambda^2-k^2)(u^2+v^2)\Big]F(u,v) = 0. \tag{9.26}$$

Alternatively, by eliminating the function F from (9.24), we obtain

$$F = -\frac{i}{\lambda-k}\frac{1}{u+iv}\Big(\frac{\partial}{\partial u}-i\frac{\partial}{\partial v}\Big)G,$$

$$\Big[\frac{\partial^2}{\partial u^2}+\frac{\partial^2}{\partial v^2}+(\lambda^2-k^2)(u^2+v^2)\Big]G((u,v) = 0. \tag{9.27}$$

The second order differential equation is the same for both cases, (9.26) and (9.27). For definiteness, let us examine the case (9.26), starting with the equation

$$\Big[\frac{\partial^2}{\partial u^2}+\frac{\partial^2}{\partial v^2}+\sigma^2(u^2+v^2)\Big]F(u,v) = 0, \quad \sigma^2 = (\lambda^2-k^2) = \epsilon^2-M^2-k^2 > 0. \tag{9.28}$$

Let $F(u,v) = U(u)V(v)$; then, instead of (9.28), we obtain

$$\Big(\frac{1}{U}\frac{d^2U}{du^2}+\sigma^2u^2\Big)+\Big(\frac{1}{V}\frac{d^2V}{dv^2}+\sigma^2v^2\Big) = 0.$$

By introducing a separating constant Λ, we get two equations in variables u and v, respectively:

$$\frac{d^2U}{du^2} + (\sigma^2 u^2 + \Lambda) U = 0, \qquad \frac{d^2V}{dv^2} + (\sigma^2 v^2 - \Lambda) V = 0.$$

For the functions U and V we have similar equations, which differ only by the sign of the separation constant Λ. For definiteness, let us consider the function $U(u)$ given by

$$\frac{d^2U}{du^2} + (\sigma^2 u^2 + \Lambda) U = 0.$$

Further, by expressing in terms of the variable $bu^2 = z$, it takes the form:

$$\Big(z\frac{d^2}{dz^2} + \frac{1}{2}\frac{d}{dz} + \sigma^2 \frac{z}{4b^2} + \frac{\Lambda}{4b}\Big) U = 0.$$

Separating the simple factor $U(z) = e^{-z/2} f(z)$, we derive

$$zf'' - zf' + \frac{1}{4} zf + \frac{1}{2}(f' - \frac{1}{2} f) + \sigma^2 \frac{z}{4b^2} f + \frac{\Lambda}{4b} f = 0.$$

Let $b^2 = -\sigma^2$ and $b = +i\sigma$, $\sigma > 0$; then we get

$$z\frac{d^2 f}{dz^2} + (\frac{1}{2} - z)\frac{df}{dz} - \frac{1 + i\Lambda/\sigma}{4} f = 0,$$

which is in fact a confluent hypergeometric equation

$$z\frac{d^2 f}{dz^2} + (c - z)\frac{df}{dz} - af = 0, \quad c = \frac{1}{2}, \quad a = \frac{1 + i\Lambda/\sigma}{4}.$$

We further use the two independent solutions

$$U_- = e^{-z/2}\Phi(a, \frac{1}{2}; z), \quad U_+ = e^{-z/2}\sqrt{z}\,\Phi(a + \frac{1}{2}, \frac{3}{2}; z).$$

Analogous results for (9.27) are obtained by the formal change $\Lambda \rightsquigarrow -\Lambda$:

$$V(v) = e^{-y/2} g(y), \quad y = +i\sigma v^2,$$
$$\frac{d^2 g}{dy^2} + \Big(\frac{1}{2} - y\Big)\frac{dg}{dy} - \frac{1 - i\Lambda/\sigma}{4} g = 0,$$
$$c' = \frac{1}{2} = c, \quad a' = \frac{1 - i\Lambda/\sigma}{4} = \frac{1}{2} - a; \tag{9.29}$$

the two independent solutions are written as follows

$$V_-(y) = e^{-y/2}\Phi(\frac{1}{2} - a, \frac{1}{2}; y), \quad V_+(y) = e^{-y/2}\sqrt{y}\,\Phi(1 - a, \frac{3}{2}; y).$$

Thus, the total functions $F(u,v)$ and $G(u,v)$ will be constructed in terms of the following solutions

$$U_-(u)=e^{-i\sigma u^2/2}\Phi\Big(a,\tfrac{1}{2};i\sigma u^2\Big),\qquad U_+(u)=e^{-i\sigma u^2/2}u\Phi\Big(a+\tfrac{1}{2},\tfrac{3}{2};i\sigma u^2\Big),$$
$$V_-(v)=e^{-i\sigma v^2/2}\Phi\Big(\tfrac{1}{2}-a,\tfrac{1}{2};i\sigma v^2\Big),\quad V_+(v)=e^{-i\sigma v^2/2}v\Phi\Big(1-a,\tfrac{3}{2};i\sigma v^2\Big);$$

$$U'_-(u)=e^{-i\sigma u^2/2}\Phi\Big(a',\tfrac{1}{2};i\sigma u^2\Big),\qquad U'_+(u)=e^{-i\sigma u^2/2}u\Phi\Big(a'+\tfrac{1}{2},\tfrac{3}{2};i\sigma u^2\Big),$$
$$V'_-(v)=e^{-i\sigma v^2/2}\Phi\Big(\tfrac{1}{2}-a',\tfrac{1}{2};i\sigma v^2\Big),\quad V'_+(v)=e^{-i\sigma v^2/2}v\Phi\Big(1-a',\tfrac{3}{2};i\sigma v^2\Big).$$

Now, we should turn to the first order equation in (9.26)

$$i(\lambda+k)G=\frac{1}{u-iv}\Big(\frac{\partial}{\partial u}+i\frac{\partial}{\partial v}\Big)F\,;$$

by using it and starting with any known $F(u,v)$, we can find an explicit form for the corresponding function $G(u,v)$.

Let us discuss in more detail some subtleties of the properties of the solutions and of their relations to the structure of spatial, vector and spinorial models. While considering the relation between parabolic and Cartesian coordinates:

$$x_1=\frac{u^2-v^2}{2},\qquad x_1=u\,v\,,$$

$$\underline{v=0\,:}\,x_1=+\frac{u^2}{2}\geq 0,\;x_2=0,\qquad \underline{u=0\,:}\,x=-\frac{v^2}{2}\leq 0,\;y=0,$$

we noted existence of two special regions.

Now let us consider which restrictions for the Dirac wave functions Ψ follow from the requirement of single-valuedness. First consider the variant $(--)$:

$$F_{--}(u,v)=U_-(u)V_-(v)=e^{-i\sigma u^2/2}\Phi(a,\frac{1}{2};i\sigma u^2)e^{-i\sigma v^2/2}\Phi(\frac{1}{2}-a,\frac{1}{2};i\sigma v^2),$$

$$i(\lambda+k)G=\frac{1}{(u-iv)}\Big(\frac{\partial}{\partial u}+i\frac{\partial}{\partial v}\Big)U_-(u)V_-(v)$$
$$=\frac{1}{(u-iv)}\Big(\frac{\partial}{\partial u}+i\frac{\partial}{\partial v}\Big)e^{-i\sigma u^2/2}\Phi(a,\frac{1}{2};i\sigma u^2)e^{-i\sigma v^2/2}\Phi(\frac{1}{2}-a,\frac{1}{2};i\sigma v^2)$$
$$=\frac{1}{(u-iv)}e^{-i\sigma u^2/2}e^{-i\sigma v^2/2}$$
$$\times\Big[\Big(-i\sigma u\Phi(a,\frac{1}{2};i\sigma u^2)+2i\sigma u\frac{d}{d(i\sigma u^2)}\Phi(a,\frac{1}{2};i\sigma u^2)\Big)\Phi(\frac{1}{2}-a,\frac{1}{2};i\sigma v^2)$$
$$+i\Phi(a,\frac{1}{2};i\sigma u^2)\Big(-i\sigma v\Phi(\frac{1}{2}-a,\frac{1}{2};i\sigma v^2)+2i\sigma v\frac{d}{d(i\sigma v^2)}\Phi(\frac{1}{2}-a,\frac{1}{2};i\sigma v^2)\Big)\Big].$$

We note that the related pair of functions, $F_{--} \rightsquigarrow G_{--}(u,v)$, are single-valued in the special region:

$$F_{--}(u=0,v) = +F_{--}(u=0,-v) \neq 0, \quad F_{--}(u,v=0) = +G_{--}(-u,v=0) \neq 0;$$
$$G_{--}(u=0,v) = +G_{--}(u=0,-v) \neq 0, \quad G_{--}(u,v=0) = +G_{--}(-u,v=0) \neq 0.$$

Let us consider the variant (++):

$$F_{++}(u,v) = U_+(u)V_+(v) = e^{-i\sigma u^2/2} u\Phi(a+\frac{1}{2},\frac{3}{2};i\sigma u^2) e^{-i\sigma v^2/2} v\Phi(1-a,\frac{3}{2};i\sigma v^2),$$

$$i(\lambda+k)G = \frac{1}{(u-iv)}\Big(\frac{\partial}{\partial u}+i\frac{\partial}{\partial v}\Big)U_+(u)V_+(v)$$
$$= \frac{1}{(u-iv)}\Big(\frac{\partial}{\partial u}+i\frac{\partial}{\partial v}\Big)e^{-i\sigma u^2/2} u\Phi(a+\frac{1}{2},\frac{3}{2};i\sigma u^2) e^{-i\sigma v^2/2} v\Phi(1-a,\frac{3}{2};i\sigma v^2)$$
$$= e^{-i\sigma u^2/2} e^{-i\sigma v^2/2} \frac{1}{(u-iv)}$$
$$\times\Big[\Big((-i\sigma u^2+1)\Phi(a+\frac{1}{2},\frac{3}{2};i\sigma u^2) + 2i\sigma u^2 \frac{d}{d(i\sigma u^2)}\Phi(a+\frac{1}{2},\frac{3}{2};i\sigma u^2)\Big) v\Phi(1-a,\frac{3}{2};i\sigma v^2)$$
$$+iu\Phi(a+\frac{1}{2},\frac{3}{2};i\sigma u^2)\Big((-i\sigma v^2+1)\Phi(1-a,\frac{3}{2};i\sigma v^2) + 2i\sigma v^2 \frac{d}{d(i\sigma v^2)}\Phi(1-a,\frac{3}{2};i\sigma v^2)\Big)\Big].$$

The related functions, $F_{++} \rightsquigarrow G_{++}(u,v)$ are single-valued in the spacial region:

$$F_{++}(u=0,+v) = +F_{++}(u=0,-v) = 0, \tag{9.30}$$
$$F_{++}(u,v=0) = +F_{++}(-u,v=0) = 0; \tag{9.31}$$
$$G_{++}(u=0,+v) = +G_{++}(u=0,-v) \neq 0, \tag{9.32}$$
$$G_{++}(u,v=0) = +G_{++}(-u,v=0) \neq 0. \tag{9.33}$$

Let us consider the variant (−+):

$$F_{-+}(u,v) = U_-(u)V_+(v) = e^{-i\sigma u^2/2}\Phi(a,\frac{1}{2};i\sigma u^2) e^{-i\sigma v^2/2}\, v\,\Phi(1-a,\frac{3}{2};i\sigma v^2),$$

$$i(\lambda+k)G = \frac{1}{(u-iv)}\Big(\frac{\partial}{\partial u}+i\frac{\partial}{\partial v}\Big)U_-(u)V_+(v)$$
$$= \frac{1}{(u-iv)}\Big(\frac{\partial}{\partial u}+i\frac{\partial}{\partial v}\Big)e^{-i\sigma u^2/2}\Phi(a,\frac{1}{2};i\sigma u^2) e^{-i\sigma v^2/2}\, v\,\Phi(1-a,\frac{3}{2};i\sigma v^2)$$
$$= e^{-i\sigma u^2/2} e^{-i\sigma v^2/2} \frac{1}{(u-iv)}$$
$$\times\Big[\Big(-i\sigma u\Phi(a,\frac{1}{2};i\sigma u^2) + 2i\sigma u \frac{d}{d(i\sigma u^2)}\Phi(a,\frac{1}{2};i\sigma u^2)\Big) v\Phi(1-a,\frac{3}{2};i\sigma v^2)$$
$$+i\Phi(a,\frac{1}{2};i\sigma u^2)\Big((-i\sigma v^2+1)\Phi(1-a,\frac{3}{2};i\sigma v^2) + 2i\sigma v^2 \frac{d}{d(i\sigma v^2)}\Phi(1-a,\frac{3}{2};i\sigma v^2)\Big)\Big].$$

The related functions, $F_{-+} \rightsquigarrow G_{-+}(u,v)$, are double-valued in the spacial region:

$$F_{-+}(u=0,+v) = -F_{-+}(u=0,-v) \neq 0, \quad F_{-+}(u,v=0) = -F_{-+}(-u,v=0) = 0,$$
$$G_{-+}(u=0,+v) = -G_{-+}(u=0,-v) \neq 0, \quad G_{-+}(u,v=0) = -G_{-+}(-u,v=0) \neq 0. \quad (9.34)$$

Let us consider the variant $(+-)$:

$$F_{+-}(u,v) = U_+(u)V_-(v) = e^{-i\sigma u^2/2} u\Phi(a+\frac{1}{2},\frac{3}{2};i\sigma u^2) e^{-i\sigma v^2/2}\Phi(\frac{1}{2}-a,\frac{1}{2};i\sigma v^2),$$

$$i(\lambda+k)G = \frac{1}{(u-iv)}\Big(\frac{\partial}{\partial u}+i\frac{\partial}{\partial v}\Big)U_+(u)V_-(v)$$

$$= \frac{1}{(u-iv)}\Big(\frac{\partial}{\partial u}+i\frac{\partial}{\partial v}\Big)e^{-i\sigma u^2/2} u\Phi(a+\frac{1}{2},\frac{3}{2};i\sigma u^2) e^{-i\sigma v^2/2}\Phi(\frac{1}{2}-a,\frac{1}{2};i\sigma v^2)$$

$$= e^{-i\sigma u^2/2} e^{-i\sigma v^2/2}\frac{1}{(u-iv)}$$

$$\times\Big[\Big((-i\sigma\, u^2+1)\,\Phi(a+\frac{1}{2},\frac{3}{2};i\sigma u^2)+2i\sigma\, u^2\,\frac{d}{d(i\sigma u^2)}\,\Phi(a+\frac{1}{2},\frac{3}{2};i\sigma u^2)\Big)\,\Phi(\frac{1}{2}-a,\frac{1}{2};i\sigma v^2)$$

$$+iu\Phi(a+\frac{1}{2},\frac{3}{2};i\sigma u^2)\Big(-i\sigma\, v\,\Phi(\frac{1}{2}-a,\frac{1}{2};i\sigma v^2)+2i\sigma\, v\,\frac{d}{d(i\sigma v^2)}\Phi(\frac{1}{2}-a,\frac{1}{2};i\sigma v^2)\Big)\Big].$$

The related functions, $F_{+-} \rightsquigarrow G_{+-}(u,v)$, are double-valued in the spacial region:

$$\begin{aligned} F_{+-}(u=0,+v) &= -F_{+-}(u=0,-v) = 0, \\ F_{+-}(+u,v=0) &= -F_{+-}(-u,v=0) \neq 0; \\ G_{+-}(u=0,+v) &= -G_{+-}(u=0,-v) \neq 0, \\ G_{+-}(+u,v=0) &= -G_{+-}(-u,v=0) \neq 0. \end{aligned} \quad (9.35)$$

When using the spinor space model, the two sets of couples (u,v) and $(-u,v)$ (or, similarly, the sets of couples (u,v) and $(u,-v)$) represent different geometrical points, so the requirement of single valuedness in the case of a spinor space does not imply that the values of the wave functions must be equal at the points (u,v) and $(-u,v)$:

$$\Psi(u,v) = \Psi((x_1,x_2)^{(1)}) \neq \Psi(-u,v) = \Psi((x_1,x_2)^{(2)}).$$

Therefore, we conclude that the solutions $F(u,v), G(u.v)$ of the types $(--)$ and $(++)$ are single-valued in the spaces with vector structure, whereas the solutions $F(u,v), G(u,v)$ of the types $(-+)$ and $(+-)$ are not single-valued in spaces with vector structure; so the solutions of these two types $(-+)$ and $(+-)$ must be discarded. However, the types of solutions $(-+)$ and $(+-)$ are valid in the space with spinor structure.

Chapter 10

Maxwell Equations in Space with Spinor Structure

10.1 Spinor Form of Maxwell Equations

To introduce spinor notations, let us start with the ordinary Dirac equation [283]

$$(i\gamma^a\partial_a - m)\Psi = 0\,, \quad \gamma^a = \begin{vmatrix} 0 & \bar{\sigma}^a \\ \sigma^a & 0 \end{vmatrix}\,, \quad \Psi = \begin{vmatrix} \xi^\alpha \\ \eta_{\dot{\alpha}} \end{vmatrix}, \{\alpha, \dot{\alpha}\} = 1,2\,; \tag{10.1}$$

$\sigma^a = (I, \sigma^j)$, $\bar{\sigma}^a = (I, -\sigma^j)$. In 2-spinor form we have

$$i\sigma^a\partial_a\xi = m\eta\,, \quad i\bar{\sigma}^a\partial_a\eta = m\xi\,. \tag{10.2}$$

It is convenient to attach spinor indices to Pauli matrices, $\sigma^a = (\sigma^a)_{\dot{\beta}\alpha}$, $\bar{\sigma}^a = (\bar{\sigma}^a)^{\beta\dot{\alpha}}$; then Eqs. (10.2) read

$$i(\sigma^a\partial_a)_{\dot{\beta}\alpha}\,\xi^\alpha = m\eta_{\dot{\beta}}\,, \quad i(\bar{\sigma}^a\partial_a)^{\beta\dot{\alpha}}\,\eta_{\dot{\alpha}} = m\xi^\beta\,. \tag{10.3}$$

The electromagnetic tensor is equivalent to a pair of symmetrical 2-rank spinors: $F_{mn} \longleftrightarrow \{\xi^{\alpha\beta}, \eta_{\dot{\alpha}\dot{\beta}}\}$; correspondingly, the eight Maxwell equations are presented as follows

$$(\sigma^a\partial_a)_{\dot{\rho}\alpha}\,\xi^{\alpha\beta} = (\sigma^b)_{\dot{\rho}\alpha}\omega^{\alpha\beta}\,J_b\,, \quad (\bar{\sigma}^a\partial_a)^{\rho\dot{\alpha}}\,\eta_{\dot{\alpha}\dot{\beta}} = (\bar{\sigma}^b)^{\rho\dot{\alpha}}\omega_{\dot{\alpha}\dot{\beta}}\,J_b\,; \tag{10.4}$$

the second equation is conjugate to the first one. In (10.4) we use spinor metrical matrices [283]

$$(\epsilon_{\alpha\beta}) = i\sigma^2\,, \quad (\epsilon^{\alpha\beta}) = -i\sigma^2\,; \quad (\epsilon_{\dot{\alpha}\dot{\beta}}) = i\sigma^2\,, \quad (\epsilon^{\dot{\alpha}\dot{\beta}}) = -i\sigma^2\,. \tag{10.5}$$

In order to prove the equivalence of the spinor form (10.4) to the ordinary Maxwell equations in vector notations, let us take into account the identities [283]

$$\begin{gathered}(\xi^{\alpha\beta}) = \Sigma^{mn}F_{mn}\sigma^2\,, \quad (\eta_{\dot{\alpha}\dot{\beta}}) = -\bar{\Sigma}^{mn}F_{mn}\sigma^2\,, \\ \Sigma^{mn} = \frac{1}{4}(\bar{\sigma}^m\sigma^n - \bar{\sigma}^n\sigma^m)\,, \; \bar{\Sigma}^{mn} = \frac{1}{4}(\sigma^m\bar{\sigma}^n - \sigma^n\bar{\sigma}^m)\,.\end{gathered} \tag{10.6}$$

Then, the equations (10.4) may be written as

$$\sigma^a \partial_a \Sigma^{mn} F_{mn} = -\sigma^b J_b , \qquad \bar{\sigma}^a \partial_a \bar{\Sigma}^{mn} F_{mn} = -\bar{\sigma}^b J_b . \tag{10.7}$$

We shall to take into account the identities

$$\Sigma^{mn} F_{mn} = \sigma^1 (F_{01} - iF_{23}) + \sigma^2 (F_{02} - iF_{31}) + \sigma^3 (F_{03} - iF_{12}) ,$$
$$\bar{\Sigma}^{mn} F_{mn} = \sigma^1 (-F_{01} - iF_{23}) + \sigma^2 (-F_{02} - iF_{31}) + \sigma^3 (-F_{03} - iF_{12}) ;$$

with the notations

$$F_{01} = -E^1 , F_{02} = -E^2 , F_{03} = -E^3 , F_{23} = B^1 , F_{31} = B^2 , F_{12} = B^3 ; \tag{10.8}$$

they read

$$\begin{aligned} \Sigma^{mn} F_{mn} &= -\sigma^1 (E^1 + iB^1) - \sigma^1 (E^2 + iB^2) - \sigma^1 (E^3 + iB^3) = -\sigma^j a_j , \\ \bar{\Sigma}^{mn} F_{mn} &= \sigma^1 (E^1 - iB^1) + \sigma^1 (E^2 - iB^2) + \sigma^1 (E^3 - iB^3) = +\sigma^j b_j , \end{aligned} \tag{10.9}$$

and

$$(\xi^{\alpha\beta}) = \left| \begin{array}{cc} -i(a_1 - ia_2) & ia_3 \\ ia_3 & +i(a_1 + ia_2) \end{array} \right| , (\eta_{\dot{\alpha}\dot{\beta}}) = \left| \begin{array}{cc} -i(b_1 - ib_2) & ib_3 \\ ib_3 & i(b_1 + ib_2) \end{array} \right| .$$

Therefore the equations (10.7) may be presented in the form

$$(\partial_0 + \sigma^l \partial_l)\, (\sigma^k a_k) = J_0 + \sigma^j J_j , \quad (\partial_0 - \sigma^l \partial_l)\, (\sigma^k b_k) = -J_0 + \sigma^j J_j ,$$

whence we derive

$$\sigma^n \partial_0 a_n + (\delta_{lk} + i\epsilon_{nlk} \sigma^n) \partial_l a_k = J_0 + \sigma^n J_n ,$$
$$\sigma^n \partial_0 b_n - (\delta_{lk} + i\epsilon_{nlk} \sigma^n) \partial_l b_k = -J_0 + \sigma^n J_n .$$

Thus, we have the four equations

$$\partial_l a_l = J_0 , \quad \partial_0 a_n + i\epsilon_{nlk} \partial_l a_k = J_n , \quad \partial_l b_l = J_0 , \quad \partial_0 b_n - i\epsilon_{nlk} \partial_l b_k = J_n ,$$

or, differently,

$$\begin{aligned} &(1) \quad \partial_l (E^l + iB^l) = J_0 , \quad (2) \quad \partial_0 (E^l + iB^l) + i\epsilon_{nlk} \partial_l (E^k + iB^k) = J_n , \\ &(1') \quad \partial_l (E^l - iB^l) = J_0 , \quad (2') \quad \partial_0 (E^l - iB^l) - i\epsilon_{nlk} \partial_l (E^k - iB^l) = J_n . \end{aligned}$$

Summing and subtracting the equations, we obtain

$$1 + 1' , \quad \partial_l E^l = J_0 , \quad 1 - 1' , \quad \partial_l B^l = 0 ,$$
$$2 + 2' , \quad \partial_0 E^n - \epsilon_{nlk} \partial_l B^k = J_k , \quad 2 - 2' , \quad \partial_0 B^n + \epsilon_{nlk} \partial_l E^k = 0 ;$$

they may be identified with the Maxwell equations in vector form

$$\text{div}\,\mathbf{E} = J^0 , \quad \text{div}\,\mathbf{B} = 0 , \quad \text{rot}\,\mathbf{B} = \partial_0 \mathbf{E} + \mathbf{J} , \quad \text{rot}\,\mathbf{E} = -\partial_0 \mathbf{B} , \tag{10.10}$$

where

$$\mathbf{E} = (E^n), \mathbf{B} = (B^n), \quad J^0 = J_0, \mathbf{J} = (J^n) = (-J_n).$$

The covariant Maxwell equations in spinor form can be constructed within the tetrad method used for generalizing the Dirac equation

$$i\sigma^\alpha(x)\left[\partial_\alpha + \Sigma_\alpha(x)\right]\xi(x) = m\eta(x), \quad i\bar{\sigma}^\alpha(x)\left[\partial_\alpha + \bar{\Sigma}_\alpha(x)\right]\eta(x) = m\xi(x). \quad (10.11)$$

So, the Maxwell equations in spinor form are generalized as follows

$$\begin{aligned} i\sigma^\alpha(x)\left[\partial_\alpha + \Sigma_\alpha(x)\otimes I + I\otimes\Sigma_\alpha(x)\right]\xi(x) &= \sigma^\beta(x)(-i\sigma^2)J_\beta(x), \\ i\bar{\sigma}^\alpha(x)\left[\partial_\alpha + \bar{\Sigma}_\alpha(x)\otimes I + I\otimes\bar{\Sigma}_\alpha(x)\right]\eta(x) &= \sigma^\beta(x)(+i\sigma^2)J_\beta(x), \end{aligned} \quad (10.12)$$

where

$$\begin{aligned} \sigma^\alpha(x) = \sigma^a e^\alpha_{(a)}(x), &\quad \bar{\sigma}^\alpha(x) = \bar{\sigma}^a e^\alpha_{(a)}(x), \\ \Sigma_\alpha(x) = \frac{1}{2}\Sigma^{ab} e^\beta_{(a)} \nabla_\alpha(e_{(b)\beta}), &\quad \bar{\Sigma}_\alpha(x) = \frac{1}{2}\bar{\Sigma}^{ab} e^\beta_{(x)} \nabla_\alpha(e_{(b)\beta}), \\ \Sigma^{ab} = \frac{1}{4}(\bar{\sigma}^a\sigma^b - \bar{\sigma}^b\sigma^a), &\quad \bar{\Sigma}^{ab} = \frac{1}{4}(\sigma^a\bar{\sigma}^b - \sigma^b\bar{\sigma}^a). \end{aligned} \quad (10.13)$$

As in the Minkowski space, the second equation in (10.12) is conjugate to the first, so it suffices to study only the first one (further we follow the case without the current $_\beta(x)$)

$$\sigma^\alpha(x)\left[\partial_\alpha + \Sigma_\alpha(x)\otimes I + I\otimes\Sigma_\alpha(x)\right]\xi(x) = 0. \quad (10.14)$$

In (10.14), the quantity $\xi(x)$ stands for a symmetric 2-rank spinor; it can be treated as a 2×2-matrix function. Equation (10.14) may be presented with the help of Ricci rotation coefficients $\gamma_{abc}(x)$:

$$\begin{aligned} &\left[\sigma^c e^\alpha_{(c)}(x)\partial_\alpha + \sigma^c\left(\frac{1}{2}\Sigma^{ab}\otimes I + I\otimes\frac{1}{2}\Sigma^{ab}\right)\gamma_{abc}(x)\right]\xi(x) = 0; \\ &\gamma_{abc} = +e_{(a)\beta;\alpha}e^\beta_{(b)}e^\alpha_{(c)}, \quad \Sigma^{ab} = \frac{1}{4}(\bar{\sigma}^a\sigma^b - \bar{\sigma}^b\sigma^a). \end{aligned} \quad (10.15)$$

10.2 Cylindrical Parabolic Coordinates

Let us construct the solutions of spinor Maxwell equations in cylindric parabolic coordinates:

$$\begin{aligned} &x_1 = \frac{u^2 - v^2}{2}, \quad x_2 = uv, \quad x_3 = z, \\ &v = +\sqrt{-x_1 + \sqrt{x_1^2 + x_2^2}}, \quad u = \pm\sqrt{+x_1 + \sqrt{x_1^2 + x_2^2}}. \end{aligned} \quad (10.16)$$

The metric of Minkowski space takes the form (let $x^\alpha = (t,u,v,z)$)

$$g_{\alpha\beta} = \begin{vmatrix} 1 & 0 & 0 & 0 \\ 0 & -(u^2+v^2) & 0 & 0 \\ 0 & 0 & -(u^2+v^2) & 0 \\ 0 & 0 & 0 & -1 \end{vmatrix}. \tag{10.17}$$

We shall further use the diagonal tetrad (recall $g_{\alpha\beta}(x) e^\alpha_{(k)} e^\beta_{(l)} = \eta_{kl}$),

$$e^\alpha_{(k)} = \begin{vmatrix} 1 & 0 & 0 & 0 \\ 0 & 1/\sqrt{u^2+v^2} & 0 & 0 \\ 0 & 0 & 1/\sqrt{u^2+v^2} & 0 \\ 0 & 0 & 0 & 1 \end{vmatrix}, e_{(k)\alpha} = \begin{vmatrix} 1 & 0 & 0 & 0 \\ 0 & -\sqrt{u^2+v^2} & 0 & 0 \\ 0 & 0 & -\sqrt{u^2+v^2} & 0 \\ 0 & 0 & 0 & -1 \end{vmatrix}.$$

In order to find the Ricci rotation coefficients, let us introduce the auxiliary quantities (see in [264]): $\lambda_{abc} = \gamma_{abc} - \gamma_{acb}$. For λ_{abc} we easily derive the following representation

$$\lambda_{abc} = \gamma_{abc} - \gamma_{acb} = (e_{(a)\alpha;\beta} - e_{(a)\beta;\alpha})\, e^\alpha_{(c)} e^\beta_{(b)}$$
$$= (\partial_\beta e_{(a)\alpha} - \Gamma^\rho_{\alpha\beta} e_{(a)\rho} - \partial_\alpha e_{(a)\beta} + \Gamma^\rho_{\beta\alpha} e_{(a)\rho})\, e^\alpha_{(c)} e^\beta_{(b)};$$

that is

$$\lambda_{abc} = [\partial_\beta e_{(a)\alpha} - \partial_\alpha e_{(a)\beta}]\, e^\alpha_{(c)} e^\beta_{(b)}; \tag{10.18}$$

according to (10.18), λ_{abc} are calculated with the use of ordinary derivatives. Besides, there exists an identity

$$\frac{1}{2}(\lambda_{abc} + \lambda_{bca} - \lambda_{cab}) \equiv \frac{1}{2}(\gamma_{abc} - \gamma_{acb} + \gamma_{bca} - \gamma_{bac} - \gamma_{cab} + \gamma_{cba}) \equiv \gamma_{abc}. \tag{10.19}$$

We need the explicit expressions for λ_{abc}. First of all, we have $\lambda_{0bc} = 0$, $\lambda_{3bc} = 0$. Further, taking in mind diagonal structure of the tetrad, we derive the formula

$$\lambda_{1bc} = e^\beta_{(b)} e^1_{(c)} \partial_\beta e_{(1)1} - e^1_{(b)} e^\beta_{(c)} \partial_\beta e_{(1)1}$$
$$= e^1_{(b)} e^1_{(c)} \partial_1 e_{(1)1} + e^2_{(b)} e^1_{(c)} \partial_2 e_{(1)1} - e^1_{(b)} e^1_{(c)} \partial_1 e_{(1)1} - e^1_{(b)} e^2_{(c)} \partial_2 e_{(1)1},$$

that is

$$\lambda_{1[bc]} = [e^2_{(b)} e^1_{(c)} - e^1_{(b)} e^2_{(c)}]\, \partial_2 e_{(1)1} = \begin{vmatrix} 0 & 0 & 0 & 0 \\ 0 & 0 & -e^1_{(1)} e^2_{(2)} & 0 \\ 0 & e^1_{(1)} e^2_{(2)} & 0 & 0 \\ 0 & 0 & 0 & 0 \end{vmatrix} \partial_2 e_{(1)1};$$

and similarly

$$\lambda_{2bc} = e^\beta_{(b)} e^2_{(c)} \partial_\beta e_{(2)2} - e^2_{(b)} e^\beta_{(c)} \partial_\beta e_{(2)2}$$
$$= e^1_{(b)} e^2_{(c)} \partial_1 e_{(2)2} + e^2_{(b)} e^2_{(c)} \partial_2 e_{(2)2} - e^2_{(b)} e^1_{(c)} \partial_1 e_{(2)2} - e^2_{(b)} e^2_{(c)} \partial_2 e_{(2)2},$$

so that

$$\lambda_{2[bc]} = [e^1_{(b)}e^2_{(c)} - e^2_{(b)}e^1_{(c)}]\,\partial_1 e_{(2)2} = \begin{vmatrix} 0 & 0 & 0 & 0 \\ 0 & 0 & e^1_{(1)}e^2_{(2)} & 0 \\ 0 & -e^1_{(1)}e^2_{(2)} & 0 & 0 \\ 0 & 0 & 0 & 0 \end{vmatrix} \partial_1 e_{(2)2}.$$

Allowing for the relations

$$e^1_{(1)} = e^2_{(2)} = 1/\sqrt{u^2+v^2}\,, \quad e_{(1)1} = e_{(2)2} = -\sqrt{u^2+v^2}\,,$$

we obtain needed formulas for coefficients $\lambda_{1[bc]}$ and $\lambda_{2[bc]}$:

$$\lambda_{1[bc]} = \frac{1}{(u^2+v^2)^{3/2}} \begin{vmatrix} 0 & 0 & 0 & 0 \\ 0 & 0 & +v & 0 \\ 0 & -v & 0 & 0 \\ 0 & 0 & 0 & 0 \end{vmatrix}, \quad \lambda_{2[bc]} = \frac{1}{(u^2+v^2)^{3/2}} \begin{vmatrix} 0 & 0 & 0 & 0 \\ 0 & 0 & -u & \\ 0 & +u & 0 & 0 \\ 0 & 0 & 0 & 0 \end{vmatrix}.$$

Thus, the only nonvanishing coefficients are:

$$\begin{aligned} \lambda_{1[12]} &= +\frac{v}{(u^2+v^2)^{3/2}}, \quad \lambda_{1[21]} = -\frac{v}{(u^2+v^2)^{3/2}}, \\ \lambda_{2[12]} &= -\frac{u}{(u^2+v^2)^{3/2}}, \quad \lambda_{1[21]} = +\frac{u}{(u^2+v^2)^{3/2}}. \end{aligned} \tag{10.20}$$

When using the formula $\gamma_{[ab]c} = \frac{1}{2}(-\lambda_{c[ab]} + \lambda_{a[bc]} - \lambda_{b[ac]})$, it is convenient to split it into four cases

$$\gamma_{[ab]0} = \frac{1}{2}(-\lambda_{0[ab]} + \lambda_{a[b0]} - \lambda_{b[a0]}), \quad \gamma_{[ab]3} = \frac{1}{2}(-\lambda_{3[ab]} + \lambda_{a[b3]} - \lambda_{b[a3]}),$$
$$\gamma_{[ab]1} = \frac{1}{2}(-\lambda_{1[ab]} + \lambda_{a[b1]} - \lambda_{b[a1]}), \quad \gamma_{[ab]2} = \frac{1}{2}(-\lambda_{2[ab]} + \lambda_{a[b2]} - \lambda_{b[a2]}).$$

In this way we find the nonvanishing Ricci coefficients

$$\gamma_{[12]1} = -\gamma_{[21]1} = -\lambda_{1[12]} = -\frac{v}{(u^2+v^2)^{3/2}},$$
$$\gamma_{[12]2} = -\gamma_{[21]2} = -\lambda_{2[12]} = +\frac{u}{(u^2+v^2)^{3/2}}.$$

Now let us turn to the Maxwell equations, using the tetrad formalism:

$$\Big[\sigma^c e^\alpha_{(c)}(x)\partial_\alpha + \sigma^c(\frac{1}{2}\Sigma^{ab}\otimes I + I\otimes\frac{1}{2}\Sigma^{ab})\gamma_{[ab]c}(x)\Big]\xi(x) = 0, \tag{10.21}$$

where

$$\Sigma^{0j} = \frac{1}{2}\sigma^j,\ \Sigma^{12} = -\frac{i}{2}\sigma^3,\ \Sigma^{23} = -\frac{i}{2}\sigma^1,\ \Sigma^{31} = -\frac{i}{2}\sigma^2.$$

Equation (10.21) takes the form

$$\Big[\sigma^0\partial_t + \sigma^3\partial_z + \sigma^1 e^1_{(1)}(x)\partial_u + \sigma^2 e^2_{(2)}(x)\partial_v$$
$$+\sigma^1(\Sigma^{12}\otimes I + I\otimes\Sigma^{12})\gamma_{[12]1} + \sigma^2(\Sigma^{12}\otimes I + I\otimes\Sigma^{12})\gamma_{[12]2}\Big]\xi(x) = 0,$$

or, differently,

$$\Big\{\sigma^0\partial_t + \sigma^3\partial_z + \frac{1}{\sqrt{u^2+v^2}}\left(\sigma^1\partial_u + \sigma^2\partial_v\right) + \frac{i/2}{(u^2+v^2)^{3/2}}$$
$$\times \left[v\sigma^1(\sigma^3\otimes I + I\otimes\sigma^3) - u\sigma^2(\sigma^3\otimes I + I\otimes\sigma^3)\right]\Big\}\xi(x) = 0. \tag{10.22}$$

We shall use the following substitution for the electromagnetic spinor $\xi(x)$:

$$\xi(x) = e^{-i\epsilon t}e^{ikz}\begin{vmatrix} f(u,v) & h(u,v) \\ h(u,v) & g(u,v) \end{vmatrix}, \tag{10.23}$$

where f, g, h stand for some functions over the variables u, v. Taking into account expressions for Pauli matrices

$$\sigma^0 = \begin{vmatrix} 1 & 0 \\ 0 & 1 \end{vmatrix}, \sigma^1 = \begin{vmatrix} 0 & 1 \\ 1 & 0 \end{vmatrix}, \sigma^2 = \begin{vmatrix} 0 & -i \\ i & 0 \end{vmatrix}, \sigma^3 = \begin{vmatrix} 1 & 0 \\ 0 & -1 \end{vmatrix},$$

from (10.22) we derive

$$-i\epsilon\begin{vmatrix} f & h \\ h & g \end{vmatrix} + ik\begin{vmatrix} f & h \\ -h & -g \end{vmatrix} + \frac{1}{\sqrt{u^2+v^2}}\begin{vmatrix} (\partial_u - i\partial_v)h & (\partial_u - i\partial_v)g \\ (\partial_u + i\partial_v)f & (\partial_u + i\partial_v)h \end{vmatrix}$$
$$+ \frac{i}{(u^2+v^2)^{3/2}}\begin{vmatrix} 0 & -(v+iu)g \\ (v-iu)f & 0 \end{vmatrix} = \begin{vmatrix} 0 & 0 \\ 0 & 0 \end{vmatrix},$$

which lead to the four equations

$$\begin{aligned}
(11) \quad & -i\epsilon f + ikf + \frac{1}{\sqrt{u^2+v^2}}(\partial_u - i\partial_v)h = 0, \\
(22) \quad & -i\epsilon g - ikg + \frac{1}{\sqrt{u^2+v^2}}(\partial_u + i\partial_v)h = 0, \\
(12) \quad & -i\epsilon h + ikh + \frac{1}{\sqrt{u^2+v^2}}(\partial_u - i\partial_v)g - \frac{i(v+iu)}{(u^2+v^2)^{3/2}}g = 0, \\
(21) \quad & -i\epsilon h - ikh + \frac{1}{\sqrt{u^2+v^2}}(\partial_u + i\partial_v)f + \frac{i(v-iu)}{(u^2+v^2)^{3/2}}f = 0.
\end{aligned} \tag{10.24}$$

They may be written differently,

$$-i\epsilon f + ikf + \frac{1}{\sqrt{u^2+v^2}}\partial_u h - \frac{i}{\sqrt{u^2+v^2}}\partial_v h = 0,$$
$$-i\epsilon g - ikg + \frac{1}{\sqrt{u^2+v^2}}\partial_u h + \frac{i}{\sqrt{u^2+v^2}}\partial_v h = 0,$$
$$-i\epsilon h + ikh + \frac{1}{\sqrt{u^2+v^2}}\left(\partial_u + \frac{u}{u^2+v^2}\right)g - \frac{i}{\sqrt{u^2+v^2}}\left(\partial_v + \frac{v}{u^2+v^2}\right)g = 0,$$
$$-i\epsilon h - ikh + \frac{1}{\sqrt{u^2+v^2}}\left(\partial_u + \frac{u}{u^2+v^2}\right)f + \frac{i}{\sqrt{u^2+v^2}}\left(\partial_v + \frac{v}{u^2+v^2}\right)f = 0.$$

Taking into account the following two identities for the function g:

$$\frac{1}{\sqrt{u^2+v^2}}\Big(\partial_u+\frac{u}{u^2+v^2}\Big)g=\frac{1}{u^2+v^2}\partial_u\sqrt{u^2+v^2}g, \quad \sqrt{u^2+v^2}g\equiv\bar{g},$$
$$\frac{1}{\sqrt{u^2+v^2}}\Big(\partial_v+\frac{v}{u^2+v^2}\Big)g=\frac{1}{u^2+v^2}\partial_v\sqrt{u^2+v^2}g, \quad \sqrt{u^2+v^2}g\equiv\bar{g},$$

and the similar ones for f:

$$\frac{1}{\sqrt{u^2+v^2}}\Big(\partial_u+\frac{u}{u^2+v^2}\Big)f=\frac{1}{u^2+v^2}\partial_u\sqrt{u^2+v^2}f, \quad \sqrt{u^2+v^2}f\equiv\bar{f},$$
$$\frac{1}{\sqrt{u^2+v^2}}\Big(\partial_v+\frac{v}{u^2+v^2}\Big)f=\frac{1}{u^2+v^2}\partial_v\sqrt{u^2+v^2}f, \quad \sqrt{u^2+v^2}f\equiv\bar{f},$$

and introducing the new functions

$$\sqrt{u^2+v^2}f\equiv\bar{f}, \quad \sqrt{u^2+v^2}g\equiv\bar{g}, \tag{10.25}$$

we can write the above system as follows

$$\begin{gathered}
-i\epsilon\bar{f}+ik\bar{f}+\partial_u h-i\partial_v h=0, \quad -i\epsilon\bar{g}-ik\bar{g}+\partial_u h+i\partial_v h=0,\\
-i\epsilon h+ikh+\frac{1}{u^2+v^2}\partial_u\bar{g}-\frac{i}{u^2+v^2}\partial_v\bar{g}=0,\\
-i\epsilon h-ikh+\frac{1}{u^2+v^2}\partial_u\bar{f}+\frac{i}{u^2+v^2}\partial_v\bar{f}=0.
\end{gathered} \tag{10.26}$$

Let us sum and subtract equations within each pair, then with notations

$$\bar{f}+\bar{g}=F, \quad \bar{f}-\bar{g}=G, \quad F+G=2\bar{f}, \quad F-G=2\bar{g}; \tag{10.27}$$

we obtain

$$\begin{gathered}
-i\epsilon F+ikG+2\partial_u h=0, \; -i\epsilon G+ikF-2i\partial_v h=0,\\
-2i\epsilon h+\frac{1}{u^2+v^2}\partial_u F+\frac{i}{u^2+v^2}\partial_v G=0,\\
2ikh-\frac{1}{u^2+v^2}\partial_u G-\frac{i}{u^2+v^2}\partial_v F=0.
\end{gathered} \tag{10.28}$$

From the first two equations we find

$$F=\frac{2}{\epsilon^2-k^2}[-i\epsilon\partial_u h-k\partial_v h], \quad G=\frac{2}{\epsilon^2-k^2}[-ik\partial_u h-\epsilon\partial_v h]; \tag{10.29}$$

substituting these into the remaining two equations, we get

$$-2i\epsilon h+\frac{1}{u^2+v^2}\frac{2}{\epsilon^2-k^2}\{-i\epsilon\partial_u^2-k\partial_{uv}^2+k\partial_{uv}^2-i\epsilon\partial_v^2\}h=0,$$
$$2ikh-\frac{1}{u^2+v^2}\frac{2}{\epsilon^2-k^2}\{-ik\partial_u^2-\epsilon\partial_{uv}^2+\epsilon\partial_{uv}^2-ik\partial_v^2\}h=0,$$

whence, after a simple regrouping of the terms, there follow two coinciding equations

$$-2i\epsilon h+\frac{1}{u^2+v^2}\frac{2}{\epsilon^2-k^2}\Big\{-i\epsilon(\partial_u^2+\partial_v^2)-k\partial_{uv}^2+k\partial_{uv}^2\Big\}h=0,$$
$$2ikh-\frac{1}{u^2+v^2}\frac{2}{\epsilon^2-k^2}\Big\{-ik(\partial_u^2+\partial_v^2)-\epsilon\partial_{uv}^2+\epsilon\partial_{uv}^2\Big\}h=0.$$

Thus, we arrive at the 2-nd order equation for the variable $h(u,v)$:

$$h+\frac{1}{u^2+v^2}\frac{1}{\epsilon^2-k^2}(\partial_u^2+\partial_v^2)h=0\,;$$

which can be presented as (let $\lambda^2=\epsilon^2-k^2>0$)

$$\Big(\frac{\partial^2}{\partial u^2}+\frac{\partial^2}{\partial v^2}+\lambda^2u^2+\lambda^2v^2\Big)h=0. \qquad (10.30)$$

In this equation, the variables may be separated by the substitution $h(u,v)=U(u)V(v)$:

$$\Big(\frac{1}{U}\frac{d^2}{dv^2}U+\lambda^2u^2\Big)+\Big(\frac{1}{V}\partial_v^2V+\lambda^2v^2\Big)=0.$$

So we find the two separate equations

$$\frac{1}{U}\partial_u^2U+\lambda^2u^2=-A \implies \Big(\frac{d^2}{du^2}+\lambda^2u^2+A\Big)U=0,$$
$$\frac{1}{V}\partial_v^2V+\lambda^2v^2=+A \implies \Big(\frac{d^2}{dv^2}+\lambda^2v^2-A\Big)V=0. \qquad (10.31)$$

Let us transform the equations (10.31) to other variables:

$$X=u^2,\quad \Big(\frac{d^2}{dX^2}+\frac{1}{2X}\frac{d}{dX}+\frac{\lambda^2}{4}+\frac{A}{4X}\Big)U=0;$$
$$Y=v^2,\quad \Big(\frac{d^2}{dY^2}+\frac{1}{2Y}\frac{d}{dY}+\frac{\lambda^2}{4}-\frac{A}{4Y}\Big)V=0.$$

We look for their solutions in the form

$$U(X)=X^{\alpha}e^{\beta X}f(X),\qquad V(Y)=Y^{\rho}e^{\sigma Y}g(Y)\,. \qquad (10.32)$$

We further obtain

$$\Big[\frac{d^2}{dX^2}+(\frac{2\alpha}{X}+\frac{1}{2X}+2\beta)\frac{d}{dX}+\frac{2\alpha\beta}{X}+\frac{\beta}{2X}+\frac{A}{4X}+\frac{\alpha(\alpha-1)}{X^2}+\frac{\alpha/2}{X^2}+\beta^2+\frac{\lambda^2}{4}\Big]f(X)=0,$$

$$\Big[\frac{d^2}{dY^2}+(\frac{2\rho}{Y}+\frac{1}{2Y}+2\sigma)\frac{d}{dY}+\frac{2\rho\sigma}{Y}+\frac{\sigma}{2Y}-\frac{A}{4Y}+\frac{\rho(\rho-1)}{Y^2}+\frac{\rho/2}{Y^2}+\sigma^2+\frac{\lambda^2}{4}\Big]g(Y)=0.$$

By imposing the restrictions $\alpha=0,+\frac{1}{2}$, $2\beta=\pm i\lambda$; $\rho=0,+\frac{1}{2}$, $2\sigma=\pm i\lambda$, we get

$$\Big[X\frac{d^2}{dX^2}+(2\alpha+\frac{1}{2}+2\beta X)\frac{d}{dX}+2\alpha\beta+\frac{\beta}{2}+\frac{A}{4}\Big]f=0,$$
$$\Big[Y\frac{d^2}{dY^2}+(2\rho+\frac{1}{2}+2\sigma Y)\frac{d}{dY}+2\rho\sigma+\frac{\sigma}{2}-\frac{A}{4}\Big]g=0.$$

Let us fix the two parameters $2\beta = -i\lambda$, $2\sigma = -i\lambda$; this yields

$$\Big[X\frac{d^2}{dX^2} + (2\alpha + 1/2 - i\lambda X)\frac{d}{dX} - i\lambda\alpha - i\lambda/4 + A/4\Big] f(X) = 0,$$
$$\Big[Y\frac{d^2}{dY^2} + (2\rho + 1/2 - i\lambda Y)\frac{d}{dY} - i\lambda\rho - i\lambda/4 - A/4\Big] g(Y) = 0.$$

After transforming the equations to the new variables

$$i\lambda X = x = i\lambda u^2, \qquad i\lambda Y = y = i\lambda v^2, \quad A/4i\lambda = \Lambda, \tag{10.33}$$

we obtain

$$\begin{gathered}\Big[x\frac{d^2}{dx^2} + (2\alpha + 1/2 - x)\frac{d}{dx} - \alpha - 1/4 + \Lambda\Big] f(x) = 0,\\ \Big[y\frac{d^2}{dy^2} + (2\rho + 1/2 - y)\frac{d}{dy} - \rho - 1/4 - \Lambda\Big] g(y) = 0.\end{gathered} \tag{10.34}$$

For definiteness, let us take the values $\alpha = 0, \rho = 0$:

$$\begin{gathered}\Big[x\frac{d^2}{dx^2} + (1/2 - x)\frac{d}{dx} - (1/4 - \Lambda)\Big] f(x) = 0,\\ U(x) = e^{x/2} f(x), \quad x = i\lambda u^2;\\ \Big[y\frac{d^2}{dy^2} + (1/2 - y)\frac{d}{dy} - (1/4 + \Lambda)\Big] g(y) = 0,\\ V(y) = e^{y/2} g(y), \quad y = i\lambda v^2.\end{gathered} \tag{10.35}$$

They both can be identified with the confluent hypergeometric equation

$$zF'' + (c - z)F' - aF = 0, \quad c = 1/2, \; a = 1/4 \pm \Lambda. \tag{10.36}$$

Two linearly independent solutions may be used [1]:

$$F_1 = \Phi(a, c; z), \quad F_2 = z^{1-c}\Phi(a - c + 1, 2 - c; z). \tag{10.37}$$

Therefore, for the two equations in (10.35), we have the respective pairs of solutions:

$$\begin{gathered}U_1 = e^{x/2}\Phi(\frac{1}{4} - \Lambda, \frac{1}{2}; x), \quad U_2 = e^{x/2}\sqrt{x}\,\Phi(\frac{3}{4} - \Lambda, \frac{3}{2} - c; x);\\ V_1 = e^{y/2}\Phi(\frac{1}{4} + \Lambda, \frac{1}{2}; y), \quad V_2 = e^{y/2}\sqrt{y}\,\Phi(\frac{3}{4} + \Lambda, \frac{3}{2}; y);\end{gathered} \tag{10.38}$$

recall that $\sqrt{x} = \sqrt{i\lambda}\,u$ and $\sqrt{y} = \sqrt{i\lambda}\,v$.

10.3 Continuity and Spinor Space Structure

Let us find the explicit form for the constructed solutions in the initial Cartesian tetrad. It is known that the 2-spinor ψ_0 in Cartesian tetrad is related to the 2-spinor

ψ in cylindric parabolic tetrad by the following local gauge transformation

$$\psi = s\psi_0, \quad s = \frac{1}{(u^2+v^2)^{1/4}} \left| \begin{array}{cc} \sqrt{u+iv} & 0 \\ 0 & \sqrt{u-iv} \end{array} \right|, \qquad s^{-1} = s^{+} = \frac{1}{(u^2+v^2)^{1/4}} \left| \begin{array}{cc} \sqrt{u-iv} & 0 \\ 0 & \sqrt{u+iv} \end{array} \right|. \tag{10.39}$$

Therefore, the 2-rank spinors relate to each other by the transformation

$$\xi = S\xi_0 = (s \otimes s)\xi_0 = s\xi_0 s\,, \; \xi_0 = S^{-1}\xi = (s^{-1} \otimes s^{-1})\xi = s^{-1}\xi s^{-1}\,; \tag{10.40}$$

so, for the electromagnetic spinor in Cartesian basis

$$\xi_0 = \frac{1}{(u^2+v^2)^{1/2}} \left| \begin{array}{cc} \sqrt{u-iv} & 0 \\ 0 & \sqrt{u+iv} \end{array} \right| \left| \begin{array}{cc} f & h \\ h & g \end{array} \right| \left| \begin{array}{cc} \sqrt{u-iv} & 0 \\ 0 & \sqrt{u+iv} \end{array} \right|$$

we get

$$\xi_0 = \left| \begin{array}{cc} \frac{(u-iv)}{\sqrt{u^2+v^2}} f & h \\ h & \frac{(u+iv)}{\sqrt{u^2+v^2}} g \end{array} \right|. \tag{10.41}$$

Evidently, primary is the function h, and the two other ones f and g are determined by h. Taking into account the expressions for f and g

$$f = \frac{1}{2\sqrt{u^2+v^2}}(F+G)\,, \qquad g = \frac{1}{2\sqrt{u^2+v^2}}(F-G)\,,$$

and allowing for the formulas

$$F = \frac{2}{\epsilon^2 - k^2}\Big(-i\epsilon\partial_u h - k\partial_v h\Big), \quad G = \frac{2}{\epsilon^2 - k^2}\Big(-ik\partial_u h - \epsilon\partial_v h\Big), \tag{10.42}$$

we derive

$$f = \frac{1}{\epsilon - k}\frac{1}{\sqrt{u^2+v^2}}\Big(-i\partial_u - \partial_v\Big)h, \quad g = \frac{1}{\epsilon + k}\frac{1}{\sqrt{u^2+v^2}}\Big(-i\partial_u + \partial_v\Big)h. \tag{10.43}$$

Further, we should take in mind the product formula $h(u,v) = U(u)V(v)$ and the two possibilities for each multiplier (see (10.38)):

$$U_1 = e^{x/2}\Phi(1/4 - \Lambda, 1/2; x)\,, \quad U_2 = e^{x/2}\sqrt{x}\,\Phi(3/4 - \Lambda, 3/2; x)\,,$$

$$V_1 = e^{y/2}\Phi(1/4 + \Lambda, 1/2; y)\,, \quad V_2 = e^{y/2}\sqrt{y}\,\Phi(3/4 + \Lambda, 3/2; y)\,,$$

where

$$x = i\lambda u^2, \quad y = i\lambda v^2 \implies \frac{\partial}{\partial u} = 2i\lambda u\frac{\partial}{\partial x}, \quad \frac{\partial}{\partial v} = 2i\lambda v\frac{\partial}{\partial y}. \tag{10.44}$$

Let us change (10.43) to the alternative form

$$f = \frac{2i\lambda}{\epsilon - k}\frac{1}{\sqrt{u^2+v^2}}(-iu\frac{\partial}{\partial x} - v\frac{\partial}{\partial y})h = A(-iu\frac{\partial}{\partial x} - v\frac{\partial}{\partial y})h\,,$$
$$g = \frac{2i\lambda}{\epsilon + k}\frac{1}{\sqrt{u^2+v^2}}(-iu\frac{\partial}{\partial x} + v\frac{\partial}{\partial y})h = B(-iu\frac{\partial}{\partial x} + v\frac{\partial}{\partial y})h\,. \tag{10.45}$$

Besides, we should take in mind the rule for differentiating the confluent hypergeometric function

$$\frac{d}{dx}\Phi(a,c;x) = \frac{a}{c}\Phi(a+1,c+1;x)\,.$$

We shall follow all the four possibilities for the function $h(u,b)$:

$$h_{11} = U_1(x)V_1(y)\,, \quad h_{12} = U_1(x)V_2(y)\,,$$
$$h_{21} = U_2(x)V_1(y)\,, \quad h_{22} = U_1(x)V_2(y)\,. \qquad (10.46)$$

Consider the variant h_{11}:

$$f_{11} = A(-iu\frac{\partial}{\partial x} - v\frac{\partial}{\partial y})\,U_1(x)V_1(y)$$

$$= A\Big\{-iue^{x/2}\Big[\frac{1}{2}\Phi(1/4-\Lambda,1/2;x) + \frac{1/4-\Lambda}{1/2}\Phi(5/4-\Lambda,3/2;x)\Big]V_1(y)$$

$$-vU_1(x)e^{y/2}\Big[\frac{1}{2}\Phi(1/4+\Lambda,1/2;y) + \frac{1/4+\Lambda}{1/2}\Phi(5/4+\Lambda,3/2;y)\Big]\Big\},$$

$$g_{11} = B(-iu\frac{\partial}{\partial x} + v\frac{\partial}{\partial y})\,U_1(x)V_1(y)$$

$$= B\Big\{-iue^{x/2}\Big[\frac{1}{2}\Phi(1/4-\Lambda,1/2;x) + \frac{1/4-\Lambda}{1/2}\Phi(5/4-\Lambda,3/2;x)\Big]V_1(y)$$

$$+vU_1(x)e^{y/2}\Big[\frac{1}{2}\Phi(1/4+\Lambda,1/2;y) + \frac{1/4+\Lambda}{1/2}\Phi(5/4+\Lambda,3/2;y)\Big]\Big\}.$$

Consider the variant h_{12}:

$$f_{12} = A(-iu\frac{\partial}{\partial x} - v\frac{\partial}{\partial y})\,U_1(x)V_2(y)$$

$$= A\Big\{-iue^{x/2}\Big[\frac{1}{2}\Phi(1/4-\Lambda,1/2;x) + \frac{1/4-\Lambda}{1/2}\Phi(5/4-\Lambda,3/2;x)\Big]V_2(y)$$

$$-vU_1(x)e^{y/2}\sqrt{y}\Big[\frac{1}{2}\Phi(3/4+\Lambda,3/2;y) + \frac{1}{2y}\Phi(3/4+\Lambda,3/2;y)$$

$$+\frac{3/4+\Lambda}{3/2}\Phi(7/4+\Lambda,5/2;y)\Big]\Big\},$$

$$g_{12} = A(-iu\frac{\partial}{\partial x} + v\,\frac{\partial}{\partial y})\,U_1(x)V_2(y)$$

$$= B\Big\{-iue^{x/2}\Big[\frac{1}{2}\Phi(1/4-\Lambda,1/2;x) + \frac{1/4-\Lambda}{1/2}\Phi(5/4-\Lambda,3/2;x)\Big]V_2(y)$$

$$+vU_1(x)e^{y/2}\sqrt{y}\Big[\frac{1}{2}\Phi(3/4+\Lambda,3/2;y) + \frac{1}{2y}\Phi(3/4+\Lambda,3/2;y)$$

$$+\frac{3/4+\Lambda}{3/2}\Phi(7/4+\Lambda,5/2;y)\Big]\Big\}.$$

Consider the variant h_{21}:

$$f_{21} = A(-iu\frac{\partial}{\partial x} - v\frac{\partial}{\partial y})\, U_2(x)V_1(y)$$

$$= A\Big\{ -iue^{x/2}\sqrt{x}\Big[\frac{1}{2}\Phi(3/4-\Lambda,3/2;x) + \frac{1}{2x}\Phi(3/4-\Lambda,3/2;x)$$

$$+\frac{3/4-\Lambda}{3/2}\Phi(7/4+\Lambda,5/2;x)\Big]V_1(y)$$

$$-vU_2(x)e^{y/2}\Big[\frac{1}{2}\Phi(1/4+\Lambda,1/2;y) + \frac{1/4+\Lambda}{1/2}\Phi(5/4+\Lambda,3/2;y)\Big]\Big\},$$

$$g_{21} = B(-iu\ \frac{\partial}{\partial x} + v\frac{\partial}{\partial y})\, U_2(x)V_1(y)$$

$$= A\Big\{ -iue^{x/2}\sqrt{x}\Big[\frac{1}{2}\Phi(3/4-\Lambda,3/2;x) + \frac{1}{2x}\Phi(3/4-\Lambda,3/2;x)$$

$$+\frac{3/4-\Lambda}{3/2}\Phi(7/4+\Lambda,5/2;x)\Big]V_1(y)$$

$$+vU_2(x)e^{y/2}\Big[\frac{1}{2}\Phi(1/4+\Lambda,1/2;y) + \frac{1/4+\Lambda}{1/2}\Phi(5/4+\Lambda,3/2;y)\Big]\Big\}.$$

Consider the variant h_{22}:

$$f_{22} = A(-iu\frac{\partial}{\partial x} - v\frac{\partial}{\partial y})\, U_2(x)V_2(y)$$

$$= A\Big\{ -iue^{x/2}\sqrt{x}\Big[\frac{1}{2}\Phi(3/4-\Lambda,3/2;x) + \frac{1}{2x}\Phi(3/4-\Lambda,3/2;x)$$

$$+\frac{3/4-\Lambda}{3/2}\Phi(7/4+\Lambda,5/2;x)\Big]V_2(y)$$

$$-vU_2(x)e^{y/2}\sqrt{y}\Big[\frac{1}{2}\Phi(3/4+\Lambda,3/2;y) + \frac{1}{2y}\Phi(3/4+\Lambda,3/2;y)$$

$$+\frac{3/4+\Lambda}{3/2}\Phi(7/4+\Lambda,5/2;y)\Big]\Big\},$$

$$g_{22} = A(-iu\frac{\partial}{\partial x} + v\frac{\partial}{\partial y})\, U_2(x)V_2(y)$$

$$= A\Big\{ -iue^{x/2}\sqrt{x}\Big[\frac{1}{2}\Phi(3/4-\Lambda,3/2;x) + \frac{1}{2x}\Phi(3/4-\Lambda,3/2;x)$$

$$+\frac{3/4-\Lambda}{3/2}\Phi(7/4+\Lambda,5/2;x)\Big]V_2(y)$$

$$+v\cdot U_2(x)e^{y/2}\sqrt{y}\Big[\frac{1}{2}\Phi(3/4+\Lambda,3/2;y) + \frac{1}{2y}\Phi(3/4+\Lambda,3/2;y)$$

$$+\frac{3/4+\Lambda}{3/2}\Phi(7/4+\Lambda,5/2;y)\Big]\Big\}.$$

We recall the formulas for parabolic cylindric coordinates

$$x = \frac{u^2 - v^2}{2}, \quad y = u\,v, \quad z = z. \tag{10.47}$$

In order to get inverse transformation, from the first equation (10.47) we exclude the variable u; then – the variable v:

$$2x = -v^2 + \frac{1}{v^2}\,y^2\,,\; 2x = +u^2 - \frac{1}{u^2}\,y^2 \;\Longrightarrow\; v^4 + 2x\,v^2 - y^2 = 0\,,\; u^4 - 2x\,u^2 - y^2 = 0\,.$$

We further find

$$v^2 = -x + \sqrt{x^2 + y^2}\,, \quad u^2 = +x + \sqrt{x^2 + y^2}\,. \tag{10.48}$$

For parameterizing of the ordinary Cartesian space (x,y,z), we shall use the formulas

$$v = +\sqrt{-x + \sqrt{x^2 + y^2}}\,, \quad u = \pm\sqrt{+x + \sqrt{x^2 + y^2}}\,. \tag{10.49}$$

The correspondence between (x,y) and (u,v) is clarified by the formulas

$$\begin{gathered} u = k\,\cos\phi\,, \quad v = k\,\sin\phi\,, \quad \phi \in [\,0,\,\pi\,]\,; \\ x = (k^2/2)\,\cos 2\phi\,, \quad y = (k^2/2)\,\sin 2\phi\,, \quad 2\phi \in [0,\,2\pi]\,. \end{gathered} \tag{10.50}$$

In fact, there exists a peculiarity in parameterizing the half-line $x > 0$ by coordinates $(u, v = 0)$.

If one takes in mind the space models with spinor structure, then the symmetry between the coordinates u and v becomes complete: they relate to the Cartesian coordinates $(x,y,z) \oplus (x,y,z)$ by the formulas

$$v = \pm\sqrt{-x + \sqrt{x^2 + y^2}}\,, \quad u = \pm\sqrt{+x + \sqrt{x^2 + y^2}}\,. \tag{10.51}$$

To the spinor space model there corresponds the domain $\tilde{G}(u,v)$, which is symmetrical with respect coordinates u and v, and without having any special identification rules for the boundary points in the plane (u,v).

This is an important matter, because the expected solutions must be single-valued in the whole space. Evidently, in different space models, we should expect different single-valued solutions.

To clarify this point, it suffices to follow the behavior of constructed solutions (see (10.41) when $v \to 0$:

$$v \to 0, \quad \xi_0 = \left| \begin{array}{cc} \frac{(u-iv)}{\sqrt{u^2+v^2}} f & h \\ h & \frac{(u+iv)}{\sqrt{u^2+v^2}} g \end{array} \right| \to \left| \begin{array}{cc} \frac{u}{\sqrt{u^2}}\, f & h \\ h & \frac{u}{\sqrt{u^2}}\, g \end{array} \right| . \tag{10.52}$$

In the vector space model, the following identity must hold

$$\xi_0(u, v = 0) = \xi_0(-u, v = 0)\,; \tag{10.53}$$

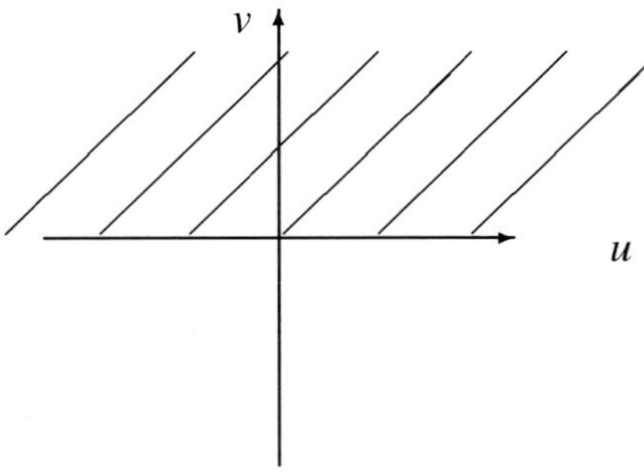

Figure 10.1: The domain $G(u,v)$ for vector space

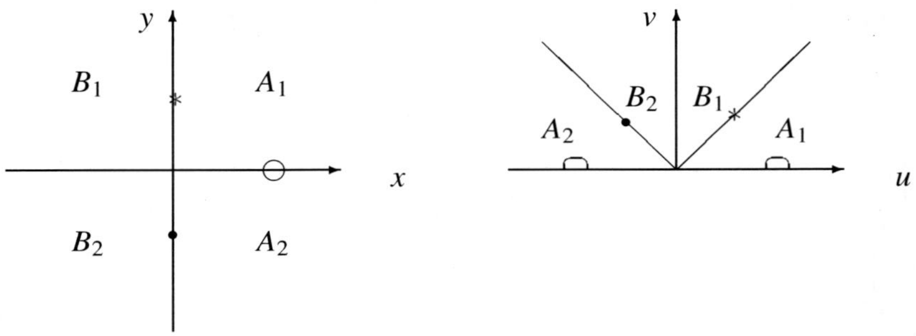

Figure 10.2: The map $G(x,y) \Longrightarrow G(u,v)$

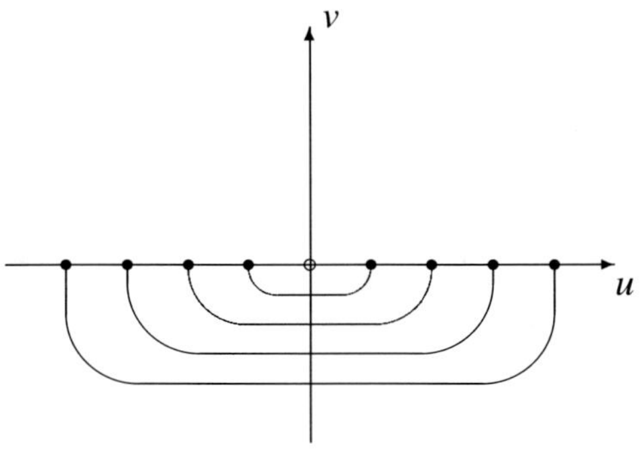

Figure 10.3: Identification for boundary points in $G(u,v)$

however in the spinor space model, this requirement may be ignored.

Allowing for the above expressions relations for h_{ij}, f_{ij}, g_{ij}, we find the following behavior of the four solutions at $v \to 0$ (the plus (+) or minus (−) means appropriateness and non-appropriateness of functions as single-valued in vector space model; see (10.52)):

$$\begin{aligned}
& f_{11} \sim u \cdot \text{function}(u^2), \qquad g_{11} \sim u \cdot \text{function}(u^2), \qquad (+)\,; \\
& f_{12} \sim \text{function}(u^2), \qquad g_{12} \sim \text{function}(u^2), \qquad (-)\,; \\
& f_{21} \sim u^2 \cdot \text{function}(u^2), \qquad g_{21} \sim u^2 \cdot \text{function}(u^2), \qquad (-)\,; \\
& f_{22} \sim u \cdot \text{function}(u^2), \qquad g_{11} \sim u \cdot \text{function}(u^2) \qquad (+)\,.
\end{aligned} \tag{10.54}$$

Similarly, from (10.46) it follows

$$\begin{aligned}
& h_{11} = U_1(x)V_1(y) \sim \text{function}(u^2), \qquad (+)\,; \\
& h_{12} = U_1(x)V_2(y) \sim 0\,, \qquad (+,-)\,; \\
& h_{21} = U_2(x)V_1(y) \sim u \cdot \text{function}(u^2)\,, \qquad (-); \\
& h_{22} = U_1(x)V_2(y) \sim 0\,, \qquad (+,-)\,.
\end{aligned} \tag{10.55}$$

Thus, we conclude that in the vector space we have single valued solutions only for variants h_{11} and h_{22}; in spinor space all four variants $h_{11}, h_{12}, h_{21}, h_{22}$ provide us with single-valued solutions.

10.4 Helicity Operator

First, let us solve a subsidiary task: find consequences of diagonalization of the helicity operator for electromagnetic spinor (related to plane waves)

$$\begin{aligned}
\Sigma = -\frac{i}{2}\Big[\partial_1(\sigma_1 \otimes I + I \otimes \sigma_1) + \partial_2(\sigma_2 \otimes I + I \otimes \sigma_2) + \partial_3(\sigma_3 \otimes I + I \otimes \sigma_3)\Big] \\
= -\frac{i}{2}\Big[\,\partial_1\Sigma_1 + \partial_2\Sigma_2 + \partial_3\Sigma_3\,\Big]\,.
\end{aligned} \tag{10.56}$$

Taking in mind the substitution for ξ:

$$\xi = e^{-i\epsilon t} e^{ik_1 x} e^{ik_2 y} e^{ik_3 z} \left| \begin{array}{cc} f & h \\ h & g \end{array} \right| ,$$

we have the following expression for Σ:

$$\Sigma = \frac{1}{2}\Big[k_1(\sigma_1\xi + \xi\tilde{\sigma}_1) + (\sigma_2\xi + \xi\tilde{\sigma}_2) + k_2 + k_3(\sigma_3\xi + \xi\tilde{\sigma}_3)\Big]$$

$$= \frac{1}{2}\left\{ k_1 \left| \begin{array}{cc} 2h & f+g \\ f+g & 2h \end{array} \right| + k_2 \left| \begin{array}{cc} -2ih & i(f-g) \\ i(f-g) & 2ih \end{array} \right| + k_3 \left| \begin{array}{cc} 2f & 0 \\ 0 & -2g \end{array} \right| \right\}.$$

Therefore, from the eigenvalue equation $\Sigma\xi = \sigma\xi$, we get the four equations

$$\begin{aligned} (k_1 - ik_2)\,h + (k_3 - \sigma)\,f = 0\,, \\ (k_1 + ik_2)\,h - (k_3 + \sigma)\,g = 0\,, \\ \frac{1}{2}(k_1 + ik_2)\,f + \frac{1}{2}(k_1 - ik_2)\,g - \sigma\,h = 0\,, \\ \frac{1}{2}(k_1 + ik_2)\,f + \frac{1}{2}(k_1 - ik_2)\,g - \sigma\,h = 0\,; \end{aligned} \tag{10.57}$$

the last two equations coincide. In (10.57), we have the system of three equations

$$\left|\begin{array}{ccc} (k_3 - \sigma) & 0 & (k_1 - ik_2) \\ 0 & -(k_3 + \sigma) & (k_1 + ik_2) \\ \frac{1}{2}(k_1 + ik_2) & \frac{1}{2}(k_1 - ik_2) & -\sigma \end{array}\right| \left|\begin{array}{c} f \\ g \\ h \end{array}\right| = 0\,. \tag{10.58}$$

For the eigenvalues σ we have a cubic equation

$$(k_3 - \sigma)(k_3 + \sigma)\sigma + \frac{1}{2}(k_1 + ik_2)(k_1 - ik_2)(k_3 + \sigma) - \frac{1}{2}(k_1 + ik_2)(k_1 - ik_2)(k_3 - \sigma) = 0\,,$$

or $(k_1^2 + k_2^2 + k_3^2 - \sigma^2) = 0$; so we obtain the three roots

$$\sigma = 0, +k, -k; \;\; k = \sqrt{k_1^2 + k_2^2 + k_3^2} = \epsilon\,. \tag{10.59}$$

Let us find the solutions of the system (10.58) at $\sigma = 0$:

$$\left|\begin{array}{ccc} k_3 & 0 & (k_1 - ik_2) \\ 0 & -k_3 & (k_1 + ik_2) \\ \frac{1}{2}(k_1 + ik_2) & \frac{1}{2}(k_1 - ik_2) & 0 \end{array}\right| \left|\begin{array}{c} f \\ g \\ h \end{array}\right| = 0\,,$$

or,

$$k_3 f + (k_1 - ik_2)h = 0\,, \; -k_3 g + (k_1 + ik_2)h = 0, \; (k_1 + ik_2)f + (k_1 - ik_2)g = 0\,;$$

with the first two equations, the third one turns to be an identity $0 \equiv 0$; therefore we get

$$\sigma = 0\,, \quad f = -\frac{k_1 - ik_2}{k_3}h\,, \quad g = +\frac{k_1 + ik_2}{k_3}h\,. \tag{10.60}$$

Let us find the solutions of the system (10.58) at $\sigma = -k$:

$$\left|\begin{array}{ccc} (k_3 + k) & 0 & (k_1 - ik_2) \\ 0 & -(k_3 - k) & (k_1 + ik_2) \\ \frac{1}{2}(k_1 + ik_2) & \frac{1}{2}(k_1 - ik_2) & k \end{array}\right| \left|\begin{array}{c} f \\ g \\ h \end{array}\right| = 0\,,$$

that is

$$(k_3 + k)f + (k_1 - ik_2)h = 0\,, \quad -(k_3 - k)g + (k_1 + ik_2)h = 0\,,$$
$$\frac{1}{2}(k_1 + ik_2)f + \frac{1}{2}(k_1 - ik_2)g + kh = 0\,;$$

with two first equations the third one turns to be an identity $0 \equiv 0$; therefore we get

$$\sigma = -k, \qquad f = -\frac{k_1 - ik_2}{k_3 + k} h, \quad g = +\frac{k_1 + ik_2}{k_3 - k} h. \tag{10.61}$$

Now let us find solutions of the system (10.58) at $\sigma = +k$:

$$\begin{vmatrix} (k_3 - k) & 0 & (k_1 - ik_2) \\ 0 & -(k_3 + k) & (k_1 + ik_2) \\ \frac{1}{2}(k_1 + ik_2) & \frac{1}{2}(k_1 - ik_2) & -k \end{vmatrix} \begin{vmatrix} f \\ g \\ h \end{vmatrix} = 0,$$

or

$$(k_3 - k) f + (k_1 - ik_2) h = 0, \quad -(k_3 + k) g + (k_1 + ik_2) h = 0,$$
$$\frac{1}{2}(k_1 + ik_2) f + \frac{1}{2}(k_1 - ik_2) g - kh = 0;$$

with two first equations the third one becomes an identity, so we arrive at

$$\sigma = +k, \quad f = -\frac{k_1 - ik_2}{k_3 - k} h, \quad g = +\frac{k_1 + ik_2}{k_3 + k} h. \tag{10.62}$$

We shall compare now (10.60)–(10.62) with the solutions of the Maxwell equations in spinor form

$$\begin{cases} (k_3 - \epsilon) f + (k_1 - ik_2) h = 0, \\ (k_1 + ik_2) f - (k_3 + \epsilon) h = 0, \end{cases} \qquad \epsilon = +k, \quad f = -\frac{k_1 - ik_2}{k_3 - k} h;$$
$$\begin{cases} (k_1 - ik_2) g + (k_3 - \epsilon) h = 0, \\ -(k_3 + \epsilon) g + (k_1 + ik_2) h = 0, \end{cases} \qquad \epsilon = +k, \quad g = +\frac{k_1 + ik_2}{k_3 + k} h. \tag{10.63}$$

We conclude that the solution of the spinor Maxwell equations is the eigenstate with $\sigma = +1$.

Now, let us consider the similar problem in cylindric parabolic coordinates. First, we transform the helicity operator Σ (see (10.56)) to cylindric parabolic coordinates and tetrad. From the eigenstate equation in Cartesian basis

$$-\frac{i}{2}[\partial_1 \Sigma_1 + \partial_2 \Sigma_2 + \partial_z \Sigma_3] \xi_0 = \sigma \xi_0,$$

it follows (see (10.39) and (10.40))

$$-\frac{i}{2}\Big[S\Sigma_1 S^{-1}(\frac{\partial}{\partial x^1} + S\frac{\partial S^{-1}}{\partial x^1}) + S\Sigma_2 S^{-1}(\frac{\partial}{\partial x^2} + S\frac{\partial S^{-1}}{\partial x^2})$$
$$+ S\Sigma_3 S^{-1}(\frac{\partial}{\partial x^2} + S\frac{\partial S^{-1}}{\partial x^3})\Big] = \sigma \xi; \tag{10.64}$$

the matrix S depends only on (x^1, x^2) (or (u, v)). We use the identities

$$S\Sigma_1 S^{-1} = (s \otimes s)[\sigma_1 \otimes I + I \otimes \sigma_1](s^{-1} \otimes s^{-1}) = s\sigma_1 s^{-1} \otimes I + I \otimes s\sigma_1 s^{-1},$$

$$S\Sigma_2 S^{-1} = (s\otimes s)[\sigma_2 \otimes I + I\otimes \sigma_2](s^{-1}\otimes s^{-1}) = s\sigma_2 s^{-1}\otimes I + I\otimes s\sigma_2 s^{-1},$$

$$S\Sigma_3 S^{-1} = (s\otimes s)[\sigma_3 \otimes I + I\otimes \sigma_3](s^{-1}\otimes s^{-1}) = s\sigma_3 s^{-1}\otimes I + I\otimes s\sigma_3 s^{-1},$$

and also the identities

$$S\frac{\partial S^{-1}}{\partial x^1} = (s\otimes s)\frac{\partial (s^{-1}\otimes s^{-1})}{\partial x^1} = s\frac{\partial s^{-1}}{\partial x^1}\otimes I + I\otimes s\frac{\partial s^{-1}}{\partial x^1},$$

$$S\frac{\partial S^{-1}}{\partial x^2} = (s\otimes s)\frac{\partial (s^{-1}\otimes s^{-1})}{\partial x^2} = s\frac{\partial s^{-1}}{\partial x^2}\otimes I + I\otimes s\frac{\partial s^{-1}}{\partial x^2}, \quad S\frac{\partial S^{-1}}{\partial x^3} = 0.$$

Taking these in mind, we obtain (see (10.64))

$$S\Sigma_1 S^{-1}(\frac{\partial}{\partial x^1} + S\frac{\partial S^{-1}}{\partial x^1})$$

$$= \left(s\sigma_1 s^{-1}\otimes I + I\otimes s\sigma_1 s^{-1}\right)\left(\frac{\partial}{\partial x^1} + s\frac{\partial s^{-1}}{\partial x^1}\otimes I + I\otimes s\frac{\partial s^{-1}}{\partial x^1}\right),$$

that is

$$S\Sigma_1 S^{-1}(\frac{\partial}{\partial x^1} + S\frac{\partial S^{-1}}{\partial x^1}) = (s\sigma_1 s^{-1}\otimes I + I\otimes s\sigma_1 s^{-1})\frac{\partial}{\partial x^1}$$

$$+s\sigma_1\frac{\partial s^{-1}}{\partial x^1}\otimes I + I\otimes s\sigma_1\frac{\partial s^{-1}}{\partial x^1} + s\sigma_1 s^{-1}\otimes s\frac{\partial s^{-1}}{\partial x^1} + s\frac{\partial s^{-1}}{\partial x^1}\otimes s\sigma_1 s^{-1}.$$

Similarly, we derive

$$S\Sigma_2 S^{-1}(\frac{\partial}{\partial x^2} + S\frac{\partial S^{-1}}{\partial x^2}) = (s\sigma_2 s^{-1}\otimes I + I\otimes s\sigma_2 s^{-1})\frac{\partial}{\partial x^2}$$

$$+s\sigma_2\frac{\partial s^{-1}}{\partial x^2}\otimes I + I\otimes s\sigma_2\frac{\partial s^{-1}}{\partial x^2} + s\sigma_2 s^{-1}\otimes s\frac{\partial s^{-1}}{\partial x^2} + s\frac{\partial s^{-1}}{\partial x^2}\otimes s\sigma_2 s^{-1},$$

and

$$S\Sigma_3 S^{-1}(\frac{\partial}{\partial x^3} + S\frac{\partial S^{-1}}{\partial x^3}) = S\Sigma_3 S^{-1}\frac{\partial}{\partial x^3} = (s\sigma_3 s^{-1}\otimes I + I\otimes s\sigma_3 s_3)\frac{\partial}{\partial x^3}.$$

Let us write down the general structure for the operator Σ:

$$\Sigma = -\frac{i}{2}\Big\{\left(s\sigma_1 s^{-1}\otimes I + I\otimes s\sigma_1 s^{-1}\right)\frac{\partial}{\partial x^1} + \left(s\sigma_2 s^{-1}\otimes I + I\otimes s\sigma_2 s^{-1}\right)\frac{\partial}{\partial x^2}$$

$$+\left(s\sigma_3 s^{-1}\otimes I + I\otimes s\sigma_3 s_3\right)\frac{\partial}{\partial x^3} + \left(s\sigma_1 s^{-1} s\frac{\partial s^{-1}}{\partial x^1}\otimes I + I\otimes s\sigma_1 s^{-1} s\frac{\partial s^{-1}}{\partial x^1}\right)$$

$$+\left(s\sigma_2 s^{-1} s\frac{\partial s^{-1}}{\partial x^2}\otimes I + s\sigma_2 s^{-1} s\frac{\partial s^{-1}}{\partial x^2}\otimes I\right)$$

$$+\left(s\sigma_1 s^{-1}\otimes s\frac{\partial s^{-1}}{\partial x^1} + s\frac{\partial s^{-1}}{\partial x^1}\otimes s\sigma_1 s^{-1}\right) + \left(s\sigma_2 s^{-1}\otimes s\frac{\partial s^{-1}}{\partial x^2} + s\frac{\partial s^{-1}}{\partial x^2}\otimes s\sigma_2 s^{-1}\right)\Big\}.$$

We readily find the formulas

$$s\sigma_3 s^{-1} = \sigma_3 \,,$$
$$s\sigma_1 s^{-1} = \frac{1}{\sqrt{u^2+v^2}} \left| \begin{array}{cc} 0 & u+iv \\ u-iv & 0 \end{array} \right| = \frac{1}{\sqrt{u^2+v^2}} (u\sigma_1 - v\sigma_2) \,,$$
$$s\sigma_2 s^{-1} = \frac{1}{\sqrt{u^2+v^2}} \left| \begin{array}{cc} 0 & v-iu \\ v+iu & 0 \end{array} \right| = \frac{1}{\sqrt{u^2+v^2}} (v\sigma_1 + u\sigma_2) \,. \qquad (10.65)$$

Taking in mind the relations

$$\frac{\partial u}{\partial x^1} = \frac{u}{u^2+v^2} \,, \quad \frac{\partial v}{\partial x^1} = \frac{-v}{u^2+v^2} \,, \quad \frac{\partial u}{\partial x^2} = \frac{v}{u^2+v^2} \,, \quad \frac{\partial v}{\partial x^2} = \frac{u}{u^2+v^2} \,,$$

we derive the formulas

$$\frac{\partial}{\partial x^1} = \frac{1}{u^2+v^2} (u\frac{\partial}{\partial u} - v\frac{\partial}{\partial v}) \,, \quad \frac{\partial}{\partial x^2} = \frac{1}{u^2+v^2} (v\frac{\partial}{\partial u} + u\frac{\partial}{\partial v}) \,. \qquad (10.66)$$

Let us consider the term

$$(s\sigma_3 s^{-1} \otimes I + I \otimes s\sigma_3 s_3)\frac{\partial}{\partial x^3} + [s\sigma_1 s^{-1} \otimes I + I \otimes s\sigma_1 s^{-1}] \,\frac{\partial}{\partial x^1}$$
$$+[s\sigma_2 s^{-1} \otimes I + I \otimes s\sigma_2 s^{-1}] \,\frac{\partial}{\partial x^2} = (\sigma_3 \otimes + I \otimes \sigma_3)\frac{\partial}{\partial x^3} + \frac{1}{\sqrt{u^2+v^2}} \frac{1}{u^2+v^2}$$
$$\times \Big\{ [(u\sigma_1 - v\sigma_2) \otimes I + I \otimes (u\sigma_1 - v\sigma_2)](u\frac{\partial}{\partial u} - v\frac{\partial}{\partial v})$$
$$+[(v\sigma_1 + u\sigma_2) \otimes I + I \otimes (v\sigma_1 + u\sigma_2)](v\frac{\partial}{\partial u} + u\frac{\partial}{\partial v}) \Big\}$$
$$= (\sigma_3 \otimes + I \otimes \sigma_3)\frac{\partial}{\partial x^3} + \frac{1}{\sqrt{u^2+v^2}} \Big[(\sigma_1 \frac{\partial}{\partial u} + \sigma_2 \frac{\partial}{\partial v}) \otimes I + I \otimes (\sigma_1 \frac{\partial}{\partial u} + \sigma_2 \frac{\partial}{\partial v}) \Big] .$$

We further consider the term

$$s\frac{\partial s^{-1}}{\partial x^1} = \frac{1}{u^2+v^2} s \Big\{ u\frac{\partial}{\partial u} - v\frac{\partial}{\partial v} \Big\} s^{-1} \frac{1}{u^2+v^2} \frac{1}{(u^2+v^2)^{1/4}} \left| \begin{array}{cc} \sqrt{u+iv} & 0 \\ 0 & \sqrt{u-iv} \end{array} \right|$$
$$\times \Big\{ -\frac{u^2}{2(u^2+v^2)(u^2+v^2)^{1/4}} \left| \begin{array}{cc} \sqrt{u-iv} & 0 \\ 0 & \sqrt{u+iv} \end{array} \right| + \frac{1}{2(u^2+v^2)^{1/4}} \left| \begin{array}{cc} \frac{u}{\sqrt{u-iv}} & 0 \\ 0 & \frac{u}{\sqrt{u+iv}} \end{array} \right|$$
$$+ \frac{v^2}{2(u^2+v^2)(u^2+v^2)^{1/4}} \left| \begin{array}{cc} \sqrt{u-iv} & 0 \\ 0 & \sqrt{u+iv} \end{array} \right| + \frac{1}{2(u^2+v^2)^{1/4}} \left| \begin{array}{cc} \frac{iv}{\sqrt{u-iv}} & 0 \\ 0 & \frac{-iv}{\sqrt{u+iv}} \end{array} \right| \Big\}$$
$$= \frac{1}{2} \frac{1}{u^2+v^2} \frac{1}{(u^2+v^2)^{1/2}} \left| \begin{array}{cc} \sqrt{u+iv} & 0 \\ 0 & \sqrt{u-iv} \end{array} \right|$$
$$\times \Big\{ -\frac{u^2}{(u^2+v^2)} \left| \begin{array}{cc} \sqrt{u-iv} & 0 \\ 0 & \sqrt{u+iv} \end{array} \right| + \left| \begin{array}{cc} \frac{u}{\sqrt{u-iv}} & 0 \\ 0 & \frac{u}{\sqrt{u+iv}} \end{array} \right|$$

$$+\frac{v^2}{(u^2+v^2)}\begin{vmatrix} \sqrt{u-iv} & 0 \\ 0 & \sqrt{u+iv} \end{vmatrix} + \begin{vmatrix} \frac{iv}{\sqrt{u-iv}} & 0 \\ 0 & \frac{-iv}{\sqrt{u+iv}} \end{vmatrix} \Big\}$$

$$= \frac{1}{2}\frac{v^2-u^2}{(u^2+v^2)^2} + \frac{1}{2}\frac{\sqrt{u+iv}\sqrt{u-iv}}{(u^2+v^2)^2}\begin{vmatrix} \frac{(u+iv)\sqrt{u+iv}}{\sqrt{u-iv}} & 0 \\ 0 & \frac{(u-iv)\sqrt{u-iv}}{\sqrt{u+iv}} \end{vmatrix}$$

$$= \frac{1}{2}\frac{v^2-u^2}{(u^2+v^2)^2} + \frac{1}{2}\frac{1}{(u^2+v^2)^2}\begin{vmatrix} (u+iv)^2 & 0 \\ 0 & (u-iv)^2 \end{vmatrix},$$

and arrive at the simple result

$$s\frac{\partial s^{-1}}{\partial x^1} = \frac{2iuv}{2(u^2+v^2)^2}\begin{vmatrix} 1 & 0 \\ 0 & -1 \end{vmatrix} = \frac{2iuv}{2(u^2+v^2)^2}\sigma_3. \tag{10.67}$$

Similarly, we consider the term

$$s\frac{\partial s^{-1}}{\partial x^2} = \frac{1}{u^2+v^2}s\Big\{v\frac{\partial}{\partial u}+u\frac{\partial}{\partial v}\Big\}s^{-1} = \frac{1}{u^2+v^2}\frac{1}{(u^2+v^2)^{1/4}}\begin{vmatrix} \sqrt{u+iv} & 0 \\ 0 & \sqrt{u-iv} \end{vmatrix}$$

$$\times\Big\{-\frac{uv}{2(u^2+v^2)(u^2+v^2)^{1/4}}\begin{vmatrix} \sqrt{u-iv} & 0 \\ 0 & \sqrt{u+iv} \end{vmatrix} + \frac{1}{2(u^2+v^2)^{1/4}}\begin{vmatrix} \frac{v}{\sqrt{u-iv}} & 0 \\ 0 & \frac{v}{\sqrt{u+iv}} \end{vmatrix}$$

$$-\frac{uv}{2(u^2+v^2)(u^2+v^2)^{1/4}}\begin{vmatrix} \sqrt{u-iv} & 0 \\ 0 & \sqrt{u+iv} \end{vmatrix} + \frac{1}{2(u^2+v^2)^{1/4}}\begin{vmatrix} \frac{-iu}{\sqrt{u-iv}} & 0 \\ 0 & \frac{+iu}{\sqrt{u+iv}} \end{vmatrix}\Big\}$$

$$= \frac{1}{2}\frac{1}{u^2+v^2}\frac{1}{(u^2+v^2)^{1/2}}\begin{vmatrix} \sqrt{u+iv} & 0 \\ 0 & \sqrt{u-iv} \end{vmatrix}$$

$$\times\Big\{-\frac{uv}{(u^2+v^2)}\begin{vmatrix} \sqrt{u-iv} & 0 \\ 0 & \sqrt{u+iv} \end{vmatrix} + \begin{vmatrix} \frac{v}{\sqrt{u-iv}} & 0 \\ 0 & \frac{v}{\sqrt{u+iv}} \end{vmatrix}$$

$$-\frac{uv}{(u^2+v^2)}\begin{vmatrix} \sqrt{u-iv} & 0 \\ 0 & \sqrt{u+iv} \end{vmatrix} + \begin{vmatrix} \frac{-iu}{\sqrt{u-iv}} & 0 \\ 0 & \frac{+iu}{\sqrt{u+iv}} \end{vmatrix}\Big\}$$

$$= -\frac{1}{2}\frac{2uv}{(u^2+v^2)^2} + \frac{1}{2}\frac{\sqrt{u+iv}\sqrt{u-iv}}{(u^2+v^2)^2}\begin{vmatrix} \frac{(v-iu)\sqrt{u+iv}}{\sqrt{u-iv}} & 0 \\ 0 & \frac{(v+iu)\sqrt{u-iv}}{\sqrt{u+iv}} \end{vmatrix}$$

$$= -\frac{1}{2}\frac{2uv}{(u^2+v^2)^2} + \frac{1}{2}\frac{1}{(u^2+v^2)^2}\begin{vmatrix} (u+iv)(v-iu) & 0 \\ 0 & (u-iv)(v+iu) \end{vmatrix},$$

so that

$$s\frac{\partial s^{-1}}{\partial x^2} = \frac{i(v^2-u^2)}{2(u^2+v^2)^2}\begin{vmatrix} 1 & 0 \\ 0 & -1 \end{vmatrix} = \frac{i(v^2-u^2)}{2(u^2+v^2)^2}\sigma_3. \tag{10.68}$$

Taking into account the formulas (10.65)–(10.68), we transform the above expression for Σ to the following form

$$\Sigma = -\frac{i}{2}\Big\{(\sigma_3\otimes+I\otimes\sigma_3)\frac{\partial}{\partial x^3}$$

$$+\frac{1}{\sqrt{u^2+v^2}}\Big[(\sigma_1\frac{\partial}{\partial u}+\sigma_2\frac{\partial}{\partial v})\otimes I+I\otimes(\sigma_1\frac{\partial}{\partial u}+\sigma_2\frac{\partial}{\partial v})\Big]$$

$$+\frac{1}{\sqrt{u^2+v^2}}\frac{2iuv}{2(u^2+v^2)^2}\Big[(u\sigma_1-v\sigma_2)\sigma_3\otimes I+I\otimes(u\sigma_1-v\sigma_2)\sigma_3\Big]$$

$$+\frac{1}{\sqrt{u^2+v^2}}\frac{i(v^2-u^2)}{2(u^2+v^2)^2}\Big[(v\sigma_1+u\sigma_2)\ \sigma_3\times I+I\otimes(v\sigma_1+u\sigma_2)\ \sigma_3\Big]$$

$$+\frac{1}{\sqrt{u^2+v^2}}\frac{2iuv}{2(u^2+v^2)^2}\Big[(u\sigma_1-v\sigma_2)\otimes\sigma_3+\sigma_3\otimes(u\sigma_1-v\sigma_2)\Big]$$

$$+\frac{1}{\sqrt{u^2+v^2}}\frac{i(v^2-u^2)}{2(u^2+v^2)^2}\Big[(v\sigma_1+u\sigma_2)\otimes\ \sigma_3+\sigma_3\otimes(v\sigma_1+u\sigma_2)\Big]\Big\}.$$

It may be written differently,

$$\Sigma=-\frac{i}{2}\Big\{(\sigma_3\otimes+I\otimes\sigma_3)\frac{\partial}{\partial x^3}$$

$$+\frac{1}{\sqrt{u^2+v^2}}\Big\{(\sigma_1\partial_u+\sigma_2\frac{\partial}{\partial v})\otimes I+I\otimes(\sigma_1\partial_u+\sigma_2\frac{\partial}{\partial v})$$

$$+\frac{2uv}{2(u^2+v^2)^2}\Big[(v\sigma_1+u\sigma_2)\otimes I+I\otimes(v\sigma_1+u\sigma_2)$$

$$+(u\sigma_1-v\sigma_2)\otimes i\sigma_3+i\sigma_3\otimes(u\sigma_1-v\sigma_2)\Big]$$

$$+\frac{v^2-u^2}{2(u^2+v^2)^2}[(-u\sigma_1+v\sigma_2)\otimes I+I\otimes(-u\sigma_1+v\sigma_2)$$

$$+(v\sigma_1+u\sigma_2)\otimes i\sigma_3+i\sigma_3\otimes(v\sigma_1+u\sigma_2)]\Big\}\Big\}.$$

Thus, the eigenvalue equation $\Sigma\xi=\sigma\xi$ leads to

$$\Big\{(\sigma_1\partial_u+\sigma_2\partial_v)\otimes I+I\otimes(\sigma_1\partial_u+\sigma_2\partial_v)$$

$$+\frac{2uv}{2(u^2+v^2)^2}\ [(v\sigma_1+u\sigma_2)\otimes I+I\otimes(v\sigma_1+u\sigma_2)]$$

$$+\frac{v^2-u^2}{2(u^2+v^2)^2}[(-u\sigma_1+v\sigma_2)\otimes I+I\otimes(-u\sigma_1+v\sigma_2)]$$

$$+\frac{2uv}{2(u^2+v^2)^2}\ [(u\sigma_1-v\sigma_2)\otimes i\sigma_3+i\sigma_3\otimes(u\sigma_1-v\sigma_2)]$$

$$+\frac{v^2-u^2}{2(u^2+v^2)^2}[(v\sigma_1+u\sigma_2)\otimes i\sigma_3+i\sigma_3\otimes(v\sigma_1+u\sigma_2)]\Big\}\xi$$

$$=\sqrt{u^2+v^2}\ (2i\sigma-ik\sigma_3\otimes-ikI\otimes\sigma_3)\ \xi. \tag{10.69}$$

By regrouping the terms within the lines 2-3, and also within the lines 3-4, we obtain a more simple form for the equation:

$$[(\sigma_1\partial_u+\sigma_2\partial_v)\otimes I+I\otimes(\sigma_1\partial_u+\sigma_2\partial_v)]\xi+\frac{1}{2(u^2+v^2)}$$

$$\times\{(u\sigma_1+v\sigma_2)\otimes I+I\otimes(u\sigma_1+v\sigma_2)+(v\sigma_1-u\sigma_2)\otimes i\sigma_3+i\sigma_3\otimes(v\sigma_1-u\sigma_2)\}\xi$$

$$=\sqrt{u^2+v^2}\,(2i\sigma-ik\sigma_3\otimes I-ikI\otimes\sigma_3)\,\xi,\quad \text{where}\quad h=\begin{vmatrix} f & h \\ h & g \end{vmatrix}.$$

First, we calculate

$$(\sigma_1\partial_u+\sigma_2\partial_v)\otimes I\begin{vmatrix} f & h \\ h & g \end{vmatrix}=\begin{vmatrix} (\partial_u-i\partial_v)h & (\partial_u-i\partial_v)g \\ (\partial_u+i\partial_v)f & (\partial_u+i\partial_v)h \end{vmatrix},$$

and

$$I\otimes(\sigma_1\partial_u+\sigma_2\frac{\partial}{\partial v})\begin{vmatrix} f & h \\ h & g \end{vmatrix}=\begin{vmatrix} (\partial_u-i\partial_v)h & (\partial_u+i\partial_v)f \\ (\partial_u-i\partial_v)g & (\partial_u+i\partial_v)h \end{vmatrix}.$$

Their sum equals to

$$(\sigma_1\partial_u+\sigma_2\frac{\partial}{\partial v})\otimes I\begin{vmatrix} f & h \\ h & g \end{vmatrix}+I\otimes(\sigma_1\partial_u+\sigma_2\frac{\partial}{\partial v})\begin{vmatrix} f & h \\ h & g \end{vmatrix}$$

$$=\begin{vmatrix} 2(\partial_u-i\partial_v)h & (\partial_u+i\partial_v)f+(\partial_u-i\partial_v)g \\ (\partial_u+i\partial_v)f+(\partial_u-i\partial_v)g & 2(\partial_u+i\partial_v)h \end{vmatrix}. \tag{10.70}$$

Then we calculates the terms

$$[(u\sigma_1+v\sigma_2)\otimes I+I\otimes(u\sigma_1+v\sigma_2)]\xi$$

$$=\begin{vmatrix} 0 & u-iv \\ u+iv & 0 \end{vmatrix}\begin{vmatrix} f & h \\ h & g \end{vmatrix}+\begin{vmatrix} f & h \\ h & g \end{vmatrix}\begin{vmatrix} 0 & u+iv \\ u-iv & 0 \end{vmatrix}$$

$$=\begin{vmatrix} 2(u-iv)h & (u+iv)f+(u-iv)g \\ (u+iv)f+(u-iv)g & 2(u+iv)h \end{vmatrix},$$

and

$$[(v\sigma_1-u\sigma_2)\otimes i\sigma_3+i\sigma_3\otimes(v\sigma_1-u\sigma_2)]\xi$$

$$=\begin{vmatrix} 0 & v+iu \\ v-iu & 0 \end{vmatrix}\begin{vmatrix} f & h \\ h & g \end{vmatrix}\begin{vmatrix} i & 0 \\ 0 & -i \end{vmatrix}+\begin{vmatrix} i & 0 \\ 0 & -i \end{vmatrix}\begin{vmatrix} f & h \\ h & g \end{vmatrix}\begin{vmatrix} 0 & v-iu \\ v+iu & 0 \end{vmatrix}$$

$$=\begin{vmatrix} 2(iv-u)h & (iv+u)f+(-iv+u)g \\ (iv+u)f+(-iv+u)g & 2(-iv-u)h \end{vmatrix};$$

their sum equals to

$$2\begin{vmatrix} 0 & (u+iv)f+(u-iv)g \\ (u+iv)f+(u-iv)g & 0 \end{vmatrix}. \tag{10.71}$$

Besides, we find

$$(2i\sigma - ik\sigma_3 \otimes -ikI \otimes \sigma_3)\,\xi = 2i \begin{vmatrix} (\sigma-k)f & \sigma h \\ \sigma h & (\sigma+k)g \end{vmatrix} . \qquad (10.72)$$

Taking into account the relations (10.70)–(10.72), we reduce the eigenvalue equation to the form

$$\begin{vmatrix} 2(\partial_u - i\partial_v)h & (\partial_u + i\partial_v)f + (\partial_u - i\partial_v)g \\ (\partial_u + i\partial_v)f + (\partial_u - i\partial_v)g & 2(\partial_u + i\partial_v)h \end{vmatrix}$$

$$+\frac{1}{(u^2+v^2)^2} \begin{vmatrix} 0 & (u+iv)f + (u-iv)g \\ (u+iv)f + (u-iv)g & 0 \end{vmatrix}$$

$$= 2i\sqrt{u^2+v^2} \begin{vmatrix} (\sigma-k)f & \sigma h \\ \sigma h & (\sigma+k)g \end{vmatrix} , \qquad (10.73)$$

whence we derive four equations.

First, let us consider the following ones:

$$2(\partial_u - i\partial_v)h = 2i\sqrt{u^2+v^2}(\sigma-k)f\,,$$

$$2(\partial_u + i\partial_v)h = 2i\sqrt{u^2+v^2}(\sigma+k)g\,,$$

or, differently,

$$(\partial_u - i\partial_v)h = i(\sigma-k)\bar{f}, \quad (\partial_u + i\partial_v)h = i(\sigma+k)\bar{g}; \qquad (10.74)$$

we recall the notations $\bar{f} = \sqrt{u^2+v^2}f$, $\bar{g} = \sqrt{u^2+v^2}g$. Let us introduce the new variables

$$\bar{f} + \bar{g} = F, \quad \bar{f} - \bar{g} = G;$$

then, summing and subtracting the equations in (10.74), we obtain the system of two linear equations with respect to F and G:

$$\sigma F - kG = -2i\partial_u h, \quad -kF + \sigma G = -2\partial_v h\,.$$

Its solution is

$$F = \frac{2}{\sigma^2 - k^2}[-i\sigma\partial_u - k\partial_v]\,h, \quad G = \frac{2}{\sigma^2-k^2}[-ik\partial_u - \sigma\partial_v]\,h. \qquad (10.75)$$

These formulas may be compared with the similar ones resulting from the Maxwell equations (10.29):

$$F = \frac{2}{\epsilon^2 - k^2}[-i\epsilon\partial_u - k\partial_v]\,h, \quad G = \frac{2}{\epsilon^2-k^2}[-ik\partial_u - \epsilon\partial_v]\,h. \qquad (10.76)$$

The relations (10.75) and (10.76) coincide, when identifying σ and ϵ.

Now consider the two remaining equations:

$$
\begin{aligned}
&(\partial_u + i\partial_v) f + (\partial_u - i\partial_v) g + \frac{1}{(u^2+v^2)}[(u+iv)f + (u-iv)g] = 2i\sqrt{u^2+v^2}\sigma h\,, \\
&(\partial_u + i\partial_v) f + (\partial_u - i\partial_v) g + \frac{1}{(u^2+v^2)}[(u+iv)f + (u-iv)g] = 2i\sqrt{u^2+v^2}\sigma h\,;
\end{aligned}
\tag{10.77}
$$

they coincide with each other. The equation from (10.77) reduces to the simpler form

$$
\frac{1}{\sqrt{u^2+v^2}}(\partial_u - \frac{u}{u^2+v^2})F
$$

$$
+\frac{i}{\sqrt{u^2+v^2}}(\partial_v - \frac{v}{u^2+v^2})G\frac{1}{(u^2+v^2)}\frac{1}{\sqrt{u^2+v^2}}[uF + ivG] = 2i\sqrt{u^2+v^2}\sigma h\,,
$$

which after regrouping the terms leads to

$$
\frac{1}{u^2+v^2}\partial_u F + \frac{i}{u^2+v^2}\partial_v G = 2i\sigma h\,. \tag{10.78}
$$

We notice that the last equation (10.78) coincides with the third equation in the system (10.28)

$$
-2i\epsilon h + \frac{1}{u^2+v^2}\partial_u F + \frac{i}{u^2+v^2}\partial_v G = 0\,,
$$

by recollecting the identity $\sigma = \omega$.

Chapter 11

Geometrization of Maxwell Electrodynamics

11.1 Optics and Lagrange Formalism

Below, the equations of motion associated with a Lagrangian inspired by relativistic optics in nonuniform moving medium are considered. The model describes optical effects in the nonuniform moving medium with special optical properties given by the self-consistent system. When using the metric restricted to the Minkowski manifold, we have established the Euler – Lagrange equations for corresponding geodesics. We have specified the general model to the special case when the metric coefficient γ linearly increases along the direction Z. The exact analytical solutions of the Euler – Lagrange equations have been constructed. The study of the obtained solutions shows that the light ray bends to the axis Z along which the effective refractive index increases.

In geometrical optics [263, 264], a significant attention is shown to the Beil metric (see [254] – [261])

$$g_{\alpha\beta}(x,y) = \varphi_{\alpha\beta}(x) + \gamma^2(x)\, y_\alpha y_\beta\,, \tag{11.1}$$

where $\gamma(x) \geq 0$ is a smooth function on the space-time M^4, and $\varphi_{\alpha\beta}(x)$ is a pseudo-Riemannian metric on M^4.

One assumes that the manifold M^4 is endowed with some local coordinates $(x^\alpha)_{\alpha=\overline{1,4}} = \left(x^1 = t, x^2, x^3, x^4\right)$, and (y^α) is the Liouville vector field on the total space of the tangent bundle TM^4; the following rule holds $y_\alpha = \varphi_{\alpha\mu} y^\mu$. Since the components of $\varphi_{\alpha\beta}(x)$ are dimensionless, the same are γy_α; so we have dimensionless combinations $[\varphi_{\alpha\beta}(x)] = 1$, $[\gamma y_\alpha] = 1$.

In this context, we restrict our study to the Minkowski manifold $\mathcal{M}^4 = \left(\mathbb{R}^4, \eta_{ij}\right)$, which has the local coordinates $(x) := \left(x^i\right)_{i=\overline{1,4}}$. The dimension of the corresponding tangent bundle $T\mathbb{R}^4$ is equal to eight, and its local coordinates are $(x,y) := (x^i, y^i)_{i=\overline{1,4}}$. Here $x^i = (x^1, x^2, x^3, x^4)$ are space-time coordinates, and $y^i = (y^1, y^2, y^3, y^4)$ is a tangent vector. It should be noted that the Latin letters i, j, k, ... run from 1 to 4. The Einstein summation convention is adopted. To avoid confusion between indices and powers, the space-time coordinates are denoted as $x^i := x^{(i)}$.

Starting with the formula (11.1), we can introduce the following metric on $T\mathbb{R}^4$:

$$g_{ij}(x,y) = \eta_{ij} + \gamma^2(x) y_i y_j \,, \tag{11.2}$$

where $\eta = (\eta_{ij}) =$ diag $(-1,1,1,1)$ is the Minkowski metric, and $y_i = \eta_{ir} y^r$. This approach is inspired by the optics framework developed in the papers [261] – [265? –267] for the dispersionless uniform moving media.

From physical point of view, geometrical optics in moving media is an interesting object for some cosmological applications [270, 271], because the effects of velocity vector field are similar to the action of gravitational or magnetic fields on charged matter waves [273, 274]. In the papers [273, 274], a special Lagrangian and metric related to the light propagation in moving isotropic dispersionless media, were studied and the gravitation-like effects for the light deflection at a vortex were demonstrated (so-called "an artificial optical black hole"); in these papers the Gordon's optical metric was used:

$$g_{ij}(x) = \eta_{ij} + \gamma^2(x) u_i(x) u_j(x) \,. \tag{11.3}$$

Commonly, the following parametrization for $\gamma^2(x)$ is applied

$$\gamma^2(x) = \frac{1}{c^2}\left(1 - \frac{1}{n^2}\right), \tag{11.4}$$

where n stands for the local refractive index of the uniform medium. It should be noted that in Gordon's optical metric, the four-velocity u_i is defined by the formula [265]

$$u_i(x) = \alpha\Big(c, -\mathbf{u}(x)\Big), \qquad \alpha = \left(1 - \frac{u^2(x)}{c^2}\right)^{-1/2},$$

and it describes the velocity of the medium. In Gordon approach, the velocity components u^i are external variables, which in general may be arbitrary.

The corresponding constitutive relations for the moving uniform media are effectively equivalent to some anisotropic medium without absorption and dispersion. In [271], the Gordon metric was extended to the complex case, to account for processes of absorption and refraction of the monochromatic light (in this case, the electromagnetic properties of the effective medium are assumed to be linear and isotropic).

In the metric (11.2), the quantity y^i designates the Liouville vector field on the total space of the tangent bundle $T\mathbb{R}^4$. The metric (11.2) may be considered as the mathematical generalization of the well known Gordon metric (11.3); however, the physical interpretation of metric (11.2) is not completely clear. In particular, the geometric structure (11.2) leads to a self-consistent system, when there arise equations for determining the variables x^i and y^i. Solving the system of differential equations related to the geometrical structure (11.2), we simultaneously should find explicit functions, x^i and y^i. After that, by the simple change of notation

$$x^i = x^i, \qquad y^i = u^i(x), \tag{11.5}$$

we obtain some solutions for Gordon system (11.3), but now the field u^i is not arbitrary, being rather determined by the generalized self-consistent model. It should be noted that, from mathematical point of view, the restriction of the metric (11.2) to the cross section $S_V : \mathbb{R}^4 \to T\mathbb{R}^4$ given by the relations (11.5) leads to the optical metric (11.3) (see [262]).

In the sequel, we will examine the special case of an anisotropic dynamical model, which is governed by the Lagrangian [259, 272]:

$$L(x,y) = \frac{1}{2}g_{ij}(x,y)y^iy^j = \frac{1}{2}(\eta_{ij}+\gamma^2 y_i y_j)y^iy^j = \frac{1}{2}\eta_{ij}y^iy^j + \frac{\gamma^2}{2}|y|^4, \tag{11.6}$$

where the following notation is used

$$|y|^2 = -(y^1)^2+(y^2)^2+(y^3)^2+(y^4)^2 = \eta_{ij}y^iy^j.$$

Assuming that the coefficient $\gamma(x)$ is invariant with respect to Lorentz transformations, we conclude that the Lagrangian (11.6) is also invariant. A similar 3-dimensional Lagrangian metric $(\eta_{ij})_{i,j=\overline{1,4}}$ is replaced with Euclidian metric $(\delta_{ij})_{i,j=\overline{1,3}}$ was studied in [260], the corresponding nonrelativistic Lagrangian being invariant with respect to the orthogonal group $O(3)$.

The Lagrangian (11.6) produces the fundamental metric

$$g_{ij}(x,y) = \frac{1}{2}\frac{\partial^2 L}{\partial y^i \partial y^j} = \sigma(x,y)\eta_{ij} + 2\gamma^2(x)y_iy_j,$$

where $\sigma(x,y) = (1/2) + \gamma^2(x)|y|^2$.

With the notation $\xi(x,y) = (1/2) + 3\gamma^2(x)|y|^2$, for the inverse matrix $[g^{-1}] = (g^{jk})$ we find

$$g^{jk}(x,y) = \frac{1}{\sigma(x,y)}\eta^{jk} - \frac{2\gamma^2(x)}{\sigma(x,y)\cdot\xi(x,y)}y^jy^k,$$

where we assume that $\sigma\cdot\xi \neq 0$. The Euler-Lagrange equations associated with the Lagrangian (11.6) can be written in the form (see [255])

$$\frac{d^2x^i}{d\tau^2} + 2G^i(x(\tau),y(\tau)) = 0, \tag{11.7}$$

where G^i is defined by the formula

$$G^i(x,y) \stackrel{def}{=} \frac{g^{ik}}{4}\left(\frac{\partial^2 L}{\partial y^k \partial x^s}y^s - \frac{\partial L}{\partial x^k}\right)$$

$$= \frac{\gamma}{\sigma}|y|^2 y^i(\gamma_s y^s) - \frac{3}{2}\frac{\gamma^3}{\sigma\xi}y^i|y|^4(\gamma_s y^s) - \frac{\gamma}{4\sigma}|y|^4\gamma_i\eta_{ii},$$

where $\gamma_s = \partial\gamma/\partial x^s$. The geometrical quantity $G^i(x,y)$ has the meaning of a semispray on the tangent space $T\mathbb{R}^4$ (see [256, 272]).

11.2 The Euler–Lagrange Equations

The Euler – Lagrange equations (11.7) for the variables

$$(y^1, y^2, y^3, y^4) = (V^{(1)}, V^{(2)}, V^{(3)}, V^{(4)}) := V, \quad V^{(i)} = \frac{dx^{(i)}}{d\tau},$$

$$v^2 = ||V||^2 = -(V^{(1)})^2 + (V^{(2)})^2 + (V^{(3)})^2 + (V^{(4)})^2,$$

lead to the following equations

$$\frac{dV^{(i)}}{d\tau} + \frac{4\gamma v^2\,(1+3\gamma^2 v^2)}{(1+2\gamma^2 v^2)\,(1+6\gamma^2 v^2)} \left(\gamma_s V^{(s)} V^{(i)}\right) - \frac{\gamma v^4}{1+2\gamma^2 v^2} \gamma_i \eta_{ii} = 0, \tag{11.8}$$

where τ is some evolution parameter. In detailed form, the equations (11.8) read as

$$\begin{aligned}
&\frac{dV^{(1)}}{d\tau} + \frac{4\gamma v^2\,(1+3\gamma^2 v^2)}{(1+2\gamma^2 v^2)\,(1+6\gamma^2 v^2)} (\gamma_s V^{(s)}) V^{(1)} + \frac{\gamma v^4}{1+2\gamma^2 v^2} \gamma_1 = 0,\\
&\frac{dV^{(2)}}{d\tau} + \frac{4\gamma v^2\,(1+3\gamma^2 v^2)}{(1+2\gamma^2 v^2)\,(1+6\gamma^2 v^2)} (\gamma_s V^{(s)}) V^{(2)} - \frac{\gamma v^4}{1+2\gamma^2 v^2} \gamma_2 = 0,\\
&\frac{dV^{(3)}}{d\tau} + \frac{4\gamma v^2\,(1+3\gamma^2 v^2)}{(1+2\gamma^2 v^2)\,(1+6\gamma^2 v^2)} (\gamma_s V^{(s)}) V^{(3)} - \frac{\gamma v^4}{1+2\gamma^2 v^2} \gamma_3 = 0,\\
&\frac{dV^{(4)}}{d\tau} + \frac{4\gamma v^2\,(1+3\gamma^2 v^2)}{(1+2\gamma^2 v^2)\,(1+6\gamma^2 v^2)} (\gamma_s V^{(s)}) V^{(4)} - \frac{\gamma v^4}{1+2\gamma^2 v^2} \gamma_4 = 0.
\end{aligned} \tag{11.9}$$

Note that when the coefficient $\gamma(x)$ is constant, $\gamma(x) = \gamma_0$, the above equations simplify

$$\frac{dV^{(i)}}{d\tau} = 0 \quad \Leftrightarrow \quad V = (V^{(1)}, V^{(2)}, V^{(3)}, V^{(4)}) = \text{constant} \Leftrightarrow$$

$$\frac{dx^{(i)}}{d\tau} = V^{(i)} \quad \Leftrightarrow \quad x(\tau) = (V^{(1)}\tau + x_0^{(1)}, V^{(2)}\tau + x_0^{(2)}, V^{(3)}\tau + x_0^{(3)}, V^{(4)}\tau + x_0^{(4)});$$

in this case, the geodesics are straight lines. Such a behavior of light rays is typical for the uniform medium with the constant refractive index.

The coefficient $\gamma(x)$ depending on the coordinates x corresponds to the nonuniform medium. We will utilize the formula (11.4) and call the quantity n as an effective refractive index (but one should not confuse n with a real refractive index of the medium).

We will consider the nonuniform medium, whose optical properties vary only along one axis, $\gamma = \gamma_4 x^{(4)} = \gamma_4 Z$; note the use of the new notations, $x^{(2)} = X$, $x^{(3)} = Y$, $x^{(4)} = Z$. The cases: $\gamma = \gamma_3 Y$ or $\gamma = \gamma_2 X$ can be treated in a similar manner. From the relation (11.4), one can find the explicit dependence of the effective refractive index on coordinate Z:

$$n^2 = \frac{1}{1 - c^2\gamma_4^2 Z^2}.$$

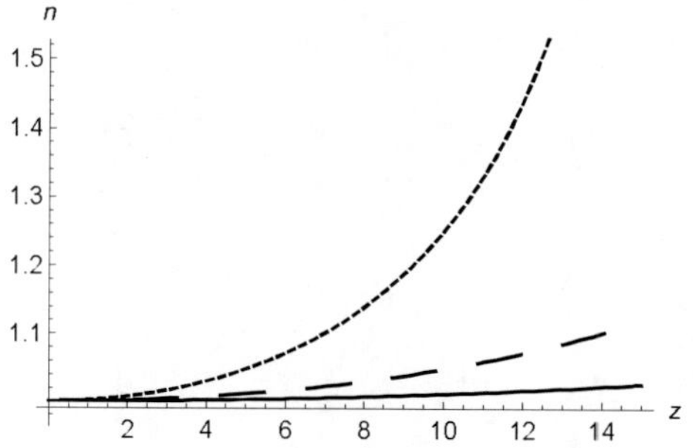

Figure 11.1: Dependence of the effective refractive index n on the coordinate Z at different values of γ_4. The following parameters are used $c = 1$; $\gamma_4 = 0.06$ (dotted line), 0.03 (dashed line), 0.01 (solid line).

The behavior of $n(Z)$ at different γ_4 is shown in Figure 11.1. For this special case $\gamma = \gamma_4 Z$, the system (11.9) takes the form

$$\frac{dV^{(1)}}{d\tau} + \frac{4\gamma v^2(1+3\gamma^2 v^2)}{(1+2\gamma^2 v^2)(1+6\gamma^2 v^2)}\gamma_4 V^{(1)}V^{(4)} = 0,$$

$$\frac{dV^{(2)}}{d\tau} + \frac{4\gamma v^2(1+3\gamma^2 v^2)}{(1+2\gamma^2 v^2)(1+6\gamma^2 v^2)}\gamma_4 V^{(2)}V^{(4)} = 0,$$

$$\frac{dV^{(3)}}{d\tau} + \frac{4\gamma v^2(1+3\gamma^2 v^2)}{(1+2\gamma^2 v^2)(1+6\gamma^2 v^2)}\gamma_4 V^{(3)}V^{(4)} = 0,$$

$$\frac{dV^{(4)}}{d\tau} + \frac{4\gamma v^2(1+3\gamma^2 v^2)}{(1+2\gamma^2 v^2)(1+6\gamma^2 v^2)}\gamma_4 (V^{(4)})^2 - \frac{\gamma v^4}{1+2\gamma^2 v^2}\gamma_4 = 0.$$

For shortness, below we use the notation $v^2 \equiv W$. From (11.10) we can derive the following equations for variables W and $V^{(4)}$:

$$\begin{aligned} &\frac{dW}{d\tau} + \frac{8\gamma(1+3\gamma^2 W)}{(1+2\gamma^2 W)(1+6\gamma^2 W)}\gamma_4 V^{(4)}W^2 - \frac{2\gamma}{1+2\gamma^2 W}\gamma_4 V^{(4)}W^2 = 0, \\ &\frac{dV^{(4)}}{d\tau} + \frac{4\gamma(1+3\gamma^2 W)}{(1+2\gamma^2 W)(1+6\gamma^2 W)}\gamma_4 (V^{(4)})^2 W - \frac{\gamma}{1+2\gamma^2 W}\gamma_4 W^2 = 0, \end{aligned} \tag{11.10}$$

which are reduced to the simpler form

$$\begin{aligned} &\frac{dW}{d\tau} + \frac{6\gamma\gamma_4}{1+6\gamma^2 W}V^{(4)}W^2 = 0, \\ &\frac{dV^{(4)}}{d\tau} + \frac{4\gamma(1+3\gamma^2 W)}{(1+2\gamma^2 W)(1+6\gamma^2 W)}\gamma_4 (V^{(4)})^2 W - \frac{\gamma}{1+2\gamma^2 W}\gamma_4 W^2 = 0. \end{aligned} \tag{11.11}$$

Taking into account the identities

$$\gamma = \gamma_4 x^{(4)}, \quad V^{(4)} = \frac{dx^{(4)}}{d\tau}, \quad x^{(4)} = Z,$$

we can write the equations (11.11) as follows

$$\frac{dW}{d\tau} + \frac{6Z\gamma_4^2}{1+6\gamma_4^2 Z^2 W} W^2 \frac{dZ}{d\tau} = 0,$$
$$\frac{d^2 Z}{d\tau^2} + \frac{4\gamma_4 Z(1+3\gamma_4^2 Z^2 W)}{(1+2\gamma_4^2 Z^2 W)(1+6\gamma_4^2 Z^2 W)} \gamma_4 \left(\frac{dZ}{d\tau}\right)^2 W - \frac{\gamma_4 Z}{1+2\gamma_4^2 Z^2 W} \gamma_4 W^2 = 0. \tag{11.12}$$

To resolve the first equation in (11.12), let us make two substitutions:

$$3\gamma_4^2 Z^2 = F_1\,, \quad \frac{1}{W} = F_2 \quad \Longrightarrow \quad F_1' = 6\gamma_4^2 Z Z'\,, \quad W' = -\frac{1}{F_2^2} F_2'\,, \tag{11.13}$$

where the derivative over τ is denoted by prime. Then the first equation in (11.12) takes the form

$$-F_2' + F_1' \frac{1}{1+2F_1/F_2} = 0.$$

Making yet another substitution $F_1/F_2 = G$, we arrive at the equation with separable variables

$$\frac{F_2'}{F_2} = \frac{G'}{1+G} \quad \Longrightarrow \quad G = c_1 F_2 - 1\,. \tag{11.14}$$

Taking in mind (11.13), from (11.14) it follows

$$Z^2 = \frac{1}{3\gamma_4^2} \frac{c_1 - W}{W^2}\,, \tag{11.15}$$

where c_1 is an arbitrary constant. From the physical point of view, the variable Z should be real, so the difference $c_1 - W$ has to be positive. At a chosen metric signature, we have the identity

$$v^2 = W = -(V^{(1)})^2 + (V^{(2)})^2 + (V^{(3)})^2 + (V^{(4)})^2\,;$$

therefore, assuming $V^{(1)} \equiv c^2$ (c is the velocity of the light in the vacuum), one concludes that $W < 0$. Because $Z^2 > 0$, the difference $c_1 - W$ has to be positive, i.e., $c_1 > W$.

From relation (11.15) it follows the expression for W:

$$W(Z) = \frac{-1 \mp \sqrt{12 c_1 \gamma_4^2 Z^2 + 1}}{6\gamma_4^2 Z^2}\,. \tag{11.16}$$

Let us introduce a new variable f defined as

$$f^2(Z) = 12 c_1 \gamma_4^2 Z^2 + 1\,. \tag{11.17}$$

Then the formula (11.16) takes the form

$$W = \frac{2c_1(f-1)}{f^2-1}\,, \tag{11.18}$$

where the negative f corresponds to the sign "$-$" in (11.16), while the positive f is defined by the sign "$+$". Because $W<0$, we conclude that $f<1$. Since Z^2 should be positive, the following restriction on f follows from the formula (11.17): at $c_1>0$ we have $f^2>1$; at $c_1<0$ we get $f^2<1$.

Substituting W from (11.16) in the second equation in (11.12), we get the nonlinear equation for the variable Z:

$$Z''+\frac{12c_1\gamma_4^2 Z}{12c_1\gamma_4^2Z^2+2\sqrt{12c_1\gamma_4^2Z^2+1}+1}Z'^2-\frac{\left(\sqrt{12c_1\gamma_4^2Z^2+1}-1\right)^2}{12\gamma_4^2Z^3\left(\sqrt{12c_1\gamma_4^2Z^2+1}+2\right)}=0; \quad (11.19)$$

in terms of the variable f we have

$$f''f+\frac{(f^3-2f-2)}{(f+2)\,(f^2-1)}\,(f')^2-\frac{12c_1^2\gamma_4^2(f-1)}{(f+1)(f-2)}=0. \quad (11.20)$$

Taking in mind the identity

$$\gamma_4^2Z^2=\gamma^2=\frac{1}{c^2}(1-\frac{1}{n^2}),$$

from (11.17) one finds

$$12(1-\frac{1}{n^2})=f^2-1 \quad \Rightarrow f^2=13-12\frac{1}{n^2}.$$

At $n=1$ we have $f=1$. The increasing of the effective refractive index n leads to the rising of the variable f; however, we should remember that the value f is restricted by the inequality $f^2<13$. Taking in mind the previously determined restrictions, one gets

$$-\sqrt{13}<f<-1 \quad \text{at} \quad c_1>0, \quad \text{and} \quad -1<f<1 \quad \text{at} \quad c_1<0. \quad (11.21)$$

Now, we will solve the equation (11.20). Because it does not contain the variable t explicitly, we can depress the equation by the substitution

$$f'\to p, \quad f''\to pp'_f, \quad \text{where} \quad p'_f=\frac{dp}{df}.$$

In this way, from (11.20) we obtain

$$-\frac{12c_1^2(f-1)\gamma_4^2}{(f+1)(f+2)}+\frac{(f^3-2f-2)\,p^2}{(f+2)\,(f^2-1)}+fpp'_f=0. \quad (11.22)$$

The last equation changes to a nonhomogeneous differential equation by means of the substitution $p^2\to K$:

$$K'_f+\frac{2\,(f^3-2f-2)}{f(f+2)\,(f^2-1)}K-\frac{24c_1^2\gamma_4^2(f-1)}{f(f+1)(f+2)}=0, \quad \text{where} \quad K'_f=dK/df. \quad (11.23)$$

Its solution in explicit form is

$$K = f'^2 = \frac{1}{f^2(f+2)^2}\left[24c_1^2\gamma_4^2\left(f^3-1\right) - c_2\left(f^2-1\right)\right]. \tag{11.24}$$

The equation (11.24) can be resolved in terms of the elliptic integrals of the second kind $E[\phi|m]$ and of the first kind $F[\phi|m]$ [275]:

$$\tau = -\frac{\sqrt{-c_2 f^2 + c_2 + \Gamma^2\left(f^3-1\right)}}{3\Gamma^2}$$

$$\times\left\{\frac{-4c_2 - 2\Gamma^2(f+5)}{f-1} - \frac{iB\left(c_2+3\Gamma^2\right)\sqrt{\left(\Gamma^2(2f+1)-c_2\right)^2 - A}}{\sqrt{f-1}\left(c_2(-f-1)+\Gamma^2\left(f^2+f+1\right)\right)}\right.$$

$$\times\left(\frac{2\left(\sqrt{A}-3c_2+3\Gamma^2\right)}{\left(2c_2-3\Gamma^2\right)}F\left[i\sinh^{-1}\left(\frac{2B}{\sqrt{f-1}}\right)\frac{-3\Gamma^2+c_2+\sqrt{A}}{-3\Gamma^2+c_2-\sqrt{A}}\right]\right.$$

$$\left.\left.+4E\left[i\sinh^{-1}\left(\frac{2B}{\sqrt{f-1}}\right)\frac{-3\Gamma^2+c_2+\sqrt{A}}{-3\Gamma^2+c_2-\sqrt{A}}\right]\right)\right\}. \tag{11.25}$$

In the above formula the following parameters are used:

$$\Gamma^2 = 24c_1^2\gamma_4^2, \quad A = \left(c_2-\Gamma^2\right)\left(c_2+3\Gamma^2\right), \quad B = \sqrt{\frac{c_2-\frac{3\Gamma^2}{2}}{\sqrt{A}+c_2-3\Gamma^2}}. \tag{11.26}$$

The constants c_2 and Γ are defined by the initial conditions for $(Z, Z' = V^{(4)})$.

Now, we can find the corresponding expressions for the velocities $V^{(1)}$, $V^{(2)}$, $V^{(3)}$ and for the coordinates $x^{(1)}$ and X, Y (see equations (11.10)). Because the first three equations in the system (11.10) have the same form, it is sufficient to solve one of them; let it be the equation for $V^{(2)} = \frac{dX}{d\tau}$. To this end, we transform the second equation in (11.10) to the variable f. By substituting expressions (11.17) and (11.16) in the second equation in (11.10) and by applying the identity

$$\frac{dV^{(2)}}{d\tau} = \frac{dV^{(2)}}{df}\frac{df}{d\tau} = \frac{dV^{(2)}}{df}f',$$

one gets

$$\frac{dV^{(2)}}{df} + \frac{1}{(f+2)}V^{(2)} = 0 \quad\Rightarrow\quad V^{(2)} = \frac{V_0^{(2)}}{f+2}. \tag{11.27}$$

Besides, we can find the coordinate X from the equation $dX/d\tau = V^{(2)}$. Having used expression (11.24) and the identity

$$\frac{dX}{d\tau} = \frac{dX}{df}\frac{df}{d\tau} = \frac{dX}{df}f',$$

we obtain

$$\frac{dX}{df} = \frac{V_0^{(2)} f}{\sqrt{\Gamma^2(f^3-1) - c_2(f^2-1)}}. \tag{11.28}$$

The solution of this equation reads

$$\begin{aligned}X(f) &= X_0 + V_0^{(2)}\Bigg[\frac{2\sqrt{\Gamma^2(f^3-1)-c_2(f^2-1)}}{\Gamma^2(f-1)} \\ &+ \frac{iB\sqrt{f-1}}{2\Gamma^2\sqrt{\Gamma^2(f^3-1)-c_2(f^2-1)}}\sqrt{(c_2-\Gamma^2(2f+1))^2 - A} \\ &\times\Bigg(\frac{\left(\sqrt{A}-3c_2+3\Gamma^2\right)F\left(i\sinh^{-1}\left(\frac{2B}{\sqrt{f-1}}\right)1-\frac{2\sqrt{A}}{3\Gamma^2-c_2+\sqrt{A}}\right)}{c_2-\frac{3\Gamma^2}{2}} \\ &+4E\left(i\sinh^{-1}\left(\frac{2B}{\sqrt{f-1}}\right)1-\frac{2\sqrt{A}}{3\Gamma^2-c_2+\sqrt{A}}\right)\Bigg)\Bigg].\end{aligned} \tag{11.29}$$

The expressions (11.29) and (11.17) define the projection of the trajectory of the light ray on the plane XZ implicitly. Its behavior depends on the parameters Γ and c_2 and is illustrated in Figure 11.2. The ray deflects onto the direction of higher values of the effective refractive index, and at some point the total internal reflection occurs. It should be emphasized that the obtained solution is not valid when $c_1 = 0$

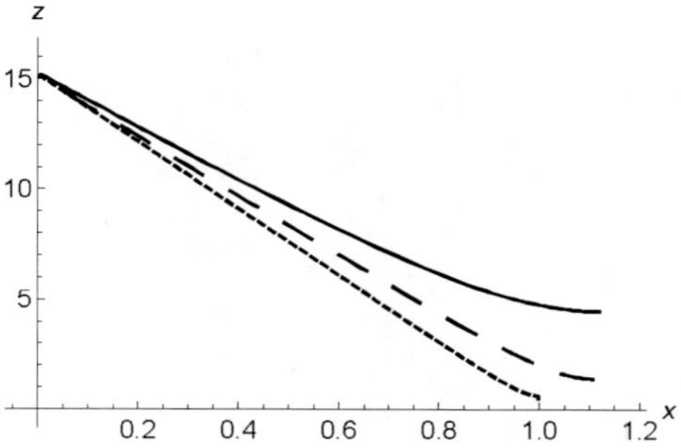

Figure 11.2: (The projects trajectories of the light ray on plane XZ at negative constant c_1. The following parameters are used: $c_1 = -0.1$; $c_2 = 1$ and $\gamma_4 = 0.06$ (dotted line), $c_2 = 0.2$ and $\gamma_4 = 0.03$ (dashed line), $c_2 = 0.02$ and $\gamma_4 = 0.01$ (solid line).

$(\Gamma = 0)$. In this case, from the formula (11.15) we readily obtain

$$Z = \frac{1}{\sqrt{3}\gamma_4}\sqrt{-\frac{1}{W}} = \frac{1}{\sqrt{3}\gamma_4 c\sqrt{1-V^2/c^2}}, \tag{11.30}$$

where $V^{(1)} = c$, $V^2 = (V^{(2)})^2 + (V^{(3)})^2 + (V^{(4)})^2$. From relation (11.30) it follows the expression for W in the form

$$W = -\frac{1}{3\gamma_4^2 Z^2}\,. \tag{11.31}$$

After substituting W from (11.31) in the second equation from (11.12), the nonlinear equation for the variable Z reduces to

$$\frac{d^2 Z}{d\tau^2} - \frac{1}{3\gamma_4^2 Z^3} = 0\,, \tag{11.32}$$

its solution reads

$$Z(\tau) = \sqrt{\frac{\tau^2}{3\gamma_4^2 Z_0^2} + \left(Z_0 + V_0^{(4)}\tau\right)^2}\,, \tag{11.33}$$

where Z_0 and $V_0^{(4)}$ stand for the coordinate Z and for the velocity Z' at the moment $t = 0$. Now, we can find the corresponding expressions for the velocities $V^{(1)}$, $V^{(2)}$, $V^{(3)}$ and the coordinates $x^{(1)}$ and X, Y. To this end, we substitute expressions (11.31) and (11.33) in equations (11.10) and take into account the identity $\gamma = \gamma_4 x^{(4)} = \gamma_4 Z$. Then one obtains

$$\frac{dV^{(1)}}{d\tau} = 0\,, \qquad \frac{dV^{(2)}}{d\tau} = 0\,, \qquad \frac{dV^{(3)}}{d\tau} = 0\,, \tag{11.34}$$

that is

$$x^{(1)} = V_0^{(1)}\tau + x_0^{(1)}\,, \quad X = V_0^{(2)}\tau + X_0\,, \qquad Y = V_0^{(3)}\tau + Y_0\,.$$

The projections of this trajectory onto the plane XZ at different γ_4 are illustrated in Fig. 11.3. The type of these trajectories is similar to the ones from the general case (see Fig. 11.2).

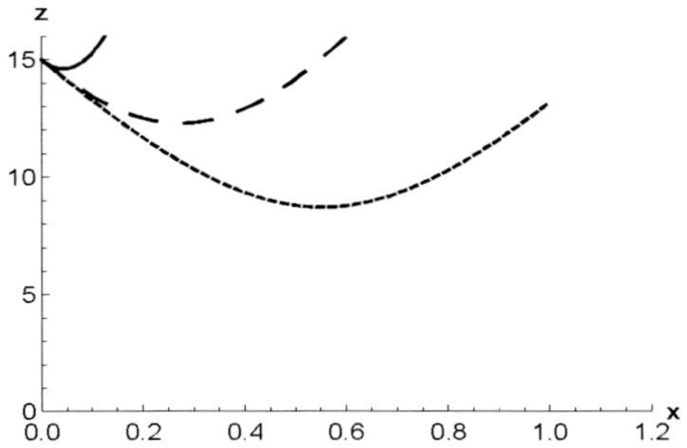

Figure 11.3: The projects trajectories of the light ray on plane XZ at zero constant c_1. The following parameters are used: $c_1 = 0$; $Z_0 = 15$; $X_0 = 0$; $V_0^{(4)} = -0.9$; $V_0^{(2)} = 0.05$; $\gamma_4 = 0.06$ (dotted line), 0.03 (dashed line), 0.01 (solid line).

Chapter 12

Finslerian Geometrization for the Problem of a Vector Particle in External Coulomb Field

12.1 Setting the Problem

Until now, the known quantum-mechanical problem of a spin 1 particle in presence of the external Coulomb field remains unsolved. The first of the three expected sub-classes of bound states was described by I.E. Tamm [283]. The lack of advances in this study relates to other two sub-classes of expected bound states. The main conclusion from I.E. Tamm's considerations consists in the statement that in two remaining sub-classes of states, the particle should fall to the center. However, the study of the nonrelativistic problem of spin 1 particle in the Coulomb field showed that there exist three correctly defined series of bound states, which are described by formulas similar to the analogue one for spinless Schrödinger particles displaced in a Coulomb field.

In [306], it was proved a possibility for some radial functions to yield different 2-nd order differential equations, instead of the expected equations of 4-th order. The present chapter is based on [307].

The present analysis starts with the the system of 4 equations for the functions $f_i(r), i = 1, \ldots 4$ derived in [307]:

$$\begin{aligned} \frac{d}{dr}f^1 &= af^3 + cf^4\,, \quad \frac{d}{dr}f^2 = df^3 + bf^4\,, \\ \frac{d}{dr}f^3 &= Af^1 + Cf^2\,, \quad \frac{d}{dr}f^4 = Df^1 + Bf^2\,, \end{aligned} \tag{12.1}$$

where we use the notations for r-dependent coefficients

$$\begin{gathered} a = 2i\nu\frac{\epsilon r+\alpha}{r}, \quad b = -\frac{\nu(\epsilon r+\alpha)}{r}, \\ A = -i\frac{\nu(\epsilon r+\alpha)}{r^3}, \quad B = -\frac{2\nu(\epsilon r+\alpha)}{r^3}, \\ c = -(2\nu^2+r^2), \quad C(r) = +i\frac{(2\nu^2+r^2)}{r^2}, \\ d = i\frac{(\epsilon r+\alpha)^2-r^2}{r^2}, \quad D = \frac{(\epsilon r+\alpha)^2-r^2}{r^4}; \end{gathered} \tag{12.2}$$

and $\nu = +\sqrt{j(j+1)/2}, j = 1,2,3,\dots$. Preserving only three independent functions c, B, D we obtain equations (the upper dot stands for the derivative d/dr)

$$\begin{gathered} \dot{f}^1 = -iBr^2 f^3 + cf^4, \quad \dot{f}^2 = ir^2 Df^3 + \frac{1}{2}Br^2 f^4, \\ \dot{f}^3 = \frac{i}{2}Bf^1 - i\frac{c}{r^2}f^2, \quad \dot{f}^4 = Df^1 + Bf^2. \end{gathered} \tag{12.3}$$

The system (12.3) may be written in matrix form

$$\dot{f}^i(r) = K^i{}_j(r) f^j(r), \qquad K = \begin{vmatrix} 0 & 0 & -iBr^2 & c \\ 0 & 0 & ir^2 D & \frac{1}{2}Br^2 \\ \frac{1}{2}iB & -icr^{-2} & 0 & 0 \\ D & B & 0 & 0 \end{vmatrix}. \tag{12.4}$$

By decomposing each complex-valued function $f_i(r)$ into the sum of real and imaginary parts:

$$f^1 = x^1 + ix^2, \quad f^2 = x^3 + ix^4, \quad f^3 = x^5 + ix^6, \quad f^4 = x^7 + ix^8, \tag{12.5}$$

from (12.4) we derive the first order equations for the eight real functions:

$$\begin{gathered} \dot{x}^1 = Br^2 x^6 + cx^7, \quad \dot{x}^2 = -Br^2 x^5 + cx^8, \\ \dot{x}^3 = -Dr^2 x^6 + \frac{Br^2}{2}x^7, \quad \dot{x}^4 = Dr^2 x^5 + \frac{Br^2}{2}x^8; \end{gathered} \tag{12.6}$$

$$\dot{x}^5 = -\frac{B}{2}x^2 + \frac{c}{r^2}x^4, \; \dot{x}^6 = \frac{B}{2}x^1 - \frac{c}{r^2}x^3, \; \dot{x}^7 = Dx^1 + Bx^3, \; \dot{x}^8 = Dx^2 + Bx^4. \tag{12.7}$$

The system (12.6)–(12.7) may be written in matrix form (let $dx^i/dr \equiv y^i(r)$)

$$y^i(r) = M^i{}_j\,(r)\,x^j(r), \tag{12.8}$$

where the matrix M is

$$M^i{}_j\,(r) = \begin{vmatrix} 0 & 0 & 0 & 0 & 0 & Br^2 & c & 0 \\ 0 & 0 & 0 & 0 & -Br^2 & 0 & 0 & c \\ 0 & 0 & 0 & 0 & 0 & -Dr^2 & \frac{1}{2}Br^2 & 0 \\ 0 & 0 & 0 & 0 & Dr^2 & 0 & 0 & \frac{1}{2}Br^2 \\ 0 & -\frac{1}{2}B & 0 & cr^{-2} & 0 & 0 & 0 & 0 \\ \frac{1}{2}B & 0 & -cr^{-2} & 0 & 0 & 0 & 0 & 0 \\ D & 0 & B & 0 & 0 & 0 & 0 & 0 \\ 0 & D & 0 & B & 0 & 0 & 0 & 0 \end{vmatrix}. \tag{12.9}$$

In order to obtain the known mathematical structure [308], let us differentiate the equation (12.8); we infer

$$\dot{y}^i(r) = \dot{M}^i{}_j(r)\,x^j(r) + M^i{}_j(r)\,y^j(r)\,. \tag{12.10}$$

Let us adhere to the conventional mathematical notation [308]:

$$\dot{y}^i(r) + 2G^i(r,x,y) = 0\,, \quad G^i(r,x,y) = -\frac{1}{2}(\dot{M}^i{}_j\,x^j + M^i{}_j\,y^j)\,. \tag{12.11}$$

The dotted matrix $\dot{M}$ is given by the formula

$$\frac{d}{dr}M = \dot{M} =$$

$$\left|\begin{array}{cccccccc} 0 & 0 & 0 & 0 & 0 & \dot{B}r^2+2Br & \dot{c} & 0 \\ 0 & 0 & 0 & 0 & -\dot{B}r^2-2Br & 0 & 0 & \dot{c} \\ 0 & 0 & 0 & 0 & 0 & -\dot{D}r^2-2rD & \frac{1}{2}\dot{B}r^2+Br & 0 \\ 0 & 0 & 0 & 0 & \dot{D}r^2+2rD & 0 & 0 & \frac{1}{2}\dot{B}r^2+Br \\ 0 & -\frac{1}{2}\dot{B} & 0 & \dot{c}r^{-2}-2cr^{-3} & 0 & 0 & 0 & 0 \\ \frac{1}{2}\dot{B} & 0 & 2cr^{-3}-\dot{c}r^{-2} & 0 & 0 & 0 & 0 & 0 \\ \dot{D} & 0 & \dot{B} & 0 & 0 & 0 & 0 & 0 \\ 0 & \dot{D} & 0 & \dot{B} & 0 & 0 & 0 & 0 \end{array}\right| .$$

Taking into account the expressions for $B(r)$ and $c(r)$, we get

$$\dot{M} =$$

$$\left|\begin{array}{cccc} 0 & 0 & 0 & 0 \\ 0 & 0 & 0 & 0 \\ 0 & 0 & 0 & 0 \\ 0 & 0 & 0 & 0 \\ 0 & -\nu(3\alpha+2\epsilon r)r^{-4} & 0 & 4\nu^2r^{-3} \\ \nu(3\alpha+2\epsilon r)r^{-4} & 0 & -4\nu^2r^{-3} & 0 \\ -2\left((\alpha+\epsilon r)(2\alpha+\epsilon r)-r^2\right)r^{-5} & 0 & 2\nu(3\alpha+2\epsilon r)r^{-4} & 0 \\ 0 & -2\left((\alpha+\epsilon r)(2\alpha+\epsilon r)-r^2\right)r^{-5} & 0 & 2\nu(3\alpha+2\epsilon r)r^{-4} \end{array}\right.$$

$$\left.\begin{array}{cccc} 0 & 2\nu\alpha r^{-2} & -2r & 0 \\ -2\nu\alpha r^{-2} & 0 & 0 & -2r \\ 0 & 2\alpha(\alpha+\epsilon r)r^{-3} & \nu\alpha r^{-2} & 0 \\ -2\alpha(\alpha+\epsilon r)r^{-3} & 0 & 0 & \nu\alpha r^{-2} \\ 0 & 0 & 0 & 0 \\ 0 & 0 & 0 & 0 \\ 0 & 0 & 0 & 0 \\ 0 & 0 & 0 & 0 \end{array}\right| . \tag{12.12}$$

In explicit form, the system of 8 equations for the functions $x_i(r)$ reads as follows

(note that it has two-part structure)

$$
\begin{aligned}
\frac{d^2x_1}{dr^2} &= \frac{2\nu\alpha}{r^2}x_6 - 2rx_7 + Br^2\frac{dx_6}{dr} + c\frac{dx_7}{dr}, \\
\frac{d^2x_2}{dr^2} &= -\frac{2\nu\alpha}{r^2}x_5 - 2rx_8 - Br^2\frac{dx_5}{dr} + c\frac{dx_8}{dr}, \\
\frac{d^2x_3}{dr^2} &= -(\dot{D}r^2 + 2Dr)x_6 + \frac{\nu\alpha}{r^2}x_7 - Dr^2\frac{dx_6}{dr} + \frac{Br^2}{2}\frac{dx_7}{dr}, \\
\frac{d^2x_4}{dr^2} &= +(\dot{D}r^2 + 2Dr)x_5 + \frac{\nu\alpha}{r^2}x_8 + Dr^2\frac{dx_5}{dr} + \frac{Br^2}{2}\frac{dx_8}{dr},
\end{aligned}
\tag{12.13}
$$

$$
\begin{aligned}
\frac{d^2x_5}{dr^2} &= -\frac{1}{2}\dot{B}x_2 + \frac{4\nu^2}{r^3}x_4 - \frac{1}{2}B\frac{dx_2}{dr} + \frac{c}{r^2}\frac{dx_4}{dr}, \\
\frac{d^2x_6}{dr^2} &= +\frac{1}{2}\dot{B}x_1 - \frac{4\nu^2}{r^3}x_3 + \frac{1}{2}B\frac{dx_1}{dr} - \frac{c}{r^2}\frac{dx_3}{dr}, \\
\frac{d^2x^7}{dr^2} &= \dot{D}x_1 + \dot{B}x_3 + D\frac{dx_1}{dr} + B\frac{dx_3}{dr}, \\
\frac{d^2x^8}{dr^2} &= \dot{D}x_2 + \dot{B}x_4 + D\frac{dx_2}{dr} + B\frac{dx_4}{dr}.
\end{aligned}
\tag{12.14}
$$

Taking into account expressions for the involved coefficients, we obtain:

$$
\begin{aligned}
\frac{d^2x_1}{dr^2} &= \frac{2\nu\alpha}{r^2}x_6 - 2rx_7 - \frac{2\nu(\epsilon r + \alpha)}{r}\frac{dx_6}{dr} - (2\nu^2 + r^2)\frac{dx_7}{dr}, \\
\frac{d^2x_2}{dr^2} &= -\frac{2\nu\alpha}{r^2}x_5 - 2rx_8 + \frac{2\nu(\epsilon r + \alpha)}{r}\frac{dx_5}{dr} - (2\nu^2 + r^2)\frac{dx_8}{dr}, \\
\frac{d^2x_3}{dr^2} &= \frac{2\alpha(\alpha + \epsilon r)}{r^3}x_6 + \frac{\nu\alpha}{r^2}x_7 - \frac{(\epsilon r + \alpha)^2 - r^2}{r^2}\frac{dx_6}{dr} - \frac{\nu(\epsilon r + \alpha)}{r}\frac{dx_7}{dr}, \\
\frac{d^2x_4}{dr^2} &= -\frac{2\alpha(\alpha + \epsilon r)}{r^3}x_5 + \frac{\nu\alpha}{r^2}x_8 + \frac{(\epsilon r + \alpha)^2 - r^2}{r^2}\frac{dx_5}{dr} - \frac{\nu(\epsilon r + \alpha)}{r}\frac{dx_8}{dr},
\end{aligned}
\tag{12.15}
$$

$$
\begin{aligned}
\frac{d^2x_5}{dr^2} &= -\frac{\nu(3\alpha + 2\epsilon r)}{r^4}x_2 + \frac{4\nu^2}{r^3}x_4 + \frac{\nu(\epsilon r + \alpha)}{r^3}\frac{dx_2}{dr} - \frac{(2\nu^2 + r^2)}{r^2}\frac{dx_4}{dr}, \\
\frac{d^2x_6}{dr^2} &= +\frac{\nu(3\alpha + 2\epsilon r)}{r^4}x_1 - \frac{4\nu^2}{r^3}x_3 - \frac{\nu(\epsilon r + \alpha)}{r^3}\frac{dx_1}{dr} + \frac{(2\nu^2 + r^2)}{r^2}\frac{dx_3}{dr},
\end{aligned}
$$

$$
\begin{aligned}
\frac{d^2x^7}{dr^2} &= -\frac{2\left((\alpha + \epsilon r)(2\alpha + \epsilon r) - r^2\right)}{r^5}x_1 + \frac{2\nu(3\alpha + 2\epsilon r)}{r^4}x_3 \\
&\quad + \frac{(\epsilon r + \alpha)^2 - r^2}{r^4}\frac{dx_1}{dr} - \frac{2\nu(\epsilon r + \alpha)}{r^3}\frac{dx_3}{dr}, \\
\frac{d^2x^8}{dr^2} &= -\frac{2\left((\alpha + \epsilon r)(2\alpha + \epsilon r) - r^2\right)}{r^5}x_2 + \frac{2\nu(3\alpha + 2\epsilon r)}{r^4}x_4 \\
&\quad + \frac{(\epsilon r + \alpha)^2 - r^2}{r^4}\frac{dx_2}{dr} - \frac{2\nu(\epsilon r + \alpha)}{r^3}\frac{dx_4}{dr}.
\end{aligned}
\tag{12.16}
$$

From geometrical point of view [308], the system of differential equations (12.15)–(12.16) is governed by the 8-vector field G^i. This field determines geometrical objects, such as a nonlinear connection, the Berwald connection, and other; therefore, some properties of the solutions of these equations may be studied by using these objects. In [308], it was proposed the apply five geometrical objects, the so called Kosambi–Cartan-Chen invariants, for studying certain properties of the solutions $\{f_i\}$.

12.2 KCC-Invariants

The first KCC-invariant (Kosambi–Cartan–Chen) ε^i is introduced by the following definition [308]

$$\varepsilon^i(r,x,y) = \frac{\partial G^i}{\partial y^j}y^j - 2G^i, \qquad \varepsilon^i = \dot{M}^i{}_j\,x^j + \frac{1}{2}M^i{}_j\,y^j, \tag{12.17}$$

whose explicit components are

$$\begin{aligned}
\varepsilon^1 &= \frac{2\nu\alpha}{r^2}x^6 - 2rx^7 + \frac{Br^2}{2}y^6 + \frac{c}{2}y^7,\\
\varepsilon^2 &= -\frac{2\nu\alpha}{r^2}x^5 - 2rx^8 - \frac{Br^2}{2}y^5 + \frac{c}{2}y^8,\\
\varepsilon^3 &= \frac{2\alpha(\alpha+\epsilon r)}{r^3}x^6 + \frac{\nu\alpha}{r^2}x^7 - \frac{Dr^2}{2}y^6 + \frac{Br^2}{4}y^7,\\
\varepsilon^4 &= -\frac{2\alpha(\alpha+\epsilon r)}{r^3}x^5 + \frac{\nu\alpha}{r^2}x^8 + \frac{Dr^2}{2}y^5 + \frac{Br^2}{4}y^8,\\
\varepsilon^5 &= -\frac{\dot{B}}{2}x^2 + \frac{4\nu^2}{r^3}x^4 - \frac{B}{4}y^2 + \frac{c}{2r^2}y^4,\\
\varepsilon^6 &= \frac{\dot{B}}{2}x^1 - \frac{4\nu^2}{r^3}x^3 + \frac{B}{4}y^1 - \frac{c}{2r^2}y^3,\\
\varepsilon^7 &= \dot{D}x^1 + \dot{B}x^3 + \frac{D}{2}y^1 + \frac{B}{2}y^3,\\
\varepsilon^8 &= \dot{D}x^2 + \dot{B}x^4 + \frac{D}{2}y^2 + \frac{B}{2}y^4;
\end{aligned} \tag{12.18}$$

or, in explicit form,

$$\begin{aligned}
\varepsilon^1 &= \frac{2\nu\alpha}{r^2}x^6 - 2rx^7 - \frac{\nu(\epsilon r+\alpha)}{r}y^6 - \frac{2\nu^2+r^2}{2}y^7,\\
\varepsilon^2 &= -\frac{2\nu\alpha}{r^2}x^5 - 2rx^8 + \frac{\nu(\epsilon r+\alpha)}{r}y^5 - \frac{2\nu^2+r^2}{2}y^8,\\
\varepsilon^3 &= \frac{2\alpha(\alpha+\epsilon r)}{r^3}x^6 + \frac{\nu\alpha}{r^2}x^7 - \frac{(\epsilon r+\alpha)^2-r^2}{2r^2}y^6 - \frac{\nu(\epsilon r+\alpha)}{2r}y^7,\\
\varepsilon^4 &= -\frac{2\alpha(\alpha+\epsilon r)}{r^3}x^5 + \frac{\nu\alpha}{r^2}x^8 + \frac{(\epsilon r+\alpha)^2-r^2}{2r^2}y^5 - \frac{\nu(\epsilon r+\alpha)}{2r}y^8,
\end{aligned} \tag{12.19}$$

$$
\begin{aligned}
\varepsilon^5 &= -\frac{\nu(3\alpha+2\epsilon r)}{r^4}x^2 + \frac{4\nu^2}{r^3}x^4 + \frac{\nu(\epsilon r+\alpha)}{2r^3}y^2 - \frac{2\nu^2+r^2}{2r^2}y^4, \\
\varepsilon^6 &= \frac{\nu(3\alpha+2\epsilon r)}{r^4}x^1 - \frac{4\nu^2}{r^3}x^3 - \frac{\nu(\epsilon r+\alpha)}{2r^3}y^1 + \frac{2\nu^2+r^2}{2r^2}y^3, \\
\varepsilon^7 &= -2\frac{(\alpha+\epsilon r)(2\alpha+\epsilon r)-r^2}{r^5}x^1 + \frac{2\nu(3\alpha+2\epsilon r)}{r^4}x^3 \\
&\quad + \frac{(\epsilon r+\alpha)^2-r^2}{2r^4}y^1 - \frac{\nu(\epsilon r+\alpha)}{r^3}y^3, \\
\varepsilon^8 &= 2\frac{(\alpha+\epsilon r)(2\alpha+\epsilon r)-r^2}{r^5}x^2 + \frac{2\nu(3\alpha+2\epsilon r)}{r^4}x^4 \\
&\quad + \frac{(\epsilon r+\alpha)^2-r^2}{2r^4}y^2 - \frac{\nu(\epsilon r+\alpha)}{r^3}y^4.
\end{aligned}
\tag{12.20}
$$

By defining the complex-valued invariant, the formulas become shorter,

$$
E^1 = \varepsilon^1 + i\varepsilon^2, \quad E^2 = \varepsilon^3 + i\varepsilon^4, \quad E^3 = \varepsilon^5 + i\varepsilon^6, \quad E^4 = \varepsilon^7 + i\varepsilon^8. \tag{12.21}
$$

The second KCC-invariant $P^i_{\ j}$ is defined by the formula [308]:

$$
P^i_{\ j} = 2\frac{\partial G^i}{\partial x^j} + 2G^s\frac{\partial^2 G^i}{\partial y^j \partial y^s} + \frac{\partial^2 G^i}{\partial y^j \partial x^s}y^s - \frac{\partial G^i}{\partial y^s}\frac{\partial G^s}{\partial y^j} - \frac{\partial^2 G^i}{\partial y^j \partial r}. \tag{12.22}
$$

The 3-rd KCC-invariant is determined by the rule

$$
R^i_{\ jk} = \frac{1}{3}\Big(\frac{\partial P^i_j}{\partial y^k} - \frac{\partial P^i_k}{\partial y^j}\Big). \tag{12.23}
$$

and it determines torsion of the Bervald connection. The 4-th KCC-invariant is the Riemann–Christoffel curvature tensor an given by the formula

$$
B^i_{\ jkl} = \frac{\delta R^i_{\ jk}}{\delta y^l}. \tag{12.24}
$$

The 5-th KCC-invariant is given by the rule

$$
D^i_{\ jkl} = \frac{\partial^3 G^i}{\partial y^j \partial y^k \partial y^l}, \tag{12.25}
$$

and it is named as Duglas tensor (or Berwald curvature tensor). Due to the linearity of the vector G^i in coordinates x^i and y^i, the 1-st and the 2-nd invariants are functions of the radial coordinate r and do not depend on x^i and y^i. Therefore, the three remaining invariants identically vanish.

12.3 Second KCC-Invariant

Using the definition (12.22), we find the 64 entries of the matrix $P^i_{\ j}$:

$$
P^i_{\ j} = 2\frac{\partial G^i}{\partial x^j} + 2G^s\frac{\partial^2 G^i}{\partial y^j \partial y^s} + \frac{\partial^2 G^i}{\partial y^j \partial x^s}y^s - \frac{\partial G^i}{\partial y^s}\frac{\partial G^s}{\partial y^j} - \frac{\partial^2 G^i}{\partial y^j \partial r}. \tag{12.26}
$$

The object ${P^i}_j$ contains 64 components and given by the matrix (we list its elements them by blocks)

$$
{P^i}_j = \left| \begin{array}{cccc}
-\frac{1}{8}\left(B^2r^2+2cD\right) & 0 & 0 & 0 \\
0 & -\frac{1}{8}\left(B^2r^2+2cD\right) & 0 & 0 \\
0 & 0 & -\frac{1}{8}(B^2r^2+2cD) & 0 \\
0 & 0 & 0 & -\frac{1}{8}(B^2r^2+2cD) \\
0 & \frac{1}{4}\dot{B} & 0 & \frac{1}{2}(2c-r\dot{c})r^{-3} \\
-\frac{1}{4}\dot{B} & 0 & \frac{1}{2}(r\dot{c}-2c)r^{-3} & 0 \\
-\frac{1}{2}\dot{D} & 0 & -\frac{1}{2}\dot{B} & 0 \\
0 & -\frac{1}{2}\dot{D} & 0 & -\frac{1}{2}\dot{B}
\end{array} \right.
$$

$$
\left. \begin{array}{cccc}
0 & -\frac{1}{2}r\left(2B+r\dot{B}\right) & -\frac{1}{2}\dot{c} & 0 \\
\frac{1}{2}r\left(2B+r\dot{B}\right) & 0 & 0 & -\frac{1}{2}\dot{c} \\
0 & \frac{1}{2}r\left(2D+r\dot{D}\right) & -\frac{1}{4}r\left(2B+r\dot{B}\right) & 0 \\
-\frac{1}{2}r\left(2D+r\dot{D}\right) & 0 & 0 & -\frac{1}{4}r\left(2B+r\dot{B}\right) \\
-\frac{1}{8}(B^2r^2+2cD) & 0 & 0 & 0 \\
0 & -\frac{1}{8}(B^2r^2+2cD) & 0 & 0 \\
0 & 0 & -\frac{1}{8}\left(B^2r^2+2cD\right) & 0 \\
0 & 0 & 0 & -\frac{1}{8}\left(B^2r^2+2cD\right)
\end{array} \right|
$$

$$
= \left| \begin{array}{cccc}
\frac{-2\nu^2+(\alpha+\epsilon r)^2-r^2}{4r^2} & 0 & 0 & 0 \\
0 & \frac{-2\nu^2+(\alpha+\epsilon r)^2-r^2}{4r^2} & 0 & 0 \\
0 & 0 & \frac{-2\nu^2+(\alpha+\epsilon r)^2-r^2}{4r^2} & 0 \\
0 & 0 & 0 & \frac{-2\nu^2+(\alpha+\epsilon r)^2-r^2}{4r^2} \\
0 & \frac{\nu(3\alpha+2\epsilon r)}{2r^4} & 0 & -\frac{2\nu^2}{r^3} \\
-\frac{\nu(3\alpha+2\epsilon r)}{2r^4} & 0 & \frac{2\nu^2}{r^3} & 0 \\
\frac{(\epsilon r+\alpha)(\epsilon r+2\alpha)-r^2}{r^5} & 0 & -\frac{\nu(3\alpha+2\epsilon r)}{r^4} & 0 \\
0 & \frac{(\epsilon r+\alpha)(\epsilon r+2\alpha)-r^2}{r^5} & 0 & -\frac{\nu(3\alpha+2\epsilon r)}{r^4}
\end{array} \right.
$$

$$
\left. \begin{array}{cccc}
0 & -\frac{\nu\alpha}{r^2} & r & 0 \\
\frac{\nu\alpha}{r^2} & 0 & 0 & r \\
0 & -\frac{\alpha(\alpha+\epsilon r)}{r^3} & -\frac{\nu\alpha}{2r^2} & 0 \\
\frac{\alpha(\alpha+\epsilon r)}{r^3} & 0 & 0 & -\frac{\nu\alpha}{2r^2} \\
\frac{-2\nu^2+(\alpha+\epsilon r)^2-r^2}{4r^2} & 0 & 0 & 0 \\
0 & \frac{-2\nu^2+(\alpha+\epsilon r)^2-r^2}{4r^2} & 0 & 0 \\
0 & 0 & \frac{-2\nu^2+(\alpha+\epsilon r)^2-r^2}{4r^2} & 0 \\
0 & 0 & 0 & \frac{-2\nu^2+(\alpha+\epsilon r)^2-r^2}{4r^2}
\end{array} \quad r \right| .
$$

In explicit form, the non-vanishing components are

$$P^1{}_1 = P^2{}_2 = P^3{}_3 = P^4{}_4 = P^5{}_5 = P^6{}_6 = P^7{}_7 = P^8{}_8 = \frac{-2\nu^2 + (\alpha + \epsilon r)^2 - r^2}{4r^2},$$
$$P^3{}_6 = -P^4{}_5 = \frac{2\nu^2}{r^3}, \quad P^5{}_2 = -P^6{}_1 = -2P^7{}_3 = -2P^8{}_4 = \frac{\nu\alpha}{r^2},$$
$$P^1{}_6 = -P^2{}_5 = -\frac{1}{2}P^3{}_7 = -\frac{1}{2}P^4{}_8 = -\frac{1}{2}\nu\frac{(3\alpha + 2\epsilon r)}{r^4},$$
$$P^1{}_7 = P^2{}_8 = \frac{((\epsilon r + \alpha)(\epsilon r + 2\alpha) - r^2)}{r^5}, \; P^7{}_1 = P^8{}_2 = r, \; P^5{}_4 = -P^6{}_3 = \frac{\alpha(\alpha + \epsilon r)}{r^3}.$$

The matrix of 2-nd invariant has the structure

$$\left| \begin{array}{cccccccc} a & . & . & . & . & \sigma & r & . \\ . & a & . & . & -\sigma & . & . & r \\ . & . & a & . & . & -\upsilon & \sigma/2 & . \\ . & . & . & a & \upsilon & . & . & \sigma/2 \\ . & -N & . & \gamma & a & . & . & . \\ N & . & -\gamma & . & . & a & . & . \\ \beta & . & N/2 & . & . & . & a & . \\ . & \beta & . & N/2 & . & . & . & a \end{array} \right|. \tag{12.27}$$

In accordance with the known statement (see [308]) a bunch of geodesic curves at the point r_0 converges (or diverges) if the real parts of all the eigenvalues of the 2-nd KCC-invariant $P^i{}_j$ are negative (or positive).

The most interesting and physically interpretable is the behavior of geodesic curves in the vicinity of the points $r = 0$ and $r = \infty$. By this reason, let us find expressions of the matrix $P^i{}_j(r)$, when $r \to 0$ and $r \to \infty$; after that we we shall can examine the asymptotical behavior of λ for the eigenvalue problem

$$P\Psi_\lambda = \lambda\Psi_\lambda\,. \tag{12.28}$$

Let $r \to 0$; then we have:

$$\begin{gathered} P^1{}_1 = P^2{}_2 = P^3{}_3 = P^4{}_4 = P^5{}_5 = P^6{}_6 = P^7{}_7 = P^8{}_8 = \frac{\alpha^1 - j(j+1)}{4} \cdot \frac{1}{r^2}, \\ P^3{}_6 = -P^4{}_5 = \frac{2\nu^2}{r^3}, \quad P^5{}_2 = -P^6{}_1 = -2P^7{}_3 = -2P^8{}_4 = \frac{\nu\alpha}{r^2}, \\ P^1{}_6 = -P^2{}_5 = -\frac{1}{2}P^3{}_7 = -\frac{1}{2}P^4{}_8 = -\frac{3\alpha\nu}{2r^4}, \\ P^1{}_7 = P^2{}_8 = \frac{2\alpha^2}{r^5}, \quad P^7{}_1 = P^8{}_2 = r, \quad P^5{}_4 = -P^6{}_3 = \frac{\alpha^2}{r^3}. \end{gathered} \tag{12.29}$$

For $r \to \infty$, we infer

$$
\begin{gathered}
P^1{}_1 = P^2{}_2 = P^3{}_3 = P^4{}_4 = P^5{}_5 = P^6{}_6 = P^7{}_7 = P^8{}_8 = \frac{\epsilon^2 - 1}{4}, \\
P^3{}_6 = -P^4{}_5 = \frac{2\nu^2}{r^3}, \quad P^5{}_2 = -P^6{}_1 = -2P^7{}_3 = -2P^8{}_4 = \frac{\nu\alpha}{r^2}, \\
P^1{}_6 = -P^2{}_5 = -\frac{1}{2}P^3{}_7 = -\frac{1}{2}P^4{}_8 = -\nu\frac{\epsilon\nu}{r^3}, \\
P^1{}_7 = P^2{}_8 = \frac{\epsilon^2 - 1}{r^3}, \qquad P^7{}_1 = P^8{}_2 = r, \quad P^5{}_4 = -P^6{}_3 = \frac{\alpha\epsilon}{r^2}.
\end{gathered}
\tag{12.30}
$$

The equation (12.28), near the point $r = 0$, gives 8 real-valued values:

$$
\begin{aligned}
\lambda_1 = \lambda_2 = \lambda_3 = \lambda_4 &= \frac{-2\sqrt{2}\alpha\nu + r(\alpha^2 - 2\nu^2)}{4r^3} \longrightarrow \frac{-2\sqrt{2}\alpha\nu}{4r^3} < 0, \\
\lambda_5 = \lambda_6 = \lambda_7 = \lambda_8 &= \frac{+2\sqrt{2}\alpha\nu + r(\alpha^2 - 2\nu^2)}{4r^3} \longrightarrow \frac{+2\sqrt{2}\alpha\nu}{4r^3} > 0.
\end{aligned}
\tag{12.31}
$$

This means that two variants are possible: some geodesic curves are convergent near $r = 0$, $(\lambda < 0)$; some geodesic curves are divergent near $r = 0$ $(\lambda > 0)$. Such a behavior correlates with existence of two solutions tending to zero, and two solutions tending to infinity.

In the region $r \to \infty$, Eq. (12.28) reduces to

$$
\frac{\left(256\alpha^2\nu^4 + r^{10}\left(-4\lambda + \epsilon^2 - 1\right)^4 - 16r^8\left(\epsilon^2 - 1\right)\left(-4\lambda + \epsilon^2 - 1\right)^2 + 512\alpha\nu^2 r^3\epsilon\right)^2}{65536\, r^{20}} = 0,
$$

and the eigenvalues are $r \longrightarrow \infty$,

$$
\begin{aligned}
\lambda_{1,2} &= \frac{\epsilon^2 - 1}{4} - \frac{\sqrt{r\left(r^3\left(\epsilon^2 - 1\right) - \sqrt{-4\alpha^2\nu^4 + r^6\left(\epsilon^2 - 1\right)^2 - 8\alpha\nu^2 r^3\epsilon}\right)}}{\sqrt{2}r^3} \longrightarrow \frac{1}{4}\left(\epsilon^2 - 1\right), \\
\lambda_{3,4} &= \frac{\epsilon^2 - 1}{4} + \frac{\sqrt{r\left(r^3\left(\epsilon^2 - 1\right) - \sqrt{-4\alpha^2\nu^4 + r^6\left(\epsilon^2 - 1\right)^2 - 8\alpha\nu^2 r^3\epsilon}\right)}}{\sqrt{2}r^3} \longrightarrow \frac{1}{4}\left(\epsilon^2 - 1\right), \\
\lambda_{5,6} &= \frac{\epsilon^2 - 1}{4} - \frac{\sqrt{r\left(r^3\left(\epsilon^2 - 1\right) + \sqrt{-4\alpha^2\nu^4 + r^6\left(\epsilon^2 - 1\right)^2 - 8\alpha\nu^2 r^3\epsilon}\right)}}{\sqrt{2}r^3} \longrightarrow \frac{1}{4}\left(\epsilon^2 - 1\right), \\
\lambda_{7,8} &= \frac{\epsilon^2 - 1}{4} + \frac{\sqrt{r\left(r^3\left(\epsilon^2 - 1\right) + \sqrt{-4\alpha^2\nu^4 + r^6\left(\epsilon^2 - 1\right)^2 - 8\alpha\nu^2 r^3\epsilon}\right)}}{\sqrt{2}r^3} \longrightarrow \frac{1}{4}\left(\epsilon^2 - 1\right).
\end{aligned}
\tag{12.32}
$$

Let us recall that to bound states there must correspond the inequality $\epsilon^2 - 1 < 0$, whereas to scattering states there must correspond the opposite inequality $\epsilon^2 - 1 > 0$. The behavior of the values (12.32) agrees with this statement. In Fig. 12.1 and Fig. 12.2 we see the behavior of the real parts of the eigenvalues depending on the radial coordinate r in the whole region $r \in (0, +\infty)$.

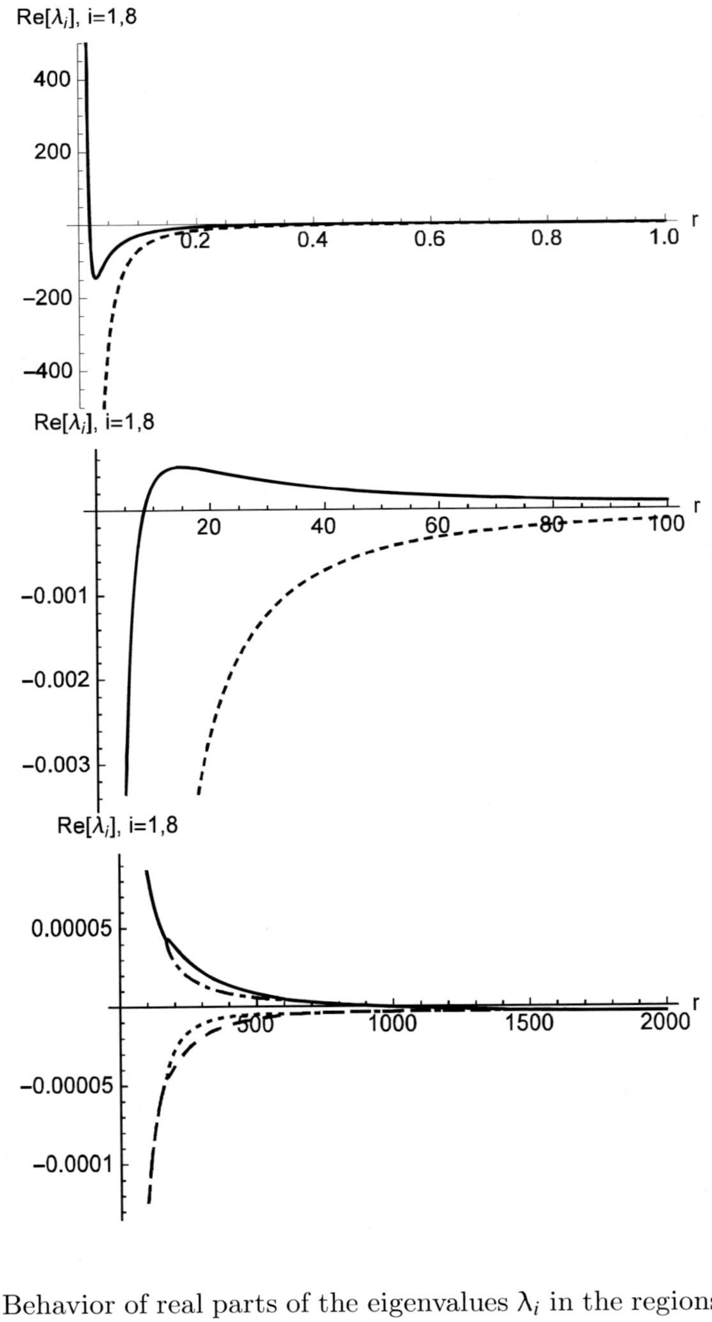

Figure 12.1: Behavior of real parts of the eigenvalues λ_i in the regions $r < 1$, $1 < r < 100$ and $r > 100$ when $\epsilon = 0,99999$, $\nu = 1$, $\alpha = 1/137$.

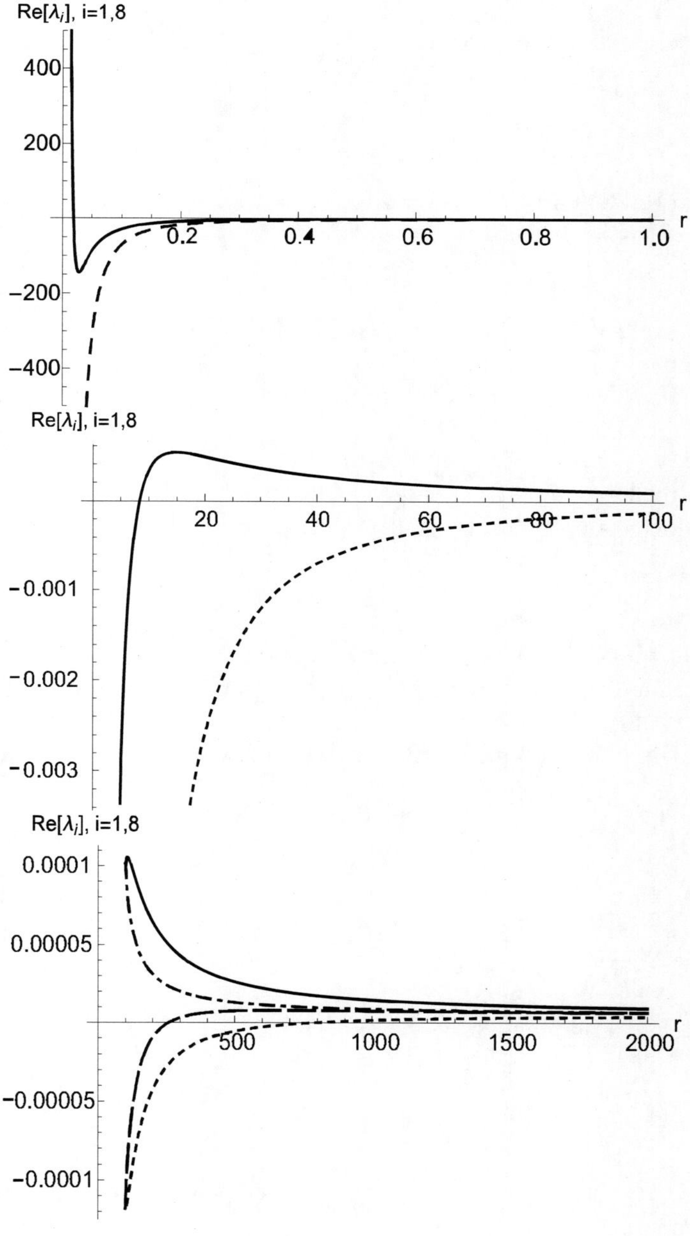

Figure 12.2: Behavior of real parts of the eigenvalues in the regions $r < 1$, $1 < r < 100$ and $r > 100$ when $\epsilon = 1{,}00001$, $\nu = 1$, $\alpha = 1/137$.

12.4 Natural Splitting 4+4

Let us introduce the complex vectors

$$z_1 = \left| \begin{array}{c} f_1 \\ f_2 \end{array} \right| , \qquad z_2 = \left| \begin{array}{c} f_3 \\ f_4 \end{array} \right| , \tag{12.33}$$

then equations (12.33) can be presented as follows

$$\dot{z}_1^i = S^i{}_j z_2^j , \qquad \dot{z}_2^i = Q^i{}_j z_1^j , \tag{12.34}$$

where S and Q are 2×2-matrices(sub-blocks of the matrix K (12.35)):

$$K = \left| \begin{array}{cc} 0 & S \\ Q & 0 \end{array} \right| , \quad S = \left| \begin{array}{cc} -iBr^2 & c \\ iDr^2 & \frac{1}{2}Br^2 \end{array} \right| , \quad Q = \left| \begin{array}{cc} \frac{1}{2}iB & -icr^{-2} \\ D & B \end{array} \right| . \tag{12.35}$$

By differentiating Eqs. in (12.34) and expressing z_1, $\dot{z}_1$ in terms of $\dot{z}_2$ and z_2 (and inversely), we derive two formally disconnected sub-systems:

$$\ddot{z}_1 = \dot{S}S^{-1}\dot{z}_1 + SQz_1 , \qquad \ddot{z}_2 = \dot{Q}Q^{-1}\dot{z}_2 + QSz_2 . \tag{12.36}$$

Due to (12.34), we conclude that having known any one of the 2-vectors, we can calculate the other 2-vector. In other words, it generally suffices to solve only one subsystem in (12.36).

Let us find the explicit forms of the matrices involved in (12.36):

$$SQ = QS = \left| \begin{array}{cc} cD + \frac{1}{2}B^2r^2 & 0 \\ 0 & cD + \frac{1}{2}B^2r^2 \end{array} \right| , \tag{12.37}$$

$$\begin{gathered} \dot{S}S^{-1} = \frac{1}{B^2r^2 + 2cD} \left| \begin{array}{cc} Br(2B + r\dot{B}) + 2D\dot{c} & 2B\dot{c} - 2c(2B + r\dot{B})r^{-1} \\ r^2(D\dot{B} - B\dot{D}) & Br(2B + r\dot{B}) + 2c(2D + r\dot{D})r^{-1} \end{array} \right| , \\ \dot{Q}Q^{-1} = \frac{1}{B^2r^2 + 2cD} \left| \begin{array}{cc} B\dot{B}r^2 - 2D(2c - r\dot{c})r^{-1} & i(c\dot{B} + B(2c - r\dot{c})r^{-1}) \\ 2ir^2(D\dot{B} - B\dot{D}) & B\dot{B}r^2 + 2c\dot{D} \end{array} \right| . \end{gathered} \tag{12.38}$$

Thus, we have two subsystems:

$$\begin{gathered} \ddot{f}_1 - \frac{1}{2}(B^2r^2 + 2cD)f_1 - \frac{2(\alpha\nu B - 2Dr^3)}{r^2(B^2r^2 + 2cD)}\dot{f}_1 + \frac{4(\alpha\nu c + Br^5)}{r^4(B^2r^2 + 2cD)}\dot{f}_2 = 0 , \\ \ddot{f}_2 - \frac{1}{2}(B^2r^2 + 2cD)f_2 + \frac{r^2(B\dot{D} - D\dot{B})}{B^2r^2 + 2cD}\dot{f}_1 - \frac{2\alpha(\nu Br^3 + 2c(\alpha + \epsilon r))}{r^5(B^2r^2 + 2cD)}\dot{f}_2 = 0 , \end{gathered} \tag{12.39}$$

and

$$\begin{gathered} \ddot{f}_3 - \frac{1}{2}(B^2r^2 + 2cD)f_3 - \frac{(B\dot{B}r^3 + 8D\nu^2)}{r(B^2r^2 + 2cD)}\dot{f}_3 - \frac{i(c\dot{B}r - 4B\nu^2)}{r(B^2r^2 + 2cD)}\dot{f}_4 = 0, \\ \ddot{f}_4 - \frac{1}{2}(B^2r^2 + 2cD)f_4 - \frac{2ir^2(\dot{B}D - B\dot{D})}{B^2r^2 + 2cD}\dot{f}_3 - \frac{(B\dot{B}r^2 + 2c\dot{D})}{B^2r^2 + 2cD}\dot{f}_4 = 0. \end{gathered} \tag{12.40}$$

Then, taking into account expressions for the coefficients $D(r), B(r), c(r)$, we derive the following equations:

$$\begin{aligned}
&\ddot{f}_1 + \frac{(-2\nu^2 - r^2 + (\alpha + r\epsilon)^2)}{r^2} f_1 \\
&- \frac{2\left(r^2\left((\alpha + r\epsilon)^2 - r^2\right) + \alpha\nu^2(\alpha + r\epsilon)\right)}{r^3\left(-2\nu^2 - r^2 + (\alpha + r\epsilon)^2\right)} \dot{f}_1 \\
&+ \frac{2\nu\left(2\alpha\nu^2 + 2r^3\epsilon + 3\alpha r^2\right)}{r^2\left(-2\nu^2 - r^2 + (\alpha + r\epsilon)^2\right)} \dot{f}_2 = 0, \\
&\ddot{f}_2 + \frac{(-2\nu^2 - r^2 + (\alpha + r\epsilon)^2)}{r^2} f_2 \\
&+ \frac{2\alpha\left(\nu^2 + r^2\right)(\alpha + r\epsilon)}{r^3\left(-2\nu^2 - r^2 + (\alpha + r\epsilon)^2\right)} \dot{f}_2 \\
&- \frac{\alpha\nu\left(r^2 + (\alpha + r\epsilon)^2\right)}{r^4\left(-2\nu^2 - r^2 + (\alpha + r\epsilon)^2\right)} \dot{f}_1 = 0,
\end{aligned} \tag{12.41}$$

or shorter,

$$\ddot{f}_1 + \alpha_{11} f_1 + \beta_{11} \dot{f}_1 + \gamma_{12} \dot{f}_2 = 0, \quad \ddot{f}_2 + \alpha_{22} f_2 + \beta_{22} \dot{f}_2 + \gamma_{21} \dot{f}_1 = 0; \tag{12.42}$$

also

$$\begin{aligned}
&\ddot{f}_3 + \frac{(-2\nu^2 - r^2 + (\alpha + r\epsilon)^2)}{r^2} f_3 \\
&- \frac{2\nu^2\left(\alpha^2 + 2r^2 + \alpha r\epsilon\right)}{r^3\left(-2\nu^2 - r^2 + (\alpha + r\epsilon)^2\right)} \dot{f}_3 - \frac{i\nu\left(2\alpha\nu^2 + 2r^3\epsilon + 3\alpha r^2\right)}{r^2\left(-2\nu^2 - r^2 + (\alpha + r\epsilon)^2\right)} \dot{f}_4 = 0, \\
&\ddot{f}_4 + \frac{(-2\nu^2 - r^2 + (\alpha + r\epsilon)^2)}{r^2} f_4 \\
&+ \frac{2\left(r^3\left(\alpha\epsilon + r\left(\epsilon^2 + 3\right)\right)\right) - \left(\nu^2 + 2r^2\right)\left(-\alpha^2 + 2r^2 - \alpha r\epsilon\right)\right)}{r^3\left(-2\nu^2 - r^2 + (\alpha + r\epsilon)^2\right)} \dot{f}_4 \\
&- \frac{2i\alpha\nu\left(r^2 + (\alpha + r\epsilon)^2\right)}{r^4\left(-2\nu^2 - r^2 + (\alpha + r\epsilon)^2\right)} \dot{f}_3 = 0,
\end{aligned} \tag{12.43}$$

or shorter,

$$\ddot{f}_3 + \alpha_{33} f_3 + \beta_{33} \dot{f}_3 + \gamma_{34} \dot{f}_4 = 0, \quad \ddot{f}_4 + \alpha_{44} f_4 + \beta_{44} \dot{f}_4 + \gamma_{43} \dot{f}_3 = 0. \tag{12.44}$$

The first KCC-invariant for the z_1-system is:

$$\begin{aligned}
E^1 &= \frac{1}{2}\left(B^2 r^2 + 2cD\right) f_1 + \frac{\left(\alpha\nu B - 2Dr^3\right)}{r^2(B^2 r^2 + 2cD)} \dot{f}_1 - \frac{2\left(\alpha\nu c + Br^5\right)}{r^4(B^2 r^2 + 2cD)} \dot{f}_2, \\
E^2 &= \frac{1}{2}\left(B^2 r^2 + 2cD\right) f_2 - \frac{r^2\left(B\dot{D} - D\dot{B}\right)}{2(B^2 r^2 + 2cD)} \dot{f}_1 + \frac{\alpha\left(\nu Br^3 + 2c(\alpha + \epsilon r)\right)}{r^5(B^2 r^2 + 2cD)} \dot{f}_2.
\end{aligned} \tag{12.45}$$

The second KCC-invariant for the z_1-system is determined by the formulas:

$$P_1{}^1 = \frac{1}{4\left(\left(B^2 r^2 + 2cD\right)^2\right.}$$

$$\times\{-2B^6r^6-4B^4r^2(3cDr^2+2)+2D(c(3r^2\dot{B}^2-8D)-8c^3D^2-24Dr^2)$$
$$+B^2(r(-3r^3\dot{B}^2-12r(r\dot{D}-4D)-8Dr)-24c^2D^2r^2+4c(2D-3r\dot{D}))$$
$$+2Br(2cDr\ddot{B}+\dot{B}(c(14D-3r\dot{D})+18Dr^2))+2B^3r^3(r\ddot{B}-2\dot{B})\},$$

$$P_1{}^2=\frac{1}{(B^2r^3+2cDr)^2}$$

$$\times\{B^3r^2(10c+10r^2)+B(cr^2(3r^2\dot{B}^2+6r\dot{D}-4D)+c^2(6r\dot{D}+8D)-12Dr^4)$$
$$+cr(\dot{B}(c(3r\dot{D}-2D)-6Dr^2)-2cDr\ddot{B})-B^2r^3(cr\ddot{B}+\dot{B}(-8c-6r^2))\},$$

$$P_2{}^1=\frac{1}{2(B^2r^2+2cD)^2}r$$

$$\times\{B^3r^2(2\dot{D}-r\ddot{D})+B(3cr\dot{D}^2-D(3r^3\dot{B}^2+2cr\ddot{D}+(2c+6r^2)\dot{D}))$$
$$+D(2cDr\ddot{B}+\dot{B}(c(2D-3r\dot{D})+6Dr^2))+B^2r^2(Dr\ddot{B}+\dot{B}(3r\dot{D}-2D))\},$$

$$P_2{}^2=-\frac{1}{4(B^2r^3+2cDr)^2}$$

$$\times\{2B^6r^8+4B^4r^4(3cDr^2+2)+2c(-3Dr^4\dot{B}^2+8c^2D^3r^2$$
$$+2c(8D^2+3r^2\dot{D}^2-2Dr(r\ddot{D}-2\dot{D})))$$
$$+B^2r^2(3r^4\dot{B}^2+24c^2D^2r^2+4c(-r^2\ddot{D}+5r\dot{D}+8D)+12r^3\dot{D})$$
$$+2Br^3(\dot{B}(c(9r\dot{D}-2D)-6Dr^2)-2cDr\ddot{B})-2B^3r^5(r\ddot{B}-2\dot{B})\}.$$

12.5 Natural Splitting, Real-Valued Representation

The equation (12.1) in natural 2+2 complex splitting may be written in real-valued form, in which there exist four linear subsystems:

$$\begin{gathered}\dot{x}^1=Br^2x^6+cx^7\,,\quad \dot{x}^3=-Dr^2x^6+\frac{Br^2}{2}x^7\,,\\ \dot{x}^2=-Br^2x^5+cx^8\,,\quad \dot{x}^4=Dr^2x^5+\frac{Br^2}{2}x^8\,,\\ \dot{x}^5=-\frac{B}{2}x^2+\frac{c}{r^2}x^4\,,\quad \dot{x}^8=Dx^2+Bx^4\,,\\ \dot{x}^6=\frac{B}{2}x^1-\frac{c}{r^2}x^3\,,\quad \dot{x}^7=Dx^1+Bx^3\,.\end{gathered}\tag{12.46}$$

Their solutions are divided into two groups:

$$\begin{gathered}x^6=\frac{2}{r^2(B^2r^2+2cD)}(\frac{Br^2}{2}\dot{x}^1-c\dot{x}^3)\,,\quad x^7=\frac{2}{r^2(B^2r^2+2cD)}(Dr^2\dot{x}^1+Br^2\dot{x}^3)\,,\\ x^1=+\frac{2r^2}{(B^2r^2+2cD)}(B\dot{x}^6+\frac{c}{r^2}\dot{x}^7)\,,\quad x^3=+\frac{2r^2}{(B^2r^2+2cD)}(-D\dot{x}^6+\frac{B}{2}\dot{x}^7)\,,\end{gathered}\tag{12.47}$$

and

$$x^5 = -\frac{2}{r^2(B^2r^2+2cD)}(\frac{Br^2}{2}\dot{x}^2 - c\dot{x}^4),\ x^8 = -\frac{2}{r^2(B^2r^2+2cD)}t(-Dr^2\dot{x}^2 - Br^2\dot{x}^4),$$
$$x^2 = \frac{2r^2}{(B^2r^2+2cD)}(B\dot{x}^5 - \frac{c}{r^2}\dot{x}^8),\ x^4 = \frac{2r^2}{(B^2r^2+2cD)}(-D\dot{x}^5 - \frac{B}{2}\dot{x}^8). \tag{12.48}$$

Let us exclude the variables x^6, x^7 in the first group (12.47):

$$\dot{x}^6 = \frac{d}{dr}\Big[\frac{2}{r^2(B^2r^2+2cD)}\Big(\frac{Br^2}{2}\frac{d}{dr}x^1 - c\frac{d}{dr}x^3\Big)\Big],$$
$$\dot{x}^7 = \frac{d}{dr}\Big[\frac{2}{r^2(B^2r^2+2cD)}\Big(Dr^2\frac{d}{dr}x^1 + Br^2\frac{d}{dr}x^3\Big)\Big], \tag{12.49}$$

so from the remaining two equations, we derive the 2-nd order system for the functions $x_1(r), x_3(r)$:

$$x^1 = +\frac{2r^2}{(B^2r^2+2cD)}\Big\{B\frac{d}{dr}\Big[\frac{2}{r^2(B^2r^2+2cD)}\Big(\frac{Br^2}{2}\frac{d}{dr}x^1 - c\frac{d}{dr}x^3\Big)\Big]$$
$$+\frac{c}{r^2}\frac{d}{dr}\Big[\frac{2}{r^2(B^2r^2+2cD)}\Big(Dr^2\frac{d}{dr}x^1 + Br^2\frac{d}{dr}x^3\Big)\Big]\Big\},$$
$$x^3 = +\frac{2r^2}{(B^2r^2+cD)}\Big\{-D\frac{d}{dr}\Big[\frac{2}{r^2(B^2r^2+2cD)}\Big(\frac{Br^2}{2}\frac{d}{dr}x^1 - c\frac{d}{dr}x^3\Big)\Big]$$
$$+\frac{B}{2}\frac{d}{dr}\Big[\frac{2}{r^2(B^2r^2+2cD)}\Big(Dr^2\frac{d}{dr}x^1 + Br^2\frac{d}{dr}x^3\Big)\Big]\Big\}. \tag{12.50}$$

Let us exclude the variables x^5, x^8 in the second group (12.48):

$$x^5 = \frac{d}{dr}\Big[-\frac{2}{r^2(B^2r^2+2cD)}\Big(\frac{Br^2}{2}\frac{d}{dr}x^2 - c\frac{d}{dr}x^4\Big)\Big],$$
$$\dot{x}^8 = \frac{d}{dr}\ \Big[-\frac{2}{r^2(B^2r^2+2cD)}\Big(-Dr^2\frac{d}{dr}x^2 - Br^2\frac{d}{dr}x^4\Big)\Big]; \tag{12.51}$$

so, by deriving the 2-nd order system for the functions $x_2(r), x_4(r)$, we infer:

$$x^2 = \frac{2r^2}{(B^2r^2+2cD)}\Big\{B\frac{d}{dr}\Big[-\frac{2}{r^2(B^2r^2+2cD)}\Big(\frac{Br^2}{2}\frac{d}{dr}x^2 - c\frac{d}{dr}x^4\Big)\Big]$$
$$-\frac{c}{r^2}\frac{d}{dr}\Big[\frac{2}{r^2(B^2r^2+2cD)}\Big(-Dr^2\frac{d}{dr}x^2 - Br^2\frac{d}{dr}x^4\Big)\Big]\Big\},$$
$$x^4 = \frac{2r^2}{(B^2r^2+2cD)}\Big\{-D\frac{d}{dr}\Big[-\frac{2}{r^2(B^2r^2+2cD)}\Big(\frac{Br^2}{2}\frac{d}{dr}x^2 - c\frac{d}{dr}x^4\Big)\Big]$$
$$-\frac{B}{2}\frac{d}{dr}\Big[-\frac{2}{r^2(B^2r^2+2cD)}\Big(-Dr^2\frac{d}{dr}x^2 - Br^2\frac{d}{dr}x^4\Big)\Big]\Big\}. \tag{12.52}$$

One might exclude from the first group the variables x_1, x_3, and derive the equations for x_6, x_7. Similarly, one might exclude from the second group the variables x_2, x_4 and derive the equations for x_5, x_8.

Let us consider the multiplier entering Eqs. (12.51) and (12.52):

$$F(r) = \frac{1}{2cD + B^2 r^2} = \frac{1}{2} \cdot \frac{r^2}{r^2 - (\epsilon r + \alpha)^2 + j(j+1)} . \quad (12.53)$$

The roots of the denominator are

$$r^2 - (\epsilon r + \alpha)^2 + j(j+1) = 0, \quad r_{1,2} = \frac{\epsilon\alpha}{1-\epsilon^2}\left[1 \pm \sqrt{1 - \frac{1-\epsilon^2}{\alpha^2\epsilon^2}[j(j+1) - \alpha^2]}\,\right]; \quad (12.54)$$

they provide us with additional singular points of the differential equations. These points are real-valued if the next inequality holds

$$1 - \epsilon^2 < \frac{\alpha^2}{j(j+1)}, \qquad \alpha = \frac{1}{137}, \quad (12.55)$$

which can be considered as a restriction for possible energy values relevant for bound states.

With the use of the notation

$$\frac{1}{2cD + B^2 r^2} = \frac{1}{2} \cdot \frac{r^2}{r^2 - (\epsilon r + \alpha)^2 + j(j+1)} = \frac{r^2}{2P}, 2cD + B^2 r^2 = \frac{2P}{r^2}, \quad (12.56)$$

the above equations may be written in a shorter form.

12.6 Projections of 8 Equations on Different Planes

Let us turn to initial system of 4 equations

$$\frac{d}{dr} f^1 = a f^3 + c f^4, \quad \frac{d}{dr} f^2 = d f^3 + b f^4,$$
$$\frac{d}{dr} f^3 = A f^1 + C f^2, \quad \frac{d}{dr} f^4 = D f^1 + B f^2, \quad (12.57)$$

where the functions $f_i(r)$ are decomposed into the sum of real and imaginary parts:

$$f^1 = x^1 + i x^2, \quad f^2 = x^3 + i x^4, \quad f^3 = x^5 + i x^6, \quad f^4 = x^7 + i x^8, \quad (12.58)$$

In 8-dimensional space let us examine 6-dimensional projection $x_1 = 0, x_2 = 0$:

$$0 = B r^2 x^6 + c x^7, \quad 0 = -B r^2 x^5 + c x^8,$$
$$\dot{x}^3 = -D r^2 x^6 + \frac{B r^2}{2} x^7, \quad \dot{x}^4 = D r^2 x^5 + \frac{B r^2}{2} x^8, \quad (12.59)$$

and

$$\dot{x}^5 = \frac{c}{r^2} x^4, \quad \dot{x}^6 = -\frac{c}{r^2} x^3, \quad \dot{x}^7 = B x^3, \quad \dot{x}^8 = B x^4 . \quad (12.60)$$

The system of 4 equations (12.59) may be considered as linear under the variables $x^5, x^6, x^7 x^8$; its solution is expressed in the terms of $\dot{x}^3, \dot{x}^4$ as follows

$$
\begin{aligned}
x^5 &= \Big(Dr^2 + \frac{Br^2}{2}\frac{Br^2}{c}\Big)^{-1}\dot{x}^4, \quad x^6 = \Big(-Dr^2 - \frac{Br^2}{2}\frac{Br^2}{c}\Big)^{-1}\dot{x}^3, \\
x^7 &= -\frac{Br^2}{c}\Big(-Dr^2 - \frac{Br^2}{2}\frac{Br^2}{c}\Big)^{-1}\dot{x}^3 \qquad x^8 = \frac{Br^2}{c}\Big(Dr^2 + \frac{Br^2}{2}\frac{Br^2}{c}\Big)^{-1}\dot{x}^4 .
\end{aligned} \tag{12.61}
$$

By substituting (12.61) into the two remaining equations (12.60), we obtain 2 equations the variable x^4:

$$
\begin{aligned}
\dot{x}^5 = \frac{c}{r^2}x^4 \quad &\Longrightarrow \quad \frac{d}{dr}\Big(Dr^2 + \frac{Br^2}{2}\frac{Br^2}{c}\Big)^{-1}\frac{d}{dr}x^4 = \frac{c}{r^2}x^4 , \\
\dot{x}^8 = Bx^4 \quad &\Longrightarrow \quad \frac{d}{dr}\frac{Br^2}{c}\Big(Dr^2 + \frac{Br^2}{2}\frac{Br^2}{c}\Big)^{-1}\frac{d}{dr}x^4 = Bx^4 ,
\end{aligned} \tag{12.62}
$$

and 2 equations the variable x^3:

$$
\begin{aligned}
\dot{x}^7 = Bx^3 \quad &\Longrightarrow \quad \frac{d}{dr}\frac{Br^2}{c}\Big(-Dr^2 - \frac{Br^2}{2}\frac{Br^2}{c}\Big)^{-1}\frac{d}{dr}x^3 = -Bx^3 , \\
\dot{x}^6 = -\frac{c}{r^2}x^3 \quad &\Longrightarrow \quad \frac{d}{dr}\Big(-Dr^2 - \frac{Br^2}{2}\frac{Br^2}{c}\Big)^{-1}\frac{d}{dr}x^3 = -\frac{c}{r^2}x^3 .
\end{aligned} \tag{12.63}
$$

Projection $x_3 = 0, x_4 = 0$:

$$
\begin{aligned}
\dot{x}^1 = Br^2 x^6 + cx^7 , \quad \dot{x}^2 = -Br^2 x^5 + cx^8 , \\
0 = -Dr^2 x^6 + \frac{Br^2}{2}x^7 , \quad 0 = Dr^2 x^5 + \frac{Br^2}{2}x^8 ,
\end{aligned} \tag{12.64}
$$

and

$$
\dot{x}^5 = -\frac{B}{2}x^2 , \quad \dot{x}^6 = \frac{B}{2}x^1 , \quad \dot{x}^7 = Dx^1 , \quad \dot{x}^8 = Dx^2 . \tag{12.65}
$$

The system of 4 equations (12.64) may be considered as linear under the variables $x^5, x^6, x^7 x^8$; its solution is expressed in the terms of $\dot{x}^1, \dot{x}^2$:

$$
\begin{aligned}
x^6 &= \Big(Br^2 + c\frac{2D}{B}\Big)^{-1}\dot{x}^1 , \quad x^7 = \frac{2D}{B}x^6 = +\frac{2D}{B}\Big(Br^2 + c\frac{2D}{B}\Big)^{-1}\dot{x}^1 , \\
x^5 &= \Big(-Br^2 - c\frac{2D}{B}\Big)^{-1}\dot{x}^2 , \qquad x_8 = -\frac{2D}{B}x^5 = -\frac{2D}{B}\Big(-Br^2 - c\frac{2D}{B}\Big)^{-1}\dot{x}^2 .
\end{aligned} \tag{12.66}
$$

By substituting (6.667.8) into the two remaining equations (12.65), we obtain 4 equations for the variables x^2, x^1:

$$
\begin{aligned}
\dot{x}^5 = -\frac{B}{2}x^2 \quad &\Longrightarrow \quad \frac{d}{dr}\Big(-Br^2 - c\frac{2D}{B}\Big)^{-1}\frac{d}{dr}x^2 = -\frac{B}{2}x^2 , \\
\dot{x}^8 = Dx^2 \quad &\Longrightarrow \quad \frac{d}{dr}\frac{2D}{B}x^5 = -\frac{d}{dr}\frac{2D}{B}\Big(-Br^2 - c\frac{2D}{B}\Big)^{-1}\frac{d}{dr}x^2 - Dx^2 ,
\end{aligned} \tag{12.67}
$$

$$\dot{x}^7 = Dx^1 \quad \Longrightarrow \quad \frac{d}{dr}\frac{2D}{B}x^6 = +\frac{d}{dr}\frac{2D}{B}\Big(Br^2 + c\frac{2D}{B}\Big)^{-1}\frac{d}{dr}x^1 = Dx^1 ,$$
$$\dot{x}^6 = \frac{B}{2}x^1 \quad \Longrightarrow \quad \frac{d}{dr}\Big(Br^2 + c\frac{2D}{B}\Big)^{-1}\frac{d}{dr}x^1 = \frac{B}{2}x^1 . \qquad (12.68)$$

Projection $x_5 = 0, x_6 = 0$:

$$\dot{x}^1 = cx^7 , \quad \dot{x}^2 = cx^8 , \quad \dot{x}^3 = \frac{Br^2}{2}x^7 , \quad \dot{x}^4 = \frac{Br^2}{2}x^8 , \qquad (12.69)$$

$$0 = -\frac{B}{2}x^2 + \frac{c}{r^2}x^4 , \quad 0 = \frac{B}{2}x^1 - \frac{c}{r^2}x^3 , \dot{x}^7 = Dx^1 + Bx^3 , \quad \dot{x}^8 = Dx^2 + Bx^4 . \qquad (12.70)$$

The system of 4 equations (12.70) may be considered as linear under the variables x^1, x^2, x^3x^4; its solution is expressed in terms of $\dot{x}^7, \dot{x}^8$ is

$$x^1 = \Big(D + B\frac{Br^2}{2c}\Big)^{-1}\dot{x}^7 , \quad x^2 = \Big(D + B\frac{Br^2}{2c}\Big)^{-1}\dot{x}^8 ,$$
$$x^3 = \frac{Br^2}{2c}x^1 = \frac{Br^2}{2c}\Big(D + B\frac{Br^2}{2c}\Big)^{-1}\dot{x}^7 , \quad x^4 = \frac{Br^2}{2c}x^2 = \frac{Br^2}{2c}\Big(D + B\frac{Br^2}{2c}\Big)^{-1}\dot{x}^8 . \qquad (12.71)$$

By substituting (12.71) into remaining equations (12.69), we obtain 4 equations for the variables x^7, x^8:

$$\dot{x}^1 = cx^7 \quad \Longrightarrow \quad \frac{d}{dr}\Big(D + B\frac{Br^2}{2c}\Big)^{-1}\frac{d}{dr}x^7 = cx^7 ,$$
$$\dot{x}^3 = \frac{Br^2}{2}x^7 \quad \Longrightarrow \quad \frac{d}{dr}\frac{Br^2}{2c}x^1 = \frac{Br^2}{2c}\Big(D + B\frac{Br^2}{2c}\Big)^{-1}\frac{d}{dr}x^7 = \frac{Br^2}{2}x^7 , \qquad (12.72)$$

$$\dot{x}^2 = cx^8 \quad \Longrightarrow \quad \frac{d}{dr}\Big(D + B\frac{Br^2}{2c}\Big)^{-1}\frac{d}{dr}x^8 = cx^8 ,$$
$$\dot{x}^4 = \frac{Br^2}{2}x^8 \quad \Longrightarrow \quad \frac{d}{dr}\frac{Br^2}{2c}x^2 = \frac{Br^2}{2c}\Big(D + B\frac{Br^2}{2c}\;\Big)^{-1}\frac{d}{dr}x^8 = \frac{Br^2}{2}x^8 . \qquad (12.73)$$

Projection $x_7 = 0, x_8 = 0$:

$$\dot{x}^1 = Br^2x^6 , \quad \dot{x}^2 = -Br^2x^5 , \quad \dot{x}^3 = -Dr^2x^6 , \quad \dot{x}^4 = Dr^2x^5 , \qquad (12.74)$$

and

$$\dot{x}^5 = -\frac{B}{2}x^2 + \frac{c}{r^2}x^4 , \quad \dot{x}^6 = \frac{B}{2}x^1 - \frac{c}{r^2}x^3 , 0 = Dx^1 + Bx^3 , \quad 0 = Dx^2 + Bx^4 . \qquad (12.75)$$

The second system (12.75) is linear under the variables x^1, x^2, x^3x^4; its solution in terms of $\dot{x}^5, \dot{x}^6$ is

$$x^3 = -\frac{D}{B}x^1 , \quad x^4 = -\frac{D}{B}x^2 ,$$
$$x^1 = \Big(\frac{B}{2} + \frac{c}{r^2}\frac{D}{B}\Big)^{-1}\dot{x}^6 , \quad x^2 = \Big(-\frac{B}{2} - \frac{c}{r^2}\frac{D}{B}\Big)^{-1}\dot{x}^5 , \qquad (12.76)$$
$$x^3 = -\frac{D}{B}\Big(\frac{B}{2} + \frac{c}{r^2}\frac{D}{B}\Big)^{-1}\dot{x}^6 , \quad x^4 = -\frac{D}{B}\Big(-\frac{B}{2} - \frac{c}{r^2}\frac{D}{B}\Big)^{-1}\dot{x}^5 .$$

By substituting (12.76) into (12.74), we obtain 4 equations for the variables x^6, x^5:

$$\dot{x}^1 = Br^2 x^6 \quad \Longrightarrow \quad \frac{d}{dr}\left(\frac{B}{2}+\frac{c}{r^2}\frac{D}{B}\right)^{-1}\frac{d}{dr}x^6 = Br^2 x^6 ,$$
$$\dot{x}^3 = -Dr^2 x^6 \quad \Longrightarrow \quad \frac{d}{dr}\frac{D}{B}\left(\frac{B}{2}+\frac{c}{r^2}\frac{D}{B}\right)^{-1}\frac{d}{dr}x^6 = Dr^2 x^6 , \qquad (12.77)$$

$$\dot{x}^2 = -Br^2 x^5 \quad \Longrightarrow \quad \frac{d}{dr}\left(-\frac{B}{2}-\frac{c}{r^2}\frac{D}{B}\right)^{-1}\frac{d}{dr}x^5 = -Br^2 x^5 ,$$
$$\dot{x}^4 = Dr^2 x^5 \quad \Longrightarrow \quad \frac{d}{dr}\frac{D}{B}\left(-\frac{B}{2}-\frac{c}{r^2}\frac{D}{B}\right)^{-1}\frac{d}{dr}x^5 = -Dr^2 x^5 . \qquad (12.78)$$

We note that in all the 2-nd order equations, there appears the same multiplier,

$$F(r) = \frac{1}{2cD+B^2r^2} ,$$

which, considering the relations

$$c = -(2\nu^2+r^2), \quad D = \frac{(\epsilon r+\alpha)^2 - r^2}{r^4}, \quad B = -\frac{2\nu(\epsilon r+\alpha)}{r^3},$$

reads

$$F(r) = \frac{1}{2cD+B^2r^2} = \frac{1}{2}\cdot\frac{r^2}{r^2-(\epsilon r+\alpha)^2+2\nu^2} . \qquad (12.79)$$

As noted above, the roots of the denominator are

$$r^2-(\epsilon r+\alpha)^2+j(j+1)=0, \quad r_{1,2} = \frac{\epsilon\alpha}{1-\epsilon^2}\left[1\pm\sqrt{1-\frac{1-\epsilon^2}{\alpha^2\epsilon^2}[j(j+1)-\alpha^2]}\,\right], \qquad (12.80)$$

and they give two singular points of the arising equations. These roots are real-valued only if

$$1-\epsilon^2 < \frac{\alpha^2}{j(j+1)}, \quad \alpha = \frac{1}{137} . \qquad (12.81)$$

Chapter 13

The Study of a Spin 1 Particle with Anomalous Magnetic Moment in the Coulomb Field

13.1 Separation of the Variables

It is known that, in the framework of the theory of relativistic wave equations, one can consider the so-called non-minimal equations, which describe particles with additional electromagnetic characteristics. In particular, the equations for spin $S = 1/2$ and $S = 1$ particles with both electric charge and anomalous magnetic moment were studied extensively [283]–[334]. In the present chapter we study the vector particle with anomalous magnetic moment in the external Coulomb field. In the case of external Coulomb field, the equation for the vector particle is too complicated, even for the case of an ordinary particle without anomalous moment. This problem has not been solved completely yet. However, in the nonrelativistic limit, the equation for an ordinary vector particle in the Coulomb field can be exactly solved. For this reason, in the present chapter we investigate the nonrelativistic problem for the particle with anomalous magnetic moment.

The initial equation has the following form (we further use the tetrad formalism)

$$\Big\{ i\beta^c \Big[i(e^{\beta}_{(c)}\partial_\beta + \frac{1}{2} j^{ab}\gamma_{abc}(x)) - e' A_c \Big] + \lambda \frac{e}{M} F_{\alpha\beta}(x) P j^{\alpha\beta}(x) - M \Big\} \Psi = 0 , \qquad (13.1)$$

where the free parameter λ is dimensionless, P is the projective operator which selects the vector component from the 10-components function

$$P = \left| \begin{array}{cc} I_4 & 0 \\ 0 & 0 \end{array} \right| , \quad M = \frac{mc}{\hbar} , \quad e' = \frac{e}{c\hbar} , \quad \Gamma = \lambda \frac{4\alpha}{M} , \quad \alpha = \frac{e^2}{\hbar c} = \frac{1}{137} . \qquad (13.2)$$

In the spherical tetrad, the equation (13.1) takes the form

$$\Big[\beta^0 (i\partial_t + \frac{\alpha}{r}) + i(\beta^3 \partial_r + \frac{1}{r}(\beta^1 j^{31} + \beta^2 j^{32})) + \frac{1}{r}\Sigma_{\theta,\phi} + \frac{\Gamma}{r^2} P j^{03} - M \Big] \Phi = 0 , \quad (13.3)$$

where the angular operator is determined by the equality:

$$\Sigma_{\theta,\phi} = i\,\beta^1 \partial_\theta + \beta^2 \frac{i\partial_\phi + i j^{12} \cos\theta}{\sin\theta} . \tag{13.4}$$

Relative to the used basis, the explicit form of the total angular momentum operator components is given by the formulas

$$j_1 = l_1 + \frac{\cos\phi}{\sin\theta} i j^{12}, \quad j_2 = l_2 + \frac{\sin\phi}{\sin\theta} i j^{12}, \quad j_3 = l_3, \quad j^{12} = \beta^1\beta^2 - \beta^2\beta^1 .$$

We shall further use the wave function and the Duffin – Kemmer matrices in cyclic representation; the matrix $i j^{12}$ has the diagonal structure

$$i j^{12} = \left| \begin{array}{cccc} 0 & 0 & 0 & 0 \\ 0 & t_3 & 0 & 0 \\ 0 & 0 & t_3 & 0 \\ 0 & 0 & 0 & t_3 \end{array} \right| , \quad t_3 = \left| \begin{array}{ccc} +1 & 0 & 0 \\ 0 & 0 & 0 \\ 0 & 0 & -1 \end{array} \right| .$$

The form of the projective operator P does not change under transition from Cartesian basis to the cyclic one.

The system of radial equations for an ordinary vector particle in the Coulomb field is well-known [334]. To obtain the generalized system for the vector particle with anomalous magnetic moment, it suffices to specify the additional term in the equation:

$$\frac{\Gamma}{r^2} P j^{03} = \frac{\Gamma}{r^2} P(\beta^0\beta^3 - \beta^3\beta^0) .P\, j^{03} = \left| \begin{array}{cccccccccc} 0 & 0 & -1 & 0 & 0 & 0 & 0 & 0 & 0 & 0 \\ 0 & 0 & 0 & 0 & 0 & 0 & 0 & 0 & 0 & 0 \\ -1 & 0 & 0 & 0 & 0 & 0 & 0 & 0 & 0 & 0 \\ 0 & 0 & 0 & 0 & 0 & 0 & 0 & 0 & 0 & 0 \\ 0 & 0 & 0 & 0 & 0 & 0 & 0 & 0 & 0 & 0 \\ 0 & 0 & 0 & 0 & 0 & 0 & 0 & 0 & 0 & 0 \\ 0 & 0 & 0 & 0 & 0 & 0 & 0 & 0 & 0 & 0 \\ 0 & 0 & 0 & 0 & 0 & 0 & 0 & 0 & 0 & 0 \\ 0 & 0 & 0 & 0 & 0 & 0 & 0 & 0 & 0 & 0 \\ 0 & 0 & 0 & 0 & 0 & 0 & 0 & 0 & 0 & 0 \end{array} \right| . \tag{13.5}$$

The structure of the 10-components wave function for the vector particle with quantum numbers ϵ, j, m is the following

$$\Psi(x) = \{f_0(x), \vec{f}(x), \vec{E}(x), \vec{H}(x)\}, \quad \Phi_0(x) = e^{-i\epsilon t} f_0(r) D_0 ,$$

$$\vec{\Phi}(x) = e^{-i\epsilon t} \left| \begin{array}{c} f_1(r) D_{-1} \\ f_2(r) D_0 \\ f_3(r) D_{+1} \end{array} \right| , \vec{E}(x) = e^{-i\epsilon t} \left| \begin{array}{c} E_1(r) D_{-1} \\ E_2(r) D_0 \\ E_3(r) D_{+1} \end{array} \right| , \vec{H}(x) = e^{-i\epsilon t} \left| \begin{array}{c} H_1(r) D_{-1} \\ H_2(r) D_0 \\ H_3(r) D_{+1} \end{array} \right| , \tag{13.6}$$

where D stands for the Wigner functions: $D_\sigma = D^j_{-m,\sigma}(\phi, \theta, 0)$, $\sigma = 0, -1, +1$.

After performing the needed calculations, one finds the system of radial equations

$$\begin{gathered}
(\frac{d}{dr}+\frac{2}{r})\,E_2-\frac{\nu}{r}\,(E_1+E_3)-\frac{\Gamma}{r^2}f_2=m\,f_0\,,\\
+i\,(\epsilon+\frac{\alpha}{r})\,E_1\;\;+i\,(\frac{d}{dr}+\frac{1}{r})\,H_1+i\frac{\nu}{r}\,H_2=mf_1\,,\\
+i\,(\epsilon+\frac{\alpha}{r})\,E_2-i\frac{\nu}{r}\,(H_1-H_3)-\frac{\Gamma}{r^2}f_0=m\,f_2\,,\\
+i\,(\epsilon+\frac{\alpha}{r})\,E_3\;\;-i\,(\frac{d}{dr}+\frac{1}{r})\,H_3-i\frac{\nu}{r}\,H_2=m\,f_3\,;\\
-i\,(\epsilon+\frac{\alpha}{r})f_1+\frac{\nu}{r}f_0=mE_1,-i(\epsilon+\frac{\alpha}{r})f_2-\frac{d}{dr}f_0=mE_2\,,\\
-i(\epsilon+\frac{\alpha}{r})f_3+\frac{\nu}{r}f_0=mE_3\,,-i\,(\frac{d}{dr}+\frac{1}{r})f_1-i\frac{\nu}{r}f_2=mH_1\,,\\
+i\frac{\nu}{r}(f_1-f_3)=mH_2\,,+i(\frac{d}{dr}+\frac{1}{r})f_3+i\frac{\nu}{r}f_2=mH_3\,.
\end{gathered} \tag{13.7}$$

We additionally diagonalize the spatial inversion operator $\hat{\Pi}$. In the Cartesian basis β^a, this operator takes the form

$$\hat{\Pi}=\left|\begin{array}{cccc}1&0&0&0\\0&-I&0&0\\0&0&-I&0\\0&0&0&+I\end{array}\right|\hat{P},\;\;\hat{P}\Psi(\vec{r})=\Psi(-\vec{r})\,. \tag{13.8}$$

After transition to the spherical basis, we obtain an alternative representation of this operator

$$\hat{\Pi}'=\left|\begin{array}{cccc}1&0&0&0\\0&\Pi_3&0&0\\0&0&\Pi_3&0\\0&0&0&-\Pi_3\end{array}\right|\hat{P},\;\Pi_3=\left|\begin{array}{ccc}0&0&-1\\0&-1&0\\-1&0&0\end{array}\right|. \tag{13.9}$$

The eigenvalue equation $\hat{\Pi}'\Psi=P\Psi$ has two types of solutions:

$$\begin{gathered}
P=(-1)^{j+1},\quad f_0=0\,,\;f_3=-f_1\,,\;f_2=0\,,\;E_3=-E_1\,,\;E_2=0\,,\;H_3=H_1\,;\\
P=(-1)^{j},\quad f_3=+f_1\,,\;E_3=+E_1\,,\;H_3=-H_1\,,\;H_2=0\,.
\end{gathered} \tag{13.10}$$

For states with $P=(-1)^{j+1}$, we obtain four equations:

$$\begin{gathered}
i\Big(\epsilon+\frac{\alpha}{r}\Big)E_1+i\Big(\frac{d}{dr}+\frac{1}{r}\Big)H_1+i\frac{\nu}{r}H_2=mf_1\,,\\
-i\Big(\epsilon+\frac{\alpha}{r}\Big)f_1=mE_1\,,\;-i\Big(\frac{d}{dr}+\frac{1}{r}\Big)f_1=mH_1\,,\;2i\frac{\nu}{r}f_1=mH_2\,.
\end{gathered} \tag{13.11}$$

Here the anomalous magnetic moment does not manifest itself in any way in the external Coulomb field. The last 4-equations system allows the following exact solution for the main function f_1:

$$\Big(\frac{d^2}{dr^2}+\frac{2}{r}\frac{d}{dr}+(\epsilon+\frac{\alpha}{r})^2-\frac{j(j+1)}{r^2}\Big)f_1=0\,. \tag{13.12}$$

The same equation arises in the theory of the scalar particle, in the presence of the Coulomb field. Its exact solutions and the corresponding energy spectrum are well-known.

For states with parity $P=(-1)^j$, we have six equations:

$$
\begin{gathered}
-(\frac{d}{dr}+\frac{2}{r})E_2-2\frac{\nu}{r}E_1-\frac{\Gamma}{r^2}f_2=mf_0\,,\; i(\epsilon+\frac{\alpha}{r})E_1+i(\frac{d}{dr}+\frac{1}{r})H_1=mf_1\,,\\
+i(\epsilon+\frac{\alpha}{r})E_2-2i\frac{\nu}{r}H_1-\frac{\Gamma}{r^2}f_0=mf_2\,,\; -i(\epsilon+\frac{\alpha}{r})f_2-\frac{d}{dr}f_0=mE_2\,,\\
-i(\epsilon+\frac{\alpha}{r})f_1+\frac{\nu}{r}f_0=mE_1\,,\; i(\frac{d}{dr}+\frac{1}{r})f_1+i\frac{\nu}{r}f_2=-mH_1\,.
\end{gathered}
\tag{13.13}
$$

13.2 The Case of Minimal $j=0$

For states with minimal $j=0$, we should use other substitutions

$$
\begin{gathered}
\Phi_0(x)=e^{-i\epsilon t}\,f_0(r)\,,\quad \vec{\Phi}(x)=e^{-i\epsilon t}\left|\begin{array}{c}0\\ f_2(r)\\ 0\end{array}\right|,\\
\vec{E}(x)=e^{-i\epsilon t}\left|\begin{array}{c}0\\ E_2(r)\\ 0\end{array}\right|,\quad \vec{H}(x)=e^{-i\epsilon t}\left|\begin{array}{c}0\\ H_2(r)\\ 0\end{array}\right|.
\end{gathered}
\tag{13.14}
$$

Here we have the following four radial equations

$$
\begin{gathered}
-\Big(\frac{d}{dr}+\frac{2}{r}\Big)E_2-\frac{\Gamma}{r^2}f_2=mf_0\,,\quad +i\Big(\epsilon+\frac{\alpha}{r}\Big)E_2-\frac{\Gamma}{r^2}f_0=mf_2\,,\\
-i\Big(\epsilon+\frac{\alpha}{r}\Big)f_2-\frac{d}{dr}f_0=mE_2\,,\; H_2=0\,.
\end{gathered}
\tag{13.15}
$$

By excluding the variable E_2, we find the system for the unknown functions f_0, f_2:

$$
\begin{gathered}
i\Big((\epsilon+\frac{\alpha}{r})\frac{d}{dr}+\frac{2\epsilon}{r}+\frac{\alpha+i\Gamma m}{r^2}\Big)f_2+\Big(\frac{d^2}{dr^2}+\frac{2}{r}\frac{d}{dr}-m^2\Big)f_0=0\,,\\
f_2=i\frac{r^2}{P(r)}\Big((\epsilon+\frac{\alpha}{r})\frac{d}{dr}-\frac{i\Gamma m}{r^2}\Big)f_0\,,
\end{gathered}
$$

where $P(r)=(\epsilon^2-m^2)r^2+2\epsilon\alpha r+\alpha^2$. The function f_2 can be eliminated,

$$
\begin{gathered}
\frac{d^2f}{dx^2}+\Big(-\frac{2x}{P}+\frac{2\alpha^2x+6E\alpha x^2+4E^2x^3}{P^2}\\
+\frac{(2E^2-2E^4)x^5+(4E\alpha-6E^3\alpha)x^4+(2\alpha^2-6E^2\alpha^2)x^3-2E\alpha^3x^2}{P^3}\Big)\frac{df}{dx}\\
+(\frac{x^2}{P}+\frac{-2ixE\gamma+\gamma^2-i\alpha\gamma}{P^2}\\
+\frac{(-2iE\gamma+2iE^3\gamma)x^3+(-2i\alpha\gamma+4iE^2\alpha\gamma)x^2+2iE\alpha^2\gamma x}{P^3})f=0\,,
\end{gathered}
$$

where the following dimensionless variables x, E and γ are used:

$$x = mr\,, \quad E = \frac{\epsilon}{m}\,, \quad \gamma = m\Gamma\,, \quad \alpha = \frac{1}{137}\,. \tag{13.16}$$

We note that the obtained equation has a complicated set of singular points. In the next section we will derive its much simpler nonrelativistic analogue.

13.3 The Non-Relativistic Approximation at $j = 0$

In the system

$$\frac{1}{m}\Big(-(\frac{d}{dr} + \frac{2}{r})E_2 - \frac{\Gamma}{r^2} f_2 \Big) = f_0\,,$$
$$+i\Big(\epsilon + \frac{\alpha}{r}\Big)E_2 - \frac{\Gamma}{r^2} f_0 = m f_2\,, \quad -i\Big(\epsilon + \frac{\alpha}{r}\Big) f_2 - \frac{d}{dr} f_0 = m E_2$$

we shall first exclude the non-dynamical variable $f_0(r)$:

$$\begin{aligned} &+i(\epsilon + \frac{\alpha}{r})E_2 - \frac{\Gamma}{mr^2}\Big(-(\frac{d}{dr} + \frac{2}{r})E_2 - \frac{\Gamma}{r^2} f_2 \Big) = m f_2\,, \\ &-i(\epsilon + \frac{\alpha}{r}) f_2 - \frac{1}{m}\frac{d}{dr}\Big(-(\frac{d}{dr} + \frac{2}{r})E_2 - \frac{\Gamma}{r^2} f_2 \Big) = m E_2\,. \end{aligned} \tag{13.17}$$

Then we introduce the the big and the small components:

$$f_2 = (B_2 + M_2)\,, \quad iE_2 = (B_2 - M_2)\,. \tag{13.18}$$

We simultaneously separate the rest energy by means of the formal substitution $\epsilon \Longrightarrow m + E$, where E is the nonrelativistic energy. This results in

$$\Big(E + \frac{\alpha}{r}\Big)(B_2 - M_2) - \frac{\Gamma}{mr^2}\Big[i(\frac{d}{dr} + \frac{2}{r})(B_2 - M_2) - \frac{\Gamma}{r^2}(B_2 + M_2) \Big] = 2mM_2\,,$$

$$\Big(E + \frac{\alpha}{r}\Big)(B_2 + M_2) - \frac{1}{m}\frac{d}{dr}\Big[-(\frac{d}{dr} + \frac{2}{r})(B_2 - M_2) - \frac{i\Gamma}{r^2}(B_2 + M_2) \Big] = -2mM_2\,.$$

To obtain the equation for the big component B_2, we sum these equations and then neglect the small component M_2:

$$2(E + \frac{\alpha}{r})B_2 - \frac{\Gamma}{mr^2}\Big(i(\frac{d}{dr} + \frac{2}{r}) - \frac{\Gamma}{r^2} \Big) B_2 + \frac{1}{m}\frac{d}{dr}\Big(\frac{d}{dr} + \frac{2}{r} + \frac{i\Gamma}{r^2} \Big) B_2 = 0.$$

Taking into account that from the physical point of view the parameter Γ is imaginary, and making the substitution $i\Gamma \Longrightarrow \Gamma$, we find (while changing the notation as $B_2(r) = R(r)$):

$$\frac{d^2R}{dr^2} + \frac{2}{r}\frac{dR}{dr} + \Big(2m(E + \frac{\alpha}{r}) - \frac{2}{r^2} - \frac{4\Gamma}{r^3} - \frac{\Gamma^2}{r^4} \Big) R = 0\,. \tag{13.19}$$

The equation (13.19) has two irregular singular points $r=0$ and $r=\infty$, both of them having rank 2. This equation belongs to the double confluent Heun type. The local Frobenius solutions in the vicinity of the point $r=0$ are constructed as

$$R = e^{Ar} r^{B} e^{\frac{C}{r}} f(r) . \tag{13.20}$$

For the function $f(r)$ we obtain the equation

$$f'' + \Big(2A + \frac{2B+2}{r} - \frac{2C}{r^2}\Big) f' + \Big(2mE + A^2 + \frac{2AB + 2m\alpha + 2A}{r}$$

$$+ \frac{B^2 + B - 2AC - 2}{r^2} + \frac{-4\Gamma - 2BC}{r^3} + \frac{-\Gamma^2 + C^2}{r^4}\Big) f = 0 .$$

Let us impose the restrictions

$$A = \pm\sqrt{-2mE} \;\; (E<0) , \quad C = \pm\Gamma , \quad B = -\frac{2\Gamma}{C} = \mp 2 ;$$

to describe the bounded states one should use the following parameters:

$$\begin{array}{llll} \Gamma > 0 , & A = -\sqrt{-2mE} , & C = -\Gamma , & B = +2 ; \\ \Gamma < 0 , & A = -\sqrt{-2mE} , & C = +\Gamma , & B = -2 . \end{array} \tag{13.21}$$

Taking into account (13.21), the previous equation reduces to

$$f'' + \Big(2A + \frac{2B+2}{r} - \frac{2C}{r^2}\Big) f' + \Big(\frac{2AB + 2m\alpha + 2A}{r} + \frac{B^2 + B - 2AC - 2}{r^2}\Big) f = 0 ;$$

we notice that the two cases which differ, depending on the sign of Γ:

$$\Gamma > 0, \quad f'' + \Big(-2\sqrt{-2mE} + \frac{6}{r} + \frac{2\Gamma}{r^2}\Big) f'$$

$$+ \Big(\frac{-6\sqrt{-2mE} + 2m\alpha}{r} + \frac{4 - 2\Gamma\sqrt{-2mE}}{r^2}\Big) f = 0 ; \tag{13.22}$$

$$\Gamma < 0, \quad f'' + \Big(-2\sqrt{-2mE} - \frac{2}{r} - \frac{2\Gamma}{r^2}\Big) f'$$

$$+ \Big(\frac{+2\sqrt{-2mE} + 2m\alpha}{r} + \frac{2\Gamma\sqrt{-2mE}}{r^2}\Big) f = 0 . \tag{13.23}$$

Both these equations have the same mathematical structure:

$$f'' + (a + \frac{a_1}{r} + \frac{a_2}{r^2}) f' + (\frac{b_1}{r} + \frac{b_2}{r^2}) f = 0 . \tag{13.24}$$

The solutions of Eq. (13.24) are constructed as power series with 3-terms recurrence relations:

$$k \geq 2, \;\; [a(k-1) + b_1]\, c_{k-1} + [k(k-1) + a_1 k + b_2]\, c_k + a_2 (k+1)\, c_{k+1} = 0 ,$$

or shortly

$$P_{k-1}c_{k-1}+P_kc_k+P_{k+1}c_{k+1}=0, \tag{13.25}$$

where

$$P_{k-1}=a(k-1)+b_1, \quad P_k=k(k-1)+a_1k+b_2, \quad P_{k+1}=a_2(k+1).$$

Dividing the relation (13.25) by $c_{k-1}k^2$ and tending $k\to\infty$:

$$\frac{1}{k^2}[a(k-1)+b_1]+\frac{1}{k^2}[k(k-1)+a_1k+b_2]\frac{c_k}{c_{k-1}}+\frac{1}{k^2}a_2(k+1)\frac{c_{k+1}}{c_k}\frac{c_k}{c_{k-1}}=0,$$

we obtain the algebraic equation that defines the convergence radius:

$$r=0, \qquad \Longrightarrow \qquad R_{\text{conv}}=\frac{1}{|r|}=\infty. \tag{13.26}$$

Let us write down the expression for the coefficients in (13.25):

$$\begin{gathered}\Gamma>0, \quad P_{k-1}=-2\sqrt{-2mE}\,(k-1)-6\sqrt{-2mE}+2m\alpha,\\ P_k=k(k-1)+6k+4-2\sqrt{-2mE}\,\Gamma, \quad P_{k+1}=2\Gamma(k+1);\end{gathered} \tag{13.27}$$

$$\begin{gathered}\Gamma<0, \quad P_{k-1}=-2\sqrt{-2mE}\,(k-1)+2\sqrt{-2mE}+2m\alpha,\\ P_k=k(k-1)-2k+2\sqrt{-2mE}\,\Gamma, \quad P_{k+1}=-2\Gamma(k+1).\end{gathered} \tag{13.28}$$

To get some quantization rule, we apply the condition to determine transcendental Heun functions:

$$\Gamma>0, \quad P_{k-1}=-2\sqrt{-2mE}(k-1)-6\sqrt{-2mE}+2m\alpha=0,$$

$$\Gamma<0, \quad p_{k-1}=-2\sqrt{-2mE}\,(k-1)+2\sqrt{-2mE}+2m\alpha\;=0.$$

From this, we find two following different formulas for the energy values:

$$\Gamma>0,\, E=-\frac{m\alpha^2}{2}\frac{1}{(k+2)^2}\; k\geq 2; \qquad \Gamma<0,\, E=-\frac{m\alpha^2}{2}\frac{1}{(k-2)^2},\; k>2. \tag{13.29}$$

These formulas seem to be physically interpretable; however, they hardly describe correct energy spectra, since they do not depend on the parameter of anomalous magnetic moment Γ.

13.4 Non-Relativistic Equations, $j=1,2,3,...$

Let us start with the relativistic equations:

$$-\Big(\frac{d}{dr}+\frac{2}{r}\Big)E_2-2\frac{\nu}{r}E_1-\frac{\Gamma}{r^2f_2}=mf_0, \quad i\Big(\epsilon+\frac{\alpha}{r}\Big)E_1+i\Big(\frac{d}{dr}+\frac{1}{r}\Big)H_1=mf_1,$$

$$i\Big(\epsilon+\frac{\alpha}{r}\Big)E_2-2i\frac{\nu}{r}H_1-\frac{\Gamma}{r^2}f_0=mf_2, \quad -i\Big(\epsilon+\frac{\alpha}{r}\Big)f_2-\frac{d}{dr}f_0=mE_2,$$

$$-i\Big(\epsilon+\frac{\alpha}{r}\Big)f_1+\frac{\nu}{r}f_0=mE_1, \quad +i\Big(\frac{d}{dr}+\frac{1}{r}\Big)f_1+i\frac{\nu}{r}f_2=-mH_1.$$

By excluding the non-dynamical variables f_0, H_1, we get four equations

$$+i(\epsilon+\frac{\alpha}{r})E_1 - i(\frac{d}{dr}+\frac{1}{r})\frac{1}{m}\Big[i(\frac{d}{dr}+\frac{1}{r})f_1 + i\frac{\nu}{r}f_2\Big] = mf_1\,,$$

$$-i(\epsilon+\frac{\alpha}{r})f_1 - \frac{\nu}{r}\frac{1}{m}\Big[(\frac{d}{dr}+\frac{2}{r})E_2 + 2\frac{\nu}{r}E_1 + \frac{\Gamma}{r^2}f_2\Big] = mE_1\,,$$

$$i(\epsilon+\frac{\alpha}{r})E_2 + 2i\frac{\nu}{r}\frac{1}{m}\Big[i(\frac{d}{dr}+\frac{1}{r})f_1 + i\frac{\nu}{r}f_2\Big]$$

$$+\frac{\Gamma}{r^2}\frac{1}{m}\Big[(\frac{d}{dr}+\frac{2}{r})E_2 + 2\frac{\nu}{r}E_1 + \frac{\Gamma}{r^2}f_2\Big] = mf_2,$$

$$-i(\epsilon+\frac{\alpha}{r})f_2 + \frac{d}{dr}\frac{1}{m}\Big[(\frac{d}{dr}+\frac{2}{r})E_2 + 2\frac{\nu}{r}E_1 + \frac{\Gamma}{r^2}f_2\Big] = mE_2\,.$$

The big and the small components are determined by the formulas

$$f_1 = (\Psi_1+\psi_1),\quad iE_1 = (\Psi_1-\psi_1),\quad f_2 = (\Psi_2+\psi_2),\quad iE_2 = (\Psi_2-\psi_2). \qquad (13.30)$$

Taking in mind Eqs. (13.30) in the previous system (also separating the rest energy by the substitution $\epsilon = (m+E)$), we derive

$$(E+\frac{\alpha}{r})(\Psi_1-\psi_1) + (\frac{d}{dr}+\frac{1}{r})\frac{1}{m}\Big[(\frac{d}{dr}+\frac{1}{r})(\Psi_1+\psi_1) + \frac{\nu}{r}(\Psi_2+\psi_2)\Big] = 2m\psi_1\,,$$

$$(E+\frac{\alpha}{r})(\Psi_1+\psi_1) - \frac{\nu}{r}\frac{1}{m}\Big[(\frac{d}{dr}+\frac{2}{r})(\Psi_2-\psi_2) + 2\frac{\nu}{r}(\Psi_1-\psi_1) + \frac{i\Gamma}{r^2}(\Psi_2+\psi_2)\Big] = -2m\psi_1\,,$$

$$(E+\frac{\alpha}{r})(\Psi_2-\psi_2) - 2\frac{\nu}{r}\frac{1}{m}\Big[(\frac{d}{dr}+\frac{1}{r})(\Psi_1+\psi_1) + \frac{\nu}{r}(\Psi_2+\psi_2)\Big] +$$

$$+\frac{\Gamma}{r^2}\frac{1}{m}\Big[-i(\frac{d}{dr}+\frac{2}{r})(\Psi_2-\psi_2) - 2i\frac{\nu}{r}(\Psi_1-\psi_1) + \frac{\Gamma}{r^2}(\Psi_2+\psi_2)\Big] = 2m\psi_2\,,$$

$$(E+\frac{\alpha}{r})(\Psi_2+\psi_2) + \frac{d}{dr}\frac{1}{m}\Big[(\frac{d}{dr}+\frac{2}{r})(\Psi_2-\psi_2) + 2\frac{\nu}{r}(\Psi_1-\psi_1) + \frac{i\Gamma}{r^2}(\Psi_2+\psi_2)\Big] = -2m\psi_2\,.$$

To get the nonrelativistic equations for the big components Ψ_1 and Ψ_2, we sum the equations within each pair and neglect the small components. In this way we obtain

$$\Big(\frac{d^2}{dr^2}+\frac{2}{r}\frac{d}{dr}+\frac{\beta-\lambda r}{r}-\frac{2\nu^2}{r^2}\Big)\Psi_1 - \nu\frac{2r+\Gamma}{r^3}\Psi_2 = 0\,,$$

$$\Big(\frac{d^2}{dr^2}+\frac{2}{r}\frac{d}{dr}+\frac{\beta-\lambda r}{r}-\frac{2\nu^2}{r^2}-\frac{2}{r^2}-\frac{4\Gamma}{r^3}-\frac{\Gamma^2}{r^4}\Big)\Psi_2 - 2\nu\frac{2r+\Gamma}{r^3}\Psi_1 = 0. \qquad (13.31)$$

In (13.31), we have performed the change $i\Gamma \to \Gamma$, and the following notations have been used:

$$2mE = -\lambda,\quad \lambda > 0,\quad 2m\alpha = \beta,\quad 2\nu^2 = j(j+1) \equiv L. \qquad (13.32)$$

From the system (13.31) we can derive the following 4-th order equation for the function $\Psi_1(r)$:

$$\frac{d^4}{dr^4}\Psi_1 + \Big[-\frac{4}{2r+\Gamma}+\frac{10}{r}\Big]\frac{d^3}{dr^3}\Psi_1$$

$$+\Big[-2\lambda+\frac{2\Gamma\beta-24}{\Gamma}\frac{1}{r}+\frac{22-2L}{r^2}-\frac{4\Gamma}{r^3}-\frac{\Gamma^2}{r^4}+\frac{48}{\Gamma(2r+\Gamma)}+\frac{8}{(2r+\Gamma)^2}\Big]\frac{d^2}{dr^2}\Psi_1$$

$$+\Big[+\frac{-8L+64-10\Gamma^2\lambda-4\Gamma\beta}{\Gamma^2 r}+\frac{4L-24+8\Gamma\beta}{\Gamma r^2}+\frac{8-6L}{r^3}-\frac{8\Gamma}{r^4}-\frac{2\Gamma^2}{r^5}$$

$$+\frac{4\Gamma^2\lambda+16L-128+8\Gamma\beta}{\Gamma^2(2r+\Gamma)}-\frac{32}{\Gamma(2r+\Gamma)^2}\Big]\frac{d}{dr}\Psi_1$$

$$+\Big[\lambda^2+\frac{16\Gamma^2\lambda+64L+32\Gamma\beta-2\beta\lambda\Gamma^3}{\Gamma^3 r}+\frac{-10\Gamma^2\lambda-24L-12\Gamma\beta+\beta^2\Gamma^2+2\lambda L\Gamma^2}{\Gamma^2 r^2}$$

$$+\frac{4\Gamma^2\lambda+4\Gamma\beta+8L-2\Gamma\beta L}{\Gamma r^3}+\frac{-4\Gamma\beta+L^2+\Gamma^2\lambda-4L}{r^4}-\frac{\Gamma^2\beta}{r^5}$$

$$+\frac{-32\Gamma^2\lambda-128L-64\Gamma\beta}{\Gamma^3(2r+\Gamma)}+\frac{-32L-8\Gamma^2\lambda-16\Gamma\beta}{\Gamma^2(2r+\Gamma)^2}\Big]\Psi_1=0.$$

The structure of equation may be briefly written as

$$\frac{d^4}{dr^4}\Psi_1+\Big[-\frac{4}{2r+\Gamma}+\frac{10}{r}\Big]\frac{d^3}{dr^3}\Psi_1$$

$$+\Big[-2\lambda+\frac{a_1}{r}+\frac{a_2}{r^2}+\frac{a_3}{r^3}+\frac{a_4}{r^4}+\frac{a_5}{2r+\Gamma}+\frac{a_6}{(2r+\Gamma)^2}\Big]\frac{d^2}{dr^2}\Psi_1$$

$$+\Big[\frac{b_1}{r}+\frac{b_2}{r^2}+\frac{b_3}{r^3}+\frac{b_4}{r^4}+\frac{b_5}{r^5}+\frac{b_6}{2r+\Gamma}+\frac{b_7}{(2r+\Gamma)^2}\Big]\frac{d}{dr}\Psi_1$$

$$+\Big[\lambda^2+\frac{c_1}{r}+\frac{c_2}{r^2}+\frac{c_3}{r^3}+\frac{c_4}{r^4}+\frac{c_5}{r^5}+\frac{c_6}{2r+\Gamma}+\frac{c_7}{(2r+\Gamma)^2}\Big]\Psi_1=0. \qquad (13.33)$$

In the neighborhood of the regular singular point $r=-\Gamma/2$, Eq. (13.33) simplifies:

$$\Big[(\frac{d^4}{dr^4}-\frac{4}{2r+\Gamma}\frac{d^3}{dr^3}+\frac{a_6}{(2r+\Gamma)^2}\frac{d^2}{dr^2}+\frac{b_7}{(2r+\Gamma)^2}\frac{d}{dr}+\frac{c_7}{(2r+\Gamma)^2}\Big]\Psi_1=0,$$

and its solutions have the form

$$\Psi_1=(2r+\Gamma)^s, \qquad s=0,\,-1,\,-3,\,-4\,. \qquad (13.34)$$

Only for $s=0$, the corresponding solution behaves regularly near the point $r=-\Gamma/2$. The point $r=0$ is an irregular singular point of the rank 2.

Therefore, the local Frobenius solutions should be searched in the form

$$\Psi_1(r)=e^{Dr}r^A e^{B/r}f(r)\,. \qquad (13.35)$$

The differential equation for the function $f(r)$ has the following structure

$$f''''+(...)f'''+(...)f''+(...)f'+(\lambda^2-2\lambda D^2+D^4+...)f=0\,. \qquad (13.36)$$

The explicit form of this equation is too complicated, so we shall omit this. We shall study the equation for physically relevant values of the parameters.

The parameter D is fixed by the requirement

$$\lambda^2 - 2\lambda D^2 + D^4 = 0, \quad (D^2-\lambda)^2 = 0, \quad D = \pm\sqrt{\lambda} = \pm\sqrt{-2mE} = \pm\sqrt{-2\epsilon}. \qquad (13.37)$$

Since the coefficient of the term $\frac{1}{r^8} f(r)$ needs to be zero, the following restriction on the parameter B is imposed:

$$B^2(a_4 + B^2) = 0 \quad \Longrightarrow \quad B_1 = 0,\, B_2 = +\Gamma,\, B_3 = -\Gamma, \qquad (13.38)$$

where we note that one of the roots is double degenerate.

Since $B_1 = 0$, the coefficient of $\frac{1}{r^7} f(r)$ is identically zero, so we impose the restriction of vanishing on the coefficient of $\frac{1}{r^6} f(r)$:

$$a_4 A^2 \Gamma + b_5 A \Gamma - a_4 A \Gamma = 0 \qquad \Longrightarrow \quad A_1 = 0,\, A_2 = -1. \qquad (13.39)$$

While $B = \pm\Gamma$, from the requirement of zero-coefficient for r^{-7}, we obtain the equation for the parameter A:

$$2a_4 A - a_3 B + b_5 + 4AB^2 - 2a_4 - 2B^2 = 0$$

$$\Longrightarrow \quad A(2a_4 + 4B^2) = a_3 B - b_5 + (2a_4 + 2B^2).$$

Taking into account that

$$a_4 = -\Gamma^2, \quad a_3 = -4\gamma, \quad b_5 = -2\Gamma^2, \quad B = \pm\Gamma,$$

one finds $2\Gamma^2 A = -4\Gamma(\pm\Gamma) + 2\Gamma^2 + 0$, or,

$$B = +\Gamma,\, A_3 = -1; \quad B = -\Gamma,\, A_4 = +3. \qquad (13.40)$$

Therefore, we have four independent solutions (we consider only the case of negative D):

$$\begin{array}{ll}
(I,II)\ D = -\sqrt{-2\epsilon}, B = 0, A_1 = 0,\ A_2 = -1, & \begin{array}{l}\Psi_1 = e^{Dr} f_1(r), \\ \Psi_2 = e^{Dr}\frac{1}{r} f_2(r);\end{array} \\
(III)\ D = -\sqrt{-2\epsilon}, B = +\Gamma, A_3 = -1, & \Psi_3 = e^{Dr}\frac{1}{r} e^{+\Gamma/r} f_3(r); \\
(IV)\ D = -\sqrt{-2\epsilon}, B = -\Gamma,\ A_4 = +3, & \Psi_4 = e^{Dr} r^3 e^{-\Gamma/r} f_4(r).
\end{array} \qquad (13.41)$$

Let us study the solution of type I:

$$D = -\sqrt{\lambda}, \quad B = 0, \quad A_1 = 0; \qquad (13.42)$$

the equation (13.36) takes the form

$$\frac{d^4}{dr^4} f + \left(P_0 + \frac{P_1}{r} + \frac{P_2}{2r+\Gamma}\right) \frac{d^3}{dr^3} f$$

$$+ \left(Q_0 + \frac{Q_1}{r} + \frac{Q_2}{r^2} + \frac{Q_3}{r^3} + \frac{Q_4}{r^4} + \frac{Q_5}{2r+\Gamma} + \frac{Q_6}{(2r+\Gamma)^2}\right) \frac{d^2}{dr^2} f$$

$$+\Big(\frac{M_1}{r}+\frac{M_2}{r^2}+\frac{M_3}{r^3}+\frac{M_4}{r^4}+\frac{M_5}{r^5}+\frac{M_6}{2r+\Gamma}+\frac{M_7}{(2r+\Gamma)^2}\Big)\frac{d}{dr}f$$

$$+\Big(\frac{N_1}{r}+\frac{N_2}{r^2}+\frac{N_3}{r^3}+\frac{N_4}{r^4}+\frac{N_5}{r^5}+\frac{N_6}{2r+\Gamma}+\frac{N_7}{(2r+\Gamma)^2}\Big)f=0. \tag{13.43}$$

The solutions for the functions are constructed as power series:

$$f=\sum_{n=0}^{\infty}d_n r^n,\quad f'=\sum_{n=1}^{\infty}n d_n r^{n-1},\quad f''=\sum_{n=2}^{\infty}n(n-1)d_n r^{n-2},$$

$$f'''=\sum_{n=3}^{\infty}n(n-1)(n-2)d_n r^{n-3},\quad f''''=\sum_{n=4}^{\infty}n(n-1)(n-2)(n-3)d_n r^{n-4}.$$

Equating the coefficients of the variable which have the same power r^n, we find the 8-terms recurrence relation:

$n\geq 6$,

$$\begin{aligned}
&[\,4N_1+2N_6\,]\,d_{n-6}\\
&+[\,4Q_0(n-5)(n-6)+(4M_1+2M_6)(n-5)+(4\Gamma N_1+4N_2+\Gamma)\,N_6+N_7)\,d_{n-5}\\
&+[\,4P_0(n-4)(n-5)(n-6)+(4Q_1+4\Gamma Q_0+2Q_5)(n-4)(n-5)\\
&+(4\Gamma M_1+4M_2+\Gamma M_6+M_7)(n-4)+(\Gamma^2 N_1+4N_3+4\Gamma N_2)\,]\,d_{n-4}\\
&[\,4(n-3)(n-4)(n-5)(n-6)+(4\Gamma P_0+4P_1+2P_2)(n-3)(n-4)(n-5)\\
&+(\Gamma Q_5+\Gamma^2 Q_0+4\Gamma Q_1+4Q_2+Q_6)(n-3)(n-4)\\
&+(4\Gamma M_2+\Gamma^2 M_1+4M_3)(n-3)+(4\Gamma N_3+\Gamma^2 N_2+4N_4)\,]\,d_{n-3}\\
&+[\,4\Gamma(n-2)(n-3)(n-4)(n-5)+(4\Gamma P_1+\Gamma^2 P_0+\Gamma P_2)(n-2)(n-3)(n-4)\\
&+(\Gamma^2 Q_1+4Q_3+4\Gamma Q_2)(n-2)(n-3)+(4M_4+4\Gamma M_3+\Gamma^2 M_2)(n-2)\\
&+(4\Gamma N_4+\Gamma^2 N_3+4N_5)\,]\,d_{n-2}\\
&+[\,\Gamma^2(n-1)(n-2)(n-3)(n-4)+\Gamma^2 P_1(n-1)(n-2)(n-3)\\
&+(4\Gamma Q_3+\Gamma^2 Q_2+4Q_4)(n-1)(n-2)+(\Gamma^2 M_3+4\Gamma M_4+4M_5)(n-1)\\
&+(\Gamma^2 N_4+4\Gamma N_5)\,]\,d_{n-1}\\
&+[\,(4\Gamma Q_4+\Gamma^2 Q_3)n(n-1)+(\Gamma^2 M_4+4\Gamma M_5)n+\Gamma^2 N_5\,]\,d_n\\
&+(n+1)\Gamma^2[\,nQ_4+M_5\,]\,d_{n+1}=0.
\end{aligned} \tag{13.44}$$

According to the Poincaré-Perrone method, we divide Eq. (13.44) by $n^4 d_{n-6}$ and tend $n\longrightarrow\infty$. In this way we arrive at an algebraic equation which defines the possible convergence radius:

$$4r^3+4\Gamma r^4+\Gamma^2 r^5=0,\quad r=0,r=-\frac{2}{\Gamma}\quad\Longrightarrow\quad R_{conv}=\frac{1}{|r|}=\infty,\quad\frac{\Gamma}{2}. \tag{13.45}$$

Now, let us investigate the solutions of the type II:

$$D=-\sqrt{\lambda}\,,\;B=0\,,\;A_2=-1\,.$$

The equation (13.36) takes the from

$$\frac{d^4}{dr^4}f+\Big(P_0'+\frac{P_1'}{r}+\frac{P_2'}{2r+\Gamma}\Big)\frac{d^3}{dr^3}f$$
$$+\Big(Q_0'+\frac{Q_1'}{r}+\frac{Q_2'}{r^2}+\frac{Q_3'}{r^3}+\frac{Q_4'}{r^4}+\frac{Q_5'}{2r+\Gamma}+\frac{Q_6'}{(2r+\Gamma)^2}\Big)\frac{d^2}{dr^2}f$$
$$+\Big(\frac{M_1'}{r}+\frac{M_2'}{r^2}+\frac{M_3'}{r^3}+\frac{M_4'}{r^4}+\frac{(M_5'=0)}{r^5}+\frac{M_6'}{2r+\Gamma}+\frac{M_7'}{(2r+\Gamma)^2}\Big)\frac{d}{dr}f$$
$$+\Big(\frac{N_1'}{r}+\frac{N_2'}{r^2}+\frac{N_3'}{r^3}+\frac{N_4'}{r^4}+\frac{N_5'}{r^5}+\frac{N_6'}{2r+\Gamma}+\frac{N_7'}{(2r+\Gamma)^2}\Big)f=0. \tag{13.46}$$

In contrast to the equation (13.43), here $M_5'=0$. The solutions of this type are not relevant for bound states, since they diverge near the point $r=0$ due to the term $r^A=r^{-1}$.

For the solution of type III, the main equation has the form:

$$\frac{d^4}{dr^4}f+\Big(P_0+\frac{P_1}{r}+\frac{P_2}{r^2}+\frac{P_3}{2r+\Gamma}\Big)\frac{d^3}{dr^3}f$$
$$+\Big(Q_0+\frac{Q_1}{r}+\frac{Q_2}{r^2}+\frac{Q_3}{r^3}+\frac{Q_4}{r^4}+\frac{Q_5}{2r+\Gamma}+\frac{Q_6}{(2r+\Gamma)^2}\Big)\frac{d^2}{dr^2}f$$
$$+\Big(\frac{M_1}{r}+\frac{M_2}{r^2}+\frac{M_3}{r^3}+\frac{M_4}{r^4}+\frac{M_5}{r^5}+\frac{M_6}{r^6}+\frac{M_7}{2r+\Gamma}+\frac{M_8}{(2r+\Gamma)^2}\Big)\frac{d}{dr}f$$
$$+\Big(\frac{N_1}{r}+\frac{N_2}{r^2}+\frac{N_3}{r^3}+\frac{N_4}{r^4}+\frac{N_5}{r^5}+\frac{N_6}{r^6}+\frac{N_7}{2r+\Gamma}+\frac{N_8}{(2r+\Gamma)^2}\Big)f=0. \tag{13.47}$$

In this case, we construct the solutions as power series with 9-terms recurrence relations:

$n=7,8,9...$

$$(2N_7+4N_1)\,d_{n-7}$$
$$+[4Q_0(n-6)(n-7)+(4M_1+2M_7)(n-6)+(4\Gamma N_1+N_8+4N_2+\Gamma N_7)]\,d_{n-6}$$
$$+[4P_0(n-5)(n-6)(n-7)+(4Q_1+4\Gamma Q_0+2Q_5)(n-5)(n-6)$$
$$+(4\Gamma M_1+\Gamma M_7+4M_2+M_8)(n-5)+(\Gamma^2N_1+4\Gamma N_2+4N_3)]\,d_{n-5}$$
$$+[4(n-4)(n-5)(n-6)(n-7)+(4P_1+2P_3+4\Gamma P_0)(n-4)(n-5)(n-6)$$
$$+(4\Gamma Q_1+\Gamma^2Q_0+\Gamma Q_5+4Q_2+Q_6)(n-4)(n-5)+(\Gamma^2M_1+4\Gamma M_2+4M_3)(n-4)$$
$$+(\Gamma^2N_2+4\Gamma N_3+4N_4)]\,d_{n-4}+[4\Gamma(n-3)(n-4)(n-5)(n-6)$$
$$+(\Gamma P_3+\Gamma^2P_0+4P_2+4\Gamma P_1)(n-3)(n-4)(n-5)+(4Q_3+4\Gamma Q_2+\Gamma^2Q_1)(n-3)(n-4)$$

$$+(4M_4+4\Gamma M_3+\Gamma^2 M_2)(n-3)+(4\Gamma N_4+4N_5+\Gamma^2 N_3)]\,d_{n-3}$$
$$+[\Gamma^2(n-2)(n-3)(n-4)(n-5)+(\Gamma^2 P_1+4P_2\Gamma)(n-2)(n-3)(n-4)$$
$$+(\Gamma^2 Q_2+4Q_4+4\Gamma Q_3)(n-2)(n-3)+(4M_5+4\Gamma M_4+\Gamma^2 M_3)(n-2)$$
$$+(4\Gamma N_5+4N_6+\Gamma^2 N_4)]\,d_{n-2}+[\Gamma^2 P_2(n-1)(n-2)(n-3)$$
$$+(4\Gamma Q_4+\Gamma^2 Q_3)(n-1)(n-2)+(4\Gamma M_5+\Gamma^2 M_4+4M_6)(n-1)$$
$$+(\Gamma^2 N_5+4\Gamma N_6)]\,d_{n-1}+[\Gamma^2 Q_4 n(n-1)$$
$$+(\Gamma^2 M_5+4\Gamma M_6)n+\Gamma^2 N_6]\,d_n+\Gamma^2 M_6(n+1)\,d_{n+1}=0. \qquad (13.48)$$

Studying convergence of the series by the Poincaré-Perrone method, we find the following possible convergence radii:

$$4r^3+4\Gamma r^4+\Gamma^2 r^5=0, \quad r=0, r=-\frac{2}{\Gamma} \quad \Longrightarrow \quad R_{conv}=\frac{1}{|r|}=\infty, \quad \frac{|\Gamma|}{2}.$$

Finally, let us consider the last case *IV* at $B=-\Gamma, A=3$. The equation (13.36) has the form:

$$\frac{d^4}{dr^4}f+\left(P_0^{'}+\frac{P_1^{'}}{r}+\frac{P_2^{'}}{r^2}+\frac{P_3^{'}}{2r+\Gamma}\right)\frac{d^3}{dr^3}f$$
$$+\left(Q_0^{'}+\frac{Q_1^{'}}{r}+\frac{Q_2^{'}}{r^2}+\frac{Q_3^{'}}{r^3}+\frac{Q_4^{'}}{r^4}+\frac{Q_5^{'}}{2r+\Gamma}+\frac{Q_6^{'}}{(2r+\Gamma)^2}\right)\frac{d^2}{dr^2}f$$
$$+\left(\frac{M_1^{'}}{r}+\frac{M_2^{'}}{r^2}+\frac{M_3^{'}}{r^3}+\frac{M_4^{'}}{r^4}+\frac{M_5^{'}}{r^5}+\frac{M_6^{'}}{r^6}+\frac{M_7^{'}}{2r+\Gamma}+\frac{M_8^{'}}{(2r+\Gamma)^2}\right)\frac{d}{dr}f$$
$$+\left(\frac{N_1^{'}}{r}+\frac{N_2^{'}}{r^2}+\frac{N_3^{'}}{r^3}+\frac{N_4^{'}}{r^4}+\frac{N_5^{'}}{r^5}+\frac{N_6^{'}}{r^6}+\frac{N_7^{'}}{2r+\Gamma}+\frac{N_8^{'}}{(2r+\Gamma)^2}\right)f=0.$$

The structure of this equation is the same as in case III, so we have formally the same recurrence relations and the same convergence radii.

13.5 KCC-Geometrical Approach to the Problem

Now we consider the problem of spin 1 particle with anomalous magnetic moment in the external Coulomb field, by applying the Kosambi–Cartan–Chern geometrical theory. This geometrical study of the relevant system of differential equations is based on the use of KCC-invariants [163–166].

In this approach, one considers a system of second order differential equations

$$\dot{y}^i(r)+2Q^i(r,x,y)=0, \qquad (13.49)$$

which corresponds to the the Euler-Lagrange equations for some differential system associated to a Lagrangian function L. In (13.49), the symbol x^i designates so called coordinates, their derivatives in the argument r are $y^i = dx^i/dr = \dot{x}^i$, and the quantities Q_i are determined through some Lagrangian L, as follows

$$Q^i = \frac{1}{4} g^{il} \Big(\frac{\partial^2 L}{\partial x^k \partial y^l} y^k - \frac{\partial L}{\partial x^l} + \frac{\partial^2 L}{\partial y^l \partial r} \Big), \qquad g_{ij} = \frac{1}{2} \frac{\partial^2 L}{\partial y^i \partial y^j}. \tag{13.50}$$

The first and the second invariants, $\varepsilon^i(r,x,y)$ and $P^i{}_j$, are respectively defined by

$$\varepsilon^i = \frac{\partial Q^i}{\partial y^j} y^j - 2Q^i,$$
$$P^i{}_j = 2\frac{\partial Q^i}{\partial x^j} + 2Q^s \frac{\partial^2 Q^i}{\partial y^j \partial y^s} - \frac{\partial^2 Q^i}{\partial y^j \partial x^s} y^s - \frac{\partial Q^i}{\partial y^s} \frac{\partial Q^s}{\partial y^j} - \frac{\partial^2 Q^i}{\partial y^j \partial r}. \tag{13.51}$$

The second invariant $P^i{}_j$ describes the Jacobi stability of the system. There is an analogy between the equations of Riemannian geodesic deviation, the ones governed by the second KCC-invariant:

$$\frac{D^2 \xi^i}{Ds^2} = R^i_{kjl} \frac{dx^k}{ds} \frac{dx^l}{ds} \xi^j = -K^i_j \xi^j, \qquad \frac{D^2 \xi^i}{Dr^2} = P^i_j \xi^j. \tag{13.52}$$

It is known that a pencil of geodesic curves which emerge from the same point r_0 converges (or diverges) if the real parts of all eigenvalues of the invariant $P^i{}_j$ are negative (or positive) ones.

We start with the radial system of two second-order differential equations (13.31) for two radial functions, for the nonrelativistic case. It should be emphasized that we follow the case of bound states, hence assuming $\nu = \sqrt{j(j+1)/2}$, $j = 1,2,3,\ldots$.

We further use the notations $x^i = \Psi_i(r)$, $y^i = (d/dr)\Psi_i(r) = \dot{\Psi}_i(r)$. We readily find the relevant quantities Q^i:

$$Q^1(r,\Psi_i,\dot{\Psi}_i) = \frac{1}{2}\Big(\frac{2}{r}\dot{\Psi}_1 + (2m\frac{\alpha + Er}{r} - \frac{2\nu^2}{r^2})\Psi_1 - \nu \frac{2r+\Gamma}{r^3}\Psi_2 \Big),$$
$$Q^2(r,\Psi_i,\dot{\Psi}_i) = \frac{1}{2}\Big(\frac{2}{r}\dot{\Psi}_2 + (2m\frac{\alpha + Er}{r} - \frac{2\nu^2}{r^2} - \frac{2}{r^2} - \frac{4\Gamma}{r^3} - \frac{\Gamma^2}{r^4})\Psi_2 - 2\nu \frac{2r+\Gamma}{r^3}\Psi_1 \Big). \tag{13.53}$$

By direct calculation, for the first invariant ε^i we find two components:

$$\varepsilon^1 = \frac{\nu \Psi_2(\Gamma + 2r)}{r^3} + \Psi_1 \Big(-2mE + \frac{2\nu^2}{r^2} - \frac{2m\alpha}{r} \Big) - \frac{\dot{\Psi}_1}{r},$$
$$\varepsilon^2 = \frac{2\nu \Psi_1(\Gamma + 2r)}{r^3} + \Psi_2 \Big(-2mE + \frac{\Gamma^2}{r^4} + \frac{4\Gamma}{r^3} + \frac{2(\nu^2+1)}{r^2} - \frac{2m\alpha}{r} \Big) - \frac{\dot{\Psi}_2}{r}; \tag{13.54}$$

$$P^i{}_j = \begin{vmatrix} 2m\frac{\alpha+Er}{r} - \frac{2\nu^2}{r^2} & -\frac{(2r+\Gamma)\nu}{r^3} \\ -\frac{2(2r+\Gamma)\nu}{r^3} & -\frac{\Gamma^2}{r^4} - \frac{4\Gamma}{r^3} + 2m\frac{\alpha+Er}{r} - \frac{2(\nu^2+1)}{r^2} \end{vmatrix}. \tag{13.55}$$

The eigenvalues Λ_1, Λ_2 of the second invariant are given by the formula

$$\Lambda_{1,2} = 2mE + \frac{1-2\nu^2}{r^2} + \frac{2m\alpha}{r} - \frac{(\Gamma+2r)^2}{2r^4} \pm \frac{\sqrt{(\Gamma^2+2r^2+4\Gamma r)^2 + 8\nu^2 r^2 (\Gamma+2r)^2}}{2r^4}. \tag{13.56}$$

Let us specify their behavior near the singular points $r=0$, $r=\infty$, and $r=-\Gamma/2$:

$$r \to 0,\ \Lambda_1 \to \frac{2m\alpha}{r} > 0,\ \Lambda_2 \to -\frac{\Gamma^2}{r^4} < 0; \qquad r \to \infty,\ \Lambda_1, \Lambda_2 \to 2mE < 0;$$
$$r \to -\frac{\Gamma}{2},\ \Lambda_1 \to 2mE - \frac{8\nu^2}{\Gamma^2} - \frac{4m\alpha}{\Gamma} < 0, \quad \Lambda_2 \to 2mE - \frac{8(\nu^2-1)}{\Gamma^2} - \frac{4m\alpha}{\Gamma} < 0. \tag{13.57}$$

The behavior of the real parts of eigenvalues near the singular points $r = 0, \infty, -\Gamma/2$ correlates with the properties of solutions near the points $r = 0, \infty, -\Gamma/2$, for quantum mechanical bound states.

The next step is to construct a Lagrangian function L for the phase space $\dot{\Psi}_i$, Ψ_i, defined by (13.53). We will search for the function L in the form

$$L = g_{ij}(r) y^i y^j + b_j(r,x) y^j, \qquad b_j(r,x) = h_{ij}(r)\, x^i. \tag{13.58}$$

By substituting (13.58) into (13.50) and assuming that the tensor g_{ij} is diagonal, $g_{12} = g_{21} = 0$, one gets

$$Q^1 = \frac{2\dot{g}_{11} y^1 + \dot{h}_{21} x^2 + \dot{h}_{11} x^1 + (h_{21} - h_{12}) y^2}{4 g_{11}},$$
$$Q^2 = \frac{2\dot{g}_{22} y^2 + \dot{h}_{22} x^2 + \dot{h}_{12} x^1 + (h_{12} - h_{21}) y^1}{4 g_{22}}. \tag{13.59}$$

After equating the particular terms in (13.53) with the corresponding terms in (13.59) (remembering also that $x^i = \Psi_i$; $y^i = \dot{\Psi}_i$), we derive the system of equations with respect to $g_{ij}(r)$ and $h_{ij}(r)$:

$$h_{21} - h_{12} = 0, \quad \frac{\dot{g}_{11}}{2g_{11}} = \frac{1}{r}, \quad \frac{\dot{g}_{22}}{2g_{22}} = \frac{1}{r}, \quad \frac{\dot{h}_{11}}{4g_{11}} = \frac{2mr(\alpha+Er) - 2\nu^2}{2r^2},$$
$$\frac{\dot{h}_{21}}{4g_{11}} = -\frac{\nu(\Gamma+2r)}{2r^3}, \quad \frac{\dot{h}_{12}}{4g_{22}} = -\frac{\nu(\Gamma+2r)}{r^3},$$
$$\frac{\dot{h}_{22}}{4g_{22}} = -\frac{\Gamma^2}{2r^4} - \frac{r\left(\nu^2 - mr(Er+\alpha) + 1\right) + 2\Gamma}{r^3}.$$

Its solution is given by the relations:

$$g_{11} = 2C_1 r^2,\ g_{22} = C_1 r^2,\ h_{12} = h_{21} = C_2 - 4\nu C_1\,(\Gamma \ln r + 2r),$$
$$h_{11} = C_2 - 4C_1\,(\frac{-2mEr^3}{3} - m\alpha r^2 + 2\nu^2 r), \tag{13.60}$$
$$h_{22} = C_2 - 2C_1\left[-\frac{2mEr^3}{3} - m\alpha r^2 - \frac{\Gamma^2}{r} + 4\Gamma \ln r + 2(\nu^2+1) r\right],$$

where C_1, C_2 are the arbitrary constants. Finally, the Lagrangian L can be written in the form

$$L = C_1 \Big[r^2\,(2\dot\Psi_1^2 + \dot\Psi_2^2) - 4(-\frac{2}{3} mEr^3 - m\alpha r^2 + 2\nu^2 r)\,\dot\Psi_1\Psi_1$$
$$-2\Big(-\frac{2mEr^3}{3} - m\alpha r^2 - \frac{\Gamma^2}{r} + 4\Gamma\ln r + 2(\nu^2+1)r \Big)\dot\Psi_2\Psi_2 \tag{13.61}$$
$$-4\nu\,(\Gamma\ln r + 2r)\,(\dot\Psi_1\Psi_2 + \dot\Psi_2\Psi_1)\Big] + C_2\Big[\Psi_1\dot\Psi_1 + \Psi_2\dot\Psi_2\Big] + C_3.$$

It can be readily checked that the last summands with C_2 and C_3 give zero at the substitution into the formula for the coefficients Q^i and, consequently, this does not change the equation of motion. Hence the Lagrangian has an arbitrariness up to the term $C_2[\Psi_1\dot\Psi_1 + \Psi_2\dot\Psi_2] + C_3$.

The task of constructing the Lagrangian function may be solved even without the restriction $b_j(r,x) = h_{ij}(r)\,x^i$. In this case, assuming that the tensor g_{ij} is diagonal ($g_{12} = g_{21} = 0$), we derive

$$Q^1 = \frac{1}{4g_{11}}\Big(2\dot g_{11} y^1 + \frac{\partial b_1}{\partial r} + (\frac{\partial b_1}{\partial x^2} - \frac{\partial b_2}{\partial x^1})y^2\Big),$$
$$Q^2 = \frac{1}{4g_{22}}\Big(2\dot g_{22} y^2 + \frac{\partial b_2}{\partial r} + (\frac{\partial b_2}{\partial x^1} - \frac{\partial b_1}{\partial x^2})y^2\Big).$$

Further, we obtain the system of equations with respect to $g_{ij}(r)$ and to $b_j(r,x)$:

$$\frac{\partial b_1}{\partial x^2} - \frac{\partial b_2}{\partial x^1} = 0, \qquad \frac{\dot g_{11}}{2g_{11}} = \frac{1}{r}, \qquad \frac{\dot g_{22}}{2g_{22}} = \frac{1}{r},$$
$$\frac{1}{4g_1}\frac{\partial b_1}{\partial r} = \frac{x^1\,\big(r(2m\alpha + 2mEr) - 2\nu^2\big)}{2r^2} - \frac{\nu x^2(\Gamma + 2r)}{2r^3},$$
$$\frac{1}{4g_2}\frac{\partial b_2}{\partial r} = -\frac{\nu x^1(\Gamma + 2r)}{r^3} - \frac{x^2\,\big(\Gamma^2 + r^2\,(2\nu^2 - 2mr(Er + \alpha) + 2) + 4\Gamma r\big)}{2r^4}.$$

Its solution is given by the formulas:

$$g_{11} = 2C_1 r^2, \quad g_{22} = C_1 r^2,$$
$$b_1 = B_1(x^1, x^2) - 4C_1\Big\{ r x^1 [2\nu^2 - \frac{2mEr^2}{3} - mr\alpha] + \nu x^2(\Gamma\ln r + 2r)\Big\},$$
$$b_2 = B_2(x^1, x^2) - 4C_1\Big\{ x^2\Big[-\frac{\Gamma^2}{2r} + r(\nu^2+1) - \frac{mEr^3}{3} - \frac{m\alpha r^2}{2} + 2\Gamma\ln r\Big]$$
$$+\nu x^1(\Gamma\ln r + 2r)\Big\},$$

where C_1 is an arbitrary constant. The two functions $B_1(x^1,x^2)$ and $B_2(x^1,x^2)$ obey the following restriction

$$\frac{\partial B_1(x^1,x^2)}{\partial x^2} - \frac{\partial B_2(x^1,x^2)}{\partial x^1} = 0. \tag{13.62}$$

From (13.62), we conclude that the 2-dimensional vector field of components B_1, B_2 can be represented as a gradient of a scalar function

$$B_1(x^1,x^2) = \frac{\partial}{\partial x^1}\varphi(x^1,x^2), \quad B_2(x^1,x^2) = \frac{\partial}{\partial x^2}\varphi(x^1,x^2), \quad B_i = \mathrm{grad}\,\varphi\,. \qquad (13.63)$$

Therefore, there exists some freedom in choosing the Lagrangian (the constant C_1 may be taken as 1):

$$L = 2r^2(y^1)^2 + r^2(y^2)^2 + \left(\frac{2\Gamma^2}{r} + \frac{4}{3}mEr^3 + 2m\alpha r^2 - 4\left(\nu^2+1\right)r - 8\Gamma\ln r\right)x^2y^2$$
$$-4\nu(\Gamma\ln r + 2r)x^1y^2 - 4\nu(\Gamma\ln r + 2r)x^2y^1$$
$$-r\left(8\nu^2 - \frac{8mEr^2}{3} - 4m\alpha r\right)x^1y^1 + y^1\frac{\partial\varphi}{\partial x^1} + y^2\frac{\partial\varphi}{\partial x^2}\,.$$

Chapter 14

Vector Particle with Electric Quadruple Moment in the Coulomb Field

14.1 Initial Equation

It is known that in the framework of relativistic wave equation theory one can propose so-called nonminimal equations which describe particles with additional electromagnetic characteristics, with spectra of spin and mass states [283]–[307]. Within this approach, in [308]–[334] the problem of spin 1 particles with additional anomalous magnetic and quadruple electric moments has been investigated. The equations were studied and solved for the particles in the external homogenous electric and magnetic fields. The equation for the vector particle in the external Coulomb field is very complicated even in the case of the ordinary particle without additional electromagnetic moments, and has not been solved in full yet. However, at nonrelativistic limit, the equation for the ordinary vector particle in the Coulomb field can be solved exactly. In the present work we study the nonrelativistic problem of a vector particle with additional quadruple moment in the external Coulomb field.

The initial equation has the following form (further we use the conventional tetrad formalism)

$$\Big\{ i\beta^c \Big[i(e^{\beta}_{(c)}\partial_\beta + \frac{1}{2} j^{ab}\gamma_{abc}(x)) - e'A_c \Big] + \lambda \frac{e}{M} F_{\alpha\beta}(x)\bar{P} j^{\alpha\beta}(x) - M \Big\} \Psi = 0; \qquad (14.1)$$

the free parameter λ related to the quadruple moment is dimensionless, $\bar{P}$ is a projective operator separating from the 10-component wave function the tensor component; we use the notations:

$$\bar{P} = \begin{vmatrix} 0 & 0 \\ 0 & I_6 \end{vmatrix}, \quad M = \frac{mc}{\hbar}, \quad e' = \frac{e}{c\hbar}, \quad \Gamma = \lambda \frac{4\alpha}{M}, \quad \alpha = \frac{e^2}{\hbar c}. \qquad (14.2)$$

In spherical tetrad, Eq. (14.1) has the form

$$\Big[\beta^0 (i\partial_t + \frac{\alpha}{r}) + i\Big(\beta^3\,\partial_r + \frac{1}{r}(\beta^1 j^{31} + \beta^2 j^{32})\Big) + \frac{1}{r}\Sigma_{\theta,\phi} + \frac{\Gamma}{r^2}\bar{P} j^{03} - M \Big] \Phi(x) = 0, \qquad (14.3)$$

where, depending on the angular variables, the operator $\Sigma_{\theta,\phi}$ is determined by the equality

$$\Sigma_{\theta,\phi} = i\,\beta^1 \partial_\theta + \beta^2 \frac{i\partial_\phi + i j^{12}\cos\theta}{\sin\theta}\,. \tag{14.4}$$

The components of the operator of the total angular momentum are given in this basis by the formulas

$$j_1 = l_1 + \frac{\cos\phi}{\sin\theta} i j^{12}, \quad j_2 = l_2 + \frac{\sin\phi}{\sin\theta} i j^{12}, \quad j_3 = l_3\,, \quad j^{12} = \beta^1\beta^2 - \beta^2\beta^1\,. \tag{14.5}$$

Below we shall apply the Duffin – Kemmer matrices in cyclic basis

$$\beta^0 = \left| \begin{array}{cccccccccc} 0 & 0 & 0 & 0 & 0 & 0 & 0 & 0 & 0 & 0 \\ 0 & 0 & 0 & 0 & +i & 0 & 0 & 0 & 0 & 0 \\ 0 & 0 & 0 & 0 & 0 & +i & 0 & 0 & 0 & 0 \\ 0 & 0 & 0 & 0 & 0 & 0 & +i & 0 & 0 & 0 \\ 0 & -i & 0 & 0 & 0 & 0 & 0 & 0 & 0 & 0 \\ 0 & 0 & -i & 0 & 0 & 0 & 0 & 0 & 0 & 0 \\ 0 & 0 & 0 & -i & 0 & 0 & 0 & 0 & 0 & 0 \\ 0 & 0 & 0 & 0 & 0 & 0 & 0 & 0 & 0 & 0 \\ 0 & 0 & 0 & 0 & 0 & 0 & 0 & 0 & 0 & 0 \\ 0 & 0 & 0 & 0 & 0 & 0 & 0 & 0 & 0 & 0 \end{array} \right|, \beta^3 = \left| \begin{array}{cccccccccc} 0 & 0 & 0 & 0 & 0 & i & 0 & 0 & 0 & 0 \\ 0 & 0 & 0 & 0 & 0 & 0 & 0 & +1 & 0 & 0 \\ 0 & 0 & 0 & 0 & 0 & 0 & 0 & 0 & 0 & 0 \\ 0 & 0 & 0 & 0 & 0 & 0 & 0 & 0 & 0 & -1 \\ 0 & 0 & 0 & 0 & 0 & 0 & 0 & 0 & 0 & 0 \\ i & 0 & 0 & 0 & 0 & 0 & 0 & 0 & 0 & 0 \\ 0 & 0 & 0 & 0 & 0 & 0 & 0 & 0 & 0 & 0 \\ 0 & -1 & 0 & 0 & 0 & 0 & 0 & 0 & 0 & 0 \\ 0 & 0 & 0 & 0 & 0 & 0 & 0 & 0 & 0 & 0 \\ 0 & 0 & 0 & +1 & 0 & 0 & 0 & 0 & 0 & 0 \end{array} \right|$$

$$\beta^1 = \frac{1}{\sqrt{2}} \left| \begin{array}{cccccccccc} 0 & 0 & 0 & 0 & -i & 0 & +i & 0 & 0 & 0 \\ 0 & 0 & 0 & 0 & 0 & 0 & 0 & 0 & +1 & 0 \\ 0 & 0 & 0 & 0 & 0 & 0 & 0 & +1 & 0 & +1 \\ 0 & 0 & 0 & 0 & 0 & 0 & 0 & 0 & +1 & 0 \\ -i & 0 & 0 & 0 & 0 & 0 & 0 & 0 & 0 & 0 \\ 0 & 0 & 0 & 0 & 0 & 0 & 0 & 0 & 0 & 0 \\ +i & 0 & 0 & 0 & 0 & 0 & 0 & 0 & 0 & 0 \\ 0 & 0 & -1 & 0 & 0 & 0 & 0 & 0 & 0 & 0 \\ 0 & -1 & 0 & -1 & 0 & 0 & 0 & 0 & 0 & 0 \\ 0 & 0 & -1 & 0 & 0 & 0 & 0 & 0 & 0 & 0 \end{array} \right|,$$

$$\beta^2 = \frac{1}{\sqrt{2}} \left| \begin{array}{cccccccccc} 0 & 0 & 0 & 0 & 1 & 0 & 1 & 0 & 0 & 0 \\ 0 & 0 & 0 & 0 & 0 & 0 & 0 & 0 & -i & 0 \\ 0 & 0 & 0 & 0 & 0 & 0 & 0 & +i & 0 & -i \\ 0 & 0 & 0 & 0 & 0 & 0 & 0 & 0 & +i & 0 \\ -1 & 0 & 0 & 0 & 0 & 0 & 0 & 0 & 0 & 0 \\ 0 & 0 & 0 & 0 & 0 & 0 & 0 & 0 & 0 & 0 \\ -1 & 0 & 0 & 0 & 0 & 0 & 0 & 0 & 0 & 0 \\ 0 & 0 & +i & 0 & 0 & 0 & 0 & 0 & 0 & 0 \\ 0 & -i & 0 & +i & 0 & 0 & 0 & 0 & 0 & 0 \\ 0 & 0 & -i & 0 & 0 & 0 & 0 & 0 & 0 & 0 \end{array} \right|.$$

In this basis the matrix $ij^{12} = i(\beta^1\beta^2 - \beta^2\beta^1)$ has the diagonal structure

$$ij^{12} = \left| \begin{array}{cccc} 0 & 0 & 0 & 0 \\ 0 & t_3 & 0 & 0 \\ 0 & 0 & t_3 & 0 \\ 0 & 0 & 0 & t_3 \end{array} \right|, \quad t_3 = \left| \begin{array}{ccc} +1 & 0 & 0 \\ 0 & 0 & 0 \\ 0 & 0 & -1 \end{array} \right|.$$

We re-express the 10×10 matrices from Cartesian basis to the cyclic basis by means of the similarity transformation

$$\Psi_{cycl} = S\Psi_{Cart}, \quad S = \begin{vmatrix} 1 & 0 & 0 & 0 \\ 0 & U & 0 & 0 \\ 0 & 0 & U & 0 \\ 0 & 0 & 0 & U \end{vmatrix}, \quad U = \begin{vmatrix} -1/\sqrt{2} & i/\sqrt{2} & 0 \\ 0 & 0 & 1 \\ +1/\sqrt{2} & i/\sqrt{2} & 0 \end{vmatrix}, \tag{14.6}$$

$$\beta^c_{cycl} = S\beta^c_{cart}S^{-1}, \quad \bar{P}_{cycl} = S\bar{P}_{cart}S^{-1} \equiv \bar{P}_{Cart}\,,$$

and we note that the the projective operator P preserves its form in this new basis. Further, all the formulas will refer to the cyclic basis.

The system of radial equations for the ordinary vector particle in Coulomb field is known (see [303, 304]). To get similar system for a particle with quadruple moment, it suffices to find the explicit form of additional term in the equation:

$$\frac{\Gamma}{r^2}\bar{P}j^{03} = \frac{\Gamma}{r^2}\bar{P}(\beta^0\beta^3 - \beta^3\beta^0)\,, \tag{14.7}$$

where

$$\beta^0\beta^3 - \beta^3\beta^0 = \begin{vmatrix} 0 & 0 & -1 & 0 & 0 & 0 & 0 & 0 & 0 & 0 \\ 0 & 0 & 0 & 0 & 0 & 0 & 0 & 0 & 0 & 0 \\ -1 & 0 & 0 & 0 & 0 & 0 & 0 & 0 & 0 & 0 \\ 0 & 0 & 0 & 0 & 0 & 0 & 0 & 0 & 0 & 0 \\ 0 & 0 & 0 & 0 & 0 & 0 & 0 & -i & 0 & 0 \\ 0 & 0 & 0 & 0 & 0 & 0 & 0 & 0 & 0 & 0 \\ 0 & 0 & 0 & 0 & 0 & 0 & 0 & 0 & 0 & +i \\ 0 & 0 & 0 & 0 & +i & 0 & 0 & 0 & 0 & 0 \\ 0 & 0 & 0 & 0 & 0 & 0 & 0 & 0 & 0 & 0 \\ 0 & 0 & 0 & 0 & 0 & 0 & -i & 0 & 0 & 0 \end{vmatrix},$$

$$\bar{P}\,j^{03} = \begin{vmatrix} 0 & 0 & 0 & 0 & 0 & 0 & 0 & 0 & 0 & 0 \\ 0 & 0 & 0 & 0 & 0 & 0 & 0 & 0 & 0 & 0 \\ 0 & 0 & 0 & 0 & 0 & 0 & 0 & 0 & 0 & 0 \\ 0 & 0 & 0 & 0 & 0 & 0 & 0 & 0 & 0 & 0 \\ 0 & 0 & 0 & 0 & 0 & 0 & 0 & -i & 0 & 0 \\ 0 & 0 & 0 & 0 & 0 & 0 & 0 & 0 & 0 & 0 \\ 0 & 0 & 0 & 0 & 0 & 0 & 0 & 0 & 0 & +i \\ 0 & 0 & 0 & 0 & +i & 0 & 0 & 0 & 0 & 0 \\ 0 & 0 & 0 & 0 & 0 & 0 & 0 & 0 & 0 & 0 \\ 0 & 0 & 0 & 0 & 0 & 0 & -i & 0 & 0 & 0 \end{vmatrix}.$$

14.2 Separating the Variables in the Relativistic Equation

The most general form of 10-component wave function with quantum numbers ϵ, j, m is given below

$$\Psi = \{\,\Psi_0, \vec{\Psi}, \vec{E}, \vec{H}\,\},\; \Psi_0 = e^{-i\epsilon t} f_0(r) D_0,\; \vec{\Psi}(x) = e^{-i\epsilon t} \begin{vmatrix} f_1(r)D_{-1} \\ f_2(r)D_0 \\ f_3(r)D_{+1} \end{vmatrix},$$
$$\vec{E} = e^{-i\epsilon t} \begin{vmatrix} E_1(r)D_{-1} \\ E_2(r)D_0 \\ E_3(r)D_{+1} \end{vmatrix},\quad \vec{H} = e^{-i\epsilon t} \begin{vmatrix} H_1(r)D_{-1} \\ H_2(r)D_0 \\ H_3(r)D_{+1} \end{vmatrix}, \tag{14.8}$$

where we use Wigner D-functions: $D_\sigma = D^j_{-m,\sigma}(\phi, \theta, 0)$, $\quad \sigma = 0, -1, +1$.

Starting with the known system of radial equations for the ordinary vector particle is (let $\nu = \sqrt{j(j+1)}/\sqrt{2}$):

$$
\begin{gathered}
-(\frac{d}{dr} + \frac{2}{r})E_2 - \frac{\nu}{r}\,(E_1 + E_3) = m f_0\,,\\
+i(\epsilon + \frac{\alpha}{r})E_1 + i(\frac{d}{dr} + \frac{1}{r})H_1 + i\frac{\nu}{r}\,H_2 = m f_1\,,\\
+i(\epsilon + \frac{\alpha}{r})E_2 - i\frac{\nu}{r}(H_1 - H_3) = m f_2\,,\\
+i(\epsilon + \frac{\alpha}{r})E_3 - i(\frac{d}{dr} + \frac{1}{r})H_3 - i\frac{\nu}{r}H_2 = m f_3\,;\\
-i(\epsilon + \frac{\alpha}{r})\Phi_1 + \frac{\nu}{r}\Phi_0 = mE_1\,, \quad -i(\epsilon + \frac{\alpha}{r})f_2 - \frac{d}{dr}f_0 = mE_2\,,\\
-i(\epsilon + \frac{\alpha}{r})f_3 + \frac{\nu}{r}f_0 = mE_3\,, \quad -i(\frac{d}{dr} + \frac{1}{r})f_1 - i\frac{\nu}{r}f_2 = mH_1\,,\\
+i\frac{\nu}{r}(f_1 - f_3) = mH_2\,, \quad +i(\frac{d}{dr} + \frac{1}{r})f_3 + i\frac{\nu}{r}f_2 = mH_3\,,
\end{gathered}
\tag{14.9}
$$

and accounting for the explicit form of additional term in the equation (the multiplier $e^{-i\epsilon t}$ is omitted)

$$
\frac{\Gamma}{r^2}\,\bar{P} j^{03}\,\Psi = \frac{\Gamma}{r^2}
\begin{vmatrix}
0 & 0 & 0 & 0 & 0 & 0 & 0 & 0 & 0 & 0\\
0 & 0 & 0 & 0 & 0 & 0 & 0 & 0 & 0 & 0\\
0 & 0 & 0 & 0 & 0 & 0 & 0 & 0 & 0 & 0\\
0 & 0 & 0 & 0 & 0 & 0 & 0 & 0 & 0 & 0\\
0 & 0 & 0 & 0 & 0 & 0 & 0 & -i & 0 & 0\\
0 & 0 & 0 & 0 & 0 & 0 & 0 & 0 & 0 & 0\\
0 & 0 & 0 & 0 & 0 & 0 & 0 & 0 & 0 & +i\\
0 & 0 & 0 & 0 & +i & 0 & 0 & 0 & 0 & 0\\
0 & 0 & 0 & 0 & 0 & 0 & 0 & 0 & 0 & 0\\
0 & 0 & 0 & 0 & 0 & 0 & -i & 0 & 0 & 0
\end{vmatrix}
\begin{vmatrix}
f_0 D_0\\ f_1 D_{-1}\\ f_2 D_0\\ f_3 D_{+1}\\ E_1 D_{-1}\\ E_2 D_0\\ E_3 D_{+1}\\ H_1 D_{-1}\\ H_2 D_0\\ H_3 D_{+1}
\end{vmatrix}
= \frac{\Gamma}{r^2}
\begin{vmatrix}
0\\ 0\\ 0\\ 0\\ -iH_1 D_{-1}\\ 0\\ iH_3 D_{+1}\\ iE_1 D_{-1}\\ 0\\ -iE_3 D_{+1}
\end{vmatrix},
$$

we derive the following system

$$
\begin{aligned}
-(\frac{d}{dr}+\frac{2}{r})E_2-\frac{\nu}{r}(E_1+E_3)=mf_0\,,\\
+i(\epsilon+\frac{\alpha}{r})E_1+i(\frac{d}{dr}+\frac{1}{r})\,H_1+i\frac{\nu}{r}H_2=mf_1\,,\\
+i(\epsilon+\frac{\alpha}{r})E_2-i\frac{\nu}{r}(H_1-H_3)=mf_2\,,\\
+i(\epsilon+\frac{\alpha}{r})E_3-i(\frac{d}{dr}+\frac{1}{r})H_3-i\frac{\nu}{r}H_2=mf_3\,;
\end{aligned}
\tag{14.10}
$$

$$
\begin{aligned}
-i(\epsilon+\frac{\alpha}{r})f_1+\frac{\nu}{r}f_0-i\frac{\Gamma}{r^2}H_1=mE_1\,,\\
-i(\epsilon+\frac{\alpha}{r})f_2-\frac{d}{dr}f_0=mE_2\,,-i(\epsilon+\frac{\alpha}{r})f_3+\frac{\nu}{r}f_0+i\frac{\Gamma}{r^2}H_3=mE_3\,,
\end{aligned}
\tag{14.11}
$$

$$
\begin{aligned}
-i(\frac{d}{dr}+\frac{1}{r})f_1-i\frac{\nu}{r}f_2+i\frac{\Gamma}{r^2}E_1=mH_1\,,\ +i\frac{\nu}{r}(f_1-f_3)=mH_2\,,\\
+i(\frac{d}{dr}+\frac{1}{r})f_3+i\frac{\nu}{r}f_2-i\frac{\Gamma}{r^2}E_3=mH_3\,.
\end{aligned}
\tag{14.12}
$$

Additionally to the operators $\vec{j}^2, j_3$, we diagonalize the operator of spatial reflection $\hat{\Pi}$. In Cartesian matrix basis β^a, this operator has the ordinary form

$$
\hat{\Pi}=\begin{vmatrix} 1 & 0 & 0 & 0\\ 0 & -I & 0 & 0\\ 0 & 0 & -I & 0\\ 0 & 0 & 0 & +I \end{vmatrix}\hat{P}\,,\ \ \hat{P}\Psi(\vec{r})=\Psi(-\vec{r})\,.
\tag{14.13}
$$

After expressing it in spherical tetrad and using the cyclic representation of the matrix β^a, we obtain

$$
\hat{\Pi}'=\begin{vmatrix} 1 & 0 & 0 & 0\\ 0 & \Pi_3 & 0 & 0\\ 0 & 0 & \Pi_3 & 0\\ 0 & 0 & 0 & -\Pi_3 \end{vmatrix}\hat{P}\,,\ \Pi_3=\begin{vmatrix} 0 & 0 & -1\\ 0 & -1 & 0\\ -1 & 0 & 0 \end{vmatrix}.
\tag{14.14}
$$

The eigenvalues equation $\hat{\Pi}'\Psi=P\Psi$ (accounting for the known property $\hat{P}D_\sigma=(-1)^jD_{-\sigma}$) yields the set of algebraic equations:

$$
\begin{aligned}
(-1)^jf_0=Pf_0\,,\ (-1)^jf_3=Pf_1\,,\ (-1)^jf_2=Pf_2\,,\ (-1)^jf_1=Pf_3\,,\\
(-1)^jE_3=PE_1\,,\ (-1)^jE_2=PE_2\,,\ (-1)^jE_1=PE_3\,,\\
(-1)^jH_3=-PH_1\,,\ (-1)^jH_2=-PH_2\,,\ (-1)^jH_1=-PH_3\,.
\end{aligned}
\tag{14.15}
$$

This system has two solutions:

$$
P=(-1)^{j+1},\quad f_0=0,\ f_3=-f_1,\ f_2=0,\quad E_3=-E_1,\ E_2=0,\ H_3=H_1\,;
\tag{14.16}
$$

$$P=(-1)^{j}, \quad f_3=+f_1, \quad E_3=+E_1,\ H_3=-H_1,\ H_2=0. \tag{14.17}$$

It easy to check that these restrictions are compatible with the above radial equations.

For the parity $P=(-1)^{j+1}$, we get

$$f_0=0,\ f_3=-f_1,\ f_2=0,\ E_3=-E_1,\ E_2=0,\ H_3=H_1,$$

$$0=0, \quad +i(\epsilon+\frac{\alpha}{r})\,E_1+i(\frac{d}{dr}+\frac{1}{r})H_1+i\frac{\nu}{r}H_2=mf_1,$$

$$0=0, \quad -i(\epsilon+\frac{\alpha}{r})E_1-i(\frac{d}{dr}+\frac{1}{r})H_3-i\frac{\nu}{r}H_2=-mf_1;$$

$$-i(\epsilon+\frac{\alpha}{r})f_1-i\frac{\Gamma}{r^2}H_1=mE_1, \quad 0=0, \quad i(\epsilon+\frac{\alpha}{r})f_1+i\frac{\Gamma}{r^2}H_1=-mE_1,$$

$$-i(\frac{d}{dr}+\frac{1}{r})f_1+i\frac{\Gamma}{r^2}E_1=mH_1,+2i\frac{\nu}{r}f_1=mH_2, \quad -i(\frac{d}{dr}+\frac{1}{r})f_1+i\frac{\Gamma}{r^2}E_1=mH_1,$$

so we have only four equations:

$$\begin{gathered}+i\Big(\epsilon+\frac{\alpha}{r}\ Big)E_1+i\Big(\frac{d}{dr}+\frac{1}{r}\Big)H_1+i\frac{\nu}{r}H_2=mf_1,\\ -i\Big(\epsilon+\frac{\alpha}{r}\Big)f_1-i\frac{\Gamma}{r^2}H_1=mE_1,-i\Big(\frac{d}{dr}+\frac{1}{r}\Big)f_1+i\frac{\Gamma}{r^2}E_1=mH_1, \quad 2i\frac{\nu}{r}f_1=mH_2.\end{gathered} \tag{14.18}$$

We exclude the variables H_1,H_2 from the first and second equations:

$$\begin{gathered}+i\Big(\epsilon+\frac{\alpha}{r}\Big)mE_1+i\Big(\frac{d}{dr}+\frac{1}{r}\Big)\Big[-i\Big(\frac{d}{dr}+\frac{1}{r}\Big)f_1+i\frac{\Gamma}{r^2}E_1\Big]-\frac{2\nu^2}{r^2}f_1=m^2f_1,\\ -i\Big(\epsilon+\frac{\alpha}{r}\Big)mf_1-i\frac{\Gamma}{r^2}\Big[-i\Big(\frac{d}{dr}+\frac{1}{r}\Big)f_1+i\frac{\Gamma}{r^2}E_1\Big]=m^2E_1.\end{gathered} \tag{14.19}$$

The second equation permits to express E_1 through f_1:

$$-i\Big(\epsilon+\frac{\alpha}{r}\Big)mf_1-\frac{\Gamma}{r^2}\Big(\frac{d}{dr}+\frac{1}{r}\Big)f_1=(m^2-\frac{\Gamma^2}{r^4})E_1; \tag{14.20}$$

consequently, the function E_1 can be excluded. Hence, we arrive at a 2-nd order equation for the main function f_1:

$$\begin{gathered}\frac{d^2f_1}{dr^2}+\Big[-\frac{2rm}{mr^2+\Gamma}-\frac{2rm}{mr^2-\Gamma}+\frac{6}{r}\Big]\frac{df_1}{dr}\\ +\Big[\frac{2\epsilon\alpha}{r}+\frac{-2\nu^2+\alpha^2+4}{r^2}+\frac{2i\Gamma\epsilon}{mr^3}+\frac{\Gamma(\Gamma m+i\alpha)}{mr^4}+\frac{2\nu^2\Gamma^2}{m^2r^6}\\ -m^2+\epsilon^2+\frac{2m(i\epsilon r+i\alpha-1)}{mr^2+\Gamma}-\frac{2m(i\epsilon r+i\alpha+1)}{mr^2-\Gamma}\Big]f_1=0.\end{gathered} \tag{14.21}$$

It has the irregular singular point $r=0$ of the rank 3, the irregular point $r=\infty$ of the rank 2, and four regular points which are determined by the roots of the equation, $(r^2-\Gamma/m)(r^2+\Gamma/m)=0$.

Let us consider the states with different parity:

$$P=(-1)^{j}, \quad f_3=+f_1, \quad E_3=+E_1,\ H_3=-H_1,\ H_2=0; \tag{14.22}$$

at such restrictions, the radial system takes the form

$$-(\frac{d}{dr}+\frac{2}{r})\,E_2-2\,\frac{\nu}{r}\,E_1=m\,f_0\,,\quad +i\,(\epsilon+\frac{\alpha}{r})\,E_1\;+i\,(\frac{d}{dr}+\frac{1}{r})\,H_1=mf_1\,,$$
$$+i\,(\epsilon+\frac{\alpha}{r})\,E_2-2i\frac{\nu}{r}\,H_1=m\,f_2\,,\quad +i\,(\epsilon+\frac{\alpha}{r})\,E_1\;+i\,(\frac{d}{dr}+\frac{1}{r})\,H_1=m\,f_1\,;$$
$$-i\,(\epsilon+\frac{\alpha}{r})\,f_1\;+\frac{\nu}{r}\,f_0-i\frac{\Gamma}{r^2}H_1=m\,E_1\,,\quad -i\,(\epsilon+\frac{\alpha}{r})\,f_2\;-\;\frac{d}{dr}\,f_0=m\,E_2\,,$$
$$-i\,(\epsilon+\frac{\alpha}{r})\,f_1\;+\frac{\nu}{r}\,f_0-i\frac{\Gamma}{r^2}H_1=m\,E_1\,,\quad -i\,(\frac{d}{dr}+\frac{1}{r})\,f_1-i\frac{\nu}{r}\,f_2+i\frac{\Gamma}{r^2}E_1=m\,H_1\,,$$
$$0=0\,,\quad +i\,(\frac{d}{dr}+\frac{1}{r})\,f_1+i\frac{\nu}{r}f_2-i\frac{\Gamma}{r^2}E_1=-m\,H_1\,;$$

so we arrive at six different equations:

$$-\Big(\frac{d}{dr}+\frac{2}{r}\Big)E_2-2\,\frac{\nu}{r}\,E_1=m\,f_0\,,\quad +i\,\Big(\epsilon+\frac{\alpha}{r}\Big)E_1+i\Big(\frac{d}{dr}+\frac{1}{r}\Big)H_1=mf_1\,,$$
$$+i\Big(\epsilon+\frac{\alpha}{r}\Big)E_2-2i\frac{\nu}{r}\,H_1=m\,f_2\,;\quad -i\Big(\epsilon+\frac{\alpha}{r}\Big)f_1+\frac{\nu}{r}\,f_0-i\frac{\Gamma}{r^2}H_1=m\,E_1\,,\qquad (14.23)$$
$$-i\,\Big(\epsilon+\frac{\alpha}{r}\Big)f_2\;-\;\frac{d}{dr}\,f_0=m\,E_2\,,\quad +i\Big(\frac{d}{dr}+\frac{1}{r}\Big)f_1+i\frac{\nu}{r}f_2-i\frac{\Gamma}{r^2}E_1=-m\,H_1\,.$$

14.3 States with Parity $P=(-1)^{j+1}$

In eqs (14.21) we introduce the new (dimensionless) variables $x=mr, E=\frac{\epsilon}{m}, \gamma=m\Gamma$, j so obtaining the simpler representation

$$\frac{d^2 f_1}{dx^2}+\Big[-\frac{2x}{x^2+\gamma}-\frac{2x}{x^2-\gamma}+\frac{6}{x}\Big]\frac{df_1}{dx}$$
$$+\Big[(E^2-1)+\frac{2E\alpha}{x}+\frac{-2\nu^2+\alpha^2+4}{x^2}+\frac{2i\gamma E}{x^3}+\frac{\gamma(\gamma+i\alpha)}{x^4}+\frac{2\nu^2\gamma^2}{x^6}$$
$$+\frac{2(iEx+i\alpha-1)}{x^2+\gamma}-\frac{2(iEx+i\alpha+1)}{x^2-\gamma}\Big]f_1=0\,.\qquad (14.24)$$

Here we have the singular points

$$x=0\,,\ \text{Rank}=3\,,\quad x=\infty\,,\ \text{Rank}=2\,,$$
$$x=-\sqrt{\gamma}\,,\ +\sqrt{\gamma}\,,\ -i\sqrt{\gamma}\,,\ +i\sqrt{\gamma},\ \text{Rank}=1\,.\qquad (14.25)$$

The equation is suitable to be represented in the symbolic form:

$$\frac{d^2 f_1}{dx^2}+\Big[-\frac{2x}{x^2+\gamma}-\frac{2x}{x^2-\gamma}+\frac{6}{x}\Big]\frac{df_1}{dx}$$
$$+\Big[E^2-1+\frac{a_1}{x}+\frac{a_2}{x^2}+\frac{a_3}{x^3}+\frac{a_4}{x^4}+\frac{a_6}{x^6}+\frac{L}{x^2+\gamma}+\frac{N}{x^2-\gamma}\Big]f_1=0\,.\qquad (14.26)$$

We construct its Frobenius type solutions near the point $x=0$, in the form

$$f_1(x)=e^{Dx}x^A e^{B/x}e^{C/x^2}F(x)\,,$$

$$\frac{d^2F}{dx^2}+\Big[-\frac{2x}{x^2+\gamma}-\frac{2x}{x^2-\gamma}+\frac{2A+6}{x}-\frac{2B}{x^2}-\frac{4C}{x^3}+2D\Big]\frac{dF}{dx}$$
$$+\Big[(E^2+D^2-1)+\frac{2AD+6D+a_1}{x}+\frac{A^2+5A-2BD+a_2}{x^2}$$
$$+\frac{-2AB-4B-4CD+a_3}{x^3}+\frac{-4AC+B^2-6C+a_4}{x^4}+\frac{4BC}{x^5}+\frac{4C^2+a_6}{x^6}$$
$$+\frac{-2A\gamma-2Bx-4C-2D\gamma x+L\gamma}{\gamma(x^2+\gamma)}+\frac{-2A\gamma+2Bx+4C-2D\gamma x+N\gamma}{\gamma(x^2-\gamma)}\Big]F=0.$$

Let us impose the restrictions:

$$\begin{gathered} E^2+D^2-1=0 \quad\Rightarrow\quad D=-\sqrt{1-E^2},+\sqrt{1-E^2},\\ 4C^2+a_6=0 \quad\Rightarrow\quad C=\frac{\delta}{2}\sqrt{-a_6}=\delta\frac{1}{2}\sqrt{2\nu^2(-\gamma^2)},\quad \delta=\pm 1,\\ 4BC=0 \quad\Longrightarrow\quad B=0,\\ -4AC+B^2-6C+a_4 \quad\Rightarrow\quad A=-\frac{3}{2}+\frac{1}{4}\frac{a_4}{C}=-\frac{3}{2}+\frac{\delta}{2}\frac{\gamma^2+i\gamma\alpha}{\sqrt{2\nu^2(-\gamma^2)}}. \end{gathered} \tag{14.27}$$

By physical considerations, the parameter γ has to be imaginary, so we should make a substitution $i\gamma \Longrightarrow \gamma$, and to have a regular behavior at $x=0$ the parameter C should be taken as negative one. To describe bound states, we will use the following expressions for the parameters:

$$\begin{gathered} A=-\frac{3}{2}-\frac{1}{2}\frac{\gamma^2+i\gamma\alpha}{\sqrt{2\nu^2(-\gamma^2)}},\qquad B=0,\\ C=-\frac{1}{2}\sqrt{2\nu^2(-\gamma^2)}<0,\quad D=-\sqrt{1-E^2}. \end{gathered} \tag{14.28}$$

The equation is simplified:

$$\frac{d^2F}{dx^2}+\Big[-\frac{2x}{x^2+\gamma}-\frac{2x}{x^2-\gamma}+\frac{2A+6}{x}-\frac{2B}{x^2}-\frac{4C}{x^3}+2D\Big]\frac{dF}{dx}$$
$$+\Big[\frac{2AD+6D+a_1}{x}+\frac{A^2+5A-2BD+a_2}{x^2}+\frac{-2AB-4B-4CD+a_3}{x^3}$$
$$+\frac{-2A\gamma-2Bx-4C-2D\gamma x+L\gamma}{\gamma(x^2+\gamma)}+\frac{-2A\gamma+2Bx+4C-2D\gamma x+N\gamma}{\gamma(x^2-\gamma)}\Big]F=0. \tag{14.29}$$

We construct the solutions for $F(x)$ as power series: $F=\sum_{n=0}^{\infty}d_nx^n$, and we get the 8-term recurrence relations:

$$\underline{k=6,7,8,9,\ldots} \qquad [2D(k-6)+(2AD+2D+a_1)]\,d_{k-6}$$
$$+\left[(k-5)(k-6)+(2+2A)(k-5)+\left(A^2+A-2BD+L+N+a_2\right)\right]d_{k-5}$$
$$+[-2B(k-4)+(-2AB-4CD+a_3)]\,d_{k-4}$$
$$+[-4C(k-3)+(8C-L\gamma+N\gamma)]\,d_{k-3}+\left[-2D\gamma^2(k-2)+\gamma^2\left(-2AD-6D-a_1\right)\right]d_{k-2}$$
$$+\left[-\gamma^2(k-1)(k-2)-2\gamma^2(3+A)(k-1)+\gamma^2\left(-A^2-5A+2BD-a_2\right)\right]d_{k-1}$$
$$+\left[2B\gamma^2k+\gamma^2(2AB+4B+4CD-a_3)\right]d_k+4C\gamma^2(k+1)d_{k+1}=0.$$

In brief, these relations can be written as

$$P_{k-6}c_{k-6}+P_{k-5}c_{k-5}+P_{k-4}c_{k-4}+P_{k-3}c_{k-3}$$
$$+P_{k-2}c_{k-2}+P_{k-1}c_{k-1}+P_k c_k+P_{k+1}c_{k+1}=0. \tag{14.30}$$

Applying the Poincaré-Perrone method, let us analyze the convergence radius of the power series. To do this, we divide the last relation by $d_{k-6}k^2$,

$$\begin{aligned}
&[2D(k-6)+(2AD+2D+a_1)]\\
&+\left[(k-5)(k-6)+(2+2A)(k-5)+\left(A^2+A-2BD+L+N+a_2\right)\right]\frac{d_{k-5}}{d_{k-6}}\\
&+[-2B(k-4)+(-2AB-4CD+a_3)]\frac{d_{k-4}}{d_{k-5}}\frac{d_{k-5}}{d_{k-6}}\\
&+[-4C(k-3)+(8C-L\gamma+N\gamma)]\frac{d_{k-3}}{d_{k-4}}\frac{d_{k-4}}{d_{k-5}}\frac{d_{k-5}}{d_{k-6}}\\
&+\left[-2D\gamma^2(k-2)+\gamma^2(-2AD-6D-a_1)\right]\frac{d_{k-2}}{d_{k-3}}\frac{d_{k-3}}{d_{k-4}}\frac{d_{k-4}}{d_{k-5}}\frac{d_{k-5}}{d_{k-6}}\\
&+\left[-\gamma^2(k-1)(k-2)-2\gamma^2(3+A)(k-1)+\gamma^2\left(-A^2-5A+2BD-a_2\right)\right]\\
&\qquad\times\frac{d_{k-1}}{d_{k-2}}\frac{d_{k-2}}{d_{k-3}}\frac{d_{k-3}}{d_{k-4}}\frac{d_{k-4}}{d_{k-5}}\frac{d_{k-5}}{d_{k-6}}\\
&+\left[2B\gamma^2k+\gamma^2(2AB+4B+4CD-a_3)\right]\frac{d_k}{d_{k-1}}\frac{d_{k-1}}{d_{k-2}}\frac{d_{k-2}}{d_{k-3}}\frac{d_{k-3}}{d_{k-4}}\frac{d_{k-4}}{d_{k-5}}\frac{d_{k-5}}{d_{k-6}}\\
&+4C\gamma^2(k+1)\frac{d_{k+1}}{d_k}\frac{d_k}{d_{k-1}}\frac{d_{k-1}}{d_{k-2}}\frac{d_{k-2}}{d_{k-3}}\frac{d_{k-3}}{d_{k-4}}\frac{d_{k-4}}{d_{k-5}}\frac{d_{k-5}}{d_{k-6}}=0,
\end{aligned}$$

also tend $k\to\infty$. As a result, we get the algebraic equation for R:

$$R-\gamma^2R^5=0 \quad\Rightarrow\quad R=0,\ \pm\frac{1}{\sqrt{\gamma}},\ \pm\frac{1}{\sqrt{-\gamma}}.$$

The modulus of the parameter R determines the possible convergence radii.

$$R=\lim_{k\to\infty}\frac{d_{k-5}}{d_{k-6}},\quad R_{\text{conv}}=\left|\frac{1}{R}\right|=|\sqrt{\gamma}|,\ +\infty. \tag{14.31}$$

As a quantization rule we use the restrictions separating the transcendent Frobenius functions (see (14.30))

$$P_{k-6}=0 \implies 2D(k-6)+2AD+2D+a_1=0,\quad k\geq 6, \tag{14.32}$$

where

$$A=-\frac{3}{2}-\frac{1}{2}\frac{\gamma^2+i\gamma\alpha}{\sqrt{2\nu^2(-\gamma^2)}},\quad D=-\sqrt{1-E^2},\quad a_1=2E\alpha.$$

Taking into account that the parameter γ is imaginary, we make the change $i\gamma\Longrightarrow\gamma$, and then

$$A=-\frac{3}{2}-\frac{1}{2}\frac{-\gamma^2+\gamma\alpha}{\sqrt{2\nu^2\gamma^2}},\quad D=-\sqrt{1-E^2},\quad a_1=2E\alpha, \tag{14.33}$$

and the transcendency condition takes the form (let $k-6=n$, $n=0,1,2,...$)

$$-\sqrt{1-E^2}\,n+\Big(\frac{3}{2}+\frac{\gamma\alpha-\gamma^2}{2\sqrt{l(l+1)\gamma^2}}\Big)\sqrt{1-E^2}-\sqrt{1-E^2}+E\alpha=0,$$

or

$$\alpha E=\sqrt{1-E^2}\Big(n-1/2-\frac{\gamma\alpha-\gamma^2}{2\sqrt{\gamma^2 l(l+1)}}\Big). \qquad (14.34)$$

Depending on the sign of the parameter γ, there arise two different equations:

$$\gamma>0,\quad \alpha E=\sqrt{1-E^2}\Big(n-1/2-\frac{\alpha-\gamma}{2\sqrt{l(l+1)}}\Big),$$
$$\gamma<0,\quad \alpha E=\sqrt{1-E^2}\Big(n-1/2+\frac{\alpha-\gamma}{2\sqrt{l(l+1)}}\Big). \qquad (14.35)$$

The structure of these equations is the same, so the energy spectra are similar, but still different:

$$\alpha E=\sqrt{1-E^2}N \quad \Longrightarrow \quad E=\frac{1}{\sqrt{1+\frac{\alpha^2}{N^2}}};$$
$$\gamma>0,\quad N=n-1/2-\frac{\alpha-\gamma}{2\sqrt{l(l+1)}};\quad \gamma<0,\quad N=n-1/2+\frac{\alpha-\gamma}{2\sqrt{l(l+1)}}. \qquad (14.36)$$

14.4 The Case of Minimal $j=0$

Let us consider the case of minimal value of the total momentum $j=0$. The relevant substitution for the corresponding wave function should be used,

$$\Phi_0(x)=e^{-i\epsilon t}f_0(r),\quad \vec{\Phi}(x)=e^{-i\epsilon t}\left|\begin{array}{c}0\\ f_2(r)\\ 0\end{array}\right|,$$
$$\vec{E}(x)=e^{-i\epsilon t}\left|\begin{array}{c}0\\ E_2(r)\\ 0\end{array}\right|,\quad \vec{H}(x)=e^{-i\epsilon t}\left|\begin{array}{c}0\\ H_2(r)\\ 0\end{array}\right|. \qquad (14.37)$$

The corresponding equations follow from the general one with the restrictions

$$\nu=0,\quad f_1=f_3=0,\quad E_1=E_3=0,\quad H_1=H_3=0. \qquad (14.38)$$

So, we get

$$-(\frac{d}{dr}+\frac{2}{r})E_2=mf_0,\quad +i(\epsilon+\frac{\alpha}{r})E_2=mf_2,\quad 0=0,\quad 0=0,$$
$$-i(\epsilon+\frac{\alpha}{r})f_2-\frac{d}{dr}f_0=mE_2,\quad 0=0,\quad 0=0,\quad 0=mH_2,\quad 0=0.$$

In the case of minimal $j=0$, the electric quadruple moment does not manifest itself:

$$-\Big(\frac{d}{dr}+\frac{2}{r}\Big)E_2 = mf_0, \quad +i\Big(\epsilon+\frac{\alpha}{r}\Big)E_2 = mf_2,$$
$$-i\Big(\epsilon+\frac{\alpha}{r}\Big)f_2 - \frac{d}{dr}f_0 = mE_2, \quad H_2 = 0. \tag{14.39}$$

After excluding the variables f_2, E_2, we obtain a 2-nd order equation for the function $f_0(r)$:

$$\frac{d^2 f_0}{dr^2} + \Big(\frac{4}{r} + \frac{-m+\epsilon}{mr-\epsilon r-\alpha} + \frac{-m-\epsilon}{mr+\epsilon r+\alpha}\Big)\frac{df_0}{dr} + \Big(\frac{2\epsilon\alpha}{r} + \frac{\alpha^2}{r^2} - m^2 + \epsilon^2\Big)f_0 = 0,$$

or, in the dimensionless variables $x = mr, E = \epsilon/m$:

$$\frac{d^2 f_0}{dx^2} + \Big(\frac{4}{x} + \frac{-1+E}{x-Ex-\alpha} + \frac{-1-E}{x+Ex+\alpha}\Big)\frac{df_0}{dx} + \Big(\frac{2E\alpha}{x} + \frac{\alpha^2}{x^2} + E^2 - 1\Big)f_0 = 0. \tag{14.40}$$

This equation has three regular singular points and one irregular point of the rank 2 at infinity.

Alternatively, by excluding the variables f_0, f_2, we get the simpler equation for the main function E_2:

$$\frac{d^2 E_2}{dr^2} + \frac{2}{r}\frac{dE_2}{dr} + \Big(\epsilon^2 - m^2 + \frac{2\epsilon\alpha}{r} + \frac{\alpha^2-2}{r^2}\Big)E_2 = 0.$$

The last equation transformed to the variable $y = 2\sqrt{m^2-\epsilon^2}r$ takes the form

$$y\frac{d^2E_2}{dy^2} + 2\frac{dE_2}{dy} + \Big(-\frac{1}{4}y + \frac{\alpha^2-2}{y} + \frac{\epsilon\alpha}{\sqrt{m^2-\epsilon^2}}\Big)E_2 = 0.$$

Its solutions are constructed in terms of confluent hypergeometric functions, according to standard procedure:

$$E_2 = y^A e^{By} F(y), \quad y\frac{d^2F}{dy^2} + (2A+2+2By)\frac{dF}{dy}$$
$$+\Big[\Big(B^2-\frac{1}{4}\Big)y + \frac{A^2+A+\alpha^2-2}{y} + 2AB + 2B + \frac{\epsilon\alpha}{\sqrt{m^2-\epsilon^2+}}\Big]F = 0.$$

By imposing on the parameters the evident restrictions

$$A = -\frac{1}{2} \pm \frac{1}{2}\sqrt{9-4\alpha^2}, \quad B = -\frac{1}{2},$$

we reduce the task to simplify the problem to a confluent hypergeometric equation

$$y\frac{d^2F}{dy^2} + (2A+2-y)\frac{dF}{dy} + \Big(-1-A+\frac{\epsilon\alpha}{\sqrt{m^2-\epsilon^2}}\Big) = 0,$$

with the parameters

$$a = 1 + A - \frac{\epsilon\alpha}{\sqrt{m^2-\epsilon^2}}, \quad c = 2A+2.$$

The quantization condition is chosen as usually

$$a = \frac{1}{2}(1+\sqrt{9-4\alpha^2}) - \frac{\epsilon\alpha}{\sqrt{m^2-\epsilon^2}} = -n\,,$$

so we can derive the formula for energy levels

$$\epsilon = \frac{m}{\sqrt{1+\alpha^2/N^2}}, \quad N = \frac{1}{2}(1+\sqrt{9-4\alpha^2})+n,\; n=0,1,2,\dots . \tag{14.41}$$

14.5 Non-Relativistic Approximation, $P=(-1)^{j+1}, j=1,2,3,\dots$

Let us perform the nonrelativistic approximation in the radial system

$$\begin{gathered} P=(-1)^{j+1}, \qquad i\Big(\epsilon+\frac{\alpha}{r}\Big)E_1 + i\Big(\frac{d}{dr}+\frac{1}{r}\Big)H_1 + i\frac{\nu}{r}H_2 = mf_1\,, \\ -i\Big(\epsilon+\frac{\alpha}{r}\Big)f_1 - i\frac{\Gamma}{r^2}H_1 = mE_1\,, -i\Big(\frac{d}{dr}+\frac{1}{r}\Big)f_1 + i\frac{\Gamma}{r^2}E_1 = mH_1\,, 2i\frac{\nu}{r}f_1 = mH_2\,. \end{gathered} \tag{14.42}$$

By using the third and the fourth equations, we exclude the non-dynamical variables H_1, H_2:

$$\begin{gathered} i(\epsilon+\frac{\alpha}{r})E_1 + \frac{1}{m}(\frac{d}{dr}+\frac{1}{r})\Big[(\frac{d}{dr}+\frac{1}{r})f_1 - \frac{\Gamma}{r^2}E_1\Big] - \frac{2\nu^2}{mr^2}f_1 = mf_1\,, \\ -i(\epsilon+\frac{\alpha}{r})f_1 - \frac{\Gamma}{mr^2}\Big[(\frac{d}{dr}+\frac{1}{r})f_1 - \frac{\Gamma}{r^2}E_1\Big] = mE_1\,. \end{gathered}$$

The big and the small components are introduced by the relations

$$f_1 = (\Psi_1+\psi_1)\,, \quad iE_1 = (\Psi_1-\psi_1)\,; \tag{14.43}$$

then the previous equations take the form (concurrently we separate the rest energy by formal change $\epsilon = m+E$, where E stands for the nonrelativistic energy)

$$\begin{gathered} (m+E+\frac{\alpha}{r})(\Psi_1-\psi_1) + \frac{1}{m}(\frac{d}{dr}+\frac{1}{r})\Big[(\frac{d}{dr}+\frac{1}{r})(\Psi_1+\psi_1) + \frac{i\Gamma}{r^2}(\Psi_1-\psi_1)\Big] \\ -\frac{2\nu^2}{mr^2}(\Psi_1+\psi_1) = m(\Psi_1+\psi_1)\,, \\ (m+E+\frac{\alpha}{r})(\Psi_1+\psi_1) - \frac{i\Gamma}{mr^2}\Big[(\frac{d}{dr}+\frac{1}{r})(\Psi_1+\psi_1) + \frac{i\Gamma}{r^2}(\Psi_1-\psi_1)\Big] = m(\Psi_1-\psi_1)\,. \end{gathered}$$

By re-grouping the terms and neglecting the small component in comparison with the big one, we derive the following two equations (recalling that $2\nu^2 = j(j+1)$)

$$\begin{gathered} \Big(E+\frac{\alpha}{r}\Big)\Psi_1 + \frac{1}{m}(\frac{d}{dr}+\frac{1}{r})\Big[(\frac{d}{dr}+\frac{1}{r})\Psi_1 + \frac{i\Gamma}{r^2}\Psi_1\Big] - \frac{j(j+1)}{mr^2}\Psi_1 = 2m\psi_1\,, \\ \Big(E+\frac{\alpha}{r}\Big)\Psi_1 - \frac{i\Gamma}{mr^2}\Big[(\frac{d}{dr}+\frac{1}{r})\Psi_1 + \frac{i\Gamma}{r^2}\Psi_1\Big] = -2m\psi_1\,; \end{gathered}$$

after summing these, we find a 2-nd order equation for the big component

$$\Big\{(\frac{d}{dr}+\frac{1}{r})[(\frac{d}{dr}+\frac{1}{r})+\frac{i\Gamma}{r^2}\Psi_1]+2m(E+\frac{\alpha}{r})-\frac{j(j+1)}{r^2}-\frac{i\Gamma}{r^2}[(\frac{d}{dr}+\frac{1}{r})+\frac{i\Gamma}{r^2}\Psi_1]\Big\}\Psi_1=0\,.$$

Making the needed change $i\Gamma \Longrightarrow \Gamma$, we obtain

$$\Big\{(\frac{d}{dr}+\frac{1}{r})(\frac{d}{dr}+\frac{1}{r}+\frac{\Gamma}{r^2})+2m(E+\frac{\alpha}{r})-\frac{j(j+1)}{r^2}-\frac{\Gamma}{r^2}(\frac{d}{dr}+\frac{1}{r}+\frac{\Gamma}{r^2})\Big\}\Psi_1=0\,.$$

The final form of the nonrelativistic radial equation is (let $\Psi_1(r)=R(r)$):

$$\Big\{\frac{d^2}{dr^2}+\frac{2}{r}\frac{d}{dr}+2m(E+\frac{\alpha}{r})-\frac{j(j+1)}{r^2}-\frac{2\Gamma}{r^3}-\frac{\Gamma^2}{r^4}\Big\}R(r)=0\,. \qquad (14.44)$$

This equation has two singular points, $r=0$ and ∞; both of them have the rank 2. Therefore, it is the double confluent Heun equation. In the vicinity of the point $r=0$, its Frobenius solutions are constructed in the form

$$R(r)=e^{Cr}r^Ae^{\frac{B}{r}}f(r)\,,\qquad \frac{d^2f}{dr^2}+\Big(\frac{2+2A}{r}-\frac{2B}{r^2}+2C\Big)\frac{df}{dr}$$
$$+\Big(\frac{2AC+2C+2m\alpha}{r}+\frac{A^2+A-2BC-j^2-j}{r^2}$$
$$+\frac{-2AB-2\Gamma}{r^3}+\frac{B^2-\Gamma^2}{r^4}+C^2+2mE\Big)f=0\,.$$

With the evident restrictions on parameters (C must be negative)

$$C=-\sqrt{-2mE}\,;\quad B=\Gamma,\ A=-1\,;\qquad B=-\Gamma,\ A=+1 \qquad (14.45)$$

the equation becomes simpler. The negative values of the parameter B correspond to the bound states. Depending on the sign of Γ, there exist two different sets of parameters:

$$\begin{aligned}&\Gamma>0\,,\quad A=+1\,,B=-\Gamma\,,C=-\sqrt{-2mE}\,;\\ &\Gamma<0\,,\quad A=-1\,,B=+\Gamma\,,\ C=-\sqrt{-2mE}\,.\end{aligned} \qquad (14.46)$$

The equation for $f(r)$ is formally the same:

$$\frac{d^2f}{dr^2}+\Big(2C+\frac{2+2A}{r}-\frac{2B}{r^2}\Big)\frac{df}{dr}$$

$$+\Big(\frac{2AC+2C+2m\alpha}{r}+\frac{A^2+A-2BC-j^2-j}{r^2}\Big)f=0\,,$$

or shortly,

$$\frac{d^2f}{dr^2}+\Big(a+\frac{a_1}{r}+\frac{a_2}{r^2}\Big)\frac{df}{dr}+\Big(\frac{b_1}{r}+\frac{b_2}{r^2}\Big)f=0\,.$$

The solutions $f(r)$ are constructed as power series: $f=\sum_{k=0}^{\infty} c_k r^k$, and we obtain the recurrence formulas:

$$k=0, \quad b_2 c_0 + a_2 c_1 = 0,$$
$$k=1, \quad b_1 c_0 + (a_1+b_2) c_1 + 2 a_2 c_2 = 0,$$
$$k=2, \quad (a+b_1) c_1 + (2+2a_1+b_2) c_2 + 3 a_2 c_3 = 0.$$

Therefore, the general formula for the 3-term recurrence relations is

$$k=1,2,3,4,..., \qquad [a(k-1)+b_1]\, c_{k-1} + [k(k-1)+a_1 k + b_2]\, c_k + a_2(k+1)\, c_{k+1} = 0, \tag{14.47}$$

or shortly $P_{k-1}c_{k-1}+P_k c_k + P_{k+1}c_{k+1}=0$, where

$$P_{k-1}=a(k-1)+b_1, \quad P_k = k(k-1)+a_1 k + b_2, \quad P_{k+1}=a_2(k+1).$$

In accordance with the Poincaré – Perrone method, we divide the relation by $k^2 c_{k-1}$ and tend $k\to\infty$:

$$\frac{1}{k^2}[a(k-1)+b_1] + \frac{1}{k^2}[k(k-1)+a_1 k + b_2]\frac{c_k}{c_{k-1}} + \frac{1}{k^2}a_2(k+1)\frac{c_{k+1}}{c_k}\frac{c_k}{c_{k-1}} = 0;$$

as a result, the algebraic equation which determines the possible convergence radius, is found:

$$r=0 \quad \Rightarrow \quad R_{\text{conv}} = \frac{1}{|r|} = \infty. \tag{14.48}$$

Let us present the explicit form of the quantities entering the recurrence relations

$$P_{k-1} = 2C(k-1) + 2AC + 2C + 2m\alpha,$$
$$P_k = k(k-1) + (2+2A)k + A^2 + A - 2BC - j^2 - j, \quad P_{k+1} = -2B(k+1),$$

and consider the transcendency condition for Heun functions:

$$P_{n-1}=0 \quad \Longrightarrow \quad C = -\frac{m\alpha}{[(k-1)+A+2]}, \tag{14.49}$$

where $C=-\sqrt{-2mE}$, and

$$\begin{aligned} &\Gamma>0, \quad A=+1, B=-\Gamma,\, C=-\sqrt{-2mE}; \\ &\Gamma<0, \quad A=-1, B=+\Gamma,\, C=-\sqrt{-2mE}. \end{aligned} \tag{14.50}$$

Then, depending on the sign of Γ, we obtain different spectra:

$$\Gamma>0, \quad R(r) = e^{-\sqrt{-2mE}r}\, r e^{\frac{\pm\Gamma}{r}} f(r), \quad E = -\frac{m\alpha^2}{2(k+2)^2}, \tag{14.51}$$

$$\Gamma<0, \quad R(r) = e^{-\sqrt{-2mE}r}\, r^{-1} e^{\frac{\pm\Gamma}{r}} f(r), \quad E = -\frac{m\alpha^2}{2k^2}. \tag{14.52}$$

The solutions of both types, at $\Gamma>0$ and $\Gamma<0$ respectively, could describe bound states as they tend to zero at $r\to 0$. However, the formulas for energy do not depend on Γ, so these spectra hardly are physical ones.

14.6 Non-Relativistic Radial Equations, the Case of $j = 1,2,3,...$

We start from Eqs. (14.23):

$$
\begin{gathered}
-\Big(\frac{d}{dr}+\frac{2}{r}\Big)E_2 - 2\frac{\nu}{r}E_1 = mf_0\,, \quad +i\Big(\epsilon+\frac{\alpha}{r}\Big)E_1 + i\Big(\frac{d}{dr}+\frac{1}{r}\Big)H_1 = mf_1\,,\\
+i\Big(\epsilon+\frac{\alpha}{r}\Big)E_2 - 2i\frac{\nu}{r}H_1 = mf_2\,, \quad -i\Big(\epsilon+\frac{\alpha}{r}\Big)f_1 + \frac{\nu}{r}f_0 - i\frac{\Gamma}{r^2}H_1 = mE_1\,, \qquad (14.53)\\
-i\Big(\epsilon+\frac{\alpha}{r}\Big)f_2 - \frac{d}{dr}f_0 = mE_2\,, \quad +i\left(\frac{d}{dr}+\frac{1}{r}\right)f_1 + i\frac{\nu}{r}f_2 - i\frac{\Gamma}{r^2}E_1 = -mH_1\,.
\end{gathered}
$$

By eliminating the variables f_0, H_1

$$
f_0 = -\frac{1}{m}\Big[\Big(\frac{d}{dr}+\frac{2}{r}\Big)E_2 + 2\frac{\nu}{r}E_1\Big], \quad H_1 = -\frac{1}{m}\Big[i\Big(\frac{d}{dr}+\frac{1}{r}\Big)f_1 + i\frac{\nu}{r}f_2 - i\frac{\Gamma}{r^2}E_1\Big],
$$

we reduce the remaining equations to the form

$$
\begin{gathered}
+i\Big(\epsilon+\frac{\alpha}{r}\Big)E_1 - \frac{i}{m}\Big(\frac{d}{dr}+\frac{1}{r}\Big)\Big[i\Big(\frac{d}{dr}+\frac{1}{r}\Big)f_1 + i\frac{\nu}{r}f_2 - i\frac{\Gamma}{r^2}E_1\Big] = mf_1\,,\\
+i\Big(\epsilon+\frac{\alpha}{r}\Big)E_2 + 2i\frac{\nu}{r}\frac{1}{m}\Big[i\Big(\frac{d}{dr}+\frac{1}{r}\Big)f_1 + i\frac{\nu}{r}f_2 - i\frac{\Gamma}{r^2}E_1\Big] = mf_2;\\
-i\Big(\epsilon+\frac{\alpha}{r}\Big)f_1 - \frac{\nu}{r}\frac{1}{m}\Big[\Big(\frac{d}{dr}+\frac{2}{r}\Big)E_2 + 2\frac{\nu}{r}E_1\Big]\\
+i\frac{\Gamma}{r^2}\frac{1}{m}\Big[i\Big(\frac{d}{dr}+\frac{1}{r}\Big)f_1 + i\frac{\nu}{r}f_2 - i\frac{\Gamma}{r^2}E_1\Big] = mE_1\,,\\
-i\Big(\epsilon+\frac{\alpha}{r}\Big)f_2 + \frac{1}{m}\frac{d}{dr}\Big[\Big(\frac{d}{dr}+\frac{2}{r}\Big)E_2 + 2\frac{\nu}{r}E_1\Big] = mE_2\,.
\end{gathered}
$$

The big and the small components are introduced by the formulas

$$
f_1 = (\Psi_1+\psi_1)\,, \quad iE_1 = (\Psi_1-\psi_1)\,, \quad f_2 = (\Psi_2+\psi_2)\,, \quad iE_2 = (\Psi_2-\psi_2)\,; \quad (14.54)
$$

then the previous equations take the form (concurrently we separate the rest energy by the substitution $\epsilon = m+E$)

$$
\begin{gathered}
\Big(m+E+\frac{\alpha}{r}\Big)(\Psi_1-\psi_1) - \frac{i}{m}\left(\frac{d}{dr}+\frac{1}{r}\right)\Big[i\Big(\frac{d}{dr}+\frac{1}{r}\Big)(\Psi_1+\psi_1)\\
+i\frac{\nu}{r}(\Psi_2+\psi_2) - \frac{\Gamma}{r^2}(\Psi_1-\psi_1)\Big] = m(\Psi_1+\psi_1)\,,\\
\Big(m+E+\frac{\alpha}{r}\Big)(\Psi_1+\psi_1) - \frac{\nu}{r}\frac{1}{m}\Big[\Big(\frac{d}{dr}+\frac{2}{r}\Big)(\Psi_2-\psi_2) + 2\frac{\nu}{r}(\Psi_1-\psi_1)\Big]\\
-\frac{\Gamma}{r^2}\frac{1}{m}\Big[i\Big(\frac{d}{dr}+\frac{1}{r}\Big)(\Psi_1+\psi_1) + i\frac{\nu}{r}(\Psi_2+\psi_2) - \frac{\Gamma}{r^2}(\Psi_1-\psi_1)\Big] = m(\Psi_1-\psi_1)\,,\\
\Big(m+E+\frac{\alpha}{r}\Big)(\Psi_2-\psi_2) + 2i\frac{\nu}{r}\frac{1}{m}\Big[i\Big(\frac{d}{dr}+\frac{1}{r}\Big)(\Psi_1+\psi_1)\\
+i\frac{\nu}{r}(\Psi_2+\psi_2) - \frac{\Gamma}{r^2}(\Psi_1-\psi_1)\Big] = m(\Psi_2+\psi_2)\,,
\end{gathered}
$$

$$\Big(m+E+\frac{\alpha}{r}\Big)(\Psi_2+\psi_2)+\frac{1}{m}\frac{d}{dr}\Big[\Big(\frac{d}{dr}+\frac{2}{r}\Big)(\Psi_2-\psi_2)+2\frac{\nu}{r}(\Psi_1-\psi_1)\Big]=m(\Psi_2-\psi_2).$$

Regrouping the terms, we obtain

$$\Big(E+\frac{\alpha}{r}\Big)(\Psi_1-\psi_1)-\frac{i}{m}\Big(\frac{d}{dr}+\frac{1}{r}\Big)\Big[i\Big(\frac{d}{dr}+\frac{1}{r}\Big)(\Psi_1+\psi_1)$$
$$+i\frac{\nu}{r}(\Psi_2+\psi_2)-\frac{\Gamma}{r^2}(\Psi_1-\psi_1)\Big]=2m\psi_1,$$
$$\Big(E+\frac{\alpha}{r}\Big)(\Psi_1+\psi_1)-\frac{\nu}{r}\frac{1}{m}\Big[\Big(\frac{d}{dr}+\frac{2}{r}\Big)(\Psi_2-\psi_2)+2\frac{\nu}{r}(\Psi_1-\psi_1)\Big]$$
$$-\frac{\Gamma}{r^2}\frac{1}{m}\Big[i\Big(\frac{d}{dr}+\frac{1}{r}\Big)(\Psi_1+\psi_1)+i\frac{\nu}{r}f_2-\frac{\Gamma}{r^2}(\Psi_1-\psi_1)\Big]=-2m\psi_1,$$
$$\Big(E+\frac{\alpha}{r}\Big)(\Psi_2-\psi_2)+2i\frac{\nu}{r}\frac{1}{m}\Big[i\Big(\frac{d}{dr}+\frac{1}{r}\Big)(\Psi_1+\psi_1)$$
$$+i\frac{\nu}{r}(\Psi_2+\psi_2)-\frac{\Gamma}{r^2}(\Psi_1-\psi_1)\Big]=2m\psi_2,$$
$$\Big(E+\frac{\alpha}{r}\Big)(\Psi_2+\psi_2)+\frac{1}{m}\frac{d}{dr}\Big[\Big(\frac{d}{dr}+\frac{2}{r}\Big)(\Psi_2-\psi_2)+2\frac{\nu}{r}(\Psi_1-\psi_1)\Big]=-2m\psi_2.$$

To derive the needed equations for the big components Ψ_1 and Ψ_2, we sum the equations in each pair and then neglect the small components in comparison with the big ones. This results in

$$2\Big(E+\frac{\alpha}{r}\Big)\Psi_1+\frac{1}{m}\Big(\frac{d}{dr}+\frac{1}{r}\Big)\Big[\Big(\frac{d}{dr}+\frac{1}{r}\Big)\Psi_1+\frac{\nu}{r}\Psi_2+i\frac{\Gamma}{r^2}\Psi_1\Big]$$
$$-\frac{\nu}{r}\frac{1}{m}\Big[\Big(\frac{d}{dr}+\frac{2}{r}\Big)\Psi_2+2\frac{\nu}{r}\Psi_1\Big]-\frac{i\Gamma}{r^2}\frac{1}{m}\Big[\Big(\frac{d}{dr}+\frac{1}{r}\Big)\Psi_1+\frac{\nu}{r}\Psi_2+\frac{i\Gamma}{r^2}\Psi_1\Big]=0,$$
$$2\Big(E+\frac{\alpha}{r}\Big)\Psi_2-2\frac{\nu}{r}\frac{1}{m}\Big[\Big(\frac{d}{dr}+\frac{1}{r}\Big)\Psi_1+\frac{\nu}{r}\Psi_2+\frac{i\Gamma}{r^2}\Psi_1\Big]$$
$$+\frac{1}{m}\frac{d}{dr}\Big[\Big(\frac{d}{dr}+\frac{2}{r}\Big)\Psi_2+2\frac{\nu}{r}\Psi_1\Big]=0.$$

Allowing that Γ is imaginary, we make the change $i\Gamma\Longrightarrow\Gamma$, so producing the system

$$2m\Big(E+\frac{\alpha}{r}\Big)\Psi_1+\Big(\frac{d}{dr}+\frac{1}{r}\Big)\Big[\Big(\frac{d}{dr}+\frac{1}{r}\Big)\Psi_1+\frac{\nu}{r}\Psi_2+\frac{\Gamma}{r^2}\Psi_1\Big]$$
$$-\frac{\nu}{r}\Big[\Big(\frac{d}{dr}+\frac{2}{r}\Big)\Psi_2+2\frac{\nu}{r}\Psi_1\Big]-\frac{\Gamma}{r^2}\Big[\Big(\frac{d}{dr}+\frac{1}{r}\Big)\Psi_1+\frac{\nu}{r}\Psi_2+\frac{\Gamma}{r^2}\Psi_1\Big]=0,$$
$$2m\Big(E+\frac{\alpha}{r}\Big)\Psi_2-2\frac{\nu}{r}\Big[\Big(\frac{d}{dr}+\frac{1}{r}\Big)\Psi_1+\frac{\nu}{r}\Psi_2+\frac{\Gamma}{r^2}\Psi_1\Big]+\frac{d}{dr}\Big[\Big(\frac{d}{dr}+\frac{2}{r}\Big)\Psi_2+2\frac{\nu}{r}\Psi_1\Big]=0.$$

We further get (recalling that $2\nu^2=j(j+1)$)

$$\Big(\frac{d^2}{dr^2}+\frac{2}{r}\frac{d}{dr}+2mE+\frac{2m\alpha}{r}-\frac{2\nu^2}{r^2}-\frac{2\Gamma}{r^3}-\frac{\Gamma^2}{r^4}\Big)\Psi_1-\nu\Big(\frac{2}{r^2}+\frac{\Gamma}{r^3}\Big)\Psi_2=0,$$
$$\Big(\frac{d^2}{dr^2}+\frac{2}{r}\frac{d}{dr}+2mE+\frac{2m\alpha}{r}-\frac{2\nu^2+2}{r^2}\Big)\Psi_2-2\nu\Big(\frac{2}{r^2}+\frac{\Gamma}{r^3}\Big)\Psi_1=0. \quad (14.55)$$

This system for two functions permits us to construct the fourth order equations for the functions $\Psi_1(r)$ and $\Psi_2(r)$. It suffices to study only one of them, for example, the one for the function Ψ_1. We will use the dimensionless variables:

$$x = rm, \quad \Gamma m = \gamma, \quad \epsilon = \frac{E}{m}; \tag{14.56}$$

as a result we get

$$\begin{aligned}
&\frac{d^4\Psi_1}{dx^4} + \left[\frac{10}{x} - \frac{4}{2x+\gamma}\right]\frac{d^3\Psi_1}{dx^3} \\
&+\left[4\epsilon + \frac{-24+4\alpha\gamma}{\gamma x} + \frac{22-4\nu^2}{x^2} - \frac{2\gamma}{x^3} - \frac{\gamma^2}{x^4} + \frac{48}{(2x+\gamma)\gamma} + \frac{8}{(2x+\gamma)^2}\right]\frac{d^2\Psi_1}{dx^2} \\
&+\left[\frac{64-16\nu^2+20\gamma^2\epsilon-8\alpha\gamma}{\gamma^2 x} + \frac{-24+8\nu^2+16\alpha\gamma}{\gamma x^2} + \frac{8-12\nu^2}{x^3}\right. \\
&\left.+\frac{-128-8\gamma^2\epsilon+16\alpha\gamma+32\nu^2}{(2x+\gamma)\gamma^2} - \frac{32}{(2x+\gamma)^2\gamma}\right]\frac{d\Psi_1}{dx} \\
&+\left[+4\epsilon^2 + \frac{128\nu^2+8\epsilon\alpha\gamma^3+64\alpha\gamma-32\gamma^2\epsilon}{x\gamma^3}\right. \\
&+\frac{-24\alpha\gamma-48\nu^2+20\gamma^2\epsilon+4\alpha^2\gamma^2-8\epsilon\nu^2\gamma^2}{\gamma^2x^2} \\
&+\frac{8\alpha\gamma-4\gamma^2\epsilon+16\nu^2-8\alpha\nu^2\gamma}{\gamma x^3} + \frac{-8\nu^2-4\alpha\gamma-2\gamma^2\epsilon+4\nu^4}{x^4} \\
&-\frac{2\gamma(-2+2\nu^2+\alpha\gamma)}{x^5} + \frac{2\gamma^2}{x^6} \\
&\left.+\frac{-128\alpha\gamma+64\gamma^2\epsilon-256\nu^2}{(2x+\gamma)\gamma^3} + \frac{-32\alpha\gamma+16\gamma^2\epsilon-64\nu^2}{(2x+\gamma)^2\gamma^2}\right]\Psi_1 = 0.
\end{aligned} \tag{14.57}$$

The symbolic structure of the equation (14.57) is written as follows (let $\Psi_1 = \Psi$)

$$\begin{aligned}
&\frac{d^4}{dx^4}\Psi_1 + \left[\frac{10}{x} - \frac{4}{2x+\gamma}\right]\frac{d^3}{dx^3}\Psi \\
&+\left[4\epsilon + \frac{a_1}{x} + \frac{a_2}{x^2} + \frac{a_3}{x^3} + \frac{a_4}{x^4} + \frac{a_5}{2x+\gamma} + \frac{a_6}{(2x+\gamma)^2}\right]\frac{d^2}{dx^2}\Psi \\
&+\left[\frac{b_1}{x} + \frac{b_2}{x^2} + \frac{b_3}{x^3} + \frac{b_4}{2x+\gamma} + \frac{b_5}{(2x+\gamma)^2}\right]\frac{d}{dx}\Psi \\
&+\left[4\epsilon^2 + \frac{c_1}{x} + \frac{c_2}{x^2} + \frac{c_3}{x^3} + \frac{c_4}{x^4} + \frac{c_5}{x^5} + \frac{c_6}{x^6} + \frac{c_7}{2x+\gamma} + \frac{c_8}{(2x+\gamma)^2}\right]\Psi = 0.
\end{aligned} \tag{14.58}$$

We will search for Frobenius type solutions in the form

$$\Psi(x) = x^A e^{Bx} e^{C/x} f(x); \tag{14.59}$$

this yields

$$
\frac{d^4 f}{dx^4} + \Big[\frac{4A+10}{x} - \frac{4C}{x^2} + 4B - \frac{4}{(2x+\gamma)}\Big]\frac{d^3 f}{dx^3}
$$

$$
+ \Big[\frac{-24C + a_1\gamma^2 - 12A\gamma + 30B\gamma^2 + 12AB\gamma^2}{\gamma^2 x} + \frac{6A^2\gamma + a_2\gamma + 12C - 12BC\gamma + 24A\gamma}{\gamma x^2}
$$

$$
+ \frac{a_3 - 12AC - 18C}{x^3} + \frac{6C^2 + a_4}{x^4}
$$

$$
+ \frac{48C + 24A\gamma - 12B\gamma^2 + a_5\gamma^2}{(2x+\gamma)\gamma^2} + 6B^2 + 4\epsilon + \frac{a_6}{(2x+\gamma)^2}\Big]\frac{d^2 f}{dx^2}
$$

$$
+ \Big[\frac{1}{\gamma^4 x}\Big(8\epsilon A\gamma^4 + 2a_1 B\gamma^4 + 24A^2\gamma^2 - 24A\gamma^2 - 96C\gamma + 96C^2 + 12AB^2\gamma^4 + 2a_5 A\gamma^3
$$

$$
+ 4a_5 C\gamma^2 + 2a_6 A\gamma^2 + 8a_6 C\gamma - 24AB\gamma^3 + 96AC\gamma + 30B^2\gamma^4 + b_1\gamma^4 - 48BC\gamma^2\Big)
$$

$$
+ \frac{1}{\gamma^3 x^2}(-8\epsilon C\gamma^3 + 2a_1 A\gamma^3 + 2a_2 B\gamma^3 - 12A^2\gamma^2 + 12A\gamma^2 + 48C\gamma - 48C^2
$$

$$
+ 48AB\gamma^3 + 12A^2B\gamma^3 - 12B^2C\gamma^3 - 2a_5 C\gamma^2 - 2a_6 C\gamma - 48AC\gamma + b_2\gamma^3 + 24BC\gamma^2)
$$

$$
+ \frac{1}{\gamma^2 x^3}(-2a_1 C\gamma^2 + 2a_2 A\gamma^2 + 2a_3 B\gamma^2 - 24C\gamma + 24C^2 - 36BC\gamma^2
$$

$$
+ 24AC\gamma + 4A^3\gamma^2 + 18A^2\gamma^2 - 22A\gamma^2 + b_3\gamma^2 - 24ABC\gamma^2)
$$

$$
+ \frac{-2a_2 C\gamma + 2a_3 A\gamma + 2a_4 B\gamma - 12C^2 - 24AC\gamma + 12BC^2\gamma - 12A^2C\gamma + 36C\gamma}{\gamma x^4}
$$

$$
+ \frac{-2a_3 C + 2a_4 A + 12AC^2 + 6C^2}{x^5} - \frac{2C\,(a_4 + 2C^2)}{x^6}
$$

$$
+ 8\epsilon B + 4B^3 + \frac{2a_6 B\gamma^2 - 4a_6 A\gamma - 8a_6 C + b_5\gamma^2}{(2x+\gamma)^2\gamma^2}
$$

$$
+ \frac{1}{(2x+\gamma)\gamma^4}(2a_5 B\gamma^4 - 48A^2\gamma^2 + 48A\gamma^2 + 192C\gamma - 192C^2 - 4a_5 A\gamma^3
$$

$$
-8a_5 C\gamma^2 - 4a_6 A\gamma^2 - 16a_6 C\gamma + 48AB\gamma^3 - 192AC\gamma - 12B^2\gamma^4 + b_4\gamma^4 + 96BC\gamma^2\Big)\Big]\frac{df}{dx}
$$

$$
+ \Big[\frac{1}{\gamma^6 x}(288AC\gamma^2 + 384C^2\gamma - 96A^2C\gamma^2 + 96BC^2\gamma^2 - 32A\gamma^3 - 16A^3\gamma^3 - 192AC^2\gamma
$$

$$
-8a_5 C^2\gamma^2 - 96BC\gamma^3 - 2a_5 A^2\gamma^4 + 2a_5 A\gamma^4 + 24a_6 C\gamma^2 + 2b_4 C\gamma^4 + 8a_5 C\gamma^3 + 4a_6 A\gamma^3
$$

$$
-24AB\gamma^4 - 24B^2C\gamma^4 - 32a_6 C^2\gamma + 24A^2B\gamma^4 - 4a_6 A^2\gamma^3 + 4b_5 C\gamma^3 - 8a_5 AC\gamma^3 + 96ABC\gamma^3
$$

$$
+4a_5 BC\gamma^4 - 24a_6 AC\gamma^2 + 8a_6 BC\gamma^3 + 2a_5 AB\gamma^5 + 2a_6 AB\gamma^4 - 128C^3 - 12AB^2\gamma^5 + 48A^2\gamma^3
$$

$$
-192C\gamma^2 + 10B^3\gamma^6 + b_1 B\gamma^6 + a_1 B^2\gamma^6 + 4AB^3\gamma^6 + b_4 A\gamma^5 + b_5 A\gamma^4 + c_1\gamma^6 + 8\epsilon AB\gamma^6)
$$

$$
+ \frac{1}{\gamma^5 x^2}(-144AC\gamma^2 - 192C^2\gamma + 48A^2C\gamma^2 - 48BC^2\gamma^2 + b_2 B\gamma^5 + 16A\gamma^3 + 8A^3\gamma^3 - 4B^3C\gamma^5
$$

$$
+b_1 A\gamma^5 + 96AC^2\gamma + 2a_1 AB\gamma^5 - 8\epsilon BC\gamma^5 + 4a_5 C^2\gamma^2 + 48BC\gamma^3 + a_5 A^2\gamma^4 - a_5 A\gamma^4
$$

$$
-8a_6 C\gamma^2 - b_4 C\gamma^4 - 4a_5 C\gamma^3 - a_6 A\gamma^3 + 12AB\gamma^4 + 12B^2C\gamma^4 + 12a_6 C^2\gamma - 12A^2B\gamma^4
$$

$$
+a_6 A^2\gamma^3 - b_5 C\gamma^3 + 4a_5 AC\gamma^3 - 48ABC\gamma^3 - 2a_5 BC\gamma^4 + 8a_6 AC\gamma^2 - 2a_6 BC\gamma^3 + 64C^3
$$

$$
+24AB^2\gamma^5 - 24A^2\gamma^3 + a_2 B^2\gamma^5 + 4\epsilon A^2\gamma^5 + 6A^2B^2\gamma^5 - 4\epsilon A\gamma^5 + 96C\gamma^2 + c_2\gamma^5)
$$

$$
+ \frac{1}{\gamma^4 x^3}(72AC\gamma^2 + 96C^2\gamma - 24A^2C\gamma^2 + 24BC^2\gamma^2 - 8A\gamma^3 - 4A^3\gamma^3 - 48AC^2\gamma - 2a_5 C^2\gamma^2
$$

$$
-24BC\gamma^3 + 2a_6 C\gamma^2 + 2a_5 C\gamma^3 - 22AB\gamma^4 - 18B^2C\gamma^4 - 4a_6 C^2\gamma + 18A^2B\gamma^4 - 2a_1 BC\gamma^4
$$

$$
+2a_2 AB\gamma^4 - 8\epsilon AC\gamma^4 - 12AB^2C\gamma^4 - 2a_5 AC\gamma^3 + 24ABC\gamma^3 - 2a_6 AC\gamma^2 - 32C^3 + 12A^2\gamma^3
$$

$$
-48C\gamma^2 + c_3\gamma^4 + 4A^3B\gamma^4 + b_2 A\gamma^4 + a_3 B^2\gamma^4 - a_1 A\gamma^4 + 8\epsilon C\gamma^4 + a_1 A^2\gamma^4 + b_3 B\gamma^4 - b_1 C\gamma^4)
$$

$$+\frac{1}{\gamma^3x^4}(-36AC\gamma^2-48C^2\gamma+12A^2C\gamma^2-12BC^2\gamma^2+14A\gamma^3+4A^3\gamma^3+24AC^2\gamma$$
$$+a_5C^2\gamma^2+36BC\gamma^3+a_6C^2\gamma-12A^2BC\gamma^3-2a_1AC\gamma^3+2a_3AB\gamma^3-2a_2BC\gamma^3$$
$$-24ABC\gamma^3+16C^3-19A^2\gamma^3+24C\gamma^2+c_4\gamma^3+A^4\gamma^3-a_2A\gamma^3+2a_1C\gamma^3$$
$$+a_4B^2\gamma^3+a_2A^2\gamma^3+4\epsilon C^2\gamma^3-b_2C\gamma^3+6B^2C^2\gamma^3+b_3A\gamma^3)$$
$$+\frac{1}{\gamma^2x^5}(46AC\gamma^2+a_1C^2\gamma^2+a_3A^2\gamma^2+2a_2C\gamma^2-a_3A\gamma^2-b_3C\gamma^2+24C^2\gamma-8C^3-6A^2C\gamma^2$$
$$+6BC^2\gamma^2-4A^3C\gamma^2-2a_3BC\gamma^2+2a_4AB\gamma^2-12AC^2\gamma-36C\gamma^2+c_5\gamma^2+12ABC^2\gamma^2-2a_2AC\gamma^2)$$
$$+\frac{a_2C^2\gamma+a_4A^2\gamma+2a_3C\gamma-a_4A\gamma+4C^3+6A^2C^2\gamma-4BC^3\gamma-2a_3AC\gamma-2a_4BC\gamma-24C^2\gamma+c_6\gamma}{\gamma x^6}$$
$$-\frac{C\left(-a_3C-2a_4+4AC^2+2a_4A-2C^2\right)}{x^7}+\frac{C^2\left(a_4+C^2\right)}{x^8}+B^4+4\epsilon^2+4\epsilon B^2$$
$$+\frac{1}{(2x+\gamma)\gamma^6}\Big(-576AC\gamma^2-768C^2\gamma+192A^2C\gamma^2-192BC^2\gamma^2+64A\gamma^3+32A^3\gamma^3+384AC^2\gamma$$
$$+16a_5C^2\gamma^2+192BC\gamma^3+4a_5A^2\gamma^4-4a_5A\gamma^4-48a_6C\gamma^2-4b_4C\gamma^4-16a_5C\gamma^3-8a_6A\gamma^3$$
$$+48AB\gamma^4+48B^2C\gamma^4+64a_6C^2\gamma-48A^2B\gamma^4+8a_6A^2\gamma^3-8b_5\gamma^3+16a_5AC\gamma^3-192ABC\gamma^3$$
$$-8a_5BC\gamma^4+48a_6AC\gamma^2-16a_6BC\gamma^3-4a_5AB\gamma^5-4a_6AB\gamma^4+256C^3$$
$$+24AB^2\gamma^5-96A^2\gamma^3+384C\gamma^2-4B^3\gamma^6+c_7\gamma^6+a_5B^2\gamma^6+b_4B\gamma^6-2b_4A\gamma^5-2b_5A\gamma^4\Big)$$
$$+\frac{1}{(2x+\gamma)^2\gamma^4}\Big(a_6B^2\gamma^4+b_5B\gamma^4+4a_6A^2\gamma^2+16a_6C^2-2b_5\gamma^3-4a_6A\gamma^2$$
$$-16a_6C\gamma-4b_5C\gamma^2+c_8\gamma^4-4a_6AB\gamma^3+16a_6AC\gamma-8a_6BC\gamma^2\Big)\Big]f=0. \quad (14.60)$$

We impose restrictions on the parameters B and C:

$$B^4+4\epsilon^2+4\epsilon B^2=0 \quad \Longrightarrow \quad B=-\sqrt{-2\epsilon},+\sqrt{-2\epsilon}, \quad (14.61)$$

$$\frac{1}{x^8}, \quad C^2\left(a_4+C^2\right)=0 \Rightarrow C=0,\ \pm\sqrt{-a_4} \Rightarrow C_1=0,\ C_2=+\gamma,\ C_3=-\gamma. \quad (14.62)$$

While $C=C_1=0$, the coefficient at $1/x^7$ vanishes, so we require the coefficient at $1/x^6$ be equal to zero as well:

$$a_2C^2\gamma+a_4A^2\gamma+2a_3C\gamma-a_4A\gamma+4C^3+6A^2C^2\gamma$$
$$-4BC^3\gamma-2a_3AC\gamma-2a_4BC\gamma-24C^2\gamma+c_6\gamma=0,$$

or

$$a_4A^2\gamma-a_4A\gamma+c_6\gamma=0 \Longrightarrow -\gamma^3A^2+\gamma^3A+2\gamma^3=0 \Longrightarrow$$

$$A_1=-1,\ A_2=+2. \quad (14.63)$$

If $C=C_2=+\gamma$, we demand the multiplier at $1/x^7$ to be equal to zero:

$$-a_3C-2a_4+4AC^2+2a_4A-2C^2=0,\ a_3=-2\gamma \Longrightarrow$$
$$2\gamma^2+2A\gamma^2=0 \quad \Longrightarrow \quad A_3=-1. \quad (14.64)$$

Let $C = C_3 = -\gamma$; then we demand the multiplier at $1/x^7$ to be equal to zero:

$$-a_3C - 2a_4 + 4AC^2 + 2a_4A - 2C^2 = 0\,, \quad a_3 = -2\gamma \implies$$

$$-2\gamma^2 + 2A\gamma^2 = 0 \implies A_4 = +1\,. \tag{14.65}$$

Thus, there are four types of solutions (below only the negative values of B are used):

$$\begin{aligned}
&I, \quad B = -\sqrt{-2\epsilon}\,, \quad C = 0\,, \quad A = -1\,, \quad \Psi = e^{Bx}\frac{1}{x}f_1(x)\,;\\
&II, \quad B = -\sqrt{-2\epsilon}\,, \quad C = 0\,, \quad A = +2\,, \quad \Psi = e^{Bx}x^2 f_2(x)\,;\\
&III, \quad B = -\sqrt{-2\epsilon}\,, \quad C = +\gamma\,, \quad A = -1\,, \quad \Psi(x) = e^{Bx}\frac{1}{x}e^{+\gamma/x}f_3(x)\,;\\
&IV, \quad B = -\sqrt{-2\epsilon}\,, \quad C_3 = -\gamma\,, \quad A = +1\,, \quad \Psi(x) = e^{Bx}xe^{-\gamma/x}f_4(x)\,.
\end{aligned} \tag{14.66}$$

Only three cases may be be suitable for describing the bound states:

$$II; \quad III \quad \text{at negative } \gamma\,; \quad IV \quad \text{at positive } \gamma\,. \tag{14.67}$$

Let us examine the variant II. The main equation takes the following form:

$$\begin{aligned}
&\frac{d^4\Phi}{dx^4} + \left[4B + \frac{18}{x} - \frac{4}{2x+\gamma}\right]\frac{d^3\Phi}{dx^3}\\
&+\Big[4\epsilon + 6B^2 + \frac{a_1\gamma + 54B\gamma - 24}{\gamma x} + \frac{a_2 + 72}{x^2} + \frac{a_3}{x^3}\\
&+\frac{a_4}{x^4} + \frac{a_5\gamma - 12B\gamma + 48}{(2x+\gamma)\gamma} + \frac{a_6}{(2x+\gamma)^2}\Big]\frac{d^2\Phi}{dx^2}\\
&+\Big[\frac{4a_5\gamma + 48 + b_1\gamma^2 + 16\epsilon\gamma^2 + 4a_6 - 48B\gamma + 2a_1B\gamma^2 + 54B^2\gamma^2}{\gamma^2 x}\\
&+\frac{-24 + 4a_1\gamma + b_2\gamma + 2a_2B\gamma + 144B\gamma}{\gamma x^2} + \frac{b_3 + 60 + 4a_2 + 2a_3B}{x^3} + \frac{4a_3 + 2a_4B}{x^4} + \frac{4a_4}{x^5}\\
&+\frac{-96 + b_4\gamma^2 - 8a_6 - 8a_5\gamma + 96B\gamma + 2a_5B\gamma^2 - 12B^2\gamma^2}{(2x+\gamma)\gamma^2} + \frac{b_5\gamma - 8a_6 + 2a_6B\gamma}{(2x+\gamma)^2\gamma}\Big]\frac{d\Phi}{dx}\\
&+\Big[\frac{1}{\gamma^3 x}(4a_5B\gamma^2 + 4a_6B\gamma + 2b_4\gamma^2 - 4a_5\gamma - 8a_6\\
&+2b_5\gamma + c_1\gamma^3 + a_1B^2\gamma^3 + b_1B\gamma^3 - 24B^2\gamma^2 + 48B\gamma + 16\epsilon B\gamma^3 + 18B^3\gamma^3)\\
&+\frac{8\epsilon\gamma^2 + 2a_5\gamma + c_2\gamma^2 + 2b_1\gamma^2 + 2a_6 + a_2B^2\gamma^2 + b_2B\gamma^2 - 24B\gamma + 4a_1B\gamma^2 + 72B^2\gamma^2}{\gamma^2 x^2}\\
&+\frac{2a_1 + 2b_2 + c_3 + a_3B^2 + b_3B + 4a_2B + 60B}{x^3} + \frac{c_4 + 2a_2 + 2b_3 + 4a_3B + a_4B^2}{x^4}\\
&+\frac{c_5 + 2a_3 + 4a_4B}{x^5} + \frac{1}{(2x+\gamma)\gamma^3}(-8a_5B\gamma^2 - 8a_6B\gamma + 8a_5\gamma + c_7\gamma^3 - 4b_5\gamma - 4b_4\gamma^2\\
&+16a_6 + a_5B^2\gamma^3 + b_4B\gamma^3 + 48B^2\gamma^2 - 4B^3\gamma^3 - 96B\gamma)\\
&+\frac{-8a_6B\gamma + c_8\gamma^2 - 4b_5\gamma + 8a_6 + a_6B^2\gamma^2 + b_5B\gamma^2}{(2x+\gamma)^2\gamma^2}\Big]\Phi = 0\,.
\end{aligned}$$

Shortly this equation can be written as

$$\frac{d^4\Phi}{dx^4}+\left[P_0+\frac{P_1}{x}+\frac{p_1}{2x+\gamma}\right]\frac{d^3\Phi}{dx^3}$$
$$+\left[Q_0+\frac{Q_1}{x}+\frac{Q_2}{x^2}+\frac{Q_3}{x^3}+\frac{Q_4}{x^4}+\frac{q_1}{2x+\gamma}+\frac{q_2}{(2x+\gamma)^2}\right]\frac{d^2\Phi}{dx^2}$$
$$+\left[\frac{M_1}{x}+\frac{M_2}{x^2}+\frac{M_3}{x^3}+\frac{M_4}{x^4}+\frac{M_5}{x^5}+\frac{m_1}{2x+\gamma}+\frac{m_2}{(2x+\gamma)^2}\right]\frac{d\Phi}{dx}$$
$$+\left[\frac{N_1}{x}+\frac{N_2}{x^2}+\frac{N_3}{x^3}+\frac{N_4}{x^4}+\frac{N_5}{x^5}+\frac{n_1}{2x+\gamma}+\frac{n_2}{(2x+\gamma)^2}\right]\Phi=0. \qquad (14.68)$$

Let us multiply Eq. (14.68) by $x^5(2x+\gamma)^2$:

$$\left[4x^7+4\gamma x^6+\gamma^2x^5\right]\frac{d^4\Phi}{dx^4}$$
$$+\left[4P_0x^7+(4P_0\gamma+4P_1+2p_1)x^6+\left(P_0\gamma^2+4P_1\gamma+p_1\gamma\right)x^5+P_1\gamma^2x^4\right]\frac{d^3\Phi}{dx^3}$$
$$+\Big[4Q_0x^7+(4Q_0\gamma+4Q_1+2q_1)x^6+\left(Q_0\gamma^2+4Q_1\gamma+4Q_2+\gamma q_1+q_2\right)x^5$$
$$+\left(Q_1\gamma^2+4Q_2\gamma+4Q_3\right)x^4+\left(Q_2\gamma^2+4Q_3\gamma+4Q_4\right)x^3+\left(Q_3\gamma^2+4Q_4\gamma\right)x^2+Q_4\gamma^2x\Big]\frac{d^2\Phi}{dx^2}$$
$$+\Big[(4M_1+2m_1)x^6+(4M_1\gamma+4M_2+m_1\gamma+m_2)x^5+\left(M_1\gamma^2+4M_2\gamma+4M_3\right)x^4$$
$$+\left(M_2\gamma^2+4M_3\gamma+4M_4\right)x^3+\left(M_3\gamma^2+4M_4\gamma+4M_5\right)x^2+\left(M_4\gamma^2+4M_5\gamma\right)x+M_5\gamma^2\Big]\frac{d\Phi}{dx}$$
$$+\Big[(4N_1+2n_1)x^6+(4N_1\gamma+4N_2+\gamma n_1+n_2)x^5+\left(N_1\gamma^2+4N_2\gamma+4N_3\right)x^4$$
$$+\left(N_2\gamma^2+4N_3\gamma+4N_4\right)x^3+\left(N_3\gamma^2+4N_4\gamma+4N_5\right)x^2+\left(N_4\gamma^2+4N_5\gamma\right)x+N_5\gamma^2\Big]\Phi=0,$$

and we search for its solution as a power series: $\Phi=\sum_{l=0}^{\infty}d_lx^l$. After calculations, we we find the 8-term recurrence relations ($k\geq 6$):

$$(4N_1+2n_1)d_{k-6}+[4Q_0(k-5)(k-6)(4M_1+2m_1)(k-5)+(4N_1\gamma+4N_2+\gamma n_1+n_2)]d_{k-5}$$
$$+[4P_0(k-4)(k-5)(k-6)+(4Q_0\gamma+4Q_1+2q_1)(k-4)(k-5)$$
$$+(4M_1\gamma+4M_2+m_1\gamma+m_2)(k-4)+\left(N_1\gamma^2+4N_2\gamma+4N_3\right)\Big]d_{k-4}$$
$$+[4(k-3)(k-4)(k-5)(k-6)+(4P_0\gamma+4P_1+2p_1)(k-3)(k-4)(k-5)$$
$$+\left(Q_0\gamma^2+4Q_1\gamma+4Q_2+\gamma q_1+q_2\right)(k-3)(k-4)$$
$$+\left(M_1\gamma^2+4M_2\gamma+4M_3\right)(k-3)+\left(N_2\gamma^2+4N_3\gamma+4N_4\right)\Big]d_{k-3}$$
$$+\Big[4\gamma(k-2)(k-3)(k-4)(k-5)+\left(P_0\gamma^2+4P_1\gamma+p_1\gamma\right)(k-2)(k-3)(k-4)$$
$$+\left(Q_1\gamma^2+4Q_2\gamma+4Q_3\right)(k-2)(k-3)$$
$$+\left(M_2\gamma^2+4M_3\gamma+4M_4\right)(k-2)+\left(N_3\gamma^2+4N_4\gamma+4N_5\right)\Big]d_{k-2}$$
$$+\Big[\gamma^2(k-1)(k-2)(k-3)(k-4)+P_1\gamma^2(k-1)(k-2)(k-3)$$
$$+\left(Q_2\gamma^2+4Q_3\gamma+4Q_4\right)(k-1)(k-2)+\left(M_3\gamma^2+4M_4\gamma+4M_5\right)(k-1)$$
$$+\left(N_4\gamma^2+4N_5\gamma\right)\Big]d_{k-1}++\Big[\left(Q_3\gamma^2+4Q_4\gamma\right)k(k-1)+\left(M_4\gamma^2+4M_5\gamma\right)k+N_5\gamma^2\Big]d_k$$
$$+\Big[Q_4\gamma^2(k+1)k+M_5\gamma^2(k+1)\Big]d_{k+1}=0,$$

or, shortly,

$$
\begin{aligned}
&D_{k-6}d_{k-6}+D_{k-5}d_{k-5}+D_{k-4}d_{k-4}+D_{k-3}d_{k-3}\\
&+D_{k-2}d_{k-2}+D_{k-1}d_{k-1}+D_k d_k+D_{k+1}d_{k+1}=0.
\end{aligned} \tag{14.69}
$$

We divide the equation (14.69) by $d_{k-6}k^4$ and tend $k\to\infty$; we consequently get an algebraic equation for the parameter R:

$$
4R^3+4\gamma R^4+\gamma^2R^5=0 \quad \Rightarrow \quad R=0,\ -\frac{2}{\gamma}.
$$

Therefore, following convergence radii are possible:

$$
R_{\text{conv}}=\frac{1}{|R|}=\frac{|\gamma|}{2},\ \infty. \tag{14.70}
$$

As a quantization rule, we use the transcendence condition. We perform this by the following 8-term recurrence relation

$$
\begin{aligned}
&D_{k-6}d_{k-6}+D_{k-5}d_{k-5}+D_{k-4}d_{k-4}+D_{k-3}d_{k-3}\\
&+D_{k-2}d_{k-2}+D_{k-1}d_{k-1}+D_k d_k+D_{k+1}d_{k+1}=0.
\end{aligned} \tag{14.71}
$$

We require vanishing the coefficient D_{k-6}, $k\geq 6$: $D_{k-6}=4N_1+2n_1=0$, where

$$
\begin{aligned}
N_1 &= \frac{1}{\gamma^3}(4a_5B\gamma^2+4a_6B\gamma+2b_4\gamma^2-4a_5\gamma-8a_6\\
&+2b_5\gamma+c_1\gamma^3+a_1B^2\gamma^3+b_1B\gamma^3-24B^2\gamma^2+48B\gamma+16\epsilon B\gamma^3+18B^3\gamma^3),\\
n_1 &= \frac{1}{\gamma^3}(-8a_5B\gamma^2-8a_6B\gamma+8a_5\gamma+c_7\gamma^3-4b_5\gamma-4b_4\gamma^2\\
&+16a_6+a_5B^2\gamma^3+b_4B\gamma^3+48B^2\gamma^2-4B^3\gamma^3-96B\gamma)=0.
\end{aligned}
$$

Taking into consideration the explicit form of N_1 and n_1, we get

$$
D_{k-6}=64B\epsilon+2b_4B+2a_5B^2+4b_1B+4a_1B^2+4c_1+64B^3+2c_7.
$$

Accounting for the expressions of $a_1, a_5, b_1, b_4, c_1, c_7$, we derive

$$
D_{k-6}=64\,(B+\frac{1}{4}\alpha)\,(B^2+2\,\epsilon)\,. \tag{14.72}
$$

Because of $B=\pm\sqrt{-2\epsilon}$, the coefficient D_{k-6} vanishes identically. This means that the structure of the power series is described by 7-term recurrence relations.

Let us study the transcendency condition for this 7-term power series, assuming that the coefficient at d_{n-5} vanishes:

$$
4Q_0(k-5)(k-6)+(4M_1+2m_1)(k-5)+(4N_1\gamma+4N_2+\gamma n_1+n_2)=0,
$$

where

$$B = -\sqrt{-2\epsilon}, \quad Q_0 = 4\epsilon + 6B^2,$$

$$M_1 = \frac{4a_5\gamma + 48 + b_1\gamma^2 + 16\epsilon\gamma^2 + 4a_6 - 48B\gamma + 2a_1B\gamma^2 + 54B^2\gamma^2}{\gamma^2},$$

$$m_1 = \frac{-96 + b_4\gamma^2 - 8a_6 - 8a_5\gamma + 96B\gamma + 2a_5B\gamma^2 - 12B^2\gamma^2}{\gamma^2},$$

$$N_1 = \frac{1}{\gamma^3}(4a_5B\gamma^2 + 4a_6B\gamma + 2b_4\gamma^2 - 4a_5\gamma - 8a_6 +$$
$$+2b_5\gamma + c_1\gamma^3 + a_1B^2\gamma^3 + b_1B\gamma^3 - 24B^2\gamma^2 + 48B\gamma + 16\epsilon B\gamma^3 + 18B^3\gamma^3),$$

$$N_2 = \frac{8\epsilon\gamma^2 + 2a_5\gamma + c_2\gamma^2 + 2b_1\gamma^2 + 2a_6 + a_2B^2\gamma^2 + b_2B\gamma^2 - 24B\gamma + 4a_1B\gamma^2 + 72B^2\gamma^2}{\gamma^2},$$

$$n_1 = \frac{1}{\gamma^3}(-8a_5B\gamma^2 - 8a_6B\gamma + 8a_5\gamma + c_7\gamma^3 - 4b_5\gamma - 4b_4\gamma^2 +$$
$$+16a_6 + a_5B^2\gamma^3 + b_4B\gamma^3 + 48B^2\gamma^2 - 4B^3\gamma^3 - 96B\gamma) = 0,$$

$$n_2 = \frac{-8a_6B\gamma + c_8\gamma^2 - 4b_5\gamma + 8a_6 + a_6B^2\gamma^2 + b_5B\gamma^2}{\gamma^2}.$$

We further obtain

$$68B^3\gamma + \left[(a_5 + 4a_1)\gamma + a_6 - 72k + 24k^2 + 4a_2\right]B^2$$
$$+ \left[(b_4 + 4b_1 + 64\epsilon)\gamma + (8a_1 + 4a_5)k + 4b_2 + b_5 - 12a_5 - 24a_1\right]B$$
$$+ (4c_1 + c_7)\gamma + 16\epsilon k^2 + (2b_4 - 112\epsilon + 4b_1)k + c_8 + 192\epsilon + 4c_2 - 6b_4 - 12b_1 = 0,$$

which, using the explicit expressions for a_1, a_2, a_5, ..., reduces to

$$68B^3\gamma + \left[48 + 16\alpha\gamma - 72k - 16\nu^2 + 24k^2\right]B^2 + \left[136\epsilon\gamma + (32k - 48)\alpha\right]B$$
$$+ (32 - 32\nu^2 + 32\alpha\gamma + 16k^2 - 48k)\epsilon + 16\alpha^2 = 0.$$

With the equality $B = -\sqrt{-2\epsilon}$) in mind, we derive a quadratic equation with respect to B, and get the following solutions:

$$\epsilon_1 = -\frac{1}{2}\frac{\alpha^2}{(k-2)^2}, \quad \epsilon_2 = -\frac{1}{2}\frac{\alpha^2}{(k-1)^2}. \tag{14.73}$$

These expressions for energy formulas do not include the quantum number j, so they are unlikely to be relevant for the correct energy spectra.

Let us examine the solutions of type *III* (14.66). In order to describe the bound states, the parameter γ must be negative. When $A = -1$, $B = -\sqrt{-2\epsilon}$, and $C = +\gamma$, the equation (14.60) takes the form:

$$\frac{d^4f}{dx^4} + \left[4B + \frac{6}{x} - \frac{4\gamma}{x^2} - \frac{4}{2x+\gamma}\right]\frac{d^3f}{dx^3}$$

$$+ \left[6B^2 + 4\epsilon + \frac{-12\gamma + a_1\gamma^2 + 18B\gamma^2}{\gamma^2 x} + \frac{-6\gamma + a_2\gamma - 12B\gamma^2}{\gamma x^2} - \frac{8\gamma}{x^3} + \frac{5\gamma^2}{x^4}\right.$$
$$\left. + \frac{24\gamma - 12B\gamma^2 + a_5\gamma^2}{(2x+\gamma)\gamma^2} + \frac{a_6}{(2x+\gamma)^2}\right]\frac{d^2f}{dx^2}$$

$$+ \left[4B^3 + 8\epsilon B + \frac{-8\epsilon\gamma^4 + 2a_1B\gamma^4 - 48\gamma^2 + 18B^2\gamma^4 + 2a_5\gamma^3 + 6a_6\gamma^2 - 24B\gamma^3 + b_1\gamma^4}{\gamma^4 x}\right.$$

$$+\frac{-8\epsilon\gamma^4-2a_1\gamma^3+2a_2B\gamma^3+24\gamma^2-12B\gamma^3-12B^2\gamma^4-2a_5\gamma^3-2a_6\gamma^2+b_2\gamma^3}{\gamma^3x^2}$$

$$+\frac{-2a_1\gamma^3-2a_2\gamma^2-16B\gamma^3+12\gamma^2+b_3\gamma^2}{\gamma^2x^3}+\frac{-2a_2\gamma^2+40\gamma^2+10B\gamma^3}{\gamma x^4}-\frac{2\gamma^3}{x^6}$$

$$+\frac{2a_5B\gamma^4+96\gamma^2-4a_5\gamma^3-12a_6\gamma^2+48B\gamma^3-12B^2\gamma^4+b_4\gamma^4}{(2x+\gamma)\gamma^4}+\frac{2a_6B\gamma^2-4a_6\gamma+b_5\gamma^2}{(2x+\gamma)^2\gamma^2}\Bigg]\frac{df}{dx}$$

$$+\Bigg[\frac{1}{\gamma^6x}\Big(-32\gamma^3+c_1\gamma^6+6B^3\gamma^6+2a_5B\gamma^5+6a_6B\gamma^4-48B\gamma^4+4a_5\gamma^4$$

$$+8a_6\gamma^3+b_4\gamma^5-12B^2\gamma^5+3b_5\gamma^4+b_1B\gamma^6+a_1B^2\gamma^6-8\epsilon B\gamma^6\Big)$$

$$+\frac{1}{\gamma^5x^2}\Big(16\gamma^3-b_1\gamma^5+c_2\gamma^5-4B^3\gamma^6-2a_5B\gamma^5-2a_6B\gamma^4+24B\gamma^4-2a_5\gamma^4-2a_6\gamma^3$$

$$-b_4\gamma^5-6B^2\gamma^5-b_5\gamma^4+b_2B\gamma^5+a_2B^2\gamma^5-8\epsilon B\gamma^6+8\epsilon\gamma^5-2a_1B\gamma^5\Big)$$

$$+\frac{1}{\gamma^4x^3}\Big(-8\gamma^3-b_1\gamma^5-b_2\gamma^4+2a_1\gamma^4+c_3\gamma^4$$

$$+12B\gamma^4+2a_5\gamma^4-8B^2\gamma^5+b_3B\gamma^4+16\epsilon\gamma^5-2a_1B\gamma^5-2a_2B\gamma^4\Big)$$

$$+\frac{-20\gamma^3-b_2\gamma^4+4a_1\gamma^4+2a_2\gamma^3-b_3\gamma^3+c_4\gamma^3+40B\gamma^4+a_5\gamma^4+a_6\gamma^3+5B^2\gamma^5+4\epsilon\gamma^5-2a_2B\gamma^4}{\gamma^3x^4}$$

$$+\frac{-60\gamma^3+a_1\gamma^4+4a_2\gamma^3-b_3\gamma^3+c_5\gamma^2}{\gamma^2x^5}+\frac{a_2\gamma^3-22\gamma^3-2B\gamma^4}{\gamma x^6}$$

$$+\frac{1}{(2x+\gamma)\gamma^6}\Big(64\gamma^3+a_5B^2\gamma^6+b_4B\gamma^6-4B^3\gamma^6-4a_5B\gamma^5-12a_6B\gamma^4+c_7\gamma^6$$

$$+96B\gamma^4-8a_5\gamma^4-16a_6\gamma^3-2b_4\gamma^5+24B^2\gamma^5-6b_5\gamma^4\Big)$$

$$+\frac{a_6B^2\gamma^4+b_5B\gamma^4-8a_6\gamma^2-2b_5\gamma^3+c_8\gamma^4-4a_6B\gamma^3}{(2x+\gamma)^2\gamma^4}\Bigg]f=0. \qquad (14.74)$$

Shortly, this equation can be written as

$$\frac{d^4f}{dx^4}+\left[P_0+\frac{P_1}{x}+\frac{P_2}{x^2}+\frac{p_1}{2x+\gamma}\right]\frac{d^3f}{dx^3}$$
$$+\left[Q_0+\frac{Q_1}{x}+\frac{Q_2}{x^2}+\frac{Q_3}{x^3}+\frac{Q_4}{x^4}+\frac{q_1}{2x+\gamma}+\frac{q_2}{(2x+\gamma)^2}\right]\frac{d^2f}{dx^2}$$
$$+\left[M_0+\frac{M_1}{x}+\frac{M_2}{x^2}+\frac{M_3}{x^3}+\frac{M_4}{x^4}+\frac{M_6}{x^6}+\frac{m_1}{2x+\gamma}+\frac{m_2}{(2x+\gamma)^2}\right]\frac{df}{dx}$$
$$+\left[\frac{N_1}{x}+\frac{N_2}{x^2}+\frac{N_3}{x^3}+\frac{N_4}{x^4}+\frac{N_5}{x^5}+\frac{N_6}{x^6}+\frac{n_1}{2x+\gamma}+\frac{n_2}{(2x+\gamma)^2}\right]f=0. \qquad (14.75)$$

Then the solutions are constructed as power series: $\Phi=\sum_{l=0}^{\infty}d_lx^l$, and we derive the following 9-term recurrence relations ($k \geq 7$):

$$
\begin{aligned}
&[4M_0(k-7)+(4N_1+2n_1)]\,d_{k-7} \\
&+[4Q_0(k-6)(k-7)+(4M_0\gamma+4M_1+2m_1)(k-6)+(4N_1\gamma+4N_2+\gamma n_1+n_2)]\,d_{k-6} \\
&+[4P_0(k-5)(k-6)(k-7)+(4Q_0\gamma+4Q_1+2q_1)(k-5)(k-6) \\
&+\left(M_0\gamma^2+4M_1\gamma+4M_2+\gamma m_1+m_2\right)(k-5)+\left(N_1\gamma^2+4N_2\gamma+4N_3\right)]d_{k-5} \\
&+[4(k-4)(k-5)(k-6)(k-7)+(4P_0\gamma+4P_1+2p_1)(k-4)(k-5)(k-6) \\
&+\left(P_0\gamma^2+4P_1\gamma+4P_2+\gamma p_1\right)(k-4)(k-5)(k-6) \\
&+\left(Q_0\gamma^2+4Q_1\gamma+4Q_2+q_1\gamma+q_2\right)(k-4)(k-5) \\
&+\left(M_1\gamma^2+4M_2\gamma+4M_3\right)(k-4)+\left(N_2\gamma^2+4N_3\gamma+4N_4\right)]d_{k-4} \\
&+[4\gamma(k-3)(k-4)(k-5)(k-6)+\left(Q_1\gamma^2+4Q_2\gamma+4Q_3\right)(k-3)(k-4) \\
&+\left(M_2\gamma^2+4M_3\gamma+4M_4\right)(k-3)+\left(N_3\gamma^2+4N_4\gamma+4N_5\right)]d_{k-3} \\
&+[\gamma^2(k-2)(k-3)(k-4)(k-5)+\left(P_1\gamma^2+4P_2\gamma\right)(k-2)(k-3)(k-4) \\
&+\left(Q_2\gamma^2+4Q_3\gamma+4Q_4\right)(k-2)(k-3)+\left(M_3\gamma^2+4M_4\gamma\right)(k-2)+\left(N_4\gamma^2+4N_5\gamma+4N_6\right)]d_{k-2} \\
&+[P_2\gamma^2(k-1)(k-2)(k-3)+\left(Q_3\gamma^2+4Q_4\gamma\right)(k-1)(k-2) \\
&+\left(M_4\gamma^2+4M_6\right)(k-1)+\left(N_5\gamma^2+4N_6\gamma\right)]d_{k-1} \\
&+[Q_4\gamma^2k(k-1)+4M_6\gamma k+N_6\gamma^2]d_k+M_6\gamma^2(k+1)d_{k+1}=0.
\end{aligned}
$$

To study the convergence of the power series, we divide the expression by $k^4 d_{k-7}$ and tend $k \to \infty$. As a result we obtain ane algebraic equation for the quantity $R=\lim_{k\to\infty}\frac{d_{k+1}}{d_k}$:

$$
4R^3+4\gamma R^4+\gamma^2R^5=0 \quad \Rightarrow R=0,\ -\frac{2}{\gamma}, \quad R_{\text{conv}}=\frac{1}{|R|}=\frac{|\gamma|}{2},\infty. \tag{14.76}
$$

The quantization rule is determined from the transcendency condition. So we demand the vanishing of the coefficient D_{k-7}:

$$
4M_0(k-7)+4N_1+2n_1=0.
$$

Taking into account the explicit form of the parameters M_0, N_1, n_1

$$
(16k-96)B^3+(2a_5+4a_1)B^2+[(32k-256)\epsilon+4b_1+2b_4]B+4c_1+2c_7=0,
$$

and we get the identity $0=0$. So, we get the 8-term recurrence relations in (14.76). Therefore, in order to find a quantization rule, the coefficient at d_{k-6} has to be zero:

$$
4Q_0(k-6)(k-7)+(4M_0\gamma+4M_1+2m_1)(k-6)+4N_1\gamma+4N_2+\gamma n_1+n_2=0
$$

or,

$$
\begin{aligned}
&(16\gamma k-92\gamma)B^3+(720+16\alpha\gamma+24k^2-264k-16\nu^2)B^2+ \\
&\qquad +((32\gamma k-184\gamma)\epsilon-176\alpha+32\alpha k)B+ \\
&\qquad +(480+32\alpha\gamma-32\nu^2-176k+16k^2)\epsilon+16\alpha^2=0.
\end{aligned} \tag{14.77}
$$

Since $B=-\sqrt{-2\epsilon}$, we get the following equation with respect to ϵ:

$$-32\,(k-5)\,(k-6)\,\epsilon+(176-32k)\,\alpha\sqrt{-2\epsilon}+16\,\alpha^2=0\,,$$

whence we derive

$$\epsilon_1=-\frac{1}{2}\,\frac{\alpha^2}{(k-5)^2}\,,\qquad \epsilon_2=-\frac{1}{2}\,\frac{\alpha^2}{(k-6)^2}\,. \tag{14.78}$$

These spectra are unlikely to be correct, as well.

Now, we consider the solutions of the 4-th type (14.66):

IV, $\Psi(x)=e^{Bx}xe^{-\gamma/x}f_4(x)$. At the chosen values of the parameters, the equation (14.60) takes the form

$$\frac{d^4f}{dx^4}+\left[4B+\frac{14}{x}+\frac{4\gamma}{x^2}-\frac{4}{2x+\gamma}\right]\frac{d^3f}{dx^3}$$

$$+\left[6B^2+4\epsilon+\frac{12\gamma+a_1\gamma^2+42B\gamma^2}{\gamma^2x}+\frac{18\gamma+a_2\gamma+12B\gamma^2}{\gamma x^2}+\frac{28\gamma}{x^3}+\frac{5\gamma^2}{x^4}\right.$$

$$\left.+\frac{-24\gamma-12B\gamma^2+a_5\gamma^2}{(2x+\gamma)\gamma^2}+\frac{a_6}{(2x+\gamma)^2}\right]\frac{d^2f}{dx^2}$$

$$+\left[4B^3+8\epsilon B+\frac{8\epsilon\gamma^4+2a_1B\gamma^4+96\gamma^2+42B^2\gamma^4-2a_5\gamma^3-6a_6\gamma^2+24B\gamma^3+b_1\gamma^4}{\gamma^4x}\right.$$

$$+\frac{8\epsilon\gamma^4+2a_1\gamma^3+2a_2B\gamma^3-48\gamma^2+36B\gamma^3+12B^2\gamma^4+2a_5\gamma^3+2a_6\gamma^2+b_2\gamma^3}{\gamma^3x^2}$$

$$+\frac{2a_1\gamma^3+2a_2\gamma^2+56B\gamma^3+24\gamma^2+b_3\gamma^2}{\gamma^2x^3}+\frac{2a_2\gamma^2-16\gamma^2+10B\gamma^3}{\gamma x^4}+\frac{12\gamma^2}{x^5}+\frac{2\gamma^3}{x^6}$$

$$\left.+\frac{2a_5B\gamma^4-192\gamma^2+4a_5\gamma^3+12a_6\gamma^2-48B\gamma^3-12B^2\gamma^4+b_4\gamma^4}{(2x+\gamma)\gamma^4}+\frac{2a_6B\gamma^2+4a_6\gamma+b_5\gamma^2}{(2x+\gamma)^2\gamma^2}\right]\frac{df}{dx}$$

$$+\left[\frac{1}{\gamma^6x}\Big(320\gamma^3-8a_5\gamma^4+96B\gamma^4-32a_6\gamma^3+12B^2\gamma^5+b_1B\gamma^6+a_1B^2\gamma^6+c_1\gamma^6\right.$$

$$+14B^3\gamma^6-b_4\gamma^5-3b_5\gamma^4-2a_5B\gamma^5-6a_6B\gamma^4+8\epsilon B\gamma^6\Big)$$

$$+\frac{1}{\gamma^5x^2}\Big(-160\gamma^3+2a_1B\gamma^5+4a_5\gamma^4-48B\gamma^4+12a_6\gamma^3+18B^2\gamma^5+b_2B\gamma^5+a_2B^2\gamma^5+$$

$$+c_2\gamma^5+4B^3\gamma^6+b_4\gamma^5+b_5\gamma^4+2a_5B\gamma^5+2a_6B\gamma^4+b_1\gamma^5+8\epsilon B\gamma^6\Big)$$

$$+\frac{80\gamma^3+2a_1B\gamma^5+2a_2B\gamma^4-2a_5\gamma^4+24B\gamma^4-4a_6\gamma^3+28B^2\gamma^5+b_3B\gamma^4+c_3\gamma^4+b_1\gamma^5+b_2\gamma^4}{\gamma^4x^3}$$

$$+\frac{-40\gamma^3+2a_2B\gamma^4+a_5\gamma^4-16B\gamma^4+a_6\gamma^3+5B^2\gamma^5+c_4\gamma^3+4\epsilon\gamma^5+b_2\gamma^4+b_3\gamma^3}{\gamma^3x^4}$$

$$+\frac{20\gamma^3+a_1\gamma^4+b_3\gamma^3+12B\gamma^4+c_5\gamma^2}{\gamma^2x^5}+\frac{a_2\gamma^3-20\gamma^3+2B\gamma^4}{\gamma x^6}$$

$$+\frac{1}{(2x+\gamma)\gamma^6}\Big(-640\gamma^3+a_5B^2\gamma^6+16a_5\gamma^4-192B\gamma^4+64a_6\gamma^3-24B^2\gamma^5$$

$$+b_4B\gamma^6-4B^3\gamma^6+c_7\gamma^6+2b_4\gamma^5+6b_5\gamma^4+4a_5B\gamma^5+12a_6B\gamma^4\Big)$$

$$\left.+\frac{a_6B^2\gamma^4+b_5B\gamma^4+16a_6\gamma^2+2b_5\gamma^3+c_8\gamma^4+4a_6B\gamma^3}{(2x+\gamma)^2\gamma^4}\right]f=0.$$

Schematically, the equation can be written as follows:

$$\frac{d^4 f}{dx^4}+\left[P_0+\frac{P_1}{x}+\frac{P_2}{x^2}+\frac{p_1}{2x+\gamma}\right]\frac{d^3 f}{dx^3}$$
$$+\left[Q_0+\frac{Q_1}{x}+\frac{Q_2}{x^2}+\frac{Q_3}{x^3}+\frac{Q_4}{x^4}+\frac{q_1}{2x+\gamma}+\frac{q_2}{(2x+\gamma)^2}\right]\frac{d^2 f}{dx^2}$$
$$+\left[M_0+\frac{M_1}{x}+\frac{M_2}{x^2}+\frac{M_3}{x^3}+\frac{M_4}{x^4}+\frac{M_5}{x^5}+\frac{M_6}{x^6}+\frac{m_1}{2x+\gamma}+\frac{m_2}{(2x+\gamma)^2}\right]\frac{df}{dx}$$
$$+\left[\frac{N_1}{x}+\frac{N_2}{x^2}+\frac{N_3}{x^3}+\frac{N_4}{x^4}+\frac{N_5}{x^5}+\frac{N_6}{x^6}+\frac{n_1}{2x+\gamma}+\frac{n_2}{(2x+\gamma)^2}\right]f=0. \qquad (14.79)$$

Its solutions are constructed as power series; we derive the 9-term recurrence relations:

$\underline{k\geq 7,}$

$$[4M_0(k-7)+(4N_1+2n_1)]d_{k-7}$$
$$+[4Q_0(k-6)(k-7)+(4M_0\gamma+4M_1+2m_1)(k-6)+(4N_1\gamma+4N_2+\gamma n_1+n_2)]d_{k-6}$$
$$+[4P_0(k-5)(k-6)(k-7)+(4Q_0\gamma+4Q_1+2q_1)(k-5)(k-6)$$
$$+(M_0\gamma^2+4M_1\gamma+4M_2+\gamma m_1+m_2)(k-5)+(N_1\gamma^2+4N_2\gamma+4N_3)]d_{k-5}$$
$$+[4(k-4)(k-5)(k-6)(k-7)+(4P_0\gamma+4P_1+2p_1)(k-4)(k-5)(k-6)$$
$$+(P_0\gamma^2+4P_1\gamma+4P_2+\gamma p_1)(k-4)(k-5)(k-6)$$
$$+(Q_0\gamma^2+4Q_1\gamma+4Q_2+q_1\gamma+q_2)(k-4)(k-5)$$
$$+(M_1\gamma^2+4M_2\gamma+4M_3)(k-4)+(N_2\gamma^2+4N_3\gamma+4N_4)]d_{k-4}$$
$$+[4\gamma(k-3)(k-4)(k-5)(k-6)+(Q_1\gamma^2+4Q_2\gamma+4Q_3)(k-3)(k-4)$$
$$+(M_2\gamma^2+4M_3\gamma+4M_4)(k-3)+(N_3\gamma^2+4N_4\gamma+4N_5)]d_{k-3}$$
$$+[\gamma^2(k-2)(k-3)(k-4)(k-5)+(P_1\gamma^2+4P_2\gamma)(k-2)(k-3)(k-4)$$
$$+(Q_2\gamma^2+4Q_3\gamma+4Q_4)(k-2)(k-3)$$
$$+(M_3\gamma^2+4M_4\gamma+4M_5)(k-2)+(N_4\gamma^2+4N_5\gamma+4N_6)]d_{k-2}$$
$$+[P_2\gamma^2(k-1)(k-2)(k-3)+(Q_3\gamma^2+4Q_4\gamma)(k-1)(k-2)$$
$$+(M_4\gamma^2+4M_5\gamma+4M_6)(k-1)+(N_5\gamma^2+4N_6\gamma)]d_{k-1}$$
$$+[Q_4\gamma^2 k(k-1)+(M_5\gamma^2+4M_6\gamma)k+N_6\gamma^2]d_k+M_6\gamma^2(k+1)d_{k+1}=0.$$

To study the convergence of the power series, we divide the expression by $k^4 d_{k-7}$ and tending $k\to\infty$, we get the algebraic equation for the convergence radius:

$$R=\lim_{k\to\infty}\frac{d_{k+1}}{d_k},\quad 4R^3+4\gamma R^4+\gamma^2 R^5=0$$
$$\Longrightarrow\ R=0,\ -\frac{2}{\gamma},\quad R_{\text{conv}}=\frac{1}{|R|}=\frac{|\gamma|}{2},\ \infty. \qquad (14.80)$$

Considering the coefficient at $D_{k-7}, k\geq 8$:

$$4M_0(k-7)+4N_1+2n_1=0,$$

and taking into account the explicit expressions of the parameters M_0, N_1, n_1

$$(16k-64)B^3+(2a_5+4a_1)B^2+[(32k-192)\epsilon+4b_1+2b_4]B+4c_1+2c_7=0,$$

we get the identity $0=0$. So, we have 8-term recurrence relations. The transcendence condition means that the coefficient at d_{k-6} vanishes:

$$4Q_0(k-6)(k-7)+(4M_0\gamma+4M_1+2m_1)(k-6)+(4N_1\gamma+4N_2+\gamma n_1+n_2)$$

or,

$$(-32k^2-384+224k)\,\epsilon-32\,(k-\frac{7}{2})\alpha\sqrt{-2\epsilon}+16\,\alpha^2=0;$$

then the quantization rule is as follows:

$$\epsilon_1=-\frac{1}{2}\frac{\alpha^2}{(k-3)^2},\qquad \epsilon_2=-\frac{1}{2}\frac{\alpha^2}{(k-4)^2}. \tag{14.81}$$

These spectra also are unlikely to be correct. All the constructed solutions are exact, but we have no reliable rules for quantization of energy levels. The transcendency condition resolve this difficulty only partly.

14.7 KCC-Geometrical Approach

In this section we will try to study the previous arising mathematical tasks, by applying the special geometric method based on the so called KCC-invariants [163] – [166]. In this approach, one considers a system of second order differential equations

$$\dot{y}^i(r)+2Q^i(r,x,y)=0, \tag{14.82}$$

which corresponds to the the Euler-Lagrange equations for some dynamical system with Lagrangian L. In (14.82), the symbol x^i designates the so called coordinates, their derivatives with respect to the argument r are $y^i=dx^i/dr=\dot{x}^i$, and the quantities Q_i are determined through some Lagrangian L, as follows

$$Q^i=\frac{1}{4}g^{il}\Big(\frac{\partial^2L}{\partial x^k\partial y^l}y^k-\frac{\partial L}{\partial x^l}+\frac{\partial^2L}{\partial y^l\partial r}\Big),\qquad g_{ij}=\frac{1}{2}\frac{\partial^2L}{\partial y^i\partial y^j}. \tag{14.83}$$

The first and the second invariants, $\varepsilon^i(r,x,y)$ and $P^i{}_j$ are introduced by the definitions

$$\varepsilon^i=\frac{\partial Q^i}{\partial y^j}y^j-2Q^i,$$
$$P^i{}_j=2\frac{\partial Q^i}{\partial x^j}+2Q^s\frac{\partial^2Q^i}{\partial y^j\partial y^s}-\frac{\partial^2Q^i}{\partial y^j\partial x^s}y^s-\frac{\partial Q^i}{\partial y^s}\frac{\partial Q^s}{\partial y^j}-\frac{\partial^2Q^i}{\partial y^j\partial r}. \tag{14.84}$$

The second invariant $P^i{}_j$ relates to the Jacobi stability of a dynamical system. There is an analogy between the equations of geodesic deviation expressed in terms of the Riemann curvature and in terms of the second KCC-invariant:

$$\frac{D^2\xi^i}{Ds^2}=R^i_{kjl}\frac{dx^k}{ds}\frac{dx^l}{ds}\xi^j=-K^i_j\xi^j,\qquad \frac{D^2\xi^i}{Dr^2}=P^i_j\xi^j. \tag{14.85}$$

It is known that a pencil of geodesic curves from the some point r_0 converges (or diverges) if the real parts of all the eigenvalues of the invariant $P^i{}_j$ are negative (or positive) ones.

We start from the system (14.55) of two second-order differential equations for two radial functions of spin 1 particle with electric quadruple moment in the external Coulomb field. Let us write it as:

$$\begin{gathered}\Big(\frac{d^2}{dr^2}+\frac{2}{r}\frac{d}{dr}+2m\frac{\alpha+Er}{r}-\frac{2\nu^2}{r^2}-\frac{2\Gamma}{r^3}-\frac{\Gamma^2}{r^4}\Big)\Psi_1(r)-\nu\frac{2r+\Gamma}{r^3}\Psi_2(r)=0,\\ \Big(\frac{d^2}{dr^2}+\frac{2}{r}\frac{d}{dr}+2m\frac{\alpha+Er}{r}-\frac{2\nu^2}{r^2}-\frac{2}{r^2}\Big)\Psi_2(r)-2\nu\frac{2r+\Gamma}{r^3}\Psi_1(r)=0.\end{gathered} \tag{14.86}$$

We shall apply the following notations $x^i=\Psi_i(r)$, $y^i=(d/dr)\Psi_i(r)=\dot{\Psi}_i(r)$. Then, by comparing the equations (14.86) and (14.82), one finds the quantities Q^i:

$$\begin{aligned}Q^1(r,\Psi_i,\dot{\Psi}_i)&=\Big(Em+\frac{\alpha m}{r}-\frac{\Gamma^2}{2r^4}-\frac{\Gamma}{r^3}-\frac{\nu^2}{r^2}\Big)\Psi_1-\nu\frac{(\Gamma+2r)}{2r^3}\Psi_2+\frac{1}{r}\dot{\Psi}_1,\\ Q^2(r,\Psi_i,\dot{\Psi}_i)&=\Big(Em+\frac{\alpha m}{r}-\frac{\nu^2}{r^2}-\frac{1}{r^2}\Big)\Psi_2-\nu\frac{(\Gamma+2r)}{r^3}\Psi_1+\frac{1}{r}\dot{\Psi}_2.\end{aligned} \tag{14.87}$$

By direct calculation according the formula (14.84), the first and the second KCC-invariants are

$$\begin{aligned}\varepsilon^1&=\Psi_1\Big(-2Em-\frac{2\alpha m}{r}+\frac{\Gamma^2}{r^4}+\frac{2\Gamma}{r^3}+\frac{2\nu^2}{r^2}\Big)+\nu\Psi_2\Big(\frac{\Gamma}{r^3}+\frac{2}{r^2}\Big)-\frac{\dot{\Psi}_1}{r},\\ \varepsilon^2&=2\Psi_2\Big(-Em-\frac{\alpha m}{r}+\frac{\nu^2+1}{r^2}\Big)+2\nu\Psi_1\Big(\frac{\Gamma}{r^3}+\frac{2}{r^2}\Big)-\frac{\dot{\Psi}_2}{r};\end{aligned} \tag{14.88}$$

$$P^i{}_j=\begin{vmatrix}-\frac{\Gamma^2}{r^4}-\frac{2\Gamma}{r^3}-\frac{2\nu^2}{r^2}+2Em+\frac{2m\alpha}{r} & -\frac{\nu(2r+\Gamma)}{r^3}\\ -\frac{2\nu(2r+\Gamma)}{r^3} & 2Em+\frac{2\alpha m}{r}-\frac{2(\nu^2+1)}{r^2}\end{vmatrix}. \tag{14.89}$$

The eigenvalues Λ_1,Λ_2 of the second invariant are given by the formulas

$$\Lambda_{1,2}=2Em+\frac{2\alpha m}{r}-\frac{\Gamma^2}{2r^4}-\frac{\Gamma}{r^3}-\frac{2\nu^2+1}{r^2}\pm\frac{\sqrt{(\Gamma^2-2r^2+2\Gamma r)^2+8\nu^2r^2(\Gamma+2r)^2}}{2r^4}.$$

Let us study the behavior of the eigenvalues Λ^i near the singular points $r=0$, $r=\infty$, and $r=-\Gamma/2$. We find out that

$$\begin{gathered}r\to 0,\quad \Lambda^1\to-\frac{2}{r^2}<0,\quad \Lambda^2\to-\frac{\Gamma^2}{r^4}<0;\\ r\to\infty,\quad \Lambda^1,\Lambda^2\to 2Em<0;\\ r\to-\frac{\Gamma}{2},\qquad \Lambda^1\to 2Em-\frac{4m\alpha}{\Gamma}-\frac{8\nu^2}{\Gamma^2}-\frac{8}{\Gamma^2}<0,\\ \Lambda^2\to 2Em-\frac{4m\alpha}{\Gamma}-\frac{8\nu^2}{\Gamma^2}<0.\end{gathered} \tag{14.90}$$

Since the real parts of all eigenvalues of the 2-nd KCC-invariant are negative, the different branches of the solution converge near the singular points $r = 0, \infty, -\Gamma/2$. This correlates with the behavior of solutions near the singular points, for bound quantum mechanical states (discrete spectra).

The third KCC-invariant

$$R^i_{jk} = \frac{1}{3}\Big(\frac{\partial P^i_j}{\partial y^k} - \frac{\partial P^i_k}{\partial y^j}\Big) \tag{14.91}$$

determines the torsion of the Berwald connection. The fourth KCC-invariant is an extension of the Riemann–Christoffel tensor

$$B^i_{jkl} = \frac{\partial R^i_{jk}}{\partial y^l}. \tag{14.92}$$

Finally, the fifth KCC-invariant extends the Duglas tensor.

$$D^i_{jkl} = \frac{\partial^3 Q^i}{\partial y^j \partial y^k \partial y^l}. \tag{14.93}$$

Since the vector field Q^i (14.87) is linear in the coordinates $x^i \equiv \Phi^i$ and $y^i \equiv \dot{\Psi}^i$, the first (14.88) and the second KCC invariants (14.89) are functions of the radial coordinate r, and do not depend on x^i and y^i, and the third, fourth and fifth invariants identically vanish.

The next step is to construct a Lagrangian function L for the phase space $\dot{\Psi}_i$, Ψ_i, such that the formulas for coefficients Q^i (14.87) hold true, and the dynamics of the system is defined by the equations (14.86). We shall search for the function L in the form

$$L = g_{ij}(r) y^i y^j + b_j(r,x) y^j. \tag{14.94}$$

Let us assume that the tensor g_{ij} is diagonal, $g_{12} = g_{21} = 0$. In this case, substituting (14.94) into (14.83) we derive

$$\begin{aligned} Q^1 &= \frac{1}{4g_{11}}\Big(2\dot{g}_{11}y^1 + \frac{\partial b_1}{\partial r} + \Big(\frac{\partial b_1}{\partial x^2} - \frac{\partial b_2}{\partial x^1}\Big)y^2\Big), \\ Q^2 &= \frac{1}{4g_{22}}\Big(2\dot{g}_{22}y^2 + \frac{\partial b_2}{\partial r} + \Big(\frac{\partial b_2}{\partial x^1} - \frac{\partial b_1}{\partial x^2}\Big)y^2\Big). \end{aligned} \tag{14.95}$$

By equating the terms from (14.87) with the corresponding terms from (14.95), we obtain the system of equations with respect to $g_{ij}(r)$ and to $b_j(r,x)$:

$$\frac{\partial b_1}{\partial x^2} - \frac{\partial b_2}{\partial x^1} = 0, \qquad \frac{\dot{g}_{11}}{2g_{11}} = \frac{1}{r}, \qquad \frac{\dot{g}_{22}}{2g_{22}} = \frac{1}{r},$$

$$\frac{1}{4g_1}\frac{\partial b_1}{\partial r} = -\frac{x^1\left(\Gamma^2 - 2r^2\left(mr(\alpha + Er) - \nu^2\right) + 2\Gamma r\right)}{2r^4} - \frac{\nu x^1(\Gamma + 2r)}{2r^3},$$

$$\frac{1}{4g_2}\frac{\partial b_2}{\partial r} = -\frac{x^2\left(-Emr^2 + \nu^2 - \alpha m r + 1\right)}{r^2} - \frac{\nu x^1(\Gamma + 2r)}{r^3}.$$

Its solution is given by the formulas

$$g_{11} = 2C_1 r^2 , \qquad g_{22} = C_1 r^2 ,$$

$$b_1 = B_1(x^1, x^2)$$

$$2C_1 \Big\{ -\frac{2}{3} Emr^3 x^1 - \alpha m r^2 x^1 - \frac{\Gamma^2 x^1}{r} + \ln r (\Gamma \nu x^2 + 2\Gamma x^1) + 2\nu r (\nu x^1 + x^2) \Big\} ,$$

$$b_2 = B_2(x^1, x^2)$$

$$-4C_1 \Big\{ -\frac{1}{3} Emr^3 x^2 - \frac{1}{2} \alpha m r^2 x^2 + \Gamma \nu x^1 \ln r + r(2\nu x^1 + \nu^2 x^2 + x^2) \Big\} ,$$

where C_1 is an arbitrary constant. The two functions $B_1(x^1, x^2)$ and $B_2(x^1, x^2)$ obey the following restriction

$$\frac{\partial B_1(x^1, x^2)}{\partial x^2} - \frac{\partial B_2(x^1, x^2)}{\partial x^1} = 0 . \tag{14.96}$$

In accordance with the known theorem, from (14.96) we conclude that this 2-dimensional vector field B_1, B_2 can be presented as a gradient of a scalar function

$$B_1(x^1, x^2) = \frac{\partial}{\partial x^1} \varphi(x^1, x^2) , \quad B_2(x^1, x^2) = \frac{\partial}{\partial x^2} \varphi(x^1, x^2) , \quad B_i = \text{grad}\, \varphi . \tag{14.97}$$

There exist some freedom in choosing the Lagrangian (the constant C_1 may be taken as 1):

$$L = 2r^2 (y^1)^2 + r^2 (y^2)^2 + 4x^1 y^1 \Big(\frac{2}{3} Emr^3 + \alpha m r^2 + \frac{\Gamma^2}{r} - 2\Gamma \ln r - 2\nu^2 r \Big)$$
$$+ \frac{2}{3} r x^2 y^2 \left(mr(3\alpha + 2Er) - 6\left(\nu^2 + 1\right)\right) - 4\nu \left(x^2 y^1 + x^1 y^2\right) (\Gamma \ln r + 2r)$$
$$+ y^1 \frac{\partial \varphi}{\partial x^1} + y^2 \frac{\partial \varphi}{\partial x^2} , \quad \varphi = \varphi(x^1, x^2) . \tag{14.98}$$

Chapter 15

Massive and Massless Fields with Spin 3/2, Solutions and Helicity Operator

15.1 Massive and Massless Spin 3/2 Fields

The theory of a spin 3/2 particle has attracted a steady interest after the seminal investigation by Pauli and Fierz [313] – [371]. Let us recall the most significant aspect of the spin 3/2 particle theory. First of all, it is the problem of choosing an initial system of equations. The most consistent approach is based on Lagrangian formalism and a correct first order equation for the multi-component wave function, based on the general theory of 1-st order relativistic wave equations. However, investigations are based on the use of 2-nd order equations. Such a choice is of prime importance when we take into account the presence of external electromagnetic (or gravitational) fields. Applying the first order approach ensures the correct solving of the problem with independent degrees of freedom in the presence of external fields (for instance, see [370], [371]). Great attention has been given in this theory to the existence of solutions which correspond to a particle moving with velocity greater than the light velocity. Finally, a separate interest has the massless case for the spin 3/2 field, when – as shown by Pauli and Fierz – there exists a specific gauge symmetry: the 4-gradient of arbitrary bispinor function $\Phi(x)$ provides us with a solution for the massless field equation. The main goal of the present chapter is to examine the problem of degrees of freedom for the massive and massless 3/2 particle, and the role of helicity operator, when constructing plane wave solutions. We shall basically use the Rarita – Schwinger formalism.

In the Rarita – Schwinger approach, the wave equation for a massive spin 3/2 particle takes the form

$$\Big(\gamma^a\partial_a + iM\Big)\Psi_c - \frac{1}{3}\Big(\gamma^b\partial_c + \gamma_c\partial^b\Big)\Psi_b + \frac{1}{3}\gamma_c\Big(\gamma^a\partial_a - iM\Big)\gamma^b\Psi_b = 0\,, \tag{15.1}$$

where Ψ_b is vector-bispinor under the action of the Lorentz group. We shall need the following formulas for Dirac matrices:

$$\gamma^a\gamma^b+\gamma^b\gamma^a=2g^{ab}, \gamma^a\gamma_a=4, \gamma^a\gamma^b\gamma^d=\gamma^a g^{bd}-\gamma^b g^{ad}+\gamma^d g^{ab}+i\gamma^5\epsilon^{abdc}\gamma_c; \quad (15.2)$$

the Levi-Civita symbol ϵ^{cabd} is specified as $\epsilon^{0123}=+1$. Starting from Eq. (15.1), one can derive additional constraints on $\Psi_a(x)$. Indeed, by multiplying Eq. (15.1) by the matrix γ^c, we get

$$\partial_b\,\Psi^b=\frac{iM}{2}\gamma_b\Psi^b, \quad (15.3)$$

which is the first constraint. Now, let us act on equation (15.1) by the operator ∂^c; further, with the use of the formulas (15.2), we obtain

$$iM\,\partial^c\Psi_c+\gamma^a\partial_a\Big(\frac{2}{3}\partial^c\Psi_c-\frac{iM}{3}\,\gamma^c\Psi_c\Big)=0, \quad (15.4)$$

which is the second constraint. Whence, with the the first condition (15.3) in mind, it follows

$$\partial^c\Psi_c=0 \quad (M\neq 0). \quad (15.5)$$

Therefore, the first constraint gives $\gamma_b\Psi^b=0$. Taking into account two last restrictions, instead of the initial equation (15.1), we arrive at the following equivalent system

$$\Big(i\gamma^a\partial_a-M\Big)\Psi_c=0, \quad \partial^c\Psi_c=0, \quad \gamma^c\Psi_c=0. \quad (15.6)$$

The case of massless particle is substantially different. From the very beginning, in Eq. (15.1) we set $M=0$:

$$\gamma^a\partial_a\Psi_c-\frac{1}{3}\Big(\gamma^b\partial_c+\gamma_c\partial^b\Big)\Psi_b+\frac{1}{3}\gamma_c\gamma^a\partial_a\gamma^b\Psi_b=0. \quad (15.7)$$

Now, the first constraint reads as

$$\partial_b\,\Psi^b=0. \quad (15.8)$$

The second condition (15.3) takes the form $\gamma^a\partial_a\partial^c\Psi_c=0$; evidently it does not add restrictions to the constraint (15.8). Therefore, for the massless case, Eq. (15.7) may be written in the split form:

$$\gamma^a\partial_a\Psi_c-\frac{1}{3}\gamma^b\partial_c\Psi_b+\frac{1}{3}\gamma_c\gamma^a\gamma^b\partial_a\Psi_b=0, \quad \partial_b\Psi^b=0. \quad (15.9)$$

We further use the wave equation (15.7). It can be transformed to a form, which emphasizes the existence of gauge solutions of gradient type. To this end, we write the equation (15.7) in matrix form,

$$\Gamma^a\partial_a\Psi=0, \quad \Psi=(\Psi_l), \quad (15.10)$$

where the 16×16 matrices Γ^a are given by the formula

$$(\Gamma^a)_k{}^l = \gamma^a \, \delta_k^l - \frac{1}{3}\gamma^l \delta_k^a - \frac{1}{3}\gamma_k g^{al} + \frac{1}{3}\gamma_k \gamma^a \gamma^l . \tag{15.11}$$

Let us make two transformations on Eq. (15.10): first we multiply it by a matrix C with nonvanishing determinant, and then we introduce new wave function by a linear transformation S:

$$\Gamma^a \implies \Gamma'^a = C\,\Gamma^a \implies \tilde{\Gamma}^a = S\,\Gamma'^a\,S^{-1}, \; \tilde{\Psi} = S\,\Psi; \tag{15.12}$$

these two transformations will be specified below. Let the matrices C and S have the structure:

$$\begin{gathered} C_a{}^b = \delta_a^b + c\gamma_a\gamma^b, \quad S_a{}^b = \delta_a^b + a\gamma_a\gamma^b, \\ (S^{-1})_a{}^b = \delta_a^b + b\gamma_a\gamma^b, \quad a+b+4ab = 0. \end{gathered} \tag{15.13}$$

where a, b, c are some numerical parameters; the constraint on a and b follows from the condition $S\,S^{-1} = I$. In accordance with (15.12) and (15.13), we first find Γ'^a:

$$(\Gamma'^a)_k{}^l = \gamma^a \, \delta_k^l - \frac{1}{3}\gamma^l \, \delta_k^a + (2c - \frac{1}{3})\, \gamma_k \, g^{al} + \frac{1}{3}\, \gamma_k \, \gamma^a \, \gamma^l ,$$

and then we find the matrices $\tilde{\Gamma}^a$:

$$\begin{aligned} (\tilde{\Gamma}^a)_k{}^l = {} & \gamma^a \delta_\rho^l \Big\{ 1 - \Big[\frac{b+1}{3} + b(\frac{2c-1}{3}(1+4a) + 2a) \Big] \Big\} \\ & + \gamma^l \delta_k^a \Big\{ \frac{2b-1}{3} + \Big[\frac{b+1}{3} + b(\frac{2c-1}{3}(1+4a) + 2a) \Big] \Big\} \\ & + \gamma_k g^{al} \Big\{ \Big[(2c-1)\frac{1+4a}{3} + 2a \Big] + \Big[\frac{b+1}{3} + b((2c-1)\frac{1+4a}{3} + 2a) \Big] \Big\} \\ & + i\gamma^5 \epsilon_k^{als} \gamma_s \Big[\frac{b+1}{3} + b((2c-1)\frac{1+4a}{3} + 2a) \Big] . \end{aligned} \tag{15.14}$$

Let us require that in $\tilde{\Gamma}^a$ all the terms except the one containing the Levy-Civita symbol vanish; this results in the following equations:

$$\begin{gathered} a+b+4ab = 0, \; 1 - \Big[\frac{b+1}{3} + b((2c-1)\frac{1+4a}{3} + 2a) \Big] = 0, \\ \frac{2b+1}{3} + \Big[\frac{b+1}{3} + b((2c-1)\frac{1+4a}{3} + 2a) \Big] = 0, \\ (1+4a)\frac{2c-1}{3} + 2a + \Big[\frac{b+1}{3} + b((2c-1)\frac{1+4a}{3} + 2a) \Big] = 0. \end{gathered}$$

It is readily verified that its solution is given as

$$a = -\frac{1}{3}, \quad b = -1, \quad c = +2. \tag{15.15}$$

Thus, we find the needed linear transformation

$$\tilde{\Psi}_k = S_k{}^l \Psi_l \,, \quad S_k{}^l = \delta_k^l - \frac{1}{3}\, \gamma_k \gamma^l \,; \tag{15.16}$$

in this new basis, the matrices $\tilde{\Gamma}^a$ have the structure

$$(\tilde{\Gamma}^a)_k{}^l = +i\, \gamma^5 \epsilon_k{}^{als} \gamma_s \,.$$

Therefore, we arrive at the equivalent form of the wave equation for the massless spin 3/2 particle:

$$(\tilde{\Gamma}^a)_l{}^k \tilde{\Psi}_l(x) = 0 \quad \Longrightarrow \quad i\, \gamma^5 \epsilon_k{}^{nal} \gamma_n \partial_a \tilde{\Psi}_l(x) = 0 \,; \tag{15.17}$$

in Eq. (15.17) we may omit the multiplier $i\gamma^5$.

From Eq. (15.17), it follows that any vector-bispinor $\tilde{\Psi}_l^{(0)}$ as the gradient of arbitrary bispinor $\varphi(x)$:

$$\tilde{\Psi}_l^{(0)}(x) = \partial_l \varphi(x) \tag{15.18}$$

provides us with some solution of Eq. (15.17). This property is called gauge symmetry in the massless spin 3/2 field theory. In the initial basis, such gradient-type solutions are presented by the formula

$$\Psi_l^{(0)}(x) = \left(\delta_l{}^k - \gamma_l \gamma^k \right) \partial_k \varphi(x) \,. \tag{15.19}$$

Note that the explicit form of Eq. (15.17) is rather cumbersome:

$$\begin{aligned} \gamma_1(\partial_2\tilde{\Psi}_3 - \partial_3\tilde{\Psi}_2) + \gamma_2(\partial_3\tilde{\Psi}_1 - \partial_1\tilde{\Psi}_3) + \gamma_3(\partial_1\tilde{\Psi}_2 - \partial_2\tilde{\Psi}_1) = 0\,, \\ -\gamma_0(\partial_2\tilde{\Psi}_3 - \partial_3\tilde{\Psi}_2) + \partial_0(\gamma_2\tilde{\Psi}_3 - \gamma_3\tilde{\Psi}_2) - (\gamma_2\partial_3 - \gamma_3\partial_2)\tilde{\Psi}_0 = 0\,, \\ -\gamma_0(\partial_3\tilde{\Psi}_1 - \partial_1\tilde{\Psi}_3) + \partial_0(\gamma_3\tilde{\Psi}_1 - \gamma_1\tilde{\Psi}_3) - (\gamma_3\partial_1 - \gamma_1\partial_3)\tilde{\Psi}_0 = 0\,, \\ -\gamma_0(\partial_1\tilde{\Psi}_2 - \partial_2\tilde{\Psi}_1) + \partial_0(\gamma_1\tilde{\Psi}_2 - \gamma_2\tilde{\Psi}_1) - (\gamma_1\partial_2 - \gamma_2\partial_1)\tilde{\Psi}_0 = 0\,. \end{aligned} \tag{15.20}$$

15.2 Separating the Variables

We start with the free massive spin 3/2 particle, which is described by the set of three equations

$$\left(i\gamma^a \partial_a - mc/\hbar \right) \Psi_l(x) = 0\,, \quad \gamma^l \Psi_l(x) = 0\,, \quad \partial_l \Psi^l(x) = 0\,, \tag{15.21}$$

where the 16-component wave function $\Psi_l(x)$ is a vector-bispinor with respect to the Lorentz group:

$$\Psi_{A(l)}(x) = \left| \begin{array}{cccc} \Psi_{1(0)}(x) & \Psi_{1(1)}(x) & \Psi_{1(2)}(x) & \Psi_{1(3)}(x) \\ \Psi_{2(0)}(x) & \Psi_{2(1)}(x) & \Psi_{2(2)}(x) & \Psi_{2(3)}(x) \\ \Psi_{3(0)}(x) & \Psi_{3(1)}(x) & \Psi_{3(2)}(x) & \Psi_{3(3)}(x) \\ \Psi_{4(0)}(x) & \Psi_{4(1)}(x) & \Psi_{4(2)}(x) & \Psi_{4(3)}(x) \end{array} \right|\,, \tag{15.22}$$

where A and (l) designate bispinor and vector indices, respectively. We shall use the Dirac matrices in spinor basis:

$$\gamma^0 = \begin{vmatrix} 0 & 0 & 1 & 0 \\ 0 & 0 & 0 & 1 \\ 1 & 0 & 0 & 0 \\ 0 & 1 & 0 & 0 \end{vmatrix}, \gamma^1 = \begin{vmatrix} 0 & 0 & 0 & -1 \\ 0 & 0 & -1 & 0 \\ 0 & 1 & 0 & 0 \\ 1 & 0 & 0 & 0 \end{vmatrix}, \gamma^2 = \begin{vmatrix} 0 & 0 & 0 & +i \\ 0 & 0 & -i & 0 \\ 0 & -i & 0 & 0 \\ +i & 0 & 0 & 0 \end{vmatrix}, \gamma^3 = \begin{vmatrix} 0 & 0 & -1 & 0 \\ 0 & 0 & 0 & 1 \\ 1 & 0 & 0 & 0 \\ 0 & -1 & 0 & 0 \end{vmatrix}.$$

The solutions of Eqs. (15.21) are searched in the form of plane waves:

$$\Psi_l(x) = e^{-i\epsilon t/\hbar} e^{+ip_k x^k} A_l, \qquad (A_l) = \begin{vmatrix} a_0 & a_1 & a_2 & a_3 \\ b_0 & b_1 & b_2 & b_3 \\ c_0 & c_1 & c_2 & c_3 \\ d_0 & d_1 & d_2 & d_3 \end{vmatrix}. \tag{15.23}$$

We shall use the quantities

$$\frac{mc}{\hbar} = M\,, \quad \frac{\epsilon}{\hbar c} = E\,, \quad \frac{p_i}{\hbar} = k_j\,.$$

From Eqs. (15.21), we get three algebraic equations (remembering that $A^0 = A_0, A^j = -A_j$)

$$\left(\gamma^0 E - \gamma^j k_j - M\right) A_l = 0, \quad \gamma^0 A_0 + \gamma^j A_j = 0, \quad E A_0 + k_j A_j = 0. \tag{15.24}$$

The first equation in (15.24) takes the form

$$\begin{vmatrix} -M & 0 & (E+k_3) & (k_1 - ik_2) \\ 0 & -M & (k_1 + ik_2) & (E - k_3) \\ (E-k_3) & -(k_1 - ik_2) & -M & 0 \\ -(k_1 + ik_2) & (E+k_3) & 0 & -M \end{vmatrix} \begin{vmatrix} a_l \\ b_l \\ c_l \\ d_l \end{vmatrix} = 0, \quad l = 0,1,2,3; \tag{15.25}$$

the determinant of the system must be equal to zero:

$$(-M^2 + E^2 - k_3^2 - k_2^2 - k_1^2)^2 = 0 \implies E^2 = M^2 + \mathbf{k}^2. \tag{15.26}$$

It is readily proved that the determinant of the 3×3-matrix vanishes too:

$$\det \begin{vmatrix} -M & 0 & (E+k_3) \\ 0 & -M & (k_1 + ik_2) \\ (E-k_3) & -(k_1 - ik_2) & -M \end{vmatrix} = -M(M^2 - E^2 + \mathbf{k}^2) = 0,$$

This means that the rank of the system (15.25) equals to 2. Therefore, in the system (15.25), only two equations are independent; let these be the first two ones

$$(E+k_3)\,c_l + (k_1 - ik_2)\,d_l = M\,a_l\,, \quad (k_1 + ik_2)\,c_l + (E - k_3)\,d_0 = M\,b_l\,. \tag{15.27}$$

The solution is (recall that $l = 0,1,2,3$)

$$c_l = \frac{E - k_3}{M}\,a_l - \frac{k_1 - ik_2}{M}\,b_l\,, \quad d_l = -\frac{k_1 + ik_2}{M}\,a_l + \frac{E + k_3}{M}\,b_l\,, \tag{15.28}$$

or shortly,

$$c_l = \alpha a_l + \beta b_l, \; d_l = \rho a_l + \gamma b_l \,. \tag{15.29}$$

The second equation in (15.24) gives

$$\begin{gathered} a_0 + a_3 = -(b_1 - ib_2)\,, \quad b_0 - b_3 = -(a_1 + ia_2)\,, \\ c_0 - c_3 = d_1 - id_2\,, \quad d_0 + d_3 = c_1 + ic_2\,. \end{gathered} \tag{15.30}$$

Taking in mind (15.29), one excludes the variables c_l and d_l in (15.30):

$$\begin{gathered} a_0 + a_3 = -(b_1 - ib_2)\,, \qquad b_0 - b_3 = -(a_1 + ia_2)\,, \\ \alpha(a_0 - a_3) + \beta(b_0 - b_3) = \rho(a_1 - ia_2) + \gamma(b_1 - ib_2)\,, \\ \rho(a_0 + a_3) + \gamma(b_0 + b_3) = \alpha(a_1 + ia_2) + \beta(b_1 + ib_2)\,. \end{gathered} \tag{15.31}$$

In (15.31) we have 4 linear constraints on 8 variables, a_l, b_l. Finally, the third equation in (15.24) yields

$$\begin{gathered} Ea_0 + k_1 a_1 + k_2 a_2 + k_3 a_3 = 0\,, \quad Eb_0 + k_1 b_1 + k_2 b_2 + k_3 b_3 = 0\,, \\ Ec_0 + k_1 c_1 + k_2 c_2 + k_3 c_3 = 0\,, \quad Ed_0 + k_1 d_1 + k_2 d_2 + k_3 d_3 = 0\,. \end{gathered} \tag{15.32}$$

We immediately note that the last two equations in (15.32), if one takes into account (15.28), become consequences of the two first equations from (15.32). Therefore, in the system (15.32) we may preserve only the two independent relations

$$Ea_0 + k_1 a_1 + k_2 a_2 + k_3 a_3 = 0\,, \quad Eb_0 + k_1 b_1 + k_2 b_2 + k_3 b_3 = 0\,. \tag{15.33}$$

Let us collect the equations (15.31) and (15.33) together:

$$\begin{gathered} a_0 + a_3 = -(b_1 - ib_2)\,, \qquad b_0 - b_3 = -(a_1 + ia_2)\,, \\ \alpha(a_0 - a_3) + \beta(b_0 - b_3) = \rho(a_1 - ia_2) + \gamma(b_1 - ib_2)\,, \\ \rho(a_0 + a_3) + \gamma(b_0 + b_3) = \alpha(a_1 + ia_2) + \beta(b_1 + ib_2)\,, \\ Ea_0 + k_1 a_1 + k_2 a_2 + k_3 a_3 = 0\,,\; Eb_0 + k_1 b_1 + k_2 b_2 + k_3 b_3 = 0\,. \end{gathered} \tag{15.34}$$

It is readily checked that from the first two equations in (15.34), there follow the last two ones. Indeed, the first leads, by exclusion method, the equations which contain respectively the variables a_l and b_l (remembering the explicit form of the coefficients $\alpha, \beta, \rho, \sigma$):

$$\begin{gathered} (E - k_3)(a_0 - a_3) + (k_1 + ik_2)(a_1 - ia_2) \\ = -(k_1 - ik_2)(a_1 + ia_2) - (E + k_3)(a_0 + a_3)\,, \\ (k_1 + ik_2)(b_1 - ib_2) + (E - k_3)(b_0 - b_3) \\ = -(E + k_3)(b_0 + b_3) - (k_1 - ik_2)(b_1 + ib_2)\,; \end{gathered} \tag{15.35}$$

they lead to

$$Ea_0 + k_1 a_1 + k_2 a_2 + k_3 a_3 = 0\,, \quad Eb_0 + k_1 b_1 + k_2 b_2 + k_3 b_3 = 0\,. \tag{15.36}$$

Therefore, in the system (15.34) we may ignore the last two equations. Turning to (15.34), with the help of the first two equations,

$$a_0 = -a_3 - (b_1 - ib_2)\,, \qquad b_0 = +b_3 - (a_1 + ia_2)\,, \tag{15.37}$$

we may exclude the variables a_0, b_0; this yields

$$\begin{aligned} k_1 a_1 + k_2 a_2 + k_3 a_3 &= +Ea_3 + E(b_1 - ib_2)\,, \\ k_1 b_1 + k_2 b_2 + k_3 b_3 &= -Eb_3 + E(a_1 + ia_2)\,. \end{aligned} \tag{15.38}$$

In (15.38), we have 2 equations for 6 variables, so there should exist 4 independent solutions.

It is readily proved that taking in mind (15.37), from (15.38) there follow Eqs. (15.36). This means that Eqs. (15.38) are equivalent to the following 4 equations

$$\begin{gathered} Ea_0 + k_1 a_1 + k_2 a_2 + k_3 a_3 = 0\,, \quad Eb_0 + k_1 b_1 + k_2 b_2 + k_3 b_3 = 0\,, \\ a_0 + a_3 = -(b_1 - ib_2)\,, \quad b_0 - b_3 = -(a_1 + ia_2)\,. \end{gathered} \tag{15.39}$$

In order to relate the 4 independent solutions of the system (15.39) to the quantum number of some physical operators, in the next section we will study the problem of eigenstates of the helicity operator for vector-bispinor plane waves.

15.3 Helicity Operator

On vector-bispinor plane waves, let us diagonalize the helicity operator Σ:

$$\Sigma = -i\,(S_1\partial_1 + S_2\partial_2 + S_3\partial_3)\,, \quad \Sigma\,\Psi(x) = \sigma\,\Psi(x)\,. \tag{15.40}$$

Taking in mind the substitution in the plane wave form, we find the following representation for the helicity operator

$$\Sigma = i\,(k_1\sigma^{23} + k_2\sigma^{31} + k_3\sigma^{12}) \otimes I + I \otimes i\,(k_1 j^{23} + k_2 j^{31} + k_3 j^{12})\,. \tag{15.41}$$

We need the expressions for the matrices:

$$\sigma^{23} = \frac{1}{2}\gamma^2\gamma^3 = -\frac{i}{2}\begin{vmatrix} 0 & 1 & 0 & 0 \\ 1 & 0 & 0 & 0 \\ 0 & 0 & 0 & 1 \\ 0 & 0 & 1 & 0 \end{vmatrix}, \quad \sigma^{31} = \frac{1}{2}\gamma^3\gamma^1 = -\frac{i}{2}\begin{vmatrix} 0 & -i & 0 & 0 \\ i & 0 & 0 & 0 \\ 0 & 0 & 0 & -i \\ 0 & 0 & i & 0 \end{vmatrix},$$

$$\sigma^{12} = \frac{1}{2}\gamma^1\gamma^2 = -\frac{i}{2}\begin{vmatrix} 1 & 0 & 0 & 0 \\ 0 & -1 & 0 & 0 \\ 0 & 0 & 1 & 0 \\ 0 & 0 & 0 & -1 \end{vmatrix}, \quad j^{23} = \delta^2_k g^{3l} - \delta^3_k g^{2l} = \begin{vmatrix} 0 & 0 & 0 & 0 \\ 0 & 0 & 0 & 0 \\ 0 & 0 & 0 & -1 \\ 0 & 0 & 1 & 0 \end{vmatrix},$$

$$j^{31} = \delta^3_k g^{1l} - \delta^1_k g^{3l} = \left| \begin{array}{cccc} 0 & 0 & 0 & 0 \\ 0 & 0 & 0 & 1 \\ 0 & 0 & 0 & 0 \\ 0 & -1 & 0 & 0 \end{array} \right|, \quad j^{12} = \delta^1_k g^{2l} - \delta^2_k g^{1l} = \left| \begin{array}{cccc} 0 & 0 & 0 & 0 \\ 0 & 0 & -1 & 0 \\ 0 & 1 & 0 & 0 \\ 0 & 0 & 0 & 0 \end{array} \right| .$$

With these in mind, for the operator Σ we find

$$\Sigma = \frac{1}{2} \left| \begin{array}{cccc} k_3 & k_1 - ik_2 & 0 & 0 \\ k_1 + ik_2 & -k_3 & 0 & 0 \\ 0 & 0 & k_3 & k_1 - ik_2 \\ 0 & 0 & k_1 + ik_2 & -k_3 \end{array} \right| \otimes I + I \otimes i \left| \begin{array}{cccc} 0 & 0 & 0 & 0 \\ 0 & 0 & -k_3 & k_2 \\ 0 & k_3 & 0 & -k_1 \\ 0 & -k_2 & k_1 & 0 \end{array} \right| . \qquad (15.42)$$

To clarify the action of Σ on the wave function, it suffices to recall the action of generators on the vector-bispinor:

$$[\delta_{AB} + \delta\omega_{12}\, i(\sigma^{12})_{AB}]\; [\delta_{ln} + \delta\omega_{12}\, i(j^{12})_{ln}]\Psi_{Bn}$$

$$= \Psi_{Al} + \delta\omega_{12} \, \{ \, i(\sigma^{12})_{AB}\Psi_{Bl} + i(j^{12})_{ln}\Psi_{An}(\tilde{j}^{12})_{nl} \, \} ,$$

where the symbol $\tilde{S}$ over S designates the transposed matrix.

Taking in mind the identity from below

$$\left| \begin{array}{cccc} a_0 & a_1 & a_2 & a_3 \\ b_0 & b_1 & b_2 & b_3 \\ c_0 & c_1 & c_2 & c_3 \\ d_0 & d_1 & d_2 & d_3 \end{array} \right| \left| \begin{array}{cccc} 0 & 0 & 0 & 0 \\ 0 & 0 & +k_3 & -k_2 \\ 0 & -k_3 & 0 & +k_1 \\ 0 & +k_2 & -k_1 & 0 \end{array} \right|$$

$$= \left| \begin{array}{cccc} 0 & (-k_3 a_2 + k_2 a_3) & (k_3 a_1 - k_1 a_3) & (-k_2 a_1 + k_1 a_2) \\ 0 & (-k_3 b_2 + k_2 b_3) & (k_3 b_1 - k_1 b_3) & (-k_2 b_1 + k_1 b_2) \\ 0 & (-k_3 c_2 + k_2 c_3) & (k_3 c_1 - k_1 c_3) & (-k_2 c_1 + k_1 c_2) \\ 0 & (-k_3 d_2 + k_2 d_3) & (k_3 d_1 - k_1 d_3) & (-k_2 d_1 + k_1 d_2) \end{array} \right| ,$$

we reduce the eigenvalue equation $\Sigma\,\Psi = \sigma\,\Psi$ to the form

$$\frac{1}{2} \left| \begin{array}{cccc} k_3 & (k_1 - ik_2) & 0 & 0 \\ (k_1 + ik_2) & -k_3 & 0 & 0 \\ 0 & 0 & k_3 & (k_1 - ik_2) \\ 0 & 0 & (k_1 + ik_2) & -k_3 \end{array} \right| \left| \begin{array}{c} a_0 \\ b_0 \\ c_0 \\ d_0 \end{array} \right| = \sigma \left| \begin{array}{c} a_0 \\ b_0 \\ c_0 \\ d_0 \end{array} \right| ,$$

$$\frac{1}{2} \left| \begin{array}{cccc} k_3 & (k_1 - ik_2) & 0 & 0 \\ (k_1 + ik_2) & -k_3 & 0 & 0 \\ 0 & 0 & k_3 & (k_1 - ik_2) \\ 0 & 0 & (k_1 + ik_2) & -k_3 \end{array} \right| \left| \begin{array}{c} a_1 \\ b_1 \\ c_1 \\ d_1 \end{array} \right| + i \left| \begin{array}{c} (k_2 a_3 - k_3 a_2) \\ (k_2 b_3 - k_3 b_2) \\ (k_2 c_3 - k_3 c_2) \\ (k_2 d_3 - k_3 d_2) \end{array} \right| = \sigma \left| \begin{array}{c} a_1 \\ b_1 \\ c_1 \\ d_1 \end{array} \right| ,$$

$$\frac{1}{2} \left| \begin{array}{cccc} k_3 & (k_1 - ik_2) & 0 & 0 \\ (k_1 + ik_2) & -k_3 & 0 & 0 \\ 0 & 0 & k_3 & (k_1 - ik_2) \\ 0 & 0 & (k_1 + ik_2) & -k_3 \end{array} \right| \left| \begin{array}{c} a_2 \\ b_2 \\ c_2 \\ d_2 \end{array} \right| + i \left| \begin{array}{c} (k_3 a_1 - k_1 a_3) \\ (k_3 b_1 - k_1 b_3) \\ (k_3 c_1 - k_1 c_3) \\ (k_3 d_1 - k_1 d_3) \end{array} \right| = \sigma \left| \begin{array}{c} a_2 \\ b_2 \\ c_2 \\ d_2 \end{array} \right| ,$$

$$\frac{1}{2} \left| \begin{array}{cccc} k_3 & (k_1 - ik_2) & 0 & 0 \\ (k_1 + ik_2) & -k_3 & 0 & 0 \\ 0 & 0 & k_3 & (k_1 - ik_2) \\ 0 & 0 & (k_1 + ik_2) & -k_3 \end{array} \right| \left| \begin{array}{c} a_3 \\ b_3 \\ c_3 \\ d_3 \end{array} \right| + i \left| \begin{array}{c} (k_1 a_2 - k_2 a_1) \\ (k_1 b_2 - k_2 b_1) \\ (k_1 c_2 - k_2 c_1) \\ (k_1 d_2 - k_2 d_1) \end{array} \right| = \sigma \left| \begin{array}{c} a_3 \\ b_3 \\ c_3 \\ d_3 \end{array} \right| ,$$

whence we derive the 16 linear equations:

$$(k_3 - 2\sigma)\, a_0 + (k_1 - ik_2)\, b_0 = 0\,,$$
$$(k_1 + ik_2)\, a_0 - (k_3 + 2\sigma)\, b_0 = 0\,,$$

$$(k_3 - 2\sigma)\, c_0 + (k_1 - ik_2)\, d_0 = 0\,,$$
$$(k_1 + ik_2)\, c_0 - (k_3 + 2\sigma)\, d_0 = 0\,,$$

$$(k_3 - 2\sigma)\, a_1 + (k_1 - ik_2)\, b_1 + 2ik_2\, a_3 - 2ik_3\, a_2 = 0\,,$$
$$(k_1 + ik_2)\, a_1 - (k_3 + 2\sigma)\, b_1 + 2ik_2\, b_3 - 2ik_3\, b_2 = 0\,,$$

$$(k_3 - 2\sigma)\, c_1 + (k_1 - ik_2)\, d_1 + 2ik_2\, c_3 - 2ik_3\, c_2 = 0\,,$$
$$(k_1 + ik_2)\, c_1 - (k_3 + 2\sigma)\, d_1 + 2ik_2\, d_3 - 2ik_3\, d_2 = 0\,,$$

$$(k_3 - 2\sigma)\, a_2 + (k_1 - ik_2)\, b_2 + 2ik_3\, a_1 - 2ik_1\, a_3 = 0\,,$$
$$(k_1 + ik_2)\, a_2 - (k_3 + 2\sigma)\, b_2 + 2ik_3\, b_1 - 2ik_1\, b_3 = 0\,,$$

$$(k_3 - 2\sigma)\, c_2 + (k_1 - ik_2)\, d_2 + 2ik_3\, c_1 - 2ik_1\, c_3 = 0\,,$$
$$(k_1 + ik_2)\, c_2 - (k_3 + 2\sigma)\, d_2 + 2ik_3\, d_1 - 2ik_1\, d_3 = 0\,,$$

$$(k_3 - 2\sigma)\, a_3 + (k_1 - ik_2)\, b_3 + 2ik_1\, a_2 - 2ik_2\, a_1 = 0\,,$$
$$(k_1 + ik_2)\, a_3 - (k_3 + 2\sigma)\, b_3 + 2ik_1 b_2 - 2ik_2 b_1 = 0\,,$$

$$(k_3 - 2\sigma)\, c_3 + (k_1 - ik_2)\, d_3 + 2ik_1\, c_2 - 2ik_2\, c_1 = 0\,,$$
$$(k_1 + ik_2)\, c_3 - (k_3 + 2\sigma)\, d_3 + 2ik_1\, d_2 - 2ik_2\, d_1 = 0\,.$$

In the whole system, we can see 4 unlinked subsystems, $16 = 2 + 2 + 6 + 6$. The first two subsystems

$$\begin{array}{ll} (k_3 - 2\sigma)\, a_0 + (k_1 - ik_2)\, b_0 = 0\,, & (k_1 + ik_2)\, a_0 - (k_3 + 2\sigma)\, b_0 = 0\,; \\ (k_3 - 2\sigma)\, c_0 + (k_1 - ik_2)\, d_0 = 0\,, & (k_1 + ik_2)\, c_0 - (k_3 + 2\sigma)\, d_0 = 0 \end{array} \qquad (15.43)$$

lead to only two eigenvalues:

$$\sigma = \pm\frac{1}{2}\sqrt{k_1^2 + k_2^2 + k_3^2} = \pm\frac{1}{2}k\,,$$
$$b_0 = \frac{\pm k - k_3}{k_1 - ik_2}\, a_0 = \frac{k_1 + ik_2}{\pm k + k_3} a_0\,, \quad d_0 = \frac{\pm k - k_3}{k_1 - ik_2} c_0 = \frac{k_1 + ik_2}{\pm k + k_3}\, c_0\,. \qquad (15.44)$$

We can easily check that formulas (15.28) lead to the expressions for c_0, d_0, which agree with (15.44):

$$c_0 = \frac{E \mp k}{M}\, a_0\,, \quad d_0 = \frac{k_1 + ik_2}{\pm k + k_3}\left\{\frac{E \mp k}{M}\, a_0\right\} \quad \Longrightarrow \quad d_0 = \frac{k_1 + ik_2}{\pm k + k_3}\, c_0\,.$$

Now, let us examine the equations for a_j, b_j:

$$\begin{aligned} &+(k_3 - 2\sigma)\, a_1 - 2ik_3\, a_2 + 2ik_2\, a_3 = -(k_1 - ik_2)\, b_1\,, \\ &+2ik_3\, a_1 + (k_3 - 2\sigma)\, a_2 - 2ik_1\, a_3 = -(k_1 - ik_2)\, b_2\,, \\ &-2ik_2\, a_1 + 2ik_1\, a_2 + (k_3 - 2\sigma)\, a_3 = -(k_1 - ik_2)\, b_3\,, \\ &-(k_3 + 2\sigma)\, b_1 - 2ik_3\, b_2 + 2ik_2\, b_3 = -(k_1 + ik_2)\, a_1\,, \\ &+2ik_3\, b_1 - (k_3 + 2\sigma)\, b_2 - 2ik_1\, b_3 = -(k_1 + ik_2)\, a_2\,, \\ &-2ik_2\, b_1 + 2ik_1\, b_2 - (k_3 + 2\sigma)\, b_3 = -(k_1 + ik_2)\, a_3\,, \end{aligned} \qquad (15.45)$$

and for c_j, d_j:

$$
\begin{aligned}
&+(k_3-2\sigma)\,c_1-2ik_3\,c_2+2ik_2\,c_3=-(k_1-ik_2)\,d_1\,,\\
&+2ik_3c_1+(k_3-2\sigma)\,c_2-2ik_1c_3=-(k_1-ik_2)\,d_2\,,\\
&-2ik_2c_1+2ik_1c_2+(k_3-2\sigma)\,c_3=-(k_1-ik_2)\,d_3\,,\\
&-(k_3+2\sigma)\,d_1-2ik_3\,d_2+2ik_2\,d_3=-(k_1+ik_2)\,c_1\,,\\
&+2ik_3d_1-(k_3+2\sigma)\,d_2-2ik_1d_3=-(k_1+ik_2)\,c_2\,,\\
&-2ik_2d_1+2ik_1d_2-(k_3+2\sigma)\,d_3=-(k_1+ik_2)\,c_3\,.
\end{aligned}
\tag{15.46}
$$

The system (15.46) coincides with (15.45); by this reason it is enough to study only the system (15.45). Let us present it differently,

$$
\begin{aligned}
&\begin{vmatrix} +(k_3-2\sigma) & -2ik_3 & +2ik_2 \\ +2ik_3 & +(k_3-2\sigma) & -2ik_1 \\ -2ik_2 & +2ik_1 & +(k_3-2\sigma) \end{vmatrix} \mathbf{a}=-(k_1-ik_2)\,\mathbf{b}\,,\\
&\begin{vmatrix} -(k_3+2\sigma) & -2ik_3 & +2ik_2 \\ +2ik_3 & -(k_3+2\sigma) & -2ik_1 \\ -2ik_2 & +2ik_1 & -(k_3+2\sigma) \end{vmatrix} \mathbf{b}=-(k_1+ik_2)\,\mathbf{a}\,,
\end{aligned}
\tag{15.47}
$$

whence we derive the separate equations for the variables **a** and **b**:

$$
\left\{ \begin{vmatrix} -(k_3+2\sigma) & -2ik_3 & +2ik_2 \\ +2ik_3 & -(k_3+2\sigma) & -2ik_1 \\ -2ik_2 & +2ik_1 & -(k_3+2\sigma) \end{vmatrix} \begin{vmatrix} (k_3-2\sigma) & -2ik_3 & +2ik_2 \\ +2ik_3 & (k_3-2\sigma) & -2ik_1 \\ -2ik_2 & +2ik_1 & (k_3-2\sigma) \end{vmatrix} - (k_1^2+k_2^2) \right\} \mathbf{a}=0\,,
$$

$$
\left\{ \begin{vmatrix} (k_3-2\sigma) & -2ik_3 & +2ik_2 \\ +2ik_3 & (k_3-2\sigma) & -2ik_1 \\ -2ik_2 & +2ik_1 & (k_3-2\sigma) \end{vmatrix} \begin{vmatrix} -(k_3+2\sigma) & -2ik_3 & +2ik_2 \\ +2ik_3 & -(k_3+2\sigma) & -2ik_1 \\ -2ik_2 & +2ik_1 & -(k_3+2\sigma) \end{vmatrix} - (k_1^2+k_2^2) \right\} \mathbf{b}=0\,.
$$

We shall solve the equation for **a**, and then find the vector **b** from (15.47).

First, let us consider the case of the simple orientation of the plane wave:

$$
(0,0,k_3),\qquad \begin{vmatrix} 4\sigma^2+3k_3^2 & 8ik_3\sigma & 0 \\ -8ik_3\sigma & 4\sigma^2+3k_3^2 & 0 \\ 0 & 0 & 4\sigma^2-k_3^2 \end{vmatrix} \begin{vmatrix} a_1 \\ a_2 \\ a_3 \end{vmatrix} = 0\,,
\tag{15.48}
$$

whence we derive the equation

$$
[(4\sigma^2+3k_3^2)^2-8^2k_3^2\sigma^2]\,(4\sigma^2-k_3^2)=0\,,
$$

that is,

$$
\sigma=-\frac{1}{2}k,-\frac{1}{2}k,\quad +\frac{1}{2}k,+\frac{1}{2}k,\quad -\frac{3}{2}k,\quad +\frac{3}{2}k\,;
\tag{15.49}
$$

we note that the roots $\pm k/2$ are doubly degenerate.

Let us find the corresponding solutions for **a**:

$$\begin{vmatrix} 4\sigma^2+3k_3^2 & 8ik_3\sigma & 0 \\ -8ik_3\sigma & 4\sigma^2+3k_3^2 & 0 \\ 0 & 0 & 4\sigma^2-k_3^2 \end{vmatrix} \begin{vmatrix} a_1 \\ a_2 \\ a_3 \end{vmatrix} = 0 .$$

At $\sigma = \pm\frac{1}{2}k_3$, we have the system

$$\begin{vmatrix} 4k_3^2 & \pm 4ik_3^2 & 0 \\ \mp 4ik_3^2 & 4k_3^2 & 0 \\ 0 & 0 & 0 \end{vmatrix} \begin{vmatrix} a_1 \\ a_2 \\ a_3 \end{vmatrix} = 0 ;$$

the rank of the matrix is equal to 1, and the solutions are

$$\sigma = -\frac{1}{2}k_3, \quad a_1 - ia_2 = 0, \quad a_3 \text{ is arbitrary} ; \tag{15.50}$$
$$\sigma = +\frac{1}{2}k_3, \quad a_1 + ia_2 = 0, \quad a_3 \text{ is arbitrary}.$$

For definiteness, let $a_3 = \pm a_1$; correspondingly, at each σ, we have two solutions:

$$\begin{aligned} \sigma &= -\frac{1}{2}k_3 , \quad \mathbf{a} = a_1(1, -i, \pm 1) ; \\ \sigma &= +\frac{1}{2}k_3 , \quad \mathbf{a} = a_1(1, +i, \pm 1) . \end{aligned} \tag{15.51}$$

Now, let $\sigma = \pm 3/2$; then the solutions are

$$\begin{aligned} &\sigma = -\frac{3}{2}k_3 , \quad \begin{vmatrix} 12k_3^2 & -12ik_3^2 & 0 \\ +12ik_3^2 & 12k_3^2 & 0 \\ 0 & 0 & 8k_3^2 \end{vmatrix} \begin{vmatrix} a_1 \\ a_2 \\ a_3 \end{vmatrix} = 0, \quad a_2 = -ia_1 , \; a_3 = 0; \\ &k = +\frac{3}{2}k_3 , \quad \begin{vmatrix} 12k_3^2 & +12ik_3^2 & 0 \\ -12ik_3^2 & 12k_3^2 & 0 \\ 0 & 0 & 8k_3^2 \end{vmatrix} \begin{vmatrix} a_1 \\ a_2 \\ a_3 \end{vmatrix} = 0, \quad a_2 = +ia_1 , \; a_3 = 0 . \end{aligned} \tag{15.52}$$

Now we turn to the case of general orientation:

$$\begin{vmatrix} 4\sigma^2 - k^2 + 4(k_2^2 + k_3^2) & +8i\sigma k_3 - 4k_1k_2 & -8i\sigma k_2 - 4k_1k_3 \\ -8i\sigma k_3 - 4k_1k_2 & 4\sigma^2 - k^2 + 4(k_1^2 + k_3^2) & 8i\sigma k_1 - 4k_2k_3 \\ 8i\sigma k_2 - 4k_1k_3 & -8i\sigma k_1 - 4k_2k_3 & 4\sigma^2 - k^2 + 4(k_1^2 + k_2^2) \end{vmatrix} \mathbf{a} = 0 . \tag{15.53}$$

It is readily proved that from the vanishing of the main determinant, there follow the yet known eigenvalues:

$$\sigma = -\frac{1}{2}k, -\frac{1}{2}k, \quad +\frac{1}{2}k, +\frac{1}{2}k, \quad -\frac{3}{2}k, +\frac{3}{2}k . \tag{15.54}$$

It is more convenient to use dimensionless quantities:

$$\frac{k_i}{k} = n_i , \quad n_in_i = 1 , \quad \frac{\sigma}{k} \Longrightarrow \sigma , \qquad \sigma = -\frac{1}{2}, +\frac{1}{2}, -\frac{1}{2}, +\frac{1}{2}, -\frac{3}{2}, +\frac{3}{2} ;$$

then the above equation (15.53) reads

$$\begin{vmatrix} 4\sigma^2-1+4(n_2^2+n_3^2) & +8i\sigma n_3-4n_1n_2 & -8i\sigma n_2-4n_1n_3 \\ -8i\sigma n_3-4n_1n_2 & 4\sigma^2-1+4(n_1^2+n_3^2) & 8i\sigma n_1-4n_2n_3 \\ 8i\sigma n_2-4n_1n_3 & -8i\sigma n_1-4n_2n_3 & 4\sigma^2-1+4(n_1^2+n_2^2) \end{vmatrix} \mathbf{a}=0. \qquad (15.55)$$

Let $\sigma=\pm 1/2$; then we have

$$\begin{aligned} (n_2^2+n_3^2)\,a_1+(\pm in_3-n_1n_2)\,a_2-(\pm in_2+n_1n_3)\,a_3=0\,, \\ -(\pm in_3+n_1n_2)\,a_1+(n_1^2+n_3^2)\,a_2+(\pm in_1-n_2n_3)\,a_3=0\,, \\ (\pm in_2-n_1n_3)\,a_1-(\pm in_1+n_2n_3)\,a_2+(n_1^2+n_2^2)\,a_3=0\,. \end{aligned} \qquad (15.56)$$

The rank of the matrix equals to 1, and only one equation is independent; let this be the third:

$$(\pm in_2-n_1n_3)\,a_1-(\pm in_1+n_2n_3)\,a_2+(1-n_3^2)\,a_3=0\,. \qquad (15.57)$$

This equation has two independent solutions (this correlates with the double multiplicity of the roots $\sigma=\pm 1/2$). The first and the most simple is $\mathbf{a}^{(1)}=(n_1,n_2,n_3)$; indeed

$$(\pm in_2-n_1n_3)\,n_1-(\pm in_1+n_2n_3)\,n_2+(1-n_3^2)\,n_3\equiv 0\,.$$

Taking in mind the structure of Eq. (15.57), the second solution should have the form of a vector product:

$$\mathbf{a}^{(2)}=\begin{vmatrix} \mathbf{e}_1 & \mathbf{e}_2 & \mathbf{e}_3 \\ n_1 & n_2 & n_3 \\ (\pm in_2-n_1n_3) & -(\pm in_1+n_2n_3) & (1-n_3^2) \end{vmatrix}$$

$$=(\pm in_1n_3+n_2)\,\mathbf{e}_1+(\pm in_2n_3-n_1)\,\mathbf{e}_2\mp i(1-n_3^2)\,\mathbf{e}_3\,.$$

Thus, there exist two independent solutions of Eq. (15.57) at $\sigma=\pm 1/2$:

$$\mathbf{a}^{(1)}=(n_1,n_2,n_3)=\mathbf{n}\,,\quad \mathbf{a}^{(2)}=(\,\pm in_1n_3+n_2;\ \pm in_2n_3-n_1;\ \mp i(1-n_3^2)\,)\,; \qquad (15.58)$$

we recall that in the case $n_1=0, n_2=0$, these formulas are not valid, and, instead, the result (15.51) should be used.

Now we study the case $\sigma=\pm 3/2$. The rank of the system (15.55) equals to 2; let us preserve two first equations:

$$\begin{aligned} (2+n_2^2+n_3^2)\,a_1+(2i\sigma n_3-n_1n_2)\,a_2=(+2i\sigma n_2+n_1n_3)\,a_3\,, \\ -(2i\sigma n_3+n_1n_2)\,a_1+(2+n_1^2+n_3^2)\,a_2=(-2i\sigma n_1+n_2n_3)\,a_3\,, \end{aligned} \qquad (15.59)$$

The determinant of this system equals to $6(1-n_3^2)$, and it vanishes at $n_3=\pm 1$, this latter case being peculiar. In the general case, the solution of the system (15.59) is

$$\begin{aligned} a_1&=\frac{(3-4\sigma^2)n_1n_3+4i\sigma n_2}{6+(3-4\sigma^2)n_3^2}\,a_3=\frac{-n_1n_3\pm in_2}{1-n_3^2}\,a_3\,,\ \text{let}\ a_3=1-n_3^2\,; \\ a_2&=\frac{(3-4\sigma^2)n_2n_3-4i\sigma n_1}{6+(3-4\sigma^2)n_3^2}\,a_3=\frac{-n_2n_3\mp in_1}{1-n_3^2}\,a_3\,,\ \text{let}\ a_3=1-n_3^2\,. \end{aligned} \qquad (15.60)$$

The choice of a_3 fixes the normalization of this solution.

For all the sets $\{a_1, a_2, a_3\}_\sigma$, one can find the related sets $\{b_1, b_2, b_3\}_\sigma$ (see (15.47)):

$$\begin{vmatrix} b_1 \\ b_2 \\ b_3 \end{vmatrix} = -\frac{1}{n_1 - in_2} \begin{vmatrix} +(n_3 - 2\sigma) & -2in_3 & +2in_2 \\ +2in_3 & +(n_3 - 2\sigma) & -2in_1 \\ -2in_2 & +2in_1 & +(n_3 - 2\sigma) \end{vmatrix} \begin{vmatrix} a_1 \\ a_2 \\ a_3 \end{vmatrix}, \tag{15.61}$$

which may be presented in vector form

$$\mathbf{b} = \frac{1}{n_1 - in_2} \left[(2\sigma - n_3)\, \mathbf{a} - 2i\, \mathbf{n} \times \mathbf{a} \right]. \tag{15.62}$$

Firstly, consider the case $\sigma = \pm 1/2$. For the case of solutions of the type $\mathbf{a}^{(1)} = \mathbf{n}$, we derive

$$\mathbf{b}^{(1)} = \frac{\pm 1 - n_3}{n_1 - in_2}\, \mathbf{a}^{(1)}. \tag{15.63}$$

For solutions of the second type

$$\mathbf{a}^{(2)} = (\pm in_1 n_3 + n_2; \pm in_2 n_3 - n_1; \mp i(1 - n_3^2)),$$

$$\mathbf{b}^{(2)} = \frac{1}{n_1 - in_2} [(\pm 1 - n_3)\, \mathbf{a}^{(2)} - 2i\, \mathbf{n} \times \mathbf{a}^{(2)}],$$

allowing for the identities

$$\begin{aligned} (\mathbf{n} \times \mathbf{a}^{(2)})_1 &= n_2 a_3^{(2)} - n_3 a_2^{(2)} = (\mp i) a_1^{(2)}, \\ (\mathbf{n} \times \mathbf{a}^{(2)})_2 &= n_3 a_1^{(2)} - n_1 a_3^{(2)} = (\mp i) a_2^{(2)}, \\ (\mathbf{n} \times \mathbf{a}^{(2)})_3 &= n_1 a_2^{(2)} - n_2 a_1^{(2)} = (\mp i) a_3^{(2)}, \end{aligned} \tag{15.64}$$

we derive

$$\mathbf{b}^{(2)} = \frac{1}{n_1 - in_2} [(\pm 1 - n_3) - 2i\,(\mp i)]\, \mathbf{a}^{(2)} = \frac{(\mp 1 - n_3)}{n_1 - in_2}\, \mathbf{a}^{(2)}. \tag{15.65}$$

It should be emphasized that in (15.63) and in (15.65), the multipliers are different.

Now consider the states with helicities $\sigma = \pm 3/2$:

$$a_1 = (-n_1 n_3 \pm in_2), \qquad a_2 = (-n_2 n_3 \mp in_1), \qquad a_3 = 1 - n_3^2,$$

$$\begin{aligned} b_1 &= \frac{1}{n_1 - n_2} \{ (\pm 3 - n_3) a_1 - 2i(n_2 a_3 - n_3 a_2) \} = \frac{\pm 1 - n_3}{n_1 - in_2} (-n_1 n_3 \pm in_2), \\ b_2 &= \frac{1}{n_1 - n_2} \{ (\pm 3 - n_3) a_2 - 2i(n_3 a_1 - n_1 a_3) \} = \frac{\pm 1 - n_3}{n_1 - in_2} (-n_2 n_3 \mp in_1), \\ b_2 &= \frac{1}{n_1 - n_2} \{ (\pm 3 - n_3) a_3 - 2i(n_1 a_2 - n_2 a_1) \} = \frac{\pm 1 - n_3}{n_1 - in_2} (1 - n_3^2), \end{aligned} \tag{15.66}$$

or shortly,

$$\mathbf{b} = \frac{\pm 1 - n_3}{n_1 - in_2}\, \mathbf{a}. \tag{15.67}$$

Let us collect the results together:

$$\sigma=\pm 1/2, \quad \{a_0,b_0,c_0,d_0,a_j^{(1)},b_j^{(1)}\}, \quad \{a_0,b_0,c_0,d_0,a_j^{(2)},b_j^{(2)}\},$$
$$b_0=\frac{\pm 1-n_3}{n_1-in_2}\,a_0, \quad d_0=\frac{\pm 1-n_3}{n_1-in_2}\,c_0; \tag{15.68}$$

$$\mathbf{a}^{(1)}=\mathbf{n}, \quad \mathbf{b}^{(1)}=\frac{\pm 1-n_3}{n_1-in_2}\,\mathbf{a}^{(1)}, \quad \mathbf{c}^{(1)}\sim\mathbf{n}, \quad \mathbf{d}^{(1)}=\frac{\pm 1-n_3}{n_1-in_2}\,\mathbf{c}^{(1)}; \tag{15.69}$$

$$\begin{gathered}
a_1^{(2)}=(\pm in_1n_3+n_2),\ a_2^{(2)}=(\pm in_2n_3-n_1),\\
a_3^{(2)}=\mp i(1-n_3^2),\ \mathbf{b}^{(2)}=\frac{\mp 1-n_3}{n_1-in_2}\,\mathbf{a}^{(2)},\\
c_1^{(2)}\sim(\pm in_1n_3+n_2),\ c_2^{(2)}\sim(\pm in_2n_3-n_1),\\
c_3^{(2)}\sim\mp i(1-n_3^2),\ \mathbf{d}^{(2)}=\frac{\mp 1-n_3}{n_1-in_2}\,\mathbf{c}^{(2)};
\end{gathered} \tag{15.70}$$

$$\begin{gathered}
\sigma=\pm 3/2, \quad a_0=0, \quad b_0=0, \quad c_0=0, \quad d_0=0,\\
a_1=(-n_1n_3\pm in_2),\ a_2=(-n_2n_3\mp in_1),\ a_3=1-n_3^2, \quad \mathbf{b}=\frac{\pm 1-n_3}{n_1-in_2}\,\mathbf{a}\\
c_1=(-n_1n_3\pm in_2),\ c_2=(-n_2n_3\mp in_1),\ c_3=1-n_3^2, \quad \mathbf{d}=\frac{\pm 1-n_3}{n_1-in_2}\,\mathbf{c}.
\end{gathered} \tag{15.71}$$

15.4 Helicity Operator and Solutions of the Wave Equation

The above study of the wave equation has provided the linear constraints

$$\begin{gathered}
Ea_0+k_1a_1+k_2a_2+k_3a_3=0, \quad Eb_0+k_1b_1+k_2b_2+k_3b_3=0,\\
a_0+a_3=-(b_1-ib_2), \quad b_0-b_3=-(a_1+ia_2),
\end{gathered} \tag{15.72}$$

and

$$c_l=\frac{E-k_3}{M}\,a_l-\frac{k_1-ik_2}{M}\,b_l, \quad d_l=-\frac{k_1+ik_2}{M}\,a_l+\frac{E+k_3}{M}\,b_l. \tag{15.73}$$

First, let us consider states with helicities $\sigma=\pm 3/2$:

$$\begin{gathered}
a_0=0, \quad b_0=0, \quad c_0=0, \quad d_0=0,\\
a_1=(-n_1n_3\pm in_2),\ a_2=(-n_2n_3\mp in_1),\ a_3=1-n_3^2, \quad \mathbf{b}=\frac{\pm 1-n_3}{n_1-in_2}\,\mathbf{a}\\
c_1\sim(-n_1n_3\pm in_2),\ c_2\sim(-n_2n_3\mp in_1),\ c_3\sim 1-n_3^2; \quad \mathbf{d}=\frac{\pm 1-n_3}{n_1-in_2}\,\mathbf{c};
\end{gathered} \tag{15.74}$$

the symbol $\sim$ reminds that $\mathbf{c}$ is fixed up to an arbitrary multiplier. The equations (15.72) take the form

$$\begin{gathered}
n_1a_1+n_2a_2+n_3a_3=0, \qquad n_1b_1+n_2b_2+n_3b_3=0,\\
a_3=-(b_1-ib_2), \qquad b_3=(a_1+ia_2),
\end{gathered} \tag{15.75}$$

Let us prove that the relations (15.75) agree with (15.74). First, we take into account (15.74) in the first equation from (15.75):

$$n_1 a_1 + n_2 a_2 + n_3 a_3 = 0 \implies$$

$$n_1(-n_1 n_3 \pm i n_2) + n_2(-n_2 n_3 \mp i n_1) + n_3(n_1^2 + n_2^2) \implies 0 = 0 .$$

We do not need re-check the second relation from (15.75).

Now, let us check relation $a_3 = -(b_1 - i b_2)$:

$$\begin{aligned} -(b_1 - i b_2) &= -\frac{\pm 1 - n_3}{n_1 - i n_2} \{(-n_1 n_3 \pm i n_2) - i(-n_2 n_3 \mp i n_1)\} \\ &= -\frac{\pm 1 - n_3}{n_1 - i n_2} \{n_1(-n_3 \mp 1) - i n_2(-n_3 \mp 1)\} \\ &= -(\pm 1 - n_3)(\mp 1 - n_3) = 1 - n_3^2 = a_3 . \end{aligned} \tag{15.76}$$

Now, let us check the relation $b_3 = (a_1 + i a_2)$:

$$(a_1 + i a_2) = (-n_1 n_3 \pm i n_2) + i(-n_2 n_3 \mp i n_1)$$

$$= -n_1 n_3 \pm i n_2 - i n_2 n_3 \pm n_1 = (n_1 + i n_2)(\pm 1 - n_3) \equiv b_3 .$$

Taking in mind (15.75), we find c_j:

$$c_j = \frac{E - k_3}{M} a_j - \frac{k_1 - i k_2}{M} \frac{\pm k - k_3}{k_1 - i k_2} a_j = \frac{E \mp k}{M} a_j . \tag{15.77}$$

Now, let us find d_j:

$$d_j = -\frac{k_1 + i k_2}{M} a_j + \frac{E + k_3}{M} \frac{\pm k - k_3}{k_1 - i k_2} \frac{k_1 + i k_2}{k_1 + i k_2} a_j$$

$$= -\frac{k_1 + i k_2}{M} a_j \pm \frac{E + k_3}{M} \frac{k \mp k_3}{(k - k_3)(k + k_3)} (k_1 + i k_2)\, a_j = \frac{k_1 + i k_2}{k_3 \pm k} \left\{ \frac{E \mp k}{M} a_j \right\},$$

that is

$$d_j = \frac{k_1 + i k_2}{k_3 \pm k} c_j = \frac{\pm k - k_3}{k_1 - i k_2} c_j ; \tag{15.78}$$

this coincides with (15.73). Thus, the states with helicities $\sigma = \pm 3/2$ actually are exact solutions of the wave equation.

Now we turn to the case when $\sigma = \pm 1/2$. Since we have a double generation, first let us consider each of them separately. The states of the first type are given by (see (15.44))

$$\begin{gathered} b_0^{(1)} = \frac{\pm 1 - n_3}{n_1 - i n_2} a_0^{(1)} = \frac{n_1 + i n_2}{\pm 1 + n_3} a_0^{(1)}, \quad d_0^{(1)} = \frac{\pm 1 - n_3}{n_1 - i n_2} c_0^{(1)} = \frac{n_1 + i n_2}{\pm 1 + n_3} c_0^{(1)}, \\ \mathbf{a}^{(1)} = \mathbf{n}, \qquad \mathbf{b}^{(1)} = \frac{\pm 1 - n_3}{n_1 - i n_2} \mathbf{a}^{(1)}, \quad \mathbf{c}^{(1)} \sim \mathbf{a}^{(1)}, \qquad \mathbf{d}^{(1)} = \frac{\pm 1 - n_3}{n_1 - i n_2} \mathbf{c}^{(1)} . \end{gathered} \tag{15.79}$$

The symbol $\sim$ may be specified with the help of the linear relations

$$\mathbf{c}^{(1)} = \frac{E-k_3}{M}\,\mathbf{a}^{(1)} - \frac{k_1-ik_2}{M}\,\mathbf{b}^{(1)} = \frac{E\mp k}{M}\,\mathbf{a}^{(1)},$$
$$\mathbf{d}^{(1)} = -\frac{k_1+ik_2}{M}\,\mathbf{a}^{(1)} + \frac{E+k_3}{M}\,\mathbf{b}^{(1)} = \frac{k_1+ik_2}{k_3\pm k}\left\{\frac{E\mp k}{M}\mathbf{a}^{(1)}\right\} = \frac{\pm 1-n_3}{n_1-in_2}\,\mathbf{c}^{(1)}. \tag{15.80}$$

We can easily see that these states do not produce solutions of the wave equation. Indeed, we have (see (15.72))

$$Ea_0^{(1)}+k_1a_1^{(1)}+k_2a_2^{(1)}+k_3a_3^{(1)}=0 \quad\Longrightarrow\quad Ea_0^{(1)}+k=0, \quad a_0^{(1)}=-\frac{k}{E}. \tag{15.81}$$

Equation $Eb_0^{(1)}+k_1b_1^{(1)}+k_2b_2^{(1)}+k_3b_3^{(1)}=0$ gives the same result. Let us check the two remaining equations (see (15.72)):

$$a_0^{(1)}+a_3^{(1)} = -(b_1^{(1)}-ib_2^{(1)}) \Longrightarrow$$

$$-\frac{k}{E}+n_3 = -\frac{\pm 1-n_3}{n_1-in_2}(n_1-in_2) = \pm 1+n_3 \Longrightarrow \frac{k}{E}=\mp 1;$$

$$b_0^{(1)}-b_3^{(1)} = -(a_1^{(1)}+ia_2^{(1)}) \Longrightarrow \frac{\pm 1-n_3}{n_1-in_2}(-\frac{k}{E}-n_3) = -(n_1+in_2) \Longrightarrow$$

$$(\pm 1-n_3)(\frac{k}{E}+n_3) = (1-n_3)(1+n_3) \quad\Longrightarrow\quad \frac{k}{E}+n_3=\pm 1+n_3, \quad \frac{k}{E}=\pm 1;$$

these two identities cannot be valid, because $E=\sqrt{k^2+M^2}$.

Now consider the states with $\sigma=\pm 1/2$, of second type. First, note the relations

$$b_0^{(2)} = \frac{\pm 1-n_3}{n_1-in_2}\,a_0^{(2)}, \quad d_0^{(2)} = \frac{\pm 1-n_3}{n_1-in_2}\,c_0^{(2)}, \quad c_0^{(2)} = \frac{E\mp k}{M}\,a_0^{(2)}. \tag{15.82}$$

Now consider the vectors $\mathbf{a}^{(2)},\mathbf{b}^{(2)},\mathbf{c}^{(2)},\mathbf{d}^{(2)}$:

$$a_1^{(2)} = \pm in_1n_3+n_2, \quad a_2^{(2)} = \pm in_2n_3-n_1,$$

$$a_3^{(2)} = \mp i(1-n_3^2), \quad \mathbf{b}^{(2)} = \frac{\mp 1-n_3}{n_1-in_2}\,\mathbf{a}^{(2)},$$

$$c_1^{(2)} \sim (\pm in_1n_3+n_2), \quad c_2^{(2)} \sim (\pm in_2n_3-n_1),$$

$$c_3^{(2)} \sim \mp i(1-n_3^2)\,), \quad \mathbf{d}^{(2)} = \frac{\mp 1-n_3}{n_1-in_2}\,\mathbf{c}^{(2)}.$$

The symbol $\sim$ may be specified with help of the linear restrictions (see (15.73)):

$$\mathbf{c}^{(2)} = \frac{E-k_3}{M}\,\mathbf{a}^{(2)} - \frac{k_1-ik_2}{M}\,\mathbf{b}^{(2)} = \frac{E\pm k}{M}\,\mathbf{a}^{(2)},$$
$$\mathbf{d}^{(2)} = -\frac{k_1+ik_2}{M}\,\mathbf{a}^{(2)} + \frac{E+k_3}{M}\,\mathbf{b}^{(2)} = \frac{\mp 1-n_3}{n_1-in_2}\,\mathbf{c}^{(2)}. \tag{15.83}$$

Let us introduce the notations:

$$\Gamma = \frac{\pm 1 - n_3}{n_1 - i n_2}, \qquad R = \frac{\mp 1 - n_3}{n_1 - i n_2}; \tag{15.84}$$

then the above formulas read shorter,

$$b_0^{(2)} = \Gamma\, a_0^{(2)}, \quad d_0^{(2)} = \Gamma\, c_0^{(a)}, \quad \mathbf{b}^{(2)} = R\, \mathbf{a}^{(2)}, \quad \mathbf{d}^{(2)} = R\, \mathbf{c}^{(2)}. \tag{15.85}$$

Now we may check the first consequence of the wave equation (see (15.72))

$$E a_0^{(2)} + k_1 a_1^{(2)} + k_2 a_2^{(2)} + k_3\,,\, a_3^{(2)} = 0; \tag{15.86}$$

it yields

$$E a_0^{(2)} + k\{\, n_1(\pm i n_1 n_3 + n_2)$$

$$+ n_2(\pm i n_2 n_3 - n_1) + n_3(\mp i)(n_1^2 + n_2^2)\,\} = 0 \implies a_0^{(2)} = 0\,.$$

Similarly, we derive

$$E b_0^{(2)} + k_1 b_1^{(2)} + k_2^{(2)} b_2 + k_3 b_3^{(2)} = 0 \quad \implies \quad b_0^{(2)} = 0\,.$$

Thus, we obtain

$$a_0^{(2)} = 0, \quad b_0^{(2)} = 0, \quad c_0^{(2)} = 0, \quad d_0^{(2)} = 0\,. \tag{15.87}$$

Let us check two remaining restrictions (see (15.72))

$$a_3^{(2)} = -(b_1^{(2)} - i b_2^{(2)})\,, \quad a_3^{(2)} = \mp i(1 - n_3^2)\,,$$

$$-(b_1^{(2)} - i b_2^{(2)}) = -R\,[\pm i n_1 n_3 + n_2 - i(\pm i n_2 n_3 - n_1)]$$

$$= \pm i(1 \pm n_3)(1 \pm n_3)\,\frac{1 - n_3^2}{(1 - n_3)(1 + n_3)} = \pm i(1 - n_3^2)\,\frac{1 \pm n_3}{1 \mp n_3} = \frac{R}{\Gamma} a_3^{(2)} \neq a_3^{(2)};$$

we conclude that the needed restriction does not hold. Similarly, we derive

$$b_3^{(2)} = a_1^{(2)} + i a_2^{(2)}, \quad b_3^{(2)} = \frac{\mp 1 - n_3}{n_1 - i n_2}(\mp i)(1 - n_3^2)$$

$$= \frac{\mp 1 - n_3}{n_1 - i n_2}(\mp i)(n_1 - i n_2)(n_1 + i n_2) = i(n_1 + i n_2)(1 \pm n_3)\,,$$

$$a_1^{(2)} + i a_2^{(2)} = \pm i n_1 n_3 + n_2 + i(\pm i n_2 n_3 - n_1)$$

$$= -i(n_1 + i n_2)(1 \mp n_3)\cdot\frac{(1 \pm n_3)}{(1 \pm n_3)} = -i(n_1 + i n_2)(1 \pm n_3)\frac{(1 \mp n_3)}{(1 \pm n_3)} = \frac{\Gamma}{R} b_3^{(2)} \neq b_3^{(2)};$$

again, the needed restriction does not hold. Therefore, each separate helicity state, of type (1) and (2), with $\sigma = \pm 1/2$, does not provide us with solutions of the wave equation.

Let us search for solutions of the wave equations as a linear combination of these states:

$$
\begin{aligned}
a_0 = Fa_0^{(1)} + Ga_0^{(2)}, \quad \mathbf{a} = F\mathbf{a}^{(1)} + G\mathbf{a}^{(2)}, \\
b_0 = F\Gamma a_0^{(1)} + G\Gamma a_0^{(2)}, \quad \mathbf{b} = F\Gamma\mathbf{a}^{(1)} + GR\mathbf{a}^{(2)};
\end{aligned}
\qquad (15.88)
$$

we remind that the normalization of the vectors $\mathbf{a}^{(1)}, \mathbf{a}^{(2)}, \mathbf{b}^{(1)}, \mathbf{b}^{(2)}$ is fixed, but the parameters $a_0^{(1)}, a_0^{(2)}$ are arbitrary.

Two first restrictions from (15.72) give

$$
\begin{aligned}
Ea_0 + \mathbf{ka} = 0 \quad &\Longrightarrow \quad E(Fa_0^{(1)} + Ga_0^{(2)}) + Fk + G\cdot 0 = 0, \\
Eb_0 + \mathbf{kb} = 0 \quad &\Longrightarrow \quad E\Gamma(Fa_0^{(1)} + Ga_0^{(2)}) + F\Gamma k + G\cdot 0 = 0,
\end{aligned}
$$

that is

$$
Fa_0^{(1)} + Ga_0^{(2)} + F\frac{k}{E} = 0 \Longrightarrow F(a_0^{(1)} + \frac{k}{E}) + Ga_0^{(2)} = 0. \qquad (15.89)
$$

Taking in mind (15.81) and (15.86), $a_0^{(1)} = -\frac{k}{E}$, $a_0^{(2)} = 0$, we conclude that these two equations do not impose any restrictions on the parameters F and G.

Now consider the two remaining equations from (15.72):

$$
a_0 + a_3 = -(b_1 - ib_2), \quad b_0 - b_3 = -(a_1 + ia_2).
$$

They are equivalent to

$$
\begin{aligned}
(Fa_0^{(1)} + Ga_0^{(2)}) + (Fa_3^{(1)} + Ga_3^{(2)}) = -\Big\{(F\Gamma a_1^{(1)} + GRa_1^{(2)}) - i(F\Gamma a_2^{(1)} + GRa_2^{(2)}))\Big\}, \\
(F\Gamma a_0^{(1)} + G\Gamma a_0^{(2)}) - (F\Gamma a_3^{(1)} + GRa_3^{(2)}) = -\Big\{(Fa_1^{(1)} + Ga_1^{(2)}) + i(Fa_2^{(1)} + Ga_2^{(2)})\Big\},
\end{aligned}
$$

which, after re-grouping the terms, lead to

$$
\begin{aligned}
F\left[(a_0^{(1)} + a_3^{(1)}) + \Gamma(a_1^{(1)} - ia_2^{(1)})\right] + G\left[(a_0^{(2)} + a_3^{(2)}) + R(a_1^{(2)} - ia_2^{(2)})\right] = 0, \\
F\left[\Gamma(a_0^{(1)} - a_3^{(1)}) + (a_1^{(1)} + ia_2^{(1)})\right] + G\left[(\Gamma a_0^{(2)} - Ra_3^{(2)}) + (a_1^{(2)} + ia_2^{(2)})\right] = 0,
\end{aligned}
\qquad (15.90)
$$

whence, taking into account $a_0^{(1)} = -k/E$ and $a_0^{(2)} = 0$, we derive the following two equations:

$$
\begin{aligned}
F\left[-k/E + a_3^{(1)} + \Gamma(a_1^{(1)} - ia_2^{(1)})\right] + G\left[a_3^{(2)} + R(a_1^{(2)} - ia_2^{(2)})\right] = 0, \\
F\left[-\Gamma k/E - \Gamma a_3^{(1)} + (a_1^{(1)} + ia_2^{(1)})\right] + G\left[-Ra_3^{(2)} + (a_1^{(2)} + ia_2^{(2)})\right] = 0.
\end{aligned}
\qquad (15.91)
$$

Now we should use the formulas

$$
\begin{aligned}
a_1^{(1)} = n_1, \quad a_2^{(1)} = n_2, \quad a_3^{(1)} = n_3, \quad \Gamma = \frac{\pm 1 - n_3}{n_1 - in_2}, \; R = \frac{\mp 1 - n_3}{n_1 - in_2}, \\
a_1^{(2)} = \pm in_1 n_3 + n_2, \quad a_2^{(2)} = \pm in_2 n_3 - n_1, \quad a_3^{(2)} = \mp i(1 - n_3^2).
\end{aligned}
$$

First, we find

$$a_3^{(1)} + \Gamma(a_1^{(1)} - ia_2^{(1)}) = n_1 + \frac{\pm 1 - n_3}{n_1 - in_2}(n_1 - in_2) = \pm 1\,,$$
$$a_3^{(2)} + R(a_1^{(2)} - ia_2^{(2)}) = \mp i[(1 - n_3(1+n_3) + (1 \pm n_3)(1 \pm n_3)] = \mp 2i(1 \pm n_3)\,;$$

therefore the first equation in (15.91) takes the form

$$(-\frac{k}{E} \pm 1)\, F \mp 2i(1 \pm n_3))\, G = 0\,. \qquad (15.92)$$

Similarly, we get

$$-\Gamma a_3^{(1)} + (a_1^{(1)} + ia_2^{(1)}) = -\frac{\pm 1 - n_3}{n_1 - in_2} n_3 + (n_1 + in_2)\frac{n_1 - in_2}{n_1 - in_2}$$
$$= \frac{1}{n_1 - in_2}[(\mp 1 + n_3)n_3 + (1 - n_3)(1 + n_3)] = \frac{1 \mp n_3}{n_1 - in_2}\,,$$
$$-Ra_3^{(2)} + (a_1^{(2)} + ia_2^{(2)}) = \pm i(\mp 1 - n_3)(n_1 + in_2) \pm i(n_1 + in_2)(n_3 \mp 1)$$
$$= \pm i(n_1 + in_2)[\mp 1 - n_3 + n_3 \mp 1] = -2i(n_1 + in_2)\,;$$

therefore, the second equation in (15.91) takes the form

$$F\Big[\mp \frac{1 \mp n_3}{n_1 - in_2}\frac{k}{E} + \frac{1 \mp n_3}{n_1 - in_2}\Big] + G\Big[-2i(n_1 + in_2)\Big] = 0\,,$$

or,

$$\mp \frac{1 \mp n_3}{n_1 - in_2}(\frac{k}{E} \mp 1)\, F - 2i(n_1 + in_2)\, G = 0\,,$$

that is

$$(\frac{k}{E} \mp 1)\, F \pm 2i(1 \pm n_3)\, G = 0\,; \qquad (15.93)$$

we note that Eq. (15.93) coincides with the first one (15.92). Thus, we have found the coefficients of the needed linear combination of the two type of states with helicity $\sigma = \pm 1/2$. These coefficients are fixed up to some total arbitrary multiplier, which relates to the normalization of solutions.

15.5 The Plane Wave Solutions in Massless Case

In this section, we turn to the massless spin 3/2 particle. The system (15.20) may be written in matrix form (by changing the notation, $\tilde{\Psi}^l(x) = \Phi^l(x)$):

$$\left|\begin{array}{cccc} 0 & (\gamma^2\partial_3 - \gamma^3\partial_2) & (\gamma^3\partial_1 - \gamma^1\partial_3) & (\gamma^1\partial_2 - \gamma^2\partial_1) \\ (\gamma^2\partial_3 - \gamma^3\partial_2) & 0 & -(\gamma^3\partial_0 + \gamma^0\partial_3) & (\gamma^2\partial_0 + \gamma^0\partial_2) \\ (\gamma^3\partial_1 - \gamma^1\partial_3) & (\gamma^3\partial_0 + \gamma^0\partial_3) & 0 & -(\gamma^1\partial_0 + \gamma^0\partial_1) \\ (\gamma^1\partial_2 - \gamma^2\partial_1) & -(\gamma^2\partial_0 + \gamma^0\partial_2) & (\gamma^1\partial_0 + \gamma^0\partial_1) & 0 \end{array}\right| \left|\begin{array}{c} \Phi^0 \\ \Phi^1 \\ \Phi^2 \\ \Phi^3 \end{array}\right| = 0\,. \qquad (15.94)$$

We will search for solutions in the form of plane waves

$$\Phi^{l}(x) = e^{ik_a x^a} A^{l}, \quad k_a x^a = k_0 x^0 + k_j x^j, \quad k_0 = -\epsilon,$$

where $A^l = (A^0, A^1, A^2, A^3)$ stand for 4-columns; then Eq. (15.94) takes the form

$$\begin{vmatrix} 0 & (\gamma^2 k_3 - \gamma^3 k_2) & (\gamma^3 k_1 - \gamma^1 k_3) & (\gamma^1 k_2 - \gamma^2 k_1) \\ (\gamma^2 k_3 - \gamma^3 k_2) & 0 & -(\gamma^3 k_0 + \gamma^0 k_3) & (\gamma^2 k_0 + \gamma^0 k_2) \\ (\gamma^3 k_1 - \gamma^1 k_3) & (\gamma^3 k_0 + \gamma^0 k_3) & 0 & -(\gamma^1 k_0 + \gamma^0 k_1) \\ (\gamma^1 k_2 - \gamma^2 k_1) & -(\gamma^2 k_0 + \gamma^0 k_2) & (\gamma^1 k_0 + \gamma^0 k_1) & 0 \end{vmatrix} \begin{vmatrix} A^0 \\ A^1 \\ A^2 \\ A^3 \end{vmatrix} = 0. \quad (15.95)$$

It is readily seen that the gradient-type solutions $\Phi_l^{(0)}(x) = \partial_l e^{ikx}\varphi$ identically satisfy this equation,

$$\begin{vmatrix} 0 & (\gamma^2 k_3 - \gamma^3 k_2) & (\gamma^3 k_1 - \gamma^1 k_3) & (\gamma^1 k_2 - \gamma^2 k_1) \\ (\gamma^2 k_3 - \gamma^3 k_2) & 0 & -(\gamma^3 k_0 + \gamma^0 k_3) & (\gamma^2 k_0 + \gamma^0 k_2) \\ (\gamma^3 k_1 - \gamma^1 k_3) & (\gamma^3 k_0 + \gamma^0 k_3) & 0 & -(\gamma^1 k_0 + \gamma^0 k_1) \\ (\gamma^1 k_2 - \gamma^2 k_1) & -(\gamma^2 k_0 + \gamma^0 k_2) & (\gamma^1 k_0 + \gamma^0 k_1) & 0 \end{vmatrix} \begin{vmatrix} k^0 \varphi \\ -k_1 \varphi \\ -k_2 \varphi \\ -k_3 \varphi \end{vmatrix} = 0.$$

It is convenient to present the system (15.95) as four equations:

$$\begin{gathered} (\gamma^2 k_3 - \gamma^3 k_2) A^1 + (\gamma^3 k_1 - \gamma^1 k_3) A^2 + (\gamma^1 k_2 - \gamma^2 k_1) A^3 = 0, \\ (\gamma^3 k_0 + \gamma^0 k_3) A^2 - (\gamma^2 k_0 + \gamma^0 k_2) A^3 = (\gamma^2 k_3 - \gamma^3 k_2) A^0, \\ -(\gamma^3 k_0 + \gamma^0 k_3) A^1 + (\gamma^1 k_0 + \gamma^0 k_1) A^3 = (\gamma^3 k_1 - \gamma^1 k_3) A^0, \\ (\gamma^2 k_0 + \gamma^0 k_2) A^1 - (\gamma^1 k_0 + \gamma^0 k_1) A^2 = (\gamma^1 k_2 - \gamma^2 k_1) A^0. \end{gathered} \quad (15.96)$$

From the first equation, allowing for the identity

$$(\gamma^1 k_2 - \gamma^2 k_1)(\gamma^1 k_2 - \gamma^2 k_1) = -k_2^2 - k_1^2,$$

we get an expression for A^3:

$$\begin{aligned} (k_1^2 + k_2^2) A^3 = {} & (\gamma^1 \gamma^2 k_2 k_3 + k_1 k_2 \gamma^2 \gamma^3 + k_3 k_1 + \gamma^3 \gamma^1 k_2^2) A^1 \\ & + (\gamma^1 \gamma^3 k_2 k_1 + k_1 k_3 \gamma^2 \gamma^1 + k_3 k_2 + \gamma^3 \gamma^2 k_1^2) A^2. \end{aligned} \quad (15.97)$$

Similarly, from the fourth equation, we get

$$\begin{aligned} -(k_1^2 + k_2^2) A^0 = {} & (\gamma^1 \gamma^2 k_0 k_2 + k_1 k_2 \gamma^0 \gamma^2 + k_1 k_0 + \gamma^1 \gamma^0 k_2^2) A^1 \\ & + (\gamma^2 \gamma^1 k_0 k_1 + k_2 k_1 \gamma^0 \gamma^1 + k_2 k_0 + \gamma^2 \gamma^0 k_1^2) A^2. \end{aligned} \quad (15.98)$$

For brevity, let us apply the notations

$$k_1 = a, \quad k_2 = b, \quad k_3 = c, \quad k_0 = d, \quad d^2 - a^2 - b^2 - c^2 = 0.$$

Then, the expressions for A^3 and A^0 read

$$\begin{aligned}(a^2+b^2)A^3 &= (bc\gamma^1\gamma^2+ab\gamma^2\gamma^3+ac+b^2\gamma^3\gamma^1)A^1\\ &\quad +(ab\gamma^1\gamma^3+ac\gamma^2\gamma^1+bc+a^2\gamma^3\gamma^2)A^2,\\ -(a^2+b^2)A^0 &= (db\gamma^1\gamma^2+ab\gamma^0\gamma^2+da+b^2\gamma^1\gamma^0)A^1\\ &\quad +(da\gamma^2\gamma^1+ab\gamma^0\gamma^1+db+a^2\gamma^2\gamma^0)A^2.\end{aligned} \tag{15.99}$$

With the help of (15.99), one can exclude the variables A^0, A^3 from the 2-nd and 3-rd equations in (15.96). So we find two equations with respect to A^1 and A^2:

$$\begin{aligned}&ab\{d\,\gamma^1\gamma^2\gamma^3+c\,\gamma^0\gamma^1\gamma^2+a\,\gamma^0\gamma^2\gamma^3+b\,\gamma^0\gamma^3\gamma^1\}A^2\\ &=b^2\{d\,\gamma^1\gamma^2\gamma^3+c\,\gamma^0\gamma^1\gamma^2+a\,\gamma^0\gamma^2\gamma^3+b\,\gamma^0\gamma^3\gamma^1\}A^1,\end{aligned} \tag{15.100}$$

$$\begin{aligned}&ab\{d\,\gamma^1\gamma^2\gamma^3+c\,\gamma^0\gamma^1\gamma^2+a\,\gamma^0\gamma^2\gamma^3+b\,\gamma^0\gamma^3\gamma^1\}A^1\\ &=a^2\{d\,\gamma^1\gamma^2\gamma^3+c\,\gamma^0\gamma^1\gamma^2+a\,\gamma^0\gamma^2\gamma^3+b\,\gamma^0\gamma^3\gamma^1\}A^2.\end{aligned} \tag{15.101}$$

With the notation

$$K=d\,\gamma^1\gamma^2\gamma^3+c\,\gamma^0\gamma^1\gamma^2+a\,\gamma^0\gamma^2\gamma^3+b\,\gamma^0\gamma^3\gamma^1, \tag{15.102}$$

the equations (15.100)–(15.101) may be written as

$$aK\,A^2=bKA^1,\; bK\,A^1=aKA^2 \quad\Longrightarrow\quad K(aA^2-bA^1)=0,\; K(bA^1-aA^2)=0.$$

In fact, here we have only one equation (further, the notation $k_a=(d,a,b,c)$ is not applied)

$$K(k_1A^2-k_2A^1)=0, \quad K=k_0\gamma^1\gamma^2\gamma^3+k_1\gamma^0\gamma^2\gamma^3+k_2\gamma^0\gamma^3\gamma^1+k_3\gamma^0\gamma^1\gamma^2. \tag{15.103}$$

Taking into account the definition

$$\gamma^5=i\gamma^0\gamma^1\gamma^2\gamma^3=\begin{vmatrix}1&0&0&0\\0&1&0&1\\0&0&-1&0\\0&0&0&-1\end{vmatrix}, \quad (\gamma^5)^2=I, \quad \gamma^5\gamma^a=-\gamma^a\gamma^5, \tag{15.104}$$

and the identities

$$\begin{aligned}&i\gamma^5\gamma^0=-\gamma^0\gamma^1\gamma^2\gamma^3\gamma^0=\gamma^1\gamma^2\gamma^3, \quad i\gamma^5\gamma^1=-\gamma^0\gamma^1\gamma^2\gamma^3\gamma^1=\gamma^0\gamma^2\gamma^3,\\ &i\gamma^5\gamma^2=-\gamma^0\gamma^1\gamma^2\gamma^3\gamma^2=\gamma^0\gamma^3\gamma^1, \quad i\gamma^5\gamma^3=-\gamma^0\gamma^1\gamma^2\gamma^3\gamma^3=\gamma^0\gamma^1\gamma^2,\end{aligned} \tag{15.105}$$

we obtain a different representation for the matrix K:

$$K=i\gamma^5(k_0\gamma^0+k_1\gamma^1+k_2\gamma^2+k_3\gamma^3). \tag{15.106}$$

Therefore, Eq. (15.103) is written as follows (the multiplier $i\gamma^5$ is omitted):

$$(k_0\gamma^0+k_1\gamma^1+k_2\gamma^2+k_3\gamma^3)A=0, \quad A=(k_1A^2-k_2A^1). \tag{15.107}$$

Equation (15.107) may be presented as (let $A=(a,b,c,d)$

$$\begin{vmatrix} 0 & 0 & k^0-k_3 & -k_1+ik_2 \\ 0 & 0 & -k_1-ik_2 & k^0+k_3 \\ k^0+k_3 & k_1-ik_2 & 0 & 0 \\ k_1+ik_2 & k^0-k_3 & 0 & 0 \end{vmatrix} \begin{vmatrix} a \\ b \\ c \\ d \end{vmatrix} = 0,$$

whence there follow two systems with respect to d,d and a,b (let $k_0=-\epsilon, \epsilon>0$):

$$(\epsilon+k_3)c+(k_1-ik_2)d=0, \quad (k_1+ik_2)c+(\epsilon-k_3)d=0;$$

$$(\epsilon-k_3)a-(k_1-ik_2)b=0, \quad -(k_1+ik_2)a+(\epsilon+k_3)b=0.$$

Their solutions are

$$b=\frac{\epsilon-k_3}{k_1-ik_2}a=\frac{k_1+ik_2}{\epsilon+k_3}a, \qquad d=-\frac{\epsilon+k_3}{k_1-ik_2}c=-\frac{k_1+ik_2}{\epsilon-k_3}c. \tag{15.108}$$

Temporarily we shall use the shortening form for the solutions (15.108):

$$b=(\epsilon-k_3)a, \quad d=-(\epsilon+k_3)c; \tag{15.109}$$

in fact we make the change of notations

$$\frac{\epsilon-k_3}{k_1-ik_2} \Longrightarrow \epsilon-k_3, \qquad \frac{\epsilon+k_3}{k_1-ik_2} \Longrightarrow \epsilon+k_3. \tag{15.110}$$

However, in the very end, these changes in notations (15.110) must be considered throughout.

Therefore, the general solution may be presented as

$$A=-k_2A^1+k_1A^2=a\begin{vmatrix} 1 \\ (\epsilon-k_3) \\ 0 \\ 0 \end{vmatrix}+c\begin{vmatrix} 0 \\ 0 \\ 1 \\ -(\epsilon+k_3) \end{vmatrix}. \tag{15.111}$$

Since the parameters a and c are independent, in (15.111) we have two linearly independent solutions. Let us fix the solutions $A_{(1)}$ and $A_{(2)}$ in the following way:

(1), $a=k_1, c=-k_2$,

$$A_{(1)}=-k_2A^1_{(1)}+k_1A^2_{(1)}=-k_2\begin{vmatrix} 0 \\ 0 \\ 1 \\ -(\epsilon+k_3) \end{vmatrix}+k_1\begin{vmatrix} 1 \\ (\epsilon-k_3) \\ 0 \\ 0 \end{vmatrix}; \tag{15.112}$$

(2), $a=-k_2, c=k_1$,

$$A_{(2)}=-k_2A^1_{(2)}+k_1A^2_{(2)}=-k_2\begin{vmatrix} 1 \\ (\epsilon-k_3) \\ 0 \\ 0 \end{vmatrix}+k_1\begin{vmatrix} 0 \\ 0 \\ 1 \\ -(\epsilon+k_3) \end{vmatrix}, \tag{15.113}$$

or differently

$$(1) \qquad A^1_{(1)} = \left| \begin{array}{c} 0 \\ 0 \\ 1 \\ -(\epsilon + k_3) \end{array} \right| , \quad A^2_{(1)} = \left| \begin{array}{c} 1 \\ (\epsilon - k_3) \\ 0 \\ 0 \end{array} \right| ; \tag{15.114}$$

$$(2) \qquad A^1_{(2)} = \left| \begin{array}{c} 1 \\ (\epsilon - k_3) \\ 0 \\ 0 \end{array} \right| , \quad A^2_{(2)} = \left| \begin{array}{c} 0 \\ 0 \\ 1 \\ -(\epsilon + k_3) \end{array} \right| . \tag{15.115}$$

Below we shall take into account the four identities

$$k_1 A^1_{(1)} + k_2 A^2_{(1)} = \left| \begin{array}{c} k_2 \\ k_2(\epsilon - k_3) \\ k_1 \\ -k_1(\epsilon + k_3) \end{array} \right| , k_1 A^1_{(2)} + k_2 A^2_{(2)} = \left| \begin{array}{c} k_1 \\ k_1(\epsilon - k_3) \\ k_2 \\ -k_2(\epsilon + k_3) \end{array} \right| , \tag{15.116}$$

$$-k_2 A^1_{(1)} + k_1 A^2_{(1)} = \left| \begin{array}{c} k_1 \\ k_1(\epsilon - k_3) \\ -k_2 \\ +k_2(\epsilon + k_3) \end{array} \right| , -k_2 A^1_{(2)} + k_1 A^2_{(2)} = \left| \begin{array}{c} -k_2 \\ -k_2(\epsilon - k_3) \\ k_1 \\ -k_1(\epsilon + k_3) \end{array} \right| . \tag{15.117}$$

The concomitant components $A^0_{(1)}, A^3_{(1)}$ and $A^0_{(2)}, A^3_{(2)}$ can be calculated according to (15.97) and (15.98); let us write them down differently:

$$(k_1^2 + k_2^2) A^3 = k_3 (k_1 A^1 + k_2 A^2) - (k_1 \gamma^2 \gamma^3 + k_2 \gamma^3 \gamma^1 + k_3 \gamma^1 \gamma^2)(-k_2 A^1 + k_1 A^2) ,$$
$$-(k_1^2 + k_2^2) A^0 = k_0 (k_1 A^1 + k_2 A^2) + (\gamma^0 \gamma^1 k_2 - \gamma^0 \gamma^2 k_1 - k_0 \gamma^1 \gamma^2)(-k_2 A^1 + k_1 A^2) ,$$

whence we get

$$(k_1^2 + k_2^2) A^3 = k_3 (k_1 A^1 + k_2 A^2)$$

$$+ i \left| \begin{array}{cccc} k_3 & k_1 - i k_2 & 0 & 0 \\ k_1 + i k_2 & -k_3 & 0 & 0 \\ 0 & 0 & k_3 & k_1 - i k_2 \\ 0 & 0 & k_1 + i k_2 & -k_3 \end{array} \right| (-k_2 A^1 + k_1 A^2) ,$$

$$-(k_1^2 + k_2^2) A^0 = k_0 (k_1 A^1 + k_2 A^2)$$

$$+ i \left| \begin{array}{cccc} k_0 & k_1 - i k_2 & 0 & 0 \\ -k_1 - i k_2 & -k_0 & 0 & 0 \\ 0 & 0 & k_0 & -k_1 + i k_2 \\ 0 & 0 & k_1 + i k_2 & -k_0 \end{array} \right| (-k_2 A^1 + k_1 A^2) .$$

We can further obtain the expressions for $A^0_{(1)}, A^3_{(1)}$ and $A^0_{(2)}, A^3_{(2)}$:

$$A^3_{(1)} = \left| \begin{array}{c} (k_1+ik_2)^{-1}[ik_3+ik_1(\epsilon-k_3)] \\ (k_1-ik_2)^{-1}[ik_1-ik_3(\epsilon-k_3)] \\ (k_1+ik_2)^{-1}[k_3+ik_2(\epsilon+k_3)] \\ (k_1-ik_2)^{-1}[-ik_2-k_3(\epsilon+k_3)] \end{array} \right|,$$

$$A^0_{(1)} = \left| \begin{array}{c} (k_1+ik_2)^{-1}[-ik_0-ik_1(\epsilon-k_3)] \\ (k_1-ik_2)^{-1}[+ik_1+ik_0(\epsilon-k_3)] \\ (k_1+ik_2)^{-1}[-k_0+ik_2(\epsilon+k_3)] \\ (k_1-ik_2)^{-1}[+ik_2+k_0(\epsilon+k_3)] \end{array} \right|, \tag{15.118}$$

$$A^3_{(2)} = \left| \begin{array}{c} (k_1+ik_2)^{-1}[+k_3-ik_2(\epsilon-k_3)] \\ (k_1-ik_2)^{-1}[-ik_2+k_3(\epsilon-k_3)] \\ (k_1+ik_2)^{-1}[+ik_3-ik_1(\epsilon+k_3)] \\ (k_1-ik_2)^{-1}[+ik_1+ik_3(\epsilon+k_3)] \end{array} \right|,$$

$$A^0_{(2)} = \left| \begin{array}{c} (k_1+ik_2)^{-1}[-k_0+ik_2(\epsilon-k_3)] \\ (k_1-ik_2)^{-1}[-ik_2-k_0(\epsilon-k_3)] \\ (k_1+ik_2)^{-1}[-ik_0-ik_1(\epsilon+k_3)] \\ (k_1-ik_2)^{-1}[-ik_1-ik_0(\epsilon+k_3)] \end{array} \right|. \tag{15.119}$$

15.6 Relation to Initial Basis

We recall the formulas which relate the two bases:

$$\Psi_l = (\delta_l^k - \gamma_l \gamma^k)\tilde{\Psi}_k, \quad \Psi_l(x) = e^{ikx} B_l, \quad \tilde{\Psi}_l(x) = e^{ikx} A_l, \quad l = 0,1,2,3. \tag{15.120}$$

In component form they read

$$\begin{aligned} B^0 &= A^0 - \gamma^0(\gamma^0 A^0 - \gamma^1 A^1 - \gamma^2 A^2 - \gamma^3 A^3), \\ B^1 &= A^1 - \gamma^1(\gamma^0 A^0 - \gamma^1 A^1 - \gamma^2 A^2 - \gamma^3 A^3), \\ B^2 &= A^2 - \gamma^2(\gamma^0 A^0 - \gamma^1 A^1 - \gamma^2 A^2 - \gamma^3 A^3), \\ B^3 &= A^3 - \gamma^3(\gamma^0 A^0 - \gamma^1 A^1 - \gamma^2 A^2 - \gamma^3 A^3). \end{aligned} \tag{15.121}$$

Let us find the blocks

$$(\gamma^0 A^0 - \gamma^1 A^1 - \gamma^2 A^2 - \gamma^3 A^3)_{(1)}$$

$$= \left| \begin{array}{c} (k_1+ik_2)^{-1}[-k_0+2ik_2(\epsilon+k_3)+k_3] - (\epsilon+k_3) \\ (k_1-ik_2)^{-1}[2ik_2+(k_0+k_3)(\epsilon+k_3)]+1 \\ (k_1+ik_2)^{-1}i[-k_0-2k_1(\epsilon-k_3)-k_3]+i(\epsilon-k_3) \\ (k_1-ik_2)^{-1}i[2k_1+(k_0-k_3)(\epsilon-k_3)]-i \end{array} \right|, \tag{15.122}$$

$$
(\gamma^0 A^0 - \gamma^1 A^1 - \gamma^2 A^2_{(2)} - \gamma^3 A^3)_{(2)}
= \left| \begin{array}{c} (k_1+ik_2)^{-1} i[-k_0 - 2k_1(\epsilon+k_3)+k_3] + i(\epsilon+k_3) \\ (k_1-ik_2)^{-1} i[-2k_1 - (k_0+k_3)(\epsilon+k_3)] + i \\ (k_1+ik_2)^{-1}[-k_0 + 2ik_2(\epsilon-k_3) - k_3] - (\epsilon-k_3) \\ (k_1-ik_2)^{-1}[-2ik_2 - (k_0-k_3)(\epsilon-k_3)] - 1 \end{array} \right| . \qquad (15.123)
$$

Now, (15.122) and (15.123) are to be taken into account in (18.46). We may split the last ones into two groups, and after the needed calculations, we get

$$
B^1_{(1)} = \left| \begin{array}{c} (k_1-ik_2)^{-1} i[+2k_1 + (k_0-k_3)(\epsilon-k_3)] - i \\ (k_1+ik_2)^{-1} i[-k_0 - 2k_1(\epsilon-k_3) - k_3] + i(\epsilon-k_3) \\ (k_1-ik_2)^{-1}[-2ik_2 - (k_0+k_3)(\epsilon+k_3)] \\ (k_1+ik_2)^{-1}[+k_0 - 2ik_2(\epsilon+k_3) - k_3] \end{array} \right| = \left| \begin{array}{c} a_1 \\ b_1 \\ c_1 \\ d_1 \end{array} \right|_{(1)},
$$

$$
B^1_{(2)} = \left| \begin{array}{c} (k_1-ik_2)^{-1}[-2ik_2 - (k_0-k_3)(\epsilon-k_3)] \\ (k_1+ik_2)^{-1}[-k_0 + 2ik_2(\epsilon-k_3) - k_3] \\ (k_1-ik_2)^{-1} i[+2k_1 + (k_0+k_3)(\epsilon+k_3)] - i \\ (k_1+ik_2)^{-1} i[+k_0 + 2k_1(\epsilon+k_3) - k_3] - i(\epsilon+k_3) \end{array} \right| = \left| \begin{array}{c} a_1 \\ b_1 \\ c_1 \\ d_1 \end{array} \right|_{(2)},
$$

$$
B^2_{(1)} = \left| \begin{array}{c} (k_1-ik_2)^{-1}[+2k_1 + (k_0-k_3)(\epsilon-k_3)] \\ (k_1+ik_2)^{-1}[+k_0 + 2k_1(\epsilon-k_3) + k_3] \\ (k_1-ik_2)^{-1}[-2k_2 + i(k_0+k_3)(\epsilon+k_3)] + i \\ (k_1+ik_2)^{-1}[+ik_0 + 2k_2(\epsilon+k_3) - ik_3] + i(\epsilon+k_3) \end{array} \right| = \left| \begin{array}{c} a_2 \\ b_2 \\ c_2 \\ d_2 \end{array} \right|_{(1)},
$$

$$
B^2_{(2)} = \left| \begin{array}{c} (k_1-ik_2)^{-1}[-2k_2 + i(k_0-k_3)(\epsilon-k_3)] + i \\ (k_1+ik_2)^{-1}[-ik_0 - 2k_2(\epsilon-k_3) - ik_3] - i(\epsilon-k_3) \\ (k_1-ik_2)^{-1}[+2k_1 + (k_0+k_3)(\epsilon+k_3)] \\ (k_1+ik_2)^{-1}[-k_0 - 2k_1(\epsilon+k_3) + k_3] \end{array} \right| = \left| \begin{array}{c} a_2 \\ b_2 \\ c_2 \\ d_2 \end{array} \right|_{(2)},
$$

$$
B^3_{(1)} = \left| \begin{array}{r} (k_1+ik_2)^{-1}[-ik_1(\epsilon-k_3) - ik_3] + i(\epsilon-k_3) \\ (k_1-ik_2)^{-1}[-ik_1 - ik_0(\epsilon-k_3)] + i \\ (k_1+ik_2)^{-1}[-ik_2(\epsilon+k_3) - k_3] + (\epsilon+k_3) \\ (k_1-ik_2)^{-1}[+ik_2 + k_0(\epsilon+k_3)] + 1 \end{array} \right| = \left| \begin{array}{c} a_3 \\ b_3 \\ c_3 \\ d_3 \end{array} \right|_{(1)},
$$

$$
B^3_{(2)} = \left| \begin{array}{r} (k_1+ik_2)^{-1}[+ik_2(\epsilon-k_3) - k_3] - (\epsilon-k_3) \\ (k_1-ik_2)^{-1}[+ik_2 + k_0(\epsilon-k_3)] + 1 \\ (k_1+ik_2)^{-1}[+ik_0 + ik_1(\epsilon+k_3)] - i(\epsilon+k_3) \\ (k_1-ik_2)^{-1}[-ik_1 - ik_0(\epsilon+k_3)] + i \end{array} \right| = \left| \begin{array}{c} a_3 \\ b_3 \\ c_3 \\ d_3 \end{array} \right|_{(2)},
$$

$$B^0_{(1)} = \left| \begin{array}{r} (k_1+ik_2)^{-1}[+ik_1(\epsilon-k_3)+ik_3]-i(\epsilon-k_3) \\ (k_1-ik_2)^{-1}[-ik_1+ik_3(\epsilon-k_3)]+i \\ (k_1+ik_2)^{-1}[-ik_2(\epsilon+k_3)-k_3]+(\epsilon+k_3) \\ (k_1-ik_2)^{-1}[-ik_2-k_3(\epsilon+k_3)]-1 \end{array} \right| = \left| \begin{array}{c} a_0 \\ b_0 \\ c_0 \\ d_0 \end{array} \right|_{(1)},$$

$$B^0_{(2)} = \left| \begin{array}{r} (k_1+ik_2)^{-1}[-ik_2(\epsilon-k_3)+k_3]+(\epsilon-k_3) \\ (k_1-ik_2)^{-1}[+ik_2-k_3(\epsilon-k_3)]+1 \\ (k_1+ik_2)^{-1}[+ik_1(\epsilon+k_3)-ik_3]-i(\epsilon+k_3) \\ (k_1-ik_2)^{-1}[+ik_1+ik_3(\epsilon+k_3)]-i \end{array} \right| = \left| \begin{array}{c} a_0 \\ b_0 \\ c_0 \\ d_0 \end{array} \right|_{(2)}. \tag{15.124}$$

These formulas (15.124) may be simplified, by taking in mind the following identities and notations

$$k_0 = -\epsilon\,,\ (\epsilon+k_3)(\epsilon-k_3) = (k_1-ik_2)(k_1+ik_2)\,, \quad \epsilon = k\,,\ d\frac{k_j}{\epsilon} = n_j\,, \quad n_jn_j = 1\,.$$

In this way we derive

$$\begin{aligned} a^{(1)}_1 &= 0,\ a^{(1)}_2 = 1\,,\ a^{(1)}_3 = -\frac{n_2}{1+n_3} - \frac{in_3}{n_1+in_2}\,,\ a^{(1)}_0 = \frac{n_2}{1+n_3} + \frac{in_3}{n_1+in_2}\,, \\ a^{(2)}_1 &= 1\,,\ a^{(2)}_2 = 0\,,\ a^{(2)}_3 = -\frac{n_1}{1+n_3} - \frac{n_3}{n_1+in_2}\,,\ a^{(2)}_0 = \frac{n_1}{1+n_3} + \frac{n_3}{n_1+in_2}\,, \end{aligned} \tag{15.125}$$

$$\begin{aligned} b^{(1)}_1 &= 0\,,\ b^{(1)}_2 = \frac{n_1+in_2}{1+n_3}\,,\ b^{(1)}_3 = \frac{n_2n_3+in_1}{(n_1-in_2)(1+n_3)}\,,\ b^{(1)}_0 = \frac{n_2+in_1n_3}{(n_1-in_2)(1+n_3)}\,, \\ b^{(2)}_1 &= \frac{n_1+in_2}{1+n_3}\,,\ b^{(2)}_2 = 0\,,\ b^{(2)}_3 = \frac{n_1n_3-in_2}{(n_1-in_2)(1+n_3)}\,,\ b^{(2)}_0 = \frac{n_1-in_2n_3}{(n_1-in_2)(1+n_3)}\,, \end{aligned} \tag{15.126}$$

$$\begin{aligned} c^{(1)}_1 &= 1\,,\ c^{(1)}_2 = 0\,,\ c^{(1)}_3 = \frac{n_1}{1-n_3} - \frac{n_3}{n_1+in_2}\,,\ c^{(1)}_0 = \frac{n_1}{1-n_3} - \frac{n_3}{n_1+in_2}\,, \\ c^{(2)}_1 &= 0\,,\ c^{(2)}_2 = 1\,,\ c^{(2)}_3 = \frac{n_2}{1-n_3} - \frac{i}{n_1+in_2}\,,\ c^{(2)}_0 = \frac{n_2}{1-n_3} - \frac{in_3}{n_1+in_2}\,, \end{aligned} \tag{15.127}$$

$$\begin{aligned} d^{(1)}_1 &= -\frac{n_1+in_2}{1-n_3}\,,\ d^{(1)}_2 = 0\,,\ d^{(1)}_3 = \frac{-n_1n_3-in_2}{(1-n_3)(n_1-in_2)}\,,\ d^{(1)}_0 = \frac{-n_1-in_2n_3}{(1-n_3)(n_1-in_2)}\,, \\ d^{(2)}_1 &= 0\,,\ d^{(2)}_2 = -\frac{n_1+in_2}{1-n_3}\,,\ d^{(2)}_3 = \frac{-n_2n_3+in_1}{(1-n_3)(n_1-in_2)}\,,\ d^{(2)}_0 = \frac{-n_2+in_1n_3}{(1-n_3)(n_1-in_2)}\,. \end{aligned} \tag{15.128}$$

15.7 Helicity Operator

We have constructed above the eigenvectors of the helicity operator for spin 3/2 particles. Let us recall the used notations and results. The vector-bispinor wave function $\Psi_l(x)$ is written as a matrix (A is a bispinor index, (l) is vector one)

$$\Psi_{A(l)}(x) = \begin{vmatrix} \Psi_{1(0)}(x) & \Psi_{1(1)}(x) & \Psi_{1(2)}(x) & \Psi_{1(3)}(x) \\ \Psi_{2(0)}(x) & \Psi_{2(1)}(x) & \Psi_{2(2)}(x) & \Psi_{2(3)}(x) \\ \Psi_{3(0)}(x) & \Psi_{3(1)}(x) & \Psi_{3(2)}(x) & \Psi_{3(3)}(x) \\ \Psi_{4(0)}(x) & \Psi_{4(1)}(x) & \Psi_{4(2)}(x) & \Psi_{4(3)}(x) \end{vmatrix}$$

$$= e^{ikx} \begin{vmatrix} a_0 & a_1 & a_2 & a_3 \\ b_0 & b_1 & b_2 & b_3 \\ c_0 & c_1 & c_2 & c_3 \\ d_0 & d_1 & d_2 & d_3 \end{vmatrix} = e^{ikx}\{A_0, A_1, A_2, A_3\}. \tag{15.129}$$

There are 4 possible eigenvalues σ for the helicity operator (the values $\pm$ are doubly degenerate)

$$\sigma = -\frac{1}{2}k, -\frac{1}{2}k, \quad +\frac{1}{2}k, +\frac{1}{2}k, \quad -\frac{3}{2}k, \quad +\frac{3}{2}k \quad k = \sqrt{k_1^2 + k_2^2 + k_2^2}; \tag{15.130}$$

the dimensionless quantities are more convenient:

$$\frac{k_i}{k} = n_i, \quad n_i n_i = 1, \quad \frac{\sigma}{k} \Longrightarrow \sigma, \qquad \sigma = -\frac{1}{2}, +\frac{1}{2}, -\frac{1}{2}, +\frac{1}{2}, -\frac{3}{2}, +\frac{3}{2}. \tag{15.131}$$

For each value from $\sigma = \pm 1/2$ there exist two eigenstates (they are marked by (I) and (II)):

$$\Psi_0^I = \begin{vmatrix} 1 \\ \frac{\pm 1 - n_3}{n_1 - n_2} \\ 1 \\ \frac{\pm 1 - n_3}{n_1 - n_2} \end{vmatrix}, \Psi_1^I = \begin{vmatrix} n_1 \\ n_1 \frac{\pm 1 - n_3}{n_1 - n_2} \\ n_1 \\ n_1 \frac{\pm 1 - n_3}{n_1 - n_2} \end{vmatrix}, \Psi_2^I = \begin{vmatrix} n_2 \\ n_2 \frac{\pm 1 - n_3}{n_1 - n_2} \\ n_2 \\ n_3 \frac{\pm 1 - n_3}{n_1 - n_2} \end{vmatrix}, \Psi_3^I = \begin{vmatrix} n_3 \\ n_3 \frac{\pm 1 - n_3}{n_1 - n_2} \\ n_3 \\ n_3 \frac{\pm 1 - n_3}{n_1 - n_2} \end{vmatrix}, \tag{15.132}$$

$$\Psi_0^{II} = \begin{vmatrix} 1 \\ \frac{\pm 1 - n_3}{n_1 - n_2} \\ 1 \\ \lambda_2' \frac{\pm 1 - n_3}{n_1 - n_2} \end{vmatrix}, \quad \Psi_1^{II} = \begin{vmatrix} (\pm i n_1 n_3 + n_2) \\ (\pm i n_1 n_3 + n_2) \frac{\mp 1 - n_3}{n_1 - n_2} \\ (\pm i n_1 n_3 + n_2) \\ (\pm i n_1 n_3 + n_2) \frac{\mp 1 - n_3}{n_1 - n_2} \end{vmatrix},$$

$$\Psi_2^{II} = \begin{vmatrix} (\pm i n_2 n_3 - n_1) \\ (\pm i n_2 n_3 - n_1) \frac{\mp 1 - n_3}{n_1 - n_2} \\ (\pm i n_2 n_3 - n_1) \\ (\pm i n_2 n_3 - n_1) \frac{\mp 1 - n_3}{n_1 - n_2} \end{vmatrix}, \quad \Psi_3^{II} = \begin{vmatrix} [\mp i(1 - n_3^2)] \\ [\mp i(1 - n_3^2)] \frac{\mp 1 - n_3}{n_1 - n_2} \\ [\mp i(1 - n_3^2)] \\ [\mp i(1 - n_3^2)] \frac{\mp 1 - n_3}{n_1 - n_2} \end{vmatrix}. \tag{15.133}$$

The eigenstates for helicities $\sigma = \pm 3/2$ are given by the formulas

$$\Psi_0^{III} = \begin{vmatrix} 0 \\ 0 \\ 0 \\ 0 \end{vmatrix}, \quad \Psi_1^{III} = \begin{vmatrix} (-n_1 n_3 \pm i n_2) \\ (-n_1 n_3 \pm i n_2) \frac{\pm 1 - n_3}{n_1 - n_2} \\ (-n_1 n_3 \pm i n_2) \\ (-n_1 n_3 \pm i n_2) \frac{\mp 1 - n_3}{n_1 - n_2}, \end{vmatrix},$$

$$\Psi_2^{III} = \begin{vmatrix} (-n_2 n_3 \mp i n_1) \\ (-n_2 n_3 \mp i n_1) \frac{\pm 1 - n_3}{n_1 - n_2} \\ (-n_2 n_3 \mp i n_1) \\ (-n_2 n_3 \mp i n_1) \frac{\mp 1 - n_3}{n_1 - n_2}, \end{vmatrix}, \quad \Psi_3^{III} = \begin{vmatrix} (1 - n_3^2) \\ (1 - n_3^2) \frac{\pm 1 - n_3}{n_1 - n_2} \\ (1 - n_3^2) \\ (1 - n_3^2) \frac{\mp 1 - n_3}{n_1 - n_2}, \end{vmatrix}, \tag{15.134}$$

We shall relate the helicity states (15.132)–(15.134) with the solutions for the massless particle:

$$B^0_{(1)} = \left| \begin{array}{c} a_0 \\ b_0 \\ c_0 \\ d_0 \end{array} \right|_{(1)} , B^1_{(1)} = \left| \begin{array}{c} a_1 \\ b_1 \\ c_1 \\ d_1 \end{array} \right|_{(1)} , B^2_{(1)} = \left| \begin{array}{c} a_2 \\ b_2 \\ c_2 \\ d_2 \end{array} \right|_{(1)} , B^3_{(1)} = \left| \begin{array}{c} a_3 \\ b_3 \\ c_3 \\ d_3 \end{array} \right|_{(1)} , \tag{15.135}$$

$$B^0_{(2)} = \left| \begin{array}{c} a_0 \\ b_0 \\ c_0 \\ d_0 \end{array} \right|_{(2)} , B^1_{(2)} = \left| \begin{array}{c} a_1 \\ b_1 \\ c_1 \\ d_1 \end{array} \right|_{(2)} , B^2_{(2)} = \left| \begin{array}{c} a_2 \\ b_2 \\ c_2 \\ d_2 \end{array} \right|_{(2)} , B^3_{(2)} = \left| \begin{array}{c} a_3 \\ b_3 \\ c_3 \\ d_3 \end{array} \right|_{(2)} . \tag{15.136}$$

For the helicities $\sigma = \pm 3/2$, the quantities a_0, b_0, c_0, d_0 are equal to zero; however, the quantities a_0, b_0, c_0, d_0 in massless solutions are not, and therefore they cannot be constructed only in terms of helicity solutions with $\sigma = \pm 3/2$. Besides, it should be emphasized that in describing helicity eigenstates, the structure $\{a_0, a_1, a_2, a_3\}$ is the main one, and it determines all the remaining quantities by means of linear relations. For this reason, it suffices to connect the quantities $\{a_0, a_1, a_2, a_3\}_{(1,2)}$ in (15.136) with the relevant variables in (15.132)–(15.134).

Let us write down the two sets of parameters for helicity states,

$$\begin{gathered} a_0^{I+} = 1, \quad a_1^{I+} = n_1, \quad a_2^{I+} = n_2, \quad a_3^{I+} = n_3, \\ a_0^{I-} = 1, \quad a_1^{I-} = n_1, \quad a_2^{I-} = n_2, \quad a_3^{I-} = n_3, \\ a_0^{II+} = 1, a_1^{II+} = in_1n_3 + n_2, a_2^{II+} = in_2n_3 - n_1, a_3^{II+} = -i(1 - n_3^2), \\ a_0^{II-} = 1, a_1^{II-} = -in_1n_3 + n_2, a_2^{II-} = -in_2n_3 - n_1, a_3^{II-} = +i(1 - n_3^2), \\ a_0^{III+} = 0, a_1^{III+} = -n_1n_3 + in_2, a_2^{III+} = -n_2n_3 - in_1, a_3^{III+} = (1 - n_3^2), \\ a_0^{III-} = 0, a_1^{III-} = -n_1n_3 - in_2, a_2^{III-} = -n_2n_3 + in_1, a_3^{III-} = (1 - n_3^2), \end{gathered} \tag{15.137}$$

and for the massless solutions

$$\begin{gathered} (1),\ a_0^{(1)} = \frac{n_2}{1+n_3} + \frac{in_3}{n_1 + in_2}, a_1^{(1)} = 0, a_2^{(1)} = 1, a_3^{(1)} = -\frac{n_2}{1+n_3} - \frac{in_3}{n_1 + in_2}, \\ (2),\ a_0^{(2)} = \frac{n_1}{1+n_3} + \frac{n_3}{n_1 + in_2}, a_1^{(2)} = 1, a_2^{(2)} = 0,\ a_3^{(2)} = -\frac{n_1}{1+n_3} - \frac{n_3}{n_1 + in_2}. \end{gathered} \tag{15.138}$$

Instead of the variables from (15.138), it is more convenient to use their linear combinations:

$$a_l^+ = a_l^{(2)} + ia_l^{(1)}, \quad a_l^- = a_l^{(2)} - ia_l^{(1)}, \tag{15.139}$$

because they seem to be simpler

$$\begin{gathered} a_0^+ = \frac{n_1 + in_2}{1+n_3}, \quad a_1^+ = 1, \quad a_2^+ = +i, \quad a_3^+ = -\frac{n_1 + in_2}{1+n_3}, \\ a_0^- = \frac{1+n_3}{n_1 + in_2}, \quad a_1^- = 1, \quad a_2^- = -i, \quad a_3^- = -\frac{1+n_3}{n_1 + in_2}. \end{gathered} \tag{15.140}$$

We search for linear expansions of the massless solutions (15.140) in terms of helicity states (the index l takes on the values $0,1,2,3$),

$$a_l^+ = \alpha a_l^{I+} + \alpha' a_l^{I-} + \beta a_l^{II+} + \beta' a_l^{II-} + \gamma a_l^{III+} + \gamma' a_l^{III-}\,,$$
$$a_l^- = \alpha a_l^{I+} + \alpha' a_l^{I-} + \beta a_l^{II+} + \beta' a_l^{II-} + \gamma a_l^{III+} + \gamma' a_l^{III-}\,; \tag{15.141}$$

in both formulas, the first two terms define only one parameter, $(\alpha+\alpha')$. Each equation provides us with four linear relations:

a_l^+-solution,

$$a_0^+ = \frac{n_1+in_2}{1+n_3} = (\alpha+\alpha') + (\beta+\beta') + \gamma\cdot 0 + \gamma'\cdot 0\,,$$
$$a_1^+ = 1 = (\alpha+\alpha')n_1 + \beta(in_1n_3+n_2)$$
$$+\beta'(-in_1n_3+n_2) + \gamma(-n_1n_3+in_2) + \gamma'(-n_1n_3-in_2)\,,$$
$$a_2^+ = i = (\alpha+\alpha')n_2 + \beta(in_2n_3-n_1) \tag{15.142}$$
$$+\beta'(-in_2n_3-n_1) + \gamma(-n_2n_3-in_1) + \gamma'(-n_2n_3+in_1)\,,$$
$$a_3^+ = -\frac{n_1+in_2}{1+n_3} = (\alpha+\alpha')n_3 + \beta(-i)(1-n_3^2)$$
$$+\beta' i(1-n_3^2) + \gamma(1-n_3^2) + \gamma'(1-n_3^2)\,;$$

a_l^--solution,

$$a_0^- = \frac{1+n_3}{n_1+in_2} = (\alpha+\alpha') + (\beta+\beta') + \gamma\cdot 0 + \gamma'\cdot 0\,,$$
$$a_1^- = 1 = (\alpha+\alpha')n_1 + \beta(in_1n_3+n_2)$$
$$+\beta'(-in_1n_3+n_2) + \gamma(-n_1n_3+in_2) + \gamma'(-n_1n_3-in_2)\,,$$
$$a_2^- = -i = (\alpha+\alpha')n_2 + \beta(in_2n_3-n_1) \tag{15.143}$$
$$+\beta'(-in_2n_3-n_1) + \gamma(-n_2n_3-in_1) + \gamma'(-n_2n_3+in_1)\,,$$
$$a_{30}^- = -\frac{1+n_3}{n_1+in_2} = (\alpha+\alpha')n_3 + \beta(-i)(1-n_3^2)$$
$$+\beta' i(1-n_3^2) + \gamma(1-n_3^2) + \gamma'(1-n_3^2)\,,$$

whence after regrouping the terms, we obtain (let $\alpha+\alpha' = \sigma$)

a_l^+-solution,

$$a_0^+ = \frac{n_1 + in_2}{1+n_3} = \sigma + (\beta + \beta'),$$

$$\begin{aligned} a_1^+ = 1 = \sigma n_1 + (\beta+\beta')n_2 + i(\beta-\beta')n_1 n_3 - (\gamma+\gamma')n_1 n_3 + i(\gamma-\gamma')n_2, \\ a_2^+ = i = \sigma n_2 - (\beta+\beta')n_1 + i(\beta-\beta')n_2 n_3 - (\gamma+\gamma')n_2 n_3 - i(\gamma-\gamma')n_1, \end{aligned} \tag{15.144}$$

$$a_3^+ = -\frac{n_1 + in_2}{1+n_3} = \sigma n_3 - i(\beta-\beta')(1-n_3^2) + (\gamma+\gamma')(1-n_3^2);$$

a_l^--solution,

$$a_0^- = \frac{1+n_3}{n_1 + in_2} = \sigma + (\beta + \beta'),$$

$$\begin{aligned} a_1^- = 1 = \sigma n_1 + (\beta+\beta')n_2 + i(\beta-\beta')n_1 n_3 - (\gamma+\gamma')n_1 n_3 + i(\gamma-\gamma')n_2, \\ a_2^- = -i = \sigma n_2 - (\beta+\beta')n_1 + i(\beta-\beta')n_2 n_3 - (\gamma+\gamma')n_2 n_3 - i(\gamma-\gamma')n_1, \end{aligned} \tag{15.145}$$

$$a_3^- = -\frac{1+n_3}{n_1 + in_2} = \sigma n_3 - i(\beta-\beta')(1-n_3^2) + (\gamma+\gamma')(1-n_3^2).$$

With the new notations

$$\sigma = x_0, \quad \beta+\beta' = x_1, \quad i(\beta-\beta') = x_2, \quad \gamma+\gamma' = x_3, \quad i(\gamma-\gamma') = x_4, \tag{15.146}$$

the above systems read (it should be notified that coordinates x_2, x_3 enter this system only as the combination $y = x_2 - x_3$):

a_l^+-solution,

$$\left| \begin{array}{cccc} 1 & 1 & 0 & 0 \\ n_1 & n_2 & n_1 n_3 & n_2 \\ n_2 & -n_1 & n_2 n_3 & -n_1 \\ n_3 & 0 & -(1-n_3^2) & 0 \end{array} \right| \left| \begin{array}{c} x_0 \\ x_1 \\ y \\ x_4 \end{array} \right| = \left| \begin{array}{c} \frac{n_1+in_2}{1+n_3} \\ 1 \\ +i \\ -\frac{n_1+in_2}{1+n_3} \end{array} \right| = \left| \begin{array}{c} a_0^+ \\ a_1^+ \\ a_2^+ \\ a_3^+ \end{array} \right|; \tag{15.147}$$

a_l^--solution,

$$\left| \begin{array}{cccc} 1 & 1 & 0 & 0 \\ n_1 & n_2 & n_1 n_3 & n_2 \\ n_2 & -n_1 & n_2 n_3 & -n_1 \\ n_3 & 0 & -(1-n_3^2) & 0 \end{array} \right| \left| \begin{array}{c} x_0 \\ x_1 \\ y \\ x_4 \end{array} \right| = \left| \begin{array}{c} \frac{n_1-in_2}{1-n_3} \\ 1 \\ -i \\ -\frac{n_1-in_2}{1+n_3} \end{array} \right| = \left| \begin{array}{c} a_0^- \\ a_1^- \\ a_2^- \\ a_3^- \end{array} \right|. \tag{15.148}$$

We have two similar systems with the same main determinant, $n_1^2 + n_1^2$; they may be readily solved – see below.

From the equations (15.147) and (15.148), it follows that the coefficients in the expansion of any massless solution in terms of the helicity solutions

$$a_l = x_0 a_l^I + \beta a_l^{II+} + \beta' a_l^{II-} + \gamma a_l^{III+} + \gamma' a_l^{III-} \tag{15.149}$$

may be presented as a linear combinations of four variables $x_0; x_1, y, x_4$; the last three them, x_0, x_1, y, x_4 determine the constraints on $\beta, \beta', \gamma, \gamma'$ (see (15.146)):

$$\begin{array}{lcl} \beta+\beta' = x_1 & & \beta+\beta' = x_1 \\ i(\beta-\beta')-(\gamma+\gamma') = y & \Longrightarrow & \beta-\beta' = -iy - i(\gamma+\gamma') \\ i(\gamma-\gamma') = x_4 & & i(\gamma+\gamma') = x_4 + 2i\gamma' \end{array}$$

which are equivalent to

$$\beta = \frac{1}{2}(x_1 - iy - x_4) - i\gamma', \ \beta = \frac{1}{2}(x_1 + iy + x_4) + i\gamma', \ \gamma = -ix_4 + \gamma'. \quad (15.150)$$

Taking into account (15.150) in the expansion (15.149), we may decompose this expansion into a sum of two terms, (the seconde one is proportional to an arbitrary parameter γ'):

$$a_l = \left\{ x_0 a_l^I + \frac{1}{2}(x_1 - iy - x_4) a_l^{II+} \frac{1}{2}(x_1 + iy + x_4) a_l^{II-} - ix_4 a_l^{III+} \right\}$$
$$-i\gamma' \left\{ a_l^{II+} - a_l^{II-} + i a_l^{III+} + i\gamma' a_l^{III-} \right\}. \quad (15.151)$$

By using the explicit form of the helicity solutions, we prove that this additional term identically vanishes. Therefore, the final result reduces to the expansion

$$a_l = \left\{ x_0 a_l^I + \frac{1}{2}(x_1 - iy - x_4) a_l^{II+} \frac{1}{2}(x_1 + iy + x_4) a_l^{II-} - ix_4 a_l^{III+} \right\}. \quad (15.152)$$

In order to find the explicit expression (15.152) in the case of special massless states a_l^+ and a_l^-, we need the solutions of the linear system (15.147) and (15.148); these solutions are

a_l^+-solution,

$$x_0 = \frac{1-n_3}{n_1 - in_2}, \quad x_1 = 0, \quad y = \frac{1}{n_1 - n_2}, \quad x_4 = -\frac{i}{n_1 - in_2}; \quad (15.153)$$

a_l^--solution,

$$x_0 = (1+n_3)\frac{1-2n_3}{n_1 + in_2}, \quad x_1 = (1+n_3)\frac{2n_3}{n_1 + in_2},$$
$$y = \frac{1+2n_3}{n_1 + in_2}, \quad x_4 = \frac{i}{n_1 + in_2} - (1+n_3)\frac{2n_3}{n_1 + in_2}. \quad (15.154)$$

Thus, we have found two massless solutions for spin 3/2 field, and found their expansions as linear combinations of the eigenstates with helicities $\sigma = \pm 1/2, \pm 3/2$.

Chapter 16

Solutions with Spherical Symmetry for a Massive Spin 3/2 Particle

16.1 System of Equations and Spherical Symmetry

In the present chapter we examine the problem of finding the exact solutions with spherical symmetry for a massive spin 3/2 particle. For simplicity, we restrict ourselves to the Minkowski space-time model. The wave equation for such a particle may be presented in the form (we assume the use of the tetrad formalism, see [139])

$$\begin{gathered} \Big[\, i\gamma^{\beta}(x)(\nabla_{\beta} + \Gamma_{\beta}(x)) - m \, \Big] \, \Psi_{\alpha} = 0 \, , \\ \gamma^{\alpha}(x)\Psi_{\alpha} = 0 \, , \;\; (\nabla_{\alpha} + \Gamma_{\alpha})\Psi^{\alpha} = 0 \, . \end{gathered} \tag{16.1}$$

The mass parameter is designated as $m = Mc/\hbar$; the wave function $\Psi_{\alpha}(x)$ behaves as a bispinor with respect to tetrad transformations, and as a generally covariant vector with respect to coordinate transformations; consider the notation $\gamma^{\beta}(x) = e^{\beta}_{(a)}(x)\gamma^{a}$; the symbol $\Gamma_{\beta}(x)$ designates the bispinor connection

$$\Gamma_{\beta}(x) = \frac{1}{2}(\sigma^{ab})_{k}^{\;\, l} e^{\beta}_{(a)}(\nabla_{\alpha} e_{(b)\beta}) \, , \qquad \sigma^{ab} = \frac{\gamma^{a}\gamma^{b} - \gamma^{b}\gamma^{a}}{4} \, . \tag{16.2}$$

it will be convenient to use below the wave function with tetrad-vector index $\Psi_{l}(x)$, which relates to the previous function $\Psi_{\alpha}(x)$ in accordance with the rule

$$\Psi_{l}(x) = e^{\beta}_{(l)}(x)\Psi_{\beta}(x) \, , \quad \Psi_{\beta}(x) = e^{(l)}_{\beta}(x)\Psi_{l}(x) \, . \tag{16.3}$$

With (16.3) in mind, the first equation in (16.1) is transformed (the symbol $;\alpha$ in $e^{(l)}_{\beta;\alpha}$ stands for the covariant derivative ∇_{α})

$$\Big\{ i\gamma^{\alpha}(x) \Big[(\partial_{\alpha} + \Gamma_{\alpha})\delta^{l}_{k} + e^{\beta}_{(k)} e^{(l)}_{\beta;\alpha} \Big] - m\delta^{l}_{k} \Big\} \Psi_{l} = 0 \, , \tag{16.4}$$

where

$$e^{\beta}_{(k)} e^{(l)}_{\beta;\alpha} = (L_{\alpha})_{k}^{\;\, l} = \frac{1}{2}(j^{ab})_{k}^{\;\, l} e^{\beta}_{(a)}(\nabla_{\alpha} e_{(b)\beta}) \, , \quad (j^{ab})_{k}^{\;\, l} = \delta^{a}_{k}\, g^{bl} - \delta^{b}_{k} g^{al} \, . \tag{16.5}$$

The generators J^{ab} for the spin-vector representation are $J^{ab} = \sigma^{ab} \otimes I + I \otimes j^{ab}$. Taking in mind (16.5), equation (16.4) is transformed to the form

$$\Big[i\gamma^{\alpha}(x)(\partial_{\alpha} + B_{\alpha}) - m\Big]\Psi = 0, \quad B_{\alpha} = \Gamma_{\alpha} \otimes I + I \otimes L_{\alpha}. \tag{16.6}$$

While using the wave function $\Psi_l(x)$ (see (16.3)), the additional constraints in (16.1) read as follows:

$$\gamma^l \Psi_l = 0, \quad \Big[e^{(l)\alpha}\partial_{\alpha} + e^{(l)\alpha}) + e^{(l)\alpha}\Gamma_{\alpha}\Big]\Psi_l = 0. \tag{16.7}$$

We will specify the equations in spherical coordinates $x^{\alpha} = (t,\, r,\, \theta,\, \phi)$:

$$dS^2 = dt^2 - dr^2 - r^2(d\theta^2 + \sin^2\theta d\phi^2). \tag{16.8}$$

Using the spherical tetrad, for $\gamma^{\alpha}(x)$ and $B_{\alpha}(x)$ we get the expressions

$$\begin{gathered} \gamma^{\alpha} = (\gamma^0, \gamma^3, \gamma^1/r, \gamma^2/r\sin\theta), \quad \Gamma_{\theta} = \sigma^{31}, \; \Gamma_{\phi} = \sin\theta\sigma^{32} + \cos\theta\sigma^{12}, \\ L_{\theta} = j^{31}, \; L_{\phi} = \sin\theta j^{32} + \cos\theta j^{12}, \quad B_{\theta} = J^{31}, \; B_{\phi} = \sin\theta J^{32} + \cos\theta J^{12}. \end{gathered} \tag{16.9}$$

Correspondingly, the equation (16.6) takes the form

$$\Big[i\gamma^0\partial_0 + i\gamma^3\partial_r + \frac{i\gamma^1 J^{31} + i\gamma^2 J^{32}}{r} + \frac{1}{r}\Sigma_{\theta,\phi} - m\Big]\Psi = 0. \tag{16.10}$$

The (θ,ϕ)-dependent operator $\Sigma_{\theta,\phi}$ is determined by the relation

$$\Sigma_{\theta,\phi} = i\gamma^1\partial_{\theta} + \gamma^2\frac{i\partial_{\phi} + (i\sigma^{12} \otimes I + I \otimes ij^{12})\cos\theta}{\sin\theta}. \tag{16.11}$$

Having in mind the identity

$$\frac{i\gamma^1 J^{31} + i\gamma^2 J^{32}}{r} = \frac{i\gamma^3}{r} + \frac{\gamma^1 \otimes T^2 - \gamma^2 \otimes T^1}{r},$$

we change the above equation to the form

$$\Big[i\gamma^0\partial_0 + i\gamma^3(\partial_r + \frac{1}{r}) + \frac{\gamma^1 \otimes T^2 - \gamma^2 \otimes T^1}{r} + \frac{1}{r}\Sigma_{\theta,\phi} - m\Big]\Psi = 0. \tag{16.12}$$

Searching for the solutions of equation (16.12) in spherically symmetric form, we diagonalize the operator of the total angular momentum. In Cartesian basis, the components of the total angular momentum are defined as follows

$$J_i^{Cart} = l_i + S_i, \quad S_1 = iJ^{23}, \quad S_2 = iJ^{31}, \quad S_3 = iJ^{12}. \tag{16.13}$$

Using the spinor representation for Dirac matrices and taking in mind the above definition for generators (16.5), we obtain

$$S_i = \frac{1}{2}\,\Sigma_i \otimes I \,+\, I \otimes T_i\,, \qquad T_i = \begin{vmatrix} 0 & 0 \\ 0 & \tau_i \end{vmatrix}. \tag{16.14}$$

Correspondingly, the wave function has two indices, of bispinor A and of 4-vector (l):

$$\Psi_{A(l)} = \left| \begin{array}{cccc} \Psi_{1(0)} & \Psi_{1(1)} & \Psi_{1(2)} & \Psi_{1(3)} \\ \Psi_{2(0)} & \Psi_{2(1)} & \Psi_{2(2)} & \Psi_{2(3)} \\ \Psi_{3(0)} & \Psi_{3(1)} & \Psi_{3(2)} & \Psi_{3(3)} \\ \Psi_{4(0)} & \Psi_{4(1)} & \Psi_{4(2)} & \Psi_{4(3)} \end{array} \right| . \tag{16.15}$$

The wave functions for the particle in Cartesian and spherical tetrads are related by θ, ϕ-dependent gauge transformations:

$$\Psi^{sph} = S\,\Psi^{Cart}, \quad S = \left| \begin{array}{cc} U_2 & 0 \\ 0 & U_2 \end{array} \right| \otimes \left| \begin{array}{cc} 1 & 0 \\ 0 & O_3 \end{array} \right| , \tag{16.16}$$

where $U_2(\theta, \phi)$ and $O_3(\theta, \phi)$ are as shown below

$$U_2 = \left| \begin{array}{cc} \cos\theta/2 e^{i\phi/2} & \sin\theta/2 e^{-i\phi/2} \\ -\sin\theta/2 e^{i\phi/2} & \cos\theta/2 e^{-i\phi/2} \end{array} \right| , (L_a{}^b) = \left| \begin{array}{cccc} 1 & 0 & 0 & 0 \\ 0 & \cos\theta\cos\phi & \cos\theta\sin\phi & -\sin\theta \\ 0 & -\sin\phi & \cos\phi & 0 \\ 0 & \sin\theta\cos\phi & \sin\theta\sin\phi & \cos\theta \end{array} \right| .$$

After performing the needed transformation $J_i^{Cart} = l_i + S_i$, $J_i = S\, J_i^{Cart}\, S^{-1}$ for the operators J_i in spherical tetrad basis, we obtain the alternative expressions [?]

$$J_1 = l_1 + S_3 \frac{\sin\phi}{\sin\theta}, \quad J_2 = l_2 + S_3 \frac{\cos\phi}{\sin\theta}, \quad J_3 = l_3 = -i\frac{\partial}{\partial\phi} . \tag{16.17}$$

We address the problem of searching the eigenvectors of the two operators $\vec{J}^2$ and J_3:

$$\vec{J}^2\, \psi(\theta,\phi) = j(j+1)\, \psi(\theta,\phi), \quad J_3\, \psi(\theta,\phi) = m\, \psi(\theta,\phi), \tag{16.18}$$

where the eigenvalues m of J_3 should not be confused with the mass parameter $m = \frac{Mc}{\hbar}$, reduces to combining them in terms of the Wigner D-functions [68]. To this end, it is convenient to operate with a diagonal matrix $S_3 = i j^{12}$; however we have

$$S_3 = \frac{1}{2} \left| \begin{array}{cccc} +1 & 0 & 0 & 0 \\ 0 & -1 & 0 & 0 \\ 0 & 0 & +1 & 0 \\ 0 & 0 & 0 & -1 \end{array} \right| \otimes I + I \otimes \left| \begin{array}{cccc} 0 & 0 & 0 & 0 \\ 0 & 0 & -i & 0 \\ 0 & +i & 0 & 0 \\ 0 & 0 & 0 & 0 \end{array} \right| .$$

So we perform the change to the cyclic basis (in the vector space) $\tilde{\Psi} = (I \otimes U)\, \Psi$:

$$\left| \begin{array}{c} \tilde{\Psi}_{(0)} \\ \tilde{\Psi}_{(1)} \\ \tilde{\Psi}_{(2)} \\ \tilde{\Psi}_{(3)} \end{array} \right| = \left| \begin{array}{cccc} 1 & 0 & 0 & 0 \\ 0 & -1/\sqrt{2} & i/\sqrt{2} & 0 \\ 0 & 0 & 0 & 1 \\ 0 & +1/\sqrt{2} & i/\sqrt{2} & 0 \end{array} \right| \left| \begin{array}{c} \Psi_{(0)} \\ \Psi_{(1)} \\ \Psi_{(2)} \\ \Psi_{(3)} \end{array} \right| , \tag{16.19}$$

$$\left| \begin{array}{c} \Psi_{(0)} \\ \Psi_{(1)} \\ \Psi_{(2)} \\ \Psi_{(3)} \end{array} \right| = \left| \begin{array}{cccc} 1 & 0 & 0 & 0 \\ 0 & -1/\sqrt{2} & 0 & 1/\sqrt{2} \\ 0 & -i/\sqrt{2} & 0 & -i/\sqrt{2} \\ 0 & 0 & 1 & 0 \end{array} \right| \left| \begin{array}{c} \tilde{\Psi}_{(0)} \\ \tilde{\Psi}_{(1)} \\ \tilde{\Psi}_{(2)} \\ \tilde{\Psi}_{(3)} \end{array} \right| . \tag{16.20}$$

This results in

$$\tilde{S}_3 = \frac{1}{2} \begin{vmatrix} +1 & 0 & 0 & 0 \\ 0 & -1 & 0 & 0 \\ 0 & 0 & +1 & 0 \\ 0 & 0 & 0 & -1 \end{vmatrix} \otimes I + I \otimes \begin{vmatrix} 0 & 0 & 0 & 0 \\ 0 & +1 & 0 & 0 \\ 0 & 0 & 0 & 0 \\ 0 & 0 & 0 & -1 \end{vmatrix} . \tag{16.21}$$

In the cyclic basis, the main equation (16.6) formally preserves its form

$$\Big[i\gamma^{\alpha}(x) \Big(\partial_{\alpha} + \Gamma_{\alpha}(x) \otimes I + I \otimes \tilde{L}_{\alpha}(x) \Big) - m \Big] \tilde{\Psi}(x) = 0 . \tag{16.22}$$

An additional constraint in the basis $\tilde{\Psi}_l(x)$ reads as follows

$$\gamma^l (U^{-1})_{lk} \tilde{\Psi}_k(x) = 0 , \tag{16.23}$$

$$\Big[e^{(l)\alpha} \partial_{\alpha} + e^{(l)\alpha}(x) + e^{(l)\alpha}(x) \Gamma_{\alpha}(x) \Big] U^{-1}_{lk} \Psi_k(x) = 0 . \tag{16.24}$$

We need below the transformed generators (see (16.10))

$$i\tilde{j}^{23} = \frac{1}{\sqrt{2}} \begin{vmatrix} 0 & 0 & 0 & 0 \\ 0 & 0 & 1 & 0 \\ 0 & 1 & 0 & 1 \\ 0 & 0 & 1 & 0 \end{vmatrix} = \tilde{T}^1 , \quad i\tilde{J}^{31} = \frac{1}{\sqrt{2}} \begin{vmatrix} 0 & 0 & 0 & 0 \\ 0 & 0 & -i & 0 \\ 0 & +i & 0 & 0-i \\ 0 & 0 & +i & 0 \end{vmatrix} = \tilde{T}^2 ,$$

$$i\tilde{j}^{12} = \begin{vmatrix} 0 & 0 & 0 & 0 \\ 0 & +1 & 0 & 0 \\ 0 & 0 & 0 & 0 \\ 0 & 0 & 0 & -1 \end{vmatrix} = \tilde{T}^3 .$$

Starting with equation (16.12), after separating the simple multiplier in the wave function $\tilde{\Psi}(x) = (\epsilon^{-i\epsilon t}/r)\, \tilde{\Phi}(r,\theta,\phi)$, we obtain

$$\Big[\gamma^0 \epsilon + i\gamma^3 \partial_r + \frac{\gamma^1 \otimes \tilde{T}_2 - \gamma^2 \otimes \tilde{T}_1}{r} + \frac{1}{r} \tilde{\Sigma}_{\theta,\phi} - m \Big] \tilde{\Phi}(r,\theta,\phi) = 0 , \tag{16.25}$$

$$\tilde{\Sigma}_{\theta,\phi} = i\gamma^1 \partial_{\theta} + \gamma^2 \frac{i\partial_{\phi} + i\tilde{S}_3 \cos\theta}{\sin\theta} . \tag{16.26}$$

There exist 16 eigenvectors of the matrix $\tilde{S}_3$; therefore we have 16 eigenvectors of the operators $\vec{J}^2$, J_3:

$$\vec{J}^2\, \psi(\theta,\phi) = j(j+1)\, \psi(\theta,\phi) , \quad J_3\, \psi(\theta,\phi) = m\, \psi(\theta,\phi) .$$

So the most general substitution for the wave function with quantum numbers j,m is (the vector-bispinor is divided in two vector-spinors):

$$\tilde{\xi}_l = \begin{vmatrix} f_0\, \delta^0_i\, D_{-1/2} + f_1\, \delta^1_i\, D_{-3/2} + f_2\, \delta^2_i\, D_{-1/2} + f_3\, \delta^3_i\, D_{+1/2} \\ g_0\, \delta^0_i\, D_{+1/2} + g_1\, \delta^1_i\, D_{-1/2} + g_2\, \delta^2_i\, D_{+1/2} + g_3\, \delta^3_i\, D_{+3/2} \end{vmatrix} ; \tag{16.27}$$

the second vector-spinor $\tilde{\eta}_i$ has a similar structure, but with other radial functions:

$$f_i(r) \Longrightarrow h_i(r) , \qquad g_i(r) \Longrightarrow \nu_i(r) . \tag{16.28}$$

It should be noted that for the minimal value $j = 1/2$, we must take a more simple initial wave function:

$$j = \frac{1}{2}, \quad f_1 = 0,\ h_1 = 0,\ g_3 = 0,\ v_3 = 0.$$

Equation (16.25) may be written in split form

$$\Big[\epsilon + i\sigma_3\partial_r + \frac{\sigma_1 \otimes \tilde{T}_2 - \sigma_2 \otimes \tilde{T}_1}{r} + \frac{1}{r}\tilde{\Sigma}_{\theta,\phi}\Big]\,\tilde{\xi} = m\tilde{\eta}, \tag{16.29}$$

$$\Big[\epsilon - i\sigma_3\partial_r - \frac{\sigma_1 \otimes \tilde{T}_2 - \sigma_2 \otimes \tilde{T}_1}{r} - \frac{1}{r}\,\tilde{\Sigma}_{\theta,\phi}\Big]\,\tilde{\eta} = m\,\tilde{\xi}, \tag{16.30}$$

where

$$\tilde{\Sigma}_{\theta,\phi} = i\sigma_1\partial_\theta + \sigma_2\frac{i\partial_\phi + (1/2\sigma_3 \otimes I + I \otimes \tilde{T}_3)\cos\theta}{\sin\theta}.$$

In addition to the operators $i\partial_t$, $\vec{J}^2$, J_3, we shall diagonalize also the operator of spacial reflection. We start with the Cartesian P-operators in bispinor and vector spaces; after transforming them to the spherical-cyclic basis, we get

$$(\tilde{\Pi}_k{}^l) = \begin{vmatrix} 1 & 0 & 0 & 0 \\ 0 & 0 & 0 & 1 \\ 0 & 0 & 1 & 0 \\ 0 & 1 & 0 & 0 \end{vmatrix}, \quad \Pi = \begin{vmatrix} 0 & 0 & 0 & -1 \\ 0 & 0 & -1 & 0 \\ 0 & -1 & 0 & 0 \\ -1 & 0 & 0 & 0 \end{vmatrix}. \tag{16.31}$$

From the eigenvalue equation

$$[\,(\Pi \otimes \tilde{\Pi}_k{}^l\,)\,\hat{P}\,]\,\tilde{\Phi}_l(r,\theta,\phi) = P\,\tilde{\Phi}_k(r,\theta,\phi),$$

we derive two eigenvalues and respective sets of restrictions on the radial functions:

$$v_0 = \delta\, f_0,\ v_1 = \delta\, f_3,\ v_2 = \delta\, f_2,\ v_3 = \delta\, f_1,$$

$$h_0 = \delta\, g_0,\ h_1 = \delta\, g_3,\ h_2 = \delta\, g_2,\ h_3 = \delta\, g_1, \tag{16.32}$$

where $\delta = +1$ corresponds to the parity $P = (-1)^{J+1}$ and $\delta = -1$ corresponds to $P = (-1)^J$.

16.2 Separating the Variables

Having performed all the needed calculations for separating the variables in equations (16.29)–(16.30), with the use of the known recurrent formulas for Wigner D-functions [68]

$$\partial_\theta D_{+1/2} = \frac{1}{2}\left(aD_{-1/2} - bD_{+3/2}\right),$$

$$[\sin^{-1}\theta(-m - \frac{1}{2}\cos\theta)]D_{+1/2} = \frac{1}{2}\left(-aD_{-1/2} - bD_{+3/2}\right),$$

$$\partial_\theta D_{-1/2} = \frac{1}{2}\left(bD_{-3/2} - aD_{+1/2}\right),$$

$$[\sin^{-1}\theta(-m+\frac{1}{2}\cos\theta)]D_{-1/2}=\frac{1}{2}\left(-bD_{-3/2}-aD_{+1/2}\right),$$

$$\partial_\theta D_{+3/2}=\frac{1}{2}\left(bD_{+1/2}-cD_{+5/2}\right),$$

$$[\sin^{-1}\theta(-m-\frac{3}{2}\cos\theta)]D_{+3/2}=\frac{1}{2}\left(-bD_{+1/2}-cD_{+5/2}\right),$$

$$\partial_\theta D_{-3/2}=\frac{1}{2}\left(cD_{-5/2}-bD_{-1/2}\right),$$

$$[\sin^{-1}\theta(-m+\frac{3}{2}\cos\theta)]D_{-3/2}=\frac{1}{2}\left(-cD_{-5/2}-bD_{-1/2}\right),$$

where

$$a=j+1/2,\quad b=\sqrt{(j-1/2)(j+3/2)},\quad c=\sqrt{(j-3/2)(j+5/2)}; \tag{16.33}$$

we find 8 radial equations:

$P=(-1)^{j+1}$,

$$\begin{gathered}
\epsilon g_0-i\frac{d}{dr}g_0-i\frac{a}{r}f_0=m\,f_0\,,\epsilon f_0+i\frac{d}{dr}f_0+i\frac{a}{r}g_0=m\,g_0\,,\\
\epsilon g_3-i\frac{d}{dr}g_3-i\frac{b}{r}f_3=m\,f_1\,,\epsilon f_3+i\frac{d}{dr}f_3+i\frac{\sqrt{2}}{r}g_2+i\frac{b}{r}g_3=m\,g_1\,,\\
\epsilon g_2-i\frac{d}{dr}g_2-i\frac{\sqrt{2}}{r}f_3-i\frac{a}{r}f_2=m\,f_2\,,\\
\epsilon f_2+i\frac{d}{dr}f_2+i\frac{\sqrt{2}}{r}g_1+i\frac{a}{r}g_2=m\,g_2\,,\\
\epsilon g_1-i\frac{d}{dr}g_1-i\frac{\sqrt{2}}{r}f_2-i\frac{b}{r}f_1=m\,f_3\,,\epsilon f_1+i\frac{d}{dr}f_1+i\frac{b}{r}g_1=m\,g_3\,;
\end{gathered} \tag{16.34}$$

to obtain the similar equations for the states with parity $P=(-1)^j$, it suffices to make the formal change m to $(-m)$.

For the minimal value $j=1/2$, we get the more simple system (formally it suffices to set $f_1=0$, $g_3=0$, $b=0$, $a=1$):

$P=(-1)^{j+1}$, $j=1/2$,

$$\begin{gathered}
(\epsilon-i\frac{d}{dr})g_0=(m+\frac{i}{r})f_0,\quad (\epsilon+i\frac{d}{dr})f_0=(m-\frac{i}{r})g_0,\\
(\epsilon+i\frac{d}{dr})f_3+i\frac{\sqrt{2}}{r}g_2=mg_1,\quad (\epsilon-i\frac{d}{dr})g_2-i\frac{\sqrt{2}}{r}f_3=(m+\frac{i}{r})f_2,\\
(\epsilon+i\frac{d}{dr})f_2+i\frac{\sqrt{2}}{r}g_1=(m-\frac{i}{r})g_2,\quad (\epsilon-i\frac{d}{dr})g_1-i\frac{\sqrt{2}}{r}f_2=mf_3.
\end{gathered} \tag{16.35}$$

16.3 Separating the Variables and Additional Constraints

Now we are to separate the variables with the additional constraints

$$\gamma^l (U^{-1})_{lk}\, \tilde{\Psi}_k(x) = 0\,,$$
$$\left[e^{(l)\alpha}\partial_\alpha + e^{(l)\alpha}_{\;\;;\alpha}(x) + e^{(l)\alpha}(x)\Gamma_\alpha(x) \right] (U^{-1})_{lk}\Psi_k(x) = 0\,. \tag{16.36}$$

After a simple calculation, from the first relation in (16.36) we obtain two algebraic constraints (which are the same for both parity values; also they are valid for the case $j = 1/2$)

$$\delta = \pm 1, \quad f_0 - \sqrt{2}g_1 + f_2 = 0, \quad g_0 + \sqrt{2}f_3 - g_2 = 0\,. \tag{16.37}$$

Now we turn to the differential constraint. Using the explicit form of the components $\tilde{\Psi}_l$:

$$\tilde{\Psi}_0 = \left| \begin{array}{c} f_0 D_{-1/2} \\ g_0 D_{+3/2} \\ h_0 D_{-1/2} \\ v_0 D_{+3/2} \end{array} \right|, \qquad \tilde{\Psi}_3 = \left| \begin{array}{c} f_2 D_{-1/2} \\ g_2 D_{+1/2} \\ h_2 D_{-1/2} \\ v_2 D_{+1/2} \end{array} \right|,$$

$$\tilde{\Psi}_1 = \frac{1}{\sqrt{2}} \left| \begin{array}{c} -f_1 D_{-3/2} + f_3 D_{+1/2} \\ -g_1 D_{-1/2} + g_3 D_{+3/2} \\ -h_1 D_{-3/2} + h_3 D_{+1/2} \\ -v_1 D_{-1/2} + v_3 D_{+3/2} \end{array} \right|, \qquad \tilde{\Psi}_2 = \frac{1}{\sqrt{2}} \left| \begin{array}{c} -if_1 D_{-3/2} - if_3 D_{+1/2} \\ -ig_1 D_{-1/2} - ig_3 D_{+3/2} \\ -ih_1 D_{-3/2} - ih_3 D_{+1/2} \\ -iv_1 D_{-1/2} - iv_3 D_{+3/2} \end{array} \right|,$$

we find the three terms in (16.36). For the first term we get (remembering the multiplier at r^{-1} in the wave function, we make the change $\partial_r \Longrightarrow \partial_r - 1/r$)

$$e^{(l)\alpha}\partial_\alpha\, \tilde{\Psi}_l = e^{(0)0}\, \partial_0 \tilde{\Psi}_0 + e^{(3)r}\, (\partial_r - \frac{1}{r})\tilde{\Psi}_3 + e^{(1)\theta}\, \partial_\theta \tilde{\Psi}_1 + e^{(2)\phi}\, \partial_\phi \tilde{\Psi}_2$$
$$= -i\epsilon \left| \begin{array}{c} f_0 D_{-1/2} \\ g_0 D_{+1/2} \\ h_0 D_{-1/2} \\ v_0 D_{+1/2} \end{array} \right| - \frac{1}{\sqrt{2}\, r}\partial_\theta \left| \begin{array}{c} -f_1 D_{-3/2} + f_3 D_{+1/2} \\ -g_1 D_{-1/2} + g_3 D_{+3/2} \\ -h_1 D_{-3/2} + h_3 D_{+1/2} \\ -v_1 D_{-1/2} + v_3\, D_{+3/2} \end{array} \right|$$
$$-(\partial_r - \frac{1}{r}) \left| \begin{array}{c} f_2 D_{-1/2} \\ g_2 D_{+1/2} \\ h_2 D_{-1/2} \\ v_2 D_{+1/2} \end{array} \right| - \frac{im}{\sqrt{2}\, r \sin\theta} \left| \begin{array}{c} -if_1 D_{-3/2} - if_3 D_{+1/2} \\ -ig_1 D_{-1/2} - ig_3 D_{+3/2} \\ -ih_1 D_{-3/2} - ih_3 D_{+1/2} \\ -iv_1 D_{-1/2} - iv_3 D_{+3/2} \end{array} \right|. \tag{16.38}$$

Taking in mind the relations

$$e^{(0)\alpha}_{\;\;;\alpha} = 0, \quad e^{(1)\alpha}_{\;\;;\alpha} = -\frac{1}{r}\, \text{ctg}\,\theta, \quad e^{(2)\alpha}_{\;\;;\alpha} = 0, \quad e^{(3)\alpha}_{\;\;;\alpha} = -\frac{2}{r},$$

we find the term $e^{(l)\alpha}_{;\alpha}\tilde{\Psi}_l$:

$$e^{(l)\alpha}_{\;\;;\alpha}\, \tilde{\Psi}_l = +\frac{\text{ctg}\,\theta}{\sqrt{2}\, r} \left| \begin{array}{c} f_1\, D_{-3/2} - f_3\, D_{+1/2} \\ g_1\, D_{-1/2} - g_3\, D_{+3/2} \\ h_1\, D_{-3/2} - h_3\, D_{+1/2} \\ v_1\, D_{-1/2} - v_3\, D_{+3/2} \end{array} \right| - \frac{2}{r} \left| \begin{array}{c} f_2\, D_{-1/2} \\ g_2\, D_{+1/2} \\ h_2\, D_{-1/2} \\ v_2\, D_{+1/2} \end{array} \right|. \tag{16.39}$$

For the term $e^{(l)\alpha}\Gamma_\alpha \tilde{\Psi}_l$, taking in mind the expressions for the connection $\Gamma_\alpha(x)$, we derive

$$e^{(l)\alpha}\Gamma_\alpha\tilde{\Psi}_l = e^{(1)\theta}\Gamma_\theta\tilde{\Psi}_1 + e^{(2)\phi}\Gamma_\phi\tilde{\Psi}_2 = -\frac{1}{r}(\sigma^{31}\tilde{\Psi}_1 + \sigma^{32}\tilde{\Psi}_2) - \frac{\text{ctg}\,\theta}{r}\sigma^{12}\tilde{\Psi}_2\,,$$

and further we find

$$e^{(l)\alpha}\Gamma_\alpha\,\tilde{\Psi}_l = -\frac{1}{\sqrt{2}\,r}\left|\begin{array}{c} g_1 D_{-1/2} \\ f_3\, D_{+1/2} \\ \nu_1 D_{-1/2} \\ h_3 D_{+1/2} \end{array}\right| + \frac{\text{ctg}\,\theta}{\sqrt{2}\,r}\left|\begin{array}{c} +\frac{1}{2} f_1 D_{-3/2} + \frac{1}{2} f_3 D_{+1/2} \\ -\frac{1}{2} g_1 D_{-1/2} - \frac{1}{2} g_3 D_{+3/2} \\ +\frac{1}{2} h_1 D_{-3/2} + \frac{1}{2} h_3 D_{+1/2} \\ -\frac{1}{2}\nu_1 D_{-1/2} - \frac{1}{2}\nu_3 D_{+3/2} \end{array}\right|. \qquad (16.40)$$

By summing (16.38), (16.39) and (16.40), we get the following form for the differential constraint in (16.36):

$$-i\epsilon\left|\begin{array}{c} f_0 D_{-1/2} \\ g_0 D_{+1/2} \\ h_0 D_{-1/2} \\ \nu_0 D_{+1/2} \end{array}\right| - (\partial_r + \frac{1}{r})\left|\begin{array}{c} f_2 D_{-1/2} \\ g_2 D_{+1/2} \\ h_2 D_{-1/2} \\ \nu_2 D_{+1/2} \end{array}\right| - \frac{1}{\sqrt{2}r}\left|\begin{array}{c} g_1 D_{-1/2} \\ f_3 D_{+1/2} \\ \nu_1 D_{-1/2} \\ h_3 D_{+1/2} \end{array}\right|$$

$$+\frac{1}{\sqrt{2}\,r}\partial_\theta\left|\begin{array}{c} f_1 D_{-3/2} - f_3 D_{+1/2} \\ g_1 D_{-1/2} - g_3 D_{+3/2} \\ h_1 D_{-3/2} - h_3 D_{+1/2} \\ \nu_1 D_{-1/2} - \nu_3 D_{+3/2} \end{array}\right| + \frac{1}{\sqrt{2}\,r}\left|\begin{array}{c} f_1 \frac{-m+3/2\cos\theta}{\sin\theta} D_{-3/2} + f_3 \frac{-m-1/2\cos\theta}{\sin\theta} D_{+1/2} \\ g_1 \frac{-m+1/2\cos\theta}{\sin\theta} D_{-1/2} + g_3 \frac{-m-3/2\cos\theta}{\sin\theta} D_{+3/2} \\ h_1 \frac{-m+3/2\cos\theta}{\sin\theta} D_{-3/2} + h_3 \frac{-m-1/2\cos\theta}{\sin\theta} D_{+3/2} \\ \nu_1 \frac{-m+1/2\cos\theta}{\sin\theta} D_{-1/2} + \nu_3 \frac{-m-3/2\cos\theta}{\sin\theta} D_{+3/2} \end{array}\right| = 0.$$

Then, transforming the last two terms with the help of the consequences of the recurrent formulas

$$\partial_\theta D_{+1/2} - \frac{-m - 1/2\,\cos\theta}{\sin\theta} D_{+1/2} = +a D_{-1/2}\,,$$
$$\partial_\theta D_{-1/2} + \frac{-m + 1/2\,\cos\theta}{\sin\theta} D_{-1/2} = -a D_{+1/2}\,,$$
$$\partial_\theta D_{+3/2} - \frac{-m - 3/2\,\cos\theta}{\sin\theta} D_{+3/2} = +b D_{+1/2}\,,$$
$$\partial_\theta D_{-3/2} + \frac{-m + 3/2\,\cos\theta}{\sin\theta} D_{-3/2} = -b D_{-1/2}\,,$$

we produce

$$-i\epsilon\left|\begin{array}{c} f_0 D_{-1/2} \\ g_0 D_{+1/2} \\ h_0 D_{-1/2} \\ \nu_0 D_{+1/2} \end{array}\right| - (\frac{d}{dr} + \frac{1}{r})\left|\begin{array}{c} f_2 D_{-1/2} \\ g_2 D_{+1/2} \\ h_2 D_{-1/2} \\ \nu_2 D_{+1/2} \end{array}\right| - \frac{1}{\sqrt{2}r}\left|\begin{array}{c} g_1 D_{-1/2} \\ f_3 D_{+1/2} \\ \nu_1 D_{-1/2} \\ h_3 D_{+1/2} \end{array}\right| - \frac{1}{\sqrt{2}\,r}\left|\begin{array}{c} (b\,f_1 + a\,f_3)\,D_{-1/2} \\ (a\,g_1 + b\,g_3)\,D_{-1/2} \\ (b\,h_1 + a\,h_3)\,D_{-1/2} \\ (a\,\nu_1 + b\,\mu_3)\,D_{+1/2} \end{array}\right|.$$

Thus, we derive four differential equations in the radial variable, which after taking into account the P-parity restrictions lead us to only two constrains (they are valid for both values of parity)

$$-i\epsilon f_0 - (\frac{d}{dr} + \frac{1}{r})f_2 - \frac{1}{\sqrt{2}\,r}g_1 - \frac{1}{\sqrt{2}\,r}(b\,f_1 + a\,f_3) = 0\,,$$
$$-i\epsilon g_0 - (\frac{d}{dr} + \frac{1}{r})g_2 - \frac{1}{\sqrt{2}\,r}f_3 - \frac{1}{\sqrt{2}\,r}(a\,g_1 + b\,g_3) = 0\,. \qquad (16.41)$$

For the minimal value $j = 1/2$, equations (16.41) become simpler

$$
\begin{aligned}
&-i\epsilon f_0 - (\frac{d}{dr} + \frac{1}{r}) f_2 - \frac{1}{\sqrt{2}r} g_1 - \frac{1}{\sqrt{2}\,r} f_3 = 0\,, \\
&-i\epsilon g_0 - (\frac{d}{dr} + \frac{1}{r}) g_2 - \frac{1}{\sqrt{2}\,r} f_3 - \frac{1}{\sqrt{2}\,r} g_1 = 0\,.
\end{aligned}
\tag{16.42}
$$

16.4 Solving Equations for Functions f_0, g_0

First we are to solve equations for the functions f_0, g_0 (see the first two equations in (16.35). Summing and subtracting these two equations, we get

$$
\Big(\frac{d}{dr} + \frac{a}{r}\Big) F_0 = i(\epsilon + m) G_0\,, \quad \Big(\frac{d}{dr} - \frac{a}{r}\Big) G_0 = i(\epsilon - m) F_0\,, \tag{16.43}
$$

where $F_0 = f_0 + g_0$, $G_0 = f_0 - g_0$. From (16.43), there follow the 2-nd order equations for the separate functions

$$
\begin{aligned}
&\Big(\frac{d^2}{dr^2} - \frac{a^2 + a}{r^2} + \epsilon^2 - m^2\Big) F_0 = 0\,,\ l = a\,; \\
&\Big(\frac{d^2}{dr^2} - \frac{(a-1)a}{r^2} + \epsilon^2 - m^2\Big) G_0 = 0\ l' = a - 1\,.
\end{aligned}
\tag{16.44}
$$

They reduce to Bessel type in the variable $x = \sqrt{\epsilon^2 - m^2} r$:

$$
\Big(\frac{d^2}{dx^2} + 1 - \frac{l(l+1)}{x^2}\Big) F_0(x) = 0\,, \quad \Big(\frac{d^2}{dx^2} + 1 - \frac{l'('l+1)}{x^2}\Big) G_0(x) = 0\,. \tag{16.45}
$$

Assume that

$$
F_0(x) = a_0 \sqrt{x}\, Z_l(x)\,, \quad G_0(x) = b_0 \sqrt{x}\, Z_{l'}(x)\,;
$$

then, the equations (16.45) yield

$$
\begin{aligned}
&Z_l'' + \frac{1}{x} Z_l' + 1 - \frac{(l+1/2)^2}{x^2} Z_l = 0\,, \quad l + 1/2 = j + 1 = p\,, \\
&Z_{l'}'' + \frac{1}{x} Z_{l'}' + 1 - \frac{(l-1/2)^2}{x^2} Z_{l'} = 0\,, \quad l' + 1/2 = j = p - 1\,.
\end{aligned}
\tag{16.46}
$$

Thus, the functions F_0 and G_0 are constructed through Bessel functions.

Let us turn back to the first order equations (16.43) for the functions F_0, G_0 written in the variable x:

$$
\begin{aligned}
&(\frac{d}{dx} + \frac{p}{x}) Z_p = \sqrt{\frac{\epsilon + m}{\epsilon - m}}\, (i\frac{b_0}{a_0})\, Z_{p-1}\,, \\
&(\frac{d}{dx} - \frac{p-1}{x}) Z_{p-1} = \sqrt{\frac{\epsilon - m}{\epsilon + m}}\, (i\frac{a_0}{b_0})\, Z_p\,.
\end{aligned}
\tag{16.47}
$$

Due to the known formulas

$$
\Big(\frac{d}{dx} + \frac{p}{z}\Big) Z_p = Z_{p-1}\,, \quad \Big(\frac{d}{dx} - \frac{p}{z}\Big) Z_p = -Z_{p+1} = 0\,, \tag{16.48}
$$

we get the relative coefficient between a_0 and b_0:

$$
\sqrt{\epsilon + m}\, b_0 = -i \sqrt{\epsilon - m}\, a_0\,. \tag{16.49}
$$

16.5 The Matrix Form of the Main System

Now, we turn to the 6 equations from (16.34):

$$(\epsilon + i\frac{d}{dr})f_1 + i\frac{b}{r}g_1 = m\, g_3\,, \quad (\epsilon - i\frac{d}{dr})g_3 - i\frac{b}{r}f_3 = m\, f_1\,,$$
$$(\epsilon + i\frac{d}{dr})f_2 + i\frac{\sqrt{2}}{r}g_1 = (m - i\frac{a}{r})g_2\,, \quad (\epsilon - i\frac{d}{dr})g_2 - i\frac{\sqrt{2}}{r}f_3 = (m + i\frac{a}{r})f_2\,, \qquad (16.50)$$
$$(\epsilon + i\frac{d}{dr})f_3 + i\frac{\sqrt{2}}{r}g_2 + i\frac{b}{r}g_3 = m\, g_1\,, \quad (\epsilon - i\frac{d}{dr})g_1 - i\frac{\sqrt{2}}{r}f_2 - i\frac{b}{r}f_1 = m\, f_3\,.$$

It is convenient to employ the following variables:

$$f_1 + g_3 = F_1\,,\ f_1 - g_3 = G_1\,,\ f_2 + g_2 = F_2\,,\ f_2 - g_2 = G_2\,, f_3 + g_1 = F_3\,,\ f_3 - g_1 = G_3\,.$$

) Summing and subtracting the equations in (16.50) within each pair, we produce

$$\frac{d}{dr}G_1 + i(m-\epsilon)F_1 = +\frac{b}{r}G_3\,, \quad \frac{d}{dr}F_1 - i(m+\epsilon)G_1 = -\frac{b}{r}F_3\,;$$
$$(\frac{d}{dr} - \frac{a}{r})G_2 + i(m-\epsilon)F_2 = +\frac{\sqrt{2}}{r}G_3\,, \quad (\frac{d}{dr} + \frac{a}{r})F_2 - i(m+\epsilon)G_2 = -\frac{\sqrt{2}}{r}F_3\,; \qquad (16.51)$$
$$\frac{d}{dr}G_3 - \frac{\sqrt{2}}{r}G_2 + i(m-\epsilon)F_3 = +\frac{b}{r}G_1\,, \quad \frac{d}{dr}F_3 + \frac{\sqrt{2}}{r}F_2 - i(m+\epsilon)G_3 = -\frac{b}{r}F_1\,.$$

This system may be presented in a matrix form

$$\frac{d}{dr}\left|\begin{array}{c} F_1 \\ F_2 \\ F_3 \\ G_1 \\ G_2 \\ G_3 \end{array}\right| = \left|\begin{array}{cccccc} 0 & 0 & -\frac{b}{r} & i(\epsilon+m) & 0 & 0 \\ 0 & -\frac{a}{r} & -\frac{\sqrt{2}}{r} & 0 & i(\epsilon+m) & 0 \\ -\frac{b}{r} & -\frac{\sqrt{2}}{r} & 0 & 0 & 0 & i(\epsilon+m) \\ i(\epsilon-m) & 0 & 0 & 0 & 0 & \frac{b}{r} \\ 0 & i(\epsilon-m) & 0 & 0 & \frac{a}{r} & \frac{\sqrt{2}}{r} \\ 0 & 0 & i(\epsilon-m) & \frac{b}{r} & \frac{\sqrt{2}}{r} & 0 \end{array}\right| \left|\begin{array}{c} F_1 \\ F_2 \\ F_3 \\ G_1 \\ G_2 \\ G_3 \end{array}\right|.$$

For the case of the minimal value $j = 1/2$ ($a = 1, b = 0, F_1 = 0, G_1 = 0$), the corresponding matrix equation becomes simpler

$$\frac{d}{dr}\left|\begin{array}{c} F_2 \\ F_3 \\ G_2 \\ G_3 \end{array}\right| = \left|\begin{array}{cccc} -\frac{1}{r} & -\frac{\sqrt{2}}{r} & i(\epsilon+m) & 0 \\ -\frac{\sqrt{2}}{r} & 0 & 0 & i(\epsilon+m) \\ i(\epsilon-m) & 0 & \frac{1}{r} & \frac{\sqrt{2}}{r} \\ 0 & i(\epsilon-m) & \frac{\sqrt{2}}{r} & 0 \end{array}\right| \left|\begin{array}{c} F_2 \\ F_3 \\ G_2 \\ G_3 \end{array}\right|. \qquad (16.52)$$

The system (16.52) may be presented in (2+2)-form (F and G are 2-component columns)

$$(\frac{d}{dr} + \frac{\alpha}{r})F = i(\epsilon+m)G\,, \quad (\frac{d}{dr} - \frac{\alpha}{r})G = i(\epsilon-m)F\,, \qquad (16.53)$$

where

$$\alpha = \left|\begin{array}{cc} 1 & \sqrt{2} \\ \sqrt{2} & 0 \end{array}\right|.$$

By applying the exclusion method, we derive the second order separate equations:

$$(\frac{d^2}{dr^2}+\epsilon^2-m^2)F=\frac{1}{r^2}(\alpha^2+\alpha)F\,,\quad (\frac{d^2}{dr^2}+\epsilon^2-m^2)G=\frac{1}{r^2}(\alpha^2-\alpha)G\,. \tag{16.54}$$

A similar approach may be applied to the case of arbitrary j. The system (16.51) is presented in a (3+3)-form, as follows

$$(\frac{d}{dr}+\frac{A}{r})F=i(\epsilon+m)G\,,\quad (\frac{d}{dr}-\frac{A}{r})G=i(\epsilon-m)F\,, \tag{16.55}$$

where

$$A=\begin{vmatrix} 0 & 0 & b \\ 0 & a & \sqrt{2} \\ b & \sqrt{2} & 0 \end{vmatrix}.$$

By applying the exclusion method, we derive the following second order separate equations:

$$(\frac{d^2}{dr^2}+\epsilon^2-m^2)F=\frac{1}{r^2}(A^2+A)F\,,\quad (\frac{d^2}{dr^2}+\epsilon^2-m^2)G=\frac{1}{r^2}(A^2-A)G\,. \tag{16.56}$$

16.6 The Case of Minimal Value $j=1/2$

For the minimal value $j=1/2$ we have two equations,

$$\begin{gathered}\Big(\frac{d^2}{dr^2}+\epsilon^2-m^2\Big)F=\frac{1}{r^2}(\alpha^2+\alpha)F\,,\ \Big(\frac{d^2}{dr^2}+\epsilon^2-m^2\Big)G=\frac{1}{r^2}(\alpha^2-\alpha)G\,,\\ \alpha=\begin{vmatrix} 1 & \sqrt{2} \\ \sqrt{2} & 0 \end{vmatrix},\quad \alpha^2=\begin{vmatrix} 3 & \sqrt{2} \\ \sqrt{2} & 2 \end{vmatrix},\\ \alpha^2+\alpha=\begin{vmatrix} 4 & 2\sqrt{2} \\ 2\sqrt{2} & 2 \end{vmatrix},\ \alpha^2-\alpha=\begin{vmatrix} 2 & 0 \\ 0 & 2 \end{vmatrix}.\end{gathered} \tag{16.57}$$

The equations for G_2, G_3 turn out to be separated:

$$\Big(\frac{d^2}{dr^2}+\epsilon^2-m^2-\frac{2}{r^2}\Big)G_2=0\,,\quad \Big(\frac{d^2}{dr^2}+\epsilon^2-m^2-\frac{2}{r^2}\Big)G_3=0\,,$$

and their solutions are

$$G_2=b_2\sqrt{x}\,Z_{3/2}(x)\,,\quad G_3=b_3\sqrt{x}\,Z_{3/2}(x)\,.$$

The system for F_2, F_3 reads

$$(\frac{d^2}{dr^2}+\epsilon^2-m^2)\begin{vmatrix} F_2 \\ F_3 \end{vmatrix}=\frac{2}{r^2}\begin{vmatrix} 2 & \sqrt{2} \\ \sqrt{2} & 1 \end{vmatrix}\begin{vmatrix} F_2 \\ F_3 \end{vmatrix},\quad \Delta F=\frac{2}{r^2}TF\,. \tag{16.58}$$

By applying a linear transformation we will reduce the mixing matrix to its diagonal form

$$\bar{F}=SF\,,\quad \Delta\bar{F}=\frac{2}{r^2}(STS^{-1})\,\bar{F}\,,\quad STS^{-1}=T_0=\begin{vmatrix} \lambda_2 & 0 \\ 0 & \lambda_3 \end{vmatrix}. \tag{16.59}$$

So we get algebraic equation

$$\begin{vmatrix} (2-\lambda_2) & \sqrt{2} \\ \sqrt{2} & (1-\lambda_2) \end{vmatrix}\begin{vmatrix} s_{11} \\ s_{12} \end{vmatrix}=0\,,\quad \begin{vmatrix} (2-\lambda_3) & \sqrt{2} \\ \sqrt{2} & (1-\lambda_3) \end{vmatrix}\begin{vmatrix} s_{21} \\ s_{22} \end{vmatrix}=0\,.$$

The solutions exist if the determinant of the system vanishes; so we find the two roots $\lambda_2 = 0$, $\lambda_3 = 3$. Further, we determine the entries of the matrix by solving the linear equations

$$2s_{11} + \sqrt{2}s_{12} = 0 \quad \Longrightarrow \quad s_{12} = -\sqrt{2}\, s_{11} \ \ (\text{let } s_{11} = 1)\,,$$

$$2s_{21} + \sqrt{2}s_{22} = 3s_{21} \quad \Longrightarrow \quad s_{21} = +\sqrt{2}\, s_{22} \ \ (\text{let } s_{22} = 1\,.$$

Thus, the needed transformation (one of many possible) is

$$S = \left| \begin{array}{cc} 1 & -\sqrt{2} \\ +\sqrt{2} & 1 \end{array} \right|, \ S^{-1} = \frac{1}{3} \left| \begin{array}{cc} 1 & +\sqrt{2} \\ -\sqrt{2} & 1 \end{array} \right|,$$
$$\bar{F} = S\,F\,, \quad \left| \begin{array}{c} \bar{F}_2 \\ \bar{F}_3 \end{array} \right| = \left| \begin{array}{cc} 1 & -\sqrt{2} \\ +\sqrt{2} & 1 \end{array} \right| \left| \begin{array}{c} F_2 \\ F_3 \end{array} \right|. \tag{16.60}$$

In this way we arrive at two separate equations for $\bar{F}_2$ and $\bar{F}_3$:

$$\Big(\frac{d^2}{dr^2} + \epsilon^2 - m^2\Big)\bar{F}_2 = 0\,, \quad \Big(\frac{d^2}{dr^2} + \epsilon^2 - m^2 - \frac{6}{r^2}\Big)\bar{F}_3 = 0\,;$$

their solutions are: $\bar{F}_2 = a_2\sqrt{x}\,Z_{1/2}$ and $\bar{F}_3 = a_3\sqrt{x}\,Z_{5/2}$.

Now we are to consider the additional constrains

$$g_1 = \frac{1}{\sqrt{2}}(f_2 + f_0)\,, \quad f_3 = \frac{1}{\sqrt{2}}(g_2 - g_0)\,,$$
$$-i\epsilon f_0 - (\frac{d}{dr} + \frac{1}{r})f_2 = \frac{1}{\sqrt{2}\,r}(g_1 + f_3) = 0\,,$$
$$-i\epsilon g_0 - (\frac{d}{dr} + \frac{1}{r})g_2 = \frac{1}{\sqrt{2}\,r}(f_3 + g_1) = 0\,. \tag{16.61}$$

In other variables

$$f_0 + g_0 = F_0\,, f_0 - g_0 = G_0\,,\ f_2 + g_2 = F_2\,, f_2 - g_2 = G_2\,,\ f_3 + g_1 = F_3\,,\ f_3 - g_1 = G_3\,,$$

they read

$$F_3 = \frac{1}{\sqrt{2}}(F_2 + G_0)\,, \quad G_3 = -\frac{1}{\sqrt{2}}(G_2 + F_0)\,,$$
$$-i\epsilon F_0 - (\frac{d}{dr} + \frac{1}{r})F_2 = \frac{1}{\sqrt{2}r}\,2F_3\,, \quad -i\epsilon G_0 - (\frac{d}{dr} + \frac{1}{r})G_2 = 0\,. \tag{16.62}$$

Taking into account the formulas

$$F_2 = \frac{1}{3}a_2\sqrt{x}\,Z_{1/2} + \frac{\sqrt{2}}{3}a_3\sqrt{x}\,Z_{5/2}\,, \quad F_0 = a_0\sqrt{x}\,Z_{3/2}\,, \quad G_0 = b_0\sqrt{x}\,Z_{1/2}\,,$$
$$F_3 = -\frac{\sqrt{2}}{3}a_2\sqrt{x}\,Z_{1/2} + \frac{a_3}{3}\sqrt{x}\,Z_{5/2}\,, \quad G_2 = b_2\sqrt{x}\,Z_{3/2}\,, \quad G_3 = b_3\sqrt{x}\,Z_{3/2}\,, \tag{16.63}$$

we can transform the equations (16.62) to Bessel form. First considering algebraic relations, they give

$$-\frac{\sqrt{2}}{3}a_2\sqrt{x}\,Z_{1/2} + \frac{a_3}{3}\sqrt{x}\,Z_{5/2}$$
$$= \frac{1}{\sqrt{2}}\Big\{\frac{1}{3}a_2\sqrt{x}\,Z_{1/2} + \frac{\sqrt{2}}{3}a_3\sqrt{x}\,Z_{5/2} + b_0\sqrt{x}\,Z_{1/2}\Big\}\,,$$

$$b_3\sqrt{x}\,Z_{3/2} = -\frac{1}{\sqrt{2}}\Big\{b_2\sqrt{x}\,Z_{3/2} + a_0\sqrt{x}\,Z_{3/2}\Big\},$$

whence there follow the linear relations for the numerical coefficients

$$-\frac{\sqrt{2}}{3}a_2 = \frac{1}{\sqrt{2}}(\frac{1}{3}a_2 + b_0)\,,\; b_3 = -\frac{1}{\sqrt{2}}(b_2 + a_0) \implies$$

$$b_0 + a_2 = 0\,, \quad a_0 + b_2 + \sqrt{2}b_3 = 0\,. \tag{16.64}$$

Now consider the differential constrains in the variable x:

$$-(\frac{d}{dx} + \frac{1}{x})(\frac{1}{3}a_2\sqrt{x}Z_{1/2} + \frac{\sqrt{2}}{3}a_3\sqrt{x}Z_{5/2})$$

$$= \frac{i\epsilon}{\sqrt{\epsilon^2 - m^2}}a_0\sqrt{x}Z_{3/2} + \frac{1}{x}\sqrt{2}(-\frac{\sqrt{2}}{3}a_2\sqrt{x}Z_{1/2} + \frac{a_3}{3}\sqrt{x}Z_{5/2})\,,$$

$$-(\frac{d}{dx} + \frac{1}{x})\; b_2\sqrt{x}\,Z_{3/2} = \frac{i\epsilon}{\sqrt{\epsilon^2 - m^2}}b_0\sqrt{x}\,Z_{1/2}\,,$$

whence, taking in mind the identity

$$\frac{d}{dx}\sqrt{x} = \sqrt{x}\,(\frac{d}{dx} + \frac{1/2}{x})\,,$$

we derive

$$-\frac{1}{3}a_2\,(\frac{d}{dx} + \frac{3/2}{x})Z_{1/2} - \frac{\sqrt{2}}{3}a_3\,(\frac{d}{dx} + \frac{3/2}{x})Z_{5/2}$$

$$= \frac{i\epsilon a_0}{\sqrt{\epsilon^2 - m^2}}Z_{3/2} - \frac{2}{3}a_2\frac{1}{x}Z_{1/2} + a_3\frac{\sqrt{2}}{3}\frac{1}{x}Z_{5/2}\,,$$

$$(\frac{d}{dx} + \frac{3/2}{x})\; b_2\;Z_{3/2} = -\frac{i\epsilon b_0}{\sqrt{\epsilon^2 - m^2}}\;Z_{1/2}\,.$$

Further, after re-grouping the terms in the first equation, we obtain

$$\begin{gathered} -\frac{1}{3}a_2\,(\frac{d}{dx} - \frac{1/2}{x})Z_{1/2} - \frac{\sqrt{2}}{3}a_3\,(\frac{d}{dx} + \frac{5/2}{x})Z_{5/2} = \frac{i\epsilon a_0}{\sqrt{\epsilon^2 - m^2}}Z_{3/2}\,, \\ (\frac{d}{dx} + \frac{3/2}{x})\; b_2\;Z_{3/2} = -\frac{i\epsilon b_0}{\sqrt{\epsilon^2 - m^2}}\;Z_{1/2}\,. \end{gathered} \tag{16.65}$$

Now, taking into account the properties of Bessel functions (16.48), we derive the linear relations for the coefficients (for convenience let us write down also the consequences of the algebraic constraints)

$$\begin{gathered} a_2 - \sqrt{2}a_3 = 3\frac{i\epsilon}{\sqrt{\epsilon^2 - m^2}}a_0\,, \quad b_2 = -\frac{i\epsilon}{\sqrt{\epsilon^2 - m^2}}b_0\,, \\ b_0 + a_2 = 0\,, \quad a_0 + b_2 + \sqrt{2}b_3 = 0\,. \end{gathered} \tag{16.66}$$

With the help of two last equations in (16.66), we exclude the variables a_0, b_0 from two first equations:

$$a_2 - \sqrt{2}a_3 = -3\frac{i\epsilon}{\sqrt{\epsilon^2 - m^2}}\;(b_2 + \sqrt{2}b_3)\,, \quad b_2 = \frac{i\epsilon}{\sqrt{\epsilon^2 - m^2}}\;a_2\,. \tag{16.67}$$

Besides, from the above established relation $\sqrt{\epsilon+m}\, b_0 = -i\sqrt{\epsilon-m}\, a_0$, it follows

$$-\sqrt{\epsilon+m}\, a_2 = +i\sqrt{\epsilon-m}\,(b_2+\sqrt{2}b_3)\,.$$

Therefore, the equations (16.67) reduce to the form

$$(2\epsilon+m)\,a_2+(\epsilon-m)\sqrt{2}a_3=0\,,\quad m b_2-\epsilon\sqrt{2}b_3=0\,. \tag{16.68}$$

Now, we are to take into account that the columns $\bar{F}(x)$ and $\bar{G}(x)$ cannot be considered as independent; instead they must obey the differential condition

$$\Big(\frac{d}{dr}+\frac{\alpha}{r}\Big)F=i(\epsilon+m)G \Longrightarrow \Big(\frac{d}{dr}+\frac{S\alpha S^{-1}}{r}\Big)\bar{F}=i(\epsilon+m)\,S\bar{G}, \tag{16.69}$$

where

$$\bar{F}=\left|\begin{array}{c} a_2\sqrt{x}\,Z_{1/2} \\ a_3\sqrt{x}\,Z_{5/2}\end{array}\right|,\quad \bar{G}=\left|\begin{array}{c} b_2\sqrt{x}\,Z_{3/2} \\ b_3\sqrt{x}\,Z_{3/2}\end{array}\right|. \tag{16.70}$$

In the variable x, Eq. (16.69) has the form

$$\Big(\frac{d}{dx}+\frac{S\alpha S^{-1}}{x}\Big)\bar{F}(x)=i\sqrt{\frac{\epsilon+m}{\epsilon-m}}\,SG(x)\,. \tag{16.71}$$

The left-hand and right-hand sides of (16.71) are

$$\left|\begin{array}{c} (d/dx-1/x)\,a_2\sqrt{x}\,Z_{1/2} \\ (d/dx+2/x)\,a_3\sqrt{x}\,Z_{5/2}\end{array}\right|,$$

$$i\sqrt{\frac{\epsilon+m}{\epsilon-m}}\,SG(x)=i\sqrt{\frac{\epsilon+m}{\epsilon-m}}\left|\begin{array}{c} (b_2-\sqrt{2}b_3)\,\sqrt{x}\,Z_{3/2} \\ (\sqrt{2}b_2+b_3)\,\sqrt{x}\,Z_{3/2}\end{array}\right|.$$

Therefore, we conclude that (16.71) gives

$$(\frac{d}{dx}-\frac{1/2}{x})a_2Z_{1/2}=i\sqrt{\frac{\epsilon+m}{\epsilon-m}}(b_2-\sqrt{2}b_3)Z_{3/2}\,,$$

$$(\frac{d}{dx}+\frac{5/2}{x})a_3Z_{5/2}=i\sqrt{\frac{\epsilon+m}{\epsilon-m}}(\sqrt{2}b_2+b_3)Z_{3/2}\,;$$

whence we derive the linear relations between coefficients:

$$a_2=-i\sqrt{\frac{\epsilon+m}{\epsilon-m}}\,(b_2-\sqrt{2}b_3)\,,\quad a_3=+i\sqrt{\frac{\epsilon+m}{\epsilon-m}}\,(\sqrt{2}b_2+b_3)\,. \tag{16.72}$$

They can be readily resolved with respect to b_2, b_3:

$$b_2=+i\sqrt{\frac{\epsilon-m}{\epsilon+m}}\frac{1}{3}\,(a_2-\sqrt{2}a_3)\,,\ b_3=-i\sqrt{\frac{\epsilon-m}{\epsilon+m}}\frac{1}{3}\,(\sqrt{2}a_2+a_3)\,. \tag{16.73}$$

Let us collect together all the independent equations for the unknown coefficients

$$\begin{gathered}(2\epsilon+m)\,a_2+(\epsilon-m)\sqrt{2}a_3=0\,,\quad m b_2-\epsilon\sqrt{2}b_3=0\,,\\ b_2=+i\sqrt{\frac{\epsilon-m}{\epsilon+m}}\frac{1}{3}\,(a_2-\sqrt{2}a_3)\,,\quad b_3=-i\sqrt{\frac{\epsilon-m}{\epsilon+m}}\frac{1}{3}\,(\sqrt{2}a_2+a_3)\,.\end{gathered} \tag{16.74}$$

We readily verify that substituting the last two equations into the second one, we obtain the first equation

$$m(a_2 - \sqrt{2}a_3) + \epsilon\sqrt{2}(\sqrt{2}a_2 + a_3) = 0 \implies (2\epsilon + m)a_2 + (\epsilon - m)\sqrt{2}a_3 = 0\,.$$

This means that in (16.74) we have only three independent relations:

$$a_2 = -\frac{\epsilon - m}{2\epsilon + m}\sqrt{2}a_3\,, \quad b_2 = -i\sqrt{\frac{\epsilon - m}{\epsilon + m}\frac{1}{3}}\,\frac{3\epsilon}{2\epsilon + m}\sqrt{2}a_3\,,$$
$$b_3 = -i\sqrt{\frac{\epsilon - m}{\epsilon + m}\frac{1}{3}}\,\frac{3m}{2\epsilon + m}a_3\,,$$

which completely determine the solutions for the case of $j = 1/2$.

16.7 Studying General Case $j = 3/2, 5/2, \dots$

The two relevant systems are

$$r^2\Big(\frac{d^2}{dr^2} + \epsilon^2 - m^2\Big)\begin{vmatrix} F_1 \\ F_2 \\ F_3 \end{vmatrix} = \begin{vmatrix} b^2 & \sqrt{2}b & b \\ \sqrt{2}b & a^2 + a + 2 & \sqrt{2}(a+1) \\ b & \sqrt{2}(a+1) & b^2 + 2 \end{vmatrix}\begin{vmatrix} F_1 \\ F_2 \\ F_3 \end{vmatrix}, \tag{16.75}$$

$$r^2\Big(\frac{d^2}{dr^2} + \epsilon^2 - m^2\Big)\begin{vmatrix} G_1 \\ G_2 \\ G_3 \end{vmatrix} = \begin{vmatrix} b^2 & \sqrt{2}b & -b \\ \sqrt{2}b & a^2 - a + 2 & \sqrt{2}(a-1) \\ -b & \sqrt{2}(a-1) & b^2 + 2 \end{vmatrix}\begin{vmatrix} G_1 \\ G_2 \\ G_3 \end{vmatrix}. \tag{16.76}$$

Symbolically, the equations (16.75)–(16.76) read

$$\Delta F = TF\,, \qquad \Delta G = T'G, \quad \Delta = r^2\Big(\frac{d^2}{dr^2} + \epsilon^2 - m^2\Big)\,. \tag{16.77}$$

We are to find the linear transformations over F and G, which change the mixing matrices T and T' to their diagonal form:

$$\bar{F} = SF, \quad \Delta\bar{F} = (STS^{-1})\,\bar{F}\,, \quad STS^{-1} = T_0 = \begin{vmatrix} \lambda_1 & 0 & 0 \\ 0 & \lambda_2 & 0 \\ 0 & 0 & \lambda_3 \end{vmatrix}; \tag{16.78}$$

$$\bar{G} = SG, \quad \Delta\bar{G} = (S'T'S'^{-1})\,\bar{G}\,, \quad S'T'S'^{-1} = T'_0 = \begin{vmatrix} \lambda'_1 & 0 & 0 \\ 0 & \lambda'_2 & 0 \\ 0 & 0 & \lambda'_3 \end{vmatrix}. \tag{16.79}$$

Let us consider the first equation $ST = T_0S$; it yields

$$\begin{vmatrix} s_{11} & s_{12} & s_{13} \\ s_{21} & s_{22} & s_{23} \\ s_{31} & s_{32} & s_{33} \end{vmatrix}\begin{vmatrix} b^2 & \sqrt{2}b & b \\ \sqrt{2}b & a^2 + a + 2 & \sqrt{2}(a+1) \\ b & \sqrt{2}(a+1) & b^2 + 2 \end{vmatrix} = \begin{vmatrix} \lambda_1 & 0 & 0 \\ 0 & \lambda_2 & 0 \\ 0 & 0 & \lambda_3 \end{vmatrix}\begin{vmatrix} s_{11} & s_{12} & s_{13} \\ s_{21} & s_{22} & s_{23} \\ s_{31} & s_{32} & s_{33} \end{vmatrix},$$

whence we obtain three sub-systems

$$
\begin{aligned}
s_{11}b^2 + s_{12}\sqrt{2}b + s_{13}b &= \lambda_1 s_{11}\,,\\
s_{11}\sqrt{2}b + s_{12}(a^2+a+2) + s_{13}\sqrt{2}(a+1) &= \lambda_1 s_{12}\,,\\
s_{11}b + s_{12}\sqrt{2}(a+1) + s_{13}(b^2+2) &= \lambda_1 s_{13}\,;
\end{aligned}
$$

$$
\begin{aligned}
s_{21}b^2 + s_{22}\sqrt{2}b + s_{23}b &= \lambda_2 s_{21}\,,\\
s_{21}\sqrt{2}b + s_{22}(a^2+a+2) + s_{23}\sqrt{2}(a+1) &= \lambda_2 s_{22}\,,\\
s_{21}b + s_{22}\sqrt{2}(a+1) + s_{23}(b^2+2) &= \lambda_2 s_{23}\,;
\end{aligned}
$$

$$
\begin{aligned}
s_{31}b^2 + s_{32}\sqrt{2}b + s_{33}b &= \lambda_3 s_{31}\,,\\
s_{31}\sqrt{2}b + s_{32}(a^2+a+2) + s_{33}\sqrt{2}(a+1) &= \lambda_3 s_{32}\,,\\
s_{31}b + s_{32}\sqrt{2}(a+1) + s_{33}(b^2+2) &= \lambda_3 s_{33}\,.
\end{aligned}
$$

Here there arise three eigenvalue problems

$$
\begin{gathered}
\left|\begin{array}{ccc} b^2-\lambda_1 & \sqrt{2}b & b\\ \sqrt{2}b & (a^2+a+2)-\lambda_1 & \sqrt{2}(a+1)\\ b & \sqrt{2}(a+1) & b^2+2-\lambda_1 \end{array}\right|\left|\begin{array}{c} s_{11}\\ s_{12}\\ s_{13}\end{array}\right| = 0\,,\\
\left|\begin{array}{ccc} b^2-\lambda_2 & \sqrt{2}b & b\\ \sqrt{2}b & (a^2+a+2)-\lambda_2 & \sqrt{2}(a+1)\\ b & \sqrt{2}(a+1) & b^2+2-\lambda_2 \end{array}\right|\left|\begin{array}{c} s_{21}\\ s_{22}\\ s_{23}\end{array}\right| = 0\,,\\
\left|\begin{array}{ccc} b^2-\lambda_3 & \sqrt{2}b & b\\ \sqrt{2}b & (a^2+a+2)-\lambda_3 & \sqrt{2}(a+1)\\ b & \sqrt{2}(a+1) & b^2+2-\lambda_3 \end{array}\right|\left|\begin{array}{c} s_{31}\\ s_{32}\\ s_{33}\end{array}\right| = 0\,.
\end{gathered}
\tag{16.80}
$$

From the very beginning, it should be noted that the lines in the matrices (16.80) can be fixed only up to arbitrary numerical factors (which correlates with the linearity of the differential equations).

Nontrivial solutions exist only if the determinant of the matrices is equal to zero:

$$
\det\left|\begin{array}{ccc} b^2-\lambda & \sqrt{2}b & b\\ \sqrt{2}b & (a^2+a+2)-\lambda & \sqrt{2}(a+1)\\ b & \sqrt{2}(a+1) & b^2+2-\lambda \end{array}\right| = 0\,, \tag{16.81}
$$

the roots are (note that one root is double degenerated)

$$
\lambda_1 = \lambda_3 = ((j-1/2)(j+1/2) = j'(j'+1)\,, \qquad j' = j-1/2\,; \tag{16.82}
$$

$$
\lambda_2 = (j+3/2)(j+5/2) = j'(j'+1)\,, \qquad j' = j+3/2\,. \tag{16.83}
$$

After performing the needed calculation, we find the explicit form for the transformation matrix S (a fixed variant from various ones; below we need also the inverse matrix)

$$S = \begin{vmatrix} \sqrt{2j+3} & 0 & -\sqrt{2j-1} \\ \sqrt{2j-1} & \sqrt{2}\sqrt{2j+3} & \sqrt{2j+3} \\ \sqrt{2}\sqrt{2j+3} & -\sqrt{2j-1} & 0 \end{vmatrix},$$

$$S^{-1} = \frac{1}{8(j+1)\sqrt{2j-1}}$$

$$\times \begin{vmatrix} \sqrt{(2j+3)(2j-1)} & (2j-1) & \sqrt{2}\sqrt{(2j+3)(2j-1)} \\ \sqrt{2}(2j+3) & \sqrt{2}\sqrt{(2j+3)(2j-1)} & -2(2j+1) \\ -6j-5 & \sqrt{(2j+3)(2j-1)} & \sqrt{2}(2j+3) \end{vmatrix}. \quad (16.84)$$

The diagonalization of the matrix T' is done by a similar scheme (in this case we shall have other roots $\lambda_i \to \lambda'_i$). The equation for $\lambda'_1, \lambda'_2, \lambda'_1$ is

$$\det \begin{vmatrix} b^2 - \lambda' & \sqrt{2}b & -b \\ \sqrt{2}b & (a^2 - a + 2) - \lambda' & \sqrt{2}(a-1) \\ -b & \sqrt{2}(a-1) & b^2 + 2 - \lambda' \end{vmatrix} = 0, \quad (16.85)$$

and the roots are (note that again one root is double degenerated)

$$\lambda'_1 = \lambda'_3 = (j+1/2)(j+3/2) = j'(j'+1), \qquad j' = j+1/2; \quad (16.86)$$

$$\lambda'_2 = (j-3/2)(j-1/2) = j'(j'+1), \qquad j' = j-3/2. \quad (16.87)$$

In this case, the needed transformation S' may be chosen in the form

$$S' = \begin{vmatrix} \sqrt{j-1/2} & 0 & -\sqrt{j+3/2} \\ \sqrt{j+3/2} & -\sqrt{2}\,\sqrt{j-1/2} & \sqrt{j-1/2} \\ \sqrt{2}\,\sqrt{j-1/2} & \sqrt{j+3/2} & 0 \end{vmatrix}, \quad (16.88)$$

$$S'^{-1} = -\frac{1}{4j\sqrt{j+3/2}}$$

$$\begin{vmatrix} -\sqrt{(j+3/2)(j-1/2)} & -(j+3/2) & -\sqrt{2}\sqrt{(j-1/2)(j+3/2)} \\ \sqrt{2}(j-1/2) & \sqrt{2}\sqrt{(j+3/2)(j-1/2)} & -(2j+1) \\ 3j+1/2 & -\sqrt{(j-1/2)(j+3/2)} & -\sqrt{2}(j-1/2) \end{vmatrix}.$$

Taking into account the expressions for the roots, we can write down the following equations for the 8 separate functions:

$$\left(\frac{d^2}{dr^2} + \epsilon^2 - m^2 - \frac{j'(j'+1)}{r^2}\right) F_0 = 0,$$
$$j' = j+1/2, \quad F_0 = a_0 f_{j+1/2} = a_0\sqrt{x} Z_{j+1};$$
$$\left(\frac{d^2}{dr^2} + \epsilon^2 - m^2 - \frac{j'(j'+1)}{r^2}\right) G_0 = 0,$$
$$j' = j-1/2, G_0 = b_0 f_{j-1/2} = b_0\sqrt{x} Z_j; \quad (16.89)$$

$$\begin{gathered}\Big(\frac{d^2}{dr^2}+\epsilon^2-m^2-\frac{j'(j'+1)}{r^2}\Big)\,\bar{F}_1=0\,,\\ j'=j-1/2\,,\quad \bar{F}_1=a_1 f_{j-1/2}=a_1\sqrt{x}Z_j\,;\\ \Big(\frac{d^2}{dr^2}+\epsilon^2-m^2-\frac{j'(j'+1)}{r^2}\Big)\,\bar{G}_1=0\,,\\ j'=j+1/2\,,\quad \bar{G}_1=b_1 f_{j+1/2}=b_1\sqrt{x}Z_{j+1}\,;\end{gathered} \tag{16.90}$$

$$\begin{gathered}\Big(\frac{d^2}{dr^2}+\epsilon^2-m^2-\frac{j'(j'+1)}{r^2}\Big)\,\bar{F}_2=0\,,\\ j'=j+3/2\,,\quad \bar{F}_2=a_2 f_{j+3/2}=a_2\sqrt{x}Z_{j+2}\,;\\ \Big(\frac{d^2}{dr^2}+\epsilon^2-m^2-\frac{j'(j'+1)}{r^2}\Big)\,\bar{G}_2=0\,,\\ j'=j-3/2\,,\quad \bar{G}_2=b_2 f_{j-3/2}=b_2\sqrt{x}Z_{j-1}\,;\end{gathered} \tag{16.91}$$

$$\begin{gathered}\Big(\frac{d^2}{dr^2}+\epsilon^2-m^2-\frac{j'(j'+1)}{r^2}\Big)\,\bar{F}_3=0\,,\\ j'=j-1/2\,,\quad \bar{F}_3=a_3 f_{j-1/2}=a_3\sqrt{x}Z_j\,;\\ \Big(\frac{d^2}{dr^2}+\epsilon^2-m^2-\frac{j'(j'+1)}{r^2}\Big)\,\bar{G}_3=0\,,\\ j'=j+1/2\,,\quad \bar{G}_3=b_3 f_{j+1/2}=b_3\sqrt{x}Z_{j+1}\,.\end{gathered} \tag{16.92}$$

They reduce to equations of Bessel type. All the solutions are determined for the present time up to arbitrary numerical factors.

16.8 Further Study of the Solutions

It should be emphasized that the parameters a_1, a_2, a_3 and b_1, b_2, b_3 cannot be considered as independent, because there exists the first-order differential equation which relates F and G:

$$(\frac{d}{dr}+\frac{A}{r})F=i(m+\epsilon)G; \tag{16.93}$$

with the use of the formulas $F=S^{-1}\bar{F}$, $G=S'^{-1}\bar{G}$, it may be presented as

$$(\frac{d}{dr}+\frac{A}{r})S^{-1}\bar{F}=i(m+\epsilon)S'^{-1}\bar{G},\quad (\frac{d}{dr}+\frac{SAS^{-1}}{r})\bar{F}=i(m+\epsilon)SS'^{-1}\bar{G}. \tag{16.94}$$

Allowing for the identity

$$SAS^{-1}=\left|\begin{array}{ccc}-(j+3/2) & 0 & \sqrt{2}\\ 0 & j+3/2 & 0\\ -\sqrt{2}(j+1/2) & 0 & j+1/2\end{array}\right|, \tag{16.95}$$

we derive

$$(\frac{d}{dr}+\frac{SAS^{-1}}{r})\left|\begin{array}{c}a_1 f_{j-1/2}\\ a_2 f_{j+3/2}\\ a_3 f_{j-1/2}\end{array}\right|=\left|\begin{array}{c}(\frac{d}{dr}-\frac{j+3/2}{r})a_1 f_{j-1/2}+\frac{\sqrt{2}}{r}a_3 f_{j-1/2}\\ (\frac{d}{dr}+\frac{j+3/2}{r})a_2 f_{j+3/2}\\ -\sqrt{2}\frac{j+1/2}{r}a_1 f_{j-1/2}+(\frac{d}{dr}+\frac{j+1/2}{r})a_3 f_{j-1/2}\end{array}\right|. \tag{16.96}$$

Taking in mind the identity

$$SS^{'-1} = \begin{vmatrix} \sqrt{2}\sqrt{\frac{j-1/2}{j+3/2}}\,\frac{j+1/2}{j} & \frac{1}{\sqrt{2}\,j} & \frac{1}{j}\sqrt{\frac{j-1/2}{j+3/2}} \\ -\sqrt{2} & 0 & 2 \\ \sqrt{\frac{j-1/2}{j+3/2}}\,\frac{j+1/2}{j} & \frac{j+1/2}{j} & \frac{1}{\sqrt{2}}\frac{1}{j}\sqrt{\frac{j-1/2}{j+3/2}} \end{vmatrix}, \qquad (16.97)$$

we derive

$$SS^{'-1}\begin{vmatrix} b_1 f_{j+1/2} \\ b_2 f_{j-3/2} \\ b_3 f_{j+1/2} \end{vmatrix} = \begin{vmatrix} \sqrt{\frac{j-1/2}{j+3/2}}(\frac{j+1/2}{j}\sqrt{2}b_1 + \frac{1}{j}b_3)f_{j+1/2} + \frac{1}{j}\frac{1}{\sqrt{2}}b_2 f_{j-3/2} \\ (-\sqrt{2}b_1 + 2b_3)f_{j+1/2} \\ \frac{1}{\sqrt{2}}\sqrt{\frac{j-1/2}{j+3/2}}(\frac{j+1/2}{j}\sqrt{2}b_1 + \frac{1}{j}b_3)f_{j+1/2} + \frac{j+1/2}{j}b_2 f_{j-3/2} \end{vmatrix}. \qquad (16.98)$$

Thus, we have three differential constraints

$$(\frac{d}{dr} + \frac{j+3/2}{r})a_2 f_{j+3/2} = i(\epsilon+m)(-\sqrt{2}b_1 + 2b_3)f_{j+1/2}, \qquad (16.99)$$

$$[(\frac{d}{dr} - \frac{j+3/2}{r})a_1 + \frac{\sqrt{2}}{r}a_3]f_{j-1/2} = i(\epsilon+m)$$
$$\times\Big[\sqrt{\frac{j-1/2}{j+3/2}}(\frac{j+1/2}{j}\sqrt{2}b_1 + \frac{1}{j}b_3)f_{j+1/2} + \frac{1}{j}\frac{1}{\sqrt{2}}b_2 f_{j-3/2}\Big], \qquad (16.100)$$

$$[-\frac{2j+1}{r}a_1 + (\frac{d}{dr} + \frac{j+1/2}{r})\sqrt{2}a_3]f_{j-1/2} = i(\epsilon+m)$$
$$\times\Big[\sqrt{\frac{j-1/2}{j+3/2}}(\frac{j+1/2}{j}\sqrt{2}b_1 + \frac{1}{j}b_3)f_{j+1/2} + \frac{2j+1}{j}\frac{1}{\sqrt{2}}b_2 f_{j-3/2}\Big]. \qquad (16.101)$$

Subtracting (16.101) from (16.100), we get

$$\Big(\frac{d}{dr} - \frac{j+3/2}{r} + \frac{2j+1}{r}\Big)a_1 f_{j-1/2} - \Big(\frac{d}{dr} + \frac{j+1/2}{r} - \frac{1}{r}\Big)\sqrt{2}a_3 f_{j-1/2}$$
$$= -i(\epsilon+m)\,2\frac{b_2}{\sqrt{2}}f_{j-3/2}.$$

Now, let us multiply Eq. (16.100) by $(2j+1)$, and from the result subtract Eq. (16.101); we obtain

$$(2j+1)\Big(\frac{d}{dr} - \frac{j+3/2}{r} + \frac{1}{r}\Big)a_1 f_{j-1/2}$$
$$-\Big(\frac{d}{dr} + \frac{j+1/2}{r} - \frac{2j+1}{r}\Big)\sqrt{2}a_3 f_{j-1/2}$$
$$= i(\epsilon+m)\sqrt{\frac{j-1/2}{j+3/2}}\,[\,(2j+1)\sqrt{2}b_1 + 2b_3\,]f_{j+1/2}.$$

Therefore, we get the three equations

$$(\frac{d}{dr} + \frac{j+3/2}{r})a_2 f_{j+3/2} = i(\epsilon+m)(-\sqrt{2}b_1 + 2b_3)f_{j+1/2},$$

$$(\frac{d}{dr}+\frac{j-1/2}{r})[a_1-\sqrt{2}a_3]f_{j-1/2}=-i(\epsilon+m)\,2\frac{b_2}{\sqrt{2}}f_{j-3/2},$$

$$[(2j+1)a_1-\sqrt{2}a_3]\,(\frac{d}{dr}-\frac{j+1/2}{r})f_{j-1/2}$$

$$=i(\epsilon+m)\sqrt{\frac{j-1/2}{j+3/2}}\,[\,(2j+1)\sqrt{2}b_1+2b_3\,]f_{j+1/2}.$$

Transforming them to Bessel form with the use of the following substitutions

$$f_{j+1/2}=\sqrt{x}Z_{j+1},\; f_{j-1/2}=\sqrt{x}Z_j,\; f_{j+3/2}=\sqrt{x}Z_{j+2},\; f_{j-3/2}=\sqrt{x}Z_{j-1},$$

we obtain

$$(\frac{d}{dx}+\frac{j+2}{x})a_2Z_{j+2}=i\sqrt{\frac{\epsilon+m}{\epsilon-m}}(-\sqrt{2}b_1+2b_3)Z_{j+1},$$

$$(\frac{d}{dx}+\frac{j}{x})[a_1-\sqrt{2}a_3]Z_j=-i\sqrt{\frac{\epsilon+m}{\epsilon-m}}\,2\frac{b_2}{\sqrt{2}}Z_{j-1},$$

$$[(2j+1)a_1-\sqrt{2}a_3](\frac{d}{dx}-\frac{j}{x})Z_j$$

$$=i\sqrt{\frac{\epsilon+m}{\epsilon-m}}\sqrt{\frac{j-1/2}{j+3/2}}\,[\,(2j+1)\sqrt{2}b_1+2b_3\,]Z_{j+1},$$

whence, by applying the known properties of Bessel functions, we arrive at the linear relations for the numerical coefficients:

$$a_2=i\sqrt{\frac{\epsilon+m}{\epsilon-m}}(-\sqrt{2}b_1+2b_3),\quad a_1-\sqrt{2}a_3=-i\sqrt{\frac{\epsilon+m}{\epsilon-m}}2\frac{b_2}{\sqrt{2}},$$
$$(2j+1)a_1-\sqrt{2}a_3=-i\sqrt{\frac{\epsilon+m}{\epsilon-m}}\sqrt{\frac{j-1/2}{j+3/2}}[(2j+1)\sqrt{2}b_1+2b_3]. \tag{16.102}$$

They may be resolved for the variables a_i and for the b_i:

$$a_1=i\sqrt{\frac{\epsilon+m}{\epsilon-m}}\Big\{-\sqrt{2}\frac{j+1/2}{j}\frac{\sqrt{4j-2}}{\sqrt{4j+6}}b_1+\frac{\sqrt{2}}{2}\frac{1}{j}b_2-\frac{\sqrt{4j-2}}{\sqrt{4j+6}}\frac{1}{j}b_3\Big\},$$

$$a_2=i\sqrt{\frac{\epsilon+m}{\epsilon-m}}\Big\{-\sqrt{2}\,b_1+2b_3\Big\}, \tag{16.103}$$

$$a_3=i\sqrt{\frac{\epsilon+m}{\epsilon-m}}\Big\{-\frac{j+1/2}{j}\frac{\sqrt{4j-2}}{\sqrt{4j+6}}b_1+\frac{1}{2}\frac{2j+1}{j}b_2-\frac{\sqrt{2}}{2}\frac{\sqrt{4j-2}}{\sqrt{4j+6}}\frac{1}{j}b_3\Big\},$$

and

$$b_1=i\sqrt{\frac{\epsilon-m}{\epsilon+m}}\Big\{\frac{\sqrt{2}}{4}\frac{2j+1}{j+1}\frac{\sqrt{4j+6}}{\sqrt{4j-2}}a_1+\frac{\sqrt{2}}{4}\frac{1}{j+1}a_2-\frac{\sqrt{4j+6}}{\sqrt{4j-2}}\frac{1}{2(j+1)}a_3\Big\},$$

$$b_2=i\sqrt{\frac{\epsilon-m}{\epsilon+m}}\Big\{\frac{\sqrt{2}}{2}\,a_1-a_3\Big\}, \tag{16.104}$$

$$b_3=i\sqrt{\frac{\epsilon-m}{\epsilon+m}}\Big\{\frac{1}{4}\frac{2j+1}{j+1}\frac{\sqrt{4j+6}}{\sqrt{4j-2}}a_1-\frac{2j+1}{4(j+1)}a_2-\sqrt{2}\frac{\sqrt{4j+6}}{\sqrt{4j-2}}\frac{1}{4(j+1)}a_3\Big\}.$$

16.9 Accounting for Algebraic and Differential Constraints

The above constraints (16.37) and (16.41) being transformed to the variables F_a, G_a will read:

$$
\begin{aligned}
& F_2 + G_0 = \sqrt{2} F_3 \,, \quad G_2 + F_0 = -\sqrt{2} G_3 \,, \\
& -i\epsilon F_0 - (\frac{d}{dr} + \frac{1}{r}) F_2 = \frac{1}{\sqrt{2} r} [b\, F_1 + (a+1)\, F_3] \,, \\
& -i\epsilon G_0 - (\frac{d}{dr} + \frac{1}{r}) G_2 = \frac{1}{\sqrt{2} r} [b\, G_1 + (a-1)\, G_3] \,.
\end{aligned}
\qquad (16.105)
$$

We are to transform Eqs. (16.105) to the variables

$$F_0, \bar{F}_1, \bar{F}_2, \bar{F}_3, \; G_0, \bar{G}_1, \bar{G}_2, \bar{G}_3.$$

With the use of the formulas $F = S^{-1} \bar{F}$, $G = S'^{-1} \bar{G}$, we get the expressions for F_i, G_i:

$$F_1 = \frac{1}{8(j+1)\sqrt{2j-1}} \Big\{ \sqrt{(2j+3)(2j-1)}\, a_1 f_{j-1/2}$$

$$+ (2j-1) a_2 f_{j+3/2} + \sqrt{(2j+3)(2j-1)} \sqrt{2} a_3 f_{j-1/2} \Big\} \,,$$

$$F_2 = \frac{1}{8(j+1)\sqrt{2j-1}} \Big\{ \sqrt{2}(2j+3) a_1 f_{j-1/2}$$

$$+ \sqrt{2} \sqrt{(2j+3)(2j-1)} a_2 f_{j+3/2} - 2(2j+1) a_3 f_{j-1/2} \Big\} \,,$$

$$F_3 = \frac{1}{8(j+1)\sqrt{2j-1}} \{ (-6j-5) a_1 f_{j-1/2}$$

$$+ \sqrt{(2j+3)(2j-1)} a_2 f_{j+3/2} + \sqrt{2}(2j+3) a_3 f_{j-1/2} \Big\} \,,$$

and

$$G_1 = -\frac{1}{4j\sqrt{j+3/2}} \{ -\sqrt{(j+3/2)(j-1/2)} b_1 f_{j+1/2} - (j+3/2) b_2 f_{j-3/2}$$

$$- \sqrt{2} \sqrt{(j-1/2)(j+3/2)} b_3 f_{j+1/2} \} \,,$$

$$G_2 = -\frac{1}{4j\sqrt{j+3/2}} \Big\{ \sqrt{2}(j-1/2) b_1 f_{j+1/2}$$

$$+ \sqrt{2} \sqrt{(j+3/2)(j-1/2)} b_2 f_{j-3/2} - (2j+1) b_3 f_{j+1/2} \Big\} \,,$$

$$G_3 = -\frac{1}{4j\sqrt{j+3/2}} \{ (3j+1/2) b_1 f_{j+1/2}$$

$$- \sqrt{(j-1/2)(j+3/2)} b_2 f_{j-3/2} - \sqrt{2}(j-1/2) b_3 f_{j+1/2} \Big\} \,.$$

It is convenient to introduce the shortening notations

$$\alpha = (\det S)^{-1} = \frac{1}{8(j+1)\sqrt{2j-1}} \,, \qquad \beta = (\det S')^{-1} = -\frac{1}{4j\sqrt{j+3/2}} \,.$$

Let us substitute the expressions for F_0, F_i and G_0, G_i:

$$F_0 = a_0 f_{j+1/2} \,,$$

$$F_1 = \alpha\Big\{\sqrt{(2j+3)(2j-1)}a_1 f_{j-1/2}$$
$$+(2j-1)a_2 f_{j+3/2} + \sqrt{(2j+3)(2j-1)}\sqrt{2}a_3 f_{j-1/2}\Big\},$$
$$F_2 = \alpha\Big\{\sqrt{2}(2j+3)a_1 f_{j-1/2}$$
$$+\sqrt{2}\sqrt{(2j+3)(2j-1)}a_2 f_{j+3/2} - 2(2j+1)a_3 f_{j-1/2}\Big\},$$
$$F_3 = \alpha\{(-6j-5)a_1 f_{j-1/2}$$
$$+\sqrt{(2j+3)(2j-1)}a_2 f_{j+3/2} + \sqrt{2}(2j+3)a_3 f_{j-1/2}\Big\},$$

and

$$G_0 = b_0 f_{j-1/2},$$
$$G_1 = \beta\Big\{-\sqrt{(j+3/2)(j-1/2)}b_1 f_{j+1/2}$$
$$-(j+3/2)b_2 f_{j-3/2} - \sqrt{2}\sqrt{(j-1/2)(j+3/2)}b_3 f_{j+1/2}\Big\},$$
$$G_2 = \beta\Big\{\sqrt{2}(j-1/2)b_1 f_{j+1/2}$$
$$+\sqrt{2}\sqrt{(j+3/2)(j-1/2)}b_2 f_{j-3/2} - (2j+1)b_3 f_{j+1/2}\Big\},$$
$$G_3 = \beta\{(3j+1/2)b_1 f_{j+1/2}$$
$$-\sqrt{(j-1/2)(j+3/2)}b_2 f_{j-3/2} - \sqrt{2}(j-1/2)b_3 f_{j+1/2}\Big\}$$

into algebraic equations in (16.105); this results in

$$\alpha\Big\{\sqrt{2}(2j+3)a_1 f_{j-1/2} + \sqrt{2}\sqrt{(2j+3)(2j-1)}a_2 f_{j+3/2}$$
$$-2(2j+1)a_3 f_{j-1/2}\Big\} + b_0 f_{j-1/2}$$
$$= \sqrt{2}\,\alpha\Big\{(-6j-5)a_1 f_{j-1/2} + \sqrt{(2j+3)(2j-1)}f_{j+3/2}a_2$$
$$+\sqrt{2}(2j+3)a_3)f_{j-1/2}\Big\},$$
$$\beta\Big\{\sqrt{2}(j-1/2)b_1 f_{j+1/2} + \sqrt{2}\sqrt{(j+3/2)(j-1/2)}b_2 f_{j-3/2}$$
$$-(2j+1)b_3 f_{j+1/2}\Big\} + a_0 f_{j+1/2}$$
$$= -\sqrt{2}\,\beta\Big\{(3j+1/2)b_1 f_{j+1/2} - \sqrt{(j-1/2)(j+3/2)}b_2 f_{j-3/2}$$
$$-\sqrt{2}(j-1/2)b_3 f_{j+1/2}\Big\}.$$

After needed re-grouping the terms, we obtain linear constraints for numerical coefficients

$$8\alpha\,(j+1)(\sqrt{2}\,a_1 - a_3) + b_0 = 0, \quad 4\beta j\,(\sqrt{2}\,b_1 - b_3) + a_0 = 0, \tag{16.106}$$

whence, allowing for the expressions of α and β, we get

$$\sqrt{2}\,a_1 - a_3 = -\sqrt{2}\sqrt{j-1/2}\,b_0, \quad \sqrt{2}\,b_1 - b_3 = +\sqrt{j+3/2}\,a_0. \tag{16.107}$$

The third and the fourth equations in (16.105) respectively give

$$-(\frac{d}{dr}+\frac{1}{r})\alpha\Big\{\sqrt{2}(2j+3)\,a_1 f_{j-1/2}$$

$$+\sqrt{2}\sqrt{(2j+3)(2j-1)}a_2 f_{j+3/2}-2(2j+1)a_3 f_{j-1/2}\Big\}$$

$$=+i\epsilon a_0 f_{j+1/2}+\frac{1}{r}\frac{j+3/2}{\sqrt{2}}\alpha\{(-6j-5)\,a_1 f_{j-1/2}$$

$$+\sqrt{(2j+3)(2j-1)}a_2 f_{j+3/2}+\sqrt{2}(2j+3)a_3 f_{j-1/2}\Big\}$$

$$+\frac{\sqrt{(j-1/2)(j+3/2)}}{\sqrt{2}r}\alpha\Big\{\sqrt{(2j+3)(2j-1)}a_1 f_{j-1/2}$$

$$+(2j-1)a_2 f_{j+3/2}+\sqrt{(2j+3)(2j-1)}\sqrt{2}a_3 f_{j-1/2}\Big\},$$

and

$$-(\frac{d}{dr}+\frac{1}{r})\beta\Big\{\sqrt{2}(j-1/2)\,b_1 f_{j+1/2}$$

$$+\sqrt{2}\sqrt{(j+3/2)(j-1/2)}\;b_2 f_{j-3/2}-(2j+1)\;b_3 f_{j+1/2}\Big\}$$

$$=+i\epsilon b_0 f_{j-1/2}+\frac{1}{r}\frac{j-1/2}{\sqrt{2}}\beta\,\{(3j+1/2)b_1 f_{j+1/2}$$

$$-\sqrt{(j-1/2)(j+3/2)}b_2 f_{j-3/2}-\sqrt{2}(j-1/2)b_3 f_{j+1/2}\Big\}$$

$$+\frac{\sqrt{(j-1/2)(j+3/2)}}{\sqrt{2}r}\Big\{-\sqrt{(j+3/2)(j-1/2)}b_1 f_{j+1/2}$$

$$-(j+3/2)b_2 f_{j-3/2}-\sqrt{2}\sqrt{(j-1/2)(j+3/2)}b_3 f_{j+1/2}\Big\}.$$

After elementary manipulation, we arrive at

$$\Big[-(j+3/2)\sqrt{2}a_1+(2j+1)a_3\Big]\Big(\frac{d}{dr}+\frac{1}{r}\Big)f_{j-1/2}$$

$$-\sqrt{2}\,\sqrt{(j+3/2)(j-1/2)}\Big(\frac{d}{dr}+\frac{j+3/2}{r}\Big)a_2 f_{j+3/2}=$$

$$=+i\epsilon\frac{a_0}{2\alpha}f_{j+1/2}+\frac{1}{r}\frac{j+3/2}{\sqrt{2}}[(-2j-3)a_1+(2j+1)\sqrt{2}a_3]\,f_{j-1/2}, \tag{16.108}$$

$$\Big[-\sqrt{2}(j-1/2)b_1+(2j+1)b_3\Big]\Big(\frac{d}{dr}+\frac{1}{r}\Big)f_{j+1/2}$$

$$-\sqrt{2}\sqrt{(j+3/2)(j-1/2)}Big(\frac{d}{dr}-\frac{j-1/2}{r}\Big)b_2 f_{j-3/2}$$

$$=+i\epsilon\;\frac{b_0}{\beta}f_{j-1/2}+\frac{1}{r}\frac{j-1/2}{\sqrt{2}}\;[(2j-1)\,b_1-(2j+1)\sqrt{2}b_3]\;f_{j+1/2}. \tag{16.109}$$

Now, we transform the last equations to Bessel form (let it be $p=j+1$):

$$f_{j+1/2}=\sqrt{x}\,Z_{j+1}=\sqrt{x}\,Z_p\,,\quad f_{j+3/2}=\sqrt{x}\,Z_{j+2}=\sqrt{x}\,Z_{p+1}\,,$$
$$f_{j-1/2}=\sqrt{x}\,Z_j=\sqrt{x}\,Z_{p-1}\,,\quad f_{j-3/2}=\sqrt{x}\,Z_{j-1}=\sqrt{x}\,Z_{p-2}\,. \tag{16.110}$$

In this way we reduce the differential constraints to the form

$$[-(j+3/2)\sqrt{2}a_1+(2j+1)a_3](\frac{d}{dx}+\frac{1}{x})\sqrt{x}Z_{p-1}$$

$$-\sqrt{2}\sqrt{(j+3/2)(j-1/2)}(\frac{d}{dx}+\frac{j+3/2}{x})a_2\sqrt{x}Z_{p+1}$$

$$=+i\Gamma\frac{a_0}{2\alpha}\sqrt{x}Z_p+\frac{1}{x}\frac{j+3/2}{\sqrt{2}}[(-2j-3)a_1+(2j+1)\sqrt{2}a_3]\sqrt{x}Z_{p-1}, \qquad (16.111)$$

$$[-\sqrt{2}(j-1/2)b_1+(2j+1)b_3](\frac{d}{dx}+\frac{1}{x})\sqrt{x}Z_p$$

$$-\sqrt{2}\sqrt{(j+3/2)(j-1/2)}(\frac{d}{dx}-\frac{j-1/2}{x})b_2\sqrt{x}Z_{p-2}$$

$$=+i\Gamma\frac{b_0}{\beta}\sqrt{x}\,Z_{p-1}+\frac{1}{x}\frac{j-1/2}{\sqrt{2}}[(2j-1)b_1-(2j+1)\sqrt{2}b_3]\sqrt{x}Z_p, \qquad (16.112)$$

where the notation $\Gamma=\epsilon/\sqrt{\epsilon^2-m^2}$ is used. Further, by applying the above commutation rule, we derive

$$[-(j+3/2)\sqrt{2}a_1+(2j+1)a_3](\frac{d}{dx}+\frac{3/2}{x})Z_{p-1}$$
$$-\sqrt{2}\sqrt{(j+3/2)(j-1/2)}(\frac{d}{dx}+\frac{j+2}{x})a_2Z_{p+1}$$
$$=+i\Gamma\frac{a_0}{2\alpha}Z_p+\frac{1}{x}\frac{j+3/2}{\sqrt{2}}[(-2j-3)a_1+(2j+1)\sqrt{2}a_3]Z_{p-1}, \qquad (16.113)$$

$$[-\sqrt{2}(j-1/2)b_1+(2j+1)b_3](\frac{d}{dx}+\frac{3/2}{x})Z_p$$
$$-\sqrt{2}\sqrt{(j+3/2)(j-1/2)}(\frac{d}{dx}-\frac{j-1}{x})b_2Z_{p-2}$$
$$=i\Gamma\frac{b_0}{\beta}Z_{p-1}+\frac{1}{x}\frac{j-1/2}{\sqrt{2}}[(2j-1)b_1-(2j+1)\sqrt{2}b_3]Z_p. \qquad (16.114)$$

The second equation can be conveniently written with the use of the primed parameter:

$$p=j+1, p-1=j, p+1=j+2, \quad p-1=p'=j, p'+1=j+1, p'-1=j-1,$$

so we arrive at the two equations

$$[-(j+3/2)\sqrt{2}a_1+(2j+1)a_3](\frac{d}{dx}-\frac{p-1}{x})\,Z_{p-1}$$
$$-\sqrt{2}\,\sqrt{(j+3/2)(j-1/2)}(\frac{d}{dx}+\frac{p+1}{x})\,a_2\,Z_{p+1}=i\Gamma\frac{a_0}{2\alpha}\,Z_p,$$

$$[-(j-1/2)\sqrt{2}b_1+(2j+1)b_3](\frac{d}{dx}+\frac{p'+1}{x})\,Z_{p'+1}$$
$$-\sqrt{2}\sqrt{(j+3/2)(j-1/2)}(\frac{d}{dx}-\frac{p'-1}{x})b_2\,Z_{p'-1}=i\Gamma\frac{b_0}{\beta}\,Z_{p'},$$

whence, by applying the known properties of Bessel functions, we derive two linear equations for the numerical coefficients (we additionally write down the above two constraints as well)

$$\begin{gathered}\sqrt{2}\,a_1 - a_3 = -\sqrt{j-1/2}\,\sqrt{2}b_0\,,\\ \sqrt{2}\,b_1 - b_3 = +\sqrt{j+3/2}\,a_0\,,\\ i\frac{\Gamma}{2\alpha}\,a_0 - (j+3/2)\,\sqrt{2}a_1\\ +\sqrt{(j+3/2)(j-1/2)}\,\sqrt{2}a_2 + (2j+1)\,a_3 = 0\,,\\ i\frac{\Gamma}{\beta}\,b_0 + (j-1/2)\,\sqrt{2}b_1\\ -\sqrt{(j+3/2)(j-1/2)}\,\sqrt{2}b_2 - (2j+1)\,b_3 = 0\,.\end{gathered} \tag{16.115}$$

Let us remind that we have to remember the previously derived conditions

$$b_0 = -i\frac{\sqrt{\epsilon-m}}{\sqrt{\epsilon+m}}\,a_0\,,\quad \frac{1}{2\alpha} = 4(j+1)\sqrt{2j-1}\,,\quad \frac{1}{\beta} = -4j\sqrt{j+3/2}\,.$$

First, with the use of

$$a_0 = \frac{\sqrt{2}}{\sqrt{j+3/2}}b_1 - \frac{1}{\sqrt{j+3/2}}b_3\,,\quad b_0 = -\frac{a_1}{\sqrt{j-1/2}} + \frac{1}{\sqrt{2}\sqrt{j-1/2}}a_3\,, \tag{16.116}$$

let us exclude the variables a_0 and b_0 from last two equations in (16.115):

$$\begin{gathered} i\frac{\Gamma}{2\alpha}\frac{1}{\sqrt{j+3/2}}(\sqrt{2}b_1 - b_3) - (j+3/2)\sqrt{2}a_1\\ +\sqrt{(j+3/2)(j-1/2)}\sqrt{2}a_2 + (2j+1)\,a_3 = 0\,,\\ i\frac{\Gamma}{\beta}\frac{1}{\sqrt{j-1/2}}(-a_1 + \frac{a_3}{\sqrt{2}}) + (j-1/2)\sqrt{2}b_1\\ -\sqrt{(j+3/2)(j-1/2)}\sqrt{2}b_2 - (2j+1)b_3 = 0\,.\end{gathered} \tag{16.117}$$

In turn, due do the known relationship $\sqrt{\epsilon-m}\,a_0 = i\sqrt{\epsilon+m}\,b_0$, we get

$$\frac{\sqrt{\epsilon-m}}{\sqrt{j+3/2}}\,(\,\sqrt{2}b_1 - b_3\,) = \frac{i\sqrt{\epsilon+m}}{\sqrt{j-1/2}}\,(\,-a_1 + \frac{a_3}{\sqrt{2}}\,)\,. \tag{16.118}$$

This identity allows us to reduce the equations in (16.117) to the form, where the first contains only the variables a_i, and the second – only the variables b_i:

$$\begin{gathered}\frac{4\epsilon(j+1)}{\epsilon-m}(a_1 - \frac{a_3}{\sqrt{2}}) - (j+\frac{3}{2})a_1 + (2j+1)\frac{a_3}{\sqrt{2}} + \sqrt{(j+\frac{3}{2})(j-\frac{1}{2})}a_2 = 0\,,\\ \frac{4\epsilon j}{\epsilon+m}(b_1 - \frac{b_3}{\sqrt{2}}) - (j-\frac{1}{2})b_1 + (2j+1)\frac{b_3}{\sqrt{2}} + \sqrt{(j+\frac{3}{2})(j-\frac{1}{2})}b_2 = 0;\end{gathered} \tag{16.119}$$

remembering that coefficient a_0, b_0 may be found by means of (16.116). The general structure of Eqs. (16.119) is

$$A_1a_1 + A_2a_2 + A_3a_3 = 0\,,\qquad B_1b_1 + B_2b_2 + B_3b_3 = 0\,, \tag{16.120}$$

where

$$A_1 = \frac{4(j+1)\epsilon - (j+3/2)(\epsilon-m)}{\epsilon-m}\,,$$

$$A_3 = -\frac{1}{\sqrt{2}}\frac{4(j+1)\epsilon-(2j+1)(\epsilon-m)}{\epsilon-m},$$

$$A_2 = \sqrt{(j-1/2)(j+3/2)},$$

$$B_1 = \frac{4\epsilon j-(j-1/2)(\epsilon+m)}{\epsilon+m},$$

$$B_3 = -\frac{1}{\sqrt{2}}\frac{4\epsilon j-(2j+1)(\epsilon+m)}{\epsilon+m},$$

$$B_2 = \sqrt{(j-1/2)(j+3/2)}.$$

Having used the above established expressions for coefficients b_i through a_i, we may transform the second linear relation in (16.120) to the variables a_i:

$$\frac{(6\epsilon j+2jm+5\epsilon+3m)}{2(\epsilon-m)}a_1 - \frac{(2\epsilon j+2jm+3\epsilon+m)}{2(\epsilon-m)}\sqrt{2}a_3$$

$$+\frac{1}{4}\sqrt{4j+6}\sqrt{4j-2}\,a_2 = 0.$$

This equation coincides with the first one in (16.120). Therefore, in (16.119) we have only one independent linear constraint

$$A_1 a_1 + A_2 a_2 + A_3 a_3 = 0. \tag{16.121}$$

We are to fix two independent solutions of the main linear constraint for the coefficients a_i. In this way, we obtain two independent solutions of the equations for the spin 3/2 particle, and evidently there exist many possibilities for that choice. For simplicity, we may set

$$I,\quad a_1 = +\sqrt{2}a_3, \frac{2j\epsilon+(\epsilon+m)}{\epsilon-m}a_1 + \sqrt{(j+3/2)(j-1/2)}\,a_2 = 0,$$

$$II,\quad a_1 = -\sqrt{2}a_3, \frac{(2j+2)(\epsilon+m)}{\epsilon-m}a_1 + \sqrt{(j+3/2)(j-1/2)}\,a_2 = 0.$$

Let us fix the quantity $a_2 = -[\sqrt{(j+3/2)(j-1/2)}]^{-1}$, then the above formulas become shorter:

$$\begin{aligned} I,\quad a_1 &= \frac{\epsilon-m}{2j\epsilon+(\epsilon+m)},\ \sqrt{2}a_3 = +a_1,\ a_2 = -\frac{1}{\sqrt{(j+3/2)(j-1/2)}},\\ II,\quad a_1 &= \frac{\epsilon-m}{(2j+2)(\epsilon+m)},\ \sqrt{2}a_3 = -a_1,\ a_2 = -\frac{1}{\sqrt{(j+3/2)(j-1/2)}}. \end{aligned} \tag{16.122}$$

Chapter 17

Massless Spin 3/2 Field, Spherical Solutions, Exclusion of the Gauge Degrees of Freedom

17.1 Massless Spin 3/2 Particle, General Theory

In the theory of spin 3/2 field, a special interest has the massless case, when – as shown by Pauli and Fierz – there exists a specific gauge symmetry: the 4-gradient of the arbitrary bispinor function provides us with a solution for the massless field equation. In this chapter, we examine the problem of spherical solutions for the 16-component system of equations describing a massless spin 3/2 particle; we specify the gauge solutions with spherical symmetry and construct exact spherical solutions which do not contain gauge components.

We start with the generally covariant equation for a massless spin 3/2 particle [139]

$$\frac{i}{2}\gamma^5 \epsilon_\rho{}^{\mu\alpha\beta}\gamma_\mu(x)\,[\nabla_\alpha + \Gamma_\alpha]\,\Psi_\beta = 0\,, \tag{17.1}$$

and recall the designations for the local Dirac matrices and bispinor connection:

$$\gamma^\beta(x) = e^\beta_{(a)}(x)\gamma^a\,, \quad \eta^{ab} e_{(a)\alpha} e_{(b)\beta} = g_{\alpha\beta}(x)\,, \quad \Gamma_\alpha(x) = \frac{1}{2}(\sigma^{nm}) e^\nu_{(n)} e_{(m)\nu;\alpha}\,.$$

The covariant Levi-Civita symbol is defined by the following formula

$$\epsilon^{\alpha\beta\rho\sigma}(x) = \epsilon^{abcd} e^\alpha_{(a)}(x) e^\beta_{(b)}(x) e^\rho_{(c)}(x) e^\sigma_{(d)}(x)\,.$$

It is readily proved that if the Riemann curvature tensor for a space-time model equals to zero, then Eq. (17.1) admits the class of gradient-type solutions

$$\Psi^G_\beta(x) = D_\beta \Psi(x)\,, \quad D_\beta = (\nabla_\beta + \Gamma_\beta)\,,$$

where Φ stands for an arbitrary bispinor field. Indeed, due to the commutator relation (see [306])

$$(D_\alpha D_\beta - D_\beta D_\alpha)\Phi = \frac{1}{2}\sigma^{\mu\nu}(x) R_{\mu\nu\alpha\beta}(x) \equiv 0\,, \tag{17.2}$$

the equation (17.1) is identically satisfied by this gradient solution

$$\frac{i}{2}\gamma^5 \epsilon_\rho{}^{\mu\alpha\beta}\gamma_\mu(x) D_\alpha D_\beta \Phi = \frac{i}{4}\gamma^5 \epsilon_\rho{}^{\mu\alpha\beta}\gamma_\mu(x)[D_\alpha D_\beta - D_\beta D_\alpha]\Phi \equiv 0\,. \tag{17.3}$$

It will be convenient to use the field function with tetrad vector index $\Psi_l(x)$, related to the initial function $\Psi_\alpha(x)$, as follows

$$\Psi_\beta(x) = e^{(l)}_\beta(x)\Psi_l(x)\,, \quad \Psi_l(x) = e^\beta_{(l)}(x)\Psi_\beta(x)\,.$$

Correspondingly, equation (17.1) is transformed to

$$\frac{i}{2}\gamma^5 \epsilon_k^{\ bcd}\gamma_b\, e^\alpha_{(c)} \left[\left(\partial_\alpha + \frac{1}{2}(\sigma^{nm})e^\nu_{(n)}e_{(m)\nu;\alpha}\right)\delta^{(l)}_{(d)} + e^\beta_{(d)}e^{(l)}_{\beta;\alpha}\right]\Psi_l = 0\,. \tag{17.4}$$

Taking in mind the structure of the bispinor connection $\Gamma_\alpha = \frac{1}{2}(\sigma^{nm})e^\nu_{(n)}e_{(m)\nu;\alpha}$ and the explicit form of vector generators $(j^{ab})_d^{\ l} = \delta^a_d\, g^{bl} - \delta^b_d g^{al}$, we introduce the vector connection

$$\frac{1}{2}(j^{ab})_d^{\ l} e^\beta_{(a)}(\nabla_\alpha e_{(b)\beta}) = e^\beta_{(d)}e^{(l)}_{\beta;\alpha} = (L_\alpha)_d^{\ l}\,. \tag{17.5}$$

Therefore, Eq. (17.4) takes on the form

$$\frac{i}{2}\gamma^5 \epsilon_k^{\ bcd}\gamma_b \left[e^\nu_{(c)}\partial_\nu\delta_d^{\ l} + e^\nu_{(c)}\Gamma_\nu\delta_d^{\ l} + e^\nu_{(c)}(L_\nu)_d^{\ l}\right]\Psi_l = 0\,. \tag{17.6}$$

The last equation may be presented with the use of the Ricci rotation coefficients:

$$\frac{i}{2}\gamma^5 \epsilon_k^{\ bcd}\gamma_b \left[e^\alpha_{(c)}\partial_\alpha + \frac{1}{2}(\sigma^{nm}\otimes I + I\otimes j^{nm})\,\gamma_{[nm]c}\right]_d^{\ l}\Psi_l = 0\,, \tag{17.7}$$

where $\Psi(x)$ stands for the matrix with two indices – for bispinor A and for vector (l):

$$\Psi_{A(l)} = \left|\begin{array}{cccc} \Psi_{10} & \Psi_{11} & \Psi_{12} & \Psi_{13} \\ \Psi_{20} & \Psi_{21} & \Psi_{22} & \Psi_{23} \\ \Psi_{30} & \Psi_{31} & \Psi_{32} & \Psi_{33} \\ \Psi_{40} & \Psi_{41} & \Psi_{4(2} & \Psi_{43} \end{array}\right|\,. \tag{17.8}$$

It is convenient to handle the short form of Eq. (17.7):

$$\frac{i}{2}\gamma^5 \epsilon_k^{\ bcd}\gamma_b (D_c)_d^{\ l}\Psi_l = 0 \quad \Longrightarrow \quad \frac{i}{2}\gamma^5 \epsilon_k^{\ bcd}\gamma_b (D_c\Psi)_d = 0\,, \tag{17.9}$$

where we use the notation

$$D_c = e^\alpha_{(c)}\partial_\alpha + \frac{1}{2}(\sigma^{nm}\otimes I + I\otimes j^{nm})\,\gamma_{[nm]c}\,. \tag{17.10}$$

When constructing the spherically symmetric solutions for a massive spin 3/2 particle, the transition from vector index in Cartesian basis Ψ to the cyclic one $\bar{\Psi}$, was used. In such a cyclic basis it we find the general substitution for the field function which relates to the diagonalization of the square and third projection of the total angular momentum:

$$\bar{\Psi} = e^{-i\epsilon t}\left|\begin{array}{cccc} f_0(r)D_{-1/2} & f_1(r)D_{-3/2} & f_2(r)D_{-1/2} & f_3(r)D_{+1/2} \\ g_0(r)D_{+1/2} & g_1(r)D_{-1/2} & g_2(r)D_{+1/2} & g_3(r)D_{+3/2} \\ h_0(r)D_{-1/2} & h_1(r)D_{-3/2} & h_2(r)D_{-1/2} & h_3(r)D_{+1/2} \\ d_0(r)D_{+1/2} & d_1(r)D_{-1/2} & d_2(r)D_{+1/2} & d_3(r)D_{+3/2} \end{array}\right|\,, \tag{17.11}$$

where the symbols D designate the Wigner functions, $D_\sigma = D^j_{-m,\sigma}(\phi,\theta,0)\,;\ j = 1/2, 3/2, 5/2, \ldots$. It should be noted that at the minimal value $j = \frac{1}{2}$, the above substitution is simplified according to the relations: $j=\frac{1}{2}$, $f_1=0$, $g_3=0$, $h_1=0$, $d_3=0$. The connection between Cartesian Ψ and cyclic $\bar{\Psi}$ bases is determined by the formulas

$$\bar{\Psi}=(I\otimes U)\Psi\,,\; U=\left|\begin{array}{cccc} 1 & 0 & 0 & 0 \\ 0 & -\frac{1}{\sqrt{2}} & \frac{i}{\sqrt{2}} & 0 \\ 0 & 0 & 0 & 1 \\ 0 & \frac{1}{\sqrt{2}} & \frac{i}{\sqrt{2}} & 0 \end{array}\right|,\Psi=(I\otimes U^{-1})\bar{\Psi},U^{-1}=\left|\begin{array}{cccc} 1 & 0 & 0 & 0 \\ 0 & -\frac{1}{\sqrt{2}} & 0 & \frac{1}{\sqrt{2}} \\ 0 & -\frac{i}{\sqrt{2}} & 0 & -\frac{i}{\sqrt{2}} \\ 0 & 0 & 1 & 0 \end{array}\right|.$$

Let us transform the equation from the Cartesian basis to the cyclic one,

$$\frac{i}{2}\gamma^5\epsilon_k{}^{bcd}\gamma_b\, U^{-1}_{dp}U_{ps}\,(D_c)_s{}^l U^{-1}_{ln}\bar{\Psi}_n=0;$$

by multiplying the last equation by U_{sk}, we obtain

$$\frac{i}{2}\gamma^5\gamma_b\left\{U_{sk}\epsilon_k{}^{bcd}U^{-1}_{dp}\right\}\left\{U_{ps}(D_c)_s{}^l U^{-1}_{ln}\right\}\bar{\Psi}_n=0.$$

With the use of the special notations

$$U_{ps}(D_c)_s{}^l U^{-1}_{ln}=(\bar{D}_c)_p{}^n\,,\quad U_{sk}\epsilon_k{}^{bcd}U^{-1}_{dp}=U_{sk}\epsilon_k{}^{bcd}U^{-1}_{dp}=\bar{\epsilon}_s{}^{bcp}\,, \tag{17.12}$$

the wave equation in cyclic basis is presented as follows

$$\frac{i}{2}\gamma^5\gamma_b\bar{\epsilon}_s{}^{bcp}(\bar{D}_c)_p{}^n\bar{\Psi}_n=0. \tag{17.13}$$

Let us specify the expression for the component of the new Levi-Civita symbols $\bar{\epsilon}_s{}^{bcp}=U_{sk}\epsilon_k{}^{bcd}U^{-1}_{dp}$. It is convenient to apply the matrix notations, $\mu^{[bc]}_{kd}=\epsilon_k{}^{bcd}$, $\bar{\mu}^{[bc]}_{sp}=\bar{\epsilon}_s{}^{bcp}$; then we have the rule

$$\bar{\mu}^{[bc]}_{sp}=U_{sk}\,\mu^{[bc]}_{kd}\,U^{-1}_{dp}\,; \tag{17.14}$$

the square brackets mark the skew-symmetry in the two indices. Further taking in mind the formulas

$$\mu^{[01]}_{sp}=\left|\begin{array}{cccc} 0&0&0&0\\0&0&0&0\\0&0&0&-1\\0&0&1&0\end{array}\right|,\quad \mu^{[02]}_{sp}=\left|\begin{array}{cccc} 0&0&0&0\\0&0&0&1\\0&0&0&0\\0&-1&0&0\end{array}\right|,\quad \mu^{[03]}_{sp}=\left|\begin{array}{cccc} 0&0&0&0\\0&0&-1&0\\0&1&0&0\\0&0&0&0\end{array}\right|,$$

$$\mu^{[23]}_{sp}=\left|\begin{array}{cccc} 0&1&0&0\\1&0&0&0\\0&0&0&0\\0&0&0&0\end{array}\right|,\quad \mu^{[31]}_{sp}=\left|\begin{array}{cccc} 0&0&1&0\\0&0&0&0\\1&0&0&0\\0&0&0&0\end{array}\right|,\quad \mu^{[12]}_{sp}=\left|\begin{array}{cccc} 0&0&0&1\\0&0&0&0\\0&0&0&0\\1&0&0&0\end{array}\right|,$$

we find the explicit form for the matrices $\bar{\mu}^{[bc]}_{sp}$:

$$\bar{\mu}^{[01]}_{sp}=\frac{1}{\sqrt{2}}\left|\begin{array}{cccc} 0&0&0&0\\0&0&-i&0\\0&-i&0&-i\\0&0&-i&0\end{array}\right|,\bar{\mu}^{[02]}_{sp}=\frac{1}{\sqrt{2}}\left|\begin{array}{cccc} 0&0&0&0\\0&0&-1&0\\0&1&0&-1\\0&0&1&0\end{array}\right|,\bar{\mu}^{[03]}_{sp}=\left|\begin{array}{cccc} 0&0&0&0\\0&-i&0&0\\0&0&0&0\\0&0&0&i\end{array}\right|,$$

$$\bar{\mu}^{[23]}_{sp}=\frac{1}{\sqrt{2}}\left|\begin{array}{cccc} 0&-1&0&1\\-1&0&0&0\\0&0&0&0\\1&0&0&0\end{array}\right|,\bar{\mu}^{[31]}_{sp}=\frac{1}{\sqrt{2}}\left|\begin{array}{cccc} 0&-i&0&-i\\i&0&0&0\\0&0&0&0\\i&0&0&0\end{array}\right|,\bar{\mu}^{[12]}_{sp}=\left|\begin{array}{cccc} 0&0&1&0\\0&0&0&0\\1&0&0&0\\0&0&0&0\end{array}\right|.$$

Let us turn to Eq. (17.13), written as

$$\frac{i}{2}\gamma^5 \bar{\mu}^{[bc]}\gamma_b \bar{D}_c \bar{\Psi} = 0\,, \tag{17.15}$$

whence we derive

$$\begin{aligned}\frac{i}{2}\gamma^5 \Big\{ \Big(\gamma^1 \otimes \bar{\mu}^{[01]} + \gamma^2 \otimes \bar{\mu}^{[02]} + \gamma^3 \otimes \bar{\mu}^{[03]}\Big)\bar{D}_0 \bar{\Psi} \\ + \Big(\gamma^0 \otimes \bar{\mu}^{[03]} + \gamma^1 \otimes \bar{\mu}^{[31]} - \gamma^2 \otimes \bar{\mu}^{[23]}\Big)\bar{D}_3 \bar{\Psi} \\ + \Big(\gamma^0 \otimes \bar{\mu}^{[01]} + \gamma^2 \otimes \bar{\mu}^{[12]} - \gamma^3 \otimes \bar{\mu}^{[31]}\Big)\bar{D}_1 \bar{\Psi} \\ + \Big(\gamma^0 \otimes \bar{\mu}^{[02]} + \gamma^3 \otimes \bar{\mu}^{[23]} - \gamma^1 \otimes \bar{\mu}^{[12]}\Big)\bar{D}_2 \bar{\Psi}\Big\} = 0\,.\end{aligned} \tag{17.16}$$

17.2 Separation of the Variables

Let us consider the equation (17.16) in spherical coordinates $x^\alpha = (t, r, \theta, \phi)$:

$$dS^2 = dt^2 - dr^2 - r^2 d\theta^2 - \sin^2\theta d\phi^2\,, \quad g_{\alpha\beta} = \left|\begin{array}{cccc} 1 & 0 & 0 & 0 \\ 0 & -1 & 0 & 0 \\ 0 & 0 & -r^2 & 0 \\ 0 & 0 & 0 & -r^2\sin^2\theta \end{array}\right|, \tag{17.17}$$

and in spherical tetrad

$$e^\alpha_{(0)} = (1,0,0,0)\,, \quad e^\alpha_{(3)} = (0,1,0,0)\,, \quad e^\alpha_{(1)} = (0,0,\frac{1}{r},0)\,, \quad e^\alpha_{(2)} = (1,0,0,\frac{1}{r\sin\theta})\,. \tag{17.18}$$

To the tetrad (17.18) there correspond the following Ricci rotation coefficients

$$\gamma_{ab0} = 0, \gamma_{ab3} = 0\,,\ \gamma_{ab1} = \left|\begin{array}{cccc} 0 & 0 & 0 & 0 \\ 0 & 0 & 0 & -\frac{1}{r} \\ 0 & 0 & 0 & 0 \\ 0 & +\frac{1}{r} & 0 & 0 \end{array}\right|, \gamma_{ab2} = \left|\begin{array}{cccc} 0 & 0 & 0 & 0 \\ 0 & 0 & +\frac{\cos\theta}{r\sin\theta} & 0 \\ 0 & -\frac{\cos\theta}{r\sin\theta} & 0 & -\frac{1}{r} \\ 0 & 0 & +\frac{1}{r} & 0 \end{array}\right|. \tag{17.19}$$

For the components of the operator $\bar{D}_c$, we obtain the following expressions

$$\begin{aligned}\bar{D}_0 = \partial_t\,, \quad \bar{D}_3 = \partial_r\,, \quad \bar{D}_1 = \frac{1}{r}\partial_\theta + \frac{1}{r}(\sigma^{31} \otimes I + I \otimes \bar{j}^{31})\,, \\ \bar{D}_2 = \frac{1}{r}(\sigma^{32} \otimes I + I \otimes \bar{j}^{32}) + \frac{1}{r}\frac{\partial_\phi + \cos\theta(\sigma^{12} \otimes I + I \otimes \bar{j}^{12})}{\sin\theta}\,.\end{aligned} \tag{17.20}$$

The general substitution for the field function was given in (17.11). In [309], the restrictions for radial functions corresponding to the diagonalization of the spatial reflection operator were found

$$d_0 = \delta f_0\,, d_1 = \delta f_3\,, d_2 = \delta f_2\,, d_3 = \delta f_1\,, h_0 = \delta g_0\,, h_1 = \delta g_3\,, h_2 = \delta g_2\,, h_3 = \delta g_1\,, \tag{17.21}$$

so we have only 8 independent functions. Below, when separating the variables, we shall apply the known recurrent formulas for Wigner functions

$$
\begin{gathered}
\partial_\theta D_{+1/2} = \frac{1}{2}(a\,D_{-1/2} - b\,D_{+3/2}), \quad \partial_\theta D_{-1/2} = \frac{1}{2}(b\,D_{-3/2} - a\,D_{+1/2}), \\
\partial_\theta D_{+3/2} = \frac{1}{2}(b\,D_{+1/2} - c\,D_{+5/2}), \quad \partial_\theta D_{-3/2} = \frac{1}{2}(c\,D_{-5/2} - b\,D_{-1/2}), \\
\frac{1}{\sin\theta}(-m - \frac{1}{2}\cos\theta)D_{+1/2} = \frac{1}{2}(-a\,D_{-1/2} - b\,D_{+3/2}), \\
\frac{1}{\sin\theta}(-m + \frac{1}{2}\cos\theta)D_{-1/2} = \frac{1}{2}(-b\,D_{-3/2} - a\,D_{+1/2}), \\
\frac{1}{\sin\theta}(-m - \frac{3}{2}\cos\theta)D_{+3/2} = \frac{1}{2}(-b\,D_{+1/2} - c\,D_{+5/2}), \\
\frac{1}{\sin\theta}(-m + \frac{3}{2}\cos\theta)D_{-3/2} = \frac{1}{2}(-c\,D_{-5/2} - b\,D_{-1/2}),
\end{gathered}
\tag{17.22}
$$

where

$$
a = j + 1/2, \quad b = \sqrt{(j-1/2)(j+3/2)}, \quad c = \sqrt{(j-3/2)(j+5/2)}.
$$

We shall also need the explicit form of the Dirac matrices (in spinor basis) and of the generators:

$$
\gamma^0 = \begin{vmatrix} 0 & 0 & 1 & 0 \\ 0 & 0 & 0 & 1 \\ 1 & 0 & 0 & 0 \\ 0 & 1 & 0 & 0 \end{vmatrix}, \gamma^1 = \begin{vmatrix} 0 & 0 & 0 & -1 \\ 0 & 0 & -1 & 0 \\ 0 & 1 & 0 & 0 \\ 1 & 0 & 0 & 0 \end{vmatrix}, \gamma^2 = \begin{vmatrix} 0 & 0 & 0 & i \\ 0 & 0 & -i & 0 \\ 0 & -i & 0 & 0 \\ i & 0 & 0 & 0 \end{vmatrix}, \gamma^3 = \begin{vmatrix} 0 & 0 & -1 & 0 \\ 0 & 0 & 0 & 1 \\ 1 & 0 & 0 & 0 \\ 0 & -1 & 0 & 0 \end{vmatrix},
$$

$$
\sigma^{12} = \frac{1}{2}\begin{vmatrix} -i & 0 & 0 & 0 \\ 0 & i & 0 & 0 \\ 0 & 0 & -i & 0 \\ 0 & 0 & 0 & i \end{vmatrix}, \sigma^{31} = \frac{1}{2}\begin{vmatrix} 0 & -1 & 0 & 0 \\ 1 & 0 & 0 & 0 \\ 0 & 0 & 0 & -1 \\ 0 & 0 & 1 & 0 \end{vmatrix}, \sigma^{32} = \frac{1}{2}\begin{vmatrix} 0 & i & 0 & 0 \\ i & 0 & 0 & 0 \\ 0 & 0 & 0 & i \\ 0 & 0 & i & 0 \end{vmatrix},
$$

$$
\bar{j}^{12} = \begin{vmatrix} 0 & 0 & 0 & 0 \\ 0 & -i & 0 & 0 \\ 0 & 0 & 0 & 0 \\ 0 & 0 & 0 & i \end{vmatrix}, \bar{j}^{31} = \frac{1}{\sqrt{2}}\begin{vmatrix} 0 & 0 & 0 & 0 \\ 0 & 0 & -1 & 0 \\ 0 & 1 & 0 & -1 \\ 0 & 0 & 1 & 0 \end{vmatrix}, \bar{j}^{23} = \frac{1}{\sqrt{2}}\begin{vmatrix} 0 & 0 & 0 & 0 \\ 0 & 0 & -i & 0 \\ 0 & -i & 0 & -i \\ 0 & 0 & -i & 0 \end{vmatrix}.
$$

Now let us turn to Eq. (17.16); in explicit form it reads (we enumerate the separate terms):

$$
\begin{aligned}
&1) \quad \frac{i}{2}\gamma^5\Big\{ - i\epsilon\Big(\gamma^1 \otimes \bar{\mu}^{[01]} + \gamma^2 \otimes \bar{\mu}^{[02]} + \gamma^3 \otimes \bar{\mu}^{[03]}\Big)\bar{\Psi} \\
&2) \quad + \Big(\gamma^0 \otimes \bar{\mu}^{[03]} + \gamma^1 \otimes \bar{\mu}^{[31]} - \gamma^2 \otimes \bar{\mu}^{[23]}\Big)\partial_r \bar{\Psi} \\
&3) \quad + \frac{1}{r}(\gamma^0\sigma^{31} \otimes \bar{\mu}^{01} + \gamma^2\sigma^{31} \otimes \bar{\mu}^{12} - \gamma^3\sigma^{31} \otimes \bar{\mu}^{31})\bar{\Psi} \\
&4) \quad + \frac{1}{r}(\gamma^0 \otimes \bar{\mu}^{01}\bar{j}^{31} + \gamma^2 \otimes \bar{\mu}^{12}\bar{j}^{31} - \gamma^3 \otimes \bar{\mu}^{31}\bar{j}^{31})\bar{\Psi} \\
&5) \quad + \frac{1}{r}\Big(\gamma^0\sigma^{32} \otimes \bar{\mu}^{02} + \gamma^3\sigma^{32} \otimes \bar{\mu}^{23} - \gamma^1\sigma^{32} \otimes \bar{\mu}^{12}\Big)\bar{\Psi} \\
&6) \quad + \frac{1}{r}\Big(\gamma^0 \otimes \bar{\mu}^{02}\bar{j}^{32} + \gamma^3 \otimes \bar{\mu}^{23}\bar{j}^{32} - \gamma^1 \otimes \bar{\mu}^{12}\bar{j}^{32}\Big)\bar{\Psi} \\
&7) \quad + \frac{1}{r}\Big(\gamma^0 \otimes \bar{\mu}^{[01]} + \gamma^2 \otimes \bar{\mu}^{[12]} - \gamma^3 \otimes \bar{\mu}^{[31]}\Big)\partial_\theta \bar{\Psi} \\
&8) \quad + \frac{1}{r}\Big(\gamma^0 \otimes \bar{\mu}^{[02]} + \gamma^3 \otimes \bar{\mu}^{[23]} - \gamma^1 \otimes \bar{\mu}^{[12]}\Big)\frac{\partial_\phi + \cos\theta(\sigma^{12} \otimes I + I \otimes \bar{j}^{12})}{\sin\theta}\bar{\Psi} = 0.
\end{aligned}
\tag{17.23}
$$

In order to ease the calculations, we use the shortening notations:

$$D_{-1/2}=A,\quad D_{+1/2}=B,\quad D_{-3/2}=C,\quad D_{+3/2}=D,\quad D_{-5/2}=K,\quad D_{+5/2}=L,$$

and then the substitution of the field function reads (the multiplier $e^{-i\epsilon t}$ is omitted):

$$\bar{\Psi}=\left|\begin{array}{cccc} f_0D_{-1/2} & f_1D_{-3/2} & f_2D_{-1/2} & f_3D_{+1/2} \\ g_0D_{+1/2} & g_1D_{-1/2} & g_2D_{+1/2} & g_3D_{+3/2} \\ h_0D_{-1/2} & h_1D_{-3/2} & h_2D_{-1/2} & h_3D_{+1/2} \\ d_0D_{+1/2} & d_1D_{-1/2} & d_2D_{+1/2} & d_3D_{+3/2} \end{array}\right| = \left|\begin{array}{cccc} f_0A & f_1C & f_2A & f_3B \\ g_0B & g_1A & g_2B & g_3D \\ h_0A & h_1C & h_2A & h_3B \\ d_0B & d_1A & d_2B & d_3D \end{array}\right|. \tag{17.24}$$

After rather long calculations, we arrive at the system of 16 radial equations

$$i\sqrt{2}\frac{d}{dr}d_1+\frac{i}{r}(h_2+\frac{3}{\sqrt{2}}d_1)+\frac{1}{\sqrt{2}r}(ibh_1-iah_3+ia\sqrt{2}d_2)=0,$$

$$i\sqrt{2}\frac{d}{dr}h_3+\frac{i}{r}(d_2+\frac{3}{\sqrt{2}}h_3)+\frac{1}{\sqrt{2}r}(-iad_1+ibd_3+ia\sqrt{2}h_2)=0,$$

$$i\sqrt{2}\frac{d}{dr}g_1+\frac{i}{r}(f_2+\frac{3}{\sqrt{2}}g_1)+\frac{1}{\sqrt{2}r}(ibf_1-iaf_3+ia\sqrt{2}g_2)=0,$$

$$i\sqrt{2}\frac{d}{dr}f_3+\frac{i}{r}(g_2+\frac{3}{\sqrt{2}}f_3)+\frac{1}{\sqrt{2}r}(-iag_1+ibg_3+ia\sqrt{2}f_2)]=0,$$

$$\epsilon h_1-i\frac{d}{dr}h_1-\frac{i}{r}h_1+\frac{1}{\sqrt{2}r}(-ibh_2+ibh_0)=0,$$

$$\epsilon(\sqrt{2}h_2-d_1)+(-i\sqrt{2}\frac{d}{dr}h_0-i\frac{d}{dr}d_1)-\frac{i}{r}(\frac{1}{\sqrt{2}}(h_0+h_2)+d_1)+\frac{1}{\sqrt{2}r}(-iad_2-iad_0)=0,$$

$$\epsilon f_1+i\frac{d}{dr}f_1+\frac{i}{r}f_1+\frac{1}{\sqrt{2}r}(ibf_2+ibf_0)=0,$$

$$\epsilon(\sqrt{2}f_2-g_1)+(-i\sqrt{2}\frac{d}{dr}f_0+i\frac{d}{dr}g_1)-\frac{i}{r}(\frac{1}{\sqrt{2}}(f_0-f_2)-g_1)+\frac{1}{\sqrt{2}r}(iag_2-iag_0)=0,$$

$$\epsilon\sqrt{2}d_1+\frac{i}{r}(h_0+\frac{1}{\sqrt{2}}d_1)+\frac{1}{\sqrt{2}r}(ibh_1-iah_3+ia\sqrt{2}d_0)=0,$$

$$\epsilon\sqrt{2}h_3+\frac{i}{r}(d_0-\frac{1}{\sqrt{2}}h_3)+\frac{1}{\sqrt{2}r}(iad_1-ibd_3+ia\sqrt{2}h_0)=0,$$

$$\epsilon\sqrt{2}g_1+\frac{i}{r}(f_0-\frac{1}{\sqrt{2}}g_1)+\frac{1}{\sqrt{2}r}(-ibf_1+iaf_3+ia\sqrt{2}g_0)=0,$$

$$\epsilon\sqrt{2}f_3+\frac{i}{r}(g_0+\frac{1}{\sqrt{2}}f_3)+\frac{1}{\sqrt{2}r}(-iag_1+ibg_3+ia\sqrt{2}f_0)=0,$$

$$\epsilon(\sqrt{2}d_2-h_3)+(-i\sqrt{2}\frac{d}{dr}d_0+i\frac{d}{dr}h_3)-\frac{i}{r}(\frac{1}{\sqrt{2}}(d_0-d_2)-h_3)+\frac{1}{\sqrt{2}r}(iah_2-iah_0)=0,$$

$$\epsilon d_3+i\frac{d}{dr}d_3+\frac{i}{r}d_3+\frac{1}{\sqrt{2}r}(ibd_2+ibd_0)=0,$$

$$\epsilon(\sqrt{2}g_2-f_3)+(-i\sqrt{2}\frac{d}{dr}g_0-i\frac{d}{dr}f_3)-\frac{i}{r}(\frac{1}{\sqrt{2}}(g_0+g_2)+f_3)+\frac{1}{\sqrt{2}r}(-iaf_2-iaf_0)=0,$$

$$\epsilon g_3-i\frac{d}{dr}g_3-\frac{i}{r}g_3+\frac{1}{\sqrt{2}r}(-ibg_2+ibg_0)=0.$$

Taking into account restrictions (17.23), we obtain only 8 different equations:

$$\sqrt{2}\frac{d}{dr}g_1+\frac{1}{r}(f_2+\frac{3}{\sqrt{2}}g_1)+\frac{1}{\sqrt{2}r}(bf_1-af_3+a\sqrt{2}g_2)=0\,,$$

$$\sqrt{2}\frac{d}{dr}f_3+\frac{1}{r}(g_2+\frac{3}{\sqrt{2}}f_3)+\frac{1}{\sqrt{2}r}(-ag_1+bg_3+a\sqrt{2}f_2)=0\,,$$

$$-i\epsilon f_1+\frac{d}{dr}f_1+\frac{1}{r}f_1+\frac{1}{\sqrt{2}r}(bf_2+bf_0)=0\,,$$

$$-i\epsilon(\sqrt{2}f_2-g_1)+(-\sqrt{2}\frac{d}{dr}f_0+\frac{d}{dr}g_1)-\frac{1}{r}(\frac{1}{\sqrt{2}}(f_0-f_2)-g_1)+\frac{1}{\sqrt{2}r}(ag_2-ag_0)=0\,,$$

$$-i\epsilon\sqrt{2}g_1+\frac{1}{r}(f_0-\frac{1}{\sqrt{2}}g_1)+\frac{1}{\sqrt{2}r}(-bf_1+af_3+a\sqrt{2}g_0)=0\,,$$

$$-i\epsilon\sqrt{2}f_3+\frac{1}{r}(g_0+\frac{1}{\sqrt{2}}f_3)+\frac{1}{\sqrt{2}r}(-ag_1+bg_3+a\sqrt{2}f_0)=0\,,$$

$$-i\epsilon(\sqrt{2}g_2-f_3)+(-\sqrt{2}\frac{d}{dr}g_0-\frac{d}{dr}f_3)-\frac{1}{r}(\frac{1}{\sqrt{2}}(g_0+g_2)+f_3)+\frac{1}{\sqrt{2}r}(-af_2-af_0)=0\,,$$

$$-i\epsilon g_3-\frac{d}{dr}g_3-\frac{1}{r}g_3+\frac{1}{\sqrt{2}r}(-bg_2+bg_0)=0\,. \qquad (17.25)$$

Let us introduce the new combinations of functions

$$F_0=f_0+g_0\,,\quad G_0=f_0-g_0\,,\quad F_1=f_1+g_3\,,\quad G_1=f_1-g_3\,,$$
$$F_2=f_2+g_2\,,\quad G_2=f_2-g_2\,,\quad F_3=f_3+g_1\,,\quad G_3=f_3-g_1\,; \qquad (17.26)$$

also to simplify equations let us separate the simple multiplier $\frac{1}{r}$ at all radial functions:

$$f\Longrightarrow\frac{1}{r}f\,,\quad \left(\frac{d}{dr}+\frac{1}{r}\right)\frac{1}{r}f\Longrightarrow\frac{1}{r}\frac{d}{dr}f\,;$$

for simplicity we retain the same designations for the new radial functions. After performing the needed calculations, we obtain the new system

$$+\sqrt{2}(\frac{d}{dr}+\frac{1}{2r}-\frac{a}{2r})F_3+(\frac{1}{r}+\frac{a}{r})F_2+\frac{b}{\sqrt{2}r}F_1=0\,,$$

$$-\sqrt{2}(\frac{d}{dr}+\frac{1}{2r}+\frac{a}{2r})G_3+(\frac{1}{r}-\frac{a}{r})G_2+\frac{b}{\sqrt{2}r}G_1=0\,,$$

$$-i\epsilon\sqrt{2}F_3+(\frac{1}{r}+\frac{a}{r})F_0+(\frac{1}{\sqrt{2}r}+\frac{a}{\sqrt{2}r})G_3-\frac{b}{\sqrt{2}r}G_1=0\,,$$

$$i\epsilon\sqrt{2}G_3+(\frac{1}{r}-\frac{a}{r})G_0+(-\frac{1}{\sqrt{2}r}+\frac{a}{\sqrt{2}r})F_3-\frac{b}{\sqrt{2}r}F_1=0\,,$$

$$-i\epsilon F_1+\frac{d}{dr}G_1+\frac{b}{\sqrt{2}r}G_2+\frac{b}{\sqrt{2}r}F_0=0\,,\qquad -i\epsilon G_1+\frac{d}{dr}F_1+\frac{b}{\sqrt{2}r}F_2+\frac{b}{\sqrt{2}r}G_0=0\,,$$

$$-i\epsilon\sqrt{2}F_2+i\epsilon F_3-\sqrt{2}(\frac{d}{dr}-\frac{1}{2r}+\frac{a}{2r})F_0-\frac{d}{dr}G_3+(\frac{1}{\sqrt{2}r}-\frac{a}{\sqrt{2}r})G_2=0\,,$$

$$-i\epsilon\sqrt{2}G_2-i\epsilon G_3-\sqrt{2}(\frac{d}{dr}-\frac{1}{2r}-\frac{a}{2r})G_0+\frac{d}{dr}F_3+(\frac{1}{\sqrt{2}r}+\frac{a}{\sqrt{2}r})F_2=0\,. \qquad (17.27)$$

In the next section we will find the explicit form for the gauge solutions of the system of equations for the massless spin 3/2 particle. The corresponding radial functions must identically satisfy the general system (17.27). Besides, the gauge solutions will further be important in solving the general radial system.

17.3 Gradient Type Solutions

Let Φ be an arbitrary spherically symmetric bispinor field. We may assume that this bispinor is a solution of the Dirac equation; however, this requirement is not necessary because – as we shall later see – only the general structure of such bispinors is substantial,

$$\Phi = e^{-i\epsilon t} \left| \begin{array}{c} K_1 D_{-1/2} \\ K_2 D_{+1/2} \\ K_3 D_{-1/2} \\ K_4 D_{+1/2} \end{array} \right| ; \qquad (17.28)$$

solutions of the gradient type are determined by the relation

$$\Psi^g_c = D_c \Phi \,, \; D_c = e^{\alpha}_{(c)} (\partial_\alpha + \Gamma_\alpha \,.$$

In this way we get

$$\Psi^g_0 = D_0 \Phi = -i\epsilon \left| \begin{array}{c} K_1 D_{-1/2} \\ K_2 D_{+1/2} \\ K_3 D_{-1/2} \\ K_4 D_{+1/2} \end{array} \right| , \quad \Psi^g_3 = D_3 \Phi \left| \begin{array}{c} K_1' D_{-1/2} \\ K_2' D_{+1/2} \\ K_3' D_{-1/2} \\ K_4' D_{+1/2} \end{array} \right| ,$$

$$\Psi^g_1 = D_1 \Phi = \Big(\frac{1}{r} \partial_\theta + \frac{1}{r} \sigma^{31} \Big) \Phi = \frac{1}{r} \left| \begin{array}{c} K_1 \partial_\theta D_{-1/2} \\ K_2 \partial_\theta D_{+1/2} \\ K_3 \partial_\theta D_{-1/2} \\ K_4 \partial_\theta D_{+1/2} \end{array} \right| + \frac{1}{2r} \left| \begin{array}{c} -K_2 D_{+1/2} \\ K_1 D_{-1/2} \\ -K_4 D_{+1/2} \\ K_3 D_{-1/2} \end{array} \right| ,$$

$$\Psi^g_2 = D_2 \Phi = \Big(\frac{1}{r} \sigma^{32} + \frac{1}{r \sin\theta} (\partial_\phi + \cos\theta \sigma^{12}) \Big) \left| \begin{array}{c} K_1 D_{-1/2} \\ K_2 D_{+1/2} \\ K_3 D_{-1/2} \\ K_4 D_{+1/2} \end{array} \right|$$

$$= \frac{i}{2r} \left| \begin{array}{c} K_2 D_{+1/2} \\ K_1 D_{-1/2} \\ K_4 D_{+1/2} \\ K_3 D_{-1/2} \end{array} \right| - \frac{i}{r} \frac{1}{\sin\theta} \left| \begin{array}{c} K_1 (-m + \cos\theta \frac{1}{2}) D_{-1/2} \\ K_2 (-m - \cos\theta \frac{1}{2}) D_{+1/2} \\ K_3 (-m + \cos\theta \frac{1}{2}) D_{-1/2} \\ K_4 (-m - \cos\theta \frac{1}{2}) D_{+1/2} \end{array} \right| .$$

Taking in mind the recurrent formulas

$$\partial_\theta D_{+1/2} = \alpha D_{-1/2} - \beta D_{+3/2} \,, \quad \partial_\theta D_{-1/2} = \beta D_{-3/2} - \alpha D_{+1/2} \,,$$

$$\frac{-m - 1/2 \cos\theta}{\sin\theta} D_{+1/2} = (-\alpha D_{-1/2} - \beta D_{+3/2}) \,,$$

$$\frac{-m + 1/2 \cos\theta}{\sin\theta} D_{-1/2} = (-\beta D_{-3/2} - \alpha D_{+1/2}) \,,$$

where $2\alpha = (j + 1/2), 2\beta = \sqrt{(j - 1/2)(j + 3/2)}$; we derive the expressions for $\Phi^g_{(1)}$ and $\Phi^g_{(2)}$:

$$\Psi^g_1 = \frac{1}{r} \left| \begin{array}{c} K_1 (\beta D_{-3/2} - \alpha D_{+1/2}) \\ K_2 (\alpha D_{-1/2} - \beta D_{+3/2}) \\ K_3 (\beta D_{-3/2} - \alpha D_{+1/2}) \\ K_4 (\alpha D_{-1/2} - \beta D_{+3/2}) \end{array} \right| + \frac{1}{2r} \left| \begin{array}{c} -K_2 D_{+1/2} \\ K_1 D_{-1/2} \\ -K_4 D_{+1/2} \\ K_3 D_{-1/2} \end{array} \right| ,$$

$$\Psi^g_2 = -\frac{i}{r} \left| \begin{array}{c} K_1 (-\beta D_{-3/2} - \alpha D_{+1/2}) \\ K_2 (-\alpha D_{-1/2} - \beta D_{+3/2}) \\ K_3 (-\beta D_{-3/2} - \alpha D_{+1/2}) \\ K_4 (-\alpha D_{-1/2} - \beta D_{+3/2}) \end{array} \right| + \frac{i}{2r} \left| \begin{array}{c} K_2 D_{+1/2} \\ K_1 D_{-1/2} \\ K_4 D_{+1/2} \\ K_3 D_{-1/2} \end{array} \right| . \qquad (17.29)$$

Let us combine the components Ψ^g_a in order to get the corresponding quantities in the cyclic basis:

$$\bar{\Psi}^g_0 = \Psi^g_0 = -i\epsilon \left| \begin{array}{c} K_1 D_{-1/2} \\ K_2 D_{+1/2} \\ K_3 D_{-1/2} \\ K_4 D_{+1/2} \end{array} \right| , \qquad \bar{\Phi}^g_2 = \Phi_3 = \left| \begin{array}{c} K'_1 D_{-1/2} \\ K'_2 D_{+1/2} \\ K'_3 D_{-1/2} \\ K'_4 D_{+1/2} \end{array} \right| ,$$

$$\bar{\Psi}^g_1 = \frac{1}{\sqrt{2}}(-\Psi^g_1 + i\Psi^g_2) = \frac{1}{r}\frac{1}{\sqrt{2}} \left| \begin{array}{c} -K_1 2\beta D_{-3/2} \\ -K_2 2\alpha D_{-1/2} \\ -K_3 2\beta D_{-3/2} \\ -K_4 2\alpha D_{-1/2} \end{array} \right| + \frac{1}{r}\frac{1}{\sqrt{2}} \left| \begin{array}{c} 0 \\ -K_1 D_{-1/2} \\ 0 \\ -K_3 D_{-1/2} \end{array} \right| , \tag{17.30}$$

$$\bar{\Psi}^g_3 = \frac{1}{\sqrt{2}}(\Psi^g_1 + i\Psi^g_2 = \frac{1}{r}\frac{1}{\sqrt{2}} \left| \begin{array}{c} -K_1 2\alpha D_{+1/2} \\ -K_2 2\beta D_{+3/2} \\ -K_3 2\alpha D_{+1/2} \\ -K_4 2\beta D_{+3/2} \end{array} \right| + \frac{1}{r}\frac{1}{\sqrt{2}} \left| \begin{array}{c} -K_2 D_{+1/2} \\ 0 \\ -K_4 D_{+1/2} \\ 0 \end{array} \right| .$$

It is known that the space reflection operator for the bispinor field permits us to divide the solutions into two types; this is reached by imposing the following restrictions

$$\delta = \pm 1, \quad K_4 = \Delta K_1, \quad K_3 = \Delta K_2 ; \tag{17.31}$$

then instead of (17.30) we have

$$\bar{\Psi}^g_0 = \left| \begin{array}{c} -i\epsilon K_1 A \\ -i\epsilon K_2 B \\ -i\epsilon \Delta K_2 A \\ -i\epsilon \Delta K_1 B \end{array} \right| , \quad \bar{\Psi}^g_1 = \frac{1}{\sqrt{2}r} \left| \begin{array}{c} -2\beta K_1 C \\ -(2\alpha K_2 + K_1) A \\ -2\beta \Delta K_2 C \\ -\Delta(2\alpha K_1 + K_2) A \end{array} \right| ,$$
$$\bar{\Psi}^g_2 = \left| \begin{array}{c} K'_1 A \\ K'_2 B \\ \Delta K'_2 A \\ \Delta K'_1 B \end{array} \right| , \quad \bar{\Psi}_3 = \frac{1}{\sqrt{2}r} \left| \begin{array}{c} -(2\alpha K_1 + K_2) B \\ -2\beta K_2 D \\ -\Delta(2\alpha K_2 + K_1) B \\ -2\beta \Delta K_1 D \end{array} \right| . \tag{17.32}$$

The general structure of this gauge solution may be symbolically presented as follows (recall the substitution for the general vector-bispinor wave functions):

$$\bar{\Psi}^g = \left| \begin{array}{cccc} \bar{f}_0 A & \bar{f}_1 C & \bar{f}_2 A & \bar{f}_3 B \\ \bar{g}_0 B & \bar{g}_1 A & \bar{g}_2 B & \bar{g}_3 D \\ \Delta\bar{g}_0 A & \Delta\bar{g}_3 C & \Delta\bar{g}_2 A & \Delta\bar{g}_1 B \\ \Delta\bar{f}_0 B & \Delta\bar{f}_3 A & \Delta\bar{f}_2 B & \Delta\bar{f}_1 D \end{array} \right| , \quad \Psi = \left| \begin{array}{cccc} f_0 A & f_1 C & f_2 A & f_3 B \\ g_0 B & g_1 A & g_2 B & g_3 D \\ \delta g_0 A & \delta g_3 C & \delta g_2 A & \delta g_1 B \\ \delta f_0 B & \delta f_3 A & \delta f_2 B & \delta f_1 D \end{array} \right| ; \tag{17.33}$$

we note that these structures coincide if $\Delta \equiv \delta$. It is convenient to write down the radial functions from (17.32) explicitly,

$$\begin{array}{ll} \bar{f}_0 = -i\epsilon K_1 , & \bar{g}_0 = -i\epsilon K_2 , \\ \bar{f}_1 = -\frac{2\beta}{\sqrt{2}r} K_1 , & \bar{g}_1 = -\frac{1}{\sqrt{2}r}(2\alpha K_2 + K_1) , \\ \bar{f}_2 = K'_1 , & \bar{g}_2 = K'_2 , \\ \bar{f}_3 = -\frac{1}{\sqrt{2}r}(2\alpha K_1 + K_2) , & \bar{g}_3 = -\frac{1}{\sqrt{2}r} 2\beta K_2 . \end{array} \tag{17.34}$$

By direct calculations, we can prove that Eqs. (17.25) are identically satisfied by the functions from (17.34). It should be emphasized that in this proof, the explicit form of the radial functions from (17.34) is not used. After translating the formulas (17.34) to the variables defined in (17.26), we obtain (remembering that in all the functions we have separated the multiplier $\frac{1}{r}$)

$$\begin{aligned}
&\bar{F}_0 = -i\epsilon(K_1+K_2), && G_0 = -i\epsilon(K_1-K_2),\\
&\bar{F}_1 = -\tfrac{2\beta}{\sqrt{2}r}(K_1+K_2), && \bar{G}_1 = -\tfrac{2\beta}{\sqrt{2}r}(K_1-K_2),\\
&\bar{F}_2 = (\tfrac{d}{dr}-\tfrac{1}{r})(K_1+K_2), && \bar{G}_2 = (\tfrac{d}{dr}-\tfrac{1}{r})(K_1-K_2),\\
&\bar{F}_3 = -\tfrac{1}{\sqrt{2}r}(2\alpha+1)(K_1+K_2), && \bar{G}_3 = -\tfrac{1}{\sqrt{2}r}(2\alpha-1)(K_1-K_2).
\end{aligned} \tag{17.35}$$

By direct calculations, we can prove that Eqs. (17.27) are identically satisfied by the functions from (17.35).

17.4 Solving the System of Radial Equations

Let us turn to the system (17.27). It is convenient to simplify the explicit form of these equations by changing the variables (for simplicity, the associated notations are preserved):

$$\begin{gathered}
\frac{F_0}{\sqrt{2}} \Rightarrow F_0, \quad \frac{G_0}{\sqrt{2}} \Rightarrow G_0, \quad \frac{F_2}{\sqrt{2}} \Rightarrow F_2, \quad \frac{G_2}{\sqrt{2}} \Rightarrow G_2,\\
F_1 \Rightarrow F_1, \quad G_1 \Rightarrow G_1, \quad F_3 \Rightarrow F_3, \quad G_3 \Rightarrow G_3.
\end{gathered} \tag{17.36}$$

So the system (17.27) takes the form

$$\begin{gathered}
(\frac{d}{dr}+\frac{1-a}{2r})F_3+\frac{1+a}{r}F_2+\frac{b}{2r}F_1=0,\\
-2i\epsilon G_2-i\epsilon G_3-2(\frac{d}{dr}-\frac{1+a}{2r})G_0+\frac{d}{dr}F_3+\frac{1+a}{r}F_2=0,\\
-(\frac{d}{dr}+\frac{1+a}{2r})G_3+\frac{1-a}{r}G_2+\frac{b}{2r}G_1=0,\\
-2i\epsilon F_2+i\epsilon F_3-2(\frac{d}{dr}-\frac{1-a}{2r})F_0-\frac{d}{dr}G_3+\frac{1-a}{r}G_2=0,\\
-i\epsilon F_1+\frac{d}{dr}G_1+\frac{b}{r}G_2+\frac{b}{r}F_0=0, \quad -i\epsilon G_1+\frac{d}{dr}F_1+\frac{b}{r}F_2+\frac{b}{r}G_0=0,\\
-i\epsilon F_3+\frac{1+a}{r}F_0+\frac{1+a}{2r}G_3-\frac{b}{2r}G_1=0, \quad i\epsilon G_3+\frac{1-a}{r}G_0-\frac{1-a}{2r}F_3-\frac{b}{2r}F_1=0.
\end{gathered}$$

Let the first equation remains the same; from the first equation we subtract the second; the third equation remains the same; from the third equation we subtract the fourth; other four equations remain the same. In this way we obtain

$$\begin{aligned}
&1)\quad (\frac{d}{dr}+\frac{1-a}{2r})F_3+\frac{1+a}{r}F_2+\frac{b}{2r}F_1=0,\\
&2)\quad 2(\frac{d}{dr}-\frac{1+a}{2r})G_0+\frac{1-a}{2r}F_3+\frac{b}{2r}F_1+2i\epsilon G_2+i\epsilon G_3=0,\\
&3)\quad -(\frac{d}{dr}+\frac{1+a}{2r})G_3+\frac{1-a}{r}G_2+\frac{b}{2r}G_1=0,\\
&4)\quad 2(\frac{d}{dr}-\frac{1-a}{2r})F_0-\frac{1+a}{2r}G_3+\frac{b}{2r}G_1+2i\epsilon F_2-i\epsilon F_3=0,
\end{aligned} \tag{17.37}$$

$$5) \quad -i\epsilon F_1 + \frac{d}{dr}G_1 + \frac{b}{r}G_2 + \frac{b}{r}F_0 = 0, \quad 6) \quad -i\epsilon G_1 + \frac{d}{dr}F_1 + \frac{b}{r}F_2 + \frac{b}{r}G_0 = 0,$$

$$7) \quad -i\epsilon F_3 + \frac{1+a}{r}F_0 + \frac{1+a}{2r}G_3 - \frac{b}{2r}G_1 = 0, \quad 8) \quad i\epsilon G_3 + \frac{1-a}{r}G_0 - \frac{1-a}{2r}F_3 - \frac{b}{2r}F_1 = 0.$$

From the equations 2) and 4), we can find the expressions for F_2 and for G_2:

$$F_2 = i\frac{(a-1)F_0 + (a-1)G_2 + r(-i\epsilon F_3 + 2F_0' + G_3')}{2r\epsilon},$$
$$G_2 = -i\frac{(a+1)G_0 + (a+1)F_2 + r(F_3' - i\epsilon G_3 - 2G_0')}{2r\epsilon}. \tag{17.38}$$

From equations 7) and 8) we can find expressions for F_3 and G_3:

$$F_3 = \frac{2(a^2-1)G_0 + (a+1)(bF_1 + 4ir\epsilon F_0) - 2ibr\epsilon G_1}{a^2 - 4r^2\epsilon^2 - 1},$$
$$G_3 = \frac{-2(a^2-1)F_0 + (a-1)(bG_1 + 4ir\epsilon G_0) + 2ibr\epsilon F_1}{a^2 - 4r^2\epsilon^2 - 1}. \tag{17.39}$$

We further substitute these four formulas into equations 1), 3), 5) and 6). In this way we obtain two pairs of equations for the variables F_0, F_1, G_0, G_1:

$$-\frac{2i(a^2-1)\epsilon(3a^2-4r^2\epsilon^2-3)}{(a^2-4r^2\epsilon^2-1)^2}F_0 + \frac{2(a^2-1)(-a^3+a+4r^2\epsilon^2)}{r(a^2-4r^2\epsilon^2-1)^2}G_0$$
$$+\frac{2br\epsilon^2(-3a^2+4r^2\epsilon^2+3)}{(a^2-4r^2\epsilon^2-1)^2}F_1 + \frac{2i(a-1)(a^2-1)b\epsilon}{(a^2-4r^2\epsilon^2-1)^2}G_1$$
$$+\frac{2(a^2-1)}{a^2-4r^2\epsilon^2-1}G_0' + \frac{2ibr\epsilon(-a^2+4r^2\epsilon^2+1)}{(a^2-4r^2\epsilon^2-1)^2}G_1' = 0,$$

$$+\frac{4br\epsilon^2(-3a^2+4r^2\epsilon^2+3)}{(a^2-4r^2\epsilon^2-1)^2}F_0 - \frac{4ib\epsilon(-a^3+a+4r^2\epsilon^2)}{(a^2-4r^2\epsilon^2-1)^2}G_0$$
$$+\left(\frac{ib^2\epsilon(a^2+4r^2\epsilon^2-1)}{(a^2-4r^2\epsilon^2-1)^2} - i\epsilon\right)F_1 + \frac{4(a-1)b^2r\epsilon^2}{(a^2-4r^2\epsilon^2-1)^2}G_1$$
$$+\frac{4ibr\epsilon(-a^2+4r^2\epsilon^2+1)}{(a^2-4r^2\epsilon^2-1)^2}G_0' + \left(\frac{b^2(-a^2+4r^2\epsilon^2+1)}{(a^2-4r^2\epsilon^2-1)^2} + 1\right)G_1' = 0;$$

and

$$\frac{2(a^2-1)(a^3-a+4r^2\epsilon^2)}{r(a^2-4r^2\epsilon^2-1)^2}F_0 - \frac{2i(a^2-1)\epsilon(3a^2-4r^2\epsilon^2-3)}{(a^2-4r^2\epsilon^2-1)^2}G_0$$
$$-\frac{2i(a-1)(a+1)^2b\epsilon}{(a^2-4r^2\epsilon^2-1)^2}F_1 + \frac{2br\epsilon^2(-3a^2+4r^2\epsilon^2+3)}{(a^2-4r^2\epsilon^2-1)^2}G_1$$
$$+\frac{2(a^2-1)}{a^2-4r^2\epsilon^2-1}F_0' - \frac{2ibr\epsilon}{a^2-4r^2\epsilon^2-1}F_1' = 0,$$

$$-\frac{4ib\epsilon(a^3-a+4r^2\epsilon^2)}{(a^2-4r^2\epsilon^2-1)^2}F_0 + \frac{4br\epsilon^2(-3a^2+4r^2\epsilon^2+3)}{(a^2-4r^2\epsilon^2-1)^2}G_0$$

$$-\frac{4(a+1)b^2r\epsilon^2}{(a^2-4r^2\epsilon^2-1)^2}F_1-\frac{i\epsilon\left(-4r^2\epsilon^2\left(2a^2+b^2-2\right)+\left(a^2-1\right)\left(a^2-b^2-1\right)+16r^4\epsilon^4\right)}{(a^2-4r^2\epsilon^2-1)^2}G_1$$

$$-\frac{4ibr\epsilon}{a^2-4r^2\epsilon^2-1}F_0'+\frac{\left(-a^2+4r^2\epsilon^2+1\right)\left(-a^2+b^2+4r^2\epsilon^2+1\right)}{(a^2-4r^2\epsilon^2-1)^2}F_1'=0.$$

These pairs may be considered as linear systems with respect to the variables F_0', F_1', and G_0', G_1'. After simple calculations, we find their solutions:

$$F_0'=\frac{1}{-a^2r+4r^3\epsilon^2+r}$$

$$\times\Big[ir\epsilon(-3a^2+3+4r^2\epsilon^2)G_0+(a^3-a+4r^2\epsilon^2)F_0-i(1+a)br\epsilon F_1-2br^2\epsilon^2G_1\Big], \qquad (17.40)$$

$$G_0'=\frac{1}{-a^2r+4r^3\epsilon^2+r}$$

$$\Big[ir\epsilon(-3a^2+3+4r^2\epsilon^2)F_0-(a^3-a-4r^2\epsilon^2)G_0-2br^2\epsilon^2F_1-i(1-a)br\epsilon G_1\Big], \qquad (17.41)$$

and

$$G_1'=i\epsilon F_1, \qquad F_1'=i\epsilon G_1. \qquad (17.42)$$

From (17.42), there follow separate equations for F_1 and for G_1, with the simple solutions

$$\delta=+1\,,\ F_1=e^{+i\epsilon r}\,,\ G_1=e^{+i\epsilon r}; \qquad \delta=-1\,,\ F_1=e^{-i\epsilon r}\,,\ G_1=-e^{-i\epsilon r}. \qquad (17.43)$$

By applying the ordinary exclusion method, from (17.40) – (17.41) we derive the second order equations

$\delta=+1$

$$-\frac{be^{ir\epsilon}(-3(a-1)a+2r\epsilon(2r\epsilon+3i))}{-3a^2+4r^2\epsilon^2+3}$$

$$-\frac{8ir^2\epsilon G_0'}{-3a^2+4r^2\epsilon^2+3}+\frac{irG_0''}{\epsilon}+\frac{iG_0\left(\left(15-7a^2\right)r^2\epsilon^2+3(a-1)^2a(a+1)+4r^4\epsilon^4\right)}{r\epsilon\left(-3a^2+4r^2\epsilon^2+3\right)}=0,$$

$$\frac{be^{ir\epsilon}(3a(a+1)-2r\epsilon(2r\epsilon+3i))}{-3a^2+4r^2\epsilon^2+3} \qquad (17.44)$$

$$-\frac{8ir^2\epsilon F_0'}{-3a^2+4r^2\epsilon^2+3}+\frac{irF_0''}{\epsilon}+\frac{iF_0\left(\left(15-7a^2\right)r^2\epsilon^2+3(a-1)a(a+1)^2+4r^4\epsilon^4\right)}{r\epsilon\left(-3a^2+4r^2\epsilon^2+3\right)}=0;$$

$\delta=-1$,

$$\frac{ib\epsilon e^{-ir\epsilon}(3(a-1)a+2r\epsilon(-2r\epsilon+3i))}{-3a^2+4r^2\epsilon^2+3}$$

$$-\frac{8r^2\epsilon^2G_0'}{-3a^2+4r^2\epsilon^2+3}+\frac{G_0\left(\left(15-7a^2\right)r^2\epsilon^2+3(a-1)^2a(a+1)+4r^4\epsilon^4\right)}{r(-3a^2+4r^2\epsilon^2+3)}+rG_0''=0,$$

$$\frac{b\epsilon e^{-ir\epsilon}(2r\epsilon(3+2ir\epsilon)-3ia(a+1))}{-3a^2+4r^2\epsilon^2+3} \qquad (17.45)$$

$$-\frac{8r^2\epsilon^2F_0'}{-3a^2+4r^2\epsilon^2+3}+\frac{F_0\left(\left(15-7a^2\right)r^2\epsilon^2+3(a-1)a(a+1)^2+4r^4\epsilon^4\right)}{r(-3a^2+4r^2\epsilon^2+3)}+rF_0''=0.$$

The equations (17.44) and (17.45) allow us, by using the relations

$$
\begin{gathered}
\delta=+1, \quad F_1=e^{+i\epsilon r}, \quad G_1=e^{+i\epsilon r}, \quad F_0=\ldots, G_0=\ldots; \\
\delta=-1, \quad F_1=e^{-i\epsilon r}, \quad G_1=-e^{-i\epsilon r}; \quad F_0=\ldots, G_0=\ldots; \\
F_3=\frac{2\left(a^2-1\right) G_0+(a+1)\left(bF_1+4ir\epsilon F_0\right)-2ibr\epsilon G_1}{a^2-4r^2\epsilon^2-1}, \\
G_3=\frac{-2\left(a^2-1\right) F_0+(a-1)\left(bG_1+4ir\epsilon G_0\right)+2ibr\epsilon F_1}{a^2-4r^2\epsilon^2-1}; \\
F_2=-G_0, \qquad G_2=-F_0
\end{gathered}
\tag{17.46}
$$

to find the concomitant functions $F_1, G_1; F_2, G_2; F_3, G_3$.

To find the general solutions of the nonhomogeneous equations for F_0 and G_0, we should know their particular solutions and the general solutions of the corresponding homogeneous equations. The particular solutions may be found by using the known gauge solutions.

Let us turn to the gauge solutions (17.35) (taking in mind the change in notations according to (17.36):

$$
\begin{array}{ll}
\bar{F}_0=-i\epsilon(K_1+K_2)\frac{1}{\sqrt{2}}, & G_0=-i\epsilon(K_1-K_2)\frac{1}{\sqrt{2}}, \\
\bar{F}_1=-\frac{b}{\sqrt{2}r}(K_1+K_2), & \bar{G}_1=-\frac{b}{\sqrt{2}r}(K_1-K_2), \\
\bar{F}_2=(\frac{d}{dr}-\frac{1}{r})(K_1+K_2)\frac{1}{\sqrt{2}}, & \bar{G}_2=(\frac{d}{dr}-\frac{1}{r})(K_1-K_2)\frac{1}{\sqrt{2}}, \\
\bar{F}_3=-\frac{1}{\sqrt{2}r}(a+1)(K_1+K_2), & \bar{G}_3=-\frac{1}{\sqrt{2}r}(a-1)(K_1-K_2);
\end{array}
\tag{17.47}
$$

The particular solutions should be determined from the following relations

$$
\begin{array}{lll}
\bar{F}_1=-\frac{b}{\sqrt{2}r}(K_1+K_2)=e^{i\delta\epsilon r} & \Longrightarrow & (K_1+K_2)=-\frac{\sqrt{2}r}{b}e^{i\delta\epsilon r}, \\
\bar{G}_1=-\frac{b}{\sqrt{2}r}(K_1-K_2)=\delta e^{i\delta\epsilon r} & \Longrightarrow & (K_1-K_2)=-\delta\frac{\sqrt{2}r}{b}e^{i\delta\epsilon r},
\end{array}
\tag{17.48}
$$

whence it follows

$$
\delta=+1,\ K_1=-\frac{r\sqrt{2}}{b}e^{+i\epsilon r},\ K_2=0; \qquad \delta=+1,\ K_1=0,\ K_2=-\frac{r\sqrt{2}}{b}e^{-i\epsilon r}. \tag{17.49}
$$

Thus we have fixed the two gauge solutions:

$$
\delta=+1, \quad
\begin{array}{ll}
\bar{F}_0=\frac{i\epsilon r}{b}e^{+i\epsilon r}, & G_0=\frac{i\epsilon r}{b}e^{+i\epsilon r}, \\
\bar{F}_1=e^{+i\epsilon r}, & \bar{G}_1=e^{+i\epsilon r}, \\
\bar{F}_2=-(\frac{d}{dr}-\frac{1}{r})\frac{r}{b}e^{+i\epsilon r}, & \bar{G}_2=-(\frac{d}{dr}-\frac{1}{r})\frac{r}{b}e^{+i\epsilon r}, \\
\bar{F}_3=\frac{a+1}{b}e^{+i\epsilon r}, & \bar{G}_3=\frac{a-1}{b}e^{+i\epsilon r};
\end{array}
\tag{17.50}
$$

$$
\delta=-1, \quad
\begin{array}{ll}
\bar{F}_0=\frac{i\epsilon r}{b}e^{-i\epsilon r}, & G_0=-\frac{i\epsilon r}{b}e^{-i\epsilon r}, \\
\bar{F}_1=e^{-i\epsilon r}, & \bar{G}_1=-e^{-i\epsilon r}, \\
\bar{F}_2=-(\frac{d}{dr}-\frac{1}{r})\frac{r}{b}e^{-i\epsilon r}, & \bar{G}_2=(\frac{d}{dr}-\frac{1}{r})\frac{r}{b}e^{-i\epsilon r}, \\
\bar{F}_3=\frac{a+1}{b}e^{-i\epsilon r}, & \bar{G}_3=-\frac{a-1}{b}e^{-i\epsilon r}.
\end{array}
\tag{17.51}
$$

By direct calculation, we can verify that the functions F_0, G_0 from (17.50) and (17.51) provide exact solutions of the equations (17.44) and (17.45). It is evident that the solutions for nonhomogeneous equations constructed by this method produce two types of solutions which do not contain gauge constituents.

17.5 Solving the Homogeneous Equations

Let us write down the homogeneous equations from (17.45) and (17.46):

$\delta = +1$

$$G_0'' - \frac{8r\epsilon^2}{(-3a^2+4r^2\epsilon^2+3)} G_0' + \frac{(15-7a^2)\,r^2\epsilon^2+3(a-1)^2a(a+1)+4r^4\epsilon^4}{r^2\,(-3a^2+4r^2\epsilon^2+3)} G_0 = 0\,,$$

$$F_0'' - \frac{8r\epsilon^2}{(-3a^2+4r^2\epsilon^2+3)} F_0'$$
$$+ \frac{(15-7a^2)\,r^2\epsilon^2+3(a-1)a(a+1)^2+4r^4\epsilon^4}{r^2\,(-3a^2+4r^2\epsilon^2+3)} F_0 = 0;$$

$\delta = -1$,

$$G_0'' - \frac{8r\epsilon^2}{(-3a^2+4r^2\epsilon^2+3)} G_0'$$
$$+ \frac{(15-7a^2)\,r^2\epsilon^2+3(a-1)^2a(a+1)+4r^4\epsilon^4}{r^2\,(-3a^2+4r^2\epsilon^2+3)} G_0 = 0\,,$$

$$F_0'' - \frac{8r\epsilon^2}{(-3a^2+4r^2\epsilon^2+3)} F_0'$$
$$+ \frac{(15-7a^2)\,r^2\epsilon^2+3(a-1)a(a+1)^2+4r^4\epsilon^4}{r^2\,(-3a^2+4r^2\epsilon^2+3)} F_0 = 0\,.$$

All the four equations contain the same second order operator, so it suffices to study only one equation:

$$\frac{d^2f}{dr^2} - \frac{8r\epsilon^2}{(-3a^2+4r^2\epsilon^2+3)}\frac{df}{dr}$$
$$+ \frac{(15-7a^2)r^2\epsilon^2+3(a-1)a(a+1)^2+4r^4\epsilon^4}{r^2(-3a^2+4r^2\epsilon^2+3)} f = 0;$$

we note the identities

$$a^2-1 = (j-\frac{1}{2})(j+\frac{3}{2}) = b^2, \quad (a-1)a(a+1)^2 = b^2a(a+1), \quad 15-7a^2 = 8-7b^2;$$

it is convenient to use the dimensionless variable $\epsilon r = x$, so we get

$$\frac{d^2f}{dx^2} - \frac{8x}{(4x^2-3b^2)}\frac{df}{dx} + \frac{4x^4+(8-7b^2)x^2+3b^2a(a+1)}{x^2(4x^2-3b^2)} f = 0\,.$$

In the vicinity of the point $x=0$, the equation simplifies

$$\frac{d^2f}{dx^2} + \frac{8x}{3b^2}\frac{df}{dx} - \frac{a(a+1)}{x^2} f = 0\,, \quad f = x^A, \quad A = a+1, -a\,.$$

Let us study the singular point $x=\infty$; in the variable $y = \frac{1}{x}$, the above equation reads

$$\frac{d^2f}{dy^2} + \frac{2}{y}\frac{df}{dy} + \frac{8}{y(4-3b^2y^2)}\frac{d}{dy} + \frac{4+(8-7b^2)y^2+3b^2a(a+1)y^4}{y^4(4-3b^2y^2)} f = 0\,,$$

so the rank of the point $x=\infty$ $(y=0)$ equals to 2.

Thus, the equation under consideration has three regular points and one irregular point $x = \infty$ of the rank 2:

$$\frac{d^2 f}{dx^2} - \frac{8x}{4(x+\sqrt{3}b/2)(x-\sqrt{3}b/2)} \frac{df}{dx}$$
$$+ \frac{4x^4 + (8-7b^2)x^2 + 3b^2 a(a+1)}{4x^2(x+\sqrt{3/2}b)(x-\sqrt{3}b/2)} f = 0;$$

the behavior of solutions near the additional two points is as shown below

$$f(x) = (x+\sqrt{3}b/2)^s, \quad f(x) = (x-\sqrt{3}b/2)^s, \; s = 0, 2;$$

the solutions of Frobenius type may be searched in the form

$$f(x) = x^A e^{Bx} e^{\frac{C}{x}} F(x);$$

the functions $F(x)$ are constructed as power series with a 6-terms recurrent relation.

So we have constructed two types of solutions for the massless spin 3/2 field, which do not contain the gauge constituents.

Chapter 18

Spin 3/2 Massless Field, Cylindric Symmetry, Eliminating the Gauge Degrees of Freedom

18.1 Separating the Variables

In the present chapter, we shall follow the problem of the degrees of freedom of the massless spin 3/2 particle described in cylindric coordinates. Two solutions which do not contain the gauge component are constructed in explicit form. We also find 4 solutions of the basic wave equation which may be identified with the gauge ones. The study is based on the use of the covariant tetrad formalism.

It is known the following form of the wave equation for the spin 3/2 particle (we generally start with the case of the massive particle; see the notations in [139])

$$\gamma^5 \epsilon_\rho^{\ \sigma\alpha\beta}(x)\gamma_\sigma (iD_\alpha - M\gamma_\alpha)\Psi_\beta = 0, \quad D_\alpha = \nabla_\alpha + \Gamma_\alpha + ieA_\alpha , \tag{18.1}$$

where $M = mc/2\hbar c$ stands for the massive parameter. It is readily proved that in the massless case $(M = 0)$ if the Riemann curvature tensor of the space-time model equals to zero, then equation (18.1) has the class of gradient-type solutions

$$\Psi^G_\beta(x) = D_\beta \Psi(x), \quad D_\beta = \nabla_\beta + \Gamma_\beta ,$$

where Φ stands for an arbitrary bispinor field. Indeed, due to the commutator relation, we have

$$(D_\alpha D_\beta - D_\beta D_\alpha)\Phi = \frac{1}{2}\sigma^{\mu\nu}(x)R_{\mu\nu\alpha\beta}(x) \equiv 0; \tag{18.2}$$

the equation (18.1) is identically satisfied by this gradient solution $R_{\mu\nu\alpha\beta}(x)$.

After the transition in the field function to the tetrad presentation in vector index $\Psi_\beta = e^{(b)}_\beta \Psi_b$, equation (18.1) takes the form

$$\gamma^5 \epsilon_k^{\ can}\gamma_c \left[i(D_a)_n^{\ l} - M\gamma_a\delta_n^{\ l} \right]\Psi_l = 0, \tag{18.3}$$

where we use the extended derivative operator

$$D_a = e^\alpha_{(a)}(\partial_\alpha + ieA_\alpha) + \frac{1}{2}(\sigma^{ps} \otimes I + I \otimes j^{ps})\gamma_{[ps]a} . \tag{18.4}$$

By introducing the six matrices $\epsilon_k^{\;can} = (\mu^{[ca]})_k^{\;n}$, we can write equation (18.3) as follows

$$\gamma^5 (\mu^{[ca]})_k^{\;n} \gamma_c \left[i(D_a)_n^{\;l} - M\gamma_a \delta_n^{\;l} \right] \Psi_l = 0 . \tag{18.5}$$

Let us allow for the first summation the formula

$$\mu^{ca}\gamma_c D_a$$
$$= \mu^{[01]}[\gamma_0 D_1\Psi - \gamma_1 D_0\Psi] + \mu^{[02]}[\gamma_0 D_2\Psi - \gamma_2 D_0] + \mu^{[03]}[\gamma_0 D_3\Psi - \gamma_3 D_0]$$
$$+\mu^{[23]}[\gamma_2 D_3\Psi - \gamma_3 D_2\Psi + \mu^{[31]}[\gamma_3 D_1\Psi - \gamma_1 D_3] + \mu^{[12]}[\gamma_1 D_2\Psi - \gamma_2 D_1] ,$$

or differently (taking in mind that the matrices $\mu^{[ca]}$ act on the vector index)

$$\mu^{[ca]}\gamma_c D_a$$
$$= \left(\gamma^1 \otimes \mu^{[01]} + \gamma^2 \otimes \mu^{[02]} + \gamma^3 \otimes \mu^{[03]}\right) D_0 + \left(\gamma^0 \otimes \mu^{[01]} + \gamma^2 \otimes \mu^{[12]} - \gamma^3 \otimes \mu^{[31]}\right) D_1$$
$$+ \left(\gamma^0 \otimes \mu^{[02]} + \gamma^3 \otimes \mu^{[23]} - \gamma^1 \otimes \mu^{[12]}\right) D_2 + \left(\gamma^0 \otimes \mu^{[03]} + \gamma^1 \otimes \mu^{[31]} - \gamma^2 \otimes \mu^{[23]}\right) D_3 ,$$

and the second summation formula

$$(\mu^{[ca]})_k^{\;n} \gamma_c \gamma_a \Psi_n \quad \Longrightarrow \quad \mu^{[ca]}\gamma_c\gamma_a$$
$$= \Big\{ (\gamma_0\gamma_1 - \gamma_1\gamma_0) \otimes \mu^{[01]} + (\gamma_0\gamma_2 - \gamma_2\gamma_0) \otimes \mu^{[02]} + (\gamma_0\gamma_3 - \gamma_3\gamma_0) \otimes \mu^{[03]}$$
$$+ (\gamma_2\gamma_3 - \gamma_3\gamma_2) \otimes \mu^{[23]} + (\gamma_3\gamma_1 - \gamma_1\gamma_3) \otimes \mu^{[31]} + (\gamma_1\gamma_2 - \gamma_2\gamma_1) \otimes \mu^{[12]} \Big\}$$
$$= \Big\{ s_{01} \otimes \mu^{[01]} + s_{02} \otimes \mu^{[02]} + s_{03} \otimes \mu^{[03]} + s_{23} \otimes \mu^{[23]} + s_{31} \otimes \mu^{[31]} + s_{12} \otimes \mu^{[12]} \Big\} .$$

Correspondingly, in detailed form, the basic equation (18.5) reads

$$\left(\gamma^1 \otimes \mu^{[01]} + \gamma^2 \otimes \mu^{[02]} + \gamma^3 \otimes \mu^{[03]}\right) D_0\Psi + \left(\gamma^0 \otimes \mu^{[01]} + \gamma^2 \otimes \mu^{[12]} - \gamma^3 \otimes \mu^{[31]}\right) D_1\Psi$$
$$+ \left(\gamma^0 \otimes \mu^{[02]} + \gamma^3 \otimes \mu^{[23]} - \gamma^1 \otimes \mu^{[12]}\right) D_2\Psi + \left(\gamma^0 \otimes \mu^{[03]} + \gamma^1 \otimes \mu^{[31]} - \gamma^2 \otimes \mu^{[23]}\right) D_3\Psi$$
$$+iM\Big\{ s_{01} \otimes \mu^{[01]} + s_{02} \otimes \mu^{[02]} + s_{03} \otimes \mu^{[03]} + s_{23} \otimes \mu^{[23]} + s_{31} \otimes \mu^{[31]} + s_{12} \otimes \mu^{[12]} \Big\}\Psi = 0 . \tag{18.6}$$

Let us consider the wave equation (for generality we take into account the presence of an external uniform magnetic field). It is convenient to use the known cylindric coordinates and the diagonal tetrad

$$dS^2 = dt^2 - dr^2 - r^2 d\phi^2 - dz^2 , \; x^\alpha = (t, r, \phi, z) , \quad A_\phi = \frac{Br^2}{2} ,$$
$$e^{\beta}_{(a)}(x) = \begin{vmatrix} 1 & 0 & 0 & 0 \\ 0 & 1 & 0 & 0 \\ 0 & 0 & 1/r & 0 \\ 0 & 0 & 0 & 1 \end{vmatrix} , \; e_{(a)\beta}(y) = \begin{vmatrix} 1 & 0 & 0 & 0 \\ 0 & -1 & 0 & 0 \\ 0 & 0 & -r & 0 \\ 0 & 0 & 0 & -1 \end{vmatrix} ; \tag{18.7}$$

the Ricci rotation coefficients are

$$\gamma_{ab0} = 0 , \quad \gamma_{ab1} = 0 , \quad \gamma_{122} = -\gamma_{212} = +\frac{1}{r} , \quad \gamma_{ab3} = 0 .$$

The components of the derivative operator D_c are as follows

$$D_0 = \partial_t\,, \quad D_1 = \partial_r\,, \quad D_2 = \frac{1}{r}(\partial_\phi + \frac{ieBr^2}{2}) + \frac{1}{r}(\sigma^{12} \otimes + I \otimes j^{12})\,, \quad D_3 = \partial_z\,. \tag{18.8}$$

Thus, the equation (18.6) takes the form

$$\Big\{ \Big(\gamma^1 \otimes \mu^{[01]} + \gamma^2 \otimes \mu^{[02]} + \gamma^3 \otimes \mu^{[03]}\Big) \partial_t \Psi + \Big(\gamma^0 \otimes \mu^{[01]} + \gamma^2 \otimes \mu^{[12]} - \gamma^3 \otimes \mu^{[31]}\Big) \partial_r \Psi$$
$$+ \Big(\gamma^0 \otimes \mu^{[02]} + \gamma^3 \otimes \mu^{[23]} - \gamma^1 \otimes \mu^{[12]}\Big) D_2 \Psi + \Big(\gamma^0 \otimes \mu^{[03]} + \gamma^1 \otimes \mu^{[31]} - \gamma^2 \otimes \mu^{[23]}\Big) \partial_z \Psi \Big\}_k$$
$$+ iM \Big\{ s_{01} \otimes \mu^{[01]} + s_{02} \otimes \mu^{[02]} + s_{03} \otimes \mu^{[03]} + s_{23} \otimes \mu^{[23]} + s_{31} \otimes \mu^{[31]} + s_{12} \otimes \mu^{[12]} \Big\} \Psi = 0\,. \tag{18.9}$$

We need the expressions for the six matrices $\mu^{[ca]}$:

$$\mu^{[01]} = \begin{vmatrix} 0 & 0 & 0 & 0 \\ 0 & 0 & 0 & 0 \\ 0 & 0 & 0 & \epsilon_2{}^{013} \\ 0 & 0 & \epsilon_3{}^{012} & 0 \end{vmatrix} = \begin{vmatrix} 0 & 0 & 0 & 0 \\ 0 & 0 & 0 & 0 \\ 0 & 0 & 0 & -1 \\ 0 & 0 & 1 & 0 \end{vmatrix}, \mu^{[02]} = \begin{vmatrix} 0 & 0 & 0 & 0 \\ 0 & 0 & 0 & \epsilon_1{}^{023} \\ 0 & 0 & 0 & 0 \\ 0 & \epsilon_3{}^{021} & 0 & 0 \end{vmatrix} = \begin{vmatrix} 0 & 0 & 0 & 0 \\ 0 & 0 & 0 & 1 \\ 0 & 0 & 0 & 0 \\ 0 & -1 & 0 & 0 \end{vmatrix},$$

$$\mu^{[03]} = \begin{vmatrix} 0 & 0 & 0 & 0 \\ 0 & 0 & \epsilon_1{}^{032} & 0 \\ 0 & \epsilon_2{}^{031} & 0 & 0 \\ 0 & 0 & 0 & 0 \end{vmatrix} = \begin{vmatrix} 0 & 0 & 0 & 0 \\ 0 & 0 & -1 & 0 \\ 0 & 1 & 0 & 0 \\ 0 & 0 & 0 & 0 \end{vmatrix}, \mu^{[23]} = \begin{vmatrix} 0 & \epsilon_0{}^{231} & 0 & 0 \\ \epsilon_1{}^{230} & 0 & 0 & 0 \\ 0 & 0 & 0 & 0 \\ 0 & 0 & 0 & 0 \end{vmatrix} = \begin{vmatrix} 0 & 1 & 0 & 0 \\ 1 & 0 & 0 & 0 \\ 0 & 0 & 0 & 0 \\ 0 & 0 & 0 & 0 \end{vmatrix},$$

$$\mu^{[31]} = \begin{vmatrix} 0 & 0 & \epsilon_0{}^{312} & 0 \\ 0 & 0 & 0 & 0 \\ \epsilon_2{}^{310} & 0 & 0 & 0 \\ 0 & 0 & 0 & 0 \end{vmatrix} = \begin{vmatrix} 0 & 0 & 1 & 0 \\ 0 & 0 & 0 & 0 \\ 1 & 0 & 0 & 0 \\ 0 & 0 & 0 & 0 \end{vmatrix}, \mu^{[12]} = \begin{vmatrix} 0 & 0 & 0 & \epsilon_0{}^{123} \\ 0 & 0 & 0 & 0 \\ 0 & 0 & 0 & 0 \\ \epsilon_3{}^{120} & 0 & 0 & 0 \end{vmatrix} = \begin{vmatrix} 0 & 0 & 0 & 1 \\ 0 & 0 & 0 & 0 \\ 0 & 0 & 0 & 0 \\ 1 & 0 & 0 & 0 \end{vmatrix}.$$

We also need the expressions for the Dirac matrices:

$$\gamma^0 = \begin{vmatrix} 0 & 0 & 1 & 0 \\ 0 & 0 & 0 & 1 \\ 1 & 0 & 0 & 0 \\ 0 & 1 & 0 & 0 \end{vmatrix}, \gamma^1 = \begin{vmatrix} 0 & 0 & 0 & -1 \\ 0 & 0 & -1 & 0 \\ 0 & 1 & 0 & 0 \\ 1 & 0 & 0 & 0 \end{vmatrix}, \gamma^2 = \begin{vmatrix} 0 & 0 & 0 & i \\ 0 & 0 & -i & 0 \\ 0 & -i & 0 & 0 \\ i & 0 & 0 & 0 \end{vmatrix}, \gamma^3 = \begin{vmatrix} 0 & 0 & -1 & 0 \\ 0 & 0 & 0 & 1 \\ 1 & 0 & 0 & 0 \\ 0 & -1 & 0 & 0 \end{vmatrix};$$

$$\sigma^{12} = \frac{1}{2} \begin{vmatrix} -i & 0 & 0 & 0 \\ 0 & i & 0 & 0 \\ 0 & 0 & -i & 0 \\ 0 & 0 & 0 & i \end{vmatrix}, \quad j^{12} = \begin{vmatrix} 0 & 0 & 0 & 0 \\ 0 & 0 & -1 & 0 \\ 0 & 1 & 0 & 0 \\ 0 & 0 & 0 & 0 \end{vmatrix},$$

$$s_{01} = -\begin{vmatrix} 0 & 2 & 0 & 0 \\ 2 & 0 & 0 & 0 \\ 0 & 0 & 0 & -2 \\ 0 & 0 & -2 & 0 \end{vmatrix}, \quad s_{02} = -\begin{vmatrix} 0 & -2i & 0 & 0 \\ 2i & 0 & 0 & 0 \\ 0 & 0 & 0 & 2i \\ 0 & 0 & -2i & 0 \end{vmatrix}, \quad s_{03} = -\begin{vmatrix} 2 & 0 & 0 & 0 \\ 0 & -2 & 0 & 0 \\ 0 & 0 & -2 & 0 \\ 0 & 0 & 0 & 2 \end{vmatrix},$$

$$s_{23} = \begin{vmatrix} 0 & -2i & 0 & 0 \\ -2i & 0 & 0 & 0 \\ 0 & 0 & 0 & -2i \\ 0 & 0 & -2i & 0 \end{vmatrix}, \quad s_{31} = \begin{vmatrix} 0 & -2 & 0 & 0 \\ 2 & 0 & 0 & 0 \\ 0 & 0 & 0 & -2 \\ 0 & 0 & 2 & 0 \end{vmatrix}, \quad s_{12} = \begin{vmatrix} -2i & 0 & 0 & 0 \\ 0 & 2i & 0 & 0 \\ 0 & 0 & -2i & 0 \\ 0 & 0 & 0 & 2i \end{vmatrix}.$$

We apply the substitution for the cylindrically symmetric wave function (assuming $\epsilon > 0$):

$$\Psi_{A(n)} = e^{-i\epsilon t} e^{im\phi} e^{ikz} \Phi_{A(n)}(r)\,, \quad [\Phi_{A(n)}] = \Phi(r) = \begin{vmatrix} f_0 & f_1 & f_2 & f_3 \\ g_0 & g_1 & g_2 & g_3 \\ h_0 & h_1 & h_2 & h_3 \\ d_0 & d_1 & d_2 & d_3 \end{vmatrix}. \tag{18.10}$$

It should be noted that in the basis of cylindric tetrad, the parameter m stands for the eigenvalues of the third projection of the total angular momentum, so m takes the half-integer values $m = \pm\frac{1}{2}, \pm\frac{3}{2}, \ldots$

Allowing for the above substitution, we reduce the equation to the form

$$-i\epsilon\left(\gamma^1 \otimes \mu^{[01]} + \gamma^2 \otimes \mu^{[02]} + \gamma^3 \otimes \mu^{[03]}\right)\Phi + \left(\gamma^0 \otimes \mu^{[01]} + \gamma^2 \otimes \mu^{[12]} - \gamma^3 \otimes \mu^{[31]}\right)\frac{d}{dr}\Phi$$
$$+\left(\gamma^0 \otimes \mu^{[02]} + \gamma^3 \otimes \mu^{[23]} - \gamma^1 \otimes \mu^{[12]}\right)D_2\Phi + ik\left(\gamma^0 \otimes \mu^{[03]} + \gamma^1 \otimes \mu^{[31]} - \gamma^2 \otimes \mu^{[23]}\right)\Phi$$
$$+iM\left\{s_{01} \otimes \mu^{[01]} + s_{02} \otimes \mu^{[02]} + s_{03} \otimes \mu^{[03]} + s_{23} \otimes \mu^{[23]} + s_{31} \otimes \mu^{[31]} + s_{12} \otimes \mu^{[12]}\right\}\Phi = 0, \tag{18.11}$$

where

$$D_2 = \frac{1}{r}(i\mu + \sigma^{12} \otimes + I \otimes j^{12}), \quad \mu(r) = m + \frac{eBr^2}{2}, \quad [\mu] = 1; \tag{18.12}$$

for brevity we shall change in the following the notation, $eB \Longrightarrow B$. The main equation may be written as

$$-i\epsilon B_0\Phi + B_1\frac{d}{dr}\Phi + B_2 D_2 \Psi + ikB_3\Phi$$
$$+iM\left\{s_{01} \otimes \mu^{[01]} + s_{02} \otimes \mu^{[02]} + s_{03} \otimes \mu^{[03]} + s_{23} \otimes \mu^{[23]} + s_{31} \otimes \mu^{[31]} + s_{12} \otimes \mu^{[12]}\right\}\Phi = 0. \tag{18.13}$$

First, we calculate the terms:

$$-i\epsilon B_0\Phi = \left|\begin{array}{cccc} 0 & +\epsilon(d_3 - ih_2) & +i\epsilon(h_1 - d_3) & +\epsilon(-d_1 + id_2) \\ 0 & +\epsilon(-h_3 + id_2) & +i\epsilon(-d_1 - h_3) & +\epsilon(h_1 + ih_2) \\ 0 & -\epsilon(g_3 - if_2) & -i\epsilon(f_1 - g_3) & -\epsilon(-g_1 + ig_2) \\ 0 & -\epsilon(-f_3 + ig_2) & -i\epsilon(-g_1 - f_3) & -\epsilon(f_1 + if_2) \end{array}\right|,$$

$$\frac{d}{dr}B_1\Phi = \left|\begin{array}{cccc} +h_2' + id_3' & 0 & h_0' - h_3' & h_2' + id_0' \\ -d_2' - ih_3' & 0 & -d_0' - d_3' & d_2' - ih_0' \\ -f_2' - ig_3' & 0 & -f_0' - f_3' & f_2' - ig_0' \\ g_2' + if_3' & 0 & g_0' - g_3' & g_2' + if_0' \end{array}\right|,$$

$$B_2 D_2 \Phi$$

$$= \frac{i}{2r}\left|\begin{array}{cccc} -(2\mu-1)h_1 + (2\mu+1)d_3 - 2ih_2 & -(2\mu-1)(h_0 - h_3) & 0 & -(2\mu-1)h_1 + (2\mu+1)d_0 - 2ih_2 \\ +(2\mu+1)d_1 + (2\mu-1)h_3 + 2id_2 & +(2\mu+1)(d_0 + d_3) & 0 & -(2\mu+1)d_1 + (2\mu-1)h_0 - 2id_2 \\ (2\mu-1)f_1 - (2\mu+1)g_3 + 2if_2 & (2\mu-1)(f_0 + f_3) & 0 & -(2\mu-1)f_1 - (2\mu+1)g_0 - 2if_2 \\ -(2\mu+1)g_1 - (2\mu-1)f_3 - 2ig_2 & -(2\mu+1)(g_0 - g_3) & 0 & -(2\mu+1)g_1 - (2\mu-1)f_0 - 2ig_2 \end{array}\right|,$$

$$ikB_3\Phi = \left|\begin{array}{cccc} k(d_1 - id_2) & k(d_0 - ih_2) & ik(-d_0 + h_1) & 0 \\ k(-h_1 - ih_2) & k(-h_0 - id_2) & ik(-h_0 + d_1) & 0 \\ -k(g_1 - ig_2) & -k(g_0 + if_2) & -ik(-g_0 - f_1) & 0 \\ -k(-f_1 - if_2) & -k(-f_0 + ig_2) & -ik(-f_0 - g_1) & 0 \end{array}\right|. \tag{18.14}$$

For the massive term, we get

$$+iM\Big\{s_{01}\otimes\mu^{[01]}+s_{02}\otimes\mu^{[02]}+s_{03}\otimes\mu^{[03]}+s_{23}\otimes\mu^{[23]}+s_{31}\otimes\mu^{[31]}+s_{12}\otimes\mu^{[12]}\Big\}\Phi$$

$$=2iM\left|\begin{array}{cccc} -i(f_3+g_1)-g_2 & -f_2-i(g_0+g_3) & f_1-g_0-g_3 & -if_0+ig_1+g_2 \\ -if_1+f_2+ig_3 & -if_0+if_3+g_2 & f_0-f_3-g_1 & -if_1+f_2+ig_0 \\ -i(d_1-id_2+h_3) & -id_0+id_3+h_2 & -d_0+d_3-h_1 & -i(d_1-id_2+h_0) \\ i(d_3-h_1)+h_2 & -d_2-i(h_0+h_3) & d_1+h_0+h_3 & i(d_0+h_1+ih_2) \end{array}\right|. \tag{18.15}$$

We collect together all the summands in the equation:

$$\left|\begin{array}{cccc} 0 & +\epsilon(d_3-ih_2) & +i\epsilon(h_1-d_3) & +\epsilon(-d_1+id_2) \\ 0 & +\epsilon(-h_3+id_2) & +i\epsilon(-d_1-h_3) & +\epsilon(h_1+ih_2) \\ 0 & -\epsilon(g_3-if_2) & -i\epsilon(f_1-g_3) & -\epsilon(-g_1+ig_2) \\ 0 & -\epsilon(-f_3+ig_2) & -i\epsilon(-g_1-f_3) & -\epsilon(f_1+if_2) \end{array}\right|$$

$$+\left|\begin{array}{cccc} +h_2'+id_3' & 0 & +h_0'-h_3' & +h_2'+id_0' \\ -d_2'-ih_3' & 0 & -d_0'-d_3' & +d_2'-ih_0' \\ -f_2'-ig_3' & 0 & -f_0'-f_3' & +f_2'-ig_0' \\ +g_2'+if_3' & 0 & +g_0'-g_3' & +g_2'+if_0' \end{array}\right|$$

$$+\frac{i}{2r}\left|\begin{array}{cccc} -(2\mu-1)h_1+(2\mu+1)d_3-2ih_2 & -(2\mu-1)(h_0-h_3) & 0 & -(2\mu-1)h_1+(2\mu+1)d_0-2ih_2 \\ +(2\mu+1)d_1+(2\mu-1)h_3+2id_2 & +(2\mu+1)(d_0+d_3) & 0 & -(2\mu+1)d_1+(2\mu-1)h_0-2id_2 \\ +(2\mu-1)f_1-(2\mu+1)g_3+2if_2 & +(2\mu-1)(f_0+f_3) & 0 & -(2\mu-1)f_1-(2\mu+1)g_0-2if_2 \\ -(2\mu+1)g_1-(2\mu-1)f_3-2ig_2 & -(2\mu+1)(g_0-g_3) & 0 & -(2\mu+1)g_1-(2\mu-1)f_0-2ig_2 \end{array}\right|$$

$$+\left|\begin{array}{cccc} +k(d_1-id_2) & +k(d_0-ih_2) & +ik(-d_0+h_1) & 0 \\ +k(-h_1-ih_2) & +k(-h_0-id_2) & +ik(-h_0+d_1) & 0 \\ -k(g_1-ig_2) & -k(g_0+if_2) & -ik(-g_0-f_1) & 0 \\ -k(-f_1-if_2) & -k(-f_0+ig_2) & -ik(-f_0-g_1) & 0 \end{array}\right|$$

$$+2M\left|\begin{array}{cccc} +f_3+g_1-ig_2 & -if_2+g_0+g_3 & i(f_1-g_0-g_3) & +f_0-g_1+ig_2 \\ -g_3+f_1+if_2 & +ig_2+f_0-f_3 & i(f_0-f_3-g_1) & -g_0+f_1+if_2 \\ +h_3+d_1-id_2 & +ih_2+d_0-d_3 & i(-d_0+d_3-h_1) & +h_0+d_1-id_2 \\ -d_3+h_1+ih_2 & -id_2+h_0+h_3 & i(d_1+h_0+h_3) & -d_0-h_1-ih_2 \end{array}\right|=0.$$

We shall further follow the massless case, so the massive term vanishes, and $\mu(r)$ changes to m.

18.2 Massless Field

We have the system

$$\begin{aligned} h_2'+id_3'+\frac{i}{2r}[-(2m-1)h_1+(2m+1)d_3-2ih_2]+k(d_1-id_2)=0\,, \\ -d_2'-ih_3'+\frac{i}{2r}[+(2m+1)d_1+(2m-1)h_3+2id_2]+k(-h_1-ih_2)=0\,, \\ -f_2'-ig_3'+\frac{i}{2r}[+(2m-1)f_1-(2m+1)g_3+2if_2]-k(g_1-ig_2)=0\,, \\ g_2'+if_3'+\frac{i}{2r}[-(2m+1)g_1-(2m-1)f_3-2ig_2]-k(-f_1-if_2)=0\,; \end{aligned} \tag{18.16}$$

$$
\begin{aligned}
&+\epsilon(d_3 - ih_2) + \frac{i}{2r}[-(2m-1)(h_0 - h_3)] + k(d_0 - ih_2) = 0\,,\\
&+\epsilon(-h_3 + id_2) + \frac{i}{2r}[(2m+1)(d_0 + d_3)] + k(-h_0 - id_2) = 0\,,\\
&-\epsilon(g_3 - if_2) + \frac{i}{2r}[(2m-1)(f_0 + f_3)] - k(g_0 + if_2) = 0\,,\\
&-\epsilon(-f_3 + ig_2) - \frac{i}{2r}[(2\mu+1)(g_0 - g_3)] - k(-f_0 + ig_2) = 0;
\end{aligned}
\qquad (18.17)
$$

$$
\begin{aligned}
&+i\epsilon(h_1 - d_3) + h_0' - h_3' + ik(-d_0 + h_1) = 0\,,\\
&+i\epsilon(-d_1 - h_3) - d_0' - d_3' + ik(-h_0 + d_1) = 0\,,\\
&-i\epsilon(f_1 - g_3) - f_0' - f_3' - ik(-g_0 - f_1) = 0\,,\\
&-i\epsilon(-g_1 - f_3) + g_0' - g_3' - ik(-f_0 - g_1) = 0;
\end{aligned}
\qquad (18.18)
$$

$$
\begin{aligned}
&+\epsilon(-d_1 + id_2) + h_2' + id_0' + \frac{i}{2r}[-(2m-1)h_1 + (2m+1)d_0 - 2ih_2] = 0\,,\\
&+\epsilon(h_1 + ih_2) + d_2' - ih_0' + \frac{i}{2r}[-(2m+1)d_1 + (2m-1)h_0 - 2id_2] = 0\,,\\
&-\epsilon(-g_1 + ig_2) + f_2' - ig_0' + \frac{i}{2r}[-(2m-1)f_1 - (2m+1)g_0 - 2if_2] = 0\,,\\
&-\epsilon(f_1 + if_2) + g_2' + if_0' + \frac{i}{2r}[-(2m+1)g_1 - (2m-1)f_0 - 2ig_2] = 0\,.
\end{aligned}
\qquad (18.19)
$$

Let us verify the correctness of the system, by substituting into it the gauge functions. The gauge solutions with cylindric symmetry are determined by the formulas

$$
\bar{\Psi}_n = D_n\Phi, \quad D_n = e_{(n)}^{\alpha}\partial_\alpha + \frac{1}{2}\sigma^{kl}\gamma_{kln}, \quad \Phi = e^{-i\epsilon x^0} e^{im\phi} e^{ikz} \begin{vmatrix} K_1(r) \\ K_2(r) \\ K_3(r) \\ K_4(r) \end{vmatrix}, \qquad (18.20)
$$

$$
D_0 = \partial_0\,, \quad D_1 = \frac{\partial}{\partial r}\,, \quad D_2 = \frac{1}{r}(\partial_\phi + \sigma^{12})\,, \quad D_3 = \partial_z\,,
$$

so they are given by the matrix (the exponential multipliers are omitted)

$$
\bar{\Psi}_n = (\bar{\Psi}_{An}) = \begin{vmatrix}
-i\epsilon K_1 & K_1' & \frac{i}{r}(m-\frac{1}{2})K_1 & ikK_1 \\
-i\epsilon K_2 & K_2' & \frac{i}{r}(m+\frac{1}{2})K_2 & ikK_2 \\
-i\epsilon K_3 & K_3' & \frac{i}{r}(m-\frac{1}{2})K_3 & ikK_3 \\
-i\epsilon K_4 & K_4' & \frac{i}{r}(m+\frac{1}{2})K_4 & ikK_4
\end{vmatrix} = \begin{vmatrix}
\bar{f}_0 & \bar{f}_1 & \bar{f}_2 & \bar{f}_3 \\
\bar{g}_0 & \bar{g}_1 & \bar{g}_2 & \bar{g}_3 \\
\bar{h}_0 & \bar{h}_1 & \bar{h}_2 & \bar{h}_3 \\
\bar{d}_0 & \bar{d}_1 & \bar{d}_2 & \bar{d}_3
\end{vmatrix}. \qquad (18.21)
$$

It is readily proved that all the 16 equations are satisfied by these functions. It should be especially noted that the choice of any particular explicit form of the separated gauge functions K_1, K_2, K_3, K_4 does not matter.

To proceed with the system, let us transform it to the new variables

$$
\begin{aligned}
&f_1+if_2=F_1, \quad f_1-if_2=F_2, \quad f_1=\frac{1}{2}(F_1+F_2), \quad f_2=\frac{1}{2i}(F_1-F_2);\\
&g_1+ig_2=G_1, \quad g_1-ig_2=G_2, \quad g_1=\frac{1}{2}(G_1+G_2), \quad g_2=\frac{1}{2i}(G_1-G_2);\\
&h_1+ih_2=H_1, \quad h_1-ih_2=H_2, \quad h_1=\frac{1}{2}(H_1+H_2), \quad h_2=\frac{1}{2i}(H_1-H_2);\\
&d_1+id_2=D_1, \quad d_1-id_2=D_2, \quad d_1=\frac{1}{2}(D_1+D_2), \quad d_2=\frac{1}{2i}(D_1-D_2);\\
&f_0+f_3=F_0, \quad f_0-f_3=F_3, \quad f_0=\frac{1}{2}(F_0+F_3), \quad f_3=\frac{1}{2}(F_0-F_3);\\
&g_0+g_3=G_0, \quad g_0-g_3=G_3, \quad g_0=\frac{1}{2}(G_0+G_3), \quad g_3=\frac{1}{2}(G_0-G_3);\\
&h_0+h_3=H_0, \quad h_0-h_3=H_3, \quad h_0=\frac{1}{2}(H_0+H_3), \quad h_3=\frac{1}{2}(H_0-H_3);\\
&d_0+d_3=D_0, \quad d_0-d_3=D_3, \quad d_0=\frac{1}{2}(D_0+D_3), \quad d_3=\frac{1}{2}(D_0-D_3);
\end{aligned}
\tag{18.22}
$$

in this way, we arrive at new equations, which may be divided into two unlinked subsystems, each of 8 equations. It is convenient to apply special the following notations for the two operators:

$$
a=\frac{d}{dr}+\frac{1+2m}{2\,r}, \qquad b=\frac{d}{dr}+\frac{1-2m}{2\,r}. \tag{18.23}
$$

First, let us consider the first subsystem:

$$
\begin{aligned}
&iaF_1+iaG_3-i(b+\frac{1}{r})F_2=(\epsilon+k)G_2,\\
&ibG_2+ibF_0-i(a+\frac{1}{r})G_1=(\epsilon-k)F_1,\\
&iaG_0=(\epsilon-k)G_2, \quad ibF_3=(\epsilon+k)F_1,\\
&i(a-\frac{1}{r})F_0=F_2(\epsilon-k), \quad i(b-\frac{1}{r})G_3=G_1(\epsilon+k),\\
&ibF_0=F_1(\epsilon-k)-G_0(\epsilon+k)+G_3(\epsilon-k),\\
&iaG_3=G_2(\epsilon+k)-F_3(\epsilon-k)+F_0(\epsilon+k).
\end{aligned}
\tag{18.24}
$$

With the help of equations 7 and 8 we can exclude the variables F_0, G_3 in equations 1 and 2:

$$
\begin{aligned}
&iaF_1-F_3(\epsilon-k)+F_0(\epsilon+k)-i(b+1/r)F_2=0,\\
&ibG_2-G_0(\epsilon+k)+G_3(\epsilon-k)-i(a+1/r)G_1=0.
\end{aligned}
$$

Further, with the help of equations 3,4,5,6 we exclude the variables F_1, F_2 and G_1, G_2:

$$
-\frac{1}{(\epsilon+k)}abF_3-F_3(\epsilon-k)+F_0(\epsilon+k)+\frac{1}{(\epsilon-k)}(b+\frac{1}{r})(a-\frac{1}{r})F_0=0,
$$

$$
-b\frac{1}{(\epsilon-k)}aG_0-G_0(\epsilon+k)+G_3(\epsilon-k)+(a+\frac{1}{r})\frac{1}{(\epsilon+k)}(b-\frac{1}{r})G_3=0,
$$

or

$$\Big[ab+\epsilon^2-k^2\Big]\frac{F_3}{\epsilon+k}=\Big[(b+\frac{1}{r})(a-\frac{1}{r})+\epsilon^2-k^2\Big]\frac{F_0}{\epsilon-k},$$
$$\Big[ba+\epsilon^2-k^2\Big]\frac{G_0}{\epsilon-k}=\Big[(a+\frac{1}{r})(b-\frac{1}{r})+\epsilon^2-k^2\Big]\frac{G_3}{\epsilon+k}.$$

Taking in mind the following relations

$$ab=(b+\frac{1}{r})(a-\frac{1}{r})=\frac{d^2}{dr^2}+\frac{1}{r}\frac{d}{dr}-\frac{(m-1/2)^2}{r^2}=\Delta_-,$$
$$ba=(a+\frac{1}{r})(b-\frac{1}{r})=\frac{d^2}{dr^2}+\frac{1}{r}\frac{d}{dr}-\frac{(m+1/2)^2}{r^2}=\Delta_+, \tag{18.25}$$

we get the two second-order equations:

$$\Big[\frac{d^2}{dr^2}+\frac{1}{r}\frac{d}{dr}-\frac{(m-1/2)^2}{r^2}\Big]\Big(\frac{F_0}{\epsilon-k}-\frac{F_3}{\epsilon+k}\Big)=0,$$
$$\Big[\frac{d^2}{dr^2}+\frac{1}{r}\frac{d}{dr}-\frac{(m+1/2)^2}{r^2}\Big]\Big(\frac{G_0}{\epsilon-k}-\frac{G_3}{\epsilon+k}\Big)=0. \tag{18.26}$$

We will impose the following constraints of two types:

$$A,\qquad \frac{1}{\epsilon+k}F_3(r)=+\frac{1}{\epsilon-k}F_0(r)=f(r),\ \frac{1}{\epsilon+k}G_3(r)=+\frac{1}{\epsilon-k}G_0(r)=g(r); \tag{18.27}$$

$$B,\qquad \frac{1}{\epsilon+k}F_3(r)=-\frac{1}{\epsilon-k}F_0(r)=f(r),\ \frac{1}{\epsilon+k}G_3(r)=-\frac{1}{\epsilon-k}G_0(r)=g(r). \tag{18.28}$$

In the case (18.27), the equations (18.26) are identically satisfied and the functions $f(r)$ and $g(r)$ may be arbitrary. We shall see below that case (18.27) corresponds to the gauge solutions.

In the case (18.28), the functions $f(r),g(r)$ obey the definite equations

$$F_0,F_3,\qquad \Big[\frac{d^2}{dr^2}+\frac{1}{r}\frac{d}{dr}+\epsilon^2-k^2-\frac{(m-1/2)^2}{r^2}\Big]f(r)=0,$$
$$G_0,G_3,\qquad \Big[\frac{d^2}{dr^2}+\frac{1}{r}\frac{d}{dr}+\epsilon^2-k^2-\frac{(m+1/2)^2}{r^2}\Big]g(r)=0, \tag{18.29}$$

and all the remaining functions can be found through the formulas

$$F_1=\frac{ib}{(\epsilon+k)}F_3=ibf,\quad F_2=\frac{i(a-1/r)}{\epsilon-k}F_0=i(a-1/r)f,$$
$$G_1=\frac{i(b-1/r)}{\epsilon+k}G_3=i(b-1/r)g,\quad G_2=\frac{ia}{\epsilon-k}G_0=iag. \tag{18.30}$$

Let us prove the consistency of restrictions (18.27) with the complete system of equations. To this end, in these equations we should take into account the constraints

$$B,\qquad \frac{1}{\epsilon+k}F_3(r)=-\frac{1}{\epsilon-k}F_0(r),\qquad \frac{1}{\epsilon+k}G_3(r)=-\frac{1}{\epsilon-k}G_0(r).$$

In this way, the equations 3,4,5,6 from (18.24) give

$$F_1 = -\frac{ib}{\epsilon - k} F_0 , \quad F_2 = \frac{i(a-1/r)}{\epsilon - k} F_0 , \quad G_1 = -\frac{i(b-1/r)}{(\epsilon - k)} G_0 , \quad G_2 = \frac{ia}{(\epsilon - k)} G_0 .$$

Let us consider the remaining four equations from (18.24):

$$iaF_1 + iaG_3 - i(b+\frac{1}{r})F_2 = (\epsilon + k)G_2 \implies$$

$$-ia\frac{ib}{\epsilon - k}F_0 - ia\frac{\epsilon + k}{\epsilon - k}G_0 - i(b+\frac{1}{r})\frac{i(a-1/r)}{\epsilon - k}F_0 = \frac{\epsilon + k}{(\epsilon - k}iaG_0 \implies$$

$$\Big[ab + (b+1/r)(a-1/r)\Big]F_0 = 2i(\epsilon + k)aG_0, \quad \implies \quad abF_0 = i(\epsilon + k)aG_0 ,$$

that is

$$bF_0 = i(\epsilon + k)G_0; \quad ibG_2 + ibF_0 - i(a+\frac{1}{r})G_1 = +(\epsilon - k)F_1 \implies$$

$$ib\frac{ia}{(\epsilon - k)}G_0 + ibF_0 + i(a+\frac{1}{r})\frac{i(b-1/r)}{(\epsilon - k)}G_0 = -(\epsilon - k)\frac{ib}{\epsilon - k}F_0 ,$$

$$\Big[ba + (a+1/r)(b-1/r)\Big]G_0 = 2i(\epsilon - k)bF_0 \quad \implies \quad baG_0 = i(\epsilon - k)bF_0 ,$$

that is

$$aG_0 = i(\epsilon - k)F_0 , \quad ibF_0 = F_1(\epsilon - k) - G_0(\epsilon + k) + G_3(\epsilon - k) \implies$$

$$ibF_0 = -ibF_0 - (\epsilon + k)G_0 - (\epsilon + k)G_0 ,$$

and, moreover,

$$ibF_0 = -(\epsilon + k)G_0 , \quad iaG_3 = G_2(\epsilon + k) - F_3(\epsilon - k) + F_0(\epsilon + k) \implies$$

$$-ia\frac{\epsilon + k}{\epsilon - k}G_0 = (\epsilon + k)\frac{ia}{(\epsilon - k)}G_0 + (\epsilon + k)F_0 + F_0(\epsilon + k) ,$$

that is $iaG_0 = -(\epsilon - k)F_0$. Thus, the restrictions (18.28) are consistent with the system (18.24), only when the following four equations hold

$$abF_0 = i(\epsilon + k)aG_0 , \quad baG_0 = i(\epsilon - k)bF_0 ,$$
$$bF_0 = i(\epsilon + k)G_0 , \quad aG_0 = i(\epsilon - k)F_0 , \tag{18.31}$$

i.e., only when the two equations are independent,

$$bF_0 = i(\epsilon + k)G_0 , \quad aG_0 = i(\epsilon - k)F_0 . \tag{18.32}$$

We take in mind that in the notations $F_0 = -(\epsilon - k)f$, $G_0 = -(\epsilon - k)g$, the multiplier before f and g may be hidden in the new notations:

$$F_0 = -(\epsilon - k)f = F , \quad G_0 = -(\epsilon - k)g = G ; \tag{18.33}$$

correspondingly, the equations (18.32) can be written as follows

$$bF = i(\epsilon + k)G, \quad aG = i(\epsilon - k)F. \tag{18.34}$$

These 1-st order equations assume the above equations (18.29):

$$\begin{array}{l} bF = i(\epsilon+k)G \\ aG = i(\epsilon-k)F \end{array} \quad \Longrightarrow \quad \begin{array}{l} abF = i(\epsilon+k)aG = -(\epsilon+k)(\epsilon-k)F \\ baG = i(\epsilon-k)bF = -(\epsilon-k)(\epsilon+k)G \end{array}. \tag{18.35}$$

The existence of the constrains (18.34) means that the system of 8 equations with restriction of the type (18.28) describes only one solution, because the constrains (18.34) permit us to fix the relative coefficient between the functions $F(r)$ and $G(r)$. To derive the consequences from the other 8 equations, let us write down and compare both subsystems: the first is

$$\begin{gathered} iaF_1 + iaG_3 - i(b+\frac{1}{r})F_2 = (\epsilon+k)G_2, \\ ibG_2 + ibF_0 - i(a+\frac{1}{r})G_1 = (\epsilon-k)F_1, \\ iaG_0 = (\epsilon-k)G_2, \qquad ibF_3 = (\epsilon+k)F_1, \\ i(a-\frac{1}{r})F_0 = F_2(\epsilon-k), \quad i(b-\frac{1}{r})G_3 = G_1(\epsilon+k), \\ ibF_0 = F_1(\epsilon-k) - G_0(\epsilon+k) + G_3(\epsilon-k), \\ iaG_3 = G_2(\epsilon+k) - F_3(\epsilon-k) + F_0(\epsilon+k), \end{gathered} \tag{18.36}$$

and the second is

$$\begin{gathered} iaH_1 - iaD_0 - i(b+1/r)H_2 = -(\epsilon-k)D_2, \\ ibD_2 - ibH_3 - i(a+1/r)D_1 = -(\epsilon+k)H_1, \\ -iaD_3 = -(\epsilon+k)D_2, \qquad -ibH_0 = -(\epsilon-k)H_1, \\ -i(a-1/r)H_3 = -(\epsilon+k)H_2, \quad -i(b-1/r)D_0 = -(\epsilon-k)D_1, \\ -ibH_3 = -H_1(\epsilon+k) - D_3(\epsilon-k) + D_0(\epsilon+k), \\ -iaD_0 = -D_2(\epsilon-k) - H_0(\epsilon+k) + H_3(\epsilon-k), \end{gathered} \tag{18.37}$$

We can notice that the equations (18.37) follow from (18.36) by means of the changes:

$$\begin{array}{l} F_1 \Rightarrow H_1, \quad F_2 \Rightarrow H_2, \quad F_0 \Rightarrow -H_3, \quad F_3 \Rightarrow -H_{03}, \quad (\epsilon+k) \Rightarrow -(\epsilon-k), \\ G_1 \Rightarrow D_1, \quad G_2 \Rightarrow D_2, \quad G_0 \Rightarrow -D_3, \quad G_3 \Rightarrow -D_0, \quad (\epsilon-k) \Rightarrow +(\epsilon+k); \end{array} \tag{18.38}$$

correspondingly we derive the rules

$$F \Longrightarrow \frac{-H_3}{-(\epsilon+k)} + \frac{H_0}{-(\epsilon-k)} = -H, \qquad G \Longrightarrow \frac{-D_3}{-(\epsilon+k)} + \frac{D_0}{-(\epsilon-k)} = -D. \tag{18.39}$$

Therefore, the final equations from the first subsystem

$$\begin{gathered} iaG = -(\epsilon-k)F, \quad (\Delta_- + \epsilon^2 - k^2)F = 0, \\ ibF = -(\epsilon+k)G, \quad (\Delta_+ + \epsilon^2 - k^2)G = 0 \end{gathered} \tag{18.40}$$

transform to those for the second subsystem

$$iaD = (\epsilon + k)H, \quad (\Delta_{-} + \epsilon^2 - k^2)H = 0,$$
$$ibH = (\epsilon - k)D, \quad (\Delta_{+} + \epsilon^2 - k^2)H = 0. \tag{18.41}$$

18.3 Gauge Solutions

Let us collect together the results from solving the system of 16 equations for the case (18.27):

$$\begin{aligned}
&F_0 = (\epsilon - k)f, && F_1 = ibf, && F_2 = i(a - 1/r)f, && F_3 = (\epsilon + k)f,\\
&G_0 = (\epsilon - k)g, && G_1 = i(b - 1/r)g, && G_2 = iag, && G_3 = (\epsilon + k)g,\\
&H_0 = (\epsilon - k)h, && H_1 = ibh, && H_2 = i(a - 1/r)h, && H_3 = (\epsilon + k)h,\\
&D_0 = (\epsilon - k)d, && D_1 = i(b - 1/r)d, && D_2 = iad, && D_3 = (\epsilon + k)d.
\end{aligned} \tag{18.42}$$

In fact we have four independent solutions, determined by the four arbitrary functions

$$(f,0,0,0), \quad (0,g,0,0), \quad (0,0,h,0, \quad (0,0,0,d).$$

Let us transform the above gauge solutions (18.21) to similar variables; we obtain

$$\begin{aligned}
&\bar{F}_0 = -i(\epsilon - k)K_1, && \bar{F}_1 = (\tfrac{d}{dr} - \tfrac{m-1/2}{r})K_1, && \bar{F}_2 = (\tfrac{d}{dr} + \tfrac{m-1/2}{r})K_1, && \bar{F}_3 = -i(\epsilon + k)K_1,\\
&\bar{G}_0 = -i(\epsilon - k)K_1, && \bar{G}_1 = (\tfrac{d}{dr} - \tfrac{m+1/2}{r})K_2, && \bar{G}_2 = (\tfrac{d}{dr} + \tfrac{m+1/2}{r})K_2, && \bar{G}_3 = -i(\epsilon + k)K_2,\\
&\bar{H}_0 = -i(\epsilon - k)K_3, && \bar{H}_1 = (\tfrac{d}{dr} - \tfrac{m-1/2}{r})K_3, && \bar{H}_2 = (\tfrac{d}{dr} + \tfrac{m-1/2}{r})K_3, && \bar{H}_3 = -i(\epsilon + k)K_3,\\
&\bar{D}_0 = -i(\epsilon - k)K_4, && \bar{D}_1 = (\tfrac{d}{dr} - \tfrac{m+1/2}{r})K_4, && \bar{D}_2 = (\tfrac{d}{dr} + \tfrac{m+1/2}{r})K_4, && \bar{D}_3 = -i(\epsilon + k)K_4.
\end{aligned}$$

We can readily see that the matrix of gauge solutions coincides with the matrix (18.42) of solutions for the case (18.27) (up to the multiplier i).

18.4 Solving the Second Order Equations, the First Order Constraints

Let us find solutions of the two equations

$$\Big[\frac{d^2}{dr^2} + \frac{1}{r}\frac{d}{dr} + \epsilon^2 - k^2 - \frac{(m+1/2)^2}{r^2}\Big]G(r) = 0,$$
$$\Big[\frac{d^2}{dr^2} + \frac{2}{r}\frac{d}{dr} + \epsilon^2 - k^2 - \frac{(m-1/2)^2}{r^2}\Big]F(r) = 0. \tag{18.43}$$

In the variable $x = \sqrt{\epsilon^2 - k^2}\, r$, they take the form of Bessel equations:

$$\Big[\frac{d^2}{dx^2} + \frac{1}{x}\frac{d}{dx} + 1 - \frac{p^2}{x^2}\Big]G(x) = 0, \qquad p = m + \frac{1}{2}, -p = -m - \frac{1}{2};$$
$$\Big[\frac{d^2}{dx^2} + \frac{1}{x}\frac{d}{dx} + 1 - \frac{s^2}{x^2}\Big]F(x) = 0, s = m - \frac{1}{2} = p - 1, -s = -m + \frac{1}{2} = -p + 1. \tag{18.44}$$

Their independent solutions are

$$G_{(1)}(x) \sim J_p(x) , \qquad G_{(2)}(x) \sim J_{-p}(x) ;$$
$$F_{(1)}(x) \sim J_s(x) = J_{p-1}(x) , \quad F_{(2)}(x) \sim J_{-s}(x) = J_{-p+1}(x) . \tag{18.45}$$

We shall further consider only the solutions which are regular at $x=0$. First, we assume positive values for m. Let us transform the first order constraints to Bessel form:

$$aG = i(\epsilon - k)F , \quad \Big(\frac{d}{dr} + \frac{m+1/2}{r}\Big)G = i(\epsilon - k)F ;$$
$$bF = i(\epsilon + k)G , \quad \Big(\frac{d}{dr} - \frac{m-1/2}{r}\Big)F = i(\epsilon + k)G .$$

The first constraint gives

$$\sqrt{\epsilon+k}\Big(\frac{d}{dx} + \frac{m+1/2}{x}\Big)G = i\sqrt{\epsilon-k}F , \quad G_{(1)} = \beta J_p , \quad F_{(1)1} = \alpha J_{p-1} \implies$$
$$\sqrt{\epsilon+k}\Big(\frac{d}{dx} + \frac{p}{x}\Big)\beta J_p = i\sqrt{\epsilon-k}\alpha J_{p-1} . \tag{18.46}$$

The second constraint gives

$$\sqrt{\epsilon-k}\Big(\frac{d}{dx} - \frac{m-1/2}{x}\Big)F = i\sqrt{\epsilon+k}\, G , \quad F_{(1)} = \alpha J_{p-1} , \quad G_{(1)} = \beta J_p \implies$$
$$\sqrt{\epsilon-k}\,\Big(\frac{d}{dx} - \frac{p-1}{x}\Big)\alpha J_{p-1} = i\sqrt{\epsilon+k}\,\beta J_p . \tag{18.47}$$

Further, taking in mind the known properties of Bessel functions

$$\Big(\frac{d}{dx} + \frac{p}{x}\Big)J_p = J_{p-1} , \quad \Big(\frac{d}{dx} - \frac{p-1}{x}\Big) J_{p-1} = -J_p ,$$

we derive two algebraic relations which determine the coefficients α and β:

$$\sqrt{\epsilon+k}\,\beta = i\sqrt{\epsilon-k}\,\alpha, \ \sqrt{\epsilon-k}\,\alpha = -i\sqrt{\epsilon+k}\,\beta \implies \alpha = \sqrt{\epsilon+k} , \ \beta = i\sqrt{\epsilon-k} .$$

Therefore, the needed solution is

$$F_{(1)}(x) = \sqrt{\epsilon+k}\, J_{p-1}(x) , \quad G_{(1)}(x) = i\sqrt{\epsilon-k}\, J_p(x) , \quad m > 0 . \tag{18.48}$$

A similar solution at negative m has the form

$$F_{(2)}(x) = \sqrt{\epsilon+k}\, J_{-p+1}(x) , \quad G_{(2)}(x) = i\sqrt{\epsilon-k}\, J_{-p}(x) , \quad m < 0 . \tag{18.49}$$

According to the above noted symmetry, we can obtain similar results for the functions $H(x)$ and $D(x)$:

$$H_{(1)}(x) = \sqrt{\epsilon-k}\, J_{p-1}(x) , \quad D_{(1)}(x) = i\sqrt{\epsilon+k}\, J_p(x) , \quad m > 0 ; \tag{18.50}$$

$$H_{(2)}(x) = \sqrt{\epsilon-k}\, J_{-p+1}(x) , \quad D_{(2)}(x) = i\sqrt{\epsilon+k}\, J_{-p}(x) , \quad m < 0 . \tag{18.51}$$

Thus, six linearly independent solutions are constructed, of which four coincide with the gauge ones, and two solutions do not contain gauge degrees of freedom.

Chapter 19

On the Matrix Equation for a Spin 2 Particle in Riemannian Space-Time, Tetrad Method

19.1 The Spin 2 Particle in Minkowski Space

After the study by Pauli and Fierz [372], the theory of massive and massless fields with spin 2 has always attracted much attention [372] –[?]. Most of the studies were performed in the framework of 2-nd order differential equations. It is known that many specific difficulties may be avoided if from the very beginning we start with 1-st order systems. Apparently, the first systematic study of the theory of spin 2 fields within the first order formalism was done by F.I. Fedorov [373]. It turns out that this description requires a field function with 3 independent components. This theory was rediscovered and improved by Regee [374]. The first order approach is based from the very beginning on the general theory of relativistic wave equations by the Gel'fand–Yaglom and the Lagrangian formalisms.

We start with the known system of the first order equations for a massive spin 2 particle [139]:

$$
\begin{gathered}
\partial^a \Phi_a = m\Phi\,, \quad \frac{1}{2}\partial_a \Phi - \frac{1}{3}\partial^b \Phi_{(ab)} = m\Phi_a\,, \\
\frac{1}{2}\Big(\partial^k \Phi_{[ka]b} + \partial^k \Phi_{[kb]a} - \frac{1}{2} g_{ab}\partial^k \Phi_{[kn]}{}^{n}\Big) \\
+\Big(\partial_a \Phi_b + \partial_b \Phi_a - \frac{1}{2} g_{ab}\partial^k \Phi_k\Big) = m\Phi_{(ab)}\,, \\
\partial_a \Phi_{(bc)} - \partial_b \Phi_{(ac)} + \frac{1}{3}\Big(g_{bc}\partial^k \Phi_{(ak)} - g_{ac}\partial^k \Phi_{(bk)}\Big) = m\Phi_{[ab]c}\,,
\end{gathered}
\tag{19.1}
$$

where the field variables are scalar, vector, symmetric 2-rank tensor, and 3-rank skew-symmetric in the two first indices tensor, $m = \mathrm{i}M$. By excluding the vector and the 3-rank tensor, we obtain the 2-nd order equations with respect to the scalar and symmetric tensor:

$$
\Phi = 0\,, \quad (\Box + M^2)\Phi_{(ab)} = 0\,, \quad \Phi_{(ab)} = \Phi_{(ba)}\,, \quad \Phi^a{}_a = 0\,, \quad \partial^k \Phi_{(ka)} = 0\,. \tag{19.2}
$$

It should be stressed out that the scalar field entering the systems (19.1) and (19.2) is all-important, because it vanishes only in absence of external fields. Moreover, this scalar variable is of crucial importance for the massless case, when the first order systems read

$$\begin{gathered}
\partial^a \Phi_a = 0\,, \frac{1}{2}\partial_a \Phi - \frac{1}{3}\partial^b \Phi_{(ab)} = \Phi_a\,, \\
\frac{1}{2}\Big(\partial^k \Phi_{[ka]b} + \partial^k \Phi_{[kb]a} - \frac{1}{2} g_{ab} \partial^k \Phi_{[kn]}^{\ \ \ n}\Big) \\
+\Big(\partial_a \Phi_b + \partial_b \Phi_a - \frac{1}{2} g_{ab} \partial^k \Phi_k\Big) = 0\,, \\
\partial_a \Phi_{(bc)} - \partial_b \Phi_{(ac)} + \frac{1}{3}\Big(g_{bc}\partial^k \Phi_{(ak)} - g_{ac}\partial^k \Phi_{(bk)}\Big) = \Phi_{[ab]c}\,.
\end{gathered} \tag{19.3}$$

By excluding from (19.3) the subsidiary variables, we derive the 2-nd order equations for the massless field:

$$\begin{gathered}
\frac{1}{2}\Box\Phi - \frac{1}{3}\partial^a\partial^b \Phi_{(ab)} = 0\,, \\
(\partial_a\partial_b + \frac{1}{2} g_{ab}\Box)\Phi - \frac{1}{4} g_{ab}\Box\Phi^n_n + \Box\Phi_{(ab)} - \partial_a\partial^n\Phi_{(nb)} - \partial_b\partial^n\Phi_{(na)} = 0\,.
\end{gathered} \tag{19.4}$$

As it was shown by Pauli and Fierz, these equations have a class of so called gauge solutions:

$$\bar{\Phi} = \partial^l L_l\,, \quad \bar{\Phi}_{(ab)} = \partial_a L_b + \partial_b L_a - \frac{1}{2} g_{ab}\partial^l L_l\,, \tag{19.5}$$

where $L_l(x)$ stands for an arbitrary 4-vector. These special states do not contribute to physically observable quantities, for instance to the energy-momentum tensor. It is a matter of simple calculation to find the concomitant tensor components:

$$\begin{gathered}
\bar{\Phi}_a = \frac{1}{3}\partial_a\partial^l L_l - \frac{1}{3}\Box L_a\,, \quad \bar{\Phi}_{[ab]c} = \partial_c(\partial_a L_b - \partial_b L_a) \\
-\frac{1}{3}(g_{cb}\partial_a - g_{ca}\partial_b)\partial^l L_l + \frac{1}{3}(g_{cb}\Box L_a - g_{ca}\Box L_b)\,.
\end{gathered} \tag{19.6}$$

19.2 Structure of the Matrices of the First Order System for a Spin 2 Field

The system (19.1) can be written in equivalent form

$$\begin{gathered}
\partial_a (G^a)_{(0)}^{\ \ k}\Phi_k = m\Phi_{(0)}\,, \\
\partial_a\Big\{\frac{1}{2}(\Delta^a)_k^{\ (0)}\Phi_{(0)} - \frac{1}{3}(K^a)_k^{\ (mn)}\Phi_{mn}\Big\} = m\Phi_k\,, \\
\partial_a\Big\{\frac{1}{2}\,(B^a)_{(cd)}^{\ \ [mn]l}\Phi_{mnl} + (\Lambda^a)_{(dc)}^{\ \ k}\Phi_k\Big\} = m\Phi_{dc}\,, \\
\partial_a\Big\{(F^a)_{[kb]c}^{\ \ (mn)}\Phi_{mn}\Big\} = m\,\Phi_{kbc}\,,
\end{gathered} \tag{19.7}$$

where we use the block matrices

$$
\begin{gathered}
(G^a)_{(0)}{}^{k} = g^{ak}, \quad (\Delta^a)_k{}^{(0)} = \delta^a_k, \\
(K^a)_k{}^{(mn)} = g^{al}\frac{1}{2}\,\delta^{(mn)}_{(lk)}, \quad (\Lambda^a)_{(dc)}{}^{k} = \delta^{(ak)}_{(dc)} - \frac{1}{2}g_{dc}g^{ak}, \\
(B^a)_{(dc)}{}^{[mn]l} = g^{ak}\Big(\frac{1}{2}\delta^{[mn]}_{[kd]}\delta^l_c + \frac{1}{2}\delta^{[mn]}_{[kc]}\delta^l_d\Big) - \frac{1}{2}g_{dc}\frac{1}{2}g^{[mn],[al]}, \\
(F^a)_{[kb]c}{}^{(mn)} = \Big(\delta^a_k\frac{1}{2}\delta^{(mn)}_{(bc)} - \delta^a_b\frac{1}{2}\delta^{(mn)}_{(kc)}\Big) + \frac{1}{3}g^{ad}\Big(\frac{1}{2}\delta^{(mn)}_{(dk)}g_{bc} - \frac{1}{2}\delta^{(mn)}_{(db)}g_{kc}\Big);
\end{gathered}
\tag{19.8}
$$

the generalized Kronecker symbols are defined by the formulas

$$
\begin{gathered}
\delta^{(mn)}_{(lk)} = \delta^m_l\,\delta^n_k + \delta^n_l\,\delta^m_k, \quad \delta^{[mn]}_{[lk]} = \delta^m_l\,\delta^n_k - \delta^n_l\,\delta^m_k, \\
g^{[mn],[al]} = g^{ma}g^{nl} - g^{ml}g^{na}, \quad g^{(mn),(al)} = g^{ma}g^{nl} + g^{ml}g^{na}.
\end{gathered}
\tag{19.9}
$$

Further, by introducing the matrices

$$
\Gamma^a = \begin{vmatrix} 0 & G^a & 0 & 0 \\ \frac{1}{2}\Delta^a & 0 & -\frac{1}{3}K^a & 0 \\ 0 & \Lambda^a & 0 & \frac{1}{2}B^a \\ 0 & 0 & F^a & 0 \end{vmatrix}, \tag{19.10}
$$

we write the system (19.7) as follows

$$
(\Gamma^a\partial_a - m)\Psi = 0, \quad \Psi = \begin{vmatrix} \Phi \\ \Phi_k \\ \Phi_{(mn)} \\ \Phi_{[mn]l} \end{vmatrix}. \tag{19.11}
$$

It is a matter of straightforward calculation to get the explicit expressions for these block matrices. To this aim we will follow only the independent components of the tensors:

$$
\Psi = \{\,\Phi;\ \Phi_l;\ \vec{f},\ \vec{c},\ \vec{d},\ f_0;\ \varphi_0,\ \varphi_1,\ \varphi_2,\ \varphi_3\,\}, \quad (\Phi_{(ab)}) = \begin{vmatrix} f_0 & d_1 & d_2 & d_3 \\ d_1 & f_1 & c_3 & c_2 \\ d_2 & c_3 & f_2 & c_1 \\ d_3 & c_2 & c_1 & f_3 \end{vmatrix},
$$

while the components of the tensor $\Phi_{[mn]k}$ are written as four skew-symmetric matrices:

$$
\begin{vmatrix} \Phi_{[01]k} \\ \Phi_{[02]k} \\ \Phi_{[03]k} \\ \Phi_{[23]k} \\ \Phi_{[31]k} \\ \Phi_{[12]k} \end{vmatrix} = \begin{vmatrix} E_{10} \\ E_{20} \\ E_{30} \\ B_{10} \\ B_{20} \\ B_{30} \end{vmatrix} = \varphi_0, \quad \begin{vmatrix} E_{11} \\ E_{21} \\ E_{31} \\ B_{11} \\ B_{21} \\ B_{31} \end{vmatrix} = \varphi_1, \quad \begin{vmatrix} E_{12} \\ E_{22} \\ E_{32} \\ B_{12} \\ B_{22} \\ B_{32} \end{vmatrix} = \varphi_2, \quad \begin{vmatrix} E_{13} \\ E_{23} \\ E_{3k} \\ B_{13} \\ B_{23} \\ B_{33} \end{vmatrix} = \varphi_3.
$$

For the involved block matrices, we find the explicit expressions, which we shall include in the Appendix.

19.3 Extension to Riemannian Space-Time Geometry

The matrix equation

$$\Big(\Gamma^a \frac{\partial}{\partial x^a} - m\Big)\Psi(x) = 0, \quad \Psi = \{H; H_1; H_2; H_3\} \tag{19.12}$$

is extended to the Riemannian space-time in accordance with the tetrad method. In a space-time with given metric, let us fix a tetrad:

$$dS^2 = g_{\alpha\beta}(x)dx^\alpha dx^\beta, \quad g_{\alpha\beta}(x) \to e_{(a)\alpha}(x),$$
$$g_{\alpha\beta}(x) = \eta^{ab} e_{(a)\alpha}(x) e_{(b)\beta}(x), \quad \eta^{ab} = \text{diag}(+1,-1,-1,-1); \tag{19.13}$$

then, the generalized equation can be written as

$$\Big[\Gamma^\alpha(x)\Big(\frac{\partial}{\partial x^\alpha} + \Sigma_\alpha(x)\Big) - m\Big]\Psi(x) = 0, \tag{19.14}$$

where the local matrices $\Gamma^\alpha(x)$ are determined with the use of the tetrad

$$\Gamma^\alpha(x) = e^\alpha_{(a)}(x)\Gamma^a = \left|\begin{array}{cccc} 0 & G^\alpha(x) & 0 & 0 \\ \frac{1}{2}\Delta^\alpha(x) & 0 & -\frac{1}{3}K^\alpha(x) & 0 \\ 0 & \Lambda^\alpha(x) & 0 & \frac{1}{2}B^\alpha(x) \\ 0 & 0 & F^\alpha(x) & 0 \end{array}\right|, \tag{19.15}$$

and the connection $\Sigma_\alpha(x)$ is defined by the following relations

$$J^{ab} = \left|\begin{array}{cccc} 0 & 0 & 0 & 0 \\ 0 & J^{ab}_1 & 0 & 0 \\ 0 & 0 & J^{ab}_2 & 0 \\ 0 & 0 & 0 & J^{ab}_3 \end{array}\right|,$$
$$\Sigma_\alpha(x) = J^{ab} e^\beta_{(a)}(x)\, e_{(b)\beta;\,\alpha}(x) = \left|\begin{array}{cccc} 0 & 0 & 0 & 0 \\ 0 & (\Sigma_1)_\alpha & 0 & 0 \\ 0 & 0 & (\Sigma_2)_\alpha & 0 \\ 0 & 0 & 0 & (\Sigma_3)_\alpha \end{array}\right|, \tag{19.16}$$

where

$$\Sigma_i(x) = J^{ab}_i\, e^\beta_{(a)}(x)\, e_{(b)\beta;\,\alpha}(x), \quad i = 1,2,3;$$

and $J^{ab}_1, J^{ab}_2, J^{ab}_3$ stand for the generators for the tensors $\Phi_k, \Phi_{(mn)}, \Phi_{[mn]l}$. The equation (19.14) can be presented by using the Ricci rotation coefficients

$$\Big[\Gamma^c\Big(e^\alpha_{(c)}(x)\frac{\partial}{\partial x^\alpha} + \frac{1}{2}J^{ab}\gamma_{abc}\Big) - m\Big]\Psi(x) = 0, \tag{19.17}$$

where we recall the definition $\gamma_{[ab]c} = -\gamma_{[ba]c} = e_{(b)\rho;\,\sigma} e^\rho_{(a)} e^\sigma_{(c)}$. In block form, (19.14) reads

$$G^\alpha(x)[\partial_\alpha + (\Sigma_1)_\alpha]H_1 = mH,$$
$$\frac{1}{2}\Delta^\alpha(x)\partial_\alpha H - \frac{1}{3}K^\alpha(x)[\partial_\alpha + (\Sigma_2)_\alpha]H_2 = mH_1,$$
$$\Lambda^\alpha(x)[\partial_\alpha + (\Sigma_1)_\alpha]H_1 + \frac{1}{2}[\partial_\alpha + (\Sigma_3)_\alpha]H_3 = mH_2,$$
$$F^\alpha(x)[\partial_\alpha + (\Sigma_2)_\alpha]H_2 = mH_3. \tag{19.18}$$

In the massless case, the system slightly changes:

$$
\begin{gathered}
G^{\alpha}(x)[\partial_{\alpha}+(\Sigma_1)_{\alpha}]H_1=0\,,\\
\frac{1}{2}\Delta^{\alpha}(x)\partial_{\alpha}H-\frac{1}{3}K^{\alpha}(x)[\partial_{\alpha}+(\Sigma_2)_{\alpha}]H_2=H_1\,,\\
\Lambda^{\alpha}(x)[\partial_{\alpha}+(\Sigma_1)_{\alpha}]H_1+\frac{1}{2}[\partial_{\alpha}+(\Sigma_3)_{\alpha}]H_3=0\,,\\
F^{\alpha}(x)[\partial_{\alpha}+(\Sigma_2)_{\alpha}]H_2=H_3\,,
\end{gathered}
\tag{19.19}
$$

but its physical content is completely different.

In particular, let us detail the tetrad representation for the gauge solutions:

$$
\bar{\Phi}=\nabla_{\alpha}L^{\alpha}(x)\quad\Longrightarrow\quad\bar{\Phi}=e^{(c)\alpha}\partial_{\alpha}L_{(c)}+e^{\alpha}_{(c);\alpha}L^{(c)}\,,\tag{19.20}
$$

$$
\begin{gathered}
\bar{\Phi}_{(\alpha\beta)}=\nabla_{\alpha}L_{\beta}+\nabla_{\beta}L_{\alpha}-\frac{1}{2}g_{\alpha\beta}(x)\nabla_{\rho}\Lambda^{\rho}\quad\Longrightarrow\\
\bar{\Phi}_{(ab)}=-\left(\gamma_{[ca]b}+\gamma_{[cb]a}\right)L^{(c)}+e^{\alpha}_{(a)}\partial_{\alpha}\Lambda_{(b)}+e^{\alpha}_{(b)}\partial_{\alpha}\Lambda_{(a)}-\frac{1}{2}g_{ab}\bar{\Phi}\,.
\end{gathered}
\tag{19.21}
$$

The concomitant components are determined by the relations

$$
\begin{gathered}
\bar{H}_1=\frac{1}{2}\Delta^{\alpha}(x)\partial_{\alpha}\bar{H}-\frac{1}{3}K^{\alpha}(x)[\partial_{\alpha}+(\Sigma_2)_{\alpha}]\bar{H}_2\,,\\
\bar{H}_3=F^{\alpha}(x)\,[\partial_{\alpha}+(\Sigma_2)_{\alpha}]\,\bar{H}_2\,.
\end{gathered}
\tag{19.22}
$$

The generalized matrix equation possess an important structural property: it is symmetric under the local Lorentz group, in accordance with the following relations

$$
\begin{gathered}
\Psi'(x)=S(x)\Psi(x),\quad S(x)\Gamma^{\alpha}(x)S^{-1}(x)=\Gamma'^{\alpha}(x)\,,\\
S(x)\,\Sigma_{\alpha}(x)\,S^{-1}(x)+S(x)\,\frac{\partial}{\partial x^{\alpha}}\,S^{-1}(x)=\Sigma'_{\alpha}\,,
\end{gathered}
\tag{19.23}
$$

where the prime indicates that these quantities are determined by the formulas (19.15)–(19.16), but with the use of the primed tetrad related to the initial one by the local Lorentz transformation

$$
e^{\sigma}_{(a')}(x)=L_a{}^{b}(x)\,e^{\sigma}_{(b)}(x)\,.\tag{19.24}
$$

This symmetry proves the correctness of the constructed equation, because for a given metric one can use many different tetrads, and all the corresponding equations are equivalent due to relations (19.23). With respect to the coordinate transformation, the field function Ψ behaves as a scalar,

$$
x^{\alpha}\to x'^{\alpha},\quad\Psi(x)=\Psi'(x')\,.\tag{19.25}
$$

19.4 The Spin 2 Field in Cylindrical Coordinates

Let us specify the main equation in cylindrical coordinates $x^\alpha = (t, r, \phi, z)$ of the Minkowski space

$$dS^2 = dt^2 - dr^2 - r^2 d\phi^2 - dz^2, \quad e^\beta_{(a)}(x) = \begin{vmatrix} 1 & 0 & 0 & 0 \\ 0 & 1 & 0 & 0 \\ 0 & 0 & 1/r & 0 \\ 0 & 0 & 0 & 1 \end{vmatrix}. \tag{19.26}$$

The non-vanishing Ricci rotation coefficients are

$$\gamma_{ab0} = 0, \quad \gamma_{ab1} = 0, \quad \gamma_{122} = -\gamma_{212} = +\frac{1}{r}, \quad \gamma_{ab3} = 0. \tag{19.27}$$

Equation (19.14) takes the form

$$\Big[\Gamma^0 \frac{\partial}{\partial t} + \Gamma^1 \frac{\partial}{\partial r} + \frac{\Gamma^2}{r}(\frac{\partial}{\partial \phi} + J^{12}) + \Gamma^3 \frac{\partial}{\partial z} - m\Big]\Psi = 0. \tag{19.28}$$

Taking in mind the expressions for separate terms:

$$\Psi = \begin{vmatrix} H & 1 \\ H_1 & 4 \\ H_2 & 10 \\ H_3 & 24 \end{vmatrix}, \quad \Gamma^0 \partial_t \Psi = \begin{vmatrix} G^0 \partial_t H_1 \\ \frac{1}{2}\Delta^0 \partial_t H - \frac{1}{3} K^0 \partial_t H_2| \\ \Lambda^0 \partial_t H_1 + \frac{1}{2} B^0 \partial_t H_3 \\ F^0 \partial_t H_2 \end{vmatrix},$$

$$\Gamma^1 \frac{\partial}{\partial r}\Psi = \begin{vmatrix} G^1 \partial_r H_1 \\ \frac{1}{2}\Delta^1 \partial_r H - \frac{1}{3} K^1 \partial_r H_2 \\ \Lambda^1 \partial_r H_1 + \frac{1}{2} B^1 \partial_r H_3 \\ F^1 \partial_r H_2 \end{vmatrix}, \Gamma^3 \partial_z \Psi = \begin{vmatrix} G^3 \partial_z H_1 \\ \frac{1}{2}\Delta^3 \partial_z H - \frac{1}{3} K^3 \partial_x H_2| \\ \Lambda^3 \partial_z H_1 + \frac{1}{2} B^3 \partial_z H_3 \\ F^3 \partial_z H_2 \end{vmatrix},$$

$$\frac{\Gamma^2}{r}(\partial_\phi + J^{12})\Psi = \frac{1}{r}\begin{vmatrix} G^2(\partial_\phi + J_1^{12}) H_1 \\ \frac{1}{2}\Delta^2 \partial_\phi H - \frac{1}{3} K^2 (\partial_\phi + J_2^{12}) H_2 \\ \Lambda^2(\partial_\phi + J_1^{12}) H_1 + \frac{1}{2} B^2 (\partial_\phi + J_3^{12}) H_3 \\ F^2 (\partial_\phi + J_2^{12}) H_2 \end{vmatrix},$$

we derive the system of equations in block form

$$G^0 \partial_t H_1 + G^1 \partial_r H_1 + \frac{1}{r} G^2 (\partial_\phi + J_1^{12}) H_1 + G^3 \partial_z H_1 = mH,$$

$$\frac{1}{2}\Delta^0 \partial_t H - \frac{1}{3} K^0 \partial_t H_2 + \frac{1}{2}\Delta^1 \partial_r H - \frac{1}{3} K^1 \partial_r H_2$$
$$+ \frac{1}{r}\Big[\frac{1}{2}\Delta^2 \partial_\phi H - \frac{1}{3} K^2 (\partial_\phi + J_2^{12}) H_2\Big] + \Big[\frac{1}{2}\Delta^3 \partial_z H - \frac{1}{3} K^3 \partial_x H_2\Big] = mH_1,$$

$$\Lambda^0 \partial_t H_1 + \frac{1}{2} B^0 \partial_t H_3 + \Lambda^1 \partial_r H_1 + \frac{1}{2} B^1 \partial_r H_3$$
$$+ \frac{1}{r}\Big[\Lambda^2 (\partial_\phi + J_1^{12}) H_1 + \frac{1}{2} B^2 (\partial_\phi + J_3^{12}) H_3\Big] + \Big[\Lambda^3 \partial_z H_1 + \frac{1}{2} B^3 \partial_z H_3\Big] = mH_2,$$

$$F^0\partial_t H_2 + F^1\partial_r H_2 + \frac{1}{r}F^2(\partial_\phi + J_2^{12})H_2 + F^3\partial_z H_2 = mH_3 .$$

The transition in these equations to the massless case is evident. Let us detail the tetrad representation for the gauge solutions:

$$\bar{\Phi} = \partial_t L_{(0)} - (\partial_r + \frac{1}{r})L_{(1)} - \frac{1}{r}\partial_\phi L_{(2)} - \partial_z L_{(3)} ;$$

$$\bar{\Phi}_{(ab)} = -\Big(\gamma_{[ca]b} + \gamma_{[cb]a}\Big)L^{(c)} + e^{\alpha}_{(a)}\partial_\alpha L_{(b)} + e^{\alpha}_{(b)}\partial_\alpha L_{(a)} - \frac{1}{2}g_{ab}\bar{\Phi} ,$$

whence it follows

$$\bar{\Phi}_{00} = 2\partial_t L_{(0)} - \frac{1}{2}[\partial_t L_{(0)} - (\partial_r + \frac{1}{r})L_{(1)} - \frac{1}{r}\partial_\phi L_{(2)} - \partial_z L_{(3)}] ,$$

$$\bar{\Phi}_{11} = 2\partial_r L_{(1)} + \frac{1}{2}[\partial_t L_{(0)} - (\partial_r + \frac{1}{r})L_{(1)} - \frac{1}{r}\partial_\phi L_{(2)} - \partial_z L_{(3)}] ,$$

$$\bar{\Phi}_{22} = \frac{2}{r}L_{(1)} + \frac{2}{r}\partial_\phi L_{(2)} + \frac{1}{2}[\partial_t L_{(0)} - (\partial_r + \frac{1}{r})L_{(1)} - \frac{1}{r}\partial_\phi L_{(2)} - \partial_z L_{(3)}] ,$$

$$\bar{\Phi}_{33} = 2\partial_z L_{(3)} + \frac{1}{2}[\partial_t L_{(0)} - (\partial_r + \frac{1}{r})L_{(1)} - \frac{1}{r}\partial_\phi L_{(2)} - \partial_z L_{(3)}] ,$$

$$\bar{\Phi}_{01} = \partial_t L_{(1)} + \partial_r L_{(0)}, \bar{\Phi}_{02} = \partial_t L_{(2)} + \frac{1}{r}\partial_\phi L_{(0)}, \bar{\Phi}_{03} = \partial_t L_{(3)} + \partial_z L_{(0)} ,$$

$$\bar{\Phi}_{23} = \frac{1}{r}\partial_\phi L_{(3)} + \partial_z L_{(2)}, \bar{\Phi}_{31} = \partial_z L_{(1)} + \partial_r L_{(3)}, \bar{\Phi}_{12} = -\frac{1}{r}L_{(2)} + \partial_r L_{(2)} + \frac{1}{r}\partial_\phi L_{(1)} .$$

The concomitant gauge components are determined by the formulas

$$\bar{H}_1 = \frac{1}{2}\Delta^\alpha(x)\partial_\alpha \bar{H} - \frac{1}{3}K^\alpha(x)\Big[\partial_\alpha + (\Sigma_2)_\alpha\Big]\bar{H}_2$$

$$= \frac{1}{2}\Big[\Delta^0\partial_t + \Delta^1\partial_3 + \Delta^2\frac{\partial_\phi}{r} + \Delta^3\partial_z\Big]\bar{H} - \frac{1}{3}\Big[K^0\partial_t + K^1\partial_r + K^2\frac{\partial_\phi + J_2^{12}}{r} + K^3\partial_z\Big]\bar{H}_2 ,$$

$$\bar{H}_3 = F^\alpha(x)\Big[\partial_\alpha + (\Sigma_2)_\alpha\Big]\bar{H}_2 = \Big[F^0\partial_t + F^1\partial_r + F^2\frac{\partial_t + J_2^{12}}{r} + F^3\partial_z\Big]\bar{H}_2 .$$

19.5 The Equation in Spherical Coordinates

Let us consider the structure of the matrix equation in spherical coordinates, $x^\alpha = (t, r, \theta, \phi)$:

$$dS^2 = dt^2 - dr^2 - r^2 d\theta^2 - r^2\sin^2\theta d\phi^2, e^{\alpha}_{(0)} = (1,0,0,0) ,$$
$$e^{\alpha}_{(3)} = (0,1,0,0), e^{\alpha}_{(1)} = (0,0,\frac{1}{r},0), e^{\alpha}_{(2)} = (1,0,0,\frac{1}{r\sin\theta}) . \qquad (19.29)$$

The corresponding Ricci rotation coefficients are

$$\gamma_{ab0} = 0, \gamma_{ab3} = 0, \gamma_{ab1} = \begin{vmatrix} 0 & 0 & 0 & 0 \\ 0 & 0 & 0 & -\frac{1}{r} \\ 0 & 0 & 0 & 0 \\ 0 & +\frac{1}{r} & 0 & 0 \end{vmatrix}, \gamma_{ab2} = \begin{vmatrix} 0 & 0 & 0 & 0 \\ 0 & 0 & +\frac{\cos\theta}{r\sin\theta} & 0 \\ 0 & -\frac{\cos\theta}{r\sin\theta} & 0 & -\frac{1}{r} \\ 0 & 0 & +\frac{1}{r} & 0 \end{vmatrix} .$$

The main equation takes the form

$$\Big\{\Gamma^0\partial_t + \Big(\Gamma^3\partial_r + \frac{\Gamma^1 J^{31} + \Gamma^2 J^{32}}{r}\Big) + \frac{1}{r}\Sigma_{\theta,\phi} - m\Big\}\Psi = 0\,, \tag{19.30}$$

where we separate the angular operator

$$\Sigma_{\theta,\phi} = \Big(\Gamma^1\partial_\theta + \Gamma^2\frac{\partial_\phi + J^{12}\cos\theta}{\sin\theta}\Big)\,. \tag{19.31}$$

Using the appropriate expressions for all the involved terms

$$\Gamma^0\partial_t\Psi = \left|\begin{array}{c} G^0\partial_t H_1 \\ \frac{1}{2}\Delta^0\partial_t H - \frac{1}{3}K^0\partial_t H_2| \\ \Lambda^0\partial_t H_1 + \frac{1}{2}B^0\partial_t H_3 \\ F^0\partial_t H_2 \end{array}\right|, \qquad \Gamma^3\partial_r\Psi = \left|\begin{array}{c} G^3\partial_r H_1 \\ \frac{1}{2}\Delta^3\partial_r H - \frac{1}{3}K^3\partial_r H_2 \\ \Lambda^3\partial_r H_1 + \frac{1}{2}B^3\partial_r H_3 \\ F^3\partial_r H_2 \end{array}\right|,$$

$$\frac{1}{r}(\Gamma^1 J^{31} + \Gamma^2 J^{32)})\Psi = \frac{1}{r}\left|\begin{array}{c} (G^1 J^{31}_1 + G^2 J^{32}_1)H_1 \\ \frac{1}{2}(\Delta^1 + \Delta^2)H - \frac{1}{3}(K^1 J^{31}_2 + K^2 J^{32}_2)H_2 \\ (\Lambda^1 J^{31}_1 + \Lambda^2 J^{32}_1)H_1 + \frac{1}{2}(B^1 J^{31}_3 + B^2 J^{32}_3)H_3 \\ (F^1 J^{31}_2 + F^2 J^{32}_2)H_2 \end{array}\right|,$$

$$\Sigma_{\theta\phi}\Psi = \left|\begin{array}{c} G^1\partial_\theta H_1 + G^2\frac{\partial_\phi + \cos\theta J^{12}_1}{\sin\theta}H_1 \\ \frac{1}{2}\Delta^1\partial_\theta H + \frac{1}{2}\Delta^2\frac{\partial_\phi + \cos\theta}{\sin\theta}H - \frac{1}{3}K^1\partial_\theta H_2 - \frac{1}{3}K^2\frac{\partial_\phi + \cos\theta J^{12}_2}{\sin\theta}H_2 \\ \Lambda^1\partial_\theta H_1 + \Lambda^2\frac{\partial_\phi + \cos\theta J^{12}_1}{\sin\theta}H_1 + \frac{1}{2}B^1\partial_\theta H_3 + \frac{1}{2}B^2\frac{\partial_\phi + \cos\theta J^{12}_3}{\sin\theta}H_3 \\ F^1\partial_\theta H_2 + F^2\frac{\partial_\phi + \cos\theta J^{12}_2}{\sin\theta}H_2 \end{array}\right|,$$

we obtain the system of equations in block form

$$G^0\frac{\partial}{\partial t}H_1 + G^3\frac{\partial}{\partial r}H_1 + \frac{G^1 J^{31}_1 + G^2 J^{32}_1}{r}H_1$$

$$+\frac{1}{r}\Big(G^1\partial_\theta + G^2\frac{\partial_\phi + \cos\theta J^{12}_1}{\sin\theta}\Big)H_1 = mH\,,$$

$$\frac{1}{2}\Delta^0\frac{\partial}{\partial t}H - \frac{1}{3}K^0\frac{\partial}{\partial t}H_2$$

$$+\frac{1}{2}\Big(\Delta^3\frac{\partial}{\partial r} + \frac{\Delta^1 + \Delta^2}{r}\Big)H - \frac{1}{3}\Big(K^3\frac{\partial}{\partial r} + \frac{K^1 J^{31}_2 + K^2 J^{32}_2}{r}\Big)H_2$$

$$+\frac{1}{r}\Big[\frac{1}{2}\Big(\Delta^1\partial_\theta + \Delta^2\frac{\partial_\phi + \cos\theta}{\sin\theta}\Big)H - \frac{1}{3}\Big(K^1\partial_\theta + K^2\frac{\partial_\phi + \cos\theta J^{12}_2}{\sin\theta}\Big)H_2\Big] = mH_1\,,$$

$$\Lambda^0\frac{\partial}{\partial t}H_1 + \frac{1}{2}B^0\frac{\partial}{\partial t}H_3$$

$$+\Big(\Lambda^3\frac{\partial}{\partial r}+\frac{\Lambda^1 J_1^{31}+\Lambda^2 J_1^{32}}{r}\Big)H_1+\frac{1}{2}\Big(B^3\frac{\partial}{\partial r}+\frac{B^1 J_3^{31}+B^2 J_3^{32}}{r}\Big)H_3$$

$$+\frac{1}{r}\Big[\Big(\Lambda^1\partial_\theta+\Lambda^2\frac{\partial_\phi+\cos\theta J_1^{12}}{\sin\theta}\Big)H_1+\frac{1}{2}\Big(B^1\partial_\theta+B^2\frac{\partial_\phi+\cos\theta J_3^{12}}{\sin\theta}\Big)H_3\Big]=mH_2\,,$$

$$F^0\frac{\partial}{\partial t}H_2+F^3\frac{\partial}{\partial r}H_2+\frac{F^1 J_2^{31}+F^2 J_2^{32}}{r}H_2+\frac{1}{r}\Big(F^1\partial_\theta+F^2\frac{\partial_\phi+\cos\theta J_2^{12}}{\sin\theta}\Big)H_2=mH_3\,.$$

The transition in these equations to the massless case is evident. Let us detail the tetrad representation for the gauge solution:

$$\bar{\Phi}=\Big[\partial_t L_{(0)}-(\partial_r+\frac{2}{r})L_{(3)}-\frac{1}{r}(\partial_\theta+\frac{\cos\theta}{\sin\theta})L_{(1)}-\frac{\partial_\phi}{r\sin\theta}L_{(2)}\Big], \qquad (19.32)$$

$$\bar{\Phi}_{(ab)}=-\Big(\gamma_{[ca]b}+\gamma_{[cb]a}\Big)L^{(c)}+\Big[e^{\alpha}_{(a)}\partial_\alpha L_{(b)}+e^{\alpha}_{(b)}\partial_\alpha L_{(a)}\Big]-\frac{1}{2}g_{ab}\bar{\Phi},$$

whence it follows

$$\bar{\Phi}_{00}=2\partial_t L_{(0)}-\frac{1}{2}\Big[\partial_t L_{(0)}-(\partial_r+\frac{2}{r})L_{(3)}-\frac{1}{r}(\partial_\theta+\frac{\cos\theta}{\sin\theta})L_{(1)}-\frac{\partial_\phi}{r\sin\theta}L_{(2)}\Big],$$

$$\bar{\Phi}_{33}=2\partial_r L_{(3)}+\frac{1}{2}\Big[\partial_t L_{(0)}-(\partial_r+\frac{2}{r})L_{(3)}-\frac{1}{r}(\partial_\theta+\frac{\cos\theta}{\sin\theta})L_{(1)}-\frac{\partial_\phi}{r\sin\theta}L_{(2)}\Big],$$

$$\bar{\Phi}_{11}=\frac{2}{r}L_{(3)}+\frac{2}{r}\partial_\theta L_{(1)}+\frac{1}{2}\Big[\partial_t L_{(0)}-(\partial_r+\frac{2}{r})L_{(3)}-\frac{1}{r}(\partial_\theta+\frac{\cos\theta}{\sin\theta})L_{(1)}-\frac{\partial_\phi}{r\sin\theta}L_{(2)}\Big],$$

$$\bar{\Phi}_{22}=2\frac{\cos\theta}{r\sin\theta}L_{(1)}+\frac{2}{r}L_{(3)}+2\frac{\partial_\phi}{r\sin\theta}L_{(2)}$$

$$+\frac{1}{2}\Big[\partial_t L_{(0)}-(\partial_r+\frac{2}{r})L_{(3)}-\frac{1}{r}(\partial_\theta+\frac{\cos\theta}{\sin\theta})L_{(1)}-\frac{\partial_\phi}{r\sin\theta}L_{(2)}\Big],$$

$$\bar{\Phi}_{01}=\partial_t L_{(1)}+\frac{1}{r}\partial_\theta L_{(0)},\bar{\Phi}_{02}=\partial_t L_{(2)}+\frac{1}{r\sin\theta}\partial_\phi L_{(0)},$$

$$\bar{\Phi}_{03}=\partial_t L_{(3)}+\partial_r L_{(0)},\bar{\Phi}_{23}=\frac{\partial_\phi}{r\sin\theta}L_{(3)}+\partial_r L_{(2)},$$

$$\bar{\Phi}_{31}=-\frac{1}{r}L_{(1)}+\partial_r L_{(1)}+\frac{1}{r}\partial_\theta L_{(3)},\quad \bar{\Phi}_{12}=-\frac{\cos\theta}{r\sin\theta}L_{(2)}+\frac{1}{r}\partial_\theta L_{(2)}+\frac{1}{r\sin\theta}L_{(1)}.$$

19.6 The Structure of the Lorentzian Generators

To handleh the matrix equation, we need the explicit expressions for the involved Lorentzian generators. For the 4-vector, the generators are given by the formulas

$$j^{12}=\begin{vmatrix}0&0&0&0\\0&0&-1&0\\0&1&0&0\\0&0&0&0\end{vmatrix},\ j^{31}=\begin{vmatrix}0&0&0&0\\0&0&0&1\\0&0&0&0\\0&-1&0&0\end{vmatrix},\ j^{23}=\begin{vmatrix}0&0&0&0\\0&0&0&0\\0&0&0&-1\\0&0&1&0\end{vmatrix},$$

$$j^{01}=\begin{vmatrix}0&1&0&0\\1&0&0&0\\0&0&0&0\\0&0&0&0\end{vmatrix},\ j^{02}=\begin{vmatrix}0&0&1&0\\0&0&0&0\\1&0&0&0\\0&0&0&0\end{vmatrix},\ j^{03}=\begin{vmatrix}0&0&0&1\\0&0&0&0\\0&0&0&0\\1&0&0&0\end{vmatrix}. \qquad (19.33)$$

The generators for the symmetric tensor are defined as

$$\Phi' = (j^{ab} \otimes I + I \otimes j^{ab})\,\Phi = j^{ab}\Phi + \Phi \tilde{j}^{ab}, \quad (\Phi_{(ab)}) = \left| \begin{array}{cccc} f_0 & d_1 & d_2 & d_3 \\ d_1 & f_1 & c_3 & c_2 \\ d_2 & c_3 & f_2 & c_1 \\ d_3 & c_2 & c_1 & f_3 \end{array} \right| .$$

Further, we derive its (10×10)-representation:

$$j_2^{01} = \left| \begin{array}{cccccccccc} \cdot & \cdot & \cdot & \cdot & \cdot & \cdot & 2 & \cdot & \cdot & \cdot \\ \cdot & \cdot & \cdot & \cdot & \cdot & \cdot & \cdot & \cdot & \cdot & \cdot \\ \cdot & \cdot & \cdot & \cdot & \cdot & \cdot & \cdot & \cdot & \cdot & \cdot \\ \cdot & \cdot & \cdot & \cdot & \cdot & \cdot & \cdot & \cdot & \cdot & \cdot \\ \cdot & \cdot & \cdot & \cdot & \cdot & \cdot & \cdot & \cdot & 1 & \cdot \\ \cdot & \cdot & \cdot & \cdot & \cdot & \cdot & \cdot & 1 & \cdot & \cdot \\ 1 & \cdot & \cdot & \cdot & \cdot & \cdot & \cdot & \cdot & \cdot & 1 \\ \cdot & \cdot & \cdot & \cdot & \cdot & 1 & \cdot & \cdot & \cdot & \cdot \\ \cdot & \cdot & \cdot & \cdot & 1 & \cdot & \cdot & \cdot & \cdot & \cdot \\ \cdot & \cdot & \cdot & \cdot & \cdot & \cdot & 2 & \cdot & \cdot & \cdot \end{array} \right| , \; j_2^{02} = \left| \begin{array}{cccccccccc} \cdot & \cdot & \cdot & \cdot & \cdot & \cdot & \cdot & \cdot & \cdot & \cdot \\ \cdot & \cdot & \cdot & \cdot & \cdot & \cdot & \cdot & 2 & \cdot & \cdot \\ \cdot & \cdot & \cdot & \cdot & \cdot & \cdot & \cdot & \cdot & \cdot & \cdot \\ \cdot & \cdot & \cdot & \cdot & \cdot & \cdot & \cdot & \cdot & 1 & \cdot \\ \cdot & \cdot & \cdot & \cdot & \cdot & \cdot & \cdot & \cdot & \cdot & \cdot \\ \cdot & \cdot & \cdot & \cdot & \cdot & \cdot & 1 & \cdot & \cdot & \cdot \\ \cdot & \cdot & \cdot & \cdot & \cdot & 1 & \cdot & \cdot & \cdot & \cdot \\ \cdot & 1 & \cdot & \cdot & \cdot & \cdot & \cdot & \cdot & \cdot & 1 \\ \cdot & \cdot & \cdot & 1 & \cdot & \cdot & \cdot & \cdot & \cdot & \cdot \\ \cdot & \cdot & \cdot & \cdot & \cdot & \cdot & \cdot & 2 & \cdot & \cdot \end{array} \right| ,$$

$$j_2^{03} = \left| \begin{array}{cccccccccc} \cdot & \cdot & \cdot & \cdot & \cdot & \cdot & \cdot & \cdot & \cdot & \cdot \\ \cdot & \cdot & \cdot & \cdot & \cdot & \cdot & \cdot & \cdot & \cdot & \cdot \\ \cdot & \cdot & \cdot & \cdot & \cdot & \cdot & \cdot & \cdot & 2 & \cdot \\ \cdot & \cdot & \cdot & \cdot & \cdot & \cdot & \cdot & 1 & \cdot & \cdot \\ \cdot & \cdot & \cdot & \cdot & \cdot & \cdot & 1 & \cdot & \cdot & \cdot \\ \cdot & \cdot & \cdot & \cdot & \cdot & \cdot & \cdot & \cdot & \cdot & \cdot \\ \cdot & \cdot & \cdot & \cdot & 1 & \cdot & \cdot & \cdot & \cdot & \cdot \\ \cdot & \cdot & \cdot & 1 & \cdot & \cdot & \cdot & \cdot & \cdot & \cdot \\ \cdot & \cdot & 1 & \cdot & \cdot & \cdot & \cdot & \cdot & \cdot & 1 \\ \cdot & \cdot & \cdot & \cdot & \cdot & \cdot & \cdot & \cdot & 2 & \cdot \end{array} \right| , \; j_2^{23} = \left| \begin{array}{cccccccccc} \cdot & \cdot & \cdot & \cdot & \cdot & \cdot & \cdot & \cdot & \cdot & \cdot \\ \cdot & \cdot & \cdot & -2 & \cdot & \cdot & \cdot & \cdot & \cdot & \cdot \\ \cdot & \cdot & \cdot & 2 & \cdot & \cdot & \cdot & \cdot & \cdot & \cdot \\ \cdot & 1 & -1 & \cdot & \cdot & \cdot & \cdot & \cdot & \cdot & \cdot \\ \cdot & \cdot & \cdot & \cdot & \cdot & 1 & \cdot & \cdot & \cdot & \cdot \\ \cdot & \cdot & \cdot & \cdot & -1 & \cdot & \cdot & \cdot & \cdot & \cdot \\ \cdot & \cdot & \cdot & \cdot & \cdot & \cdot & \cdot & \cdot & \cdot & \cdot \\ \cdot & \cdot & \cdot & \cdot & \cdot & \cdot & \cdot & \cdot & -1 & \cdot \\ \cdot & \cdot & \cdot & \cdot & \cdot & \cdot & \cdot & 1 & \cdot & \cdot \\ \cdot & \cdot & \cdot & \cdot & \cdot & \cdot & \cdot & \cdot & \cdot & \cdot \end{array} \right| ,$$

$$j_2^{31} = \left| \begin{array}{cccccccccc} \cdot & \cdot & \cdot & \cdot & 2 & \cdot & \cdot & \cdot & \cdot & \cdot \\ \cdot & \cdot & \cdot & \cdot & \cdot & \cdot & \cdot & \cdot & \cdot & \cdot \\ \cdot & \cdot & \cdot & \cdot & -2 & \cdot & \cdot & \cdot & \cdot & \cdot \\ \cdot & \cdot & \cdot & \cdot & \cdot & -1 & \cdot & \cdot & \cdot & \cdot \\ -1 & \cdot & 1 & \cdot & \cdot & \cdot & \cdot & \cdot & \cdot & \cdot \\ \cdot & \cdot & \cdot & 1 & \cdot & \cdot & \cdot & \cdot & \cdot & \cdot \\ \cdot & \cdot & \cdot & \cdot & \cdot & \cdot & \cdot & \cdot & 1 & \cdot \\ \cdot & \cdot & \cdot & \cdot & \cdot & \cdot & \cdot & \cdot & \cdot & \cdot \\ \cdot & \cdot & \cdot & \cdot & \cdot & \cdot & -1 & \cdot & \cdot & \cdot \\ \cdot & \cdot & \cdot & \cdot & \cdot & \cdot & \cdot & \cdot & \cdot & \cdot \end{array} \right| ,$$

$$j_2^{12} = \left| \begin{array}{cccccccccc} \cdot & \cdot & \cdot & \cdot & \cdot & -2 & \cdot & \cdot & \cdot & \cdot \\ \cdot & \cdot & \cdot & \cdot & \cdot & 2 & \cdot & \cdot & \cdot & \cdot \\ \cdot & \cdot & \cdot & \cdot & \cdot & \cdot & \cdot & \cdot & \cdot & \cdot \\ \cdot & \cdot & \cdot & \cdot & 1 & \cdot & \cdot & \cdot & \cdot & \cdot \\ \cdot & \cdot & \cdot & -1 & \cdot & \cdot & \cdot & \cdot & \cdot & \cdot \\ 1 & -1 & \cdot & \cdot & \cdot & \cdot & \cdot & \cdot & \cdot & \cdot \\ \cdot & \cdot & \cdot & \cdot & \cdot & \cdot & \cdot & -1 & \cdot & \cdot \\ \cdot & \cdot & \cdot & \cdot & \cdot & \cdot & 1 & \cdot & \cdot & \cdot \\ \cdot & \cdot & \cdot & \cdot & \cdot & \cdot & \cdot & \cdot & \cdot & \cdot \\ \cdot & \cdot & \cdot & \cdot & \cdot & \cdot & \cdot & \cdot & \cdot & \cdot \end{array} \right| .$$

The 3-rank tensor may be presented by means of the four skew-symmetric matrices $\Phi_{[ps]t} = \varphi_t$, $t = 0,1,2,3$; so the generators act in accordance with the formula

$$\varphi'_l = j^{ab}\varphi_l + \varphi_l \tilde{j}^{ab} + (j^{ab})_l{}^t \varphi_t \,. \tag{19.34}$$

Taking into account the intermediate relations

$$ab = 12, \quad \begin{array}{l} \varphi'_0 = j^{12}\varphi_0 + \varphi_0 \tilde{j}^{12} , \\ \varphi'_1 = j^{12}\varphi_1 + \varphi_1 \tilde{j}^{12} - \varphi_2 , \\ \varphi'_2 = j^{12}\varphi_2 + \varphi_2 \tilde{j}^{12} + \varphi_1 , \\ \varphi'_3 = j^{12}\varphi_3 + \varphi_3 \tilde{j}^{12} ; \end{array} \qquad ab = 31, \quad \begin{array}{l} \varphi'_0 = j^{31}\varphi_0 + \varphi_0 \tilde{j}^{31} , \\ \varphi'_1 = j^{31}\varphi_1 + \varphi_1 \tilde{j}^{31} + \varphi_3 , \\ \varphi'_2 = j^{31}\varphi_2 + \varphi_2 \tilde{j}^{31} , \\ \varphi'_3 = j^{31}\varphi_3 + \varphi_3 \tilde{j}^{31} - \varphi_1 ; \end{array}$$

$$ab = 23, \quad \begin{array}{l} \varphi'_0 = j^{23}\varphi_0 + \varphi_0 \tilde{j}^{23} , \\ \varphi'_1 = j^{23}\varphi_1 + \varphi_1 \tilde{j}^{23} , \\ \varphi'_2 = j^{23}\varphi_2 + \varphi_2 \tilde{j}^{23} - \varphi_3 , \\ \varphi'_3 = j^{23}\varphi_3 + \varphi_3 \tilde{j}^{23} + \varphi_2 ; \end{array} \qquad ab = 01, \quad \begin{array}{l} \varphi'_0 = j^{01}\varphi_0 + \varphi_0 \tilde{j}^{01} + \varphi_1 , \\ \varphi'_1 = j^{01}\varphi_1 + \varphi_1 \tilde{j}^{01} + \varphi_0 , \\ \varphi'_2 = j^{01}\varphi_2 + \varphi_2 \tilde{j}^{01} , \\ \varphi'_3 = j^{01}\varphi_3 + \varphi_3 \tilde{j}^{01} ; \end{array}$$

$$ab = 02, \quad \begin{array}{l} \varphi'_0 = j^{02}\varphi_0 + \varphi_0 \tilde{j}^{02} + \varphi_2 , \\ \varphi'_1 = j^{02}\varphi_1 + \varphi_1 \tilde{j}^{02} , \\ \varphi'_2 = j^{02}\varphi_2 + \varphi_2 \tilde{j}^{02} + \varphi_0 , \\ \varphi'_3 = j^{02}\varphi_3 + \varphi_3 \tilde{j}^{02} ; \end{array} \qquad ab = 03, \quad \begin{array}{l} \varphi'_0 = j^{03}\varphi_0 + \varphi_0 \tilde{j}^{03} + \varphi_3 , \\ \varphi'_1 = j^{03}\varphi_1 + \varphi_1 \tilde{j}^{03} , \\ \varphi'_2 = j^{03}\varphi_2 + \varphi_2 \tilde{j}^{03} , \\ \varphi'_3 = j^{03}\varphi_3 + \varphi_3 \tilde{j}^{03} + \varphi_0 . \end{array}$$

with

$$\varphi_c = \left| \begin{array}{cccc} 0 & E_{1c} & E_{2c} & E_{3c} \\ -E_{1c} & 0 & B_{3c} & -B_{2c} \\ -E_{2c} & -B_{3c} & 0 & B_{1c} \\ -E_{3c} & B_{2c} & -B_{1c} & 0 \end{array} \right| , \qquad c = 0,1,2,3 ,$$

we can find the explicit form of the six (24×24) generators; since they are rather cumbersome, we describe them in the Appendix.

19.7 Relativistic Invariance, Additional Checking

In order to verify expressions for all the blocks in the matrices

$$(\Gamma^a \partial_a - m)\Phi = 0 , \quad \Gamma^a = \left| \begin{array}{cccc} 0 & G^a & 0 & 0 \\ \frac{1}{2}\Delta^a & 0 & -\frac{1}{3}K^a & 0 \\ 0 & \Lambda^a & 0 & \frac{1}{2}B^a \\ 0 & 0 & F^a & 0 \end{array} \right| ,$$

and the expressions for the Loretzian generators, let us examine the relativistic invariance condition

$$\Psi' = S\Psi , \quad \Psi = S^{-1}\Psi' , \quad (S\Gamma^a S^{-1}\partial_a - m)\Phi' = 0 ,$$

whence we derive

$$(\Gamma^b \partial'_b - m)\,\Phi' = 0, \quad S\Gamma^a S^{-1} = \Gamma^b L_b{}^a .$$

By specifying this for the infinitesimal transformations

$$(1+\omega_{mn}J^{mn})\Gamma^a(1-\omega_{kl}J^{kl}) = \Gamma^b\,[\delta^a_b + \omega_{ps}(j^{ps})_b{}^a],$$

we derive the needed relation

$$J^{mn}\Gamma^a - \Gamma^a J^{mn} = \Gamma^b\,(j^{mn})_b{}^a . \tag{19.35}$$

In detailed form, this reads (the indices 1,2,3, related to the tensors of 1, 2, and 3-rd ranks are omitted here):

$$\begin{array}{lll}
J^{23}\Gamma^0 - \Gamma^0 J^{23} = 0, & J^{31}\Gamma^0 - \Gamma^0 J^{31} = 0, & J^{12}\Gamma^0 - \Gamma^0 J^{12} = 0, \\
J^{23}\Gamma^1 - \Gamma^1 J^{23} = 0, & J^{31}\Gamma^1 - \Gamma^1 J^{31} = -\Gamma^3, & J^{12}\Gamma^1 - \Gamma^1 J^{12} = \Gamma^2, \\
J^{23}\Gamma^2 - \Gamma^2 J^{23} = \Gamma^3, & J^{31}\Gamma^2 - \Gamma^2 J^{31} = 0, & J^{12}\Gamma^2 - \Gamma^2 J^{12} = \Gamma^1, \\
J^{23}\Gamma^3 - \Gamma^3 J^{23} = -\Gamma^2, & J^{31}\Gamma^3 - \Gamma^3 J^{31} = \Gamma^1, & J^{12}\Gamma^3 - \Gamma^3 J^{12} = 0,
\end{array}$$

$$\begin{array}{lll}
J^{01}\Gamma^0 - \Gamma^0 J^{01} = \Gamma_1, & J^{02}\Gamma^0 - \Gamma^0 J^{02} = \Gamma_2, & J^{03}\Gamma^0 - \Gamma^0 J^{03} = \Gamma_3, \\
J^{01}\Gamma^1 - \Gamma^1 J^{01} = \Gamma_0, & J^{02}\Gamma^1 - \Gamma^1 J^{02} = 0, & J^{03}\Gamma^1 - \Gamma^1 J^{03} = 0, \\
J^{01}\Gamma^2 - \Gamma^2 J^{01} = 0, & J^{02}\Gamma^2 - \Gamma^2 J^{02} = \Gamma_0, & J^{03}\Gamma^2 - \Gamma^2 J^{03} = 0, \\
J^{01}\Gamma^3 - \Gamma^3 J^{01} = 0, & J^{02}\Gamma^3 - \Gamma^3 J^{02} = 0, & J^{03}\Gamma^3 - \Gamma^3 J^{03} = \Gamma_0.
\end{array}$$

By considering the block structure of the generators J^{mn} and of the matrices Γ^a:

$$J^{mn} = \left| \begin{array}{cccc} 0 & 0 & 0 & 0 \\ 0 & j_1^{mn} & 0 & 0 \\ 0 & 0 & j_2^{mn} & 0 \\ 0 & 0 & 0 & j_3^{mn} \end{array} \right|, \quad \Gamma^a = \left| \begin{array}{cccc} 0 & G^a & 0 & 0 \\ \frac{1}{2}\Delta^a & 0 & -\frac{1}{3}K^a & 0 \\ 0 & \Lambda^a & 0 & \frac{1}{2}B^a \\ 0 & 0 & F^a & 0 \end{array} \right|,$$

we derive 144 constraints (which are divided into six groups):

$mn = 23$,

$$-G^0 j_1^{23} = 0, \quad -G^1 j_1^{23} = 0, \quad -G^2 j_1^{23} = G^3, \quad -G^3 j_1^{23} = -G^2,$$

$$j_1^{23}\Delta^0 = 0, \quad j_1^{23}\Delta^1 = 0, \quad j_1^{23}\Delta^2 = \Delta^3, \quad j_1^{23}\Delta^3 = -\Delta^2,$$

$$-j_1^{23}K^0 + K^0 j_2^{23} = 0, -j_1^{23}K^1 + K^1 j_2^{23} = 0, -j_1^{23}K^2 + K^2 j_2^{23} = -K^3, -j_1^{23}K^3 + K^3 j_2^{23} = K^2,$$

$$j_2^{23}\Lambda^0 - \Lambda^0 j_1^{23} = 0, j_2^{23}\Lambda^1 - \Lambda^1 j_1^{23} = 0, j_2^{23}\Lambda^2 - \Lambda^2 j_1^{23} = \Lambda^3, j_2^{23}\Lambda^3 - \Lambda^3 j_1^{23} = -\Lambda^2 g^{33},$$

$$j_2^{23}B^0 - B^0 j_3^{23} = 0, j_2^{23}B^1 - B^1 j_3^{23} = 0, j_2^{23}B^2 - B^2 j_3^{23} = B^3, j_2^{23}B^3 - B^3 j_3^{23} = -B^2,$$

$$j_3^{23}F^0 - F^0 j_2^{23} = 0, j_3^{23}F^1 - F^1 j_2^{23} = 0, j_3^{23}F^2 - F^2 j_2^{23} = F^3, j_3^{23}F^3 - F^3 j_2^{23} = -F^2;$$

$mn = 31$,

$$-G^0 j_1^{31} = 0, \quad -G^1 j_1^{31} = -G^3, \quad -G^2 j_1^{31} = 0, \quad -G^3 j_1^{31} = G^1,$$

$$j_1^{31}\Delta^0 = 0, \quad j_1^{31}\Delta^1 = -\Delta^3, \quad j_1^{31}\Delta^2 = 0, \quad j_1^{31}\Delta^3 = \Delta^1,$$

$$-j_1^{31}K^0+K^0j_2^{31}=0, -j_1^{31}K^1+K^1j_2^{31}=K^3, -j_1^{31}K^2+K^2j_2^{31}=0, -j_1^{31}K^3+K^3j_2^{31}=-K^1,$$
$$j_2^{31}\Lambda^0-\Lambda^0j_1^{31}=0, j_2^{31}\Lambda^1-\Lambda^1j_1^{31}=-\Lambda^3, j_2^{31}\Lambda^2-\Lambda^2j_1^{31}=0, j_2^{31}\Lambda^3-\Lambda^3j_1^{31}=\Lambda^1,$$
$$j_2^{31}B^0-B^0j_3^{31}=0, j_2^{31}B^1-B^1j_3^{31}=-B^3, j_2^{31}B^2-B^2j_3^{31}=0, j_2^{31}B^3-B^3j_3^{31}=B^1,$$
$$j_3^{31}F^0-F^0j_2^{31}=0, j_3^{31}F^1-F^1j_2^{31}=-F^3, j_3^{31}F^2-F^2j_2^{31}=0, j_3^{31}F^3-F^3j_2^{31}=F^1;$$

$mn=12,$

$$-G^0j_1^{12}=0, \quad -G^1j_1^{12}=G^2, \quad -G^2j_1^{12}=-G^1, \quad -G^3j_1^{12}=0,$$
$$j_1^{12}\Delta^0=0, \quad j_1^{12}\Delta^1=\Delta^2, \quad j_1^{12}\Delta^2=-\Delta^1, \quad j_1^{12}\Delta^3=0,$$
$$-j_1^{12}K^0+K^0j_2^{12}=0, -j_1^{12}K^1+K^1j_2^{12}=-K^2, -j_1^{12}K^2+K^2j_2^{12}=K^1, -j_1^{12}K^3+K^3j_2^{12}=0,$$
$$j_2^{12}\Lambda^0-\Lambda^0j_1^{12}=0, j_2^{12}\Lambda^1-\Lambda^1j_1^{12}=\Lambda^2, j_2^{12}\Lambda^2-\Lambda^2j_1^{12}=-\Lambda^1, j_2^{12}\Lambda^3-\Lambda^3j_1^{12}=0,$$
$$j_2^{12}B^0-B^0j_3^{12}=0, j_2^{12}B^1-B^1j_3^{12}=B^2, j_2^{12}B^2-B^2j_3^{12}=-B^1, j_2^{12}B^3-B^3j_3^{12}=0,$$
$$j_3^{12}F^0-F^0j_2^{12}=0, j_3^{12}F^1-F^1j_2^{12}=F^2, j_3^{12}F^2-F^2j_2^{12}=-F^1, j_3^{12}F^3-F^3j_2^{12}=0;$$

$mn=01,$

$$-G^0j_1^{01}=G^1, \quad -G^1j_1^{01}=G^0, \quad -G^2j_1^{01}=0, \quad -G^3j_1^{01}=0,$$
$$j_1^{01}\Delta^0=\Delta^1, \quad j_1^{01}\Delta^1=\Delta^0, \quad j_1^{01}\Delta^2=0, \quad j_1^{01}\Delta^3=0,$$
$$-j_1^{01}K^0+K^0j_2^{01}=-K^1, -j_1^{01}K^1+K^1j_2^{01}=-K^0, -j_1^{01}K^2+K^2j_2^{01}=0, -j_1^{01}K^3+K^3j_2^{01}=0,$$
$$j_2^{01}\Lambda^0-\Lambda^0j_1^{01}=\Lambda^1, j_2^{01}\Lambda^1-\Lambda^1j_1^{01}=\Lambda^0, j_2^{01}\Lambda^2-\Lambda^2j_1^{01}=0, j_2^{01}\Lambda^3-\Lambda^3j_1^{01}=0,$$
$$j_2^{01}B^0-B^0j_3^{01}=B^1, j_2^{01}B^1-B^1j_3^{01}=B^0, j_2^{01}B^2-B^2j_3^{01}=0, j_2^{01}B^3-B^3j_3^{01}=0,$$
$$j_3^{01}F^0-F^0j_2^{01}=F^1, j_3^{01}F^1-F^1j_2^{01}=F^0, j_3^{01}F^2-F^2j_2^{01}=0, j_3^{01}F^3-F^3j_2^{01}=0;$$

$mn=02,$

$$-G^0j_1^{02}=G^2g, \quad -G^1j_1^{02}=0, \quad -G^2j_1^{02}=G^0, \quad -G^3j_1^{02}=0,$$
$$j_1^{02}\Delta^0=\Delta^2, \quad j_1^{02}\Delta^1=0, \quad j_1^{02}\Delta^2=\Delta^0, \quad j_1^{02}\Delta^3=0,$$
$$-j_1^{02}K^0+K^0j_2^{02}=-K^2, -j_1^{02}K^1+K^1j_2^{02}=0, -j_1^{02}K^2+K^2j_2^{02}=-K^0, -j_1^{02}K^3+K^3j_2^{02}=0,$$
$$j_2^{02}\Lambda^0-\Lambda^0j_1^{02}=\Lambda^2, j_2^{02}\Lambda^1-\Lambda^1j_1^{02}=0, j_2^{02}\Lambda^2-\Lambda^2j_1^{02}=\Lambda^0, j_2^{02}\Lambda^3-\Lambda^3j_1^{02}=0,$$
$$j_2^{02}B^0-B^0j_3^{02}=B^2, j_2^{02}B^1-B^1j_3^{02}=0, j_2^{02}B^2-B^2j_3^{02}=B^0, j_2^{02}B^3-B^3j_3^{02}=0,$$
$$j_3^{02}F^0-F^0j_2^{02}=F^2, j_3^{02}F^1-F^1j_2^{02}=0, j_3^{02}F^2-F^2j_2^{02}=F^0, j_3^{02}F^3-F^3j_2^{02}=0;$$

$mn=03,$

$$-G^0j_1^{03}=G^3, \quad -G^1j_1^{03}=0, \quad -G^2j_1^{03}=0, \quad -G^3j_1^{03}=G^0,$$
$$j_1^{03}\Delta^0=\Delta^3, \quad j_1^{03}\Delta^1=0, \quad j_1^{03}\Delta^2=0, \quad j_1^{03}\Delta^3=\Delta^0,$$
$$-j_1^{03}K^0+K^0j_2^{03}=-K^3, -j_1^{03}K^1+K^1j_2^{03}=0, -j_1^{03}K^2+K^2j_2^{03}=0, -j_1^{03}K^3+K^3j_2^{03}=-K^0,$$
$$j_2^{03}\Lambda^0-\Lambda^0j_1^{03}=\Lambda^3, j_2^{03}\Lambda^1-\Lambda^1j_1^{03}=0, j_2^{03}\Lambda^2-\Lambda^2j_1^{03}=0, j_2^{03}\Lambda^3-\Lambda^3j_1^{03}=\Lambda^0,$$
$$j_2^{03}B^0-B^0j_3^{03}=B^3, j_2^{03}B^1-B^1j_3^{03}=0, j_2^{03}B^2-B^2j_3^{03}=0, j_2^{03}B^3-B^3j_3^{03}=B^0,$$
$$j_3^{03}F^0-F^0j_2^{03}=F^3, j_3^{03}F^1-F^1j_2^{03}=0, j_3^{03}F^2-F^2j_2^{03}=0, j_3^{03}F^3-F^3j_2^{03}=F^0.$$

Taking into account the explicit form of all the matrix blocks and all the generators, we can prove that all these equations are satisfied. This indicates that the expressions for all the involved matrices are correct.

19.8 Matrix Blocks in the Theory of Spin 2 Particle

$$G^0 = (\ +1 \quad 0 \quad 0 \quad 0\ ), \quad G^1 = (\ 0 \quad -1 \quad 0 \quad 0\ ),$$

$$G^2 = (\ 0 \quad 0 \quad -1 \quad 0\ ), \quad G^3 = (\ 0 \quad 0 \quad 0 \quad -1\ );$$

$$(\Delta^a)_{4\times 1}, \quad \Delta^0 = \begin{vmatrix} 1 \\ 0 \\ 0 \\ 0 \end{vmatrix}, \quad \Delta^1 = \begin{vmatrix} 0 \\ 1 \\ 0 \\ 0 \end{vmatrix}, \quad \Delta^2 = \begin{vmatrix} 0 \\ 0 \\ 1 \\ 0 \end{vmatrix}, \quad \Delta^3 = \begin{vmatrix} 0 \\ 0 \\ 0 \\ 1 \end{vmatrix};$$

$$K^0 = \begin{vmatrix} 0 & 0 & 0 & 0 & 0 & 0 & 0 & 0 & 0 & 1 \\ 0 & 0 & 0 & 0 & 0 & 0 & 1 & 0 & 0 & 0 \\ 0 & 0 & 0 & 0 & 0 & 0 & 0 & 1 & 0 & 0 \\ 0 & 0 & 0 & 0 & 0 & 0 & 0 & 0 & 1 & 0 \end{vmatrix},$$

$$K^1 = \begin{vmatrix} 0 & 0 & 0 & 0 & 0 & 0 & -1 & 0 & 0 & 0 \\ -1 & 0 & 0 & 0 & 0 & 0 & 0 & 0 & 0 & 0 \\ 0 & 0 & 0 & 0 & 0 & -1 & 0 & 0 & 0 & 0 \\ 0 & 0 & 0 & 0 & -1 & 0 & 0 & 0 & 0 & 0 \end{vmatrix},$$

$$K^2 = \begin{vmatrix} 0 & 0 & 0 & 0 & 0 & 0 & 0 & -1 & 0 & 0 \\ 0 & 0 & 0 & 0 & 0 & -1 & 0 & 0 & 0 & 0 \\ 0 & -1 & 0 & 0 & 0 & 0 & 0 & 0 & 0 & 0 \\ 0 & 0 & 0 & -1 & 0 & 0 & 0 & 0 & 0 & 0 \end{vmatrix},$$

$$K^3 = \begin{vmatrix} 0 & 0 & 0 & 0 & 0 & 0 & 0 & 0 & -1 & 0 \\ 0 & 0 & 0 & 0 & -1 & 0 & 0 & 0 & 0 & 0 \\ 0 & 0 & 0 & -1 & 0 & 0 & 0 & 0 & 0 & 0 \\ 0 & 0 & -1 & 0 & 0 & 0 & 0 & 0 & 0 & 0 \end{vmatrix};$$

$$\Lambda^0 = \begin{vmatrix} \frac{1}{2} & 0 & 0 & 0 \\ \frac{1}{2} & 0 & 0 & 0 \\ \frac{1}{2} & 0 & 0 & 0 \\ 0 & 0 & 0 & 0 \\ 0 & 0 & 0 & 0 \\ 0 & 0 & 0 & 0 \\ 0 & 1 & 0 & 0 \\ 0 & 0 & 1 & 0 \\ 0 & 0 & 0 & 1 \\ \frac{3}{2} & 0 & 0 & 0 \end{vmatrix}, \quad \Lambda^1 = \begin{vmatrix} 0 & \frac{3}{2} & 0 & 0 \\ 0 & -\frac{1}{2} & 0 & 0 \\ 0 & -\frac{1}{2} & 0 & 0 \\ 0 & 0 & 0 & 0 \\ 0 & 0 & 0 & 1 \\ 0 & 0 & 1 & 0 \\ 1 & 0 & 0 & 0 \\ 0 & 0 & 0 & 0 \\ 0 & 0 & 0 & 0 \\ 0 & \frac{1}{2} & 0 & 0 \end{vmatrix},$$

$$\Lambda^2 = \begin{vmatrix} 0 & 0 & -\frac{1}{2} & 0 \\ 0 & 0 & \frac{3}{2} & 0 \\ 0 & 0 & -\frac{1}{2} & 0 \\ 0 & 0 & 0 & 1 \\ 0 & 0 & 0 & 0 \\ 0 & 1 & 0 & 0 \\ 0 & 0 & 0 & 0 \\ 1 & 0 & 0 & 0 \\ 0 & 0 & 0 & 0 \\ 0 & 0 & \frac{1}{2} & 0 \end{vmatrix}, \quad \Lambda^3 = \begin{vmatrix} 0 & 0 & 0 & -\frac{1}{2} \\ 0 & 0 & 0 & -\frac{1}{2} \\ 0 & 0 & 0 & \frac{3}{2} \\ 0 & 0 & 1 & 0 \\ 0 & 1 & 0 & 0 \\ 0 & 0 & 0 & 0 \\ 0 & 0 & 0 & 0 \\ 0 & 0 & 0 & 0 \\ 1 & 0 & 0 & 0 \\ 0 & 0 & 0 & \frac{1}{2} \end{vmatrix};$$

$$B^0 = \begin{vmatrix}
\cdot & \cdot & \cdot & \cdot & \cdot & \cdot & \frac{3}{2} & \cdot & \cdot & \cdot & \cdot & \cdot & \cdot & -\frac{1}{2} & \cdot & \cdot & \cdot & \cdot & \cdot & \cdot & -\frac{1}{2} & \cdot & \cdot & \cdot \\
\cdot & \cdot & \cdot & \cdot & \cdot & \cdot & -\frac{1}{2} & \cdot & \cdot & \cdot & \cdot & \cdot & \cdot & \frac{3}{2} & \cdot & \cdot & \cdot & \cdot & \cdot & \cdot & -\frac{1}{2} & \cdot & \cdot & \cdot \\
\cdot & \cdot & \cdot & \cdot & \cdot & \cdot & -\frac{1}{2} & \cdot & \cdot & \cdot & \cdot & \cdot & \cdot & -\frac{1}{2} & \cdot & \cdot & \cdot & \cdot & \cdot & \cdot & \frac{3}{2} & \cdot & \cdot & \cdot \\
\cdot & \cdot & \cdot & \cdot & \cdot & \cdot & \cdot & \cdot & \cdot & \cdot & \cdot & \cdot & \cdot & \cdot & 1 & \cdot & \cdot & \cdot & \cdot & 1 & \cdot & \cdot & \cdot & \cdot \\
\cdot & \cdot & \cdot & \cdot & \cdot & \cdot & \cdot & \cdot & 1 & \cdot & \cdot & \cdot & \cdot & \cdot & \cdot & \cdot & \cdot & \cdot & 1 & \cdot & \cdot & \cdot & \cdot & \cdot \\
\cdot & \cdot & \cdot & \cdot & \cdot & \cdot & \cdot & 1 & \cdot & \cdot & \cdot & \cdot & 1 & \cdot & \cdot & \cdot & \cdot & \cdot & \cdot & \cdot & \cdot & \cdot & \cdot & \cdot \\
1 & \cdot & \cdot & \cdot & \cdot & \cdot & \cdot & \cdot & \cdot & \cdot & \cdot & \cdot & \cdot & \cdot & \cdot & \cdot & \cdot & \cdot & \cdot & \cdot & \cdot & \cdot & \cdot & \cdot \\
\cdot & 1 & \cdot & \cdot & \cdot & \cdot & \cdot & \cdot & \cdot & \cdot & \cdot & \cdot & \cdot & \cdot & \cdot & \cdot & \cdot & \cdot & \cdot & \cdot & \cdot & \cdot & \cdot & \cdot \\
\cdot & \cdot & 1 & \cdot & \cdot & \cdot & \cdot & \cdot & \cdot & \cdot & \cdot & \cdot & \cdot & \cdot & \cdot & \cdot & \cdot & \cdot & \cdot & \cdot & \cdot & \cdot & \cdot & \cdot \\
\cdot & \cdot & \cdot & \cdot & \cdot & \cdot & \frac{1}{2} & \cdot & \cdot & \cdot & \cdot & \cdot & \cdot & \frac{1}{2} & \cdot & \cdot & \cdot & \cdot & \cdot & \cdot & \frac{1}{2} & \cdot & \cdot & \cdot
\end{vmatrix},$$

$$B^1 = \begin{vmatrix}
\frac{1}{2} & \cdot & \cdot & \cdot & \cdot & \cdot & \cdot & \cdot & \cdot & \cdot & \cdot & \cdot & \cdot & \cdot & \cdot & \cdot & \cdot & \frac{1}{2} & \cdot & \cdot & \cdot & \cdot & -\frac{1}{2} & \cdot \\
\frac{1}{2} & \cdot & \cdot & \cdot & \cdot & \cdot & \cdot & \cdot & \cdot & \cdot & \cdot & \cdot & \cdot & \cdot & \cdot & \cdot & \cdot & -\frac{3}{2} & \cdot & \cdot & \cdot & \cdot & -\frac{1}{2} & \cdot \\
\frac{1}{2} & \cdot & \cdot & \cdot & \cdot & \cdot & \cdot & \cdot & \cdot & \cdot & \cdot & \cdot & \cdot & \cdot & \cdot & \cdot & \cdot & \frac{1}{2} & \cdot & \cdot & \cdot & \cdot & \frac{3}{2} & \cdot \\
\cdot & \cdot & \cdot & \cdot & \cdot & \cdot & \cdot & \cdot & \cdot & \cdot & \cdot & \cdot & \cdot & \cdot & \cdot & \cdot & 1 & \cdot & \cdot & \cdot & \cdot & \cdot & \cdot & -1 \\
\cdot & \cdot & \cdot & \cdot & \cdot & \cdot & \cdot & \cdot & \cdot & \cdot & 1 & \cdot & \cdot & \cdot & \cdot & \cdot & \cdot & \cdot & \cdot & \cdot & \cdot & \cdot & \cdot & \cdot \\
\cdot & \cdot & \cdot & \cdot & \cdot & \cdot & \cdot & \cdot & \cdot & \cdot & \cdot & -1 & \cdot & \cdot & \cdot & \cdot & \cdot & \cdot & \cdot & \cdot & \cdot & \cdot & \cdot & \cdot \\
\cdot & \cdot & \cdot & \cdot & \cdot & \cdot & 1 & \cdot & \cdot & \cdot & \cdot & \cdot & \cdot & \cdot & \cdot & \cdot & \cdot & \cdot & \cdot & \cdot & \cdot & \cdot & \cdot & \cdot \\
\cdot & \cdot & \cdot & \cdot & \cdot & -1 & \cdot & \cdot & \cdot & \cdot & \cdot & \cdot & 1 & \cdot & \cdot & \cdot & \cdot & \cdot & \cdot & \cdot & \cdot & \cdot & \cdot & \cdot \\
\cdot & \cdot & \cdot & \cdot & 1 & \cdot & \cdot & \cdot & \cdot & \cdot & \cdot & \cdot & \cdot & \cdot & \cdot & \cdot & \cdot & \cdot & 1 & \cdot & \cdot & \cdot & \cdot & \cdot \\
\frac{3}{2} & \cdot & \cdot & \cdot & \cdot & \cdot & \cdot & \cdot & \cdot & \cdot & \cdot & \cdot & \cdot & \cdot & \cdot & \cdot & \cdot & -\frac{1}{2} & \cdot & \cdot & \cdot & \cdot & \frac{1}{2} & \cdot
\end{vmatrix},$$

$$B^2 = \begin{vmatrix}
\cdot & \frac{1}{2} & \cdot & \cdot & \cdot & \cdot & \cdot & \cdot & \cdot & \cdot & \cdot & \frac{3}{2} & \cdot & \cdot & \cdot & \cdot & \cdot & \cdot & \cdot & \cdot & \cdot & \frac{1}{2} & \cdot & \cdot \\
\cdot & \frac{1}{2} & \cdot & \cdot & \cdot & \cdot & \cdot & \cdot & \cdot & \cdot & \cdot & -\frac{1}{2} & \cdot & \cdot & \cdot & \cdot & \cdot & \cdot & \cdot & \cdot & \cdot & \frac{1}{2} & \cdot & \cdot \\
\cdot & \frac{1}{2} & \cdot & \cdot & \cdot & \cdot & \cdot & \cdot & \cdot & \cdot & \cdot & -\frac{1}{2} & \cdot & \cdot & \cdot & \cdot & \cdot & \cdot & \cdot & \cdot & \cdot & -\frac{3}{2} & \cdot & \cdot \\
\cdot & \cdot & \cdot & \cdot & \cdot & \cdot & \cdot & \cdot & \cdot & \cdot & \cdot & \cdot & \cdot & \cdot & \cdot & -1 & \cdot & \cdot & \cdot & \cdot & \cdot & \cdot & \cdot & \cdot \\
\cdot & \cdot & \cdot & \cdot & \cdot & \cdot & \cdot & \cdot & \cdot & -1 & \cdot & \cdot & \cdot & \cdot & \cdot & \cdot & \cdot & \cdot & \cdot & \cdot & \cdot & \cdot & \cdot & 1 \\
\cdot & \cdot & \cdot & \cdot & \cdot & \cdot & \cdot & \cdot & \cdot & \cdot & \cdot & \cdot & \cdot & \cdot & \cdot & \cdot & \cdot & 1 & \cdot & \cdot & \cdot & \cdot & \cdot & \cdot \\
\cdot & \cdot & \cdot & \cdot & \cdot & 1 & \cdot & 1 & \cdot & \cdot & \cdot & \cdot & \cdot & \cdot & \cdot & \cdot & \cdot & \cdot & \cdot & \cdot & \cdot & \cdot & \cdot & \cdot \\
\cdot & \cdot & \cdot & \cdot & \cdot & \cdot & \cdot & \cdot & \cdot & \cdot & \cdot & \cdot & \cdot & 1 & \cdot & \cdot & \cdot & \cdot & \cdot & \cdot & \cdot & \cdot & \cdot & \cdot \\
\cdot & \cdot & \cdot & -1 & \cdot & \cdot & \cdot & \cdot & \cdot & \cdot & \cdot & \cdot & \cdot & \cdot & \cdot & \cdot & \cdot & \cdot & \cdot & 1 & \cdot & \cdot & \cdot & \cdot \\
\cdot & \frac{3}{2} & \cdot & \cdot & \cdot & \cdot & \cdot & \cdot & \cdot & \cdot & \cdot & \frac{1}{2} & \cdot & \cdot & \cdot & \cdot & \cdot & \cdot & \cdot & \cdot & \cdot & -\frac{1}{2} & \cdot & \cdot
\end{vmatrix},$$

$$B^3 = \begin{vmatrix}
\cdot & \cdot & \frac{1}{2} & \cdot & \cdot & \cdot & \cdot & \cdot & \cdot & \cdot & -\frac{3}{2} & \cdot & \cdot & \cdot & \cdot & -\frac{1}{2} & \cdot & \cdot & \cdot & \cdot & \cdot & \cdot & \cdot & \cdot \\
\cdot & \cdot & \frac{1}{2} & \cdot & \cdot & \cdot & \cdot & \cdot & \cdot & \cdot & \frac{1}{2} & \cdot & \cdot & \cdot & \cdot & \frac{3}{2} & \cdot & \cdot & \cdot & \cdot & \cdot & \cdot & \cdot & \cdot \\
\cdot & \cdot & \frac{1}{2} & \cdot & \cdot & \cdot & \cdot & \cdot & \cdot & \cdot & \frac{1}{2} & \cdot & \cdot & \cdot & \cdot & -\frac{1}{2} & \cdot & \cdot & \cdot & \cdot & \cdot & \cdot & \cdot & \cdot \\
\cdot & \cdot & \cdot & \cdot & \cdot & \cdot & \cdot & \cdot & \cdot & \cdot & \cdot & \cdot & \cdot & \cdot & \cdot & \cdot & \cdot & \cdot & \cdot & \cdot & \cdot & 1 & \cdot & \cdot \\
\cdot & \cdot & \cdot & \cdot & \cdot & \cdot & \cdot & \cdot & \cdot & \cdot & \cdot & \cdot & \cdot & \cdot & \cdot & \cdot & \cdot & \cdot & \cdot & \cdot & \cdot & \cdot & -1 & \cdot \\
\cdot & \cdot & \cdot & \cdot & \cdot & \cdot & \cdot & \cdot & \cdot & 1 & \cdot & \cdot & \cdot & \cdot & \cdot & \cdot & -1 & \cdot & \cdot & \cdot & \cdot & \cdot & \cdot & \cdot \\
\cdot & \cdot & \cdot & \cdot & -1 & \cdot & \cdot & \cdot & 1 & \cdot & \cdot & \cdot & \cdot & \cdot & \cdot & \cdot & \cdot & \cdot & \cdot & \cdot & \cdot & \cdot & \cdot & \cdot \\
\cdot & \cdot & \cdot & 1 & \cdot & \cdot & \cdot & \cdot & \cdot & \cdot & \cdot & \cdot & \cdot & \cdot & 1 & \cdot & \cdot & \cdot & \cdot & \cdot & \cdot & \cdot & \cdot & \cdot \\
\cdot & \cdot & \cdot & \cdot & \cdot & \cdot & \cdot & \cdot & \cdot & \cdot & \cdot & \cdot & \cdot & \cdot & \cdot & \cdot & \cdot & \cdot & \cdot & \cdot & 1 & \cdot & \cdot & \cdot \\
\cdot & \cdot & \frac{3}{2} & \cdot & \cdot & \cdot & \cdot & \cdot & \cdot & \cdot & -\frac{1}{2} & \cdot & \cdot & \cdot & \cdot & \frac{1}{2} & \cdot & \cdot & \cdot & \cdot & \cdot & \cdot & \cdot & \cdot
\end{vmatrix};$$

$$F^0 = \begin{vmatrix}
\cdot & \cdot & \cdot & \cdot & \cdot & \cdot & \frac{2}{3} & \cdot & \cdot & \cdot \\
\cdot & \cdot & \cdot & \cdot & \cdot & \cdot & \cdot & \frac{2}{3} & \cdot & \cdot \\
\cdot & \cdot & \cdot & \cdot & \cdot & \cdot & \cdot & \cdot & \frac{2}{3} & \cdot \\
\cdot & \cdot & \cdot & \cdot & \cdot & \cdot & \cdot & \cdot & \cdot & \cdot \\
\cdot & \cdot & \cdot & \cdot & \cdot & \cdot & \cdot & \cdot & \cdot & \cdot \\
\cdot & \cdot & \cdot & \cdot & \cdot & \cdot & \cdot & \cdot & \cdot & \cdot \\
1 & \cdot & \cdot & \cdot & \cdot & \cdot & \cdot & \cdot & \cdot & -\frac{1}{3} \\
\cdot & \cdot & \cdot & \cdot & \cdot & 1 & \cdot & \cdot & \cdot & \cdot \\
\cdot & \cdot & \cdot & \cdot & 1 & \cdot & \cdot & \cdot & \cdot & \cdot \\
\cdot & \cdot & \cdot & \cdot & \cdot & \cdot & \cdot & \cdot & \cdot & \cdot \\
\cdot & \cdot & \cdot & \cdot & \cdot & \cdot & \cdot & \cdot & -\frac{1}{3} & \cdot \\
\cdot & \cdot & \cdot & \cdot & \cdot & \cdot & \cdot & \frac{1}{3} & \cdot & \cdot \\
\cdot & \cdot & \cdot & \cdot & \cdot & 1 & \cdot & \cdot & \cdot & \cdot \\
\cdot & 1 & \cdot & \cdot & \cdot & \cdot & \cdot & \cdot & \cdot & -\frac{1}{3} \\
\cdot & \cdot & \cdot & 1 & \cdot & \cdot & \cdot & \cdot & \cdot & \cdot \\
\cdot & \cdot & \cdot & \cdot & \cdot & \cdot & \cdot & \cdot & \frac{1}{3} & \cdot \\
\cdot & \cdot & \cdot & \cdot & \cdot & \cdot & \cdot & \cdot & \cdot & \cdot \\
\cdot & \cdot & \cdot & \cdot & \cdot & \cdot & -\frac{1}{3} & \cdot & \cdot & \cdot \\
\cdot & \cdot & \cdot & \cdot & 1 & \cdot & \cdot & \cdot & \cdot & \cdot \\
\cdot & \cdot & \cdot & 1 & \cdot & \cdot & \cdot & \cdot & \cdot & \cdot \\
\cdot & \cdot & 1 & \cdot & \cdot & \cdot & \cdot & \cdot & \cdot & -\frac{1}{3} \\
\cdot & \cdot & \cdot & \cdot & \cdot & \cdot & \cdot & -\frac{1}{3} & \cdot & \cdot \\
\cdot & \cdot & \cdot & \cdot & \cdot & \cdot & \frac{1}{3} & \cdot & \cdot & \cdot \\
\cdot & \cdot & \cdot & \cdot & \cdot & \cdot & \cdot & \cdot & \cdot & \cdot
\end{vmatrix},$$

$$F^{1} = \left|\begin{array}{cccccccccc}
1/3 & . & . & . & . & . & . & . & . & -1 \\
. & . & . & . & . & 1/3 & . & . & . & . \\
. & . & . & . & 1/3 & . & . & . & . & . \\
. & . & . & . & . & . & . & . & . & . \\
. & . & . & . & . & . & . & . & -1 & . \\
. & . & . & . & . & . & . & 1 & . & . \\
. & . & . & . & . & . & -2/3 & . & . & . \\
. & . & . & . & . & . & . & . & . & . \\
. & . & . & . & . & . & . & . & . & . \\
. & . & . & . & . & . & . & . & . & . \\
. & . & . & . & -2/3 & . & . & . & . & . \\
. & . & . & . & . & 2/3 & . & . & . & . \\
. & . & . & . & . & . & . & -1 & . & . \\
. & . & . & . & . & . & 1/3 & . & . & . \\
. & . & . & . & . & . & . & . & . & . \\
. & . & . & . & -1/3 & . & . & . & . & . \\
. & . & . & -1 & . & . & . & . & . & . \\
1/3 & 1 & . & . & . & . & . & . & . & . \\
. & . & . & . & . & . & . & . & -1 & . \\
. & . & . & . & . & . & . & . & . & . \\
. & . & . & . & . & . & 1/3 & . & . & . \\
. & . & . & . & . & 1/3 & . & . & . & . \\
-1/3 & . & -1 & . & . & . & . & . & . & . \\
. & . & . & 1 & . & . & . & . & . & .
\end{array}\right|,$$

$$F^{2} = \left|\begin{array}{cccccccccc}
. & . & . & . & . & 1/3 & . & . & . & . \\
. & 1/3 & . & . & . & . & . & . & . & -1 \\
. & . & . & 1/3 & . & . & . & . & . & . \\
. & . & . & . & . & . & . & . & 1 & . \\
. & . & . & . & . & . & . & . & . & . \\
. & . & . & . & . & . & -1 & . & . & . \\
. & . & . & . & . & . & . & 1/3 & . & . \\
. & . & . & . & . & . & -1 & . & . & . \\
. & . & . & . & . & . & . & . & . & . \\
. & . & . & . & 1 & . & . & . & . & . \\
. & . & . & 1/3 & . & . & . & . & . & . \\
-1 & -1/3 & . & . & . & . & . & . & . & . \\
. & . & . & . & . & . & . & . & . & . \\
. & . & . & . & . & . & . & -2/3 & . & . \\
. & . & . & . & . & . & . & . & . & . \\
. & . & . & 2/3 & . & . & . & . & . & . \\
. & . & . & . & . & . & . & . & . & . \\
. & . & . & . & . & -2/3 & . & . & . & . \\
. & . & . & . & . & . & . & . & . & . \\
. & . & . & . & . & . & . & . & -1 & . \\
. & . & . & . & . & . & . & 1/3 & . & . \\
. & 1/3 & 1 & . & . & . & . & . & . & . \\
. & . & . & . & . & -1/3 & . & . & . & . \\
. & . & . & . & -1 & . & . & . & . & .
\end{array}\right|,$$

$$F^{3} = \left|\begin{array}{cccccccccc}
. & . & . & . & 1/3 & . & . & . & . & . \\
. & . & . & 1/3 & . & . & . & . & . & . \\
. & . & 1/3 & . & . & . & . & . & . & -1 \\
. & . & . & . & . & . & . & -1 & . & . \\
. & . & . & . & . & . & 1 & . & . & . \\
. & . & . & . & . & . & . & . & . & . \\
. & . & . & . & . & . & . & . & 1/3 & . \\
. & . & . & . & . & . & . & . & . & . \\
. & . & . & . & . & . & -1 & . & . & . \\
. & . & . & . & . & -1 & . & . & . & . \\
1 & . & 1/3 & . & . & . & . & . & . & . \\
. & . & . & -1/3 & . & . & . & . & . & . \\
. & . & . & . & . & . & . & . & . & . \\
. & . & . & . & . & . & . & . & 1/3 & . \\
. & . & . & . & . & . & . & -1 & . & . \\
. & -1 & -1/3 & . & . & . & . & . & . & . \\
. & . & . & . & . & 1 & . & . & . & . \\
. & . & . & . & 1/3 & . & . & . & . & . \\
. & . & . & . & . & . & . & . & . & . \\
. & . & . & . & . & . & . & . & . & . \\
. & . & . & . & . & . & . & . & -2/3 & . \\
. & . & . & -2/3 & . & . & . & . & . & . \\
. & . & . & . & 2/3 & . & . & . & . & . \\
. & . & . & . & . & . & . & . & . & .
\end{array}\right|;$$

$$J_3^{12} =$$

$$
\left|\begin{array}{cccccccccccccccccccccccc}
\cdot & -1 & \cdot & \cdot & \cdot & \cdot & \cdot & \cdot & \cdot & \cdot & \cdot & \cdot & \cdot & \cdot & \cdot & \cdot & \cdot & \cdot & \cdot & \cdot & \cdot & \cdot & \cdot & \cdot \\
1 & \cdot & \cdot & \cdot & \cdot & \cdot & \cdot & \cdot & \cdot & \cdot & \cdot & \cdot & \cdot & \cdot & \cdot & \cdot & \cdot & \cdot & \cdot & \cdot & \cdot & \cdot & \cdot & \cdot \\
\cdot & \cdot & \cdot & \cdot & \cdot & \cdot & \cdot & \cdot & \cdot & \cdot & \cdot & \cdot & \cdot & \cdot & \cdot & \cdot & \cdot & \cdot & \cdot & \cdot & \cdot & \cdot & \cdot & \cdot \\
\cdot & \cdot & \cdot & \cdot & -1 & \cdot & \cdot & \cdot & \cdot & \cdot & \cdot & \cdot & \cdot & \cdot & \cdot & \cdot & \cdot & \cdot & \cdot & \cdot & \cdot & \cdot & \cdot & \cdot \\
\cdot & \cdot & \cdot & 1 & \cdot & \cdot & \cdot & \cdot & \cdot & \cdot & \cdot & \cdot & \cdot & \cdot & \cdot & \cdot & \cdot & \cdot & \cdot & \cdot & \cdot & \cdot & \cdot & \cdot \\
\cdot & \cdot & \cdot & \cdot & \cdot & \cdot & \cdot & \cdot & \cdot & \cdot & \cdot & \cdot & \cdot & \cdot & \cdot & \cdot & \cdot & \cdot & \cdot & \cdot & \cdot & \cdot & \cdot & \cdot \\
\cdot & \cdot & \cdot & \cdot & \cdot & \cdot & \cdot & -1 & \cdot & \cdot & \cdot & \cdot & -1 & \cdot & \cdot & \cdot & \cdot & \cdot & \cdot & \cdot & \cdot & \cdot & \cdot & \cdot \\
\cdot & \cdot & \cdot & \cdot & \cdot & \cdot & 1 & \cdot & \cdot & \cdot & \cdot & \cdot & \cdot & -1 & \cdot & \cdot & \cdot & \cdot & \cdot & \cdot & \cdot & \cdot & \cdot & \cdot \\
\cdot & \cdot & \cdot & \cdot & \cdot & \cdot & \cdot & \cdot & \cdot & \cdot & \cdot & \cdot & \cdot & \cdot & -1 & \cdot & \cdot & \cdot & \cdot & \cdot & \cdot & \cdot & \cdot & \cdot \\
\cdot & \cdot & \cdot & \cdot & \cdot & \cdot & \cdot & \cdot & \cdot & \cdot & -1 & \cdot & \cdot & \cdot & \cdot & -1 & \cdot & \cdot & \cdot & \cdot & \cdot & \cdot & \cdot & \cdot \\
\cdot & \cdot & \cdot & \cdot & \cdot & \cdot & \cdot & \cdot & \cdot & 1 & \cdot & \cdot & \cdot & \cdot & \cdot & \cdot & -1 & \cdot & \cdot & \cdot & \cdot & \cdot & \cdot & \cdot \\
\cdot & \cdot & \cdot & \cdot & \cdot & \cdot & \cdot & \cdot & \cdot & \cdot & \cdot & \cdot & \cdot & \cdot & \cdot & \cdot & \cdot & -1 & \cdot & \cdot & \cdot & \cdot & \cdot & \cdot \\
\cdot & \cdot & \cdot & \cdot & \cdot & \cdot & 1 & \cdot & \cdot & \cdot & \cdot & \cdot & \cdot & -1 & \cdot & \cdot & \cdot & \cdot & \cdot & \cdot & \cdot & \cdot & \cdot & \cdot \\
\cdot & \cdot & \cdot & \cdot & \cdot & \cdot & \cdot & 1 & \cdot & \cdot & \cdot & \cdot & 1 & \cdot & \cdot & \cdot & \cdot & \cdot & \cdot & \cdot & \cdot & \cdot & \cdot & \cdot \\
\cdot & \cdot & \cdot & \cdot & \cdot & \cdot & \cdot & \cdot & 1 & \cdot & \cdot & \cdot & \cdot & \cdot & \cdot & \cdot & \cdot & \cdot & \cdot & \cdot & \cdot & \cdot & \cdot & \cdot \\
\cdot & \cdot & \cdot & \cdot & \cdot & \cdot & \cdot & \cdot & \cdot & 1 & \cdot & \cdot & \cdot & \cdot & \cdot & \cdot & -1 & \cdot & \cdot & \cdot & \cdot & \cdot & \cdot & \cdot \\
\cdot & \cdot & \cdot & \cdot & \cdot & \cdot & \cdot & \cdot & \cdot & \cdot & 1 & \cdot & \cdot & \cdot & \cdot & 1 & \cdot & \cdot & \cdot & \cdot & \cdot & \cdot & \cdot & \cdot \\
\cdot & \cdot & \cdot & \cdot & \cdot & \cdot & \cdot & \cdot & \cdot & \cdot & \cdot & 1 & \cdot & \cdot & \cdot & \cdot & \cdot & \cdot & \cdot & \cdot & \cdot & \cdot & \cdot & \cdot \\
\cdot & \cdot & \cdot & \cdot & \cdot & \cdot & \cdot & \cdot & \cdot & \cdot & \cdot & \cdot & \cdot & \cdot & \cdot & \cdot & \cdot & \cdot & \cdot & -1 & \cdot & \cdot & \cdot & \cdot \\
\cdot & \cdot & \cdot & \cdot & \cdot & \cdot & \cdot & \cdot & \cdot & \cdot & \cdot & \cdot & \cdot & \cdot & \cdot & \cdot & \cdot & \cdot & 1 & \cdot & \cdot & \cdot & \cdot & \cdot \\
\cdot & \cdot & \cdot & \cdot & \cdot & \cdot & \cdot & \cdot & \cdot & \cdot & \cdot & \cdot & \cdot & \cdot & \cdot & \cdot & \cdot & \cdot & \cdot & \cdot & \cdot & \cdot & \cdot & \cdot \\
\cdot & \cdot & \cdot & \cdot & \cdot & \cdot & \cdot & \cdot & \cdot & \cdot & \cdot & \cdot & \cdot & \cdot & \cdot & \cdot & \cdot & \cdot & \cdot & \cdot & \cdot & \cdot & -1 & \cdot \\
\cdot & \cdot & \cdot & \cdot & \cdot & \cdot & \cdot & \cdot & \cdot & \cdot & \cdot & \cdot & \cdot & \cdot & \cdot & \cdot & \cdot & \cdot & \cdot & \cdot & \cdot & 1 & \cdot & \cdot \\
\cdot & \cdot & \cdot & \cdot & \cdot & \cdot & \cdot & \cdot & \cdot & \cdot & \cdot & \cdot & \cdot & \cdot & \cdot & \cdot & \cdot & \cdot & \cdot & \cdot & \cdot & \cdot & \cdot & \cdot
\end{array}\right|,
$$

$$J_3^{31} =$$

$$
\left|\begin{array}{cccccccccccccccccccccccc}
\cdot & \cdot & 1 & \cdot & \cdot & \cdot & \cdot & \cdot & \cdot & \cdot & \cdot & \cdot & \cdot & \cdot & \cdot & \cdot & \cdot & \cdot & \cdot & \cdot & \cdot & \cdot & \cdot & \cdot \\
\cdot & \cdot & \cdot & \cdot & \cdot & \cdot & \cdot & \cdot & \cdot & \cdot & \cdot & \cdot & \cdot & \cdot & \cdot & \cdot & \cdot & \cdot & \cdot & \cdot & \cdot & \cdot & \cdot & \cdot \\
-1 & \cdot & \cdot & \cdot & \cdot & \cdot & \cdot & \cdot & \cdot & \cdot & \cdot & \cdot & \cdot & \cdot & \cdot & \cdot & \cdot & \cdot & \cdot & \cdot & \cdot & \cdot & \cdot & \cdot \\
\cdot & \cdot & \cdot & \cdot & \cdot & 1 & \cdot & \cdot & \cdot & \cdot & \cdot & \cdot & \cdot & \cdot & \cdot & \cdot & \cdot & \cdot & \cdot & \cdot & \cdot & \cdot & \cdot & \cdot \\
\cdot & \cdot & \cdot & \cdot & \cdot & \cdot & \cdot & \cdot & \cdot & \cdot & \cdot & \cdot & \cdot & \cdot & \cdot & \cdot & \cdot & \cdot & \cdot & \cdot & \cdot & \cdot & \cdot & \cdot \\
\cdot & \cdot & \cdot & -1 & \cdot & \cdot & \cdot & \cdot & \cdot & \cdot & \cdot & \cdot & \cdot & \cdot & \cdot & \cdot & \cdot & \cdot & \cdot & \cdot & \cdot & \cdot & \cdot & \cdot \\
\cdot & \cdot & \cdot & \cdot & \cdot & \cdot & \cdot & \cdot & 1 & \cdot & \cdot & \cdot & \cdot & \cdot & \cdot & \cdot & \cdot & \cdot & 1 & \cdot & \cdot & \cdot & \cdot & \cdot \\
\cdot & \cdot & \cdot & \cdot & \cdot & \cdot & \cdot & \cdot & \cdot & \cdot & \cdot & \cdot & \cdot & \cdot & \cdot & \cdot & \cdot & \cdot & \cdot & 1 & \cdot & \cdot & \cdot & \cdot \\
\cdot & \cdot & \cdot & \cdot & \cdot & \cdot & -1 & \cdot & \cdot & \cdot & \cdot & \cdot & \cdot & \cdot & \cdot & \cdot & \cdot & \cdot & \cdot & \cdot & 1 & \cdot & \cdot & \cdot \\
\cdot & \cdot & \cdot & \cdot & \cdot & \cdot & \cdot & \cdot & \cdot & \cdot & \cdot & 1 & \cdot & \cdot & \cdot & \cdot & \cdot & \cdot & \cdot & \cdot & \cdot & 1 & \cdot & \cdot \\
\cdot & \cdot & \cdot & \cdot & \cdot & \cdot & \cdot & \cdot & \cdot & \cdot & \cdot & \cdot & \cdot & \cdot & \cdot & \cdot & \cdot & \cdot & \cdot & \cdot & \cdot & \cdot & 1 & \cdot \\
\cdot & \cdot & \cdot & \cdot & \cdot & \cdot & \cdot & \cdot & \cdot & -1 & \cdot & \cdot & \cdot & \cdot & \cdot & \cdot & \cdot & \cdot & \cdot & \cdot & \cdot & \cdot & \cdot & 1 \\
\cdot & \cdot & \cdot & \cdot & \cdot & \cdot & \cdot & \cdot & \cdot & \cdot & \cdot & \cdot & \cdot & \cdot & 1 & \cdot & \cdot & \cdot & \cdot & \cdot & \cdot & \cdot & \cdot & \cdot \\
\cdot & \cdot & \cdot & \cdot & \cdot & \cdot & \cdot & \cdot & \cdot & \cdot & \cdot & \cdot & \cdot & \cdot & \cdot & \cdot & \cdot & \cdot & \cdot & \cdot & \cdot & \cdot & \cdot & \cdot \\
\cdot & \cdot & \cdot & \cdot & \cdot & \cdot & \cdot & \cdot & \cdot & \cdot & \cdot & \cdot & -1 & \cdot & \cdot & \cdot & \cdot & \cdot & \cdot & \cdot & \cdot & \cdot & \cdot & \cdot \\
\cdot & \cdot & \cdot & \cdot & \cdot & \cdot & \cdot & \cdot & \cdot & \cdot & \cdot & \cdot & \cdot & \cdot & \cdot & \cdot & \cdot & 1 & \cdot & \cdot & \cdot & \cdot & \cdot & \cdot \\
\cdot & \cdot & \cdot & \cdot & \cdot & \cdot & \cdot & \cdot & \cdot & \cdot & \cdot & \cdot & \cdot & \cdot & \cdot & \cdot & \cdot & \cdot & \cdot & \cdot & \cdot & \cdot & \cdot & \cdot \\
\cdot & \cdot & \cdot & \cdot & \cdot & \cdot & \cdot & \cdot & \cdot & \cdot & \cdot & \cdot & \cdot & \cdot & \cdot & -1 & \cdot & \cdot & \cdot & \cdot & \cdot & \cdot & \cdot & \cdot \\
\cdot & \cdot & \cdot & \cdot & \cdot & \cdot & -1 & \cdot & \cdot & \cdot & \cdot & \cdot & \cdot & \cdot & \cdot & \cdot & \cdot & \cdot & \cdot & \cdot & 1 & \cdot & \cdot & \cdot \\
\cdot & \cdot & \cdot & \cdot & \cdot & \cdot & \cdot & -1 & \cdot & \cdot & \cdot & \cdot & \cdot & \cdot & \cdot & \cdot & \cdot & \cdot & \cdot & \cdot & \cdot & \cdot & \cdot & \cdot \\
\cdot & \cdot & \cdot & \cdot & \cdot & \cdot & \cdot & \cdot & -1 & \cdot & \cdot & \cdot & \cdot & \cdot & \cdot & \cdot & \cdot & \cdot & -1 & \cdot & \cdot & \cdot & \cdot & \cdot \\
\cdot & \cdot & \cdot & \cdot & \cdot & \cdot & \cdot & \cdot & \cdot & -1 & \cdot & \cdot & \cdot & \cdot & \cdot & \cdot & \cdot & \cdot & \cdot & \cdot & \cdot & \cdot & \cdot & 1 \\
\cdot & \cdot & \cdot & \cdot & \cdot & \cdot & \cdot & \cdot & \cdot & \cdot & -1 & \cdot & \cdot & \cdot & \cdot & \cdot & \cdot & \cdot & \cdot & \cdot & \cdot & \cdot & \cdot & \cdot \\
\cdot & \cdot & \cdot & \cdot & \cdot & \cdot & \cdot & \cdot & \cdot & \cdot & \cdot & -1 & \cdot & \cdot & \cdot & \cdot & \cdot & \cdot & \cdot & \cdot & \cdot & -1 & \cdot & \cdot
\end{array}\right|,
$$

$$J_3^{23} =$$

$$
\left|\begin{array}{cccccccccccccccccccccccc}
\cdot & \cdot & \cdot & \cdot & \cdot & \cdot & \cdot & \cdot & \cdot & \cdot & \cdot & \cdot & \cdot & \cdot & \cdot & \cdot & \cdot & \cdot & \cdot & \cdot & \cdot & \cdot & \cdot & \cdot \\
\cdot & \cdot & -1 & \cdot & \cdot & \cdot & \cdot & \cdot & \cdot & \cdot & \cdot & \cdot & \cdot & \cdot & \cdot & \cdot & \cdot & \cdot & \cdot & \cdot & \cdot & \cdot & \cdot & \cdot \\
\cdot & 1 & \cdot & \cdot & \cdot & \cdot & \cdot & \cdot & \cdot & \cdot & \cdot & \cdot & \cdot & \cdot & \cdot & \cdot & \cdot & \cdot & \cdot & \cdot & \cdot & \cdot & \cdot & \cdot \\
\cdot & \cdot & \cdot & \cdot & \cdot & \cdot & \cdot & \cdot & \cdot & \cdot & \cdot & \cdot & \cdot & \cdot & \cdot & \cdot & \cdot & \cdot & \cdot & \cdot & \cdot & \cdot & \cdot & \cdot \\
\cdot & \cdot & \cdot & \cdot & \cdot & -1 & \cdot & \cdot & \cdot & \cdot & \cdot & \cdot & \cdot & \cdot & \cdot & \cdot & \cdot & \cdot & \cdot & \cdot & \cdot & \cdot & \cdot & \cdot \\
\cdot & \cdot & \cdot & \cdot & 1 & \cdot & \cdot & \cdot & \cdot & \cdot & \cdot & \cdot & \cdot & \cdot & \cdot & \cdot & \cdot & \cdot & \cdot & \cdot & \cdot & \cdot & \cdot & \cdot \\
\cdot & \cdot & \cdot & \cdot & \cdot & \cdot & \cdot & \cdot & \cdot & \cdot & \cdot & \cdot & \cdot & \cdot & \cdot & \cdot & \cdot & \cdot & \cdot & \cdot & \cdot & \cdot & \cdot & \cdot \\
\cdot & \cdot & \cdot & \cdot & \cdot & \cdot & \cdot & \cdot & -1 & \cdot & \cdot & \cdot & \cdot & \cdot & \cdot & \cdot & \cdot & \cdot & \cdot & \cdot & \cdot & \cdot & \cdot & \cdot \\
\cdot & \cdot & \cdot & \cdot & \cdot & \cdot & \cdot & 1 & \cdot & \cdot & \cdot & \cdot & \cdot & \cdot & \cdot & \cdot & \cdot & \cdot & \cdot & \cdot & \cdot & \cdot & \cdot & \cdot \\
\cdot & \cdot & \cdot & \cdot & \cdot & \cdot & \cdot & \cdot & \cdot & \cdot & \cdot & \cdot & \cdot & \cdot & \cdot & \cdot & \cdot & \cdot & \cdot & \cdot & \cdot & \cdot & \cdot & \cdot \\
\cdot & \cdot & \cdot & \cdot & \cdot & \cdot & \cdot & \cdot & \cdot & \cdot & \cdot & -1 & \cdot & \cdot & \cdot & \cdot & \cdot & \cdot & \cdot & \cdot & \cdot & \cdot & \cdot & \cdot \\
\cdot & \cdot & \cdot & \cdot & \cdot & \cdot & \cdot & \cdot & \cdot & \cdot & 1 & \cdot & \cdot & \cdot & \cdot & \cdot & \cdot & \cdot & \cdot & \cdot & \cdot & \cdot & \cdot & \cdot \\
\cdot & \cdot & \cdot & \cdot & \cdot & \cdot & \cdot & \cdot & \cdot & \cdot & \cdot & \cdot & \cdot & \cdot & \cdot & \cdot & \cdot & \cdot & -1 & \cdot & \cdot & \cdot & \cdot & \cdot \\
\cdot & \cdot & \cdot & \cdot & \cdot & \cdot & \cdot & \cdot & \cdot & \cdot & \cdot & \cdot & \cdot & \cdot & -1 & \cdot & \cdot & \cdot & \cdot & -1 & \cdot & \cdot & \cdot & \cdot \\
\cdot & \cdot & \cdot & \cdot & \cdot & \cdot & \cdot & \cdot & \cdot & \cdot & \cdot & \cdot & \cdot & 1 & \cdot & \cdot & \cdot & \cdot & \cdot & \cdot & -1 & \cdot & \cdot & \cdot \\
\cdot & \cdot & \cdot & \cdot & \cdot & \cdot & \cdot & \cdot & \cdot & \cdot & \cdot & \cdot & \cdot & \cdot & \cdot & \cdot & \cdot & \cdot & \cdot & \cdot & \cdot & -1 & \cdot & \cdot \\
\cdot & \cdot & \cdot & \cdot & \cdot & \cdot & \cdot & \cdot & \cdot & \cdot & \cdot & \cdot & \cdot & \cdot & \cdot & \cdot & \cdot & -1 & \cdot & \cdot & \cdot & \cdot & -1 & \cdot \\
\cdot & \cdot & \cdot & \cdot & \cdot & \cdot & \cdot & \cdot & \cdot & \cdot & \cdot & \cdot & \cdot & \cdot & \cdot & \cdot & 1 & \cdot & \cdot & \cdot & \cdot & \cdot & \cdot & -1 \\
\cdot & \cdot & \cdot & \cdot & \cdot & \cdot & \cdot & \cdot & \cdot & \cdot & \cdot & \cdot & 1 & \cdot & \cdot & \cdot & \cdot & \cdot & \cdot & \cdot & \cdot & \cdot & \cdot & \cdot \\
\cdot & \cdot & \cdot & \cdot & \cdot & \cdot & \cdot & \cdot & \cdot & \cdot & \cdot & \cdot & \cdot & 1 & \cdot & \cdot & \cdot & \cdot & \cdot & \cdot & -1 & \cdot & \cdot & \cdot \\
\cdot & \cdot & \cdot & \cdot & \cdot & \cdot & \cdot & \cdot & \cdot & \cdot & \cdot & \cdot & \cdot & \cdot & 1 & \cdot & \cdot & \cdot & \cdot & 1 & \cdot & \cdot & \cdot & \cdot \\
\cdot & \cdot & \cdot & \cdot & \cdot & \cdot & \cdot & \cdot & \cdot & \cdot & \cdot & \cdot & \cdot & \cdot & \cdot & 1 & \cdot & \cdot & \cdot & \cdot & \cdot & \cdot & \cdot & \cdot \\
\cdot & \cdot & \cdot & \cdot & \cdot & \cdot & \cdot & \cdot & \cdot & \cdot & \cdot & \cdot & \cdot & \cdot & \cdot & \cdot & 1 & \cdot & \cdot & \cdot & \cdot & \cdot & \cdot & -1 \\
\cdot & \cdot & \cdot & \cdot & \cdot & \cdot & \cdot & \cdot & \cdot & \cdot & \cdot & \cdot & \cdot & \cdot & \cdot & \cdot & \cdot & 1 & \cdot & \cdot & \cdot & \cdot & 1 & \cdot
\end{array}\right|,
$$

$$J_3^{01} =$$

$$
\left| \begin{array}{cccccccccccccccccccccccc}
\cdot & \cdot & \cdot & \cdot & \cdot & \cdot & -1 & \cdot & \cdot & \cdot & \cdot & \cdot & \cdot & \cdot & \cdot & \cdot & \cdot & \cdot & \cdot & \cdot & \cdot & \cdot & \cdot & \cdot \\
\cdot & \cdot & \cdot & \cdot & \cdot & 1 & \cdot & -1 & \cdot & \cdot & \cdot & \cdot & \cdot & \cdot & \cdot & \cdot & \cdot & \cdot & \cdot & \cdot & \cdot & \cdot & \cdot & \cdot \\
\cdot & \cdot & \cdot & \cdot & -1 & \cdot & \cdot & \cdot & -1 & \cdot & \cdot & \cdot & \cdot & \cdot & \cdot & \cdot & \cdot & \cdot & \cdot & \cdot & \cdot & \cdot & \cdot & \cdot \\
\cdot & \cdot & \cdot & \cdot & \cdot & \cdot & \cdot & \cdot & \cdot & -1 & \cdot & \cdot & \cdot & \cdot & \cdot & \cdot & \cdot & \cdot & \cdot & \cdot & \cdot & \cdot & \cdot & \cdot \\
\cdot & \cdot & -1 & \cdot & \cdot & \cdot & \cdot & \cdot & \cdot & \cdot & -1 & \cdot & \cdot & \cdot & \cdot & \cdot & \cdot & \cdot & \cdot & \cdot & \cdot & \cdot & \cdot & \cdot \\
\cdot & 1 & \cdot & \cdot & \cdot & \cdot & \cdot & \cdot & \cdot & \cdot & \cdot & -1 & \cdot & \cdot & \cdot & \cdot & \cdot & \cdot & \cdot & \cdot & \cdot & \cdot & \cdot & \cdot \\
-1 & \cdot & \cdot & \cdot & \cdot & \cdot & \cdot & \cdot & \cdot & \cdot & \cdot & \cdot & \cdot & \cdot & \cdot & \cdot & \cdot & \cdot & \cdot & \cdot & \cdot & \cdot & \cdot & \cdot \\
\cdot & -1 & \cdot & \cdot & \cdot & \cdot & \cdot & \cdot & \cdot & \cdot & \cdot & 1 & \cdot & \cdot & \cdot & \cdot & \cdot & \cdot & \cdot & \cdot & \cdot & \cdot & \cdot & \cdot \\
\cdot & \cdot & -1 & \cdot & \cdot & \cdot & \cdot & \cdot & \cdot & \cdot & -1 & \cdot & \cdot & \cdot & \cdot & \cdot & \cdot & \cdot & \cdot & \cdot & \cdot & \cdot & \cdot & \cdot \\
\cdot & \cdot & \cdot & -1 & \cdot & \cdot & \cdot & \cdot & \cdot & \cdot & \cdot & \cdot & \cdot & \cdot & \cdot & \cdot & \cdot & \cdot & \cdot & \cdot & \cdot & \cdot & \cdot & \cdot \\
\cdot & \cdot & \cdot & \cdot & -1 & \cdot & \cdot & \cdot & -1 & \cdot & \cdot & \cdot & \cdot & \cdot & \cdot & \cdot & \cdot & \cdot & \cdot & \cdot & \cdot & \cdot & \cdot & \cdot \\
\cdot & \cdot & \cdot & \cdot & \cdot & -1 & \cdot & 1 & \cdot & \cdot & \cdot & \cdot & \cdot & \cdot & \cdot & \cdot & \cdot & \cdot & \cdot & \cdot & \cdot & \cdot & \cdot & \cdot \\
\cdot & \cdot & \cdot & \cdot & \cdot & \cdot & \cdot & \cdot & \cdot & \cdot & \cdot & \cdot & \cdot & \cdot & \cdot & \cdot & \cdot & \cdot & \cdot & \cdot & \cdot & \cdot & \cdot & \cdot \\
\cdot & \cdot & \cdot & \cdot & \cdot & \cdot & \cdot & \cdot & \cdot & \cdot & \cdot & \cdot & \cdot & \cdot & \cdot & \cdot & \cdot & 1 & \cdot & \cdot & \cdot & \cdot & \cdot & \cdot \\
\cdot & \cdot & \cdot & \cdot & \cdot & \cdot & \cdot & \cdot & \cdot & \cdot & \cdot & \cdot & \cdot & \cdot & \cdot & \cdot & -1 & \cdot & \cdot & \cdot & \cdot & \cdot & \cdot & \cdot \\
\cdot & \cdot & \cdot & \cdot & \cdot & \cdot & \cdot & \cdot & \cdot & \cdot & \cdot & \cdot & \cdot & \cdot & \cdot & \cdot & \cdot & \cdot & \cdot & \cdot & \cdot & \cdot & \cdot & \cdot \\
\cdot & \cdot & \cdot & \cdot & \cdot & \cdot & \cdot & \cdot & \cdot & \cdot & \cdot & \cdot & \cdot & \cdot & -1 & \cdot & \cdot & \cdot & \cdot & \cdot & \cdot & \cdot & \cdot & \cdot \\
\cdot & \cdot & \cdot & \cdot & \cdot & \cdot & \cdot & \cdot & \cdot & \cdot & \cdot & \cdot & \cdot & 1 & \cdot & \cdot & \cdot & \cdot & \cdot & \cdot & \cdot & \cdot & \cdot & \cdot \\
\cdot & \cdot & \cdot & \cdot & \cdot & \cdot & \cdot & \cdot & \cdot & \cdot & \cdot & \cdot & \cdot & \cdot & \cdot & \cdot & \cdot & \cdot & \cdot & \cdot & \cdot & \cdot & \cdot & \cdot \\
\cdot & \cdot & \cdot & \cdot & \cdot & \cdot & \cdot & \cdot & \cdot & \cdot & \cdot & \cdot & \cdot & \cdot & \cdot & \cdot & \cdot & \cdot & \cdot & \cdot & \cdot & \cdot & \cdot & 1 \\
\cdot & \cdot & \cdot & \cdot & \cdot & \cdot & \cdot & \cdot & \cdot & \cdot & \cdot & \cdot & \cdot & \cdot & \cdot & \cdot & \cdot & \cdot & \cdot & \cdot & \cdot & \cdot & -1 & \cdot \\
\cdot & \cdot & \cdot & \cdot & \cdot & \cdot & \cdot & \cdot & \cdot & \cdot & \cdot & \cdot & \cdot & \cdot & \cdot & \cdot & \cdot & \cdot & \cdot & \cdot & \cdot & \cdot & \cdot & \cdot \\
\cdot & \cdot & \cdot & \cdot & \cdot & \cdot & \cdot & \cdot & \cdot & \cdot & \cdot & \cdot & \cdot & \cdot & \cdot & \cdot & \cdot & \cdot & \cdot & \cdot & -1 & \cdot & \cdot & \cdot \\
\cdot & \cdot & \cdot & \cdot & \cdot & \cdot & \cdot & \cdot & \cdot & \cdot & \cdot & \cdot & \cdot & \cdot & \cdot & \cdot & \cdot & \cdot & \cdot & 1 & \cdot & \cdot & \cdot & \cdot
\end{array} \right|,
$$

$$J_3^{02} =$$

$$
\left| \begin{array}{cccccccccccccccccccccccc}
\cdot & \cdot & \cdot & \cdot & \cdot & -1 & \cdot & \cdot & \cdot & \cdot & \cdot & \cdot & -1 & \cdot & \cdot & \cdot & \cdot & \cdot & \cdot & \cdot & \cdot & \cdot & \cdot & \cdot \\
\cdot & \cdot & \cdot & \cdot & \cdot & \cdot & \cdot & \cdot & \cdot & \cdot & \cdot & \cdot & \cdot & -1 & \cdot & \cdot & \cdot & \cdot & \cdot & \cdot & \cdot & \cdot & \cdot & \cdot \\
\cdot & \cdot & \cdot & 1 & \cdot & \cdot & \cdot & \cdot & \cdot & \cdot & \cdot & \cdot & \cdot & \cdot & -1 & \cdot & \cdot & \cdot & \cdot & \cdot & \cdot & \cdot & \cdot & \cdot \\
\cdot & \cdot & 1 & \cdot & \cdot & \cdot & \cdot & \cdot & \cdot & \cdot & \cdot & \cdot & \cdot & \cdot & \cdot & -1 & \cdot & \cdot & \cdot & \cdot & \cdot & \cdot & \cdot & \cdot \\
\cdot & \cdot & \cdot & \cdot & \cdot & \cdot & \cdot & \cdot & \cdot & \cdot & \cdot & \cdot & \cdot & \cdot & \cdot & \cdot & -1 & \cdot & \cdot & \cdot & \cdot & \cdot & \cdot & \cdot \\
-1 & \cdot & \cdot & \cdot & \cdot & \cdot & \cdot & \cdot & \cdot & \cdot & \cdot & \cdot & \cdot & \cdot & \cdot & \cdot & \cdot & -1 & \cdot & \cdot & \cdot & \cdot & \cdot & \cdot \\
\cdot & \cdot & \cdot & \cdot & \cdot & \cdot & \cdot & \cdot & \cdot & \cdot & \cdot & -1 & \cdot & \cdot & \cdot & \cdot & \cdot & \cdot & \cdot & \cdot & \cdot & \cdot & \cdot & \cdot \\
\cdot & \cdot & \cdot & \cdot & \cdot & \cdot & \cdot & \cdot & \cdot & \cdot & \cdot & \cdot & \cdot & \cdot & \cdot & \cdot & \cdot & \cdot & \cdot & \cdot & \cdot & \cdot & \cdot & \cdot \\
\cdot & \cdot & \cdot & \cdot & \cdot & \cdot & \cdot & \cdot & \cdot & 1 & \cdot & \cdot & \cdot & \cdot & \cdot & \cdot & \cdot & \cdot & \cdot & \cdot & \cdot & \cdot & \cdot & \cdot \\
\cdot & \cdot & \cdot & \cdot & \cdot & \cdot & \cdot & \cdot & 1 & \cdot & \cdot & \cdot & \cdot & \cdot & \cdot & \cdot & \cdot & \cdot & \cdot & \cdot & \cdot & \cdot & \cdot & \cdot \\
\cdot & \cdot & \cdot & \cdot & \cdot & \cdot & \cdot & \cdot & \cdot & \cdot & \cdot & \cdot & \cdot & \cdot & \cdot & \cdot & \cdot & \cdot & \cdot & \cdot & \cdot & \cdot & \cdot & \cdot \\
\cdot & \cdot & \cdot & \cdot & \cdot & \cdot & -1 & \cdot & \cdot & \cdot & \cdot & \cdot & \cdot & \cdot & \cdot & \cdot & \cdot & \cdot & \cdot & \cdot & \cdot & \cdot & \cdot & \cdot \\
-1 & \cdot & \cdot & \cdot & \cdot & \cdot & \cdot & \cdot & \cdot & \cdot & \cdot & \cdot & \cdot & \cdot & \cdot & \cdot & \cdot & -1 & \cdot & \cdot & \cdot & \cdot & \cdot & \cdot \\
\cdot & -1 & \cdot & \cdot & \cdot & \cdot & \cdot & \cdot & \cdot & \cdot & \cdot & \cdot & \cdot & \cdot & \cdot & \cdot & \cdot & \cdot & \cdot & \cdot & \cdot & \cdot & \cdot & \cdot \\
\cdot & \cdot & -1 & \cdot & \cdot & \cdot & \cdot & \cdot & \cdot & \cdot & \cdot & \cdot & \cdot & \cdot & \cdot & 1 & \cdot & \cdot & \cdot & \cdot & \cdot & \cdot & \cdot & \cdot \\
\cdot & \cdot & \cdot & -1 & \cdot & \cdot & \cdot & \cdot & \cdot & \cdot & \cdot & \cdot & \cdot & \cdot & 1 & \cdot & \cdot & \cdot & \cdot & \cdot & \cdot & \cdot & \cdot & \cdot \\
\cdot & \cdot & \cdot & \cdot & -1 & \cdot & \cdot & \cdot & \cdot & \cdot & \cdot & \cdot & \cdot & \cdot & \cdot & \cdot & \cdot & \cdot & \cdot & \cdot & \cdot & \cdot & \cdot & \cdot \\
\cdot & \cdot & \cdot & \cdot & \cdot & -1 & \cdot & \cdot & \cdot & \cdot & \cdot & \cdot & -1 & \cdot & \cdot & \cdot & \cdot & \cdot & \cdot & \cdot & \cdot & \cdot & \cdot & \cdot \\
\cdot & \cdot & \cdot & \cdot & \cdot & \cdot & \cdot & \cdot & \cdot & \cdot & \cdot & \cdot & \cdot & \cdot & \cdot & \cdot & \cdot & \cdot & \cdot & \cdot & \cdot & \cdot & \cdot & -1 \\
\cdot & \cdot & \cdot & \cdot & \cdot & \cdot & \cdot & \cdot & \cdot & \cdot & \cdot & \cdot & \cdot & \cdot & \cdot & \cdot & \cdot & \cdot & \cdot & \cdot & \cdot & \cdot & \cdot & \cdot \\
\cdot & \cdot & \cdot & \cdot & \cdot & \cdot & \cdot & \cdot & \cdot & \cdot & \cdot & \cdot & \cdot & \cdot & \cdot & \cdot & \cdot & \cdot & \cdot & \cdot & \cdot & 1 & \cdot & \cdot \\
\cdot & \cdot & \cdot & \cdot & \cdot & \cdot & \cdot & \cdot & \cdot & \cdot & \cdot & \cdot & \cdot & \cdot & \cdot & \cdot & \cdot & \cdot & \cdot & \cdot & 1 & \cdot & \cdot & \cdot \\
\cdot & \cdot & \cdot & \cdot & \cdot & \cdot & \cdot & \cdot & \cdot & \cdot & \cdot & \cdot & \cdot & \cdot & \cdot & \cdot & \cdot & \cdot & \cdot & \cdot & \cdot & \cdot & \cdot & \cdot \\
\cdot & \cdot & \cdot & \cdot & \cdot & \cdot & \cdot & \cdot & \cdot & \cdot & \cdot & \cdot & \cdot & \cdot & \cdot & \cdot & \cdot & \cdot & -1 & \cdot & \cdot & \cdot & \cdot & \cdot
\end{array} \right|,
$$

$$J_3^{03} =$$

$$
\left| \begin{array}{cccccccccccccccccccccccc}
\cdot & \cdot & \cdot & \cdot & 1 & \cdot & \cdot & \cdot & \cdot & \cdot & \cdot & \cdot & \cdot & \cdot & \cdot & \cdot & \cdot & \cdot & -1 & \cdot & \cdot & \cdot & \cdot & \cdot \\
\cdot & \cdot & \cdot & -1 & \cdot & \cdot & \cdot & \cdot & \cdot & \cdot & \cdot & \cdot & \cdot & \cdot & \cdot & \cdot & \cdot & \cdot & \cdot & -1 & \cdot & \cdot & \cdot & \cdot \\
\cdot & \cdot & \cdot & \cdot & \cdot & \cdot & \cdot & \cdot & \cdot & \cdot & \cdot & \cdot & \cdot & \cdot & \cdot & \cdot & \cdot & \cdot & \cdot & \cdot & -1 & \cdot & \cdot & \cdot \\
\cdot & -1 & \cdot & \cdot & \cdot & \cdot & \cdot & \cdot & \cdot & \cdot & \cdot & \cdot & \cdot & \cdot & \cdot & \cdot & \cdot & \cdot & \cdot & \cdot & \cdot & -1 & \cdot & \cdot \\
1 & \cdot & \cdot & \cdot & \cdot & \cdot & \cdot & \cdot & \cdot & \cdot & \cdot & \cdot & \cdot & \cdot & \cdot & \cdot & \cdot & \cdot & \cdot & \cdot & \cdot & \cdot & -1 & \cdot \\
\cdot & \cdot & \cdot & \cdot & \cdot & \cdot & \cdot & \cdot & \cdot & \cdot & \cdot & \cdot & \cdot & \cdot & \cdot & \cdot & \cdot & \cdot & \cdot & \cdot & \cdot & \cdot & \cdot & -1 \\
\cdot & \cdot & \cdot & \cdot & \cdot & \cdot & \cdot & \cdot & \cdot & \cdot & 1 & \cdot & \cdot & \cdot & \cdot & \cdot & \cdot & \cdot & \cdot & \cdot & \cdot & \cdot & \cdot & \cdot \\
\cdot & \cdot & \cdot & \cdot & \cdot & \cdot & \cdot & \cdot & \cdot & -1 & \cdot & \cdot & \cdot & \cdot & \cdot & \cdot & \cdot & \cdot & \cdot & \cdot & \cdot & \cdot & \cdot & \cdot \\
\cdot & \cdot & \cdot & \cdot & \cdot & \cdot & \cdot & \cdot & \cdot & \cdot & \cdot & \cdot & \cdot & \cdot & \cdot & \cdot & \cdot & \cdot & \cdot & \cdot & \cdot & \cdot & \cdot & \cdot \\
\cdot & \cdot & \cdot & \cdot & \cdot & \cdot & \cdot & -1 & \cdot & \cdot & \cdot & \cdot & \cdot & \cdot & \cdot & \cdot & \cdot & \cdot & \cdot & \cdot & \cdot & \cdot & \cdot & \cdot \\
\cdot & \cdot & \cdot & \cdot & \cdot & \cdot & 1 & \cdot & \cdot & \cdot & \cdot & \cdot & \cdot & \cdot & \cdot & \cdot & \cdot & \cdot & \cdot & \cdot & \cdot & \cdot & \cdot & \cdot \\
\cdot & \cdot & \cdot & \cdot & \cdot & \cdot & \cdot & \cdot & \cdot & \cdot & \cdot & \cdot & \cdot & \cdot & \cdot & \cdot & \cdot & \cdot & \cdot & \cdot & \cdot & \cdot & \cdot & \cdot \\
\cdot & \cdot & \cdot & \cdot & \cdot & \cdot & \cdot & \cdot & \cdot & \cdot & \cdot & \cdot & \cdot & \cdot & \cdot & \cdot & 1 & \cdot & \cdot & \cdot & \cdot & \cdot & \cdot & \cdot \\
\cdot & \cdot & \cdot & \cdot & \cdot & \cdot & \cdot & \cdot & \cdot & \cdot & \cdot & \cdot & \cdot & \cdot & \cdot & -1 & \cdot & \cdot & \cdot & \cdot & \cdot & \cdot & \cdot & \cdot \\
\cdot & \cdot & \cdot & \cdot & \cdot & \cdot & \cdot & \cdot & \cdot & \cdot & \cdot & \cdot & \cdot & \cdot & \cdot & \cdot & \cdot & \cdot & \cdot & \cdot & \cdot & \cdot & \cdot & \cdot \\
\cdot & \cdot & \cdot & \cdot & \cdot & \cdot & \cdot & \cdot & \cdot & \cdot & \cdot & \cdot & \cdot & -1 & \cdot & \cdot & \cdot & \cdot & \cdot & \cdot & \cdot & \cdot & \cdot & \cdot \\
\cdot & \cdot & \cdot & \cdot & \cdot & \cdot & \cdot & \cdot & \cdot & \cdot & \cdot & \cdot & 1 & \cdot & \cdot & \cdot & \cdot & \cdot & \cdot & \cdot & \cdot & \cdot & \cdot & \cdot \\
\cdot & \cdot & \cdot & \cdot & \cdot & \cdot & \cdot & \cdot & \cdot & \cdot & \cdot & \cdot & \cdot & \cdot & \cdot & \cdot & \cdot & \cdot & \cdot & \cdot & \cdot & \cdot & \cdot & \cdot \\
-1 & \cdot & \cdot & \cdot & \cdot & \cdot & \cdot & \cdot & \cdot & \cdot & \cdot & \cdot & \cdot & \cdot & \cdot & \cdot & \cdot & \cdot & \cdot & \cdot & \cdot & \cdot & 1 & \cdot \\
\cdot & -1 & \cdot & \cdot & \cdot & \cdot & \cdot & \cdot & \cdot & \cdot & \cdot & \cdot & \cdot & \cdot & \cdot & \cdot & \cdot & \cdot & \cdot & \cdot & \cdot & -1 & \cdot & \cdot \\
\cdot & \cdot & -1 & \cdot & \cdot & \cdot & \cdot & \cdot & \cdot & \cdot & \cdot & \cdot & \cdot & \cdot & \cdot & \cdot & \cdot & \cdot & \cdot & \cdot & \cdot & \cdot & \cdot & \cdot \\
\cdot & \cdot & \cdot & -1 & \cdot & \cdot & \cdot & \cdot & \cdot & \cdot & \cdot & \cdot & \cdot & \cdot & \cdot & \cdot & \cdot & \cdot & \cdot & -1 & \cdot & \cdot & \cdot & \cdot \\
\cdot & \cdot & \cdot & \cdot & -1 & \cdot & \cdot & \cdot & \cdot & \cdot & \cdot & \cdot & \cdot & \cdot & \cdot & \cdot & \cdot & \cdot & 1 & \cdot & \cdot & \cdot & \cdot & \cdot \\
\cdot & \cdot & \cdot & \cdot & \cdot & -1 & \cdot & \cdot & \cdot & \cdot & \cdot & \cdot & \cdot & \cdot & \cdot & \cdot & \cdot & \cdot & \cdot & \cdot & \cdot & \cdot & \cdot & \cdot
\end{array} \right|.
$$

Conclusion

Let us formulate shortly the final results.

In Chapter 1, the well-known quantum mechanical problem of a spin 1/2 particle in external Coulomb potential was studied from the point of view of possible applications of the Heun function theory. It was shown, that in addition to the standard way to solve the problem in terms of the confluent hypergeometric functions, there are several other opportunities, which rest on applying the confluent Heun functions. All the ways to study this problem lead us to the same energy spectrum, which confirms their correctness.

In Chapter 2, we have studied the scalar and spinor particles in 2-dimensional spaces of constant curvature, hyperbolic Lobachevsky and spherical Riemann planes, in presence of the uniform magnetic field. The Schrödinger particle on the background of the 2-dimensional space of constant positive curvature S_2 was considered in the external magnetic field. By analogy with the case of hyperbolic plane H_2, where quasi-Cartesian coordinates exist with realization of H_2 as the Poincaré half-plane, a specific system of quasi-Cartesian coordinates (x, y) in S_2 was introduced. It turns out that we get a real 2-dimensional space only if these two coordinates are complex and obey an additional constraint. The Schrödinger equation was solved with the use of the method of separation of the variables in both coordinate systems, cylindric and quasi-Cartesian, and the energy spectrum is the same.

In Chapter 3, the quantum mechanical problem for the hydrogen atom was studied for de Sitter and anti de Sitter spaces. In both models, the problem was reduced to the general Heun equation. The energy spectrum for the atom in the de Sitter space has turned out to be quasi-stationary, we have derived an approximate expression for energy levels and estimate the probability of decay of the atom. In in the anti de Sitter model, the n atom should be a stable system, however the corresponding energy spectrum was not found. Short consideration of similar problem for a Dirac spin 1/2 particle for both models was presented.

In Chapter 4, for a spin 0 particle in spherical coordinates for expanding and oscillating de Sitter models, we have developed the non-relativistic approach. In Chapter 5, this analysis was extended to the case of the Dirac particle.

In Chapter 6, we have examined the influence of the Lobachevsky geometry of on particles with spins $0, 1/2, 1$; this geometry effectively acts as an ideal mirror distributed in the space. Since the Lobachevsky space is used in certain cosmological models, this property means that in such models it is necessary to take into account the effect of the presence of a "cosmological mirror"; it should effectively lead to a redistribution of the particle density in the space. We have generalized this analysis for the oscillating de Sitter universe, the effect of the complete reflection of the particles from the effective potential barrier, is preserved.

In Chapter 7, we have studied the isotopic doublet of Dirac fermions in the presence of the non-Abelian monopole field. A special attention was given to the problem of the nonrelativistic approximation in the theory of isotopic doublet in the non-Abelian field. This analysis was detailed for the Bogomol'nyi – Prasad – Sommerfeld momopole solution. We have extended the whole analysis to spaces of constant curvature, Lobachevsky and Riemman models. Besides, we have applied a special geometric method based on the geometrical Kosambi – Cartan – Chen invariants to study the behavior of solutions of the relevant differential equations from the point of view of the Jacobi stability.

In Chapter 8, we have performed the general mathematical study of the tunneling effect for the Dirac particle through the potential barrier generated by the Schwarzschild geometry. Our study was based on the use of Frobenius solutions of the relevant second order differential equations. We have constructed their solutions in explicit form and proved that the involved power series are convergent in all the physical region of the variable $r \in (1, +\infty)$. The results for the tunneling effect significantly differ for two situations: when the particle falls on the barrier from within, and when the particle falls from outside.

In Chapter 9, within the framework of applications of spinor theory to Quantum Mechanics, we have studied the role of spinor space structure in classifying solutions of scalar and spinor equations specified for the cylindric parabolic coordinates. The emphasis was put on doubling the set of space points, so we get an extended space model. In such a space, instead of the 2π-rotation, the 4π-rotation is considered, which transfers the space into itself. We have constructed solutions of four types. The solutions of types $(--)$ and $(++)$ are single-valued in the spaces with vector structure, whereas the solutions of types $(-+)$ and $(+-)$ are not single-valued in spaces with vector structure; so the solutions of these two types $(-+)$ and $(+-)$ must be discarded. However, all the four solutions are valid in the space with spinor structure.

In Chapter 10, we have extended the above analysis to the Maxwell theory. The Maxwell equations in any Riemannian space-time bas presented in spinor form on the base of tetrad method, when the Maxwell field is described by a local 2-nd rank symmetrical spinor. This general covariant equation was specified in cylindrical parabolic coordinates and corresponding diagonal tetrad. After separating the variables, we have derived the system of four 1-st order differential equations in partial derivatives for three functions, depending on two parabolic coordinates. The mathematical task was reduced to one 2-nd order equation in partial derivative for the main function, which determines all the remaining functions. Its solutions were constructed in terms of the confluent hypergeometric functions. We have studied the properties of the four types of constructed solutions – they must be continuous and single-valued – in the context of vector and spinor space model. It was shown that in the space with vector structure, only two variants provide us with correct solutions; in the spinor space, all the four variants are appropriate.

In Chapter 11, the equations of motion associated with a Lagrangian – inspired by relativistic optics in a nonuniform moving medium – are studied. The model describes optical effects in the nonuniform moving medium with special optical properties. We have established the Euler – Lagrange equations for the corresponding geodesics. We have specified the general model to the special case when the metric coefficient γ linearly increases along the direction Z. The exact analytical solutions of the Euler – Lagrange equations have been constructed.

In Chapter 12, the quantum mechanical problem for the vector particle was studied. With the use of the space reflection operator, the radial system of ten equations is split into independent subsystems, consisting of 4 and 6 equations, respectively. The last one reduces to a system of 4 linked first order differential equations for the complex radial functions $f^i(r)$, $i = 1,\ldots,4$. We investigate this system by using the tools of the Jacobi stability theory, namely, the Kosambi – Cartan – Chen theory. It has been shown that the second KCC-invariant is a function of only the radial coordinate r. In accordance with the general theory, a pencil of geodesic curves from the point r_0 converges (or diverges) if the real parts of all eigenvalues of the 2-nd KCC-invariant $Pi^i{}_j$ are negative (or positive). We determine the expressions for the matrix $P^i{}_j(r)$ for $r \to 0$ and $r \to \infty$, and examine the asymptotic behavior of the eigenvalue problem $P\Psi = \lambda\Psi$. The established that the behavior of eigenvalues correlates with the existence of two solutions which may be associated with the bound states of a particle in the Coulomb field.

In Chapter 13, we have studied the generalized Duffin – Kemmer – Petiau equation for the spin 1 particle with anomalous magnetic moment in presence of the Coulomb field. The main accent was given to states with the parity $P = (-1)^j$. In order to simplify the problem, the transition to the nonrelativistic approximation has been performed. The states with minimal $j = 0$ are described by a 2-nd order equation of the double confluent Heun type. By imposing the known transcendency condition, we have derived a quantization rule, which leads to the energy spectrum of the structure $E = -\text{const}/n^2$, however it does not depend on the anomalous magnetic moment. For states with $j = 1, 2, \ldots$, after performing the nonrelativistic approximation we have obtained a system of two 2-nd order linked differential equations for radial functions. This system gives a 4-th order equation for one function. Its Frobenius solutions were constructed, and the convergence of the involved power series with 8- and 9-terms recurrence relations, was studied. All the constructed solutions are exact, but any quantization rules for energies of bound states are not known yet. To study this problem, we have additionally applied the known geometrical method based on the use of geometrical KCC-invariants.

In Chapter 14, similar analysis was performed for a spin 1 particle with quadruple moment in the Coulomb field.

In Chapter 15, the solutions in the form of plane waves for a massive spin 3/2 particle, were found. The complete system of exact solutions for a massive spin 3/2 particle in momentum-helicity basis is constructed. Similar problems are studied for the massless case. The initial wave equation for vector bispinor $\Psi_a(x)$, which describes a massless spin 3/2 particle, is transformed to a new basis $\tilde{\Psi}_a(x)$, in which the gauge symmetry of the theory becomes evident: there exist solutions in the form of 4-gradient of an arbitrary bispinor $\tilde{\Psi}^0_a(x) = \partial_a \Psi(x)$. In this new basis, two independent solutions are explicitly constructed, which do not contain gauge constituents.

In Chapter 16, the wave equation for a spin 3/2 particle was investigated in spherical coordinates. The complete equation is split into the main equation, and two additional constraints. There are constructed the solutions, for which 4 operators are diagonalized: they correspond to the quantum numbers $\{\epsilon, j, m, P\}$. After separating the variables, we derive the main system of 8 radial first order equations and additional 2 algebraic and 2 differential constraints. The solutions of the radial equations are constructed as linear combinations of a number of Bessel functions. With the use of the known properties of the Bessel functions, the system of differential equations is transformed to an algebraic constraint for three numerical coefficients a_1, a_2, a_3. Its solutions may be chosen in various ways by resolving the simple condition $A_1 a_1 + A_2 a_2 + A_3 a_3 = 0$, where the coefficients A_i are expressed through the quantum numbers ϵ, j. Thus, at the fixed quantum numbers $\{\epsilon, j, m, P\}$ there exists a double-degeneration of the quantum states.

In Chapter 17, the system of equations describing a massless spin 3/2 field has been studied in spherical coordinates of the Minkowski space. It is proved that the general system reduces to two couples of independent 2-nd order nonhomogeneous differential equations; their particular solutions may be found with the use of gauge solutions of a special form. The corresponding homogeneous equations turn out to have the same form, and have three regular singularities and an irregular one, of the rank 2. The Frobenius type solutions for this equation have been specified, and the structure in power series expansion have been studied. The six remaining radial functions may be readily found by the use of simple algebraic relations. By this method, we have constructed two types of solutions for the massless spin 3/2 field, which do not contain the gauge constituents.

In Chapter 18, we have extended the first order system in matrix form for spin 2 particle to pseudo-Riemannian space-time models, by applying the tetrad method. All the intrinsic constraints on the involved tensors $\Psi(x) = \{\Phi, \Phi_c, \Phi_{(ab)}, \Phi_{[ab]c}\}$ are contained in the structure of the basic matrices. The introduced tetrad matrix equation is specified in cylindrical and spherical coordinates for the flat Minkowski space. The case of massless field is separately addressed, and we focus in this theory on a detailed matrix representation of the gauge symmetry.

Bibliography

1. Bateman, H.; and Erdelyi, A., *Higher Transcendental Functions*, Vol. 1. McGraw-Hill, New York. (1953)

2. Heun, K., Zur Theorie der Riemann'schen Functionen Zweiter Ordnung mit Verzweigungspunkten. *Math. Ann.* 1889. **33**. 161–179.

3. Ronveaux, A., (Editor). *Heun's Differential Equation.*, Oxford University Press, 1995.

4. Slavyanov, S.Yu.; Lay, W., Special functions. *A unified theory based on singularities.* Oxford University Press, Oxford, 2000.

5. Epstein, P.S., The Stark effect from the point of view of Schrodinger quantum theory. *Phys. Rev.* 1926. **28**. no 4. 695–710.

6. Baber, W.G.; Hasse, H.R., The two centre problem in wave mechanics. *Proc. Cambridge Philos. Soc.* 1935. **31**. 564–581.

7. Manning, M.F., Exact solutions of the Schrodinger equation. *Phys. Rev.* 1935. **48**. 161–164.

8. Regge,T.; Wheeler, J.A., Stability of a Schwarzschild singularity. *Phys. Rev.* 1957. **108**. 1063–1069.

9. Lamieux, A.; Bose, A.K., Construction de potentiels pour lesquelles l'equation de Schrodinger est soluble. *Ann. Inst. Henri Poincare.* 1969. **A10**. 259–270.

10. Dhurandhar,S.V.; Vishveshwara, C.V.; Cohen, J.M., Electromagnetic, neutrino and gravitational fields in the Kasner space-time with rotational symmetry. *Class. Quantum Grav.* 1984. **1**. 61–69.

11. Chaudhuri, R.N.; Mukherjee, S., On the $\mu x^2 + \lambda x^4 + \eta x^6$ interaction. *J. Phys.* 1984. **A17**. 3327–3334.

12. Léauté, B.; Marcilhacy, G.; On the Schrödinger equations of rotating harmonic, three-dimensional and doubly anharmonic oscillators and a class of confinement potentials in connection with the biconfluent Heun differential equation. *J. Phys.* 1986. **A19**. 3527–3533.

13. Leaver, E.L., Solutions to a generalized spheroidal wave equation: Teukolsky equations in general relativity, and the two-center problem in molecular quantum mechanics. *J. Math. Phys.* 1986. **27**. 1238–1265.

14. Pons, R.; Marcilhacy, G., Exact solutions for electromagnetic, neutrino and gravitational fields in Kasner spacetime. *Class. Quantum Grav.* 1987. **4**. 171–179.

15. Ray, P.P.; Mahata, K., Bounded states of the potential $V(r) = \frac{Ze^2}{r+\beta}$. *J. Phys.* 1989. **A22**. no 15. 3161–3166.

16. Costa, I.; Derruelle,N.; Novello, M.; Svaiter, N.F., Quantum fields in cosmological spacetimes: a soluble example. *Class. Quantum Grav.* 1989. **6**. 1893–1907.

17. Exton, H., The exact solution of two new types of Schrödinger equation. *J. Phys.* 1995. **A28**. 6739–6741.

18. Suzuki, H.; Takasugi, E.; Umetsu, H., Perturbations of Kerr-de Sitter black holes and Heun's equations. *Progr. Theoret. Phys.* 1998. **100**. 491–505.

19. Truong, T.T.; Bazzali, D., Exact low-lying states of two interacting equally charged particles in a magnetic field. *Phys. Lett.* 2000. **A269**. 186–193.

20. Ishkhanyan, A.M.; Suominen, K.-A., Analytic treatment of the polariton problem for a smooth interface. *J. Phys.* 2001. **A34**. L591–L598.

21. Ishkhanyan, A.M.; Suominen, K.-A. Solutions of the two-level problem in terms of biconfluent Heun functions. *J. Phys.* 2001. **A34**. 6301–6306.

22. Ralko, A.; Truong, T.T., Heun Functions and the energy spectrum of a charged particle on a sphere under magnetic field and Coulomb force. *J. Phys.* 2002. **A35**. 9573–9584.

23. Wu, S.Q.; Cai, X., Massive complex scalar field in a Kerr-Sen black hole background: exact solution of wave equation and Hawking radiation. *J. Math. Phys.* 2003. **44**. 1084–1988.

24. Fiziev, P.P., Exact solutions of Regge-Wheeler equation and quasi-normal modes of compact objects. *Class. Quantum Grav.* 2006. **23**. 2447–2468.

25. T. Oota,; Yukinori Yasui, Toric Sasaki-Einstein manifolds and Heun equation. *Nucl. Phys.* 2006. **B742**. 275–294.

26. Petroff, D., Slowly rotating homogeneous stars and the Heun equation. *Class. Quant. Grav.* 2007. **24**. 1055–1068.

27. Fiziev, P.P., Exact solutions of Regge-Wheeler equation. *J. Phys. Conf. Ser.* 2007. **66**. Paper 012016.

28. Birkandan, T.; Hortacsu, M., Examples of Heun and Mathieu functions as solutions of wave equations in curved spaces. *J. Phys.* 2007. **A40**. 1105–1116.

29. Bouaziz, D.; Bawin, M., Regularization of the singular inverse square potential in quantum mechanics with a minimal length. *Phys. Rev.* 2007. **A76**. Paper 032112.

30. Giachetti, R.; Sorace, E., States of the Dirac equation in confining potentials. *Phys. Rev. Lett.* 2008. **101**. Paper 190401.

31. Al-Badawi, A.; Sakalli, I., Solution of the Dirac equation in the rotating Bertotti-Robinson spacetime. *J. Math. Phys.* 2008. **49**. Paper 052501.

32. Ishkhanyan, A.M.; Sokhoyan,R.; Joulakian, B.; Suominen, K.-A., Rosen-Zener model in cold molecule formation. *Opt. Commun.* 2009. **282**. 218–226.

33. Ishkhanyan, A.; Sokhoyan,R.; Suominen, K.-A.; Leroy, C.; Jauslin H.-R. Quadratic-nonlinear Landau-Zener transition for association of an atomic Bose-Einstein condensate with inter-particle elastic interactions included. *Eur. Phys. J.* 2010. **D56**. 421–429.

34. Zaveri, V.H., Quarkonium and hydrogen spectra with spin-dependent relativistic wave equation. *Pramana J. Phys.* 2010. **75**. 579–598.

35. Esposito, G.; Roychowdhury, R., On the complete analytic structure of the massive gravitino propagator in four-dimensional de Sitter space. *Gen. Rel. Grav.* 2010. **42**. 1221–1238.

36. Arda, A.; Aydogdu, O.; Sever, R., Scattering of Woods-Saxon potential in Schrödinger equation. *J. Phys.* 2010. **A43**. no 42. Paper 42520.

37. Hall, R.L.; Nasser S.; Sen, K.D., Soft-core Coulomb potential and Heun's differential equation. *J. Math. Phys.* 2010. **51**. paper 022107.

38. Borissov, R.S.; Fiziev, P.P., Exact solutions of Teukolsky master equation with continuous spectrum. *Bulg. J. Phys.* 2010. **37**. 065–089.

39. Fiziev, P.P., Classes of exact solutions to the Teukolsky master equations. *Class. Quantum Grav.* 2010. **27**. Paper 135001.

40. Sesma, J., The generalized quantum isotonic oscillator. *J. Phys.* 2010. **A43**. paper 185303.

41. Vieira, H.S.; Bezerra, V.B.; Muniz, C.R., Exact solutions of the Klein-Gordon equation in the Kerr-Newman background and Hawking radiation. *Ann. Phys.* 2014. **350**. 14–28.

42. Fiziev, P.; Staicova, D., Application of the confluent Heun functions for finding the quasinormal modes of nonrotating black holes. *Phys. Rev.* 2011. **D84**. Paper 127502.

43. Kisel,V.V.; Krylov, G.G.; Ovsiyuk, E.M.; Amirfakchrian, M.; Red'kov, V.M., Wave functions of the particle with polarizability in Coulomb field. *Nonlinear Dynamics and Applications*. 2011. **18**. 168–179.

44. Cunha, M.S.; Christiansen, H.R., Confluent Heun functions in gauge theories on thick braneworlds. *Phys. Rev.* 2011. **D84**. Paper 085002.

45. Figueiredo Medeiros, E.R.; Bezerra de Mello, E.R., Relativistic quantum dynamics of a charged particle in cosmic string spacetime in the presence of magnetic field and scalar potential. *Eur. Phys. J.* 2012. **C72**. Paper 2051.

46. Ovsiyuk, E.M.; Veko, O.V.; Amirfakchrian, M., On Schrödinger equation with potential $U = -\alpha r^{-1} + \beta r + kr^2$ and the biconfluent Heun functions theory. *Nonlinear Phenomena in Complex Systems*. 2012. **15**. no 2. 163–170.

47. Ovsiyuk,E.M.; Veko, O.V.; Kisel, V.V.; Red'kov, V.M., New problems of quantum mechanics and Heun equation. *Naucno − tekchnicheskie vedomosti St. Petersburg University. Ser. Fiz. − mat*. 2012. no 1(141). 137–145.

48. Momeni,D.; Yerzhanov, K.; Myrzakulov, R., Quantized Black Hole and Heun function. *Can. J. Phys.* 2012. **90**. 877–881.

138. Ovsiyuk, E.M., On solutions of Maxwell equations in the space-time of Schwarzschild black hole. *Nonlinear Phenomena in Complex Systems*. 2012. **15**. 81–91.

50. Red'kov, V.M.; Ovsiyuk, E.M., *Quantum Mechanics in Spaces of Constant Curvature*. Nova Science Publishers. Inc., New York, 2012.

51. Downing,C.A., On a solution of the Schrödinger equation with a hyperbolic double-well potential. *J. Math. Phys.* 2013. **54**. Paper 07210.

52. Batic,D.; Williams, R.; Nowakowski, M., Potentials of the Heun class. *J. Phys.* 2013. **A46**. Paper 245204.

53. Hamzavi, M.; Rajabi, A.A., Scalar-vector-pseudoscalar Cornell potential for a spin-1/2 particle under spin and pseudospin symmetries: 1+1 dimensions. *Ann. Phys.* 2013. **334**. no 9. 316–320.

54. Hortacsu, M., *Heun Functions and their uses in Physics*. pp. 23–39 in: Proceedings of the 13th Regional Conference on Mathematical Physics, Antalya, Turkey, October 27–31, 2010, Edited by Ugur Camci and Ibrahim Semiz. World Scientific (2013); arXiv:1101.0471 [math-ph].

55. Osherov, V.I.; Ushakov, V.G., Stark problem in terms of the Stokes multipliers for the triconfluent Heun equation. *Phys. Rev.* 2013. **A88**. Paper 053414.

56. Ovsiyuk, E.M; Florea,O.; Chichurin, A.; Red'kov, V.M., Electromagnetic field in Schwarzschild black hole background. Analytical treatment and numerical simulation. *Procs. 7 − th Int. Workshop on Comp. Alg. Sys. in Teaching and Research.* September 22–25, 2013, Siedlce, Poland, Publ. Collegium Mazovia, IV, 1 (2013), 204–214.

57. Ovsiyuk,E.; Veko, O.; Kazmerchuk,K.; Kisel, V.; Red'kov, V., Quantum mechanics of a spin 1 particle in the magnetic monopole potential, in spaces of Euclid, Lobachevsky, and Riemann: nonrelativistic approximation. *Ukr. Phys. J.* 2013. **58**. 1073–1083.

58. Shahverdyan, T.A.; Mogilevtsev, D.S.; Ishkhanyan, A.M.; Red'kov, V.M., Complete-return spectrum for a generalized Rosen-Zener two-state term-crossing model. *Nonlinear Phenomena in Complex Systems*. 2013. **16**. 86–92.

59. Rajabi, A.A.; Hamzavi, M., Relativistic effect of external magnetic and Aharonov-Bohm fields on the unequal scalar and vector Cornell model. *Eur. Phys. Plus*. 2013. **128**. Paper 5.

60. Caruso, F.; Martins, J.; Oguri, V., Solving a two-electron quantum dot model in terms of polynomial solutions of a biconfluent Heun equation. *Ann. Phys.* 2014. **347**. 130–140.

61. Hoff da Silva, J.M. ; Pereira, S.H., Exact solutions to Elko spinors in spatially flat Friedmann-Robertson-Walker spacetimes. *Journal of Cosmology and Astroparticle Physics*. 2014. **3**. Paper 009.

62. Karwoswki J. ; Witek, H.A., Biconfluent Heun equation in quantum chemistry: Harmonium and related systems. *Theor. Chem. Accounts*. 2014. **133**. Paper 1494.

63. Chuchurin, A.V.; Redkov V.M., Quantum mechanical scalar particle with polarisability in the Coulomb field, analytical and numerical consideration. *Studia i Materialy*. 2013. **6**. 73–89.

64. Ovsiyuk, E.M.; Kisel, V.V.; Red'kov, V.M., *Maxwell Electrodynamics and Boson Fields in Space of Constant Curvature*. Nova Science Publishers Inc., New York, 2014.

65. Red'kov, V.M. ; Chichurin, A.V., A symbolic-numerical method for solving the differential equation describing the states of polarizable particle in Coulomb potential. *Programming and Computer Software*. 2014. **40**. no 2. 86–92.

66. *Sofia University, The Heun Project* : Heun functions, their generalizations and applications. http://theheunproject.org

67. Ovsiyuk, E.M.; Koral'kov, A.D.; Chichurin, A.V.; Red'kov, V.M., Dirac particle in the Coulomb field on the background of hyperbolic Lobachevsky model. *Nonlinear Phenomena in Complex Systems*. 2021. **24**. no 3. 260–271.

68. Varshalovich, D.A.; Moskalev, A.N.; Hersonskiy, V.K. *Quantum Theory of Angular Momentum*. Nauka, Leningrad, 1975 (in Russian).

69. Landau, L.D.; Lifshitz. E.M., *Quantum Mechanics*. Volume 3 of the Course of Theoretical Physics. Pergamon Press, 1965.

70. Berestetzkiy, V.B.; Lifshitz,E.M.; Pitaevskiy, L.B.; *Quantum Electrodynamics*. Moscow, Nauka, 1989 (in Russian).

71. Rabi, I.I., Das freie Electron in Homogenen Magnetfeld nach der Diraschen Theorie. *Z. Phys.* 1928. **49**. 507–511.

72. Landau, L., Diamagnetismus der Metalle. *Z. Phys.* 1930. **64**. 629–637.

73. Plesset, M.S., Relativistic wave mechanics of the electron deflected by magnetic field. *Phys. Rev.* 1931. **12**. 1728–1731.

74. Comtet, A.; Houston, P.J., Effective action on the hyperbolic plane in a constant external field. *J. Math. Phys.* 1985. **26**. no 1. 185–191.

75. Comtet. A., On the Landau levels on the hyperbolic plane. *Ann. of Phys.* 1987. **173**. 185–209.

76. Klauder,J.R.; Onofri, E., Landau levels and geometric quantization. *Int. J. Mod. Phys.* 1989. **A4**. 3939–3949.

77. Dunne, G.V., Hilbert space for charged particles in perpendicular magnetic fields. *Ann. of Physs.* 1992. **215**. 233–263.

78. Alimohammadi, M.; Shafei Deh Abad, A., Quantum group symmetry of the quantum Hall effect on the non-flat surfaces. *J. Phys.* 1996. **A29**. 559–564.

79. Onofri, E., Landau levels on a torus. *Int. J. Theor. Phys.* 2001. **40**. no 2. P. 537–549.

80. Drukker,N.; Fiol,B.; Simón, J., Gödel-type Universes and the Landau problem. *Journal of Cosmology and Astroparticle Physics*. 2004. **0410**. Paper 012.

81. Bogush, A.A.; Red'kov, V.M.; Krylov, G.G., Schrödinger particle in magnetic and electric fields in Lobachevsky and Riemann spaces. *Nonlinear Phenomena in Complex Systems*. 2008. **11**. no 4. 403–421.

82. Kudryashov, V.V. ; Kurochkin, Yu.A.; Ovsiyuk, E.M.; Red'kov, V.M., Motion Caused by Magnetic Field in Lobachevsky Space. *AIP Conference; Proceedings*. 2010. **1205**. 108 – 111. The Sun, the Stars, the Universe and General Relativity: International Conference in honor of Ya.B. Zeldovich's 95th anniversary. Minsk, Belarus, 20–0 23 April, 2009. ed. R. Ruffini, G. Vereschagin.

83. Kisel, V.V.; Ovsiyuk, E.M.; Red'kov, V.M., On solutions of the Duffin–Kemmer equation for spin 1 particle in the uniform magnetic field. *Doklady of the National Academy of Sciences of Belarus*. 2010. **54**. no 4. 64–71.

84. Kisel, V.V.; Ovsiyuk, E.M.; Red'kov, V.M.; Tokarevskaya,N.G., Exact solutions for a quantum-mechanical particle with spin 1 and additional intrinsic characteristic in a homogeneous magnetic field. *Acta Physica Polonica*. 2010. **B41**. no 11. 2347–2363.

85. Kudryashov, V.V.; Kurochkin, Yu.A.; Ovsiyuk, E.M.; Red'kov, V.M., Classical particle in presence of magnetic field. hyperbolic Lobachevsky and spherical Riemann models. *SIGMA*. 2010. **6**, Paper 004 [34 pages].

86. Ovsiyuk, E.M.; Kisel, V.V.; Krylov, G.G.; Red'kov, *Quantum Mechanics in Uniform Magnetic Field, New Problems*. Mozyr State University Publ. Belarus. 2011. 232 p. (in Russian)

87. Ovsiyuk, E.M., Solutions of the Dirac equation in analoges of the uniform magnetic and electric fields in spherica Riemann space. *Proceedings of the National Academy of Sciences of Belarus. Physics and Mathematics series*. 2011. no 4. 85–92.

88. Ovsiyuk, E.M.; Kisel, V.V.; Red'kov, V.M., On a Dirac particle in an uniform magnetic field in 3-dimensional spaces of constant negative curvature curvature. *Nonlinear Phenomena in Complex Systems*. 2012. **15**. no 1. 41–55.

89. Kisel. V.V,; Ovsiyuk, E.N.; Veko, O.V.; Red'kov, V.M., Quantum mechanics of the vecror particle in magnetic field on 4-dimensinal sphere. *Naucno– tekchnicheskie vedomosti St.Petersburg University. Ser. Fiz. – mat*. 2012. **1**(**141**). 128–137.

90. Kurochkin, Yu.A.; Otchik, V.S.; Ovsiyuk, E.M., Magnetic field in the Lobachevsky space and related integrable systems. *Physics of Atomic Nuclei*. 2012. **75**. no. 10. 1245–1249.

91. Red'kov, V.M.; Ovsiyuk, E.M.; Ishkhanyan, A.M., Particle in the magnetic field: 2-dimensional Riemann spherical space and complex analogue of the Poincaré half-plane. *Doklady of the National Academy of Sciences of Belarus*. 2013. **57**. no 1. 55–62.

92. Ovsiyuk, E.M. ; Voynova, Ya.A.; Kisel, V.V.; Balan, V.; Red'kov, V.M., *Techniques of projective operators used to construct solutions for a spin* 1 *particle with anomalous magnetic moment in the external magnetic field*. Chapter in: Quaternions: Theory and Applications. Ed. Sandra Griffin. – Nova Science Publishers, Inc. USA, 2017. – P. 11–46.

93. Ovsiyuk, E.M.; Kisel, V.V,; Red'kov, V.M., *Spin* 1/2 *particle with anomalous magnetic moment in presence of external magnetic field, exact solutions*. Chapter in: Relativity, Gravitation, Cosmology: Beyond Foundations. Ed. V.V. Dvoeglazov. Nova Science Publishers, Inc. New York, 2019. 65–80.

94. Olevskiy, M.N., 3-orthogonal soorinate systems in spaces of constant curvature, in which equation $\Delta_2 U + \lambda U = 0$ allows for the complete separation of the variables. *Math. Collection*. 1950. **27**. 379–426 (in Russian).

95. Ovsiyuk, E.M., *Exactly Solvable Problems of the Quantum Mechanics and Classical Field Theory in Spaces with Non – Euclidean Geometry*. Minsk, High Scool Publishing, 2013.

96. Ovsiyuk, E.M.; Veko, O.V.; Voinova, Ya.A.; Kisel, V.V.; Red'kov, V.M., *Quantum Mechanics of Particles with Spin in External Magnetic Field*. Belarussian Science. Publ. Minsk, 2017. (in Russian).

97. Dirac, P.A.M., The electron wave equation in the de Sitter space. *Ann. Math.* 1935. **36**. 657–669.

98. Hawking, S.W.; Ellis, G.F.R.; *The Large Scale Structure of Space – Time*. Cambridge University Press, 1973.

99. Gibbons, G.W., Anti-de-Sitter space-time and its uses. in: Mathematical and quantum aspects of relativity and cosmology (Pythagoreon, 1998), *Lecture Notes in Physics*. Springer-Verlag. 2000. **537**. 102–142.

100. Dirac, P.A.M., Wave equations in conformal space. *Ann. of Math.* 1936. **37**. 429–442.

101. Schrödinger, E., The proper vibrations of the expanding universe. *Physica*. 1939. **6**. 899–912.

102. Schrödinger, E., General theory of relativity and wave mechanics. *Wiss. en Natuurkund*. 1940. **10**. 2–9.

103. Goto, K., Wave equations in de Sitter space. *Progr. Theor. Phys.* 1951. **6**. 1013–1014.

104. Nachtmann, O., Quantum theory in de-Sitter space. *Commun. Math. Phys.* 1967. **6**. 1–16.

105. Chernikov, N.A.; Tagirov, E.A., Quantum theory of scalar field in de Sitter space-time. *Ann. Inst. Henri Poincare*. 1968. **IX**. 109–141.

106. Börner, G.; Dürr, H.P., Classical and quantum theory in de Sitter space. *Nuovo Cimento*. 1969. **A64**. 669–713.

107. Fushchych, W.L.; Krivsky, I.Yu., On representations of the inhomogeneous de Sitter group and equations in five-dimensional Minkowski space. *Nucl. Phys.* 1969. **B14**. 573–585.

108. Börner, G.; Dürr, H.P., Classical and quantum fields in de Sitter space. *Nuovo Cimento*. 1969. **A64**. 669–714.

109. Castagnino, M., Champs spinoriels en Relativité generale; le cas particulier de l'espace-temps de De Sitter et les équations d'ond pour les spins éleves. *Ann. Inst. Henri Poincare*. 1972. **A16**. 293–341.

110. Tagirov, E.A., Consequences of field quantization in de Sitter type cosmological models. *Ann. Phys.* 1973. **76**. 561–579.

111. Riordan, F., Solutions of the Dirac equation in finite de Sitter space. *Nuovo Cimento*. 1974. **B20**. 309–325.

112. Candelas, P.; Raine, D.J., General-relativistic quantum field theory: an exactly soluble model. *Phys. Rev.* 1975. **D12**. 965–974.

113. Schomblond, Ch.; Spindel P., Propagateurs des champs spinoriels et vectoriels dans l'univers de de Sitter. *Bull. Cl. Sci., V.Ser., Acad. R. Belg.* 1976. **62**. 124–134.

114. Hawking, S.W.; Gibbons, G.W., Cosmological event horizons, thermodynamics, and particle creation. *Phys. Rev.* 1977. **D15**. 2738–2751.

115. Avis, S.J.; Isham, C.J.; Storey, D., Quantum field theory in anti-de Sitter space-time. *Phys. Rev.* 1978. **D18**. no 10. 3565–3576.

116. Lohiya, D.; Panchapakesan, N., Massless scalar field in a de Sitter universe and its thermal flux. *J. Phys.* 1978. **A11**. 1963–1968.

117. Lohiya, D.; Panchapakesan, N., Particle emission in the de Sitter universe for massless fields with spin. *J. Phys.* 1979. **A12**. 533–539.

118. Hawking, S.; Page, D., Thermodynamics of black holes in anti-de Sitter space. *Commun. Math. Phys.* 1983. **87**. 577–588.

119. Otchik, V.S., On the Hawking radiation of spin 1/2 particles in the de Sitter space-time. *Class. Quantum Crav.* 1985. **2**. 539–543.

120. Motolla, F., Particle creation in de Sitter space. *Phys. Rev.* 1985. **D31**. 754–766.

121. Otchik, V.S.; Red'kov, V.M., *Quantum – mechanival Kepler problem in spaces of constant curvature.* Preprint no 298, Institute of Physics. AN BSSR, 1986. 49 pages (in Russian).

122. Mishima, T.; Nakayama, A., Particle production in de Sitter spacetime. *Progr. Theor. Phys.* 1987. **77**. 218–222.

123. Sánchez, N., Quantum field theory and elliptic interpretation of de Sitter space-time. *Nucl. Phys.* 1987. **B294**. 1111–1137.

124. Polarski, D., The scalar wave equation on static de Sitter and anti-de Sitter spaces. *Class. Quantum Grav.* 1989. **6**. 893–900.

125. Bros, J.; Gazeau, J.P.; Moschella, U., Quantum field theory in the de Sitter Universe. *Phys. Rev. Lett.* 1994. **73**. no 13. 1746–1749.

126. Suzuki, H.; Takasugi, E., Absorption probability of de Sitter horizon for massless fields with spin. *Mod. Phys. Lett.* 1996. **A11**. 431–436.

127. Cotaescu, I.I., Normalized energy eigenspinors of the Dirac field on anti-de Sitter spacetime. *Phys. Rev.* 1999. **D60**. Paper 124006.

128. Garidi, T., Huguet, E., Renaud, J., De Sitter waves and the zero curvature limit comments. *Phys. Rev.* 2003. **D67**. Paper 124028.

129. Moradi, S., Rouhani, S., Takook, M.V., Discrete symmetries for spinor field in de Sitter space. *Phys. Lett.* 2005. **B613**. 74–82.

130. Bachelot, A., The Dirac equation on the anti-de-Sitter universe. L'équation de Dirac sur l'univers anti de Sitter. *Comptes Rendus Mathematique*. 2007. **345**. no 8. 435–440.

131. Ovsiyuk, E.M.; Red'kov, V.M., Spherical waves of spin 1 particle in anti de Sitter space-time. *Acta Physica Polonica*. 2010. **B41**. no 6. 1247–1276.

132. Red'kov, V.M.; Ovsiyuk, E.M., On exact solutions for quantum particles with spin S = 0,1/2, 1 and de Sitter event horizon. *Ricerche di matematica*. 2011. **60**. no 1. 57–88.

133. Ovsiyuk, E.M.; Veko, O.V., Spin 1/2 particle in presence of the Abelian monopole on the background of anti de Sitter space-time. The case $j = j_{min}$. *Doklady of the National Academy of Sciences of Belarus*. 2011. **55**. no 6. 49–55 (in Russian).

134. Red'kov, V.M.; Ovsiyuk, E.M., On exact solutions for quantum particles with spin $S = 0, 1/2, 1$ and de Sitter event horizon. *Ricerche di Matematica*. 2011. **60**. no 1. 57–88.

135. Ovsiyuk, E.M.; Veko, O.V., Spin 1/2 particle in presence ob thre Abelian monopole on thebackground of anti de Sitter space-time. The case $j > j_{min}$. *Doklady of the National Academy of Sciences of Belarus*. 2011 **56**. no 1. 43–49 (in Russian).

137. Red'kov, V.M.; Ovsiyuk, E.M.; Krylov, G.G., Hawking Radiation in de Sitter Space: Calculation of the Reflection Coefficient for Quantum Particles. *Vestnik RUDN. Ser. Mat. – Inf. – Fiz.* 2012. no 4. 153–169.

137. Red'kov, V.; Ovsiyuk, E.; Veko, O., Spin 1/2 particle in the field of Dirac string on the background of de Sitter space–time. *Uzhhorod University Scientific Herald. Series Physics*. 2012. no 32. 141–150.

138. Ovsiyuk, E,M., Nonrelativistic particle in 2-dimensional Lobachevsky space, cylindrical coordinates and Poincaré half-plane. *Doklady of the National Academy of Sciences of Belarus*. 2012. **56**, no 5. 29–35.

139. Red'kov, V.M., *Fields in Riemannian space and the Lorentz group*. Publishing House: Belarusian Science, Minsk, 2009 (in Russian).

140. Red'kov, V.M., *Tetrad formalism, spherical symmetry and Schrodinger basis*. Publishing House: Belarusian Science, Minsk, 2011 (in Russian).

141. Veko, O.V.; Vlasii, N.D.; Ovsiyuk, E.M.; Red'kov, V.M.; Sitenko, Yu.A., Electromagnetic field on de Sitter expanding universe: Majorana–Oppenheimer formalism, exact solutions in non-static coordinates. *Nonlinear Phenomena in Complex Systems*. 2014. **17**. no. 1. 17–39.

142. Veko, O.V.; Ovsiyuk, E.M.; Red'kov, V.M., Dirac particle in presence of a magnetic charge in de Sitter Universe: exact solutions and transparency of the cosmological horizon. *Nonlinear Phenomena in Complex Systems*. 2014. **17**. no 4. 461–663.

143. Ovsiyuk, E.M.; Veko, O.V.; Red'kov, V.M., On modeling media with the property of ideal mirror with respect to the light and spin 1/2 particle. *Proceedings of the National Academy of Sciences of Belarus. Physics and Mathematics Series*. 2015. no 1. 76–85 (in Russian).

144. Ovsiyuk, E.M.; Veko, O.V.; Voinova, Ya.A.; Kisel, V.V.; Red'kov, V.M., On reflectiong spin 1/2 particles by "geometrical medium" of the Lobachevsky space, in presence of electric field. *Veskik of Brest University. Series* 4. *Physics and Matematics*. 2016. no 2. 21–31 (in Russian).

145. Veko, O.V.; Ovsiyuk,E.M.; Balan, V.; Red'kov, V.M., Bolyai-Lobachevsky geometrical simulation of a media acting as an ideal mirror on the particles. *Proceedings of Balkan Society of Geometers*. 2017. **24**. 74–90.

146. 't Hooft, G., Monopoles in unified gauge theories. *Nucl. Phys*. 1974. **B79**. no 2. 276–284.

147. Polyakov, A.M., Spectrum of the particles in quantum field theory. *JETP Let*. 1974. **20**. no 6. 430–433 (in Russian).

148. Julia B., Zee A. Poles with both magnetic and electric charges in non-Abelian gauge theory. *Phys. Rev*. 1975. **D11**. no 8. 2227–2232.

149. Bais, F.A.; Russel, R.J., Magnetic-monopole solution of non-Abelian gauge theory in curved space-time. *Phys. Rev*. 1975. **D11**. no 10. 2692–2695.

150. Swank, J.H.; Swank, L.J.; Dereli, T., Fermions in Yang – Mills potentials. *Phys. Rev*. 1975. **D12**. no 4. 1096–1102.

151. Jackiw, R.; Rebbi, C., Solitons with fermion number 1/2. *Phys. Rev*. 1976. **D13**. no 12. 3398–3409.

152. Jackiw, R.; Rebbi, C., Spin from isospin in a gauge theory. *Phys. Rev. Lett*. 1976. **36**. no 19. 1116–1119.

153. Hasenfratz P.; 't Hooft, G., Fermion-boson puzzle in a gauge theory. *Phys. Rev. Lett*. 1976. **36**. no 19. 1119–1122.

154. Proxvatilov, E.V.; Franke, V.A., Fermions in 't Hooft – Polyakov field. Yadernaya Fizika. 1976. **24**. no 4. 856–860 (in Russian).

155. Prasad, M.K.; Sommerfield C.M., Exact classical solution of the 't Hooft monopole and Julia – Zee dyon. *Phys. Rev. Lett*. 1975. **35**. no 12. 760–762.

156. Bogomol'nyi, E.B., Stability of classical solutions. *Yadernaya Fizika*. 1976. **24**. 861–870 (in Russian).

157. Fedorov, F.I., *The Lorentz group*. Moscow, Science. 1979 (in Russian).

158. Bogush, A.A., *Introduction to Field Theory of Elementary Particles*. Minsk, Science and Technics. 1981 (in Russian).

159. Red'kov, V.M., *The doublet of Dirac fermions in the field of the non–Abelian monopole, isotopic chiral symmetry, and parity selection rules*. 1–58; quant-ph/9901011.

160. Red'kov, V.M., *On intrinsic structure of wave functions of fermion triplet in external monopole field*. 1–40; quant-ph/9902034.

161. Red'kov, V.M., *Monopole BPS – solutions of the Yang – Mills equations in spaces of Euclid , Riemann, and Lobachevsky*. 1–8; hep-th/0306060.

162. Ovsiyuk, E.M.; Red'ko A.N.; Kisel, V.V.; Red'kov, V.M., Isotopic doublet of Dirac prticles in presence of the non-Abelian monopole, Pauli appriximation. *Problems ofPhysics, Mathematics, and Technics*. 2016. **3**(**28**). 13–22 (in Russian).

163. Atanasiu, Gh.; Balan, V.; Brinzei, N.; Rahula M., *Differential Geometry of the Second Order and Applications : Miron – Atanasiu Theory*. Moscow, 2010. 256 p. (In Russian)

164. Antonelli, P.L., *Equivalence Problem for Systems of Second Order Ordinary Differential Equations*. Encyclopedia of Mathematics, Kluwer Academic Publishers, Dordrecht, 2000.

165. Antonelli, P.L.; Bucataru, I., *KCC Theory of Systems of Second Order DifferentialEquations*. Chapter in: Handbook of Finsler Geometry. Eds. P.L. Antonelli. Springer, 2003. 1–66.

166. Antonelli, P.L.; Bucataru, I., New results about the geometric invariants in KCC-theory. *An. St. Univ. Al.I.Cuza. Iasi. Mat. N.S.* 2001. **47**. 405–420.

167. Regge, T.; Wheeler, John A., Stability of a Schwarzschild singularity. *Phys. Rev.* 1957. **108**. 1063–1069.

168. Wheeler, John A., Geons. *Phys. Rev.* 1955. **97**. 511–536.

169. Brill, D.R.; Wheeler, John A., Interaction of neutrinos and gravitational fields. *Rev. Mod. Phys.* 1957. **29**. 465–479.

170. Schwarzschild, K., Über das Gravitationsfeld eines Massenpunktes nach der Einsteinschen Theorie. *Sitzungsber. Preuss. Akad. Wissen. Phys. Math.* 1916. K1. 189–196.

171. Chandrasekhar, S., *The Mathematical Theory of Black Holes*. Oxford University Press, Oxford, 1983.

172. Bardeen, J.M.; Press, W.H., Radiation fields in the Schwarzschild background. *J. Math. Phys.* 1973. **14**. 7–19.

173. Hawking, S.W., Black hole explosions. *Nature*. 1974. **248**. no 5443. 30–31.

174. Hawking, S.W., Particle creation by black holes. *Commun. Math. Phys.* 1975. **43**. 199–220.

175. Page, D.N., Particle emission rates from a black hole: massless particles from an uncharged, non-rotating hole. *Phys. Rev.* 1976. **D13**. 198–206.

176. Frolov, V.P., Physical effects in graviational field of black holes. *FIAN Proceedings*. 1986. **169**. 3–131 (in Russian).

177. Gal'tsov, D.V., *Particles and Fields in Vicinity of Black Holes*. Moscow, 1986 (in Russian).

178. Smoller, J,; Chunjing Xie. Asymptotic behavior of massless Dirac waves in Schwarzschild geometry. *Annales Henri Poincare*. 2012. **13**. no 4. 943–989.

179. Antonelli, P.L.; Bevilacqua, L.; Rutz, S.F., Theories and models in symbiogenesis. *Nonlinear Analysis*. 2003. **4**. 743–753.

180. Zecca, A., Spin 1/2 bound states in Schwarzschild geometry. *Adv. Studies Theor. Phys.* 2007. **1**. 271–279.

181. Darboux, G., Sur la sphére de rayon nul et sur la théorie du déplacement d'une figure invariable. Bull. des Sciences Math. 1905. **9**. série 2. [Ce memoire eat un résumé des leçons que lauteur a faites à la Sorbonne en 1900 et 1904.]; Leçons sur la théorie des surfaces et les application géometriques du calcul infinitesimal. Paris, 1914.

182. Cartan, E., Les groups projectifs qui ne laissent invariante aucune multiplicité plane. *Bull. Soc. Math. France*. 1913. **41**. 53–96.

183. Cartan, E., *La theorie des Spineurs. I, II*. Actualites Sci. et Ind. 1938. **643**; Actualites Sci. et Ind. 1938.

184. Weyl, H., Elektron und Gravitation. *Zeitschrift fur Physik*. 1929. **56**. 330–352.

185. Weyl, H., Gravitation and the electron. *Proc. Nat. Acad. Sci. Amer.* 1929. **15**. 323–334.

186. Juvet, G., Opérateurs de Dirac et équations de Maxwell. *Comm. Math. Helv.* 1930. **2**. 225–235.

187. Hopf, H., Über die Abbildungen der dreidimensionalen Sphäre auf die Kugelfläche. *Math. Annalen*. 1931. **104**. 637–665; Reprinted in: *Selecta Heinz Hopf*. 38–63. Springer-Verlag, 1964.

188. De Broglie, L., Sur une analogie entre l'électron de Dirac et l'onde électromagnétique. *C.R. Acad. Sci. Paris.* 1932. **195**. 536–537; Sur le champ électromagnétique de l'onde lumineuse. *C.R. Acad. Sci. Paris.*1932. **195**. 862–864.

189. Laporte, O.; Uhlenbeck, G.E., Application of spinor analysis to the Maxwell and Dirac equations. *Phys. Rev.* 1931. **37**. no 11. 1380–1397.

190. Oppenheimer, J.R., Note on light quanta and the electromagnetic field. *Phys. Rev.* 1931. **38**. 725.

191. Majorana, E., Scientific Papers, unpublished, deposited at the "Domus Galileana". Pisa, quaderno 2. 101/1; quaderno 3. 11, 160; quaderno 15. 16; quaderno 17. 83, 159.

192. Weyl, H., *The theory of groups and quantum mechanics.* Translated by H.P. Robertson, London, Dover Publications, Inc., 1931.

193. Weyl, H., Geometrie und Physik. *Die Naturwissenschaften.* 1931. **19**. 49–58.

194. Van der Vaerden, B.L., *Die Gruppentheoretische Methode in der Quantenmechanik.* Berlin, 1932.

195. Einstein, A.; Mayer, V., Semivektoren und Spinoren. *Sitz. Ber. Preuss. Akad. Wiss. Berlin. Phys. – Math. Kl.* 1932. 522–550.

196. Mie, G., The geometry of spinors. *Annalen der Physik.* 1933. **17**. no 5. 465–500.

197. Veblen, O., Geometry of two-component spinors. *Proc. Nat. Akad. Sci.* 1933. **19**. 462–474; Geometry of four-component spinors. *Proc. Nat. Akad. Sci.* 1933. **19**. 503–517; Spinors in projective relativity. *Proc. Nat. Akad. Sci.* 1933. **19**. 979–999.

198. Einstein, A.; Mayer, W., Die Diracgleichungen für Semivektoren. *Proc. Akad. Wet. Amsterdam.* 1933. **36**. 497–516.

199. Einstein, A.; Mayer, W., Spaltung der Natürlichsten Feldgleichungen für Semi-Vektoren in Spinor-Gleichungen von Diracschen Tipus. *Proc. Akad. Wet. Amsterdam.* 1933. **36**. 615–619.

200. Schouten, J.A., Zur generellen Feldtheorie. Semi-Vektoren und Spin-raum. *Zeit. Phys.* 1933. **84**. 92–111.

201. Guth, E., Semivektoren, Spinoren und Quaternionen. *Anz. Akad. Wiss. Wien.* 1933. **70**. 200–207.

202. Bargmann, V., Über den Zusammenhang zwischen Semivektoren und Spinoren und die Reduktion der Diracgleichungen für Semivektoren. *Helv. Phys. Acta.* 1934. **7**. 57–82.

203. De Broglie, L.; Winter, M.J., Sur le spin du photon. *C.R. Acad. Sci. Paris*. 1934. **199**. 813–816.

204. Einstein, A.; Mayer, W., Darstellung der Semi-Vektoren als gewöhnliche Vektoren von Besonderem Differentiations Charakter. *Ann. of Math*. 1934. **35**. no 1. 104–110.

205. Infeld, L., Dirac's equation in general relativity theory. *Acta Phys. Polon*. 1934. **3**. no 3. 1.

206. Halpern, O.; Heller, G., On the Dirac electron in a gravitational field. *Phys. Rev*. 1935. **48**. no 5. 434–438.

207. Dirac, P.A.M., Relativistic wave equations. *Proc. Roy. Soc. London*. 1936. **A155**. 447–459.

208. Petiau, G., Contribution *à* la théorie des equations d'Ondes corpusculaires. Thesis, Univ. Paris, 1936; *Mem. Acad. Sci. Roy. Belgique*. 1936. **16**. no 2. 1–116.

209. Rumer, Yu.B., *Spinor analysis*. Moscow, 1936.

210. Dirac, P.A.M., Complex variables in quantum mechanics. *Proc. Roy. Soc. London*. 1937. **A160**. no 900. 48–59.

211. Racah, G., Sulla simmetria tra particelle e antiparticelle. *Nuovo Cim*. 1937. **14**. 322–328.

212. Whittaker, E.T., On the relations of the tensor-calculus to the spinor-calculus. *Proc. Roy. Soc. London. A*. 1937. **158**. 38–46.

213. Scherzer, O., The imaginary unit in the Dirac equation. *Ann. der Phys*. 1938. **33**. no 7. 593–595.

214. Dirac, P.A.M., La theorie de l'electron et du champ électromagnetique. *Ann. Inst. H. Poincare*. 1939. **9**. 13–49.

215. De Broglie, L., Champs réels et champs complexes en théorie électromagnétique quantique du rayonnement. *C.R. Acad. Sci. Paris*. 1940. **211**. 41–44.

216. De Wet, J.S., On spinor equations for particles with arbitrary spin and rest mass zero. *Phys. Rev*. 1940. **58**. no 3. 236–242.

217. Tonnelat, M.A., Sur la théorie du photon dans un espace de Riemann. *Ann. Phys. N.Y.* 1941. **15**. 144.

218. Milnor, J., Spin structure on manifolds. *L'Enseignement Math*. 1963. **9**. 198–203.

219. Kustaanheimo, P.; Stiefel, E., Perturbation theory of Kepler motion based on spinor regularization, *Journ. f. Reine Angew. Math., Berlin*. 1965. **218**. 204–219.

220. Aharonov, Y.; Susskind, L., Observability of the sign change of spinors under 2π rotations. *Phys. Rev*. 1967. **158**. 1237–1238.

221. Bernstein, H.J., Spin precession during interferometry of fermions and the phase factor associated with rotations through 2π radians. *Phys. Rev. Lett.* 1967. **18**. 1102–1103.

225. Geroch, R., Spinor structure of space-time in General relativity. *Int. J. Math. Phys.* 1968. **9**. 1739–1744.

223. Penrose,R., *Structure of Space – Time*. New York-Amsterdam. W.A. Benjamin Inc., 1968.

224. Held,A.; Newman, E.T.; Posadas, R. The Lorentz group and the sphere. *J. Math. Phys.* 1970. **11**. 3145–3154.

225. Geroch, R., Spinor structure of space-time in General Relativity II. *J. Math. Phys.* 1970. **11**. 343–348.

226. Borchers,H.J.; Hegerfeldt, G.C., The structure of space-time transformations, *Commun. Math. Phys.* 1972. **28**. 259–266.

227. Hehl, F.W.; von der Heyde, P., Spin and the structure of spacetime. *Ann. Inst. H. Poincare. A.* 1973. **19**. 179–196.

228. Klein, A.G.; Opat, G.I., Observability of 2π rotations: a proposed experiment. *Phys. Rev.* 1975. **D11**. 523–528.

229. Kurdgelaidze, D.F., Spinor geometry. *Izvestia Vuzov. Fizika.* 1977. **2**. 7–12 (in Russian).

230. Sternberg, S., On the role of field theories in our physical concept of geometry, *Lect. Notes Math.* 1978. **76**. 1–80.

231. Hartung, R.W., Pauli principle in Euclidean geometry. *Amer. J. Phys.* 1979. **47**. 900–910.

232. Freund Peter, G.O., Spin structures and gauge theory. *Lect. Notes Phys.* 1979. **116**. 308–310.

233. Ishikawa, H., On differentiation with respect to spinors. *Progr. Theor. Phys.* 1980. **63**. 2145–2147.

234. Sommers, P., Space spinors. *J. Math. Phys.* 1980. **21**. 2567–2571.

235. Polubarinov, I.V., *Quantum mechanics and Hopf lbundles.* In: Group Theoretical Methods in Physics. **2**. Procs. Intern. Seminar, Zvenogorod, 24–26 November 1982, 3–10.

236. Penrose, R.; Rindler, W., *Spinors and Space – Time. VolumeI : Two – spinor calculus and relativistic fields.* Cambridge University Press, 1984.

237. Bugajska, K., Spinors as fundamental objects. *J. Math. Phys.* 1985. **26**. 588–592.

238. Bugajska, K., Internal structure of fermions. *J. Math. Phys.* 1985. **26**. 1111–1117.

239. Bugajska, K., Spinors and space-times. *J. Math. Phys.*1986. **27**. 853–858.

240. Biedenharn, L.C.; Braden, H.W.; Truini, P.; van Dam, H., Relativistic wavefunctions on spinor space. *J. Phys.* 1988. **A21**. 3593–3610.

241. Red'kov, V.M., On spinor structure of pseudo-Riemannian space-time and global continuity property for fermion wave functions. *Vesti AN BSSR. Ser. fiz. − mat.* 1994. no 3. 49–55 (in Russian).

242. Red'kov, V.M., On spinor *P*-oriented model of a 3-dimensional Euclidean space. *Vesti AN BSSR. Ser. fiz. − mat.* 1995. no 3. 56–62 (in Russian).

243. Red'kov, V.M., On connection between assumption of spinor geometry and the concept of a fermion intrinsic parity. *Vesti AN BSSR. Ser.fiz. − mat.* 1996. no 1. 36–43 (in Russian).

244. Red'kov, V.M., *P*-orientation and spacial spinors. *Vesti AN BSSR. Ser. fiz. − mat.* 2000. no 2. 76–80 (in Russian).

245. Red'kov, V.M., Linear representations of spinor coverings of the Lorentz group and intrinsic space-time parity of a fermion. *Vesti AN BSSR. Ser. fiz. − mat.* 2003. no 2. 70–76 (in Russian).

246. Red'kov, V.M., Geometry of 3-spaces with spinor structure, Nonlinear Phenomena in Complex Systems. 2004. **7**. no 2. 106–128.

247. Red'kov, V.M., Spinor structure of *P*-oriented space, Kustaanheimo-Stiefel and Hopf bundle, connection between formalisms. *Nonlinear Phenomena in Complex Systems*. 2005. **8**. no 3. 222–239.

248. Red'kov, V.M., *Space with spinor structure and analytical properties of the solutions of Klein − Fock and Schrodinger equations in cylindric parabolic coordinates.* arXiv:hep-th/0612081

249. Veko, O.V.; Ovsiyuk, E.M.; Oana, A.; Neagu, M.; Balan, V.; Red'kov, V.M., *Spinor Structures in Geometry and Physics*. Nova Science Publ. Inc., New York, 2015.

250. Ovsiyuk, E.M.; Red'ko, A.N.; Balan, V.; Red'kov, V.M., The Dirac equation in parabolic cylindric coordinates and possible effects of the spinor structures in quantum physics. *Applied Sciences*. 2016. **18**. 84–107.

251. Veko, O.V.; Ovsiyuk, E.M.; Kisel, V.V.; Red'kov, V.M., *Spinor Methods in the Group Theory and Polarization Optics*. Belarusian Science. Minsk, 2019. (in Russian).

252. Ivashkevich, A.V.; Ovsiyuk, E.M.; Kisel, V.V.; Red'kov, V.M., *Spinor Maxwell Equations in Riemannian Space – Time and Modeling Constitutive Equations*. Chapter in: Understanding Quaternions. Eds. Du Peng, Haibao Hu, Dong Ding, and Zhouyue Li. Nova Science Publ. Inc., New York. 2020. 105–150.

253. Ivashkevich, A.V.; Ovsiyuk, E.M.; Kisel, V.V.; Red'kov, V.M., Spinor Maxwell equations in Riemannian space-time and the geometrical modeling of constitutive relations in electrodynamics. *Materials Physics and Mechanics*. 2020. **45**. no 1. 104–131.

254. Anastasiei, M.; Shimada, H., The Beil metrics associated to a Finsler space. *Balkan J. Geom. Appl*. 1998. **3**. no 2. 1–16.

255. Balan, V., Synge-Beil and Riemann-Jacobi jet structures with applications to physics. *Int. J. Math. Math. Sci.* 2003. **27**. 1693–1702.

256. Balan,V.; Neagu, M., *Jet Single – Time Lagrange Geometry and its Applications*. John Wiley & Sons, Inc., Hoboken, New Jersey, 2011.

257. Beil, R.G., Comparison of unified field theories. *Tensor. N.S.* 1995. **56**. 175–183.

258. Neagu, M., *Riemann – Lagrange geometry for relativistic multi – time optics*. Semin. Mech. – Differ. Dyn. Syst., West Univ. Timişoara. Romania. 2004. **87**. 1–16.

259. Neagu, M.; Oană, A., An anisotropic geometrical approach for extended relativistic dynamics. *Bull. Transilvania University of Bracsov. Romania. SeriesIII : Math. Inf. Phys.* 2016. **9**. no 1. 91–96.

260. Neagu, M.; Oană, A.; Red'kov, V.M., An anisotropic geometrical approach for nonrelativistic extended dynamics. *Ricerche di Mat*. 2013. **62**. no 2. 323–340.

261. Szász, A., Beil metrics in complex Finsler geometry. 2015. *Balkan J. of Geom. Appl*. 2015. **20**. no 2. 72–83.

262. Miron, R.; Kawaguchi, T., Relativistic geometrical optics. *Int. J. Theor. Phys.* 1991. **30**. no 11. 1521–1543.

263. Landau, L.D.; Lifshitz, E.M., *Physique Theoretique. 1. Mecanique.* (Editions Mir, Moscou, 1982) (in French).

264. Landau, L.D.; Lifshitz, E.M., *Physique Theoretique. 2.Theorie des Champs.* (Editions Mir, Moscou, 1989) (in French).

265. Gordon, W., Zur Lichtfortpflanzung nach der Relativitätstheorie, Ann. d. Phys. 1923. **72**. 421–456.

266. Mandelstam, L.I.; Tamm, I.E., Elektrodynamik der anisotropen Medien und der speziallen Relativitatstheorie. *Math. Annalen.* 1925. **95**. 154–160.

267. Synge, J.L., Relativity: the General Theory. North-Holland Publishing Company. Amsterdam, 1960.

268. Tamm, I.E., Electrodinamika anizotropnoi sredy v spetsialnonoi teorii otnositelnosti, *Zh.R,F,Kh.O, Fiz. dep.* 1924. **56**. no 2-3. 248–262 (in Russian).

269. Tamm, I.E., Kristallooptika teorii otnositelbnosti v svyazi s geometriei bikvadratichnoi formuy, *Zh.R,F,Kh.O, Fiz. dep.* 1925. **54**. no 3-4. 209–240. (in Russian).

270. Bittencourt, E.; Pereira, J.P.; Smolyaninov, I.I.; Smolyaninova, V.N., The flexibility of optical metric. 2016. *Class. Quantum Grav.* **33**. Paper 165008.

271. Chen, B.; Kantowski, R., Including absorption in Gordon's optical metri. 2009. *Phys. Rev.* **D79**. Paper 104007.

272. Miron, R.; Anastasiei, M., *The Geometry of Lagrange Spaces : Theory and Applications.* Kluwer Academic Publishers. Dordrecht, 1994.

273. Leonhardt, U.; Piwnicki, P., Optics of nonuniformly moving media. *Phys. Rev.* 1999. **A60**. 4301.

274. Leonhardt, U.; Philbin, T.G., General relativity in electrical engineering. *New Journal of Physics* . 2006. **8**. 247.

275. Byrd, P.F.; Friedman, M.D., *Handbook of Elliptic Integrals for Engineers and Scientists.* 2nd ed., Springer-Verlag, New York, 1971.

276. Proca, A., Sur les equations fondamentales des particules élémentaires. *C.R. Acad. Sci. Paris.* 1936. **202**. 1490–1492.

277. Proca, A., Théorie non relativiste des particules a spin entier. *Journ. Phys. Radium.* 1938. **9**. 61–66.

278. Duffin, R.Y., On the characteristic matrices of covariant systems. *Phys. Rev.* 1938. **54**. no 12. 1114–1114.

279. Kemmer, N., The particle aspect of meson theory. *Proc. Roy. Soc. London.* 1939. **A73**. 91–116.

280. Bhabha, H.J., Classical theory of meson. *Proc. Roy. Soc. London.* 1939. **A172**. no 5. 384-409.

281. Belinfante, F.J., The undor equation of the meson field. *Physica.* 1939. **6**. 870.

282. Belinfante, F.J., Spin of Mesons. *Physica.* 1939. **6**. 887–898.

283. Tamm, I.E., Motion of mesons in electromagnetic fields. *Doklady Akad. Nauk USSR.* 1940. **29**. 551.

284. Schrödinger, E., Pentads, tetrads, and triads of meson matrices. *Proc. Roy. Irish. Acad. A.* 1943. **48**. 135–146.

285. Schrödinger, E., Systematics of meson matrices. *Proc. Roy. Irish. Acad. A.* 1943. **49**. 29–42.

286. Kemmer, N., The algebra of meson matrices. *Proc. Camb. Phil. Soc.* 1943. **39**. 189–196.

287. Harish-Chandra, The correspondence between the particle and wave aspects of the meson and the photon. *Proc. Roy. Soc. London.* 1946. **A186**. 502–525.

288. Fujiwara, I., On the Duffin-Kemmer algebra. *Progr. Theor. Phys.* 1953. **10**. no 6. 589–616.

289. Kemmer, N., On the theory of particles of spin 1. *Helv. Phys. Acta.* 1960. **33**. no 8. 829–838.

290. Da Silveira, A., Kemmer wave equation in Riemann space. *J. Math. Phys.* 1960. **1**. no 6. 489–491.

291. Casanova, G., Particules neutre de spin 1. *C.R. Acad. Sci. Paris. A.* 1969. **268**. 673–676.

292. Fedorov, F.I.; Bogush, A.A., On the properties of Duffin–Kemmer matrices. *Doklady Natsionalnoi Akademii Nauk Belarusi.* 1962. **6**. no 2. 81–85 (in Russian).

293. Majumdar, S.D., Wave equations in curved space-time. *Phys. Rev.* 1962. **126**. no 6. 2227–2230.

294. Young, J.A.; Bludman, S.A., Electromagnetic properties of a charged vector meson. *Phys. Rev.* 1963. **131**. no 5. 2326–2334.

295. Sankaranarayanan A., Good R.H., Jr., Spin-one wave equation. Nuovo Cimento. 1965. **36**. no 4. 1503–1315.

296. Gupta, S., Wave equation of charged particle of spin 1, the Kemmer equation. *Indian J. Phys.* 1969. **43**. no 2. 92–105.

297. Goldman, T.; Tsai, W.; Yildiz, A., Consistency of spin one theory. *Phys. Rev.* 1972. **D5**. no 8. 1926–1930.

298. Fedorov, F.I., Recurrence relations for Duffin–Kemmer matrices. *Proceedings of the National academy of sciences of Belarus. Phys. – Math. series.* 1972. no 5. 73–78 (in Russian).

299. Lord, E.A., Six-dimensional formulation of meson equations. *Int. J. Theor. Phys.* 1972. **5**. no 4-6. 339–348.

300. Shamaly, A.; Capri, A.Z., First-order wave equations for integral spin. *Nuovo Cimento.* 1971. **B2**. no 2. 235–253.

301. Shamaly, A.; Capri, A.Z., Unified theories for massive spin 1 fields. *Can. J. Phys.* 1973. **51**. no 14. 1467–1470.

302. Kisel, V.V., Electrical polarizability of particles with spin 1 in the theory of relativistic wave equations. *Proceedings of the National Academy of Sciences of Belarus. Phys. – Math. series.* 1982. no 3. 73–78 (in Russian).

303. Kisel, V.V.; Ovsiyuk, E.M.; Red'kov, V.M., On the wave functions and energy spectrum for a spin 1 particle in external Coulomb field. *Nonlinear Phenomena in Complex Systems.* 2010. **13**. no. 4. 352–367.

304. Kisel, V.V.; Red'kov, V.M.; Ovsiyuk, E.M., Wave functions and energy spectrum for spin 1 particle in Coulomb field. *Doklady Natsionalnoi Akademii Nauk Belarusi.* 2011. **55**. no 1. 50–55.

307. Ovsiyuk, E.M.; Veko, O.V.; Voynova,Ya.A.; Koral'kov,A.D.; Kisel, V.V.; Red'kov, V.M., On describing bound states for a spin 1 particle in the external Coulomb field. *Balkan Society of Geometers Proceedings.* 2018. **25**. 59–78.

306. Corben, H.C.; Schwinger, J., The electromagnetic properties of mesotrons. *Phys. Rev.* 1940. **58**. 953.

307. Shamaly, A.; Capri, A.Z., Unified theories for massive spin 1 fields. *Canad. J. Phys.* 1973. **51**. no 14. 1467–1470.

308. Ovsiyuk, E.M.; Voynova, Ya.A.; Kisel, V.V.; Balan, V.; Red'kov, V.M., *Spin* 1 *Particle with Anomalous Magnetic Moment in the External Uniform Electric Field.* In: S. Griffin (Eds.) Quaternions: Theory and Applications. Nova Science Publ. Inc., New York, 2017. 47–84.

309. Ovsiyuk, E.M.; Voynova, Ya.A.; Kisel, V.V.; Balan, V.; Red'kov, V.M., *Techniques of Projective Operators Used to Construct Solutions for a Spin* 1 *Particle with Anomalous Magnetic Moment in the External Magnetic Field.* In: S. Griffin (Eds.) Quaternions: Theory and Applications. Nova Science Publ. Inc., New York, 2017. 11–46.

310. Kisel, V.M.; Voynova, Ya.A.; Ovsiyuk, E.M.; Balan, V.; Red'kov, V.M., Spin 1 particle with anomalous magnetic moment in the external uniform magnetic field. *Nonlinear Phenomena in Complex Systems.* 2017. **20**. no 1. 21–39.

311. Ovsiyuk, E.M.; Voynova, Ya.A.; Kisel, V.V.; Balan, V.; Red'kov, V.M., Spin 1 particle with anomalous magnetic moment in the external uniform electric field. *Nonlinear Phenomena in Complex Systems.* 2018. **21**. no 1. 1–20.

333. Kisel, V.V.; Ovsiyuk, E.M.; Voynova, Ya.A.; Red'kov, V.M., Quantum mechanics of spin 1 particle with quadrupole moment in external uniform magetic field. *Problems of Physics, Mathematics, and Thechnics.* 2017. **32**(**3**). 18–27 (in Russian).

313. Fierz, M., Über die relativistische theorie Kraftefreier Teilchen mit beliebigem Spin. *Helv. Phys. Acta.* 1939. **12**. 3–37.

314. Pauli, W.; Fierz, M., Über relativistische Feldleichungen von Teilchen mit beliebigem Spin im elektromagnetishen Feld. *Helv. Phys. Acta*. 1939. **12**. 297–300.

315. Fierz, M.; Pauli, W., On relativistic wave equations for particles of arbitrary spin in an electromagnetic field. *Proc. Roy. Soc. London*. 1939. **A173**. 211–232.

316. Fierz, M., Über den Drehimpuls von Teilehen mit Ruhemasse null und beliebigem Spin. *Helv. Phys. Acta*. 1940. **13**. 45–60.

317. Rarita, W.; Schwinger, J., On a theory of particles with half-integral spin. *Phys. Rev.* 1941. **60**. no 1. 61–64.

318. Ginzburg, V.L., To the theory of particles of spin 3/2. *Zh. Eksp. Teor. Fiz.* 1942. **12**. 425–442 (in Russian).

319. Ginzburg, V.L.; Smorodinsky, Ya.A., On wave equations for particles with variable spin. *Zh. Eksp. Teor. Fiz.* 1943. **13**. 274 (in Russian).

320. Davydov, A.S., Wave equations of a particle having spin 3/2 in absence of field. *Zh. Eksp. Teor. Fiz.* 1943. **13**. no 9-10. 313–319 (in Russian).

321. Bhabha, H.J.; Harish-Chandra., On the theory of point particles. *Proc. Roy. Soc. London*. 1944. **A183**. 134–141.

322. Bhabha, H.J., Relativistic wave equations for the proton. *Proc. Indian Acad. Sci.* 1945. **A21**. 241–264.

323. Bhabha, H.J., Relativistic wave equations for elementary particles. *Rev. Mod. Phys.* 1945. **17**. no 2-3. 200–215.

324. Bhabha ,H.J., The theory of the elementary particles. *Rep. Progr. Phys.* 1946. **10**. 253–271.

325. Tamm, I.E.; Davydov, A.S., To the theory of particles with spin 3/2. Zh. Eksp. Teor. Fiz. 1947. **17**. 427 (in Russian).

326. Gel'fand, I.M.; Yaglom, A.V., General relativistic invariant equations and infinitely dimensional representations of the Lorentz group. *Zh. Eksp. Theor. Fiz.* 1948. **18**. no 8. 703–733 (in Russian).

327. Fradkin, E.E., To the theory of particles with higher spins. *Zh. Eksp. Teor. Fiz.* 1950. **20**. no. 1. 27–38 (in Russian).

328. Fedorov, F.I., On minimal polynomials for the matrices of the relativistic wave equations. *Doklady AN USSR*. 1951. **79**. no 5. 787–790 (in Russian).

329. Bhabha, H.J., On a class of relativistic wave-equations of spin 3/2 *Proceedings of the Indian Academy of Sciences*. 1951. **A4**. no 6. 335–354.

330. Fedorov, F.I., Generalized relativistic wave equations. *Doklady AN USSR*. 1952. **82**. no 1. 37–40 (in Russian).

331. Belinfante, F., Intrinsic magnetic of elementary particles of spin 3/2. *Phys. Rev.* 1953. **92**. 997.

332. Corson, E.M., *Introduction to tensors, spinors and relativistic wave equations*. London, Blackie and Son, 1953.

333. Petras, M., A contribution of the theory of the Pauli-Fierz's equations a particle with spin 3/2. *Czech. J. Phys.* 1955. **5**. 169–170.

334. Petras, M., A note to Bhabha's equation for a particle with maximum spin 3/2. *Czech. J. Phys.* 1955. **5**. 418–419.

335. Fainberg, V.Ya., To the interaction theory of the particles of the higher spins with electromagnetic and meson fields. *Trudy FIAN USSR*. 1955. **6**. 269–332 (in Russian).

336. Moldauer, P.A.; Case, K.M., Properties of half-integral spin Dirac–Fierz–Pauli particles. *Phys. Rev.* 1956. **102**. no 1. 279–285.

337. Fradkin, E.E., Algebra of matrices in the theory of spin 3/2 particles. *Doklady AN USSR*.1957. **115**. no 5. 907–910 (in Russian).

338. Izmaylov, S.V.; Fradkin, E.E., On different forms of equations for spin 3/2 particle. *Uchenye Zapiski Leningrad University named after A.I. Gertsen*. 1958. **141**. 69–95 (in Russian).

339. Johnson, K.; Sudarshan, E.C.G., Inconsistency of the local field theory of charged spin 3/2 particles. *Ann. Phys.* 1961. **13**. no 1. 126–145.

340. Shelepin, L.A., Covariant theory of relativictic wave equations. *Nucl. Phys.* 1962. **33**. no 4. 580–593.

341. Weawer D.L.; Fradkin D.M., Symmetric spinor theory for any spin. *Nuovo Cimento*. 1965. **37**. no 2. 440–449.

342. Bender, C.M.; McCoy, B.M., Peculiarities of a free massless spin-3/2 field theory. *Phys. Rev.* 1966. **148**. no 4. 1375–1380.

343. Velo, G.; Zwanziger, D., Propagation and quantization of Rarita-Schwinger waves in an external electromagnetic potential. *Phys. Rev.* 1969. **186**. no 5. 1337–1341.

344. Velo, G.; Zwanziger, D., Noncausality and other defects of interaction Lagrangians for particles with spin one and higher. *Phys. Rev.* 1969. **188**. no 5. 2218–2222.

345. Aurilia, A.; Umezawa, H., Theory of high spin fields. *Phys. Rev.* 1969. **182**. no 5. 1682–1694.

346. Hagen, C.R., New inconsistencies in the quantization of spin-3/2 fields. *Phys. Rev.* 1971. **D4**. no 8. 2204–2208.

362. Pletjuchov, V.A.; Strazhev, V.I., To the theory of spin-3/2 particles. *Izvestiya Vuzov. Fizika*. 1985. **28**. no 1. 91–96 (in Russian).

363. Darkhosh, T., Is there a solution to the Rarita–Schwinger wave equation in the presence of an external electromagnetic field? *Phys. Rev.* 1985. **D32**. no 12. 3251–3255.

364. Cox, W., On the Lagrangian and Hamiltonian constraint algorithms for the Rarita–Schwinger field coupled to an external electromagnetic field. *J. Phys.* 1989. **A22**. no 10. 1599–1608.

365. Red'kov, V.M., On degrees of freedom for spin 3/2 field in de Sitter world and separation of the variables in the static coordinates. *Proceedings of the National academy of sciences of Belarus. Phys. – Math. series.* 1992. no 6. 50–56 (in Russian).

366. Deser, S.; Waldron, A.; Pascalutsa, V., Massive spin-3/2 electrodynamics. *Phys. Rev.* 2000. **D62**. Paper 105031.

367. Deser, S.; Waldron, A., Inconsistencies of massive charged gravitating higher spins. *Nucl. Phys.* 2002. **B631**. 369–387.

368. Pascalutsa, V., Correspondence of consistent and inconsistent spin-3/2 couplings via the equivalence theorem. *Phys. Lett.* 2001. **B503**. 85–90.

369. Napsuciale, M.; Kirchbach, M.; Rodriguez, S., Spin 3/2 Beyond the Rarita–Schwinger Framework. *Eur. Phys. J.* 2006. **A29**. 289–306.

370. Kisel, V.V.; Ovsiyuk, E.M.; Veko, O.V.; Voynova, Y.A.; Balan, V.; Red'kov, V.M., *Elementary Particles with InternalStructure in External Fields. I. General Theory.* Nova Science Publ. Inc., New York, 2018.

371. Ovsiyuk, E.M.; Kisel, V.V.; Veko, O.V.; Voynova, Y.A.; Balan, V.; Red'kov, V.M., *Elementary Particles with Internal Structure in External Fields. II. PhysicalProblems.* Nova Science Publ. Inc., New York, 2018.

372. Fierz,M.; Pauli, W., On relativistic wave equations for particles of arbitrary spin in an electromagnetic field. *Proc. Roy. Soc. London.* 1939. **A173**. 211-232.

373. Fedorov, F.I., To the theory of particles with spin 2. *Proceedings of Belarus State University. Ser. Phys. – Math.* 1951. **12**. 156–173 (in Russian).

374. T. Regge. On properties of the particle with spin 2. *Nuovo Cimento.* **5**, no. 2, 325–326 (1957).

375. Buchdahl, H.A., On the compatibility of relativistic wave equations for particles of higher spin in the presence of a gravitational field. *Nuovo Cimento.* 1958. **10**. 96–103.

376. Buchdahl, H.A., On the compatibility of relativistic wave equations in Riemann spaces. *Nuovo Cimento.* 1962. **25**. 486-496.

377. Krylov, B.V.; Fedorov, F.I., Equations of the first order for graviton. *Doklady of the National Academy of Sciences of Belarus*. 1967. **11**. no 8. 681–684.

378. Bogush, A.A.; Krylov, B.V.; Fedorov, F.I., On matrices of the equations for spin 2 particles. *Proceedings of the National Academy of Sciences of Belarus.Physics and Mathematics Series*. 1968. no 1. 74–81 (in Russian).

379. Fedorov, F.I., Equations of the first order for gravitational field. *Doklady of the Academy of Sciences of USSR*. 1968. **179**. no 4. 802–805 (in Russian).

380. Hagen, C.R., Minimal electromagnetic coupling of spin-two fields. *Phys. Rev.* 1972. **D6**. no 4. 984–987.

381. Velo, G., Anomalous behavior of a massive spin two charged particle in an external electromagnetic field. *Nucl. Phys.* 1972. **B3**. 389–401.

382. Krylov, B.V., On the systems of the first order for graviton. *Proceedings of the National Academy of Sciences of Belarus, Physics and Mathematics Series*. 1972. no 6. 82–89.

383. Fedorov, F.I.; Kirilov, A.A., The first order equations for graviational field in vacuum. *Acta Physica Polonica*. 1976. **B7**. no 3. 161–167.

384. Mathews, P.M.; Seetharaman, M.; Prabhakaran, J., Inconsistencies in the symmetric tensor field description of spin-2 particles in an external homogeneous magnetic field. *Phys. Rev.* 1976. **D14**. no 4. 1021–1031.

385. Kobayashi, M.; Shamaly A., Minimal electromagnetic coupling for massive spin-two fields. *Phys. Rev.* 1978. **D17**. no 8. 2179–2181.

386. Cox, W., First-order formulation of massive spin-2 field theories. *J. Phys.* 1982. **A15**. 253–268.

387. Loide, R.K., On conformally covariant spin-3/2 and spin-2 equations. *J. Phys.* 1986. **A19**. no 5. 827–829.

388. Kisel, V.V., On relativistic wave equations for a spin 2 particle. *Proceedings of the National Academy of Sciences of Belarus. Physics and Mathematics Series*. 1986. no 5. 94–99.

389. Kisel, V.V., On relativistic wave equations for a massive particle with spin 2. *Proceedings of the National academy of sciences of Belarus. Phys. – Math. series*. 1986. no 5. 94–99 (in Russian).

390. Bogush, A.A.; Kisel, V.V., On the description of the anomalous magnetic moment of a massive particle with spin 2 in the theory of relativistic wave equations. *Izvestiya Vuzov. Fizika*. 1988. **31**. no 3. 11–16 (in Russian).

391. Bogush,A.A.; Kisel, V.V.; Tokarevskaya, N.G.; Red'kov, V.M., On equations for a spin 2 particle in external electromagnetic and gravitational fields. *Proceedings of the National Academy of Sciences of Belarus, Physics and Mathematics Series*. 2003. no 1. 62–67.

392. Red'kov, V.M.; Tokarevskaya, N.G.; Kisel, V.V., Graviton in a curved space-time background and gauge symmetry. *Nonlinear Phenomena in Complex Systems*. 2003. **6**. no 3. 772–778.

393. Kisel, V.V.; Ovsiyuk, E.V.; Veko, O.V.; Red'kov, V.M., Contribution of gauge degrees of freedom in the energy-momentum tensor of the massless spin 2 field. *Proceedings of the National Academy of Sciences of Belarus, Physics and Mathematics Series*. 2015. no 2. 58–63 (in Russian).

394. Kisel, V.V.; Ovsiyuk, E.V.; Veko, O.V.; Red'kov, V.M., Nonrelativistic approximation in the theory of a spin two particle. *Doklady of the National Academy of Sciences of Belarus*. 2015. **59**. no 3. 21–27.

396. Ivashkevich, A.; Buryy, A.; Ovsiyuk, E.; Balan, V.; Kisel, V. ; Red'kov, V., On the matrix equation for a spin 2 particle in pseudo-Riemannian space-time, tetrad method. *Proceedings of Balkan Society of Geometers*. 2021. **28**. 45–66.

396. Dudko, I.G.; Semenyuk, O.A.; Kisel, V.V ; Red'kov, V.M., Spin 2 particle with anomalous magnetic moment in Riemann space-time, restriction to massless case, gauge symmetry. *Nonlinear Phenomena in Complex Systems*. 2022. **25**. no 2. 286–296.

397. Ivashkevich, A.V.; Buryy, A.V.; Ovsiyuk, Chichurin; A.S.; Red'kov V.M., Generalized helicity operator for a spin 2 particle in the presence of the external uniform magnetic field. *Nonlinear Phenomena in Complex Systems*. 2022. **25**. no 2. 104–121.

398. Krylova, N.G.; Ovsiyuk, E.M.; Balan, V.; Red'kov, V.M., Non-Abelian monopole problem, the method of geometrical KCC-invariants. *Proceedings of Balkan Society of Geometers*. 2022. **29**. 48 – 67.

399. Ovsiyuk, E.M.; Krylova, N.G.; Balan, V.; Red'kov, V.M., The doublet of Dirac particles in the non-Abelian monopole field: Pauli approximation. *Nonlinear Phenomena in Complex Systems*. 2022. **25**. no 2. 136–158.

400. Neagu, M.; Krylova, N.; Ovsiyuk, E.; Red'kov, V., Optics in anisotropic inhomogeneous media and Lagrange geometry. *International Journal of Geometric Methods in Modern Physics*. 2022. **19**. no 10. Paper 2250152.

Index

A

analytical treatment, 161

C

conformal coordinates, 15
constant curvature, 15, 17, 19, 21, 23, 25, 27, 29, 128, 140, 400, 401
continuity, 201
Coulomb problem, 1, 3, 5, 7, 9, 11, 13, 33, 34, 37, 39
cylindric coordinates, 15, 17, 183, 185, 205, 371, 372
cylindric parabolic coordinates, xiv, 177, 179, 181, 183, 185, 187, 189, 191, 209, 402
cylindric symmetry, 373, 375, 376, 377, 379, 381
cylindrical parabolic coordinates, 195, 402

D

Dirac particle, xiii, xiv, 3, 20, 111, 113, 128, 177, 179, 181, 183, 185, 187, 189, 191, 401, 402
Dirac theories, 169

E

equation in spherical coordinates, 389
Euclidean space, xi, xiii, 128, 136, 138, 142, 144
Euler--Lagrange equations, 220
external Coulomb field, 3, 12, 227, 247, 249, 259, 265

F

Finslerian geometrization, 227, 229, 231, 233, 235, 237, 239, 241, 243, 245
first order constraints, 381

G

gauge degrees of freedom, xi, 381
gauge solutions, xv, 298, 355, 361, 367, 368, 376, 380, 386, 388, 403
general Heun equation, xiii, 1, 34, 37
geometry, xiii, 43, 95, 111, 131, 132, 385, 401
gradient type solutions, 362

H

helicity operator, xv, 95, 96, 185, 207, 209, 297, 303, 310, 322, 323
Heun functions, xiii, 1, 3, 5, 6, 7, 8, 9, 11, 12, 13, 253, 278, 400
homogeneous equations, xv, 368, 403
hydrogen atom, xi, xiii, 31, 33, 34, 35, 37, 39, 41, 43, 401
hypergeometricand partially, 1

I

independent Majorana components, 109
initial equation, 146, 247, 265, 298

K

Klein-Gordon-Fock equation, 31, 37, 55, 179

L

Landau problem, 17, 22
Lobachevsky plane, 15
Lobachevsky space, xiii, 16, 20, 95, 106, 128, 132, 135, 136, 140, 401

M

magnetic field, xiii, 15, 16, 17, 20, 23, 27, 218, 265, 372, 400
Majorana particle, 109, 151, 153, 155, 156, 157, 159, 161, 163, 165, 167, 169, 171, 173, 175
Majorana spinor fields, 152
Massive and Massless Fields with Spin 3/2, 299, 301, 303, 305, 307, 309, 311, 313, 315, 317, 319, 321, 323, 325, 327
massless field, xvi, 297, 355, 375, 383, 384, 404
massless spin 3/2 field, xv, 297, 300, 355, 357, 359, 361, 363, 365, 367, 369, 403
massless spin 3/2 particle, xv, 315, 355, 361
minimal value, 274, 333, 334, 337, 338, 339, 356

N

natural splitting, 236, 241

P

Pauli approximation, xi, 111
Pauli equation, 78, 81, 90, 111, 112, 119

Q

qualitative study, 38, 39, 158

R

radial equation, xv, 2, 31, 34, 37, 40, 53, 54, 67, 77, 118, 120, 130, 155, 248, 249, 250, 267, 268, 277, 279, 334, 360, 364, 403
relativistic invariance, additional checking, 392
Riemann geometry, 45
Riemannian space, xi, xv, 23, 109, 113, 131, 136, 139, 383, 385, 402, 403

S

scalar particle, xiii, 17, 45, 47, 49, 51, 53, 55, 57, 59, 61, 63, 250
second order equations, 97, 98, 102, 158, 380
separating the variables (separation of the variables), 27, 31, 119, 170, 172, 179, 333, 358, 400, 402, 403
spherical coordinates, xiii, xv, xvi, 1, 65, 67, 69, 71, 73, 75, 77, 79, 81, 83, 85, 87, 89, 91, 93, 330, 358, 389, 401, 403, 404
spherical solutions, 355, 357, 359, 361, 363, 365, 367, 369
spherical symmetry for a massive spin 3/2 particle, 329
spin 1/2 particle, 1, 3, 5, 7, 9, 11, 13, 15, 17, 19, 21, 23, 25, 27, 29, 40, 65, 67, 69, 71, 73, 75, 77, 79, 81, 82, 83, 85, 87, 89, 91, 93, 95, 97, 99, 101, 103, 105, 107, 109, 153
spin 2 particle in Minkowski space, xv
spin 3/2 massless field, 371, 373, 375, 377, 379, 381
spinor space structure, xiv, 177, 201, 402
structure of the power series, 286

T

Tetrad method, 383
time variable, 98, 108

V

variables, xv, 2, 15, 21, 25, 28, 29, 31, 52, 56, 62, 68, 80, 91, 95, 98, 111, 116, 137, 142, 154, 180, 186, 188, 198, 200, 201, 218, 220, 221, 222, 241, 242, 243, 244, 245, 247, 251, 254, 266, 268, 270, 271, 275, 276, 279, 281, 300, 302, 303, 306, 316, 324, 326, 333, 335, 338, 340, 341, 348, 349, 353, 354, 358, 364, 365, 366, 371, 376, 377, 380, 383, 384
vector particle, xiv, 227, 229, 231, 233, 235, 237, 239, 241, 243, 245, 247, 248, 265, 267, 268, 269, 271, 273, 275, 277, 279, 281, 283, 285, 287, 289, 291, 293, 295, 402

W

wave equation, xv, 34, 40, 116, 151, 153, 177, 247, 265, 297, 298, 300, 310, 311, 312, 313, 329, 357, 371, 372, 383, 403

About the Authors

Dr. Alina V. Ivashkevich
Researcher
Department "Fundamental Interactions and Astrophysics"
B.I. Stepanov Institute of Physics NAS of Belarus, Belarus
Email: ivashkevich.alina@yandex.by

Dr. Nina G. Krylova
Assistant Professor
Physical Department
Belarusian State University, Belarus
Email: nina-kr@tut.by

Dr. Elena M. Ovsiyuk
Assistant Professor
Head of Department of theoretical Physics and Applied Informatics
Mozyr State Pedagogical University, Belarus
Email: e.ovsiyuk@mail.ru

Dr. Vasiliy V. Kisel
Assistant Professor Belarusian State University of Informatics and Radio-Electronics, Belarus
Email: vasiliy-bspu@mail.ru

Dr. Vladimir Balan
Department Mathematics-Informatics
University Politehnica of Bucharest, Romania
Email: vladimir.balan@upb.ro

Dr. Viktor M. Red'kov
Researcher
B.I. Stepanov Institute of Physics
Department "Fundamental Interactions and Astrophysics"
NAS of Belarus, Belarus
Email: v.redkov@ifanbel.bas-net.by

347. Baisya, H.L., On the Rarita–Schwinger equation for the vector-bispinor field. *Nucl. Phys.* 1971. **B29**. no 1. 104–124.

348. Singh, L.P.S., Noncausal propagation of classical Rarita-Schwinger waves. *Phys. Rev.* 1973. **D7**. no 4. 1256–1258.

349. Hortacsu, M., Demonstration of noncausality for the Rarita–Schwinger equation. *Phys. Rev.* 1974. **D9**. no 4. 928–930.

350. Seetharaman, M.; Prabhakaran, J.; Mathews, P.M., Rarita–Schwinger particles in homogeneous magnetic field and inconsistencies of spin 3/2 theories. *Phys. Rev.* 1975. **D12**. no 2. 458–466.

351. Seetharaman, M.; Prabhakaran, J.; Mathews, P.M., Causality and indefiniteness of charge in spin 3/2 field theories. *J. Phys.* 1975. **A8**. no 4. 560–565.

352. Cox, W., Algebras for causal external electromagnetic interaction in higher-spin theories. *J. Phys.* 1976. **A9**. 1025–1033.

353. Wightman, A.S., Invariant wave equations: general theory and applications to the external field problem. *Lecture Notes in Physics*. 1978. **73**. 1–101.

354. Garding, L., Mathematics of invariant wave equations. *Lect. Notes in Physics*. 1978. **73**. 102–164.

355. Capri, A.Z.; Kobes, R.L., Further problems in spin 3/2 field theories. *Phys. Rev.* 1980. **D22**. 1967–1978.

356. Aurilia, A.; Kobayashi, M.; Takahashi, Y., Remarks on the constraint structure and the quantization of the Rarita–Schwinger field. *Phys. Rev.* 1980. **D22**. no 6. 1368–1374.

357. Hagen, C.R.; Singh, L.P.S., Search for consistent interactions of the Rarita–Schwinger field. *Phys. Rev.* 1982. **D26**. no 2. 393–398.

358. Mathews, P.M.; Vijayalakshmi, B., On inequivalent classes unique-mass-spin relativistic wave equations involving repeated irreducible representations with arbitrary multiplicities. *J. Math. Phys.* 1984. **25**. no 4. 1080–1087.

359. Bogush, A.A.; Kisel, V.V., Equation for a particle with spin-3/2, which has an anomalous magnetic moment. *Izvestiya Vuzov. Fizika*. 1984. **27**. no 1. 23–27 (in Russian).

360. Bogush, A.A.; Kisel, V.V.; Fedorov, F.I., On the interpretation of additional components of wave functions in the electromagnetic interaction. *Doklady Natsionalnoi Akademii Nauk Belarusi*. 1984. **277**. no 2. 343–346 (in Russian).

361. Loide, R.K., Equations for a vector-bispinor. *J. Phys.* 1984. **A17**. no 12. 2535–2550.